ANNALS OF THE NEW YORK ACADEMY OF SCIENCES

Volume 461
March 31, 1986

FIRST ASPEN WINTER PHYSICS CONFERENCE[a]

Editor

MARTIN M. BLOCK

Scientific Advisory Committee

J. BALLAM, G. CASSIDAY, D. CLINE, J. CRONIN, P. DARRIULAT, M. L. GOLDBERGER, C. QUIGG, H. SCHOPPER, R. SCHWITTERS, F. SCIULLI, and P. SÖDING

Organizing Committee

M. M. BLOCK, L. DURAND, S. MESHKOV, D. PINES, M. RUDERMAN, D. SCHRAMM, L. M. SIMMONS, and F. ZACHARIASEN

CONTENTS

[a]The papers in this volume were presented at the First Aspen Winter Physics Conference Series, which was held at the Aspen Center for Physics in Aspen, Colorado, on January 6–19, 1985.

Part II. Electroweak Interactions

Introductory Remarks

MARTIN M. BLOCK

Department of Physics and Astronomy
Northwestern University
Evanston, Illinois 60201

The Aspen Winter Physics Conference Series was inaugurated as an annual sequence in January of 1985. The first week's conference was held on January 6–12, 1985 and was entitled, "Collider Physics at Ultrahigh Energies"; the second was held on January 13–19, 1985 and was entitled, "Electroweak Interactions."

Approximately 65 attendees from all over the world were present each week for a lively and stimulating program that was held high in the Rocky Mountains in a splendid winter setting.

In order for the local community to profit from the presence of the conferences, a popular public lecture was held each week, with the first given by Murray Gell-Mann and the second by James Cronin.

The Winter Conference Series, sponsored by the Aspen Center for Physics, was made possible by generous contributions from the Aspen Foundation, the U.S. Department of Energy, the National Science Foundation, and Northwestern University. We wish to thank the New York Academy of Sciences, and in particular Mr. Stefan Malmoli, for their help in editing and publishing these proceedings.

I would further like to take this opportunity to thank those people whose efforts made the conferences a success. In particular, I am grateful to the Scientific Advisory Committee: J. Ballam (SLAC), G. Cassiday (University of Utah), D. Cline (University of Wisconsin), J. Cronin (University of Chicago), P. Darriulat (CERN), M. L. Goldberger (CALTECH), C. Quigg (Fermilab), H. Schopper (CERN), R. Schwitters (Harvard University), F. Sciulli (Columbia University), and P. Söding (DESY). I also would like to acknowledge, for its aid and advice, the labors of love performed by the Organizing Committee, consisting of M. Block (Northwestern University), L. Durand (University of Wisconsin), S. Meshkov (National Bureau of Standards), D. Pines (University of Illinois), M. Ruderman (Columbia University), D. Schramm (University of Chicago), L. M. Simmons (Los Alamos National Lab.), and F. Zachariasen (CALTECH), which I chaired.

Experimental Future

Some Sociological Comments on High Energy Physics

LEON M. LEDERMAN

Fermi National Accelerator Laboratory
Batavia, Illinois 60510

In order to project into the experimental future, we need some historical perspective. We should be aware of trends (e.g., vectors) so that we can see where a "natural" evolution of HEP will take us and we should be sensitive to problems that, although minor in the 1960s and manageable in the 1970s, have reached near crises states in the 1980s and will become the death knell of the subject in the 1990s. Although I started this review with an idea of seeing how we were doing with our ability to measure in space and time resolution, I became trapped into sociology, so I must then also discuss this as well.

My assignment is awkward because all of the present and future experiments anywhere close to reality will be discussed in this program of speakers. What then can I add?

Accelerators? About 30 months ago and about five miles from here in *Snowmass '82,* I stated my conviction that the scientific need for observing a ~1 TeV mass scale was overwhelming, that the technology was in hand, that the cost was moderate, and that we could not afford to dabble in intermediate steps. I would like to claim here that this talk, which coined the name, "Machine-in-the-Desert" or "Desertron," was the opening shot in what was to become the goal of HEP for the next decade. After an enormous "Sturm und Drang," we have SSC. It is much too early to say whether or not this was a foolhardy decision, but if we fail, it won't be because we aren't unified and convinced that this is the right scientific step. The more we look at the technology, the better it looks. The more we look at the science, the more it is the right step. I think we are committed to SSC, but we must be prepared for some setbacks and disappointments. However, we have a reasonable program to keep us going and if the SSC is under construction by the time LEP and HERA come on (1991–92), we'll be very busy building the machine and building the detectors that are to use it. Therefore, I want to talk history now and since I'm not a scholar, but a student, I largely will use my own experience. It is also well known that, at my age, looking back is more pleasant than looking ahead.

Let me show you a number of figures, the first of which are pages from some miscellaneous *Physical Review* papers published in the ancient days when data was recorded by quill pens on fine parchment. FIGURE 1 is from research done on the Nevis cyclotron, which in 1950 was the world's highest energy machine at 400 MeV. The device was a Wilson Cloud Chamber working in a magnetic field of 5000 gauss. A total of 4000 photographs yielded the data, which was scanned and analyzed by two students and one assistant professor (me). FIGURE 2 is research done at BNL, the Cosmotron. Here, 5000 pictures in a "giant" 36″ cloud chamber were measured by one

Reprinted from THE PHYSICAL REVIEW, Vol. 86, No. 1, 17–21, April 1, 1952
Printed in U. S. A.

Scattering and Absorption of Pi-Mesons in Carbon*

H. BYFIELD, J. KESSLER, AND L. M. LEDERMAN
Columbia University, New York, New York

(Received December 3, 1951)

FIGURE 1. This is from an article on research done on the Nevis cyclotron at Columbia University in 1950.

postdoctorate (Willy Chinwosky), one graduate student (Ken Lande), and one associate professor (me).

In 1957, we are back at Nevis with four plastic scintillation counters. Again, one student, one adjunct professor, and your author did the work (FIGURE 3). FIGURE 4 is an early (1961) AGS experiment and we are up to seven authors with two graduate students. By 1967, Brookhaven experiments were getting a bit complicated with spark chambers and hodoscopes, and ten scientists, including three students, were required to do the work (FIGURE 5). By now, the experiment took a year to build and almost a year to run. The AGS lost its energy leadership by 1971 to ISR and the experiments got even more serious. Thus, our 1973 experiment (FIGURE 6) has fifteen authors and six students. I recall this being the first time I became alarmed at the size of the group and the first time I felt more like the conductor of an orchestra than a toiler in the garden of science.

Now, in FIGURE 7, we have an early ISR experiment with 14 authors, which is very modest for European groups. Note only one graduate student (who is from the U.S.). In FIGURE 8, I finally acknowledge that other physics gets done and this represents the 1974 revolution and the ψ of J/ψ fame—35 authors and who knows how many students. Although I was initially outraged at what our field was coming to, I hired Bob Hollebeek from that group and got to know other youths in the Mark-I team, e.g., Roy Schwitters. Of course, they had a marvelous experience and the conclusion was that if the group is well lead and makes a great discovery, the group size isn't necessarily disastrous.

Reprinted from THE PHYSICAL REVIEW, Vol. 105, No. 6, 1925–1927, March 15, 1957
Printed in U. S. A.

Report on Long-Lived K^0 Mesons*

K. LANDE AND L. M. LEDERMAN, *Columbia University, New York, New York*

AND

W. CHINOWSKY, *Brookhaven National Laboratory, Upton, New York*

(Received February 4, 1957)

FIGURE 2. This is from an article on research done at BNL, the Cosmotron, in 1956.

Reprinted from THE PHYSICAL REVIEW, Vol. 105, No. 4, 1415–1417, February 15, 1957
Printed in U. S. A.

Observations of the Failure of Conservation of Parity and Charge Conjugation in Meson Decays: the Magnetic Moment of the Free Muon*

RICHARD L. GARWIN,† LEON M. LEDERMAN, AND MARCEL WEINRICH

Physics Department, Nevis Cyclotron Laboratories, Columbia University, Irvington-on-Hudson, New York, New York

(Received January 15, 1957)

FIGURE 3. This is from an article on research done at the Nevis cyclotron in 1957.

VOLUME 9, NUMBER 1 PHYSICAL REVIEW LETTERS JULY 1, 1962

OBSERVATION OF HIGH-ENERGY NEUTRINO REACTIONS AND THE EXISTENCE OF TWO KINDS OF NEUTRINOS*

G. Danby, J-M. Gaillard, K. Goulianos, L. M. Lederman, N. Mistry, M. Schwartz,† and J. Steinberger†

Columbia University, New York, New York and Brookhaven National Laboratory, Upton, New York
(Received June 15, 1962)

FIGURE 4. This is from an article on an early AGS experiment done in 1961.

Reprinted from THE PHYSICAL REVIEW, Vol. 165, No. 5, 1449–1465, 25 January 1968
Printed in U. S. A.

Muon-Proton Elastic Scattering at High Momentum Transfers*

R. W. ELLSWORTH,† A. C. MELISSINOS, J. H. TINLOT,‡ H. VON BRIESEN, JR.,§ AND T. YAMANOUCHI

Department of Physics and Astronomy, The University of Rochester, Rochester, New York

AND

L. M. LEDERMAN AND M. J. TANNENBAUM||

Nevis Laboratories, Columbia University, New York, New York

AND

R. L. COOL AND A. MASCHKE

Brookhaven National Laboratory, Upton, New York

(Received 25 July 1967)

FIGURE 5. This is from an article on a later AGS experiment (1967).

VOLUME 33, NUMBER 9 PHYSICAL REVIEW LETTERS 26 AUGUST 1974

Muon-Deuteron Deep Inelastic Scattering*

I. J. Kim, A. Entenberg, H. Jöstlein,† K. Königsmann, I. Kostoulas,‡ A. C. Melissinos, E. Aslanides,§ L. Lederman, P. Limon,‖ M. May, P. Rapp, H. Gittleson, T. Kirk,¶ M. Murtagh,** M. J. Tannenbaum,†† J. Sculli,‡‡ T. White,§§ and T. Yamanouchi

Department of Physics and Astronomy, University of Rochester, Rochester, New York 14627, and Department of Physics, Columbia University, New York, New York 10027, and Department of Physics, Harvard University, Cambridge, Massachusetts 02138, and National Accelerator Laboratory, Batavia, Illinois 60510

(Received 28 May 1974)

FIGURE 6. This is from an article on another AGS experiment (1973).

Volume 46B, number 3 PHYSICS LETTERS 1 October 1973

OBSERVATION OF π^0 MESONS WITH LARGE TRANSVERSE MOMENTUM IN HIGH-ENERGY PROTON-PROTON COLLISIONS

F.W. BÜSSER*[1], L. CAMILLERI, L. Di LELLA, G. GLADDING*[2], A. PLACCI, B.G. POPE, A.M. SMITH, J.K. YOH*[3] and E. ZAVATTINI

CERN, Geneva, Switzerland

B.J. BLUMENFELD*[4] and L.M. LEDERMAN*[5]

Columbia University[*6], *N.Y., USA*

and

R.L. COOL*[5], L. LITT and S.L. SEGLER

Rockefeller University[*7], *N.Y., USA*

Received 10 August 1973

FIGURE 7. This is from an article on an early IRS experiment (1973).

VOLUME 33, NUMBER 23 PHYSICAL REVIEW LETTERS 2 DECEMBER 1974

Discovery of a Narrow Resonance in e^+e^- Annihilation*

J.-E. Augustin,† A. M. Boyarski, M. Breidenbach, F. Bulos, J. T. Dakin, G. J. Feldman, G. E. Fischer, D. Fryberger, G. Hanson, B. Jean-Marie,† R. R. Larsen, V. Lüth, H. L. Lynch, D. Lyon, C. C. Morehouse, J. M. Paterson, M. L. Perl, B. Richter, P. Rapidis, R. F. Schwitters, W. M. Tanenbaum, and F. Vannucci‡

Stanford Linear Accelerator Center, Stanford University, Stanford, California 94305

and

G. S. Abrams, D. Briggs, W. Chinowsky, C. E. Friedberg, G. Goldhaber, R. J. Hollebeek, J. A. Kadyk, B. Lulu, F. Pierre,§ G. H. Trilling, J. S. Whitaker, J. Wiss, and J. E. Zipse

Lawrence Berkeley Laboratory and Department of Physics, University of California, Berkeley, California 94720

(Received 13 November 1974)

FIGURE 8. This is from an article on research done at SLAC in 1974.

VOLUME 39, NUMBER 5 PHYSICAL REVIEW LETTERS 1 AUGUST 1977

Observation of a Dimuon Resonance at 9.5 GeV in 400-GeV Proton-Nucleus Collisions

S. W. Herb, D. C. Hom, L. M. Lederman, J. C. Sens,[a] H. D. Snyder, and J. K. Yoh
Columbia University, New York, New York 10027

and

J. A. Appel, B. C. Brown, C. N. Brown, W. R. Innes, K. Ueno, and T. Yamanouchi
Fermi National Accelerator Laboratory, Batavia, Illinois 60510

and

A. S. Ito, H. Jöstlein, D. M. Kaplan, and R. D. Kephart
State University of New York at Stony Brook, Stony Brook, New York 11974
(Received 1 July 1977)

FIGURE 9. This is from an article on research done at Fermilab (FNAL) in 1977.

Volume 83B, number 2 PHYSICS LETTERS 7 May 1979

PROPERTIES OF HADRON FINAL STATES IN e^+e^- ANNIHILATION AT 13 GeV AND 17 GeV CENTER OF MASS ENERGIES

TASSO-Collaboration

R. BRANDELIK, W. BRAUNSCHWEIG, K. GATHER, B. JAAX [1], V. KADANSKY, K. LÜBELSMEYER, H.-U. MARTYN, G. PEISE, J. RIMKUS, H.G. SANDER, D. SCHMITZ, A. SCHULTZ VON DRATZIG, D. TRINES, W. WALLRAFF
I. Physikalisches Institut der RWTH Aachen, Germany [2]

H. BOERNER, H.M. FISCHER, H. HARTMANN, E. HILGER, W. HILLEN, G. KNOP, W. KORBACH, B. LÖHR, F. ROTH, W. RÜHMER, R. WEDEMEYER, N. WERMES, M. WOLLSTADT,
Physikalisches Institut der Universität Bonn, Germany [2]

R. BÜHRING, D. HEYLAND, H. HULTSCHIG, P. JOOS, W. KOCH, U. KÖTZ, H. KOWALSKI, A. LADAGE, D. LÜKE, H.L. LYNCH, G. MIKENBERG [3], D. NOTZ, J. PYRLIK, R. RIETHMÜLLER, M. SCHLIWA, P. SÖDING, B.H. WIIK, G. WOLF
Deutsches-Elektronen-Synchrotron DESY, Hamburg, Germany

R. FOHRMANN, G. POELZ, J. RINGEL, O. RÖMER, R. RÜSCH, P. SCHMÜSER
II. Institut für Experimentalphysik der Universität Hamburg, Germany [2]

D.M. BINNIE, P.J. DORNAN, N.A. DOWNIE, D.A. GARBUTT, W.G. JONES, S.L. LLOYD, D. PANDOULAS, C. YOUNGMAN
Department of Physics, Imperial College, London, England [4]

R.J. BARLOW, R.J. CASHMORE, J. ILLINGWORTH, M. OGG, G.L. SALMON
Department of Nuclear Physics, Oxford University, England [4]

K.W. BELL, W. CHINOWSKY [5], B. FOSTER, J.C. HART, J. PROUDFOOT, D.R. QUARRIE, D.H. SAXON, P.L. WOODWORTH
Rutherford Laboratory, Chilton, England [4]

Y. EISENBERG, U. KARSHON, E. KOGAN, D. REVEL, E. RONAT, A. SHAPIRA,
Weizmann Institute, Rehovot, Israel [6]

and

J. FREEMAN, P. LECOMTE, T. MEYER, SAU LAN WU and G. ZOBERNIG
Department of Physics, University of Wisconsin, Madison, WI, USA [7]

Received 12 March 1979

FIGURE 10. This is from an article on a PETRA experiment in 1978.

Volume 122B, number 1 PHYSICS LETTERS 24 February 1983

EXPERIMENTAL OBSERVATION OF ISOLATED LARGE TRANSVERSE ENERGY ELECTRONS WITH ASSOCIATED MISSING ENERGY AT $\sqrt{s}$ = 540 GeV

UA1 Collaboration, CERN, Geneva, Switzerland

G. ARNISON [j], A. ASTBURY [j], B. AUBERT [b], C. BACCI [i], G. BAUER [l], A. BÉZAGUET [d], R. BÖCK [d], T.J.V. BOWCOCK [f], M. CALVETTI [d], T. CARROLL [d], P. CATZ [b], P. CENNINI [d], S. CENTRO [d], F. CERADINI [d], S. CITTOLIN [d], D. CLINE [l], C. COCHET [k], J. COLAS [b], M. CORDEN [c], D. DALLMAN [d], M. DeBEER [k], M. DELLA NEGRA [b], M. DEMOULIN [d], D. DENEGRI [k], A. Di CIACCIO [i], D. DiBITONTO [d], L. DOBRZYNSKI [g], J.D. DOWELL [c], M. EDWARDS [c], K. EGGERT [a], E. EISENHANDLER [f], N. ELLIS [d], P. ERHARD [a], H. FAISSNER [a], G. FONTAINE [g], R. FREY [h], R. FRÜHWIRTH [l], J. GARVEY [c], S. GEER [g], C. GHESQUIÈRE [g], P. GHEZ [b], K.L. GIBONI [a], W.R. GIBSON [f], Y. GIRAUD-HÉRAUD [g], A. GIVERNAUD [k], A. GONIDEC [b], G. GRAYER [j], P. GUTIERREZ [h], T. HANSL-KOZANECKA [a], W.J. HAYNES [j], L.O. HERTZBERGER [2], C. HODGES [h], D. HOFFMANN [a], H. HOFFMANN [d], D.J. HOLTHUIZEN [2], R.J. HOMER [c], A. HONMA [l], W. JANK [d], G. JORAT [d], P.I.P. KALMUS [f], V. KARIMÄKI [e], R. KEELER [f], I. KENYON [c], A. KERNAN [h], R. KINNUNEN [e], H. KOWALSKI [d], W. KOZANECKI [h], D. KRYN [d], F. LACAVA [d], J.-P. LAUGIER [k], J.-P. LEES [b], H. LEHMANN [a], K. LEUCHS [a], A. LÉVÊQUE [k], D. LINGLIN [b], E. LOCCI [k], M. LORET [k], J.-J. MALOSSE [k], T. MARKIEWICZ [d], G. MAURIN [d], T. McMAHON [c], J.-P. MENDIBURU [g], M.-N. MINARD [b], M. MORICCA [i], H. MUIRHEAD [d], F. MULLER [d], A.K. NANDI [j], L. NAUMANN [d], A. NORTON [d], A. ORKIN-LECOURTOIS [g], L. PAOLUZI [i], G. PETRUCCI [d], G. PIANO MORTARI [i], M. PIMIÄ [e], A. PLACCI [d], E. RADERMACHER [a], J. RANSDELL [h], H. REITHLER [a], J.-P. REVOL [d], J. RICH [k], M. RIJSSENBEEK [d], C. ROBERTS [j], J. ROHLF [d], P. ROSSI [d], C. RUBBIA [d], B. SADOULET [d], G. SAJOT [g], G. SALVI [f], G. SALVINI [i], J. SASS [k], J. SAUDRAIX [k], A. SAVOY-NAVARRO [k], D. SCHINZEL [f], W. SCOTT [j], T.P. SHAH [j], M. SPIRO [k], J. STRAUSS [l], K. SUMOROK [c], F. SZONCSO [l], D. SMITH [h], C. TAO [d], G. THOMPSON [f], J. TIMMER [d], E. TSCHESLOG [a], J. TUOMINIEMI [e], S. Van der MEER [d], J.-P. VIALLE [d], J. VRANA [g], V. VUILLEMIN [d], H.D. WAHL [l], P. WATKINS [c], J. WILSON [c], Y.G. XIE [d], M. YVERT [b] and E. ZURFLUH [d]

Aachen [a]–Annecy (LAPP) [b]–Birmingham [c]–CERN [d]–Helsinki [e]–Queen Mary College, London [f]–Paris (Coll. de France) [g]–Riverside [h]–Rome [i]–Rutherford Appleton Lab. [j]–Saclay (CEN) [k]–Vienna [l] Collaboration

Received 23 January 1983

FIGURE 11. This is from an article on research done at the CERN S$\bar{p}$pS collider in 1983.

Fermilab in 1977 (FIGURE 9) had 16 collaborators and a pretty standard collection of hardware. One of the interesting things about 1977 was that 12 or so of the 16 collaborators were full-time and thus the experiment got lots of close attention. A disease of our times is the fragmented time that people put into experiments, but let us continue. FIGURE 10 is a PETRA experiment, with 83 names. Europe has always led in the size of groups. This was a research that contributed to the discovery of gluons. Now, in 1983, we have the historic *W* discovery of UA1 with 134 authors (FIGURE 11), only two of whom were awarded the Nobel Prize (well-deserved though). My final figure here (FIGURE 12) is from a CDF report and this (still growing?) research has 209 names. We also know that some LEP detectors have more than 300 people.

Therefore, we see a clear evolution to ever larger groups, especially in the collider domain. This trend is, of course, not generated by insecurity or the need for togetherness. It is essential for putting together apparatus of unprecedented complexity. The fixed target experiments are growing too, but at a much slower rate. However, there are discussions at Fermilab of a very major fixed target facility to "really do" heavy quark physics.

For the convenience of scholars of history, I prepared TABLE 1, which compactly summarizes the history I have sampled. I recognize, somewhat arbitrarily, four periods that have strong overlaps, but which are distinguished by energy, machines, techniques, and physics trends. The point, of course, is that as we moved to the more subtle, higher energy physics goals, the experiments became complicated and the group structure grew.

Another way to look at the trends is to look at TABLE 2, which is due to Panofsky and which lists the number of frontier facilities over the past 35 years. The implications are pretty obvious.

Now, I'd like to project these trends into the future. The next figure (FIGURE 13) is from *Snowmass '84* and indicates the data handling stages for an SSC detector. It gives the cost estimates for the three types of 4π detection considered there. A lot of

EXPERIMENTERS ON COLLABORATION E-741

ROBERT E. DIEBOLD, NORMAN P. HILL, EDWARD W. MAY
LAWRENCE J. MODULMAN, NICKOLAS SOLOMEY, DAVID G. UNDERWOOD
ROBERT G. WAGNER, ARTHUR B. WICKLUND
ARGONNE NATIONAL LABORATORY

JAMES R. BENSINGER, MILCIADES CONTRERAS, LUC M. DEMORTIER
STEVE GALLO, LAWRENCE E. KIRSCH, FRANK LOMANNO
BRUCE MAGANSON, RICHARD POSTER, LEE J. SPENCER
SWLOMIT TAREM
BRANDEIS UNIVERSITY

DAN AMIDEI, MYRON CAMPBELL, HENRY J. FRISCH
CARLA GROSSO-PILCHER, TONY M. LISS, GEORGE REDLINGER
MELVYN J. SHOCHET, FREDERICK D. SNIDER, SUNIL SOMALNAR
YEONG DONG TSAI
UNIVERSITY OF CHICAGO

MUZAFFER ATAC, J. PETER BERGE, MORRIS E. BINKLEY
ALFRED E. BRENNER, JOHN TERRY CARROLL, THOMAS L. COLLINS
CHRISTOPHER T. DAY, THOMAS P. DROEGE, JOHN E. ELIAS
JAMES E. FREEMAN, IRWIN GAINES, DAVID R. HANSSEN
HANS B. JENSEN, DAVID E. JOHNSON, RICHARD W. KADEL
HANS KAUTZKY, ROBERT D. KEPHART, C. NEUMANN-HOLMES
TOKIO K. OHSKA, JAMES P. PATRICK, R.A. PERCHONOK
DAVID ROBERT QUARRIE, M. SHIBATA, DENNIS THERIOT
ALVIN V. TOLLESTRUP, KATHLEEN J. TURNER, ROBERT L. WAGNER
RYUJI YAMADA, JOHN YOE
FERMI NATIONAL ACCELERATOR LABORATORY

MINATO KAWAGUCHI
FURUI UNIVERSITY (JAPAN)

CRAIG A. BLOCKER, GEORGE W. BRANDENBURG, DAVID W. BROWN
ROBERT M. CAREY, MARK W. EATON, EDWARD KEARNS
THOMAS J. PHILLIPS, THEODOR P. SCHAAD, ROY F. SCHWITTERS
RICHARD ST. DENIS, WILLIAM A. WORSTELL
HARVARD UNIVERSITY

DAVID DERRER, LOUISA EMMONS, MICHAEL SIVERTZ
HAVERFORD COLLEGE

YASUSHI MURAKI
ICRR, TOKYO UNIVERSITY (JAPAN)

GIULIO ASCOLI, SELCUK CIHANGIR, ROBERT W. DOWNING
STEVEN M. ERREDE, ELAND E. HOLLOWAY, HOHANNES KEUTELIAN
ULRICH E. KRUSE, ROBERT D. SARD, DAVID A. SMITH
UNIVERSITY OF ILLINOIS

S. BARTALOCCI, SERGIO BERTOLUCCI, MARIO BIAGBOLI
MARCO CORDELLI, BRUNO DULACH, GIANNI FONTANA
PAOLO GIROMINI, ALESSANDRO MENGUCCI, SILVIA MIOSSI
MASSIMO PALLOTTA, ANDREA SANSONI, MASSINO SANTONI
GIANCARLO SENSOLINI
INFN, FRASCATI (ITALY)

FRANCO BEDESCHI, STEFANO BELFORTE, GIORGIO BELLETTINI
LUCA BERRETTA, RUGGERO BERTANI, FILIPPO BOSI
L. BOSISIO, EZIO CARBONI, RINO CASTALDI
FRANCO CERVELLI, MARIA CURATOLO, MARIO DELCOLLETTO
ANDREA DELCOLLETTO, RAFFAELE DELFABBRO, ANDREA DISACCO
ALESSANDRO DIVIRGLIO, BELLISARIO ESPOSITO, GIUSEPPE FAVATI
ETTORE FOCARDI, FABRIZIO GAGLIARDI, STEFANO GALBOTTI
GIANNI GENNARO, PAOLA GIANNETTI, MARCELLO A. GIORGI
MARCELLO GIVOLETTI, MARIO IACOPONI, ALDO MENZIONE
GIANFRANCO PAGANI, DIEGO PASSUELLO, DOMENICO PISTONI
FAUSTO PROFETI, LUCIANO RISTORI, ROBERTO RUBERTI
ANGELO SCRIBANO, ARNALDO STEPANINI, ALDO TAZZIOLI
GUIDO TOWELLI, RAFFAZLE TRIPICCIONE
INFN, UNIVERISTY OF PISA (ITALY)YASUO ARAI, Y. FUKUI, SHYOJI MIKAMO
MASANORI MISHIMA, KAZUO OGAMA
KEK—NAT'L LAB FOR HIGH ENERGY PHYSICS (JAPAN)

ROBERT E. BERINGER, WILLIAM C. CARITHERS, WILLIAM CHINOWSKY
MELISSA FRANKLIN, MICHAEL GOLD, BRADLY HUBBARD
JAMES SIEGRIST
LAWRENCE BERKELEY LABORATORY

MARK R. ALLEN, DANIEL A. BAUER, MATTHEW A. BERSHADY
PATRICIA CAVENEE, DAVID F. CONNOR, JOHN W. COOPER
MARSHALL MILLER, M. NEUCOMER, RICHARD VAN ERG
B. WILLIAMS
PENNSYLVANIA, UNIVERSITY OF

VIRGIL E. BARNES, WILLIAM BURLEY, KEITH CHADWICK
JAMES P. CRAIG, ANGELA DIVIRGILIO, ARTHUR F. GARFINKEL
STEPHEN KUHLMANN, ALVIN T. LAASANEN
PURDUE UNIVERSITY

THOMAS J. CHAPIN, KONSTANTIN GOULIANOS, SEBASTIAN N. WHITE
ROCKEFELLER UNIVERSITY

ANDREW F. BERETVAS, THOMAS J. DEVLIN, LELAND N. SCOTT
TUNGSHENG YANG
RUTGERS UNIVERSITY

SHIGEHARA KOBAYASHI, A. MURAKAMI
SAGA UNIVERSITY (JAPAN)

PETER M. MCINTYRE, THOMAS MEYER, MICHAEL J. SHEPKO
JOHN M. THANE, ROBERT C. WEBB
TEXAS A & M UNIVERSITY

YUZO ASANO, YOSHIO MAYOSHIDE, MASAHITO ITO
TERUKI KAMON, YUKIKO KIKUCHI, S. KIM
RUMITAKA KONDO, MIKA MASUZAWA, SHIGE MIYASHITA
MITOSHI MIYATA, SHIGEKI MORI, FUMIMIDE SATO
MASAKI SEKIGUCHI, YOSHIMOBU TAKAIWA, KOJI TAKIKAWA
AKIHIRO YAMASHITA, K. YASUOKA
TSUKUBA, UNIVERSITY OF, IBARAKI (JAPAN)

JOHN H. ARCHIBALD, DAVID B. CLINE, JEFFREY E. ELLSWORTH
MIKE FRANKOWSKI, ROBERT M. HANDLER, RICHARD J. LOVELESS
KAREN MATTISON, LEE G. PONDROM, DONALD D. REEDER
JEAN MILWE RHOADES, MARLEIGH C. SHEAFF, RONALD E. SMITH
THOMAS R. WINCH
UNIVERSITY OF WISCONSIN

FIGURE 12. This is from a CDF report (Fermilab colliding detector, 1986).

TABLE 1. History of HEP

Period	Accelerators	Detectors	Particles	Forces E.C.
1950–1963 I	Synchrocyclotrons (p) Betatrons (e) Synchrotrons (e,p) Cosmotron Bevatron 100MeV–6GeV	Cloud Chambers Bubble Chambers Spark Chambers Scint. Counters Emulsions	(1) e, μ, ν_e, ν_μ (2) $\pi^\pm, K^\pm, \bar{K}^0, R^0, \Lambda, \Sigma, \Xi$, (3) $\Delta, \eta, \sigma, \phi, \omega, \cdots$ (4) γ, W?	S) Isospin, Strangeness, S-Matrix, Regge W) V-A, 2ν's, CVC, P, C, ν-Scattering, W Search EM) Precision: Gauge Theory
1960–1973 II	+AGS, PS (p) SLAC(e) 20–30 GeV	PWC Drift Chambers HBC Pb-Glass $100/Channel	(1) u, d, s, Quarks (2) e, μ, ν_e, ν_μ, Leptons (3) γ, W?	S) Baryon Spectroscopy, Quark Dynamics via D.I.S., Dileptons W) CP Violation, ν D.I.S., Electroweak & Higgs Symmetry Breaking
1970–1982 III	Fermilab SpS Spear, Doris CESR, (e^+e^-) ISR (pp) 200–2000 GeV	Calorimetry Computerization $20/Channel	(1) $u\ c\ t\ ?\ d\ s\ b$ (2) $e\ \mu\ \tau\ \nu_e\ \nu_\mu\ \nu_\tau$? (3) γ, Gluons W? Z?	S) Large p_T Physics, Gauge Theory Revival, QCD W) Weak Natural Current EM) Electroweak + W, Z Masses
1980–1992 IV	Tevatron (p) Sp$\bar{p}$S (p$\bar{p}$) PEP, PETRA ($e^+ e^-$) (LEP ($e^+ e^-$) SLC ($e^+ e^-$) HERA) (e p) 400 GeV–20 TeV	Hermetic Calorimetry Holography Vertex Detectors $1–10/Channel	(1) $u\ c\ t\ d\ s\ b$ (2) $e\ \mu\ \tau\ \nu_e\ \nu_\mu\ \nu_\tau$ (3) $\gamma\ W^\pm\ Z^0$ Gluons	S) QCD, W-EM) } → GUTS, Particle/Cosmology Supersymmetry, Technicolor, Prequarks, Higgs Sector
>1982	SSC "800,000TeV"	?	?	?

TABLE 2. Particle Physics Accelerator Facilities at the Energy Frontier[a]

1948	1965	1980	2000
Princeton	Argonne	Brookhaven	
Berkeley	Berkeley	Cornell	
Cal Tech	Brookhaven	Fermilab	
Carnegie-Mellon	Cal Tech	SLAC	
Carnegie Inst.	CEA		
Chicago	Cornell		
Columbia	PPA		
Cornell	Stanford		
Harvard			
Illinois			
Indiana			
Maryland			
Michigan			
MIT			
Ohio State			
Penn			
Purdue			
Rochester			
Stanford			
Washington U.			
U of Washington			

[a]Universities have, until recently, been the training grounds for accelerator builders and facility directors. Accelerator technology has now become much more sophisticated and individuals have become specialized as accelerator physicists and particle physicists. It is difficult to generate accelerator leaders and experts as the number of accelerators decreases and they become divorced from the universities.

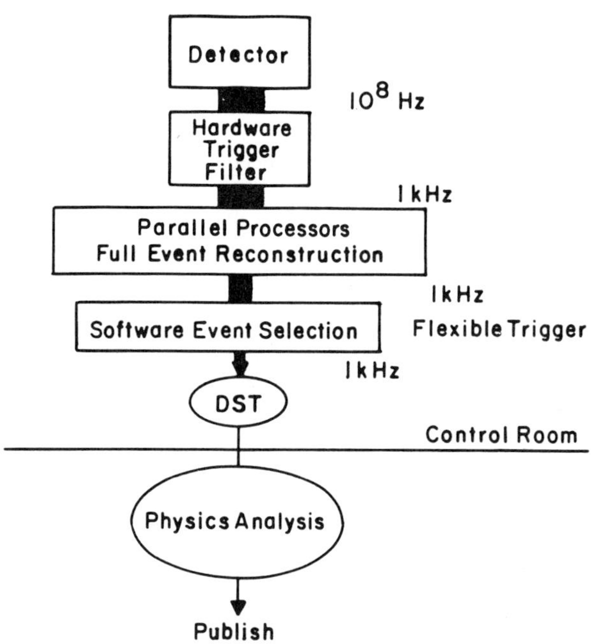

FIGURE 13. This indicates the data-handling stages of a SSC detector. This effort, as now visualized, is the scale of building Fermilab or SPS. It will take about five to six years, 500 physicists, and a new sociology. The R & D start must be soon, so as to make it cheaper and better.

very good work went into these designs, but like the SSC itself, they are extrapolations of presently understood detectors with no startling new ideas. The results, though, are staggering. If we scale the manpower effort from CDF, UA1, etc., we'll need groups of 400–500 physicists to build each of these detectors. Here, we have already made use of modularity, etc. (e.g., CDF has 200 people for a $60M detector). SSC will need, say, 450 people for a $240M detector. Small wonder that the Snowmass workers concluded that "only two" such detectors are required. Now is this inevitable? Can we make better use of industry so that only 100 physicists can manage? How does anyone marshall 450 high energy physicists into a disciplined team that designs, assembles, operates, and analyzes data from the behemoth? In the decade of the 1980s, we find ourselves in a transition from groups of fifty or so to groups of many hundreds. Here are some of the problems:

(1) How do we give the participants some of the joy and satisfaction that must have been obvious to you from my charts of the 1950s and 1960s? (Not every group will be headed by Burt Richter and discover the ψ.)
(2) What mechanisms can be devised to give proper credit and recognition to the young people in the group? (The standard "joke" is for the speaker to flash a list of authors. I once wisecracked that some author lists are made up by a Monte Carlo program.)
(3) How can we keep universities involved? Will the professor of mathematics vote for the promotion of Assistant Professor Smith, only one of the famous team of 477 scientists?
(4) How do we preserve options for these colleagues who will simply not work in large groups?
(5) Finally, how do we accelerate progress in instrumentation so that the projection we have been making turns out to be all wrong!? (In 1996, Professor Walter Mitty discovers the Higgs holding a special test tube near the interaction region.)

TABLE 3 is a matrix in which the vertical columns illustrate the subdivisions of a typical detector (not the special test tube). Presumably, each column is an assignment to one of the institutions in the mammoth collaboration. The horizontal rows are intended to illustrate the physics topics that will come out of a sample of data. The physics topics may often make more use of one of the columns and this physics subject could well be assigned to the group that made, say, the muon counters. In any case, physics analysis jobs are also parceled out to smaller groups. Perhaps even the publication based upon the analysis could be signed by the people who worked on the analysis, with the addendum, ". . . and the UPG collaboration." Since there are many different pieces of physics out of a collider facility, everyone should be highlight-authored eventually. One also has to worry about the people who are so busy keeping the apparatus working that they do no analyses, but in general this is a scheme for more meaningful publication credit. Of course, I personally will work on the test tube.

My final remarks have to do with the options for smaller efforts. Here, we must pay attention to fixed target research, as long as this is scientifically justifiable. TeV II, by 1993, will be a mature program and the AGS will be vulnerable, but if good physics is being produced, it will be important to keep these going with whatever upgrades are needed. Also, it is important to point out that if we do things right, the SSC should

TABLE 3. Collider Effort/Physics Matrix[a]

	Vertex Detector	Central Tracker	Magnet Coil	Muon Ident.	Electron Ident.	Calorimeter Inner	Calorimeter Outer	Electronics Readout	Computer On-line
Jets	0	2	3	0	0	2	3	1	0
Dijets									
W's									
Z's	1	2	2	2	2	3	4	3	2
Charm									
Bottom									
Muon Pairs									
Electron Pairs									
New Phenom. w/Jets									
Multipl. Distri.									
Higgs Things									

[a]See text for explanation.

provide many opportunities for smaller ventures. Remember that S$\bar{p}$pS and TeV I were afterthoughts imposed on an existing accelerator. As an illustration, I reproduce my contribution to *Snowmass '84* in FIGURE 14. Here, in one interaction region, three or four smaller experiments—fixed target types—can be accommodated.

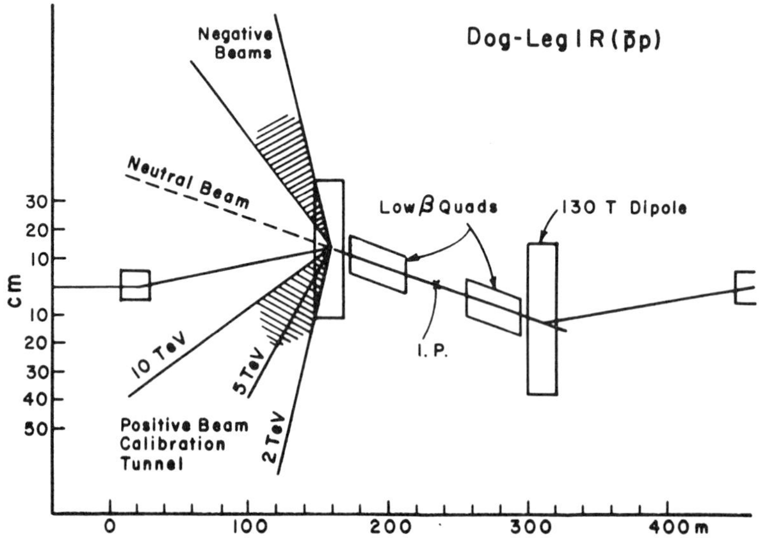

FIGURE 14. A five function interaction region: (1) A 4π calorimeter surrounds the interaction point; (2) Calibration beams are available and useful in physics; (3) Neutral beams have good, ≥ 1 TeV, neutrino flux; (4) Neutral beams may be triggered by the 4π detector; (5) No background for neighbors.

In conclusion, I have put down in this lecture some of the concerns that many of us must be feeling, if only subliminally. My attempts at resolving the problems we face are pathetic. However, the upbeat finale is that the problems are part of the process of a tremendous thrust into the unknown, an adventure that will require all of our imagination and all of our energies.

High p_T Jets from $\bar{p}p$ Collisions at $\sqrt{s} = 540$ GeV in UA2

Presented by JOHN RENNER HANSEN

UA2 Collaboration[a]
CERN CH-1211
Geneva 23, Switzerland

INTRODUCTION

The recent unambiguous identification of jets in hadronic collisions[1–4] at the CERN $\bar{p}p$ Collider and at the ISR has revived the interest in this field after the first observation of large transverse momentum (p_T) processes in early ISR experiments.[5] The prediction that the hard scattering of hadron constituents should result in the production of two hadronic jets having the same large transverse momenta as the scattered partons[6,7] has been verified in a direct way at the CERN $\bar{p}p$ Collider with the observation of a dominant two-jet structure in events depositing very large transverse energy into the central calorimeters of the UA1[3] or UA2[1] experiments. In fact, one of the most outstanding benefits of the very successful operation of the CERN $\bar{p}p$ Collider[8] at a center-of-mass energy, $\sqrt{s} = 540$ GeV, is the possibility for detailed measurements of hadronic jet production and their fragmentation properties[9–11] in an energy domain where hard processes can be separated clearly from the soft hadronic interactions.[12]

APPARATUS

The UA2 detector (FIGURE 1) has been described in detail elsewhere.[13,14] At the center of the UA2 apparatus, a vertex detector consisting of both cylindric proportional chambers and drift chambers measures particle trajectories in a region without magnetic field. The vertex detector is surrounded by a highly segmented electromagnetic and hadronic calorimeter (the central calorimeter), which covers polar angles, $40° < \theta < 140°$ $(-1 < \eta < 1)$. The forward regions ($20° < \theta < 37.5°$ and $142.5° < \theta < 160°$) are each instrumented with twelve toroidal magnet sectors that are followed by drift chambers, multitube proportional chambers, and electromagnetic calorimeters.

The present results have been obtained mainly with the central calorimeter. It is segmented into 240 cells, each covering 15° in ϕ and 10° in θ, and is built in a tower structure pointing to the center of the interaction region. The cells are segmented longitudinally into a 17 radiation length thick electromagnetic compartment (lead-scintillator) followed by two hadronic compartments (iron-scintillator) of two absorp-

[a]This UA2 Collaboration includes Bern, CERN, Copenhagen (NBI), Orsay (LAL), Pavia, and Saclay (CEN).

tion lengths each. Following an initial energy calibration of all cells in a 10 GeV electron, muon, and pion beam from the CERN PS, the calorimeter response has been tracked with a light flasher system, with measurements of the direct current induced by a ^{60}Co radioactive source, and with measurements of the average energy flow into each cell for unbiased $\bar{p}p$ collisions. The systematic uncertainty in the energy calibration for the data discussed here is less than $\pm 1.5\%$ for the electromagnetic calorimeter and less than $\pm 3.5\%$ for the hadronic one.

The response of the calorimeter to electrons, single hadrons, and multihadrons (produced in a target in front of the calorimeter) has been measured at the CERN PS and SPS machines using beams from 1 to 70 GeV. The energy resolution for electrons is measured to be $\sigma_E/E = 0.14/\sqrt{E}$ (E in GeV), whereas for hadrons, it varies from 32% at 1 GeV to 11% at 70 GeV, which is approximately proportional to $E^{-1/4}$.

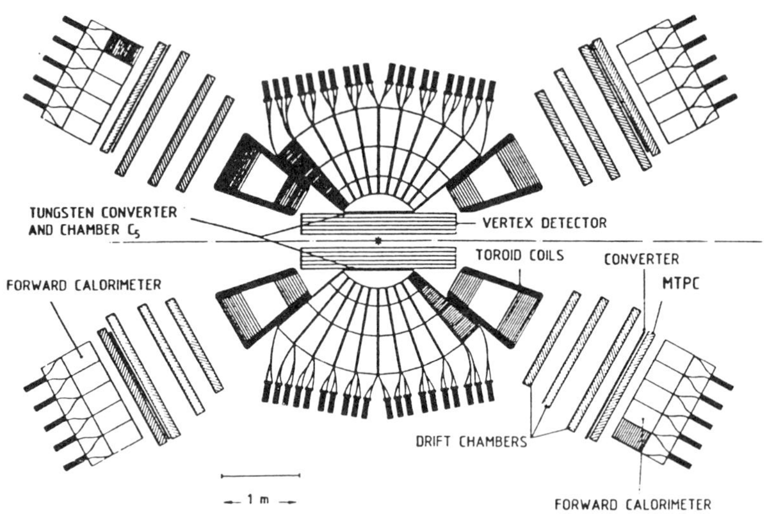

FIGURE 1. Schematic detector assembly showing UA2 in a longitudinal cut parallel to the beams.

Details of the construction, calibration, and performance of the calorimeter are reported in reference 14.

DATA TAKING AND REDUCTION

The data presented in this paper were recorded during the 1983 CERN $\bar{p}p$ Collider run that used trigger selecting $\bar{p}p$ collisions that deposited a large amount of total transverse energy (ΣE_T) into the central calorimeter. In addition, the gains of the photomultipliers were adjusted such that their response was proportional to transverse energy and their signals were linearly added. The sum was required to exceed a threshold, which was normally at about 40 GeV except for special low threshold data runs.

The background from sources other than $\bar{p}p$ collisions was suppressed at the trigger level. This was done by requiring a coincidence of the ΣE_T condition with two signals ("minimum bias" trigger) obtained from scintillator arrays covering an angular range of $0.44° < \theta < 2.84°$ on both sides of the collision region.[15] The loss of genuinely large ΣE_T $\bar{p}p$ events due to this requirement was measured in special runs where this condition was removed. For $\Sigma E_T > 40$ GeV, the loss is consistent with zero and <5% at a 90% confidence level, independent of ΣE_T. Background from cosmic rays was also found to be negligible.

A sample of "minimum bias" events was recorded simultaneously with the ΣE_T data to provide a measurement of the luminosity. Data with $\Sigma E_T > 40$ GeV were recorded for a total integrated luminosity ($\int \mathcal{L}dt = 112\ \text{nb}^{-1}$). An uncertainty of ±20% was estimated for $\int \mathcal{L}dt$ from the observed fluctuations during different running conditions and from the overall uncertainty in the cross section accepted by the "minimum bias" trigger.[15]

The full data sample was used in the following analysis for $\Sigma E_T > 60$ GeV; smaller samples and special runs were used for lower thresholds.[16]

The events collected with the ΣE_T trigger included a small (<10%) background contamination. Beam halo particles could, therefore, either satisfy the triggering condition directly or appear as an accidental overlap with a "minimum bias" $\bar{p}p$ interaction. Furthermore, the background events also exhibited a characteristic pattern in the detector different from that of large ΣE_T $\bar{p}p$ events. They were rejected from the data sample if any one of the following conditions was satisfied:

(i) if they were associated with an early signal in the small angle scintillator arrays;
(ii) if they had an abnormally large total transverse energy fraction in the hadronic compartments (more than 60% in the second one or more than 95% in both together);
(iii) if they contained only one cluster (as defined below) with more than 90% of its energy in the hadronic compartments.

These requirements reduced the background contamination in the data sample to <5%, independent of ΣE_T. The loss of good events introduced by the cuts was negligible. This has been verified by applying the time-of-flight requirement on "minimum bias" events and by using test-beam data for the energy deposition criteria.

TRANSVERSE ENERGY DISTRIBUTION

The response of the electromagnetic calorimeter compartments to the energy deposited by photons (or electrons) differs from that of hadrons by typically 20%, depending on the energy. We have adopted a constant set of weighting factors (1.18, 1.00, 1.06) for the three compartment energies (electromagnetic, and the first and second hadronic) that optimizes the resolution and linearity of the calorimeter response to high energy hadrons, which is described in detail in reference 14. The total hadronic energy in a cell of the calorimeter is measured as the weighted sum of the energies in the three compartments. In addition, each compartment contributing to the sum must exceed 150 MeV, which is well above pedestal fluctuations.

The observed ΣE_T distribution, normalized using the integrated luminosity, is presented in FIGURE 2 for two pseudorapidity intervals ($-1 < \eta < 1$ and $-0.54 < \eta < 0.54$). The transverse energies are summed over all cells in the full azimuthal acceptance, $\Delta\phi = 360°$, under the approximation that the event vertex is located at the center of the detector. The event vertex has been reconstructed for a sample of the events used for the cross-sectional measurement. Its distribution along the $\bar{p}p$ beam

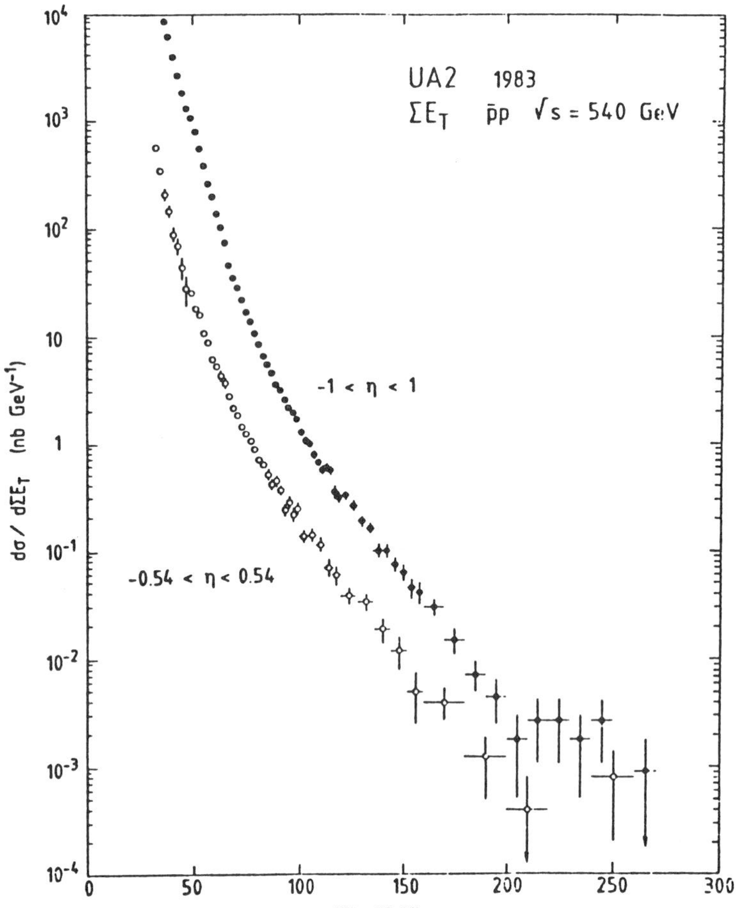

FIGURE 2. Observed total transverse energy, ΣE_T, distributions for two pseudorapidity (η) intervals.

axis is well centered in the detector and has a r.m.s. spread of 10 cm. No acceptance correction has been applied to the data shown in FIGURE 2 since we estimate that the uncertainty in the energy scale of ΣE_T due to systematic effects (150 MeV minimum compartment energy requirement, calibration errors, neglecting event vertex position, etc.) is less than $\pm 10\%$.

The ΣE_T distributions of FIGURE 2 also show a clear departure from exponential when ΣE_T exceeds ~60 GeV. This corresponds to the transverse energy where the two-jet production cross section begins to dominate over the soft hadronic interactions, as explained below.

TWO-JET DOMINANCE

In order to investigate the pattern of energy distribution in the events, a straightforward clustering algorithm has been adopted that takes advantage of the fine granularity of the calorimeter segmentation. All cells that share a common side and have a cell energy (E_{cell}) > 400 MeV are joined into a cluster. Clusters having two or more local maxima separated by a valley deeper than 5 GeV are split. In each event, the clusters are ranked in order of decreasing transverse energies, denoted by $E_T^1 > E_T^2 > E_T^3 \ldots$ The clusters contain typically three cells for $E_T = 2$ GeV and ten cells for $E_T = 40$ GeV.

The fractions, $h_1 = E_T^1/\Sigma E_T$ and $h_2 = (E_T^1 + E_T^2)/\Sigma E_T$, describe the transverse energy accounted for in each event by the cluster (respectively, the two clusters) with the largest transverse energy. An event containing only two jets of equal transverse energy would have $h_1 = 0.5$ and $h_2 = 1$ in an ideal detector. The mean values of h_1 and h_2 are shown in FIGURE 3a as a function of ΣE_T. Their behavior illustrates the emergence of a dominant two-jet structure at large ΣE_T, as previously reported.[1,9] This effect becomes even more pronounced at the largest values of ΣE_T reached in this experiment. The dependence on ΣE_T of the mean values of the ratios, $r_{21} = E_T^2/E_T^1$ and $r_{32} = E_T^3/E_T^2$, are displayed in FIGURE 3b. They show the same effect: as ΣE_T increases, r_{21} approaches about 0.9, while r_{32} decreases below 0.1. Finally, a very large fraction of the total transverse energy of large ΣE_T events is shared on average by two clusters only.

CHARGE PARTICLE MULTIPLICITY IN JETS

The charged particle multiplicity is measured by the vertex detector (see FIGURE 1) whose angular coverage is much larger than that of the central calorimeter: it extends down to $\Theta = 20°$ and over the full azimuthal range. As a result, the charged particle multiplicity measurement is almost free of acceptance corrections. The main uncertainties in this measurement arise from two sources:

(i) the efficiency of the vertex detector and of the associated pattern recognition algorithm, especially in regions of high particle density;

(ii) the procedure that defines which of the observed tracks are to be associated with the jet and which are not.

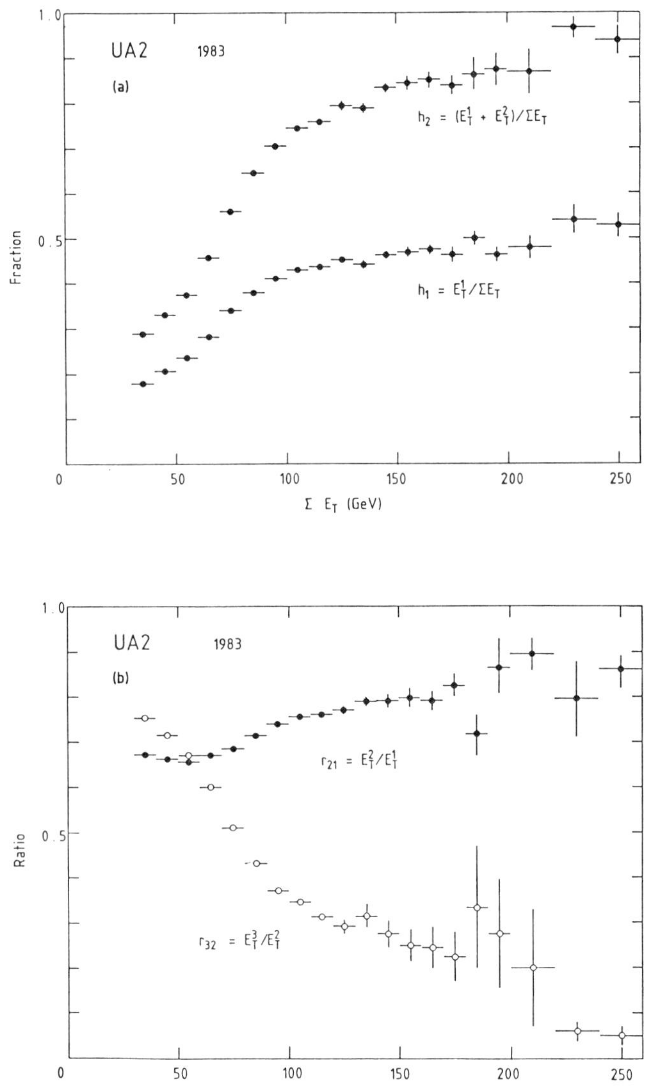

FIGURE 3. (a) The fractions, h_1 and h_2, of the total transverse energy, ΣE_T, contained in the cluster, and the two clusters, respectively, having the largest ΣE_T are displayed versus ΣE_T. (b) Dependence on ΣE_T of the ratios, $r_{21} = E_T^2/E_T^1$ and $r_{32} = E_T^3/E_T^2$ (see text).

In order to address these problems, our analysis is restricted to the transverse plane, where the track reconstruction efficiency is highest and where the particles not associated with the jets are expected to contribute a uniform azimuthal distribution.

Events are considered in which two clusters with an invariant mass in excess of 40 GeV/c^2 and an azimuthal separation greater than 150° are observed with $|\eta| < 0.7$.

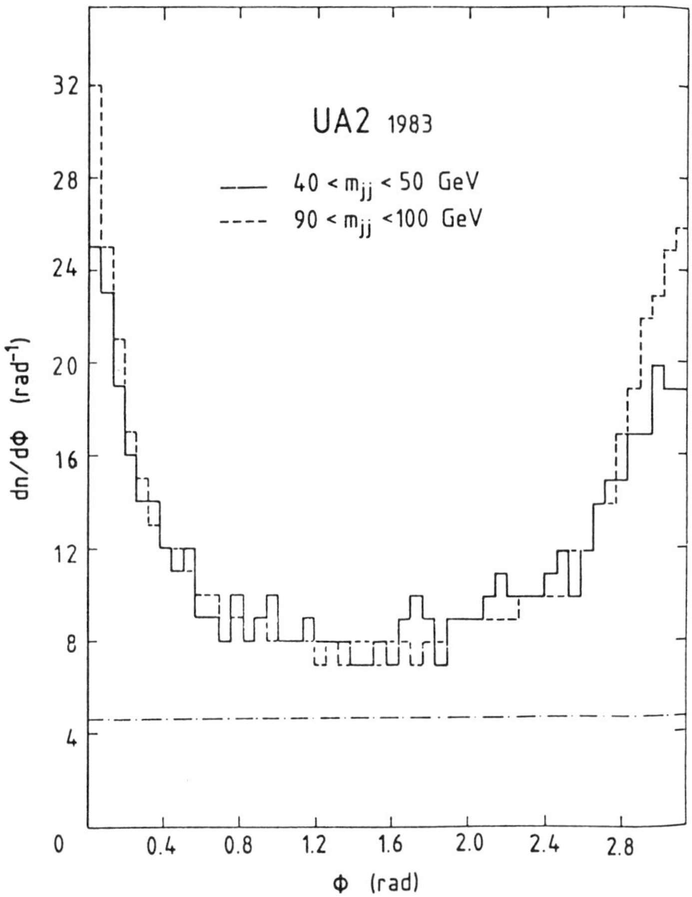

FIGURE 4. Azimuthal separation, $\Delta\phi$, between the energy cluster centroid of the leading jet and all charged transverse tracks, normalized to *n*, which is the number of events. The dashed-dotted line represents the level measured in "minimum bias" events.

$\rho(\Delta\phi)$, the distribution of the azimuthal separation ($\Delta\phi$) between each track in the vertex detector and the centroid of the highest transverse energy cluster, is shown in FIGURE 4 for two intervals of m_{jj}, which is the invariant mass associated with the two highest energy clusters. Clear peaks are seen at $\Delta\phi = 0$ and $\Delta\phi = \pi$, which is expected for two-jet events. The distribution is more peaked for the larger m_{jj} slice, showing that higher energy jets are more collimated. The distribution in $\Delta\phi$ is not symmetric about $\Delta\phi = \pi/2$. This is partly explained by the fact that the two jets are not exactly back-to-back in the transverse plane and partly by the fact that a broad high multiplicity jet will tend to deposit less energy in a leading calorimeter cluster than a

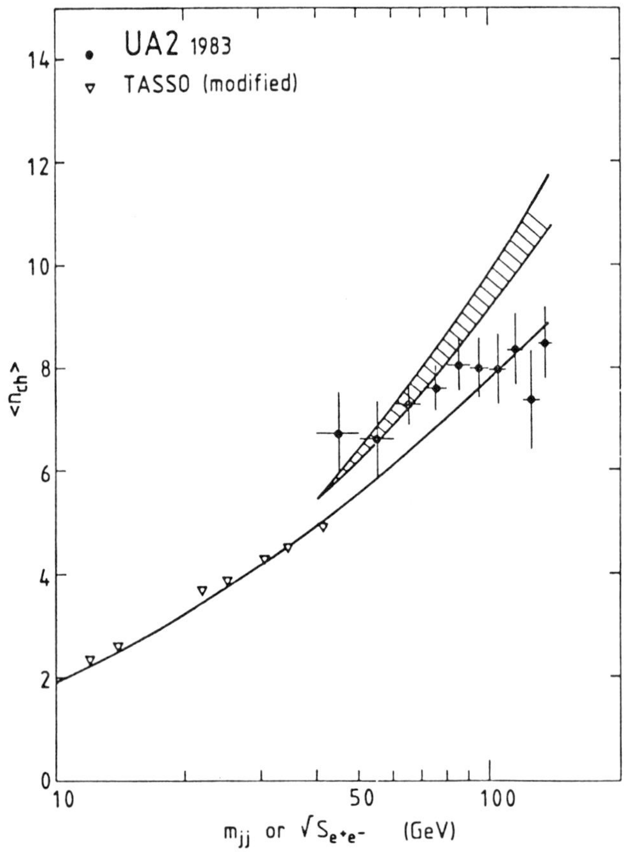

FIGURE 5. Mean charged "jet core" multiplicities, n_{ch}^{jet} (see text for definition), as a function of $\sqrt{s_{e+e-}}$ for e^+e^- data and of the invariant two-jet mass, m_{jj}, for $\bar{p}p$. The solid curve is the prediction of the model of reference 20 for quark jets and the shaded area represents the predictions for gluon jets. The uncertainty in the prediction arises from the range of parameters for which the quark jets fit the e^+e^- data.

more collimated lower multiplicity jet. Therefore, in selecting the higher transverse energy cluster, there is a bias favoring the selection of the narrower and lower multiplicity jet.

The transverse track finding efficiency of the vertex detector was measured in the angular range covered by the toroid spectrometer by comparing the two detectors for field off data. The efficiency was found to be 94.8% ± 0.8% and the fraction of spurious tracks was 21%. The loss of tracks from the finite two-track resolution was estimated from the distribution of the azimuthal separation between pairs of neighboring tracks and was evaluated to be in the range from 16% to 31%, depending on local multiplicity. The data was also corrected for γ conversions and π^0 Dalitz decays. As a check on this procedure, the mean charge multiplicity was measured for "minimum bias" events and compared with results obtained by the UA5 collaboration[17] in the same pseudorapidity range ($|\eta| < 2$). The result, 13.4 ± 0.3, is in good agreement with the UA5 measurement, 13.45 ± 0.15.

As illustrated in FIGURE 4, the track density is approximately constant over a large $\Delta\phi$ range around $\Delta\phi = \pi/2$, where it takes a value, ρ_0. On the average, ρ_0 is measured to be twice as large as in "minimum bias" events.

Furthermore, particles not associated with the jets are assumed to make a uniform contribution, $\lambda\rho_0$, to $\rho(\Delta\phi)$. This assumption was checked by measuring the track density in the forward spectrometers for events containing a jet near $\Theta = 90°$. The distribution of $\Delta\phi$, the azimuthal separation, between the forward track and the highest transverse energy cluster is flat.

A priori λ may take any value between 0 and 1; lacking knowledge of λ, we evaluate a lower bound of the mean track multiplicity in jets, $\langle n_{ch} \rangle$, by setting $\lambda = 1$. The values obtained, as a function of m_{jj}, are displayed in FIGURE 5. To avoid the bias caused by the asymmetric $\Delta\phi$ distribution discussed above, we define the jet multiplicity as the averaged multiplicity of the two jets. The results are compared with e^+e^- data,[18] which are treated the same way as the collider data.

FIGURE 5 also indicates a larger charged particle multiplicity in jet events from the collider than in similar events observed in e^+e^- collisions at lower energies. Even if λ were equal to 1, which would imply that jet fragments do not populate the region of $\Delta\phi = \pi/2$, the charged particle multiplicity of the jets observed in our data would be at least as large as that of the extrapolated e^+e^- data. These results are, therefore, consistent with expectations based on QCD calculations, which predict that at relatively low $x_T = 2E_T/\sqrt{s}$, gluon jets having higher multiplicity than quark jets[20,21] tend to dominate[22] in $\bar{p}p$ collisions.

ENERGY FLOW

To measure the transverse energy flow around the jet axis, a clean sample of two-jet events was selected as follows:

(i) The two highest transverse energy clusters in the event have transverse energies, E_T^1 and E_T^2, exceeding 15 GeV.
(ii) The ratio, E_T^2/E_T^1, exceeds 1/2.
(iii) Their azimuthal separation, $\Delta\phi^{jj}$, exceeds 140°.

To reduce the energy leaking outside the calorimeter acceptance, only clusters having a pseudorapidity, $|\eta|$, less than 0.3 (central clusters) are considered in this analysis. The distribution of the azimuthal transverse energy density, $dE_T/d\Delta\phi$, integrated over a rapidity interval of ±0.7 units around this central cluster is shown in FIGURE 6 for three different intervals of the cluster transverse energy (E_T^j). The peaking in the energy flow is much stronger than in the charged particle flow of FIGURE 4 (note the logarithmic scale in FIGURE 6). This means that the particles at wide angles to the jet axis are soft and make only a small contribution to the energy flow. The peak at $\Delta\phi \simeq 0$ is narrower than the peak at $\Delta\phi = \pi$ (not shown in the figure) because of the finite two-jet transverse momentum for the charged particle flow (see section entitled CHARGE PARTICLE MULTIPLICITY IN JETS) and because the second jet may partially escape detection in the central calorimeter. For this reason only, the region, $0 < \Delta\phi < \pi/2$, is used in the following analysis. The data are not corrected for the effects of calorimeter granularity and resolution. The main effect arises from the finite cell size ($\Delta\phi \times \Delta\theta = 15° \times 10°$), which tends to smear out the peaks. To compare the data with different fragmentation models, several Monte Carlo calculations were made. All details of the detector response were simulated for samples of events generated according to these

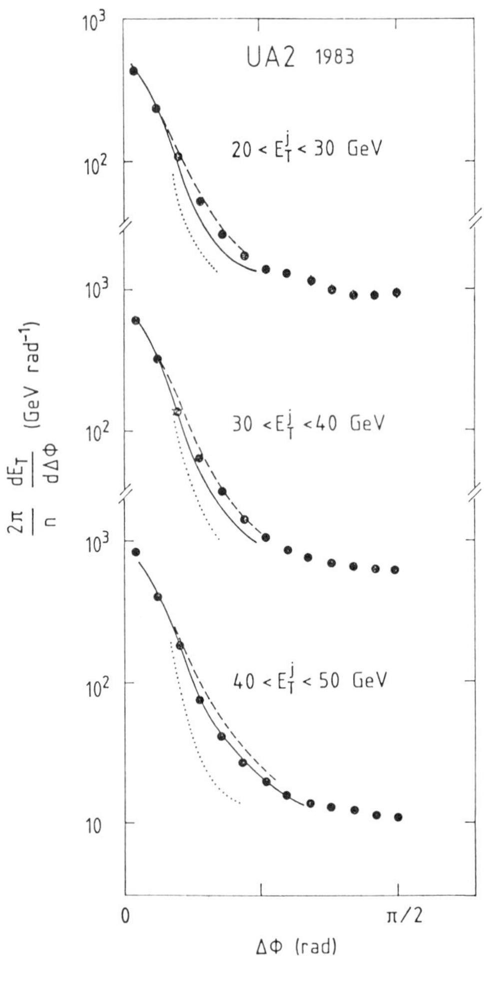

FIGURE 6. Distribution of the azimuthal transverse energy density, $dE_T/d\Delta\phi$, where $\Delta\phi$ is measured with respect to the centroid of the central cluster ($|\eta| < 0.3$) for three ranges of E_T^j: (a) $20 < E_T^j < 30$ GeV; (b) $30 < E_T^j < 40$ GeV; (c) $40 < E_T^j < 50$ GeV. The dotted curve represents the predictions of the Field-Feynman fragmentation model.[18] The dashed (solid) curve represents the predictions of the model of reference 19 for gluon (quark) jets (see text for details).

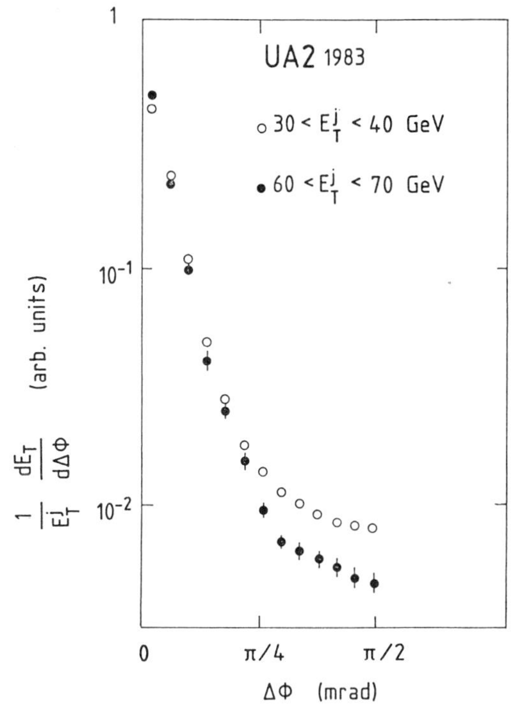

FIGURE 7. Distribution of the normalized transverse energy density, $1/E^j_T\, dE_T/d\Delta\phi$, for two intervals of the highest cluster transverse energy, E^j_T.

models. The Monte Carlo event samples were submitted to the same trigger and analysis procedure as the real data. The corresponding distributions are shown as smooth curves in FIGURE 3. The Field-Feynman fragmentation model,[20] with a r.m.s. transverse momentum of particles with respect to the jet axis, q_T, set equal to 350 MeV, produces jets that are much narrower than the data. The data can be better reproduced by models that explicitly include the effects of gluon bremsstrahlung. The data are bracketed by the predictions for $q\bar{q}$ and gg, which are expected for a mixture of quark and gluon jets as described by the model of reference 19.

In order to study the dependence of jet aperture upon energy, the normalized transverse energy flow, $(1/E^j_T)\, dE_T/d\Delta\phi$, is plotted for different slices of E^j_T in FIGURE 7. A small, but significant shrinking in the width of the jet is observed as its energy increases, although the effect of the jet's shrinking is reduced by the effects of the finite calorimeter granularity.

INCLUSIVE JET PRODUCTION CROSS SECTIONS

For a measurement of the inclusive jet production cross sections,

$$\bar{p}p \rightarrow \text{jet} + \text{anything} \quad (1)$$

$$\bar{p}p \rightarrow \text{jet}_l + \text{jet}_2 + \text{anything}, \quad (2)$$

events are selected containing at least one cluster with E_T exceeding the ΣE_T threshold for reaction (**1**) and with $E_T^1 + E_T^2$ larger than the ΣE_T threshold (and E_T^1, $E_T^2 > 10$ GeV) for reaction (**2**), after inclusion of nearby clusters as explained below. The jet direction and transverse energy are measured both from the centroid and the energy of the associated clusters, with respect to the event vertex.

The individual clusters are taken to be massless[7] in the following procedure, equating, therefore, p_T and E_T. Actually, by attributing a zero-mass momentum vector ($|\bar{p}| = E$) to each cell that participates in a cluster, one observes a mean invariant "cluster mass" of 9 GeV for clusters having $E_T > 30$ GeV. This value results from the combined effects of shower and cell sizes and of actual jet masses, and is in agreement with a Monte Carlo simulation[23] described below. For jets having $p_T > 30$ GeV, this calculation predicts a mean ratio, $|\bar{p}_T| / E_T = 0.86$, where $|\bar{p}_T|$ and E_T are summed over all generated final state particles.

In order to partly account for the final state gluon radiation effects as reported and discussed in reference 9, the jet momenta are defined by adding to the measured momentum vector of the selected cluster those of all clusters having an $E_T > 3$ GeV and a separation by an angle, ω, from the selected cluster momentum vector such that $\cos\omega > 0.2$. About 20% of the jets with $p_T > 30$ GeV include at least one such nearby cluster with $E_T > 3$ GeV. Note that in this case, the jet acquires a mass and p_T becomes different from E_T. The inclusion of these nearby clusters increases the observed transverse momentum distribution (p_T) for jets by about 15%, which is roughly independent of p_T.

The evaluation of the cross sections for reactions (**1**) and (**2**) from these events requires the measurement of the integrated luminosity for the normalization and the determination of the detector acceptance, which includes the effects of the energy smearing on the steeply falling p_T spectra. A Monte Carlo simulation of the detector is used to calculate the acceptance. The Monte Carlo events are processed through the same analysis chain as the data. The simulation reproduces the details of the event configuration and of the detector response, using the same weighting factors for the three calorimeter compartment energies as for the data. In particular, the distribution of cluster size and both hadronic fragmentation and shower development are adequately described.[9] The comparison of the transverse momentum (p_T) and the two-jet invariant mass (m_{jj}) distributions of the reconstructed Monte Carlo events with those of the initial partons used as input provides the acceptance functions, $\epsilon(p_T)$ and $\epsilon(m_{jj})$, by which the observed cross sections must be divided to obtain the corrected jet and two-jet cross sections. The function, $\epsilon(p_T)$, varies from about 0.8 to 1.0 over the range of $30 < p_T < 150$ GeV, and $\epsilon(m_{jj})$, from about 0.5 to 0.8 over the range of $50 < m_{jj} < 250$ GeV. The uncertainty due to the model dependence of the acceptance calculation has been studied by using different independent models,[9,23] by changing the details of the jet fragmentation, and by varying relevant analysis parameters (such as the accepted rapidity range, the parameters of the clustering algorithm, the weighting factors for the calorimeter compartment energies, etc.). The systematic uncertainty on ϵ is estimated to be $\pm 35\%$.

The cross section for inclusive jet production, $d^2\sigma/dp_Td\eta$, at $\eta = 0$ is shown in FIGURE 8a, and the cross section for two-jet production, $d\sigma/dm_{jj}$, with both jets in the interval, $-0.85 < \eta < 0.85$, is displayed in FIGURE 8b. The quoted errors include the statistical errors and an energy-dependent systematic uncertainty on the acceptance

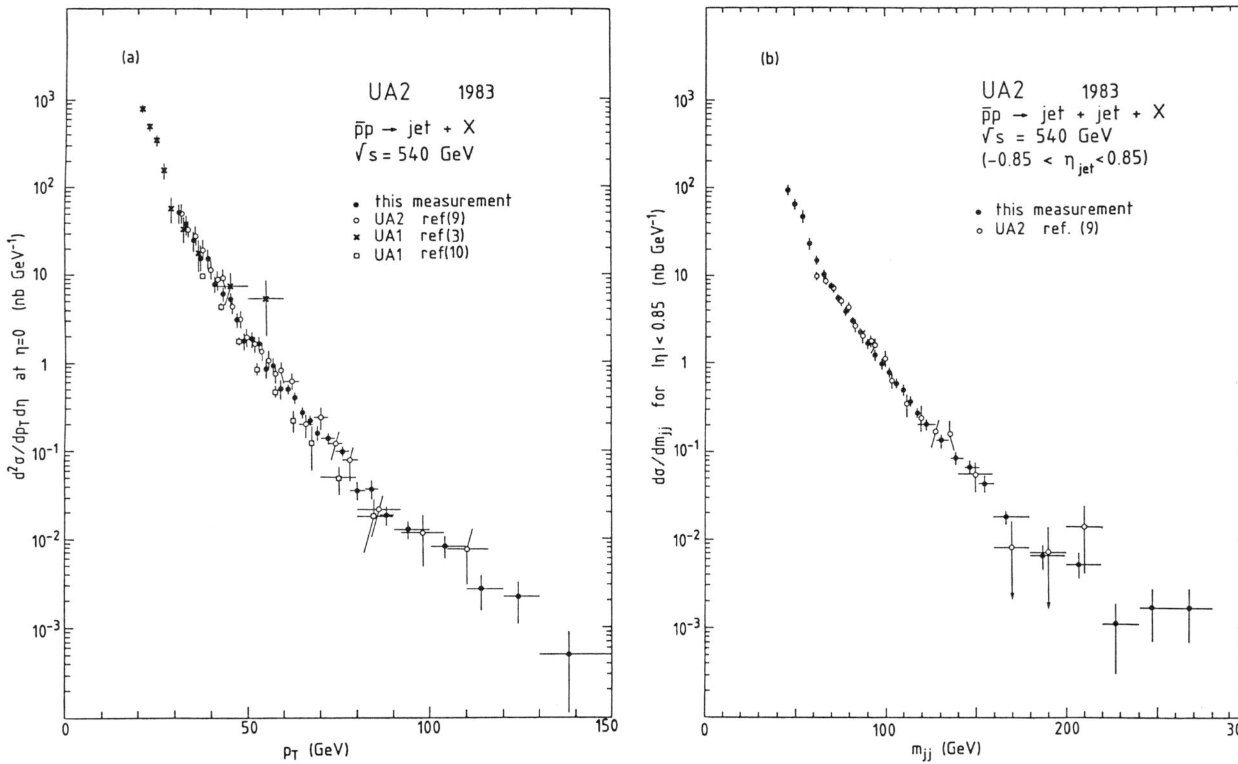

FIGURE 8. (a) Inclusive jet production cross section. The additional systematic uncertainty is ±45%. Previous measurements are from references 3, 9, and 10. (b) Two-jet production cross section with both jets in the pseudorapidity range, $-0.85 < \eta < 0.85$. The additional systematic uncertainty is ±45%. The previous measurement is from reference 9.

functions. Further uncertainties have to be considered in addition to the systematic error on the acceptance (±35%). The normalization is affected by the ±20% systematic uncertainty in the knowledge of the integrated luminosity, and the systematic error on the calorimeter energy calibration also contributes another ±20%. Therefore, the overall systematic uncertainty on the cross sections for the reactions (**1**) and (**2**) is estimated to be ±45% after adding the three contributions in quadrature.

Also shown in FIGURES 8a and 8b are the cross sections obtained previously by the same experiment[9] with a much smaller data sample (from the 1982 running period). For these data, an additional systematic error of ±40% has been quoted. The two measurements are in very good agreement. One should note, though, that the systematic uncertainties of the two results have mostly common origins and the two evaluations are, therefore, not independent. The inclusive jet production cross section reported by the UA1 experiment[3,10] is also shown in FIGURE 8a. The systematic uncertainty for this measurement is given in reference 10 as a factor of 1.65. Even though the UA1 data points are about a factor of 1.5 to 2 below the present measurement, both experiments are consistent in the region of overlap within systematic errors. In addition, a preliminary analysis of the data collected at the end of 1984, with $\sqrt{s} = 630$ GeV, shows the expected p_T dependent increase in cross section, which is due to the increase in $\sqrt{s}$.

COMPARISON WITH QCD CALCULATIONS

The measurements of the inclusive jet production cross sections, reactions (**1**) and (**2**), can be compared directly with QCD predictions. In principle, such comparisons do not require any assumption on the fragmentation mechanism; to the extent that the jets are correctly identified, they correspond directly to the hard scattered partons.[7] The observations[1,3,9,10] of the predicted increase of the yield of jets having $p_T > 20$ GeV at the CERN $\bar{p}p$ Collider with respect to top ISR energy (by about four orders of magnitude)[22] is a remarkable success of the QCD parton picture.

Several QCD calculations have since been reported. In the following, the data are compared with those of references 24–26. Two major uncertainties affect the theoretical predictions:[27] scale ambiguities (choice of Q^2) in the leading logarithm calculations and, related to it, the choice of the Λ parameter in the strong coupling constant, $\alpha_s(Q^2)$, and the parametrization used for the parton densities (structure functions) in the incoming p and $\bar{p}$. Furthermore, higher order (in α_s) contributions to the jet production cross sections are not fully included. The influence of these theoretical ambiguities has been studied, for example, in references 24–26. The calculations ignore fragmentation effects and assume massless partons.

A comparison between data and a range of these predictions is indicated in FIGURES 9a and 9b as a shaded band.[28] The solid curve in FIGURE 9a corresponds to the prediction of reference 22 and almost coincides with the results of references 24 and 25 (not separately shown) for the theoretical assumptions listed in reference 29. The latter results are also displayed in FIGURE 9b as a solid curve. The data are at a level comparable to the QCD predictions.

It has been suggested that a possible substructure of quarks and leptons would manifest itself as a new contact interaction that would be visible at large momentum

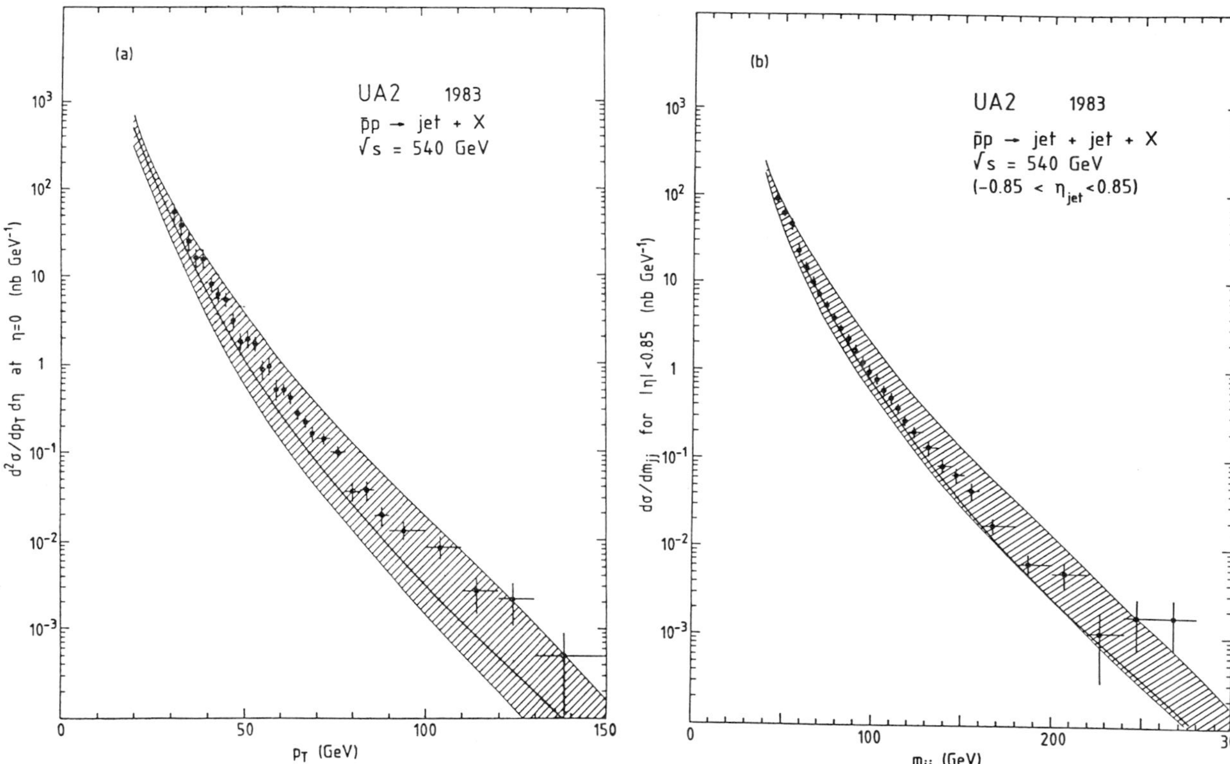

FIGURE 9. (a) Comparison of the inclusive jet production cross section with QCD calculations (see text). The additional systematic uncertainty on the data is $\pm45\%$. (b) Comparison of the two-jet production cross section with both jets in the pseudorapidity range, $-0.85 < \eta < 0.85$, with QCD calculations (see text). The additional systematic uncertainty on the data is $\pm45\%$.

transfers.[30] The cross sections for reactions (**1**) and (**2**) are expected to deviate at large p_T or m_{jj} from the QCD behavior, depending on the energy scale, ζ, which characterizes the strength of this new interaction (and the physical size of the composite states). For example, for $\zeta = 200$ GeV, the cross section of reaction (**1**) is expected to be one order of magnitude larger than the QCD prediction ($\zeta = \infty$) at $p_T > 100$ GeV (FIGURE 9a). A determination of ζ from the data is, in principle, possible, but the uncertainties attached to the QCD calculations limit its accuracy. By comparing the measured cross section for reaction (**1**) to the QCD calculations of references 24 and 29, we find a fair agreement ($\chi^2 = 45$ for 29 degrees of freedom) if we multiply the QCD cross sections by a p_T independent scale factor of 1.9. The remaining deviation can be expressed in terms of a lower limit on ζ by using the calculation of reference 31. This method, which is model dependent, gives $\zeta > 275$ GeV (95% C.L.).

SEARCH FOR STRUCTURE IN THE JET MASS SPECTRUM

The decay of W and Z vector bosons into quarks are expected to show up as a bump in the two-jet mass spectrum in the region, 70 to 100 GeV/c^2, but with a huge background from standard QCD processes. In order to improve the mass resolution, therefore, cuts used in the following sections are more strict than those used in the other sections of this paper.

For an event to enter the mass distribution, it is required to have at least two clusters, each with a transverse energy in excess of 10 GeV. The centroid of the biggest cluster must have $|\eta| < 0.7$ and all other clusters considered in the mass determination must have $|\eta| < 0.85$. Other clusters are merged with one of the two highest E_T clusters when they satisfy the criteria described earlier ($E_T > 3$ GeV, $\cos\omega > 0.2$). Therefore, this reduces the event to two main jets with a total transverse energy of ΔE_T^{cl} and a remaining event having a transverse energy of $\widetilde{\Sigma E}_T = \Sigma E_T - \Sigma E_T^{cl}$. Events having $\widetilde{\Sigma E}_T > 15$ GeV, for which the definition of the two-jet system may be ambiguous, are rejected. We also reject events having more than 6 GeV of transverse energy measured in the F/B calorimeters (where the acceptance is reduced).

In each retained event, we increase the transverse momentum vector of the weakest of the two jets to balance that of the strongest. Monte Carlo calculations show that this procedure partially compensates for losses (heavy flavor decays into undetected leptons and soft gluons at large angles to the jets) and for measurement errors. Therefore, this results in an improved resolution and a more accurate mass determination. In the W region, the mass resolution is of the order of 10 GeV/c^2. FIGURE 10 shows the final spectrum after strict cuts and p_T balance have been applied.

The QCD background, assumed to have a form, $\alpha e^{\beta\times m} + \gamma e^{\delta\times m}$, was fitted in the mass region, 40 GeV to 68 GeV (low threshold), 56 GeV to 62 GeV, and 102 GeV to 148 GeV (high threshold) (excluding the region, 66 GeV to 98 GeV), thus, resulting in a $\chi^2 = 56/67$ d.f.

Although the spectrum is well reproduced by the exponential QCD background distribution, we tried to superimpose two Gaussian distributions with a fixed width of 10 GeV/c^2 so as to describe the contribution from W and Z decays into $q\bar{q}$. The two Gaussians are linked together through the standard model by requiring the number of Z's relative to the number of W's (n_Z/n_W) to be 0.41 and the $m_Z = 1.15\ m_W$.

Therefore, the new fit (FIGURE 11) gives $\chi^2 = 56/70$ d.f. and $m_W = 75 \pm 8$ GeV/c^2. From the value and error obtained in the fit for the W contribution, we deduced, after acceptance corrections, the $W^{\pm}$ production cross section: $\sigma_{W^+} + \sigma_{W^-} = 4.8 + 3.5 - 3.0$ nb, which is in good agreement with the cross section obtained in the electron analysis.

A further search for mass bumps in the region above 100 GeV/c^2 using the same method, but this time using only one Gaussian and having the mass and width be free parameters, resulted in an excess of 69 ± 17 events above background ($\chi^2 = 44/65$ d.f.) at $m = 147 \pm 3$ GeV/c^2 and $\sigma = 13$ GeV/c^2 (FIGURE 12).

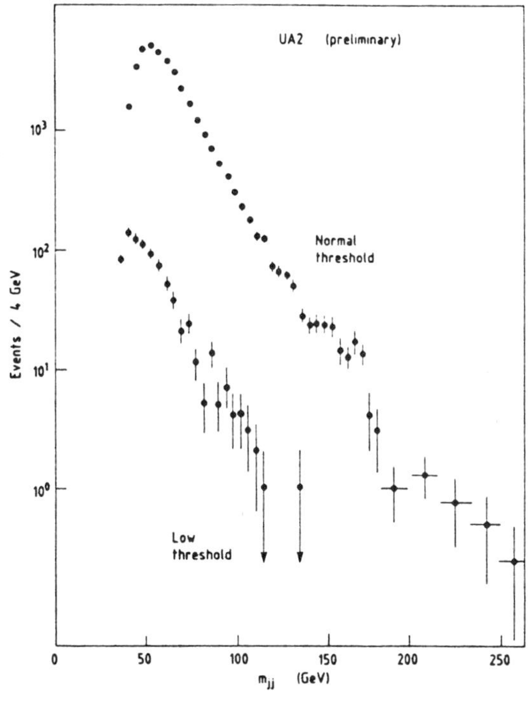

FIGURE 10. Two-jet mass spectrum after strict cuts and p_T balance have been applied.

A preliminary analysis of the 310 nb^{-1} collected in 1984 does not show a similar structure in the jet-jet mass spectrum above 120 GeV/c^2, but more work has to be done before any conclusion can be made.

DYNAMICAL DESCRIPTION OF THE HARD SCATTERING PROCESS

The factorization property of the parton-parton cross sections for any definite initial state parton pair can be generalized[32] in terms of an overall structure function,

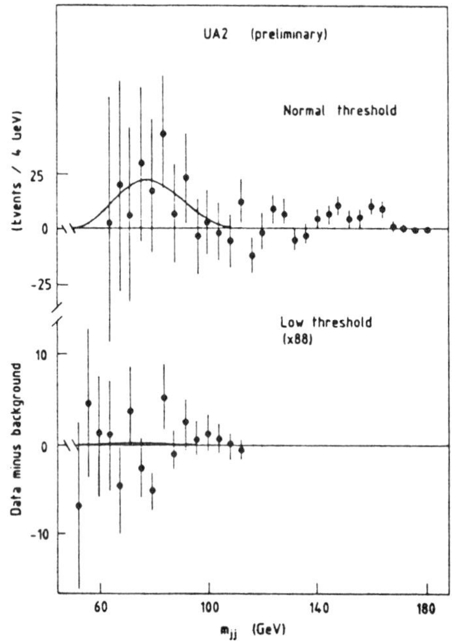

FIGURE 11. Two-jet mass spectrum after QCD background subtraction. The curve indicates the contribution from the two Gaussians, representing *W* and *Z* decays into quarks, to the fit.

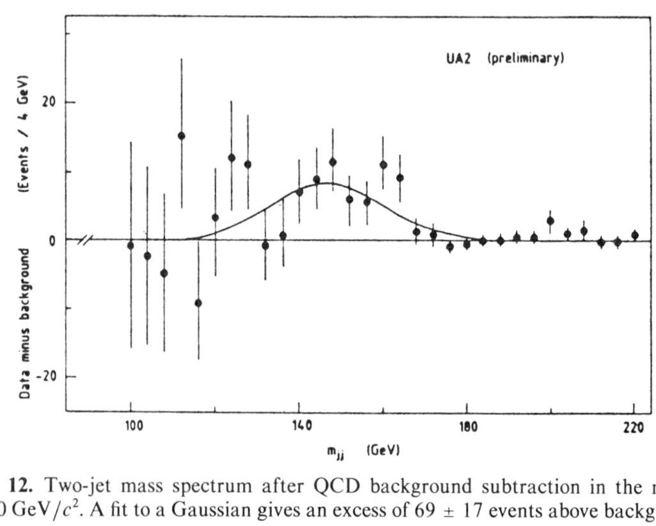

FIGURE 12. Two-jet mass spectrum after QCD background subtraction in the mass region above 100 GeV/c^2. A fit to a Gaussian gives an excess of 69 ± 17 events above background.

$F(x)$ (sum of all the quark and gluon densities), and an effective parton-parton cross section, $d\sigma/d\cos\Theta^*$, obtained from a weighted average over all the elementary subprocesses,[33]

$$\frac{d^3\sigma}{dx_1 dx_2 d\cos\Theta^*} = \frac{S(x_1, x_2)\, d\sigma}{x_1 x_2\, d\cos\Theta^*} = \frac{F(x_1)F(x_2) d\sigma}{x_1 x_2\, d\cos\Theta^*}, \tag{3}$$

where $x_1(x_2)$ represents the fraction of the longitudinal momentum carried by the interacting parton in the proton (antiproton). This analysis has been done with $\alpha_s = 12\pi/23\ell n(Q^2/\Lambda^2)$, where $Q^2 = (p_T)^2$ and $\Lambda = 0.2$ GeV. The results are rather insensitive to these definitions. As an example, a change of a factor of two in Λ corresponds to a variation of less than 10% in $F(x)$. Equation (**3**) is valid to the extent that different subprocesses have a similar Θ^* behavior and that the Q^2 dependence in both the structure functions and in $\alpha_s(Q^2)$ is small. Equation (**3**) does not include higher order QCD corrections, which are usually described by a multiplicative factor, $K < 2$,[34] and give rise to a nonzero transverse momentum for the two-jet system. In the following analysis, these higher order effects will be neglected ($K = 1$). The approximation of a small transverse momentum for the two-jet system, $p_T^{jj} \ll m^{jj}$, which is necessary to define Θ^* without kinematical ambiguities,[35] has been shown to be reasonable in the previous section. This requirement is ensured by retaining the events of the data sample, defined in the section entitled DATA TAKING AND REDUCTION, only if they have $p_T^{jj} < 20$ GeV and $\Delta\phi^{jj} > 140°$.

From the experimental point of view, the jet 4-vectors, p_1 and p_2, are computed from the cluster with the highest transverse energies according to the following procedure:

(1) The energy of the two highest clusters is corrected to take into account the different calorimeter response for hadrons impinging near the cell boundaries and the loss of energy due to the calorimeter nonlinearity and the cluster algorithm.[14] The correction varies from 18% to 1% for jet energies between 20 and 80 GeV, while the dependence on position is less than 10%;

(2) The momentum of any other cluster having $E_T > 3$ GeV and being separated by an angle, ω, from the first (second) cluster such that $\cos\omega > 0.2$ is vectorially added to the first (second) cluster momentum. As in the previous section, this correction is meant to take into account the effects of final state gluon bremsstrahlung.

The values of x_1, x_2, and $\cos\Theta^*$ are computed for each event following the convention of reference 35. The distributions, $S(x_1, x_2)$ and $d\sigma/d\cos\Theta^*$, are derived separately, according to the factorization hypothesis (**3**) that is described in detail in the following sections.

ANGULAR DISTRIBUTION

The $d\sigma/d\cos\Theta^*$ distribution is computed by considering the event distribution in the $\cos\Theta^*$-β^{jj} plane, where β^{jj} is the velocity of the two-jet system in the laboratory. The calorimeter acceptance is uniform in a well-defined domain in this plane, which depends, for each event, on the two-jet mass and on the interaction position along the beam line. A fiducial domain in the region, cos $\Theta^* < 0.6$, can be defined where the acceptance is complete. Different slices of β^{jj} are considered for different regions of cos Θ^*. The overall angular distribution is evaluated by normalizing the different cos Θ^* distributions to the same area in the interval where they overlap. The overall normalization, though, is not defined by this method. Instead, it is chosen by arbitrarily setting the data point at cos $\Theta^* = 0$ equal to 1. FIGURE 13 shows the result. The systematic uncertainties that depend on the relative normalization of different regions in the cosΘ^*-β^{jj} plane are included in the errors. The systematic uncertainties due to the cos Θ^* resolution ($\leq$0.04) and to the definitions of the kinematical variables adopted in this analysis are estimated to be <5%. In the region of overlap, the data agree well with the results of the UA1 Collaboration[36] to within the statistical and systematic errors.

The observed angular distribution can be compared with the predictions of different theories. A Rutherford-like scattering process (t-channel exchange of a vector boson) gives rise to a distribution that is proportional to $[\sin(\Theta^*/2)]^{-n}$, with $n = 4$, whereas scalar gluon theories give rise to distributions with $n \leq 2$. The data do not agree over the full cos Θ^* range with any such simple parametrization. However, the region of cos $\Theta^* \geq 0.3$ can be fitted with $n = 3.94 \pm 0.18$. This clearly shows the dominance of processes with a t-channel exchange of a vector boson in the forward

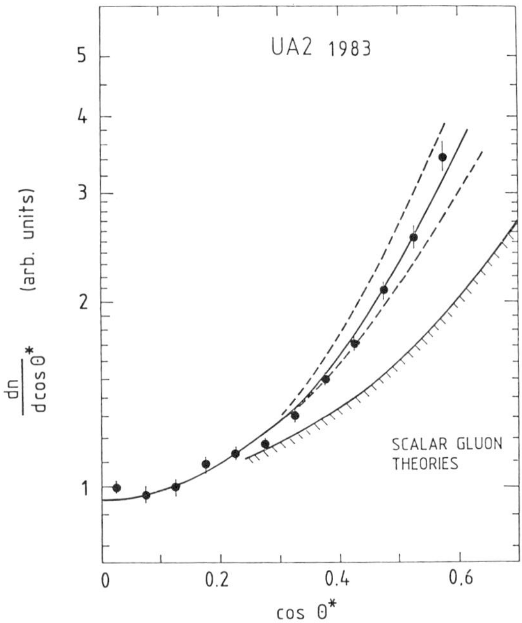

FIGURE 13. Distribution of the scattering angle in the two-jet center of mass. The normalization is defined setting the value at cos $\Theta^* = 0$ equal to 1. All the different QCD subprocesses[33] (except the pure s-channel one, $q_1 q_1 \rightarrow q_2 q_2$), separately normalized to the data, lie in the area between the two dashed curves. The full line is the overall QCD prediction discussed in the text, normalized to the data. Also, the region corresponding to scalar gluon theories[37] is shown.

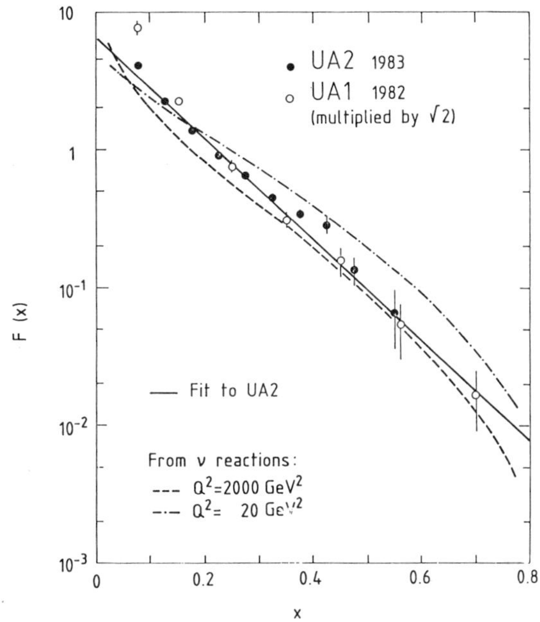

FIGURE 14. Effective structure function. The full points are the results of this experiment and the open points are derived from reference 36. The full line represents an exponential fit to the data, while the dashed lines are computed from neutrino deep inelastic scattering,[38] taking into account scaling violation effects.

region and rules out any scalar gluon theory (see reference 37 for other experimental results on the gluon spin). Therefore, the data are well reproduced by a sum of different processes like that predicted by QCD.[33] Much simpler models also agree with the data (for example, the angular distribution due only to *s*- and *t*-channel exchanges of a vector boson by spin 1/2 fermions agrees with the experimental points).

The observed distribution is compared in FIGURE 13 with the QCD prediction for the dominant parton-parton subprocesses,[33] which lie in the area between the two dashed lines. The statistical and systematic errors are such that the relative importance of the different subprocesses cannot be measured. An overall QCD prediction, obtained from a sum of all the subprocesses with each weighted by a factor depending on the structure functions (derived from reference 38), gives good agreement. This curve is shown as a full line normalized to the data.

STRUCTURE FUNCTION

The function, $S(x_1, x_2)$, defined in equation (3), is computed by weighting each event with an acceptance factor, w, which accounts for the probability to observe the event:

$$w^{-1} = \int_0^{C_{max}} (d\sigma/d\cos\Theta^*)_{QCD}\, d\cos\Theta^* \quad (4)$$

where C_{max} (function of the observed values of x_1 and x_2) is the maximum detectable value of cos Θ^* in the fiducial region and $(d\sigma/d\cos\Theta^*)_{QCD}$ is the QCD parameterization of the angular distribution. The effective structure function, $F(x)$, is evaluated according to equation (**3**) for ten bins of x between 0.05 and 0.60. The effects of the resolution in x and of the variation of the acceptance as a function of the vertex position have been taken into account. One should note that, if the usual definition of the higher order correction factor, K, were adopted, the result of the measurement would be the product, $\sqrt{K} \cdot F(x)$. FIGURE 14 show the values with their statistical errors. An empirical fit (shown as a full line in FIGURE 14) of the form, $F(x) = A \exp(-\alpha x)$, gives $A = 6.2 \pm 0.1$ and $\alpha = 8.3 \pm 0.1$. The systematic errors, which are due to the resolution, the definition of the kinematics, and the uncertainty of the luminosity, are ±30%.

Evidence for the factorization of the cross section (**3**) is illustrated in FIGURE 15. The independence of the angular distribution of x_1 and x_2 is tested separately from the factorization of $X(x_1, x_2)$ into $F(x_1) \cdot F(x_2)$. FIGURE 15a shows the value of the ratio, $(d\sigma/d\cos\Theta^*)_a/(d\sigma/d\cos\Theta^*)_b$, as a function of cos Θ^*. The a and b refer to different mass slices and, therefore, to different values of x_1 and x_2. A constant value of this ratio corresponds to the independence of $d\sigma/d\cos\Theta^*$ of mass and, as a consequence, to the validity of the first part of the factorization hypothesis. FIGURE 15b shows the ratio, $S(x_1, x_2)/S(x_1, x_2 + \Delta x)$, as a function of x_1 for different intervals of x_2 and a fixed value of $\Delta x = 0.05$. The second part of the factorization test is verified if the different x_2 intervals lie on constant lines. If, moreover, the structure function, $F(x)$, is well

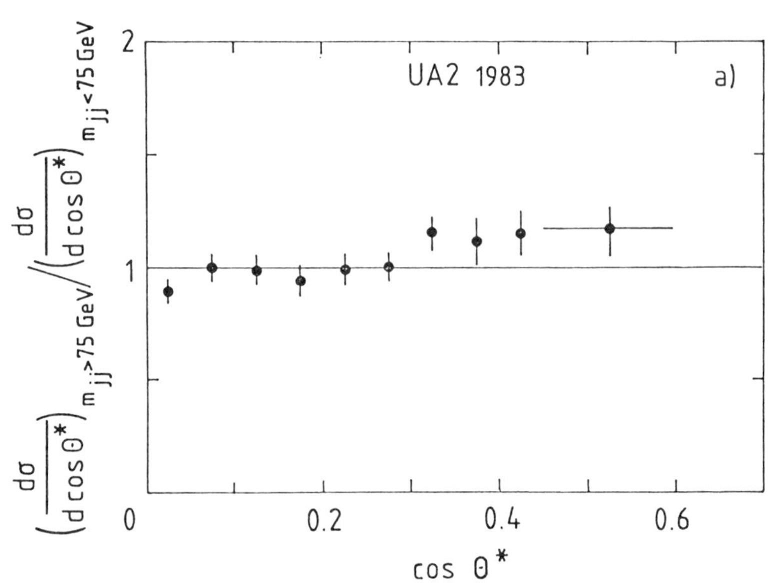

FIGURE 15a. Ratio between the angular distribution of the sample of events with $m_{jj} > 75$ GeV and the events with $m_{jj} < 75$ GeV. The horizontal lines correspond to the slope of $F(x)$.

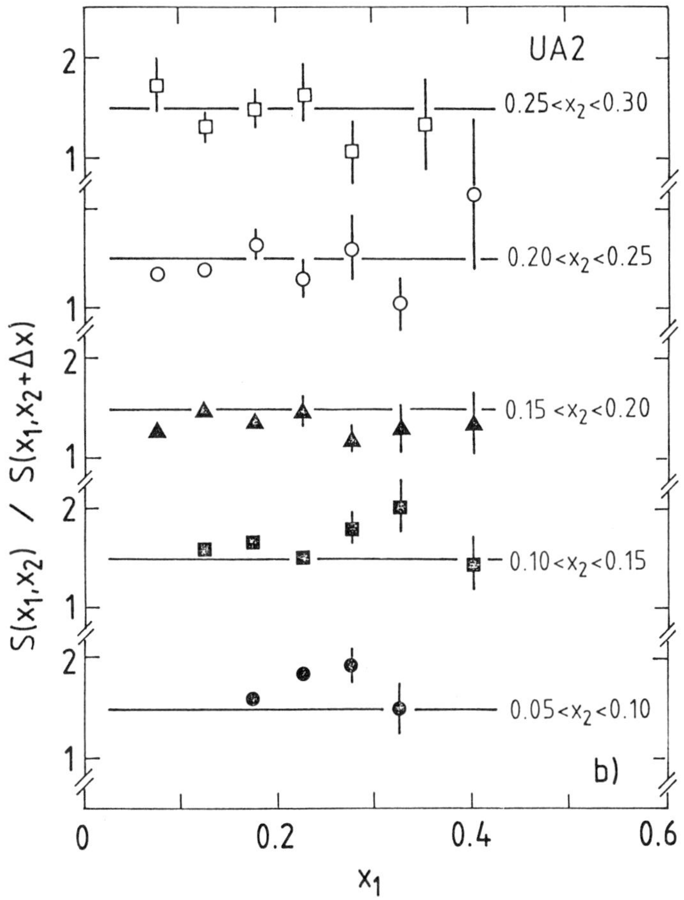

FIGURE 15b. Ratio, $S(x_1, x_2)/S(x_1, x_2 + \Delta x)$ (see eq. (**1**) in the text), for different x_2 slices and $\Delta x = 0.05$. The horizontal lines correspond to the slope of $F(x)$.

approximated by an exponential, the various lines coincide because the ratio is independent of x_2.

As a further test of the experimental method, the single- and two-jet cross sections have been computed from the measured structure function, the angular distribution, and the two-jet transverse momentum distributions. The agreement with the published results[1,9] is good within $\pm 20\%$.

The structure function agrees with the one reported in reference 36 in the region, $x \geq 0.10$, to within the statistical and systematic uncertainties. Since that analysis assumed $K = 2$, those points (shown in FIGURE 14) have been multiplied by $\sqrt{2}$.

The parton density of reference 38, extrapolated to this experiment ($\langle Q^2 \rangle \approx 2000$

GeV^2), is shown in FIGURE 14 as a dashed line. This structure function has been calculated assuming $F(x) = g(x) + 4/9\ q(x)$, where $g(x)$ is the gluon structure function and $q(x)$ is the sum over all the quark and antiquark densities. Part of the overall scale factor (~1.4) may be attributed to higher order corrections not taken into account in the computation of $F(x)$. The two distributions agree well in shape, both in the region of small x, where $F(x)$ is dominated by the gluon density, and of high x, where mainly valence quarks are present.

CONCLUSION

Jets produced at the S$\bar{p}$pS collider, at $\sqrt{s} = 530$ GeV, have been studied in detail. Cross section, structure function, and fragmentation properties are observed to be in perfect agreement with expectations from QCD and lower energy extrapolations.

NOTES AND REFERENCES

1. BANNER, M. *et al.* UA2 COLLABORATION. 1982. Phys. Lett. **1188:** 203; BAGNAIA, P. *et al.* UA2 COLLABORATION. 1984. Phys. Lett. **1388:** 430.
2. ÅKESSON, T. *et al.* AXIAL FIELD SPECTROMETER COLLABORATION. 1982. Phys. Lett. **118B:** 185 and 193.
3. ARNISON, G. *et al.* UA1 COLLABORATION. 1983. Phys. Lett. **123B:** 115.
4. ANGELIS, A. L. S. *et al.* COR COLLABORATION. 1983. Phys. Lett. **126B:** 132.
5. For reviews, see, for example: HANSEN, K. & P. HOYER, Eds. 1979. Jets in high energy collisions. Phys. Scr. **19;** DARRIULAT, P. 1980. Ann. Rev. Nucl. Part. Sci. **30:** 159.
6. FEYNMAN, R. P. 1972. Photon Hadron Interactions. Benjamin. New York; BERMAN, S. M. & M. JACOB. 1970. Phys. Rev. Lett. **25:** 1683; BERMAN, S. M., J. D. BJORKEN & J. B. KOGUT. 1971. Phys. Rev. **4D:** 3388.
7. BJORKEN, J. D. 1973. Phys. Rev. **D8:** 4098.
8. THE STAFF OF THE CERN $\bar{p}p$ PROJECT. 1981. Phys. Lett. **107B:** 231.
9. BAGNAIA, P. *et al.* UA2 COLLABORATION. 1983. Z. Phys. **C20:** 117.
10. ARNISON, G. *et al.* UA1 COLLABORATION. 1983. Phys. Lett. **132B:** 214.
11. ARNISON, G. *et al.* UA1 COLLABORATION. 1983. Phys. Lett. **132B:** 223 and CERN-EP/83-198.
12. ÅKESSON, T. & H. BENGTSSON. 1983. Phys. Lett. **120B:** 133.
13. MANSOULIÉ, B. 1983. The UA2 apparatus at the CERN $\bar{p}p$ Collider. *In* Proceedings of the Third Moriond Workshop on $\bar{p}p$ Physics, p. 609. Editions Frontières. France; DIALINAS, M. *et al.* 1983. The vertex detector of the UA2 experiment. LAL-RT/83-14, ORSAY. CONTA, C. *et al.* 1983. The system of forward-backward drift chambers in the UA2 detector. CERN-EP/83-176. Submitted to Nucl. Instrum. Methods; BORER, K. *et al.* 1983. Multitube proportional chambers for the localization of electromagnetic showers in the UA2 detector. CERN-EP/83-177. Submitted to Nucl. Instrum. Methods.
14. BEER, A. *et al.*, The central calorimeter of the UA2 experiment at the CERN $\bar{p}p$ Collider. CERN-EP/83–175. Submitted to Nucl. Instrum. Methods.
15. The cross section seen by the "minimum bias" trigger counters is 43.9 ± 3.5 mb as measured by: BATTISON, R. *et al.* UA4 COLLABORATION. 1982. Phys. Lett. **117B:** 126; SANGUINETTI, G. 1983. *In* Proceedings of the Third Moriond Workshop on $\bar{p}p$ Physics, p. 25. Editions Frontières. France.
16. The integrated luminosities for the data used at low thresholds are $\int \mathcal{L}dt = 0.56\ nb^{-1}$ for ΣE_T from 30 to 40 GeV, 5.8 nb^{-1} from 40 to 50 GeV, and 12.3 nb^{-1} from 50 to 60 GeV.
17. ALPGARD, K. *et al.* UA5 COLLABORATION. 1982. Phys. Lett. **115:** 71.
18. ALTHOFF, M., *et al.* TASSO COLLABORATION. 1983. DESY 83–010.

19. MARCHESINI, G. & B. R. WEBBER. 1984. Nucl. Phys. **B238:** 1; WEBBER, B. R. Nucl. Phys. **B238:** 492.
20. FIELD, R. D. & R. P. FEYNMAN. 1978. Nucl. Phys. **B136:** 1.
21. STERMAN, G. & S. WEINBERG. 1977. Phys. Rev. Lett. **39:** 1436; SHIZUYA, K. & S-H. H. TYE. 1978. Phys. Rev. Lett. **41:** 787; EINHORN, M. B. & B. G. WEEKS. 1978. Nucl. Phys. **B146:** 445.
22. HORGAN, R. & M. JACOB. 1981. Nucl. Phys. **B179:** 441.
23. PAIGE, F. & S. PROTOPOPESCU. 1981. ISAJET programme. BNL report 31987.
24. HUMPERT, B. Two-jet production in $\bar{p}p$ collisions. Private communication and in press.
25. ANTONIOU, N. G. *et al.* 1983. Hadron and jet production at Collider and ISR. McGill University preprint and Phys. Lett. **128B:** 257.
26. KUNSZT, K. & E. PIETARINEN. 1983. Phys. Lett. **132B:** 453.
27. ALTARELLI, G. QCD at the Collider. Ref. Th. 3733-CERN. *In* The Proceedings of the International School of Subnuclear Physics, Erice, Italy, 3–14 August 1983. To be published.
28. The upper bounds in FIGURES 9a and 9b correspond to a calculation of reference 24 using $\Lambda = 0.5$ GeV, $Q^2 = 2\, s\, t\, \hat{u}/(s^2 + t^2 + \hat{u}^2)$, and the structure functions (*SF*) of BAIER, R. *et al.* 1979. Z. Phys. **C2:** 265. The lower bound in FIGURE 9a is from reference 26 with $\Lambda = 0.5$ GeV, $Q^2 = 4p_T^2$, and *SF* from OWENS, J. F. and E. REYA. 1978. Phys. Rev. **D17:** 3003. The lower bound in FIGURE 9b uses the same *SF* and is from reference 24.
29. The calculations of reference 22 use $Q^2 = p_T^2$, $\Lambda = 0.5$ GeV, and *SF* of the type from OWENS, J. F. *et al.* 1978. Phys. Rev. **D18:** 1501. The curve shown from reference 24 uses the *SF* of ABRAMOWICZ, H. *et al.* 1982. Z. Phys. **C13:** 199, and $\Lambda = 0.5$ GeV and Q^2 as given in 28. The curve from reference 25 is obtained with $\Lambda = 0.2$ GeV, $Q^2 = 2\, p_T^2$, and *SF* from GLÜCK, M. *et al.* 1982. Z. Phys. **C13:** 119.
30. EICHTEN, E. *et al.* 1983. Phys. Rev. Lett. **50:** 811; ABOLINS, M. *et al.* Proceedings of the DPF Summer Study on Elementary Particle Physics, Snowmass, Colorado, p. 274.
31. HUMPERT, B. Feb. 1984. Jet-jet production in the multi-TeV range. Ref. TH.3817-CERN.
32. COHEN-TANNOUDJI, G. *et al.* 1983. Phys. Rev. **D28:** 1628; COMBRIDGE, B. L. & C. J. MAXWELL. 1983. RL-83-095; HALZEN, F. & P. HOYER. 1983. Phys. Lett. **130B:** 326.
33. COMBRIDGE, B. L., J. KRIPFGANZ & J. RANFT. 1977. Phys. Lett. **70B:** 234. See also: CUTLER, R. & D. SIVERS. 1978. Phys. Rev. **D17:** 196.
34. FURMAN, M. 1982. Nucl. Phys. **B197:** 413.
35. COLLINS, J. C. & D. E. SOPER. 1977. Phys. Rev. **D16:** 2219.
36. ARNISON, G. *et al.* 1984. Phys. Lett. **136B:** 294.
37. SÖDING, P. & G. WOLF. 1981. Ann. Rev. Nucl. Part. Sci. **31:** 231; DRIJARD, D. *et al.* 1983. Phys. Lett. **121B:** 433.
38. ABRAMOWICZ, H. *et al.* CDHS COLLABORATION. 1982. Z. Phys. **C12:** 289; 1982. Z. Phys. **C13:** 199; 1983. Z. Phys. **C17:** 283; EISELE, F. Private communication; See also the corresponding result from BERGSMA, F. *et al.* CHARM COLLABORATION. 1983. Phys. Lett. **123B:** 269 and preprint WA18 int./84–29. Submitted to the XXII International Conference on High Energy Physics, Leipzig, 1984.

The Phenomenology of High Energy Multijets in Hadron-Hadron Collisions[a]

R.D. FIELD

Particle Theory Group
Department of Physics
University of Florida
Gainesville, Florida 32611

INTRODUCTION

In this talk, I am not going to argue the merits or shortcomings of any of the existing QCD Monte Carlo models for hadron-hadron collisions since none of the present-day models should be taken too seriously. They all contain a little QCD perturbation theory and a lot of phenomenology, and none of the existing hadron-hadron Monte Carlo models give precisely the same predictions for experiment.[1,2] Instead, I would like to take this opportunity to compare the results of one particular QCD Monte Carlo model for hadron-hadron collisions with the predictions of the "naive 4-jet" parton model.

In the naive parton model, high transverse momentum particles are produced by a single, hard, large-angle elastic scattering of partons; one from the target hadron and one from the beam hadron. A large p_T event consists of four jets as shown in FIGURE 1. The two outgoing partons fragment into two jets with roughly equal and opposite transverse momentum. The other two jets have small transverse momentum and result from the breakup of the beam and target hadrons. One can account for this breakup by assigning the "hole," which is left in the beam and target hadron when the scattered parton is removed, the remaining momentum and allowing it also to fragment into a jet. In the naive parton model, a high p_T event is essentially a superposition of two large p_T "e^+e^-" type jets on top of a low p_T "minimum bias" background. The "minimum bias" background is assumed to look like a low p_T hadron-hadron collision, but at a lower CM energy since the two high p_T jets removed some energy from the collision.

The QCD Monte Carlo model I will use is similar to the naive parton model, but it includes some important QCD corrections. In addition to the $2 \rightarrow 2$ parton-parton hard scattering subprocesses, complicated $2 \rightarrow N$ subprocesses are estimated by including the noncollinear emission of gluons off both the initial- and final-state partons in the "leading pole" approximation.[3–8] Final-state partons have timelike invariant masses with the radiated partons being kinematically constrained to have invariant masses less than their parents, thus giving a difference that is converted into the transverse momentum of the emitted partons. The radiated partons themselves radiate more partons until all invariant masses have been degraded to some cutoff mass, μ_c^A, thus producing a "parton shower" as illustrated in FIGURE 2. Initial partons

[a]This work was supported in part by the U.S. Department of Energy under contract no. DE-AS-05-81-ER40008.

also form a shower, but in this case the partons have spacelike invariant masses. The initial partons are evolved from an initial invariant mass of $-(\mu_c^B)^2$ to a maximum (negative) invariant mass of $Q^2 = -4\hat{p}_T^2$, where $\hat{p}_T$ is the transverse momentum of the hard parton-parton subprocess.[9]

To compare with experiment, one must have a model for the way the outgoing partons turn into hadrons. In addition, in hadron-hadron collisions one must know how a proton with one parton removed turns into hadrons (beam and target jets). For e^+e^- annihilations, the approach adopted by Stephen Wolfram and myself[10] involved keeping track of the color "strings." All gluons were forcibly split into $q\bar{q}$ pairs and

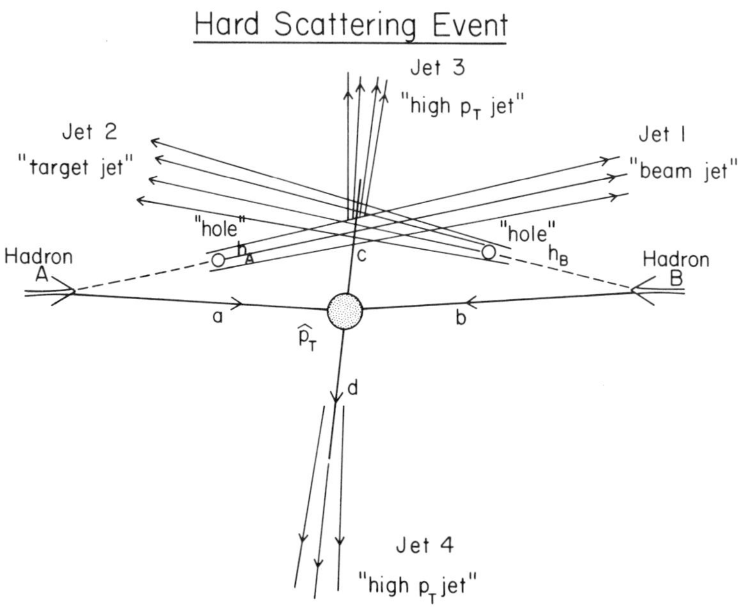

FIGURE 1. Illustration of the naive parton model in which high transverse momentum particles are produced by a single, hard, large-angle elastic scattering of partons; one from the target hadron and one from the beam hadron. The large p_T event consists of four jets; two with large transverse momentum resulting from the fragmentation of the outgoing partons and two jets with small transverse momentum resulting from the breakup of the beam and target hadrons.

color singlet clusters were formed with a distribution of invariant mass. Then, they were allowed to "decay" isotropically in their rest frame according to a simple phase-space prescription. Geoffrey Fox and I[11] used the phase-space method in hadron-hadron collisions for cluster masses less than 3 GeV. Larger mass clusters were parameterized by back-to-back Field-Feynman jets[12] in the color string CM frame.

In the analysis presented here, I do not keep track of the color strings. Instead, each parton is allowed to fragment independently into hadrons in the hadron-hadron CM frame according to the Field-Feynman (FF-jet) prescription. This is the fragmentation scheme adopted by ISAJET[9] and by Odorico.[4] I do not believe that this is a correct fragmentation procedure; however, for the present I have nothing better to suggest. In addition, the "hole" that remains after a parton is knocked out of the beam or target hadron is assigned the full remaining momentum, which is then also allowed to fragment according to the FF-jet prescription. The invariant mass cutoffs, μ_c^A and μ_c^B, are equal to 2 GeV and the QCD perturbative parameter, Λ, is equal to 0.2 GeV.

With this hadronization scheme, I can keep track of where the outgoing hadrons come from. As is illustrated in FIGURE 2, type 1 hadrons are those arising from initial partons, A_1, A_2, . . . , etc., plus the hadrons arising from the fragmentation of the "hole," h_A. Similarly, type 2 hadrons arise from the initial partons, B_1, B_2, . . . , etc., plus the "hole," h_B. Outgoing partons, C_i and D_i, fragment into hadrons of type 3 and type 4, respectively. I have been careful not to call, for example, type 3 hadrons a "jet." Type 3 hadrons may form several "jets" or no "jets," or type 3 and type 1 hadrons may conspire to form one "jet." The definition of a "jet" is at the discretion of the experimenter.

Admittedly, the parton-shower Monte Carlo model is a crude attempt to reproduce nature.[13] However, it does contain, in an approximate manner, many of the features expected from QCD. As the momentum transfer of the hard parton-parton scattering subprocess, Q, increases, the QCD parton-shower model deviates more and more from

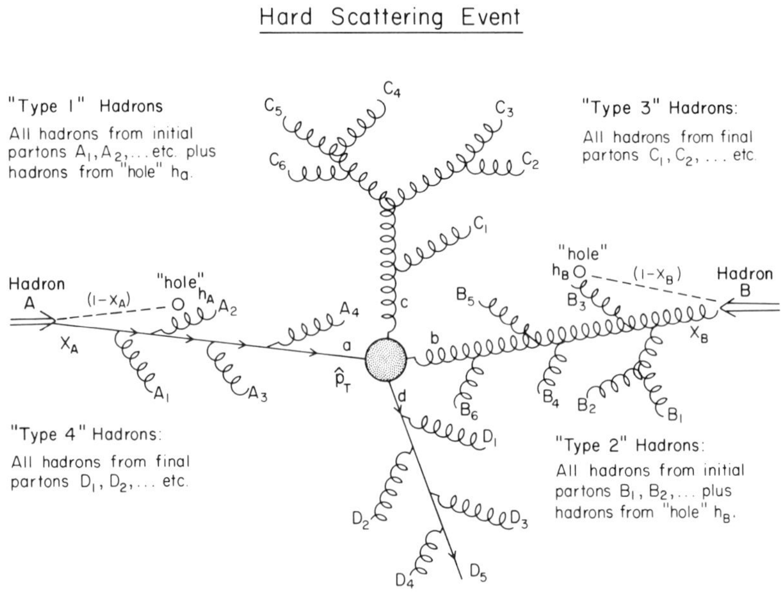

FIGURE 2. Illustration of a hard scattering QCD parton-shower Monte Carlo event that includes, in an approximate manner, the emission of gluons off both the initial- and final-state partons. Hadrons are labeled according to where they originate, with type 1 hadrons arising from the fragmentation of the initial partons, A_1, A_2, . . . , etc., plus the fragmentation of the "hole," h_A (which is assigned fractional momentum, $1 - x_A$). Similarly, type 2 hadrons arise from the fragmentation of the initial partons, B_1, B_2, . . . , etc., plus "hole," h_B. Type 3 and type 4 hadrons arise from the fragmentation of outgoing partons, C_i and D_i, respectively, and $\hat{p}_T$ is the transverse momentum of the hard constituent scattering, $ab \to cd$.

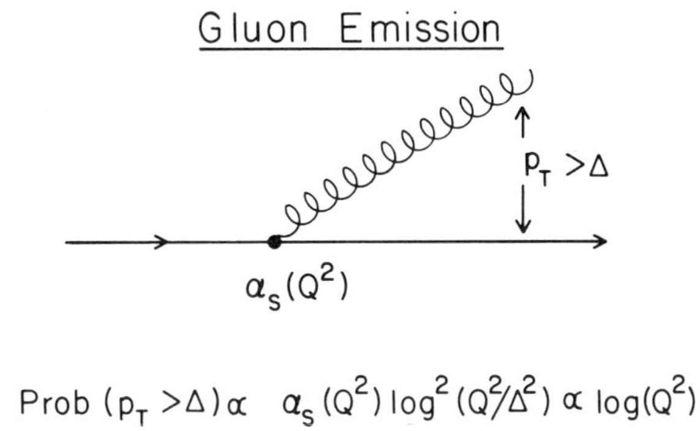

FIGURE 3. Illustration of gluon radiation. The probability of emitting a hard gluon (one with relative p_T greater than Δ) increases like $\log(Q^2)$. This causes the QCD parton-shower Monte Carlo model (FIGURE 2) to deviate more and more from the naive parton model (FIGURE 1) as the momentum transfer, $\hat{p}_T$, of the parton-parton subprocess, $ab \rightarrow cd$, increases.

the naive parton model. This is due to the fact that the probability of radiating a hard gluon off an initial- or final-state parton increases logarithmically as Q increases (see FIGURE 3). In addition, in the QCD parton-shower approach there are dynamical correlations between the outgoing high p_T jets (hadrons of type 3 and 4) and the low p_T "background" jets (hadrons of type 1 and 2). Large p_T outgoing jets mean that a large Q^2 subprocess occurred, which in turn stimulates a large amount of gluon bremsstrahlung in the initial state. In the QCD approach, it is incorrect to consider a high p_T event as a superposition of two high p_T jets on top of a "minimum bias" background. The great majority of "minimum bias" events contain no hard scattering, whereas essentially all large p_T events occur as the result of a hard parton-parton scattering.

UA2 CLUSTER ANALYSIS

In order to compare with data from UA2, events are analyzed by dividing the solid angle for pseudorapidity in the range of $|\eta| < 1$ into 240 cells, with each cell having size, $\Delta\theta\Delta\phi = 10° \times 15°$. The global transverse energy of an event is the sum of the transverse energies of all the cells. Clusters of cells are formed by including in a "cluster" all cells with a common side. However, cells with transverse energy less than $E_T(\text{min}) = 0.4$ GeV are ignored. The energy of a cell is computed, not by adding the energy of each hadron in the cell, but by using $E(\text{cell}) = E_T(\text{cell})/\sin(\theta)$, where θ is computed from the knowledge of the position of the center of the cell. A cell is considered as a massless "particle" of energy, $E(\text{cell})$, and with a direction given by the position of the center of the cell. Clusters have an energy given by the sum of the energy of each cell in the cluster and a momentum, p_{cl}, given by the vector sum of the momentums of each cell. The invariant mass of a cluster is simply $M_{cl} = (E_{cl}^2 - p_{cl}^2)^{(1/2)}$.[14] Finally, clusters are ordered according to the total E_T of all the cells in the cluster, with cluster #1 having the highest total transverse energy.[15]

FIGURE 4 shows the number of cells, on the average, with transverse energy greater

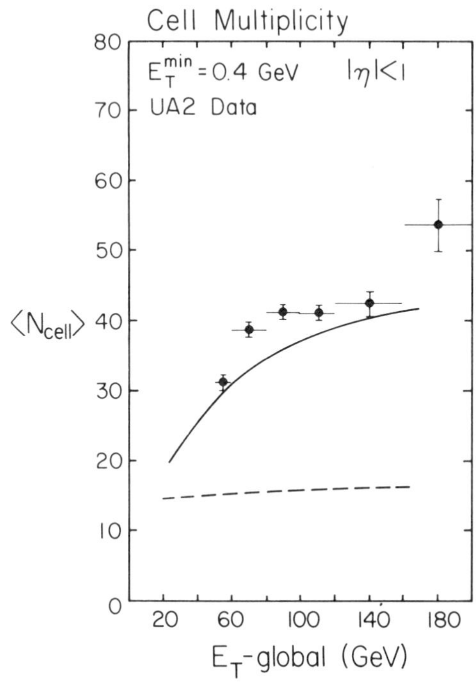

FIGURE 4. Average number of cells with $E_T(\text{min}) > 0.4$ GeV for cells of size, $\Delta\theta\Delta\phi = 10° \times 15°$, $|\eta| < 1$ (240 cells), versus the global E_T resulting from the QCD parton-shower Monte Carlo model (solid curve) and the naive parton model (dashed curve) for $\bar{p}p$ collisions at $W = 540$ GeV. The data are from UA2.[16]

than $E_T(\text{min}) = 0.4$ GeV for global E_T in the range of 20 to 200 GeV with $|\eta| < 1$. Even in this simple observable, one can see a big difference between the naive parton model and the QCD parton-shower Monte Carlo. Because of gluon bremsstrahlung, considerably more cells are populated in the QCD approach than in the naive parton model. In addition, in the QCD approach, the average multiplicity of cells with $E_T > 0.4$ GeV increases quite rapidly with the global E_T of the trigger, whereas in the naive parton model, cell multiplicity is almost independent of the global trigger E_T. In the QCD approach, large global E_T can be produced in a variety of ways. An event can have two large p_T jets resulting in a large global E_T as in the naive parton model or, on the other hand, one can produce a lot of gluon bremsstrahlung resulting in many jets, each with moderate p_T summing to give a large global E_T. The UA2 data[16] show qualitative support for the QCD approach over the naive parton model.

A crucial difference between the naive parton model and QCD is illustrated in FIGURE 5, where the average number of clusters (with $E_T(\text{each}) > 10$ GeV) is plotted versus the global E_T in the event. In the naive parton model, the multiplicity of clusters with $E_T(\text{each}) > 10$ GeV quickly becomes two and remains two as the global E_T increases. In the naive parton model, there are only two large E_T jets. QCD, on the other hand, predicts that the average number of clusters will increase beyond two. As the global E_T increases, there becomes an increasing probability of finding three, four, etc., clusters, each of which has E_T(each) > 10 GeV. This is true no matter what one chooses for the value of E_T(each). Again, the UA2 data[16] show qualitative support for the QCD approach over the naive parton model. It is important to realize that had the UA2 data in FIGURE 5 not risen above two as the global E_T increased, QCD would have been in trouble. However, given that it does rise above two, precise comparisons with the QCD parton-shower Monte Carlo are not significant.

The rich topological structure of QCD events cannot be seen by merely studying the average number of clusters as in FIGURE 5. More information is contained in FIGURE 6, which shows the probability of finding N clusters in a given event, each of which has E_T(each) > 10 GeV. For example, for global E_T in the range, $100 < E_T < 120$ GeV, the QCD parton-shower Monte Carlo model predicts a 35% probability of finding three clusters with E_T(each) < 10 GeV and a 4% probability of finding four. The naive parton model gives essentially a 100% chance of finding two clusters in every event. Although the agreement is not perfect, the UA2 data[16] strongly favor the QCD approach. In this same global E_T bin, UA2 find three clusters 28% of the time and four clusters about 4% of the time.

Another way to display the cluster probabilities is to plot the percentage of events with N clusters, each with E_T(each) > 10 GeV, versus the global trigger E_T, as is done in FIGURE 7. At low global E_T, 100% of the events contain zero clusters with

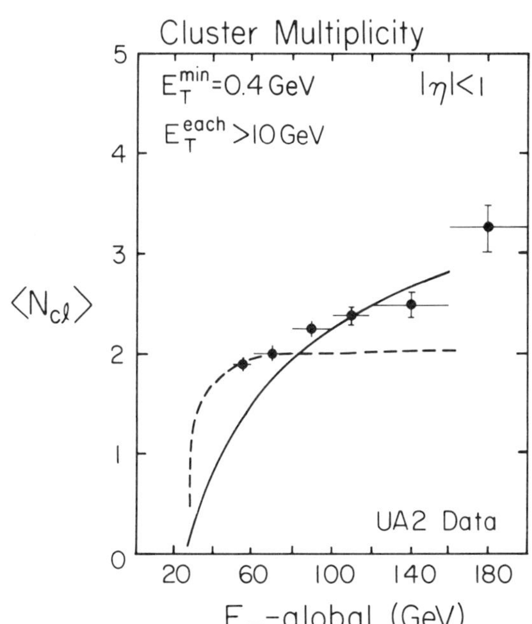

FIGURE 5. Average number of clusters, each of which has E_T(each) > 10 GeV, for cells of size, $\Delta\theta\Delta\phi = 10° \times 15°$, with $|\eta| < 1$ (240 cells) and E_T(min) = 0.4 GeV, versus the global E_T resulting from the QCD parton-shower Monte Carlo model (solid curve) and the naive parton model (dashed curve) for $\bar{p}p$ collisions at $W = 540$ GeV. The data are from UA2.[16]

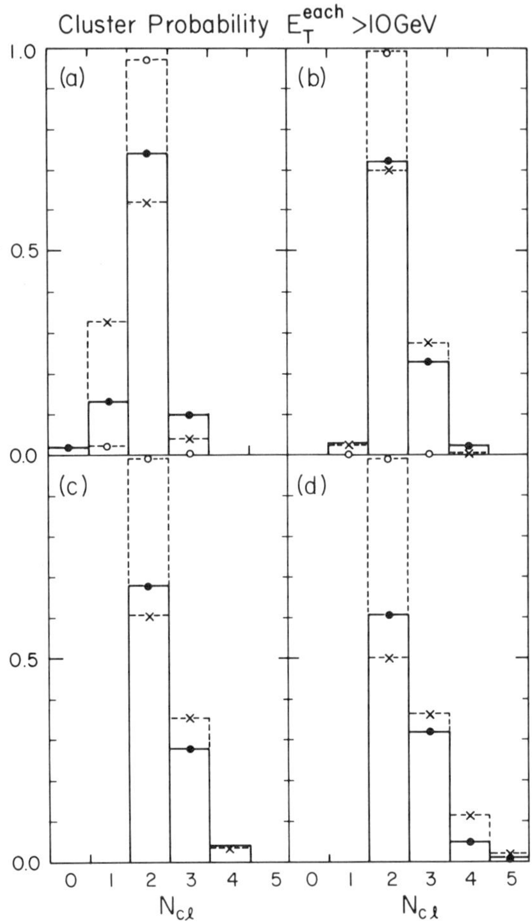

FIGURE 6. Probability of finding in a given event N clusters, each of which has E_T(each) > 10 GeV, for cells of size, $\Delta\theta\Delta\phi = 10° \times 15°$, with $|\eta| < 1$ (240 cells), resulting from the QCD parton-shower Monte Carlo model (dashed lines and crosses) and the naive parton model (dashed lines and open circles) for $\bar{p}p$ collisions at $W = 540$ GeV with global E_T values in the range: (a) $60 < E_T < 80$ GeV, (b) $80 < E_T < 100$ GeV, (c) $100 < E_T < 120$ GeV, (d) $120 < E_T < 160$ GeV. The data (solid lines and solid dots) are from UA2.[16]

E_T(each) > 10 GeV. As the global E_T increases, the number of events with just one cluster increases to about 50% and then decreases. Similarly, the percentage of events with two clusters with E_T(each) > 10 GeV increases to a maximum of about 70% and then decreases as the number of events with three and four clusters increases. At global E_T values above 150 GeV, one would expect three and four cluster events to dominate over two cluster events. In QCD, multiple cluster topologies will begin to dominate at large global E_T values regardless of the minimum E_T(each) chosen. This is in sharp contrast to the naive parton model in which the percentage of events with two clusters increases rapidly to 100% and remains 100% as the global E_T is increased. The UA2 data[16] shown in FIGURE 7 are in good qualitative agreement with the QCD parton-shower predictions. FIGURE 6 and FIGURE 7 do show, however, a tendency at large global E_T for the QCD parton-shower Monte Carlo to overestimate the number of events with three and four clusters and underestimate the number of two cluster events.

FIGURE 7 shows the percentage of events with three and four clusters with E_T(each) > 10 GeV. However, it does not show how the E_T of each cluster is distributed. Typically in multicluster events, two of the clusters will contain large amounts of E_T, with the third and fourth cluster carrying much less E_T. This can be seen in FIGURE 8, where cluster ratios are plotted versus the global E_T trigger. At global E_T of 150 GeV, the second cluster produced by the QCD Monte Carlo carries, on the average, 80% of the transverse energy of the first cluster, and the third cluster has, on the average, only about 30% of the transverse energy of the second. This should be contrasted with the naive parton model, in which one would expect at E_T(global) = 150 GeV the third cluster to carry, on the average, only about 2% of the transverse

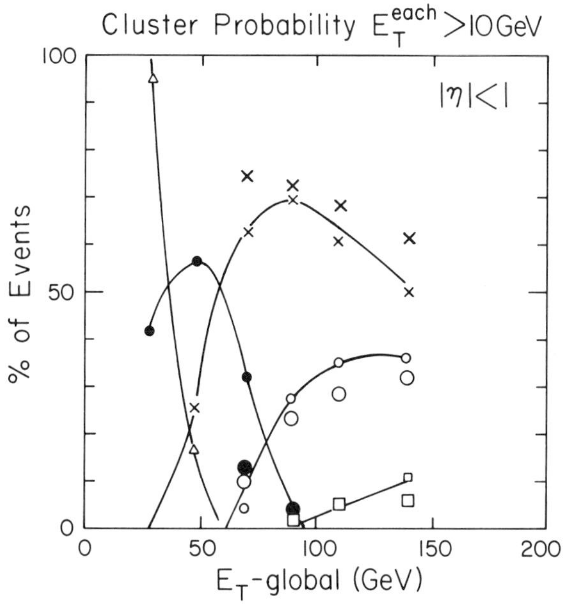

FIGURE 7. Percentage of events with N clusters, each of which has E_T(each) > 10 Gev, for cells of size, $\Delta\theta\Delta\phi$ = 10° × 15°, with $|\eta| < 1$ (240 cells), versus global E_T resulting from the QCD parton-shower Monte Carlo model (solid curves) for $\bar{p}p$ collisions at W = 540 GeV (N = 0 is indicated by open triangles; N = 1 by solid dots; N = 2 by crosses; N = 3 by open circles; N = 4 by open squares). The data are from UA2[16] and are the same as in FIGURE 6 (N = 1 is indicated by large solid circles; N = 2 by large crosses; N = 3 by large open circles; N = 4 by large open squares).

energy of the second cluster. The UA2 data[16] show qualitative agreement with the QCD parton-shower Monte Carlo model; however, at large global E_T values, the QCD model predicts values of R_{32} that are somewhat larger than the UA2 data points.

FIGURE 9 shows the fraction of the global transverse energy carried by the leading cluster, h_1, and by the first two clusters, h_{12}, versus E_T(global). Here, agreement with the UA2 data is not so good. At large E_T(global), the data show more transverse energy being carried by the first two clusters than predicted by the QCD parton-shower Monte Carlo. However, the data are still a long way from the expectations of

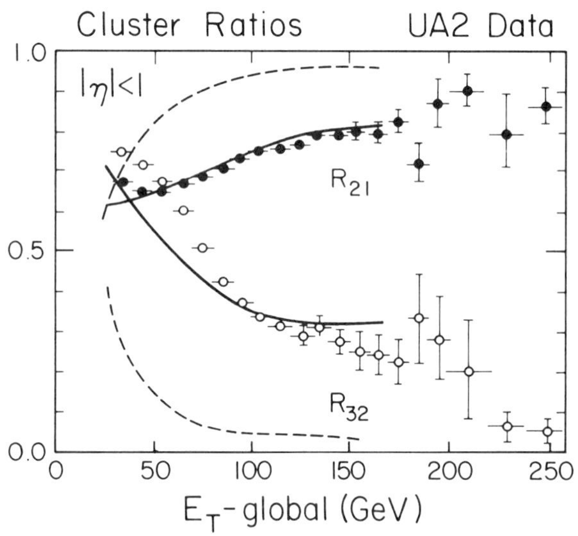

FIGURE 8. Average ratio of transverse energy carried by the first two clusters, $R_{21} = E_T(2)/E_T(1)$, and by the second two clusters, $R_{32} = E_T(3)/E_T(2)$, versus global E_T with $\eta < 1$ resulting from the QCD parton-shower Monte Carlo model (solid curves) and the naive parton model (dashed curves) for $\bar{p}p$ collisions at $W = 540$ GeV. The data are from UA2.[16]

the naive parton model. There is also disagreement at low global E_T, where the QCD Monte Carlo predicts too much transverse energy for the leading and the first two clusters. The QCD Monte Carlo predictions for h_1 and h_{12} lie above the UA2 data at low global E_T and below the data for large global E_T. I must admit that I do not understand the disagreement at low E_T.

The mean values presented in FIGURE 8 and FIGURE 9 are a bit misleading since, event-by-event, there are large deviations from these means values both in the UA2 data[16] and in the QCD parton-shower Monte Carlo. For example, in the global E_T bin, $80 < E_T < 90$ GeV, 22% of the UA2 events have $h_{12} < 0.5$ even though the mean value of h_{12} is about 0.7.[17] Typically, in QCD, there are large event-by-event fluctuations, so one rarely finds a single event that looks like the average of all the events.

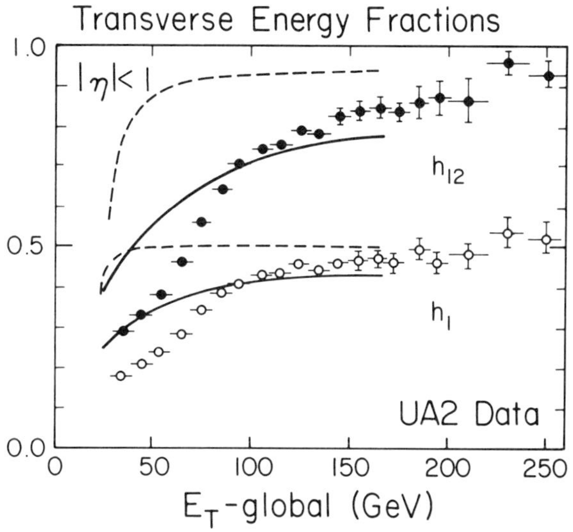

FIGURE 9. Average fraction of global transverse energy carried by the leading cluster, $h_1 = E_T(1)/E_T(\text{global})$, and by the top two clusters, $h_{12} = (E_T(1) + E_T(2))/E_T(\text{global})$, versus global E_T with $|\eta| < 1$ resulting from the QCD parton-shower Monte Carlo model (solid curves) and the naive parton model (dashed curves) for $\bar{p}p$ collisions at $W = 540$ GeV. The data are from UA2.[16]

TABLE 1. Top Three Clusters and Top Three UA1-Jets for a Global E_T Trigger, $150 < E_T$ (global) < 175 GeV, $|\eta| < 4$, Resulting from the QCD Parton-Shower Monte Carlo Model with Cell Size, $\Delta\eta\Delta\phi = 0.2 \times 15°$ (960 cells)[a]

			CLUSTERS[b]		
#	$\langle E_T$ (GeV)$\rangle$	$\langle R\rangle$	$\langle N$ (had)$\rangle$	$\langle N$ (cell)$\rangle$	$\langle M$ (GeV)$\rangle$
1	48.8 ± 9.7	0.903	20.4 ± 9.9	10.1 ± 5.3	11.5 ± 6.2
2	38.6	0.854	14.9 ± 7.0	7.4 ± 3.7	7.5 ± 4.2
3	16.0	0.524	7.3 ± 4.4	4.5 ± 2.5	2.4 ± 2.1
			UA1-JETS[c]		
#	$\langle E_T$(GeV)$\rangle$	$\langle R\rangle$	$\langle N$(had)$\rangle$	$\langle N$(cell)$\rangle$	$\langle M$(GeV)$\rangle$
1	52.4 ± 9.3	0.883	28.5 ± 9.9	18.8 ± 6.8	16.4 ± 5.3
2	43.7	0.820	27.1 ± 8.7	17.6 ± 5.9	14.6 ± 5.2
3	19.6	0.468	20.0 ± 7.5	13.5 ± 8.1	8.5 ± 3.7

[a]The ± sign refers, not to the error, but to the standard deviations from the mean, $\pm\,\sigma$, and R refers to the fraction of the transverse energy, E_T, arising from hadrons of type 3 and 4 (see FIGURE 2). For this global E_T bin, $\langle E_T(\text{global})\rangle = 158$ GeV, $\langle N(\text{cell})\rangle = 132 + 25$, $\langle N(\text{had})\rangle = 165 \pm 30$, $\langle N(\text{cell}), E_T > 0.4 \text{ GeV})\rangle = 70 \pm 13$, and $\langle \hat{W}\rangle = 121 \pm 30$ GeV, where $\hat{W}$ is the parton-parton hard scattering CM energy. M_{12} and M_{123} are the invariant masses of the top two and top three clusters (or jets), respectively.

[b]$\langle N(\text{cluster})\rangle = 42.6 \pm 8.5$, $\langle M_{12} - \hat{W}\rangle = -22.9 \pm 23.5$ GeV, $\langle M_{123} - \hat{W}\rangle = 1.7 \pm 28.1$ GeV. % events with N clusters with $E_T(\text{each}) > 15$ GeV: ($N = 2$ is 60.3%; $N = 3$ is 36.7%; $N = 4$ is 2.9%).

[c]$\langle N(\text{jet})\rangle = 3.6 \pm 1.0$, $\langle M_{12} - \hat{W}\rangle = -9.2 \pm 24.9$ GeV, $\langle M_{123} - \hat{W}\rangle = 24.5 \pm 36.0$ GeV. % events with N jets with $E_T(\text{each}) > 15$ GeV: ($N = 2$ is 34%; $N = 3$ is 46.6%; $N = 4$ is 19.2%).

UA1 ANALYSIS

For this analysis, I divide the phase-space into cells of size, $\Delta\eta\Delta\phi = 0.2 \times 15°$, and extend the pseudorapidity range to $|\eta| < 4$ (960 cells). Transverse energy clusters are defined as in the previous section with $E_T(\min) = 0.4$ GeV. In addition, "UA1-jets" are formed using the UA1 jet algorithm.[18] Here, one first considers the "hot" cells (those with transverse energy greater than 2.5 GeV). Hot cells are combined to form a "jet" if they lie within a "distance," $d = (\Delta\eta^2 + \Delta\phi^2)^{1/2} = 1$, from each other, with the jet direction being the vector sum of the momenta of each cell in the jet. Cold cells (those with $E_T < 2.5$ GeV) are added to a jet if $d < 1$ or if the angle of the cold cell relative to the jet is less than 45° and the relative p_T is less than 1 GeV. UA1-jets are not the same as clusters. This can be clearly seen in TABLE 1, which shows the top three clusters and the top three UA1-jets resulting from the QCD parton-shower Monte Carlo for a global E_T trigger in the range of $150 < E_T(\text{global}) < 175$ GeV and $\eta < 4$. For this global E_T bin, the Monte Carlo yields, on the average, 132 cells with nonzero transverse energy and an average particle multiplicity of 165. UA1-jets have, on the average, larger E_T and larger mass, and contain more hadrons than do UA1 clusters. They also obtain a larger fraction of their total E_T from hadrons of type 1 and type 2.

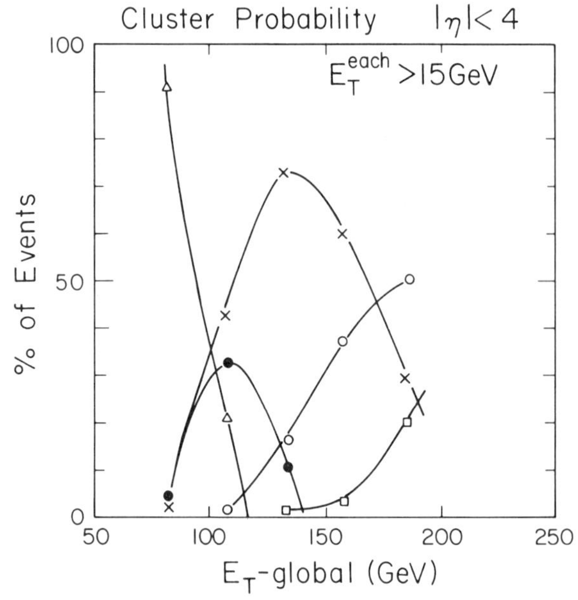

FIGURE 10. Percentage of events with N clusters, each with $E_T(\text{each}) > 15$ GeV, for cells of size, $\Delta\eta\Delta\phi = 0.2 \times 15°$, with $|\eta| < 4$ (960 cells), versus global E_T resulting from the QCD parton-shower Monte Carlo model (solid curves) for $\bar{p}p$ collisions at $W = 540$ GeV ($N = 0$ is indicated by open triangles; $N = 1$ by solid dots; $N = 2$ by crosses; $N = 3$ by open circles; $N = 4$ by open squares).

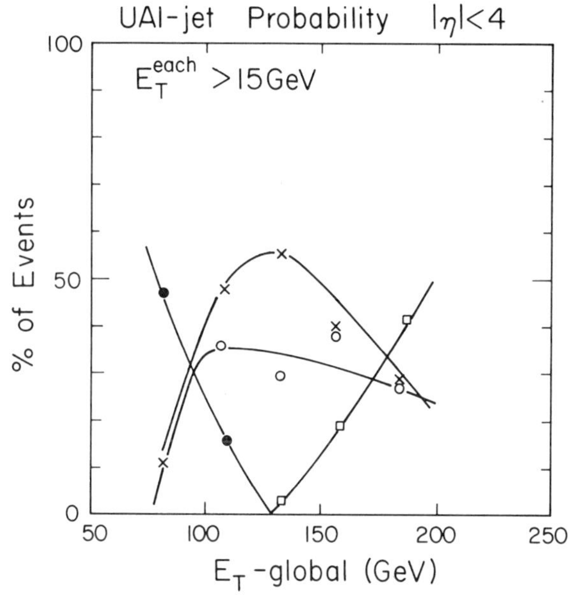

FIGURE 11. Percentage of events with N "UA1-jets," each with E_T(each) > 15 GeV, for cells of size, $\Delta\eta\Delta\phi = 0.2 \times 15°$, with $|\eta| < 4$ (960 cells), versus global E_T resulting from the QCD parton-shower Monte Carlo model (solid curves) for $\bar{p}p$ collisions at $W = 540$ GeV ($N = 1$ is indicated by solid dots; $N = 2$ by crosses; $N = 3$ by open circles; $N = 4$ by open squares).

For this global E_T bin, the Monte Carlo produced, on the average, 42.6 clusters per event, but only 3.6 UA1-jets. This is because each UA1-jet must contain at least one hot cell with $E_T > 2.5$ GeV, while a cluster only needs to contain cells with $E_T(\min) >$ 0.4 GeV.[19]

TABLE 1 also gives the probability of an event containing N clusters, each with E_T(each) > 15 GeV, resulting from the Monte Carlo for the global E_T range of 150 < E_T(global) < 175 GeV. Events with global E_T in this range contain two clusters with E_T(each) > 15 GeV 60.3% of the time, three clusters 36.7% of the time, and four clusters 2.9% of the time. On the other hand, events in this global E_T range contain two UA1-jets with E_T(each) > 15 GeV 34% of the time, three jets 46.6% of the time, and four jets 19.2% of the time. This can be seen more clearly in FIGURE 10 and FIGURE 11, which show the percentage of events with N clusters and with N UA1-jets, respectively, each of which has a transverse energy of E_T(each) > 15 GeV, versus the global E_T with $|\eta| < 4$ resulting from the QCD parton-shower Monte Carlo. Both these figures are similar to FIGURE 7, with multicluster and multijet topologies dominating at large global E_T values. For example, at global E_T values of 200 GeV, it is more likely that an event from the QCD Monte Carlo contains four UA1-jets with E_T(each) > 15 GeV than for it to contain just two jets.

Of course, the number of jets in a given event depends on the rapidity interval one examines. FIGURE 12 shows the predicted percentage of events with N UA1-jets each

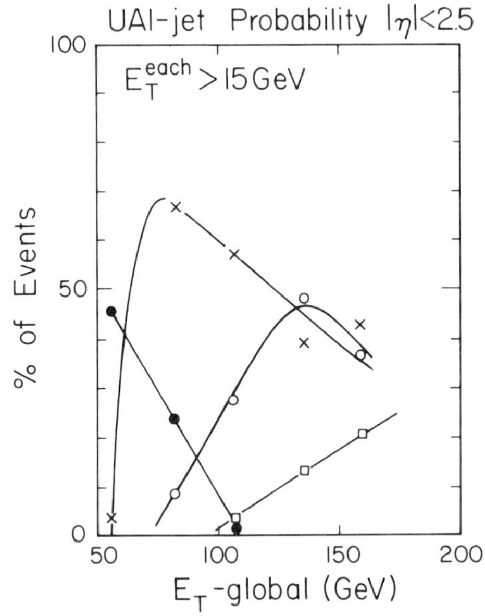

FIGURE 12. Same as FIGURE 11, but for the smaller pseudorapidity range, $|\eta| < 2.5$.

having E_T(each) > 15 GeV for the rapidity interval of $|\eta| < 2.5$. In this smaller rapidity interval, there are fewer events with three and four jets because some jets are lost outside the detection region.

The leading UA1-jet in TABLE 1 has, on the average, a transverse energy of about 52 GeV. There is, of course, another way to arrive at a sample of 52 GeV jets. One can "trigger" on 52 GeV jets rather than on global E_T as was done in TABLE 1. TABLE 2 shows the top three UA1-jets resulting from the QCD Monte Carlo for a bin in which

TABLE 2. Top Three UA1-Jets for Jet Trigger, $50 < E_T(\text{jet}) < 60$ GeV, $|\eta| < 4$, Resulting from the QCD Parton-Shower Monte Carlo Model with Cell Size, $\Delta\eta\Delta\phi = 0.2 \times 15°$ (960 cells)[a]

	UA1-JETS[b]				
#	$\langle E_T(\text{GeV})\rangle$	$\langle R\rangle$	$\langle N(\text{had})\rangle$	$\langle N(\text{cell})\rangle$	$\langle M(\text{Gev})\rangle$
1	54.1 ± 2.8 GeV	0.917	27.7 ± 10	18.5 ± 7.2	16.7 ± 5.2
2	45.3	0.857	26.3 ± 8.5	16.9 ± 5.8	14.5 ± 4.6
3	12.3	0.452	18.2 ± 7.4	9.6 ± 8.7	7.7 ± 3.8

[a]The ± sign refers, not to the error, but to the standard deviations from the mean, $\pm\sigma$, and R refers to the fraction of the transverse energy, E_T, arising from hadrons of type 3 and 4 (see FIGURE 2). For this E_T(jet) bin, $\langle E_T(\text{global})\rangle = 146 \pm 17$ GeV, $\langle N(\text{had})\rangle = 142 \pm 32$, $\langle N(\text{cell})\rangle = 106 \pm 23$, and $\langle \hat{W}\rangle = 118 \pm 28$ GeV, where $\hat{W}$ is the parton-parton hard scattering CM energy. M_{12} and M_{123} are the invariant masses of the top two and top three jets, respectively.

[b]$\langle N(\text{jet})\rangle = 3.0 \pm 1.0$, $\langle M_{12} - \hat{W}\rangle = -2.0 \pm 24.3$ GeV, $\langle M_{123} - \hat{W}\rangle = 20.4 \pm 34$ GeV. % events with N jets with E_T(each) > 15 GeV: ($N = 2$ is 61.9%; $N = 3$ is 30.6%; $N = 4$ is 7.4%).

the top jet is in the range of $50 < E_T(\text{jet}) < 60$ GeV and $|\eta| < 4$. The leading jets in TABLE 1 and TABLE 2 have, on the average, almost the same E_T (52 GeV compared to 54 GeV); however, the overall event topologies are somewhat different. In TABLE 2, the average global E_T is about 146 GeV and the average particle multiplicity is 142, whereas in TABLE 1, they are 159 GeV and 165, respectively. For the same leading jet E_T, a jet trigger results in less global E_T than a global E_T trigger. This can be seen quite clearly in FIGURE 13, which shows the leading jet E_T plotted versus the global E_T for both a jet trigger and a global E_T trigger.

You may feel that this is an obvious point—"one gets what one asks for"; however, the degree to which "one gets what one asks for" is a measure of QCD. In the naive

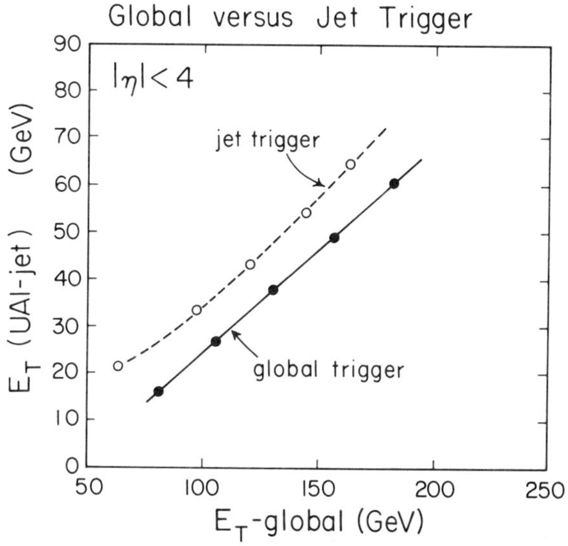

FIGURE 13. Average "UA1-jet" transverse energy versus the average global transverse energy resulting from the QCD parton-shower Monte Carlo model for a jet trigger (dashed curve) and a global E_T trigger (solid curve) for $\bar{p}p$ collisions at $W = 540$ GeV with $|\eta| < 4$.

parton model, there is only one underlying topology, four jets, and no matter what one asks for, one sees four jets. (For the naive parton model, the two curves in FIGURE 13 essentially coincide.) In QCD, one faces a multitude of possible event topologies and the choice of trigger (or even the manner in which one bins the data) preferentially selects certain topologies over others. For example, in TABLE 2, 62% of the events contain two UA1-jets each with $E_T(\text{each}) > 15$ GeV, 31% of the events contain three jets, and only 7% contain four jets. This is quite different from the values of 34%, 47%, and 19%, respectively, found in TABLE 1, even though the leading jet transverse energy is, on the average, about the same. FIGURE 14 shows the percent of events with N UA1-jets, each with $E_T(\text{each}) > 15$ GeV, versus the leading jet E_T resulting from the QCD Monte Carlo. The results in FIGURE 14 are different from those in FIGURE 12, where the multijet probabilities were plotted versus global E_T. There are considerably less three- and four-jet events and more two-jet events resulting from the Monte Carlo in FIGURE 14 than in FIGURE 12. The UA1 data[18] are in qualitative agreement with the Monte Carlo; however, as with the UA2 analysis, the data show less three-jet and more two-jet events than the Monte Carlo predicts.

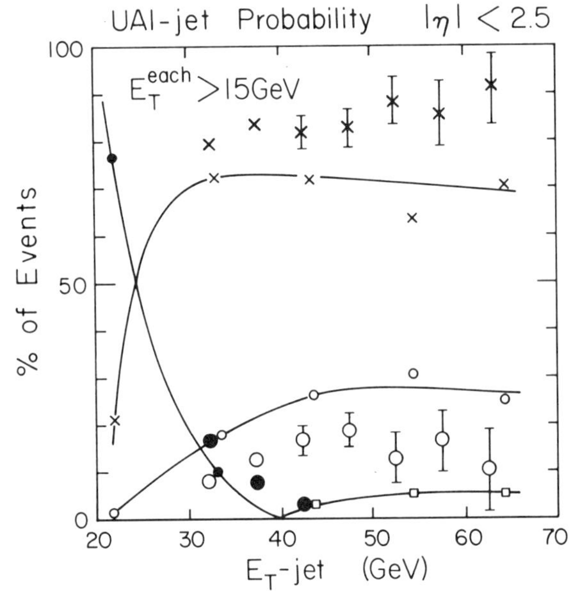

FIGURE 14. Percentage of events with N "UA1-jets," each with E_T(each) > 15 GeV, for cells of size, $\Delta\eta\Delta\phi = 0.2 \times 15°$, with $|\eta| < 2.5$, versus the transverse energy of the leading UA1-jet, E_T(jet), resulting from the QCD parton-shower Monte Carlo model (solid curves) for $\bar{p}p$ collisions at $W = 540$ GeV ($N = 1$ is indicated by solid dots; $N = 2$ by crosses; $N = 3$ by open circles; $N = 4$ by open squares). The data are from UA1[18] ($N = 1$ is indicated by large solid circles; $N = 2$ by large crosses; $N = 3$ by large open circles).

SUMMARY AND CONCLUSIONS

Data from UA1 and UA2 on the production of multijets with large transverse energy are in good qualitative agreement with the expectations of QCD and deviate greatly from the predictions of the naive parton model. The CERN $\bar{p}p$ collider data show strong qualitative evidence for QCD. This evidence is, in my opinion, as strong, if not stronger, than the evidence from data on e^+e^- annihilations or deep inelastic lepton-hadron scattering.

In all instances, the data from UA1 and UA2 are in qualitative agreement with the QCD parton-shower Monte Carlo model. However, there is a consistent trend for the model to overestimate the frequency of three- and four-jet topologies relative to two-jet topologies. This can be seen in FIGURE 7, where the Monte Carlo predicts a larger percentage of events with three and four clusters (each with E_T(each) > 10 GeV) than is seen in the UA2 data. It can also be seen in FIGURE 14, where the Monte Carlo predicts a larger fraction of events with three jets (each with E_T(each) > 15 GeV) than is seen in the UA1 data. Precise agreement with the QCD parton-shower Monte Carlo is, however, not to be expected. In fact, if the data perfectly fit the Monte Carlo predictions, it would not be any more significant than the qualitative agreement found here. The real significance is in the disagreement with the naive parton model predictions and the fact that this disagreement is always in the direction predicted by the QCD Monte Carlo. Now that I have performed this analysis, I could tune parameters in the Monte Carlo to fit the data better. In particular, to get slightly less three- and four-jet topologies, I could decrease the QCD parameter, Λ, or I could increase the cutoffs, μ_c^A and μ_c^B, or both.[20]

Certainly the QCD parton-shower Monte Carlo model contains many ad hoc guesses, some of which are surely wrong. Therefore, one cannot take detailed comparisons with data too seriously and because of the uncertainties at both the parton and hadronization level, one cannot expect to fit experiment perfectly. Nevertheless, the model does contain all the features expected from QCD and in performing this analysis I have learned some important lessons which I would now like to summarize.

(1) Hadrons of type 1 and 2 are often referred to as the "background" in a high p_T event. However, these hadrons play an integral role in the overall high E_T event topology. There are dynamical correlations between these hadrons and the hadrons of type 3 and 4 (see FIGURE 2). A high E_T event is not a superposition of two high p_T e^+e^--like jets on top of a "minimum bias" background.

(2) In QCD, mean values can be quite misleading. You rarely find a single event that looks like the average of all the events. There are large event-by-event fluctuations. Often you hear that QCD jets become "fat" as the energy increases. What this means is that if you average all the events, the average becomes "fat." On an event-by-event basis, the "fatter" average is arising from multiple jet topologies that are being included in the average. In TABLES 1 and 2, I have included some standard deviations from the mean as well as the mean values themselves. For example, TABLE 1 shows large fluctuations in the energy of a cluster (or jet) and in the number of hadrons in a cluster (or jet).

(3) UA1-jets are not the same as UA2 clusters. Both are hadron-hadron scattering observables. (The UA1-jet is a complicated observable.) Personally, I prefer clusters over "jet-finding" algorithms. On the other hand, the UA1 group has been extremely successful in doing "jet" spectroscopy and hence, their algorithm cannot be criticized. I do believe, however, that in the future we will move more in the direction of "pattern recognition" and away from "jet finding." In order to study the rich topological structures expected from QCD, one needs to examine and identify patterns of transverse energy deposition. This can be done by defining a transverse energy grid and looking at the cluster patterns.

(4) It is quite often said that at high E_T one sees "clean two-jet" events at the CERN collider or that "two-jet" events dominate at large E_T. Whether events become "cleaner" or "dirtier" at large E_T depends on your point of view. In fact, events become both "cleaner" and "dirtier" at large E_T. If you count all

jets with E_T greater than some fixed amount, for example, E_T(each) > 15 GeV as in FIGURE 11, then multijet events dominate at large E_T. Events with three, four, five, etc., jets will become more likely than two-jet events. On the other hand, two of these jets will carry, on the average, the dominant amount of global E_T as seen in FIGURE 9. The point to remember is that although the leading two jets (or clusters) carry the dominant amount of global E_T, on an event-by-event basis, the remaining E_T is not distributed uniformly in phase-space. The remaining E_T is also clustered into jets.

ACKNOWLEDGMENTS

I would like to thank Peter Jenni and the UA2 collaboration for providing me with (unpublished) data on cluster probabilities. In addition, I am grateful to J. Hansen, Jim Rohlf, and V. Vuillemin for useful discussions concerning the UA1 and UA2 data. Finally, I congratulate Martin Block and all the Aspen Crew for making the First Aspen Winter Meeting interesting, stimulating, and fun. I hope there will be many more to come.

NOTES AND REFERENCES

1. FIELD, R. D. Jets and event topologies at the SSC. Contribution to the 1984 Snowmass Summer Study on the Design and Utilization of the Superconducting Super Collider, Snowmass, Colorado, June 23–July 13, 1984.
2. HUSTON, J. A comparison of the predictions of ISAJET and FIELDAJET for jet production at the SSC. Contribution to the 1984 Snowmass Summer Study on the Design and Utilization of the Superconducting Super Collider, Snowmass, Colorado, June 23–July 13, 1984.
3. FOX, G. C. & S. WOLFRAM. 1980. Nucl. Phys. **B168:** 285; FOX, G. C. & S. WOLFRAM. 1979. A gallimaufry of e^+e^- annihilation shapes. Caltech preprint CALT-68-723. Unpublished; FOX, G. C. 1981. Lectures presented at the 1981 SLAC Summer School, CALT–68–863; FOX, G. C. & R. L. KELLY. 1981. Proton-antiproton collider physics. Caltech preprint CALT-68-890 and AIP Conference Proceedings No. 85; FIELD, R. D., G. C. FOX & R. L. KELLY. 1982. Phys. Lett. **119B:** 439.
4. ODORICO, R. 1982. Nucl. Phys. **B199:** 189; 1982. Phys. Lett. **118B:** 151; 1983. Nucl. Phys. **B288:** 381.
5. FIELD, R. D. 1981. The production of partons and hadrons in e^+e^- annihilations—Quark and gluon jet models. UFTP-81-12 and Proceedings of the Conference on Perturbative QCD, Florida State University, AIP Conference Proceedings No. 74, Particle and Fields Subseries No. 24; FIELD, R. D. 1982. Jet formation in QCD. Invited talk presented at the XIII International Symposium on Multiparticle Dynamics, Volendam, The Netherlands, June 1982, UFTP-82-28.
6. MARCHESINI, G. & B. R. WEBBER. 1983. CERN preprint TH-3525; WEBBER, B. R. 1983. CERN preprint TH-3669.
7. PAIGE, F. E. & S. D. PROTOPROPESCU. 1980. BNL-29777.
8. GOTTSCHALK, T. 1985. A backwards evolution Monte Carlo model for initial state parton showers. Caltech preprint CALT-68-1241.
9. To leading order, there is an ambiguity in the choice of the energy scale, Q. All choices that increase with the parton-parton transverse momentum, $\hat{p}_T$, are equivalent. This choice saves computer time.
10. FIELD, R. D. & S. WOLFRAM. 1983. Nucl. Phys. **B213:** 65.
11. FIELD, R. D. & G. C. FOX. 1982. QCD Monte Carlo models for e^+e^- annihilations and

hadron-hadron collisions. Invited talk presented at the Europhysics Study Conference on Jet Dynamics in Quark and Lepton Interactions, Erice, Sicily, CALT-68-965 and UFTP-82-30.

12. FIELD, R. D. & R. P. FEYNMAN. 1977. Phys. Rev. **D15:** 2590; 1978. Nucl. Phys. **B138:** 1.
13. Even at the parton level, there are theoretical uncertainties and inaccuracies in the Monte Carlo models. For example, for *2* → *3* subprocesses, the "leading pole" approximation can differ from the exact result by as much as a factor of two in certain regions of phase-space. See ELLIS, R. K. & J. OWENS. Contribution to the 1984 Snowmass Summer Study on the Design and Utilization of the Superconducting Super Collider, Snowmass, Colorado, June 23–July 13, 1984.
14. This type of transverse energy grid and cluster recognition is the heart of the CDF trigger and has been discussed by FRISCH, H. & M. SCHOCKET in CDF trigger notes.
15. The transverse energy grid and clusters discussed here are precisely those defined by UA2 (reference 16) except that I do not "split" clusters with a "valley" of E_T greater than 5 GeV into smaller clusters.
16. BAGNAIA, P. *et al.* UA2 COLLABORATION. Measurement of very large transverse momentum jet production at the CERN $\bar{\text{p}}$p Collider. CERN preprint CERN-EP/84/12.
17. FIELD, R. D. 1984. Jet topologies in hadron-hadron collisions. Invited talk presented at the 4th Topical Workshop on Proton Antiproton Collider Physics, Bern, Switzerland, UFTP-84-3.
18. ARNISON, G. *et al.* UA1 COLLABORATION. Hadronic jet production at the CERN proton-antiproton collider. CERN preprint CERN-EP/83-118.
19. Recently, the UA1 collaboration have lowered the criterion for a "hot cell" to 1.0 GeV rather than 2.5 GeV. This greatly increases the number of UA1-jets per event, but does not affect as much the number of jets with E_T(each) > 15 GeV since most of these jets have at least one cell with $E_T > 2.5$ GeV.
20. Observables such as the probability of finding N clusters (or jets) each with E_T(each) > 15 GeV do depend sensitively on the cutoffs, μ_c^A and μ_c^B. For example, if one chooses very large values for the cutoffs, then no additional parton radiation is produced and one reproduces the naive four-jet parton model.

CDF at Fermilab: A Progress Report

A. V. TOLLESTRUP

CDF Collaboration[a]

Fermi National Accelerator Laboratory[b]

Batavia, Illinois 60510

INTRODUCTION

The $\bar{p}p$ colliding beam facility at Fermilab is rapidly nearing completion and will be commissioned in 1986. The machine will provide center-of-mass energies near 2 TeV at a luminosity of 10^{30} cm^{-2} sec^{-1}. The first detector facility at B0, CDF, will be described briefly in this report. A second collision area at D0 is described elsewhere in these proceedings. In addition, other more specialized experiments, including an experiment to measure the total and elastic scattering cross section and an experiment to search for effects of a quark-gluon plasma, are planned for other collision areas.

BRIEF DETECTOR DESCRIPTION

FIGURE 1 shows a perspective drawing of the detector, and a cross section is shown in FIGURE 2. The detector components can be described best by considering what instrumentation is being constructed for each angular interval.

(1) *Inside the Beam Pipe*
The region inside the beam pipe is being fitted with Si detectors to make measurements of the elastic scattering cross sections and to implement a trigger for diffractive processes. There are four stations—two on either side of the IR. The nearest stations are in front of the low β quadrupoles, while the two farthest are actually in the machine lattice where they are used in conjunction with a lattice dipole to form a very high resolution spectrometer for very high momentum forward particles.

(2) *Beam Pipe to 2°*
This region will be instrumented in the future.

(3) *2° to 10°*
This region is covered by a vertex TPC (VTPC) and a forward drift chamber using a "Bicycle Wheel" type construction with planes of radial sense wires. Behind these chambers are the forward EM and hadron calorimeters. The forward muon system consisting of magnetized iron toroids and drift chambers covers this region and extends out to 20°.

[a]Appendix 1 gives a list of the members and their institutions.

[b]Operated by the Universities Research Association under contract with the U.S. Department of Energy.

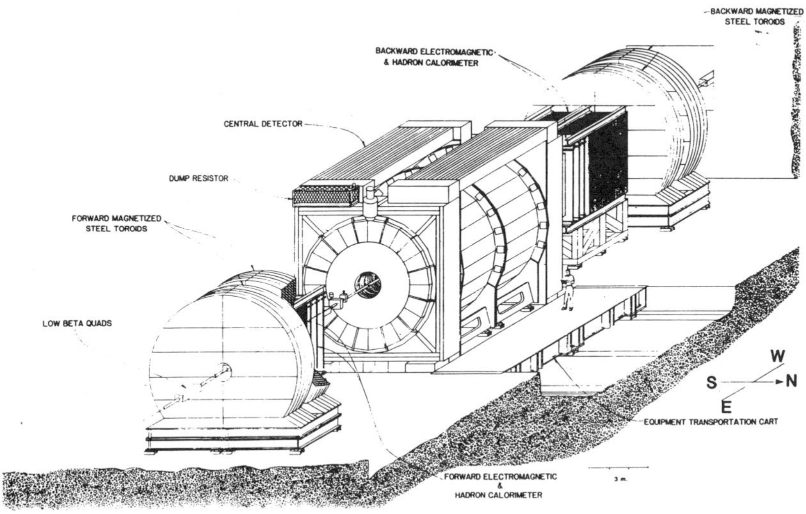

FIGURE 1. A perspective drawing of the detector.

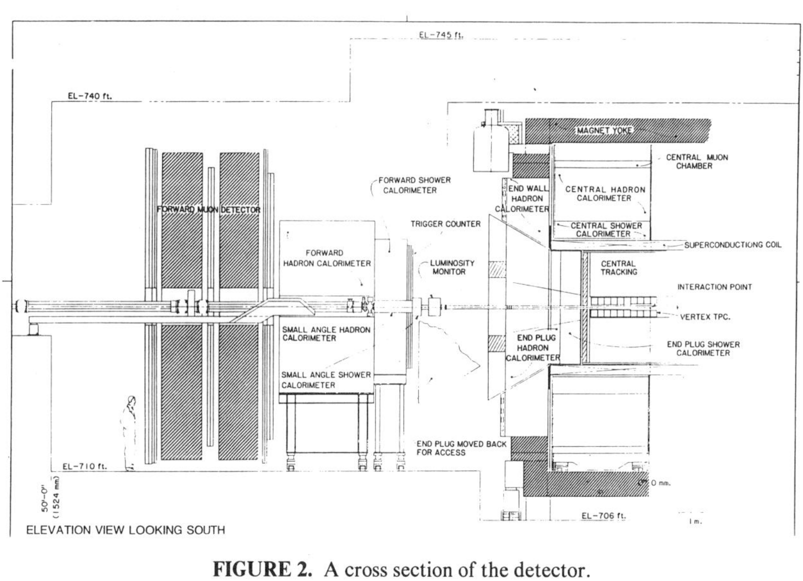

FIGURE 2. A cross section of the detector.

TABLE 1A. EM Calorimeters

	Forward		End Plug			Central
Angular range	2° to 10°		10° to 37°			37° to 90°
Size of towers in η	(18) $\Delta\eta \sim 0.1$		(15) $0.05 \leq \Delta\eta \leq 0.1$			(10) $0.087 \leq \Delta\eta \leq 0.13$
Angular size in ϕ	$\Delta\eta = 5°$		$\Delta\phi = 5°$			$\Delta\phi = 15°$
Material	Pb-gas tubes		Pb-gas tubes			Pb-scintillator
Number of layers	6	24	5	20	13	20
R.l. each layer	0.85 X_0	0.85 X_0	0.56 X_0	0.56 X_0	0.56 X_0	0.95 X_0
Layers in section	5 X_0	20.5 X_0	2.8 X_0	11.26 X_0	7.32 X_0	19 X_0
Total X_0	25.5 X_0		22.4 X_0			19 X_0
$\delta E/E$	$\sim 30\%/\sqrt{E}$		$\sim 24\%/\sqrt{E}$			$14\%/\sqrt{E}$
Position resolution	1–5 mm		1–2 mm			1.5–3 mm

TABLE 1B. Hadron Calorimeters

	Forward	End Plug	End Wall	Central
Angular range	2° to 10°	10° to 30°	30° to 45°	~45° to 90°
Construction	2 in. Fe + gas tubes	2 in. Fe + gas tubes	2 in. Fe + scint	1 in. Fe + scint
Tower size in η	$\Delta\eta = 0.1$	$\Delta\eta = 0.09$	$0.079 \leq \Delta\eta \leq 0.117$	$0.1 \leq \Delta\eta \leq 0.15$
Tower size in ϕ	$\Delta\phi = 5°$	$\Delta\phi = 5°$	$\Delta\phi = 15°$	15°
$\delta E/E$		~14% at 50 GeV	~14% at 50 GeV	$70\%/\sqrt{E}$

(4) *10° to 30°*
The end plugs of the magnet carry the EM and hadron calorimeters that service this region. Tracking is accomplished jointly by means of the VTPC and the central tracking chamber (CTC).

(5) *30° to 90°*
Starting on axis, we first come to the VTPC, which gives good track information in the RZ plane. Next the CTC in conjunction with the 1.5 T magnetic field allows for precision measurement of individual tracks. There are five superlayers of cells with axial wires and four superlayers whose wires are skew with respect to the axis and are used to give stereo information. At the outside of the CTC and just inside the coil, an axial set of tubes in the avalanche mode gives z determination on tracks by means of current division to a precision of $\sim\pm 3$ mm. This helps in track reconstruction. After passing through the coil $\sim 1\ X_0$ and $0.5\ \lambda_0$, one finds the central EM and hadron calorimetry. A strip chamber is located at approximately EM shower maximum to aid in the identifying of EM showers and the measuring of their positions to ~3 mm. Finally, on the outside of the hadron calorimeter, there is located a set of tubes in the avalanche mode that are instrumented with TDCs and where current division occurs to provide z and ϕ information on the muons that succeed in penetrating the calorimeter.

TABLES 1A and 1B show the characteristics of the various calorimeter elements of the detector. In general, the size of the projective calorimeter towers is such that a

TABLE 2A. VTPC

Drift distance (axial): 150 mm
Sense wires/octant: 16 (22) for $R = 3\text{–}1/2$ (2)
Pads/octant: 24 in three rows of 5; 8; 11 pads
Sense wire spacing: 5.5 mm
Sense wire/pad separation: 10 mm
Pad size: 12 × 45 mm
Gas: Argon/Methane—80 percent: 20 percent
Drift field: 200 V/cm
Total number of sense wires: 2176
Total number of pads: 2448
Average radiation length: 1.5 percent
Resolution in Z: 200–350 μ
Resolution in $R\phi$: 250 μ (isolated tracks)

TABLE 2B. CTC

Number of superlayers: 9
Stereo angle: 0°; 3°; 0°; −3°; 0°; +3°; 0°; −3°; 0°
Sense wires/layer: 12; 6; 12; 6; 12; 6; 12; 6; 12
Number supercells/superlayers: 30; 42; 48; 60; 72; 84; 96; 108; 120
Maximum drift distance: 35 mm
Sense wire spacing: 10 mm in plane of wires
Electric field at center of drift space: 1000 V/cm
Radius innermost sense wires: 309 mm
Radius outermost sense wires: 1320 mm
Length of wires: 3214 mm
Total number of sense wires: 6156
Total number of wires: ~36,000
Axial load on endplates: 17 tons
Gas—Argon:Ethane:Alcohol: 49.65%:49.65%:0.7%
Resolution: $\leq 200\ \mu$/wire
Double track resolution: ≤ 5 mm or 100 nsec
Z resolution: ~4 mm
Momentum resolution: $\Delta p_t/p_t \leq 0.002\ p_t$ in GeV at 90°

QCD jet will span several, thus allowing an accurate determination of the energy flow vector. In addition, the EM strip chambers have elements that are comparable in size to an EM shower in order to help with the identification of electrons and photons.

The characteristics of the tracking elements of the detector are given in TABLES 2A, B, C, and D. The system is comprised of two drift chamber systems—the CTC and the FTC. These provide track information from 2° to 178°. The vertex TDC provides complementary information to these chambers near the origin. It will also allow multiple interactions to be rapidly identified. The Si detectors in the beam pipe will cover very small angles, thereby providing a means to identify diffractive events.

The characteristics of the Tevatron are such that with a $\beta^* = 1$ meter, the beam σ is about 60 μ. With such a small interaction transverse cross section, it is anticipated that a very high resolution vertex detector will be extremely useful for identifying long-lived particles. For this reason, a silicon strip detector that fits inside the VTPC is being developed by the Pisa group. The technology is challenging as it must fit in a very confined space and operate at high rates in a location where there is always the possibility of severe radiation damage by accidental beam loss.

The solenoid was manufactured by Hitachi in Japan and is 3 m in diameter and 5 m long. The field of 1.5 T is produced by a NbTi superconductor stabilized by a sheath of extruded very pure aluminum. The conductor is cooled and supported by means of an

TABLE 2C. FTC

Angular coverage: 2°–10°
Inner radius: 12.5 cm
Outer radius: 72.5 cm
Cell: 5° in ϕ or 7272 sectors
Sense wires/sector: 20
ϕ accuracy: $\sigma_{rms} = 130\ \mu$
Total TDC: 3020
Total ADC: 1728

external cylinder of aluminum with tubing that carries liquid He welded to its surface. This configuration was achieved by first winding the coil on an internal mandrel. This structure was then insulated with glass-epoxy, cured, and then ground to a precision round cylinder. The outer support cylinder was heated to expand it radially by 3 mm and was then slipped over the coil assembly after which the inner mandrel was removed. The coil has been tested without a yoke to 60 percent of the operating current and is extremely stable and well behaved. Full scale tests in the yoke will take place in early 1985.

At present, the detector is being rapidly assembled. The central detector calorimetry is nearly complete and is being calibrated in test beams. The magnet yoke is complete and the solenoid is ready for test. The tracking chambers are being manufactured and the VTPC will be ready for preliminary beam tests in the summer of 1985. The spring of 1986 will see all components ready for test with the beam in preparation for the first physics run in the fall of that year.

TABLE 2D. Si Strip Vertex Detector

Length: 67 cm, 4 sections
Layers: 4
Thickness: (1) 200 μ
(2) 250 μ
(3) 250 μ
(4) 250 μ
Electrode spacing: (1) 200 μm
(2) 250 μm
(3) 250 μm
(4) 250 μm
Spatial resolution in $R\phi$: ~20 μm on all layers
Number of ϕ electrodes: (1) 936
(2) 1248
(3) 1740
(4) 2232
Total ϕ electrodes: 6156

APPENDIX 1
CDF COLLABORATION

Argonne National Laboratory: R. Diebold, E. May, L. Nodulman, J. Proudfoot, J. Schlereth, P. Schoessow, N. Solomey, D. Underwood, R. Wagner & A. B. Wicklund.

Brandeis University: J. Bensinger, L. DeMortier, L. Kirsch, W. Lewis, R. Poster, L. Spencer & S. Tarem.

University of Chicago: M. Campbell, H. Frisch, C. Grosso-Pilcher, T. Liss, G. Redlinger, M. Shochet, R. Snider, S. Somalwar & Y. D. Tsay.

Fermi National Accelerator Laboratory: M. Atac, E. Barsotti, P. Berge, M. Binkley, A. Brenner, T. Carroll, C. Day, T. F. Droege, J. Elias, B. Foster, J. Freeman, I. Gaines, D. Hanssen, H. Jensen, D. Johnson, R. Kadel, H. Kautzky, R. Kephart, C. Newman-Holmes, J. Patrick, R. Perchonok, D. Quarrie, D. Theriot, A. Tollerstrup, K. Turner, C. Van Ingen, R. Vidal, R. Wagner & J. Yoh.

INFN–Frascati, Italy: S. Bartolucci, S. Bertolucci, M. Cordelli, M. Curatolo, P. Esposito, P. Giromini, M. Pallotta & A. Sansoni.

Fukui University, Japan: M. Kawaguchi.

Harvard University: C. Blocker, G. Brandenburg, D. Brown, R. Carey, M. Eaton, E. Kearns, R. Schwitters & R. St. Denis.

University of Illinois: G. Ascoli, S. Cihangir, R. Downing, S. Errede, L. Holloway, I. Karlinger, H. Keutelian, U. Kruse, R. Sard, D. Smith & T. Westhusing.

KEK, Japan: Y. Fukui, S. Mikamo & M. Mishina.

Lawrence Berkeley Laboratory: W. Carithers, W. Chinowsky, R. Ely, M. Franklin, J. Siegrist & E. Wood.

University of Pennsylvania: D. Bauer, D. Connor, J. Cooper, S. Hahn, N. Lockyer, M. Miller, M. Newcomer, T. Rohlay, R. VanBerg & H. H. Williams.

INFN–University of Pisa, Italy: F. Bedeschi, G. Bellettini, L. Bosisio, F. Cervelli, R. Del Fabbro, M. Dell'Orso, A. Di Virgilio, E. Focardi, S. Galeotti, P. Giannetti, M. A. Giorgi, A. Menzione, D. Passuello, L. Ristori, A. Scribano, A. Stefanini & G. Toneli.

Purdue University: V. Barnes, K. Chadwick, R. Christian, A. Garfinkel, S. Kuhlmann, A. Laasanen & J. Simmons.

Rockefeller University: S. Belforte, T. Chapin, K. Goulianos, S. White & C. Young.

Rutgers University: A. Beretvas, A. Caracappa, T. Devlin, K. Krueger, L. Scott, T. Watts & T. Yang.

Saga University, Japan: A. Murakami.

Texas A&M University: A. Giordana, P. McIntyre, T. Meyer, M. Shepko & R. Webb.

ICRR, Tokyo University, Japan: Y. Muraki.

Tsukuba University, Japan: Y. Asano, Y. Hayashide, M. Ito, T. Kamon, Y. Kikuchi, S. Kim, K. Kondo, S. Miyashita, H. Miyata, S. Mori, Y. Morita, H. Sekiguchi, M. Shibata, Y. Takaiwa, K. Takikawa, A. Yamashita & K. Yasuoka.

University of Wisconsin: D. Cline, R. Handler, R. Loveless, L. Pondrom, D. Reeder, J. Rhoades & M. Sheaff.

The DØ Detector at the Tevatron Collider[a]

MARIS A. ABOLINS

Department of Physics and Astronomy
Michigan State University
East Lansing, Michigan 48824

INTRODUCTION

The idea of a major detector in the DØ area of the Tevatron Collider was conceived in the summer of 1983. The Fermilab Program Advisory Committee suggested to Director Lederman that he form a substantial collaboration to design and build a second detector for the Tevatron Collider to complement CDF. He invited Professor Paul Grannis of the State University of New York at Stony Brook to form such a collaboration with the aim of constructing in a timely way a powerful detector complementary to CDF in its capabilities. The group, whose current membership and institutional affiliations are given in Appendix A, was assembled that summer and fall, and the first DØ Design Report,[1] which specified the detector parameters and capabilities, was finished by December 1983. This report has been refined and improved and has been issued as a new version[2] in December of 1984. Most of the technical content of this paper is derived from that Design Report.

In the first section of this paper, I discuss some of the physics that motivated our design and summarize the reasons for building a new detector at the Tevatron Collider at this time. In the second section, I outline the basic capabilities of the detector. The main components of the detector are discussed in the third section, the electronics and data acquisition system in the fourth, and finally a possible construction schedule is presented in the fifth.

RATIONALE FOR THE DETECTOR

There are a number of reasons for building a second large detector at the Tevatron Collider:

(1) First of all, it is simply a good idea to have more than one detector at a major facility. More physics can be extracted, and in case of a major breakdown of a single detector, the program does not grind to a halt. Two detectors with complementary capabilities working side by side can have a stimulating effect on one another. The example of UA1 and UA2 is well known to everyone as is the Cornell experience with CLEO and CUSB.

(2) A detector designed today can profit from the CERN experience. The lessons of having 4π calorimetric coverage with optimal resolution can be implemented.

(3) The DØ detector is a truly modern detector embodying many of the principles

[a]This work was supported by National Science Foundation grant no. PHY-8117256.

thought to be needed for detectors at the SSC. It is worth noting that the 1984 Snowmass SSC summer study arrived at a number of possible detector designs that incorporated the essential features of the DØ design.[3]

DESIGN CONSIDERATIONS

A rational detector design should be based firmly on the physics measurements that it is expected to make. In the case of DØ, this is not a simple problem as the construction schedule is long and the field is evolving rapidly by virtue of the measurements at CERN and the expected turn-on of CDF in the next two years. Despite these difficulties, we have attempted to assess what the topics of interest are likely to be four years from now. We imagine that the physics will divide itself into two general classes:

(1) Precision measurements of the parameters of the standard model;
(2) Measurements of new phenomena.

We estimate that by 1988, our earliest possible turn-on date, CERN will have accumulated sufficient running time to produce about 200 events of $Z^0 \rightarrow e^+e^-$, as well as about 2000 events of $W^\pm \rightarrow e^\pm\nu$. With these kinds of statistics, most of the features of W and Z production and decay will be well established. Two-jet physics will have been studied out to a transverse momentum (p_T) of $\approx$200 GeV/c and the top quark from $W \rightarrow t\bar{b}$ will have been well studied. We note, however, that with an assumed luminosity of 10^{30} cm^{-2} sec^{-1}, a four month run at 50% efficiency at the Tevatron will yield 1500 $Z^0 \rightarrow e^+e^-$ and 15,000 $W^\pm \rightarrow e^\pm\nu$. Thus, on the basis of statistics and on the superior performance of the detector, certain topics dealing with the standard model should still be of interest at that late date. These include:

(1) Precision measurements of the masses of the Z and W to extract the Weinberg angle from $\sin\theta_W = 38.5/m_W$ and the parameter, $\rho = m_W^2/(m_Z^2\cos^2\theta_W)$.
(2) A precision measurement of the width of the Z^0 to place limits on the number of neutrinos.
(3) Measurements of the decay branching ratios of the Z^0 and $W^\pm$.
(4) Search for narrow states decaying electromagnetically.
(5) A study of quark decays of $W^\pm$ and Z^0.
(6) A study of jet production to extract α_s.

In the area of new phenomena there is a plethora of possibilities, and the last few years have seen an enormous amount of theoretical activity concerning conceivable new physics to be expected at collider energies. These have been summarized in many excellent reviews[4] and I shall list only some of the more discussed possibilities. These include:

(1) New heavy vector bosons.
(2) New heavy leptons.
(3) Families of supersymmetric particles.
(4) New heavy quarks.
(5) Technicolor particles.
(6) Quark compositeness.

Although there is no experimental evidence for any of these items, there are a number of tantalizing hints[5] from CERN that would indicate that all is not well with the Standard Model. Unraveling this new physics will probably be the most important task for the collider and its detectors.

An appraisal of the CERN experience and a careful analysis of the "signatures" of the expected physics suggest that our detector must measure well the following "particles":

(1) Quarks and gluons in their manifestations as jets.
(2) Charged leptons, i.e., electrons and muons.
(3) Neutrinos and other neutral noninteracting particles. This, of course, implies a measurement of the missing transverse energy.

Moreover, it is clear that the performance of the detector must be understood sufficiently well so that genuine new physics is not confused with the artifacts of the tails of resolution functions and misunderstood efficiencies or acceptances.

With these considerations in mind, we then make the following design choices and compromises:

(1) Uranium-Liquid Argon (ULA) calorimetry throughout the detector for optimum energy resolution, for symmetric response to hadronic and electromagnetic (EM) energy, for its compactness, and for its inherent stability.
(2) A high degree of lateral and longitudinal segmentation for optimum e/π separation and flexibility in measuring electrons and muons near the cores of jets.
(3) No central magnetic field. This simplifies the problem of making the acceptance uniform, makes it easier to limit cracks and dead areas, and ameliorates the central tracking problem.
(4) Transition Radiation Detectors (TRDs) for electron identification to compensate for the absence of central field.
(5) Magnetized iron toroids outside the calorimetry for muon momentum tagging.
(6) Complete calorimeter coverage down to 1° from the beam line.

FIGURE 1 shows a perspective cutaway drawing of the detector indicating its major components. FIGURE 2 is a plan view displaying more clearly the overall construction as well as the major individual subsystems: the central calorimeter (CC), the endcap calorimeters (EC), and the plug calorimeter (PC). Also pictured are the central muon drift tube detectors, CMA,CMB, and CMC, sandwiching the central iron toroid, CF. The corresponding end muon detectors are labeled EMA, EMB, and EMC, and the end toroid, EF. Note also the movable cart on which the detector rides for transportation into and out of the collision area.

DETECTOR COMPONENTS

Calorimeter

There are five separate calorimeter modules—the central, the two endcaps, and the two plugs. Their essential features and parameters are summarized in TABLE 1. It is useful here to emphasize a few points.

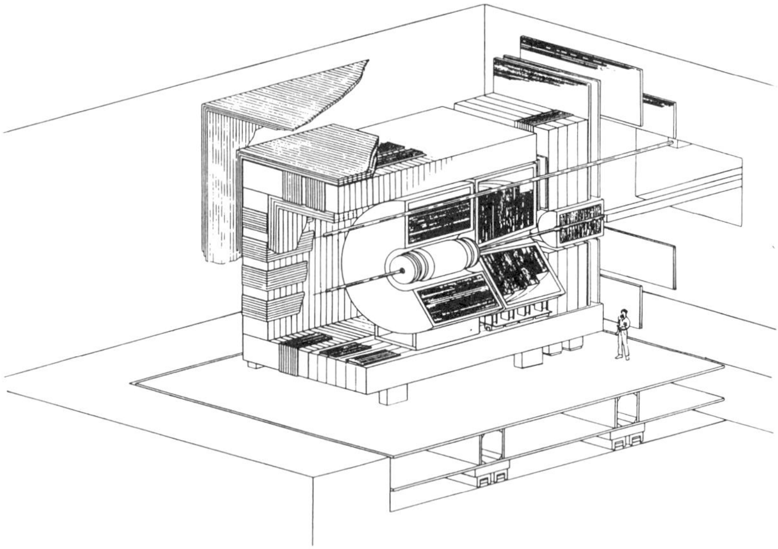

FIGURE 1. Perspective view of the DØ detector.

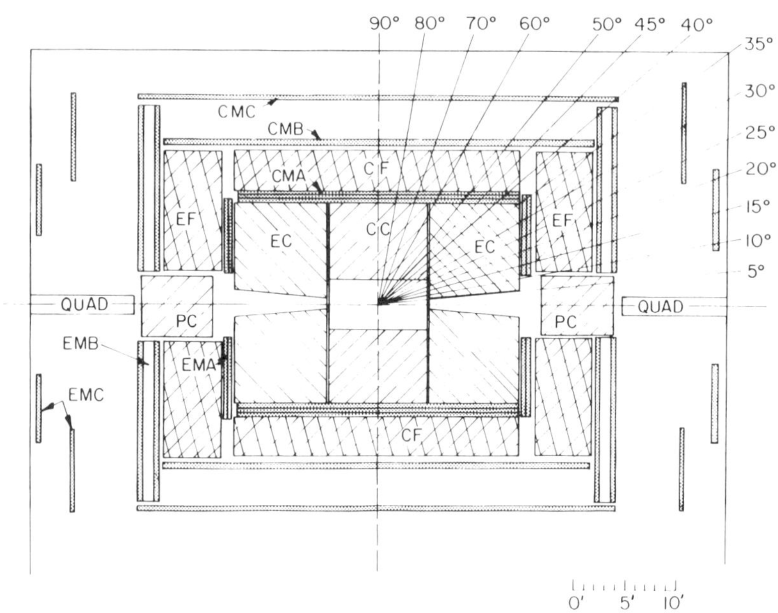

FIGURE 2. Plan view of the detector.

The absorbing material is uranium and copper with varying plate thicknesses. In the central region, there are three separate sampling structures divided radially. The first is 20 radiation lengths or 1 absorption length (λ_0) of 2 mm uranium plates for EM shower detection. The second is the hadronic section with 5 mm uranium plates totaling about 4 λ_0. The last is the leakage section of 1.8 λ_0 in copper plates. In the endcaps, the space restrictions are not as demanding so the uranium plates alternate with equal thickness copper plates throughout. In the PC, the spatial constraints are severe so all 9 λ_0 are of 6 mm thick uranium plates.

In the CC (EC), there are 32 (16) azimuthal modules. These are divided into three separate radial pieces that are rotated half a module width to distribute the cracks.

Each gap in the calorimeter contains readout boards for the collection of ionization from adjacent argon gaps. Copper clad G-10 is used with the pad structure corresponding to the projective towers etched onto the boards. In all the three types of calorimeters, CC, EC, and PC, the minimum transverse segmentation given in TABLE 1 is further subdivided into four smaller sections in the EM compartment corresponding to the shower peak for optimum shower position measurement.

TABLE 1. Calorimeter Performance Specifications and Parameters

Energy Resolution			
EM		$12\%/\sqrt{E}$ (GeV)	
Hadron		$40\%/\sqrt{E}$ (GeV)	
Segmentation ($\Delta\Phi \times \Delta\eta$)	*CC*	*EC*	*PC*
Transverse	$.1 \times .1$	$.1 \times .1$	$.1 \times .1$
Longitudinal	4 + 4	4 + 3	6
Position Resolution (EM)		8 mm/$\sqrt{E}$ (GeV)	
Solid Angle Coverage		$-4.7 \leq \eta \leq 4.7$	
Channel Count	*Towers*	*Channels*	
CC	1536	13,952	
EC(each)	1536	12,192	
PC(each)	2352	2112	
Coverage	*z*	*r*	
CC	≤ 113	$75 \leq r \leq 222$	
EC	$142 \leq z \leq 340$	≤ 207	
PC	$500 \leq z \leq 650$	≤ 65	

The expected performance of the calorimeter has been modeled by Monte Carlo and some of these results are shown in FIGURE 3, where we see plotted a number of cross sections for interesting processes versus the missing transverse component of the momentum (Mp_T).

Muon Spectrometer

The muon detection and measurement system in DØ consists of a large amount of absorber, the calorimeter, and three sets of position measuring proportion drift tubes

(PDTs) sandwiching the solid iron toroids. The essential features of this system are summarized in TABLE 2. The coverage goes down to $\theta = 11°$ with respect to the beams. The field in the CF is about 2 T and in EF, it varies from 2 T near the inner radius to 1.6 T at the outer. The total amount of material traversed by a particle is sufficiently large to make the probability of hadronic punch-through negligible. At 90°, the calorimeter is 6.9 λ_0 and the toroids are another 6.4 λ_0 for a total of 13.3 λ_0. At the minimum angle, the calorimeter is 8.9 λ_0 and the iron is 9.1 λ_0 for a total of 18 λ_0.

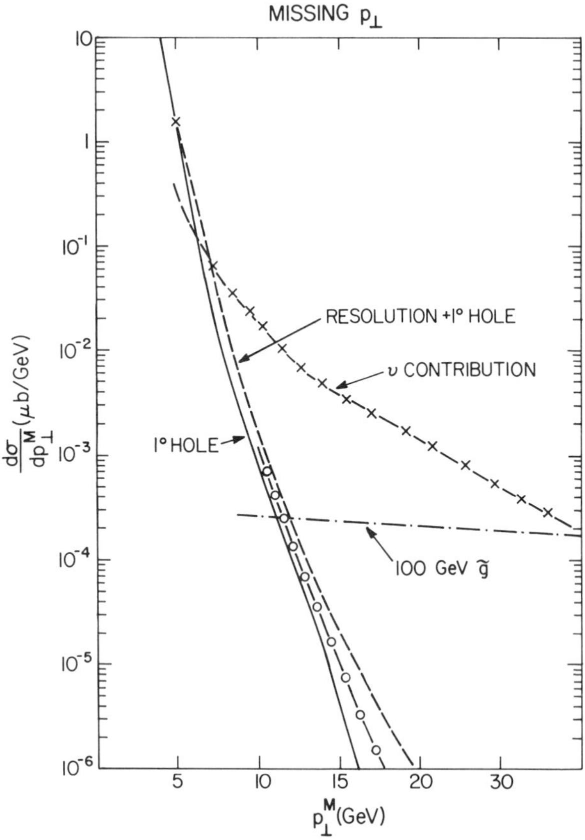

FIGURE 3. Cross section versus Mp_T for several interesting processes.

The muon trajectories are measured with planes of PDTs arranged in three layers, A, B, and C (see FIGURE 2), with the wires parallel to the magnetic field lines. The individual tubes have a cross section of 10×2.5 cm and a length of 600 cm. The orthogonal, transverse coordinates are measured by two series of sawtooth pad detectors. Their arrangement relative to the wires is indicated in FIGURE 4. There are a

TABLE 2. Muon System Performance Specifications

Momentum Resolution in Multiple Scattering Limit		≈20%	
Maximum p_T for Sign Determination (3σ)		≈300 GeV/c	
Toroids	z	r_{min}	r_{max}
Central	$z \leqq 425$	308	410
End(each)	$447 \leqq z \leqq 559$	86	410
Number of Drift Tubes		12,720	
Readouts per Channel		5	

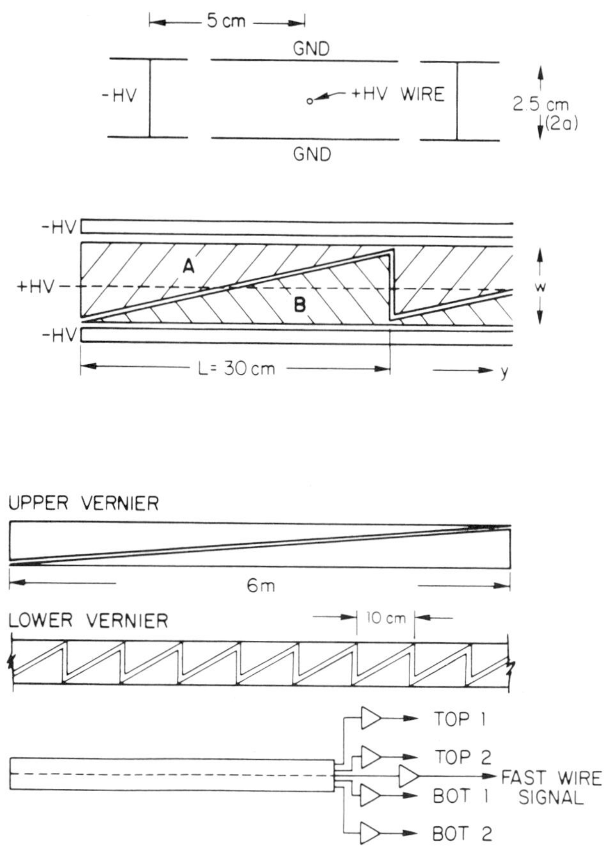

FIGURE 4. The muon detection system.

TABLE 3. Tracking System Specifications

Drift Chambers				
Resolution in $r\Delta\Phi$	200 μm			
Resolution in z	5 mm			
Two track separation	5 mm in $r\Delta\Phi$			
	5 cm in z			
Rejection of two overlapped tracks by dE/dx	50 to 1			
Solid angle coverage	$-3.5 \leqq \eta \leqq 3.5$			
	Position Coverage		Number of Wires	
	r	z	Sense	Delay
Vertex Detector	$3 \leqq r \leqq 10$	$-40 \leqq z \leqq 40$		
Central Drift(each)	$40 \leqq r \leqq 70$	$-74 \leqq z \leqq 74$	768	256
End Drift(each)	$r \leqq 65$	$80 \leqq z \leqq 137$	784	224

total of five readouts per cell, which would imply a large number of channels to be read out, but since the occupation number of the cells is expected to be small, the total number of cables needed is reduced significantly by multiplexing signals by 32 and 64.

Tracking System

The current design goals and parameters for the tracking system are summarized in TABLE 3. A cross-sectional view of the tracking system sandwiching the TRDs is shown in FIGURE 5. Note that there are modules for a vertex detector, a central drift detector, and two sets of end drift detectors. The arrangement of the central tracker in

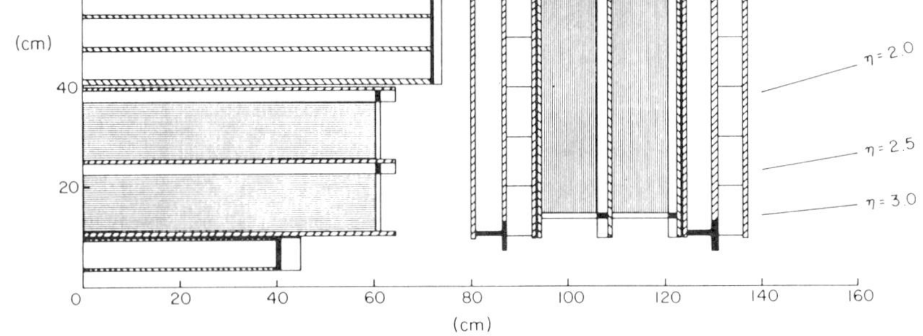

FIGURE 5. A cross-sectional view of the tracking system.

a view perpendicular to the beam is shown in FIGURE 6, where we can see the individual cells with multiple sense wires. The assembly of the endcap detectors of the individual cells is shown in FIGURE 7.

Both position and pulse-height measurements are recorded in order to reject electron-positron pairs from γ conversions. In addition to the six sense wires in each cell, there are two delay lines at the inner and outer radii of each cell oriented parallel to the wires. A timing measurement on the delay lines can locate a track to 5 mm in the coordinate along the beam, whereas the azimuthal position from the drift measurement should have an uncertainty of less than 200 μm. Note that alternate drift wires are slightly offset relative to one another for ambiguity resolution.

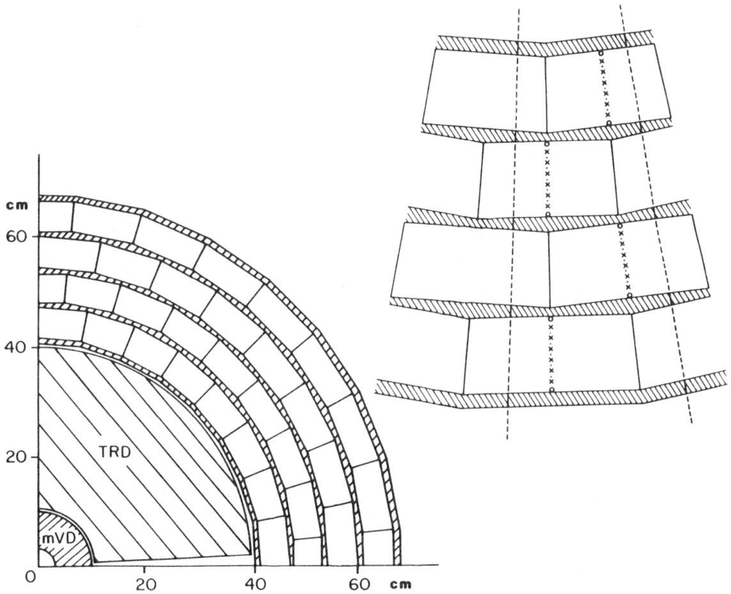

FIGURE 6. The tracking system viewed perpendicular to the beam.

Transition Radiation Detector

The TRD shown in FIGURE 5 consists of three separate detectors: a barrel central TRD and two planar endcap TRDs. Each of the detectors is constructed of two separate units of radiators and detectors. The radiator is made of 640 30-μm thick lithium plates separated by 150-μm gaps. The X-ray detector, separated from the radiator by a thin window, is a xenon-filled proportional wire chamber divided into conversion and multiplication zones as shown in FIGURE 8. The X rays produced in the radiator interact in the one cm thick conversion zone via the photoeffect to produce localized charge clusters, which then drift under the influence of an electric field toward the multiplication zone to produce localized avalanches on the wires. To

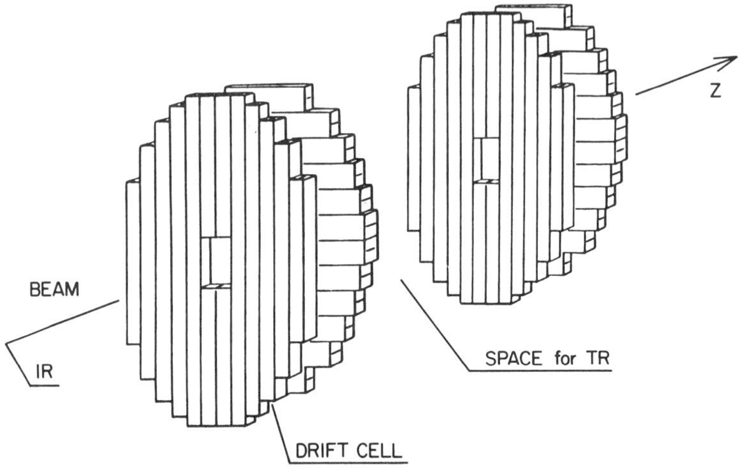

FIGURE 7. Construction of endcap tracking.

combat the background due to inevitable δ-rays, it is advantageous to count clusters of energy above some minimum threshold (~3 keV) since this appears to be a better technique than a measurement of the total energy in discriminating against the δ-ray background.[6] Estimates of hadron/electron rejection with realistic assumptions give a factor of ≈50 for our detector.

ELECTRONICS AND COMPUTING

Front End

The advantages of the liquid argon detection technique, its linearity, uniformity, and stability, can only be realized if proper care is taken to protect the minute signals

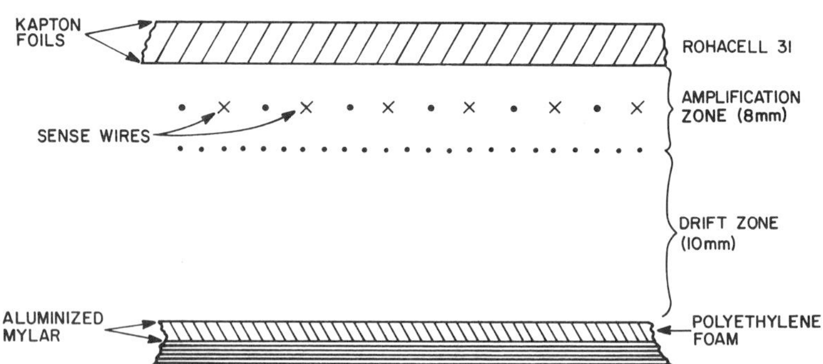

FIGURE 8. The TRD sections.

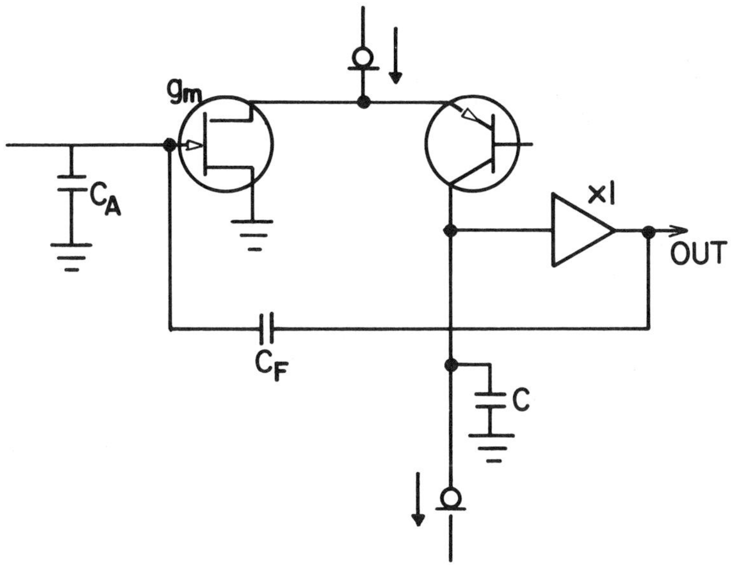

FIGURE 9. The front end amplifier for the calorimeter.

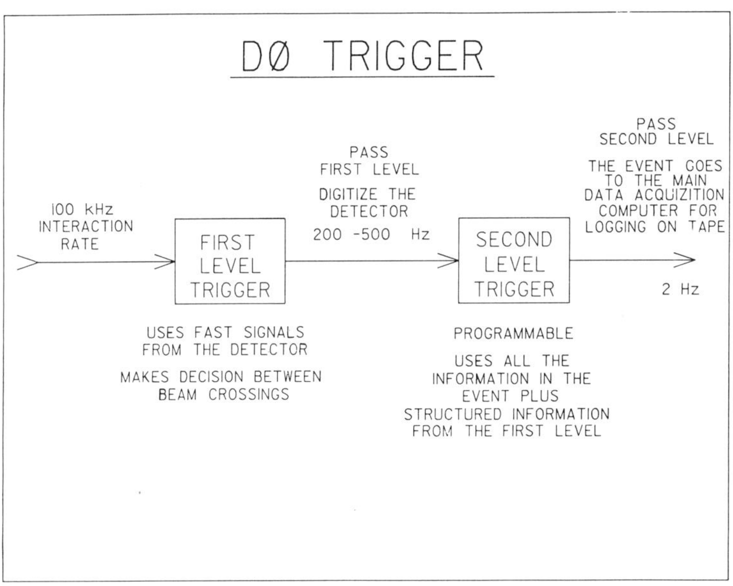

FIGURE 10. Schematic organization of trigger system.

against all forms of noise and if electronics with appropriate noise characteristics are used. The signals, expressed in numbers of electron charges, will range from 5×10^4 for muons traversing different parts of the detector, all the way to $\approx 6 \times 10^8$ electrons for a 500 GeV shower. This implies a dynamic range of 10^5 if 10% accuracy is desired.

The basic circuit for each calorimeter channel is a low-noise charge sensitive preamplifier shown in FIGURE 9. A junction field effect transistor (JFET) is used as the input stage because of its optimum signal to noise characteristics. For a 10 nf capacitor (C_A) across its input, the equivalent noise of 41,000 electrons has been measured. This is a number consistent with our expectations based on the parameters of the devices.

An essential feature of the electronics takes advantage of the periodic nature of the beam crossings (minimum time between bunches ≈ 3.5 μsec). A baseline subtractor (BLS) circuit is used to sample the signal just before crossing and just after. The signals are subtracted and the resulting difference is converted to a DC voltage proportional to the collected charge.

For reasons of noise suppression, the preamps will reside close to the detector in well-shielded enclosures. This will also be true of the BLS circuits. The resulting $\approx$44,000 analog signals will be time-multiplexed to the counting house for input to 14 bit analog to digital converters (ADCs). These ADCs will perform a conversion in ≈ 4 μsec and only a few hundred will be needed if we allow a total conversion time of 700 μsec.

Trigger

At nominal luminosity (10^{30} cm^{-2} sec^{-1}), the interaction rate at the collider will be ≈ 50 kHz. The problem confronting the trigger is to reduce this rate to a manageable tape-writing rate of ≈ 1–2 Hz without sacrificing interesting physics and introducing intractable biases. Our solution to this problem is a multilevel trigger shown schematically in FIGURE 10. In addition to the two levels of trigger indicated, there is an earlier "pre-trigger," which uses scintillators covering the plug calorimeters to signal the presence of a valid collision and from timing, infers its interaction point in the diamond. This also serves the important function of rejecting beam-gas interactions and provides accurate timing signals for later stages of the trigger.

The first level trigger is designed to be deadtime-less. That is, it performs all of its functions in the time between beam crossings. A diagram of the overall structure of this device is shown in FIGURE 11 and a list of the inputs is summarized in TABLE 4. Note that the granularity of the trigger is coarser than that built into the basic structure of the detector. The optimum size of the trigger cells was determined by Monte Carlo.

The calorimeter inputs to the trigger are used to compute the Mp_T and the total transverse energy (E_T). They are used to find clusters of energy, or jets, in both the hadronic and EM sections of the detector. A number of thresholds can be set and the number of clusters found with various thresholds can be used as inputs to further levels of the trigger. All of this information is latched in the "intermediate latch" for use in the computer controlled "AND-OR" matrix. Other inputs, for example from the muon system or from the TRDs, can be used in correlation with the calorimetric information to generate a final trigger. As an example of a trigger based on the calorimeter alone, FIGURE 12 shows the rate as function of Mp_T. Our target rate of 200 Hz at nominal luminosity is achieved with an Mp_T threshold of ≈ 5 GeV/c.

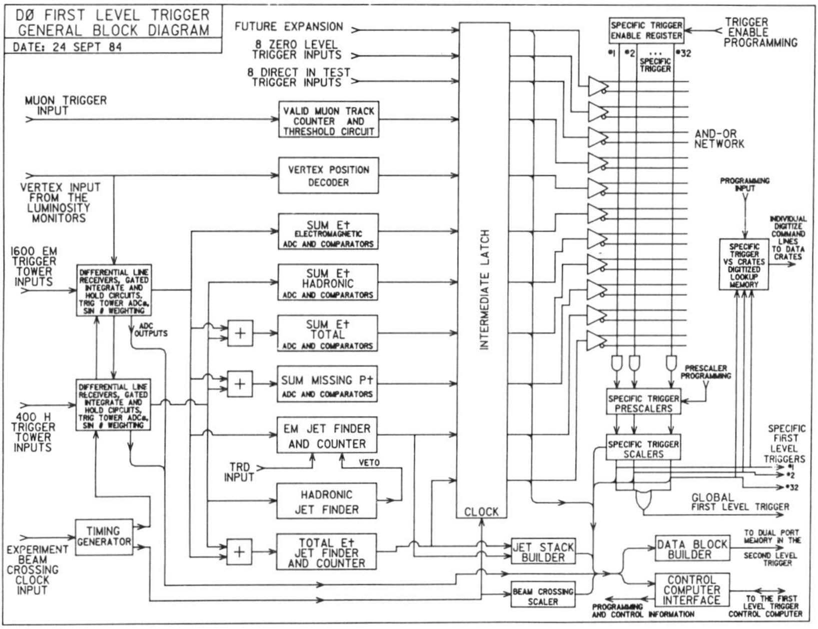

FIGURE 11. The Level 1 trigger structure.

A schematic indication of the operation of the muon trigger is given in FIGURE 13. Correlation of hits in planes "B" and "C" can be used to impose a minimum p_T cut on the muon and to require that it emanate from the interaction diamond. An indication of the problems facing this trigger is given in FIGURE 14, which shows the muon rates from all sources outside the detector. We can see that the problem is largely confined to small angles. A trigger limiting p_T to ≈ 3 GeV/c reduces this rate to a manageable 0.2 Hz. The same muon trigger is used also to suppress cosmic rays. The area of the detector is large, resulting in a substantial (≈ 88 kHz) rate of muons on layer "C." Requiring a hit in "A" as well as the "B" and "C" correlation, already mentioned, lowers this rate to about 300 Hz. Including the non-bend view and the information

TABLE 4. Inputs to Level 1 Trigger

(1) 1600 Electromagnetic Trigger Towers
(2) 400 Hadronic Trigger Towers
(3) Event Vertex Location in z
(4) Muon Trigger Data
(5) Test Data
(6) Vetos from the Zero Level Trigger
(7) Experiment Beam Crossing Clock
(8) TRD Trigger Data
(9) Other Trigger Data

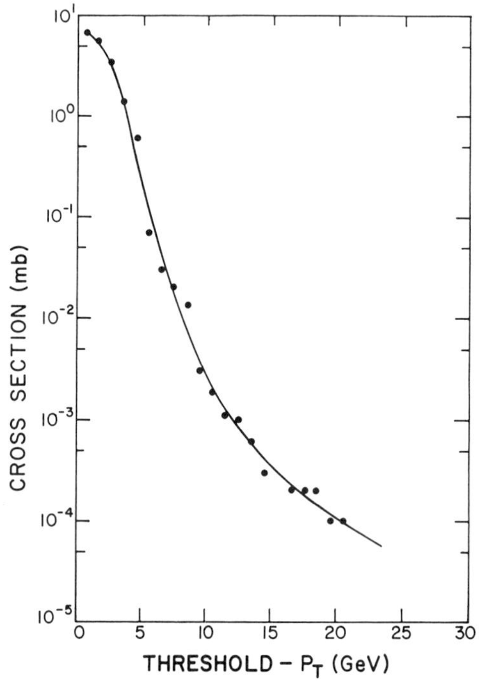

FIGURE 12. The trigger rate as a function of Mp_T.

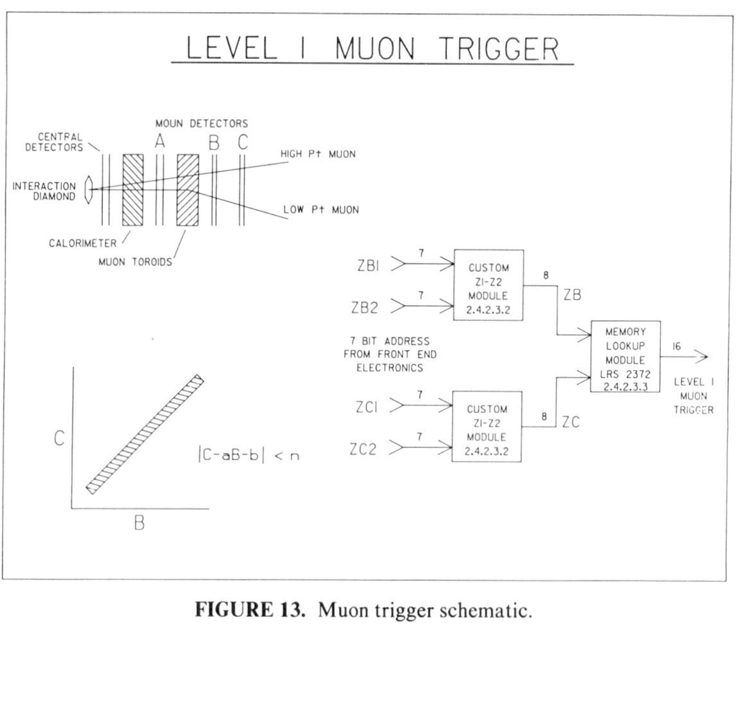

FIGURE 13. Muon trigger schematic.

from adjacent offset cells to refine the timing, we can achieve a rate of ≦ 0.5 Hz, which is tolerable.

The TRD trigger signals are derived from the data acquisition hardware in a manner indicated in FIGURE 15a. Energy clusters exceeding a preset threshold cause the counter to advance, yielding a count proportional to the total energy above threshold if care is taken to shape the pulses appropriately. These individual channels

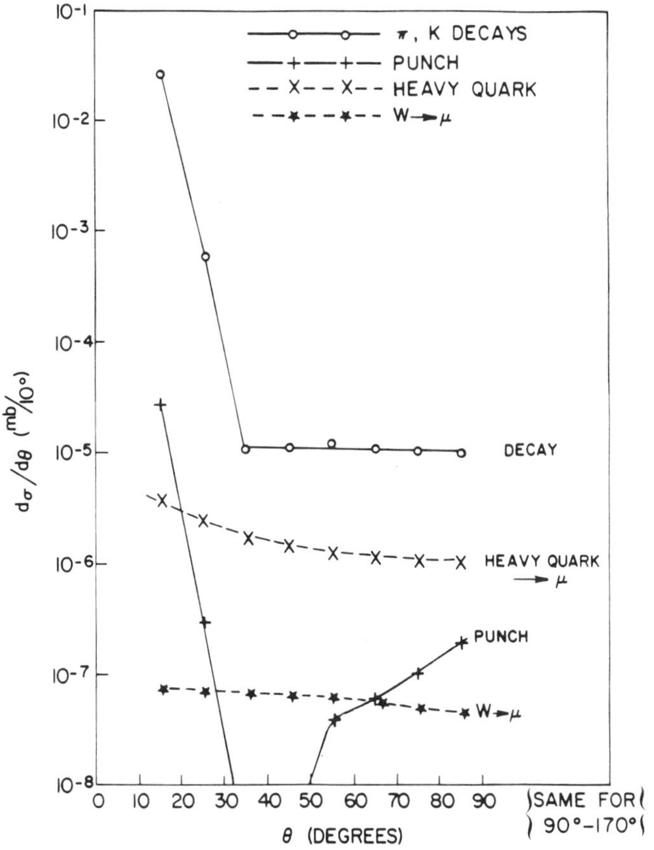

FIGURE 14. Muon rates after the steel from all sources.

are then summed (FIGURE 15b) for corresponding TRD sections to form coincidences in the azimuthal angle. Preliminary Monte Carlo calculations indicate that a π/e rejection factor of 10–20 is achievable at the trigger level.

The Level 2 trigger comprised of a large number (≈50) of fast microprocessors, currently thought to be DEC μVAX-II, is organized as shown in FIGURE 16. An acquisition supervisor, itself a fast microprocessor, directs the data flow to individual

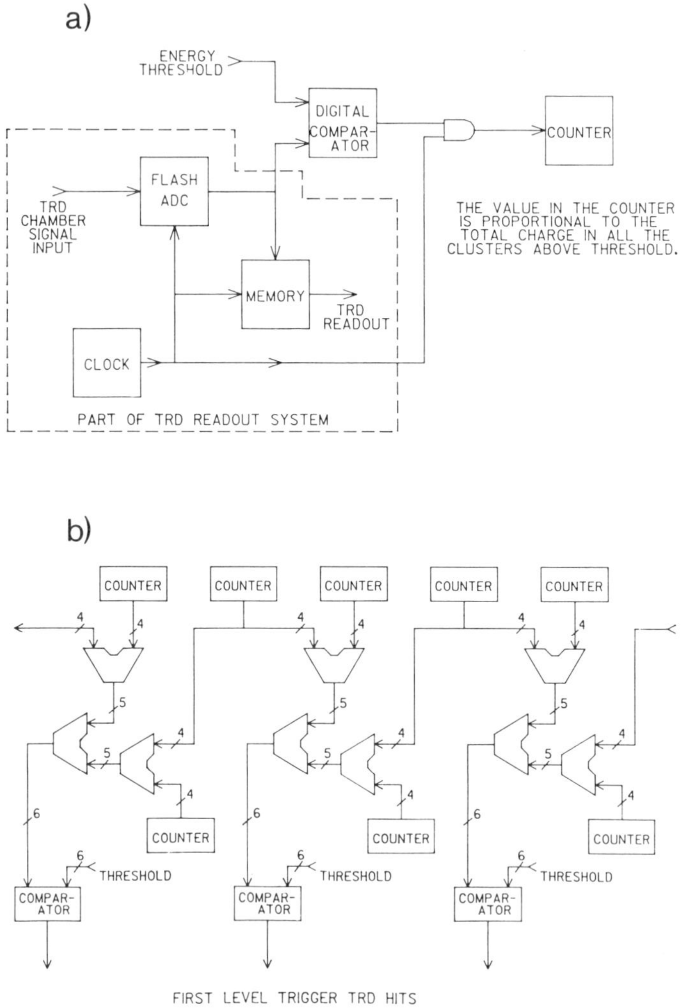

FIGURE 15. (a) TRD scaler comparator; (b) TRD adder comparator.

processors that digest data in parallel. Each processor analyzes a complete event using as input the whole data buffer from the data acquisition system as well as a substantial amount of information derived from the Level 1 trigger. On the average, each processor must examine and reject two to four events a second to achieve our desired tape-writing rate.

The system has provisions for assigning separate processors to individual parts of the detector or to distinct triggers. These features will be convenient and necessary during the setup and debugging phase.

Data Acquisition

The layout of the data acquisition system is shown in FIGURE 17. The main data acquisition computer will be a VAX-8600 or equivalent, backed up by a smaller VAX, possibly a 750. It is imagined that up to 20 interactive users will wish to use the system at any given time. We anticipate that the system will have a substantial amount (five gigabytes) of disk storage, at least four 6250 bpi tape drives, and/or an optical disk system for archival recording of data. In a normal data taking mode, we imagine a series of parallel data streams with events categorized on the basis of trigger and filter information.

TIMETABLE

A possible TeV-I running schedule is the following:

October–December 1986,
July–November 1987,
July–November 1988,
July–November 1989,
July–November 1990.

It is clear that the first two periods will not have DØ participation. In FIGURE 18, I show a construction schedule that is limited only by our collaboration's ability to design, fabricate, and assemble equipment. This is the "technology limited" schedule as it has no funding limitations built in. From it, we conclude that the DØ detector could be ready for the 1988 run.

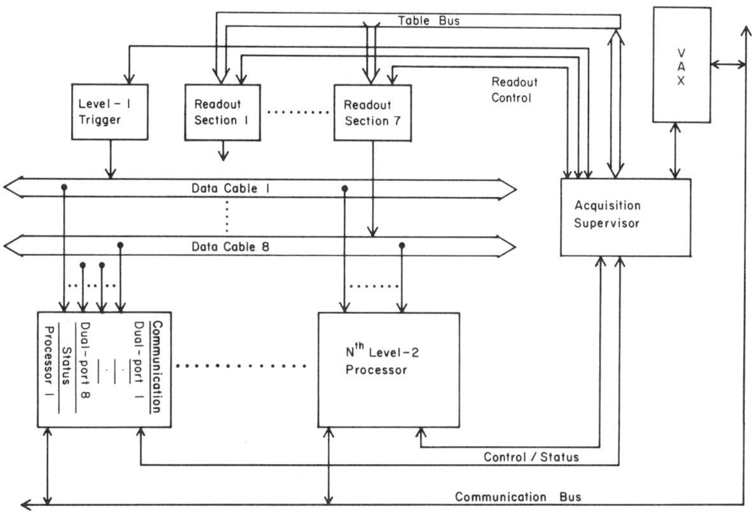

FIGURE 16. The Level 2 trigger organization.

In November of 1984, the DØ detector passed an important milestone: a formal review from the Department of Energy. An important result of this review was an official DOE cost estimate for the DØ project of \$34M with a contingency of \$14.3M, both in FY85 dollars. Current funding scenarios from DOE for Fermilab do not show this kind of money available in the appropriate time sequence to permit us to achieve a 1988 start-up date. I will not analyze any "realistic" funding plans as anything, but the

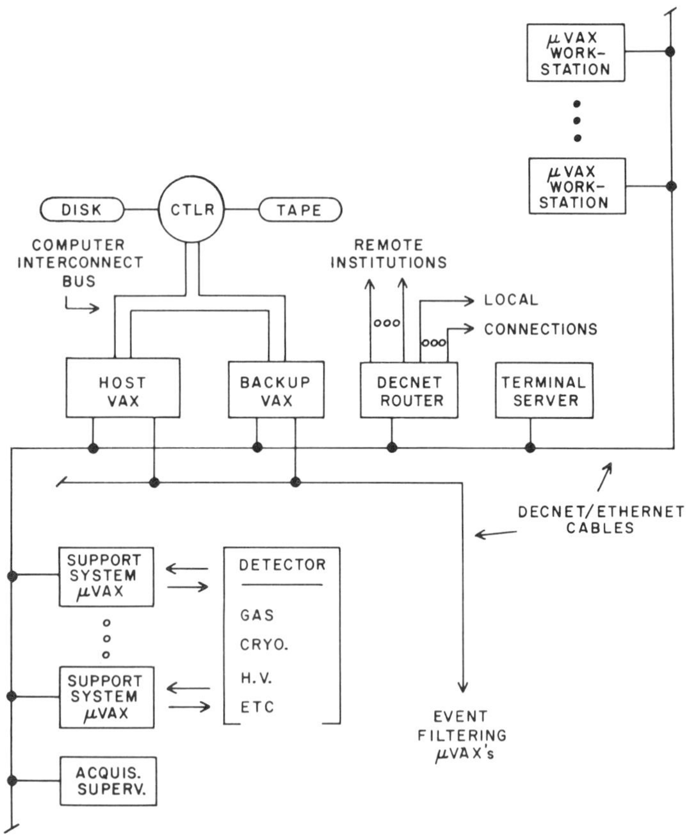

FIGURE 17. Data acquisition organization.

"technology limited" plan does not make sense from the viewpoint of optimum and effective utilization of TeV-I. I will end with a quote from Fermilab Director Leon M. Lederman at the DOE review:

> "It is hard to believe that the HEPAP priority study in June of 1985 will not recommend acceleration of DØ."

D0 MASTER SCHEDULE III Nov.16, 1984		FY85	FY86	FY87	FY88	FY89	FY90	FY91
		ONDJFMAMJJAS	ONDJFMAMJJAS	ONDJFMAMJJAS	ONDJFMAMJJAS	ONDJFMAMJJAS	ONDJFMAMJJAS	ONDJFMAMJJAS
ACCELERATOR OPERATION (o=OFF,F=FT,X=CB)		oooFFFFFXooo	oooooXXFFFFF	XXFFFFF				
2.1	Central Detectors							
2.1.1	Vertex Chamber	oooooooooooo	============	=======I				
2.1.2	Central Drift Chamber	ooooooo=====	============	=======I				
2.1.3	Forward Drift Chambers	ooooooo=====	============	=======I				
2.1.4	Transition Radiation Detectors	ooooooo=====	============	=======I				
2.2	Calorimetry							
2.2.1	Central Calorimeter	oooooooooooo	============	======IIIIII				
2.2.2	Endcap Calorimeters	oooooooooooo	============	============	IIIIII			
2.2.3	Plug Calorimeters	oo	oooooooooo==	============	====II			
2.3	Muon System							
2.3.1	Muon Toroids		oooooo======	==IIIII				
2.3.2	Proportional Drift Tubes	oooooooooooo	============	========IIII	III			
2.4	Electronics							
2.4.1	Front-end Electronics	oooooooooo==	============	============	======			
2.4.2	Fast Trigger System	oooooooooooo	============	============	======			
2.4.3	Filter & Data Acquisition System	oooooooo====	============	IIIIIIIIIIII				
2.4.4	Computers	oooooooooooo	ooo	IIIIIIIIIIII	IIII			
2.5	Associated Equipment							
2.5.1	Cryogenic Support		ooo	IIIIII				
2.5.2	Support & Moving System		oooooo======	==IIIII				
2.5.3	D0 Equipment		oooooo======	==IIIIIIIIII	II			
2.5.4	Cable Handler		oooooo======	==IIIII				
2.6	Related Efforts							
2.6.1	Installation			IIIIIIIIIIII	IIIIII			
2.6.2	Administration							

o R&D Work and Design
= Fabrication
I Installation and Testing

FIGURE 18. A construction timetable.

REFERENCES

1. DESIGN REPORT. 1983. The DØ Experiment at the Fermilab Antiproton-Proton Collider, October 1983. Fermilab Publication.
2. DESIGN REPORT. 1984. The DØ Experiment at the Fermilab Antiproton-Proton Collider, October 1984. Fermilab Publication.
3. FELDMAN, G. J. *et al.* 1984. 4π Detectors. SLAC-PUB-3460, October 1984.
4. DISCUSSIONS. *In* Proceedings of the 1982 DPF Summer Study on Elementary Particle Physics and Future Facilities. R. Donaldson, R. Gustafson & F. Paige, Eds.
5. RUBBIA, C. 1984. The Santa Fe DPF Meeting, October 1984.
6. LUDLAM, T. *et al.* 1981. Nucl. Instrum. Methods **180:** 413; FABJAN, C. W. *et al.* 1983. CERN preprint CERN-EP/83-45.

APPENDIX A
The DØ Collaboration

Brookhaven National Laboratory: S. Aronson, B. Gibbard, H. Gordon, W. Guryn, S. Kahn, M. Month, M. Murtagh, S. Protopopescu, I. Stumer & P. Yamin.

Brown University: D. Cutts, J. Hoftun, R. Lanou & R. Zeller.

Columbia University: P. Franzini, D. Son, P. M. Tuts & S. Youssef.

Fermi National Laboratory: C. Brown, W. Cooper, B. Cox, R. Dixon, G. Dugan, D. Eartly, H. Fenker, D. Finley, D. Green, H. Haggerty, S. Hansen, A. Ito, H. Jostlein, P. Koehler, E. Malamud, P. Martin, P. Mazur, J. McCarthy, R. Raja, R. Smith, M. Takasaki & R. Yamada.

Florida State University: H. Goldman.

Lawrence Berkeley Laboratory: A. Clark, L. Kerth, S. Loken, M. Strovink & W. Wenzel.

University of Maryland: S. Kunori & P. Rapp.

Michigan State University: M. Abolins, R. Brock, D. Edmunds, J. Linnemann, D. Owen, B. Pope, S. Stampke, M. Tartaglia & H. Weerts.

Northwestern University: D. Buchholz & B. Gobbi.

University of Pennsylvania: E. Gardella, W. Kononenko, W. Selove, G. Theodosiou & R. Van Berg.

University of Rochester: T. Ferbel, F. Lobkowicz & P. Slattery.

CEN Saclay: Y. Ducros, R. Hubbard, P. Mangeot, B. Mansoulie, J. Teiger, H. Zaccone & A. Zylberstejn.

State University of New York at Stony Brook: M. Adams, R. Engelman, G. Finocchiaro, P. Grannis, D. Hedin, J. Kirz, J. Lee-Franzini, S. Linn, D. Lloyd-Owen, M. Marx, R. McCarthy, L. Romero & R. D. Schamberger.

The Top Quark[a]

V. BARGER

Department of Physics
University of Wisconsin
Madison, Wisconsin 53706

INTRODUCTION

The discovery of the $W^{\pm}$ and Z bosons at the CERN $\bar{p}p$ collider[1-5] marked the beginning of a new era in particle physics in which the search for new elementary particles can be made at high energy hadron colliders. It is serendipitous that the first collider runs also provided evidence[6] for the top quark in electron + missing p_T + two-jet events, presumed to come from $W \rightarrow t\bar{b}$ and $t \rightarrow be\nu$ decays. The strategies[6-8] that allowed the identification of the top quark signal in the complicated debris of high energy $\bar{p}p$ annihilations are reviewed in this report.

The t quark is theoretically needed for renormalization in $SU(2) \times U(1)$ to cancel triangle anomalies.[9,10] Also, without a t quark, one linear combination of $-1/3$ charge quarks, b'_L, would be assigned instead as a weak isosinglet, in which case the GIM cancellation mechanism would be upset and flavor-changing neutral current couplings such as $Zb\bar{s}$ would exist. This would give neutral current b-quark (and B-meson) decays with a predicted lower limit[11,12] relative to the charged-current $b \rightarrow ce\nu$ decay. The predicted bound, $\Gamma(b \rightarrow se^+e^-)/\Gamma(b \rightarrow ce\bar{\nu}) \geq 0.12$, is found to be violated by experiments at CESR (which obtained a ratio < 0.046 with 99.9% confidence);[13] this indicates that the top quark exists.

Provided that its mass is less than 60 GeV, top is expected to be produced at the $\bar{p}p$ collider with appreciable rates from weak boson decays and $t\bar{t}$ hadroproduction.[7,8,14-26] The problem is to separate the signal from severe backgrounds. Although charged leptons from semileptonic decays provide a tag for heavy flavors, the single lepton spectrum from top does not stand out clearly from charm and bottom contributions over any kinematic range. Hence, an early top discovery was not anticipated.[14-18]

This pessimistic perspective changed soon after the W discovery when it was realized[7,8] that some of the principal elements of the $W \rightarrow e\nu$ search[1-4] could be adapted to a t-quark search.

INSIGHTS FROM $W \rightarrow e\nu$

In zeroth order QCD, the W^+ is produced by the fusion subprocess, $u\bar{d} \rightarrow W^+$, with the W momentum along the beam axis. The transverse momenta (p_T) of the W decay products are then as if the W had decayed at rest. The electron and neutrino are

[a]This research was supported in part by the University of Wisconsin Research Committee with funds granted by the Wisconsin Alumni Research Foundation, and in part by the Department of Energy under contract no. DE-AC02-76ER00881.

back-to-back in the transverse plane. Each has a p_T distribution,

$$dN/dp_T \sim [1 - (p_T/\tfrac{1}{2}M_W)^2]^{-1/2}\,[2 - (p_T/\tfrac{1}{2}M_W)^2], \tag{2.1}$$

with the "Jacobian" peak of the two-body decay at the maximum value $p_T = \tfrac{1}{2}M_W$ determined by the W mass. The measurement of missing p_T (often denoted by $\not{p}_T$) was crucial to the W discovery. The presence of joint Jacobian peaks in the $p_T(e)$ and $\not{p}_T$ distributions conclusively established the two-body $W \to e\nu$ decay signal.[1-5]

Gluon radiation gives the W some transverse momentum that smears these sharp Jacobian peaks and gives additional hadron jets. It was found that instead of the $p_T(e)$ and $\not{p}_T$ variables, a transverse mass variable[1,27,28] could be used whose distribution was essentially independent of QCD smearing effects.[28] The transverse mass for a particle, c, (such as the electron) and a missing transverse momentum, $\not{p}_T$, is defined by

$$[M_T(c\not{p}_T)]^2 = (c_T^0 + \not{p}_T)^2 - (\vec{c}_T + \vec{\not{p}}_T)^2, \tag{2.2}$$

where $c^0 = (\,|\vec{c}_T|^2 + m_c^2)^{1/2}$ and $\not{p}_T = |\vec{\not{p}}_T|$. The transverse mass distribution for $W \to e\nu$ has the upper bound,

$$M_T(e\not{p}_T) \leq M_W, \tag{2.3}$$

with a Jacobian peak at the upper endpoint. The only significant smearing of the M_T distribution is the Breit-Wigner distribution of the W mass (determined by its decay width, Γ_W) and the resolution of the $\not{p}_T$ measurements.

Out of the $W \to e\nu$ analysis came several insights that could be applied to a t-quark search in $W \to t\bar{b}$ because of basic similarities. The semileptonic $t \to be\nu$ decay gives an electron tag and $\not{p}_T$. The $\bar{b}$ jet has a Jacobian p_T distribution whose peak value depends on m_t. Transverse masses could be used to yield kinematic constraints related to the mass scales, M_W and m_t. The applications of these methods to the t search are addressed subsequently.

TOP FROM W DECAY

Rate

The number of $W \to t\bar{b}$ events relative to the number of $W \to e\nu$ events is determined by the partial width ratio of

$$\Gamma(W \to t\bar{b})/\Gamma(W \to e\nu) \simeq 3[1 - (3/2)m_t^2/M_W^2 + (1/2)m_t^6/M_W^6], \tag{3.1}$$

where m_b^2/M_W^2 is neglected. In the spectator heavy quark decay model, a fraction, $B(t \to be\nu) \approx 1/9$, will have an electron tag and the same fraction, a muon tag. Thus for $m_t = 40$ GeV, we anticipate a ratio of

$$\sigma(W \to t\bar{b};\, t \to be\nu)/\sigma(W \to e\nu) \approx 2/10 \tag{3.2}$$

before acceptance factors are taken into account. The energetic b and $\bar{b}$ quarks materialize as narrow jets of hadrons.

Realistic calculations of rates and distributions take into account the following considerations:

(i) *Electron Acceptance.* An experimental electron identification requirement is typically $p_T(e) \geq 15$ GeV (for muons, $p_T(\mu) \geq 12$ GeV in the t search).

(ii) *Electron Isolation.* To suppress the large backgrounds from $b\bar{b}$ and $c\bar{c}$ production, the electron is required to be isolated, with only limited hadronic activity in a cone about the electron direction. Typically, such an isolation requirement is[6]

$$\Sigma\, p_T\,(\text{hadrons}) < 3 \text{ GeV within } [(\Delta\eta)^2 + (\Delta\phi)^2]^{1/2} < 0.7 \tag{3.3}$$

where $\Delta\eta$ is the rapidity difference and $\Delta\phi$ is the azimuthal angle difference from the electron to the hadron. With isolation imposed, the trigger electron almost always comes from $t \to be\nu$, rather than from the other semileptonic options in the $t \to b \to c \to s$ cascade or from the semileptonic modes of the accompanying $\bar{b}$ quark.

(iii) *Jet Algorithm.* A jet algorithm similar to that used by the UA1 collaboration is employed in which partons with

$$\Delta R = [(\Delta\eta)^2 + (\Delta\phi)^2]^{1/2} < 1 \tag{3.4}$$

are coalesced. A jet recognition threshold of order,

$$p_T(j) > 10 \text{ GeV}, \tag{3.5}$$

is specified.

(iv) K *Factor.* A semiempirical cross section enhancement factor, $K = 2$, is included in σ_W, which is needed to reproduce the observed $W \to e\nu$ rate. (QCD calculations predict[29] a smaller value of $K \approx 1.3$.)

(v) W_T *Smearing.* The effects of transverse W motion from $q\bar{q} \to W +$ gluons and $qg \to qW$ subprocesses are taken into account through smearing with the W_T distribution measured[1-3] in $W \to e\nu$ events.

(vi) *Multiple Neutrinos.* The missing p_T from neutrinos associated with unidentified charged leptons in multiple semileptonic decays is included.[7,23] Also, the $\not{p}_T$ results are given gaussian experimental uncertainties with a root mean square error of 4 GeV.

(vii) *Fragmentation.* The fragmentation of the heavy quarks to spinless hadrons is described by the model of Peterson *et al.*,[30] in which the heavy Q fragments to a meson, $M(Q\bar{q})$, of fractional three-momentum, $z = p_M/p_Q$, in the subprocess center-of-mass frame with a probability distribution of

$$dD(z)/dz \propto z^{-1}[1 - z^{-1} - \epsilon(1 - z)^{-1}]^{-2}, \tag{3.6}$$

where $\epsilon = 0.15(m_c/m_Q)^2$. This representation fits existing fragmentation data on b and c quarks. For the t case, the z distribution approaches a delta function near $z = 1$ in accord with qualitative arguments that the velocities of the heavy meson and its parent heavy quark are about equal.

(viii) *Heavy-Quark Decays.* The heavy-quark decays are described by bare-quark V-A matrix elements. The charged lepton spectra from b and c decays are known to be well described by this approximation.

Applying the jet algorithm for overlapping partons, the final states of

$$W \to t\bar{b} \to e\nu b\bar{b} \to e + \not{p}_T + n \text{ jets} \tag{3.7}$$

are obtained, with mainly one- or two-jet topologies. The cross section predictions with

$K = 2$ and $m_t = 40$ GeV are as follows:

$\sqrt{s}$ (GeV)	$\sigma(e\not{p}_T j_1)$	$\sigma(e\not{p}_T j_1 j_2)$
540	21 pb	14 pb
630	28	18

(3.8)

The corresponding numbers of events can be estimated by multiplying by the integrated luminosity (of order, 0.14 pb^{-1} at $\sqrt{s} = 540$ GeV and 0.25 pb^{-1} at $\sqrt{s} = 630$ GeV) and the lepton detection efficiency (of order, 0.5). The expected numbers of electron events from $W \rightarrow t\bar{b}$ for $m_t = 40$ GeV are:

$\sqrt{s}$ (GeV)	$N(e\not{p}_T j_1)$	$N(e\not{p}_T j_1 j_2)$
540	1	1
630	3	2

(3.9)

We mainly concentrate here on the two-jet topology, for which the UA1 signal is reported. At $\sqrt{s} = 540$ GeV, UA1 finds three electron and three muon events in the two-jet category.

Two-Jet Events

Jacobian Peak of Fast $\bar{\text{b}}$ *Jet*

A distinguishing characteristic of two-body $W \rightarrow t\bar{b}$ decay is the Jacobian distribution of the $\bar{b}$ jet,[7,8] whose peak occurs at

$$p_T(\bar{b})_{peak} \simeq \tfrac{1}{2}M_W - \tfrac{1}{2}\frac{m_t^2}{M_W}. \qquad \textbf{(3.10)}$$

The full formula including m_b can be found in reference 31. The dependence of the $p_T(\bar{b})$ distribution on the t mass is illustrated in FIGURE 1. For m_t not too large, the fast

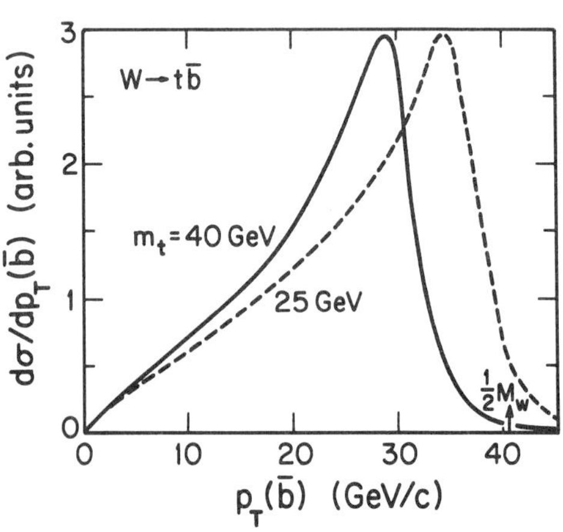

FIGURE 1. Transverse momentum distribution of the $\bar{b}$ jet in $W \rightarrow t\bar{b}$ decays; the location of the Jacobian peak depends on the t mass.

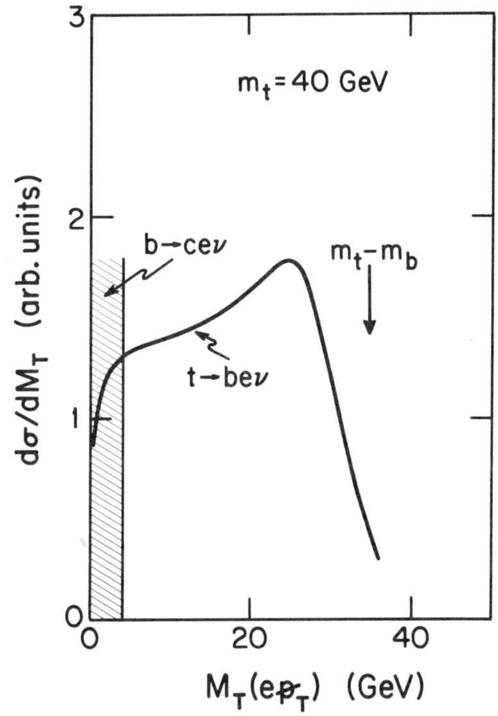

FIGURE 2. Transverse mass of $e\nu$ in $W \rightarrow t\bar{b}$ with $t \rightarrow be\nu$ decay.

jet (j_l) is almost always the $\bar{b}$ jet (92% of the time for $m_t = 40$ GeV),[31] so this association is justified. For comparisons with distributions in other transverse variables, it is more convenient to use, instead of $p_T(j_l)$, the variable, $M_T(j_l)$, defined in the $m_b \rightarrow 0$ limit by

$$M_T^2(j_l) = M_W^2 - 2M_W p_T(j_l) \tag{3.11}$$

for which the Jacobian peak occurs at

$$M_T(j_l) = m_t. \tag{3.12}$$

Transverse Mass of eν

The transverse mass of the electron and $\not{p}_T$ from $t \rightarrow be\nu$ decay has the kinematic upper bound,[7]

$$M_T(e\not{p}_T) \leq m_t - m_b. \tag{3.13}$$

Since M_T from the decay of a lighter quark such as $b \rightarrow ce\nu$ is limited to the range, $M_T(e\not{p}_T) < m_b - m_c \approx 4$ GeV, a value of M_T above 4 GeV is a signal for a heavier particle such as top, provided that $\not{p}_T$ is accurately measured and comes from the same decay as the electron. FIGURE 2 shows the expected $M_T(e\not{p}_T)$ distribution in $W \rightarrow t\bar{b}$ with $t \rightarrow be\nu$ decay.

t *Cluster Transverse Mass*

In $W \to t\bar{b}$ events, the slower jet (j_2) is almost always the b jet from the $t \to be\nu$ decay. If we treat (be) as a cluster, c, with three momentum and mass-squared

$$\vec{c} = \vec{b} + \vec{e} \qquad m_c^2 = (b + e)^2, \tag{3.14}$$

then the "cluster transverse mass," $M_T(c\not{p}_T)$, treats the three-body semileptonic decay as an effective two-body decay, $t \to c + \nu$. The M_T distribution then has a sharp Jacobian peak[7] at its upper endpoint,

$$M_T(be, \not{p}_T) \leq m_t. \tag{3.15}$$

W *Cluster Transverse Mass*

Generalizing the clustering technique, all decay products of the W can be included by taking a cluster

$$c = b + e + \bar{b} \tag{3.16}$$

so that there is effectively a two-body decay, $W \to c + \nu$. This M_T distribution has a Jacobian peak at the upper endpoint,

$$M_T(be\bar{b}, \not{p}_T) \leq M_W. \tag{3.17}$$

Invariant Mass Distributions

In events where $\not{p}_T$ is small, the invariant masses, $M(be)$ and $M(be\bar{b})$, contain essentially the same information as the transverse masses. The peak values of these distributions occur at:

$$M(be : \text{peak}) \simeq \frac{3}{4} m_t$$

$$M(be\bar{b} : \text{peak}) \simeq \frac{7}{8} M_w. \tag{3.18}$$

Thus, the invariant mass distribution of the electron and slow jet is an additional useful measure of m_t. All the above mass distributions require the b and $\bar{b}$ jets to be correctly identified; nonpeaking background contributions will come from misidentified jets.

Comparisons with Data

A good diagnostic for events of $W \to t\bar{b}$ origin is that the peak regions of all M_T and M distributions discussed above should be simultaneously populated. As shown in FIGURE 3, this expectation is fulfilled by the three electron and three muon candidate events of UA1.

$p_T(e)$ Normal to $\bar{b}$ Jet

Electrons from $t \rightarrow be\nu$ decay appear at wide angles to the axis defined by $\vec{p}_T(\bar{b})$, which is approximately $-\vec{p}_T(t)$. The component p_N of p_{eT} normal to the $\bar{b}$ jet axis in the transverse plane provides another measure of m_t. The upper endpoint of the p_N distribution is[7]

$$p_N(e_T) \leq \tfrac{1}{2}m_t - \tfrac{1}{2}\frac{m_b^2}{m_t}. \tag{3.19}$$

This broad range of the p_N distribution thereby provides another measure of m_t.

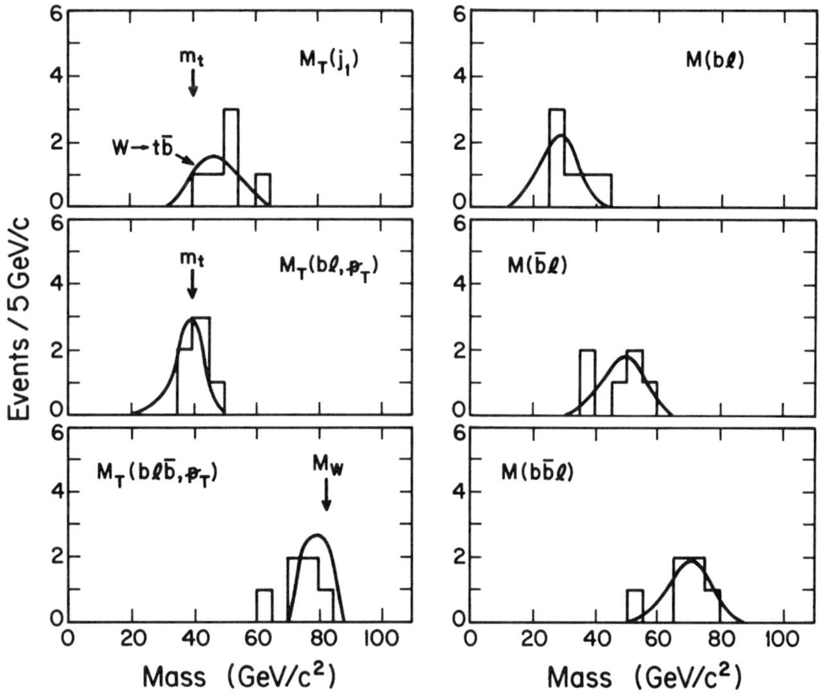

FIGURE 3. Expectations for transverse and invariant mass distributions in $e\not{p}_T j_1 j_2$ events from $W \rightarrow t\bar{b}$ and $t(40) \rightarrow be\nu$ compared with UA1 t-candidate events.

One-Jet Events

One-jet events with an isolated electron provide another interesting eventual signal for top from $W \rightarrow t\bar{b}$. The electron and the jet are expected to be nearly back-to-back. The $M_T(e\not{p}_T)$, $M(ej)$, and $M_T(ej, \not{p}_T)$ distributions for these events are rather similar to those for two-jet events of $W \rightarrow t\bar{b}$ origin, with j being the fast jet.

TOP HADROPRODUCTION

The lowest order QCD heavy-quark production subprocesses,

$$q\bar{q}, gg \rightarrow t\bar{t}, \tag{4.1}$$

are expected to be another major source of top production. The calculated $\bar{p}p$ cross section has an uncertainty of at least a factor of two from α_s^2 and a possible K factor. FIGURE 4 compares $W \rightarrow t\bar{b}$, $Z \rightarrow t\bar{t}$, and $t\bar{t}$ hadroproduction cross sections at $\sqrt{s} = 630$ GeV as a function of the t mass, assuming $K = 2$ for all mechanisms. For $m_t \approx 40$ GeV, the $t\bar{t}$ fusion cross section is about a factor of two below that of $W \rightarrow t\bar{b}$.

The topology of $t\bar{t}$ events is expected to be more complicated than $W \rightarrow t\bar{b}$. Applying the jet algorithm in the $t\bar{t}$ case, significant cross sections for one-, two-, and three-jet events with an electron and $\not{p}_T$ are obtained. For $m_t = 40$ GeV and $K = 2$, the results are as follows:

$\sqrt{s}$	$\sigma(e\not{p}_T j_1)$	$\sigma(e\not{p}_T j_1 j_2)$	$\sigma(e\not{p}_T j_1 j_2 j_3)$	
540 GeV	9 pb	20 pb	10 pb	(4.2)
630 GeV	11	29	13	

The two-jet cross sections from $t\bar{t}$ and $W \rightarrow t\bar{b}$ have similar values. The $t\bar{t}$ distributions in the two-jet category are compared with those from $W \leftarrow t\bar{b}$ in FIGURE 5. The two sources cover similar ranges of transverse and invariant masses, which allow for some possible confusion of the two signals; however, the distributions from $t\bar{t}$ are broader and do not have the distinctive peaks near m_t and M_W.

BACKGROUNDS FROM *b* AND *c* QUARKS

Cross sections for hadroproduction of heavy-quark pairs, $Q\bar{Q}$, via the QCD fusion mechanism decrease sharply as the quark mass, m_Q, increases. Consequently, the $c\bar{c}$ and $b\bar{b}$ contributions are enormously larger than $t\bar{t}$ or $W \rightarrow t\bar{b}$. This serious background must be suppressed in order to find the top signal.

At high $p_T(e)$, the $b\bar{b}$ contribution dominates over $c\bar{c}$ because the $b \rightarrow B$ fragmentation is harder than $c \rightarrow D$ and the V-A electron spectrum from $b \rightarrow e$ decay is harder than that from $c \rightarrow e$; there is also a secondary $b \rightarrow c \rightarrow e$ contribution, but this is relatively soft. FIGURE 6 compares $b\bar{b}$, $c\bar{c}$ fusion expectations (for $K = 2$) with the inclusive $p_T(\mu)$ spectrum measured[6] in the UA1 experiment at $\sqrt{s} = 540$ GeV, which may contain some background contamination from pion and kaon decays. There is general agreement between the theoretical and experimental results.

FIGURE 7 compares the $p_T(e)$ spectra from $q\bar{q}$, $gg \rightarrow b\bar{b}$, $t\bar{t}$ and $W \rightarrow t\bar{b}$ sources. There is no kinematic range where single leptons expected from t decay stand out clearly from the b, c background. This situation is completely changed, however, once an appropriate isolation requirement is imposed on the charged lepton. The $b \rightarrow e$ contribution for $p_T(e) > 10$ GeV is essentially eliminated, with little effect on the top signals. This is easy to understand from the kinematics of heavy-quark decays. The large energy release in t decays leads to a wide dispersal of the decay products, and these leptons stand a good chance of being well-isolated from the accompanying

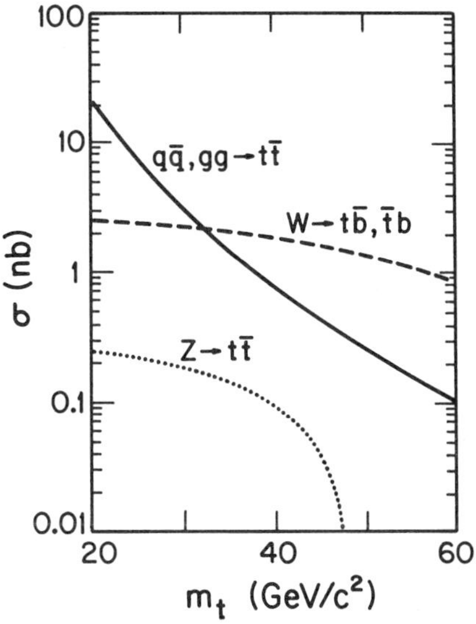

FIGURE 4. Cross sections for electroweak and hadroproduction of t quarks in $\bar{p}p$ collisions at $\sqrt{s} = 630$ GeV versus the t mass. ($K = 2$ is assumed in all cases.)

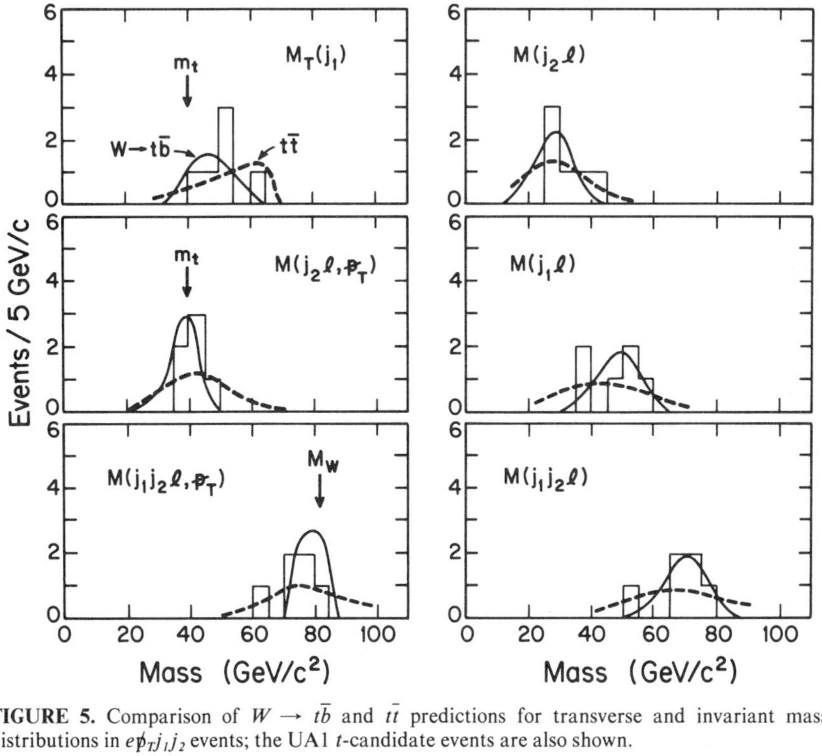

FIGURE 5. Comparison of $W \rightarrow t\bar{b}$ and $t\bar{t}$ predictions for transverse and invariant mass distributions in $e\not{p}_T j_1 j_2$ events; the UA1 t-candidate events are also shown.

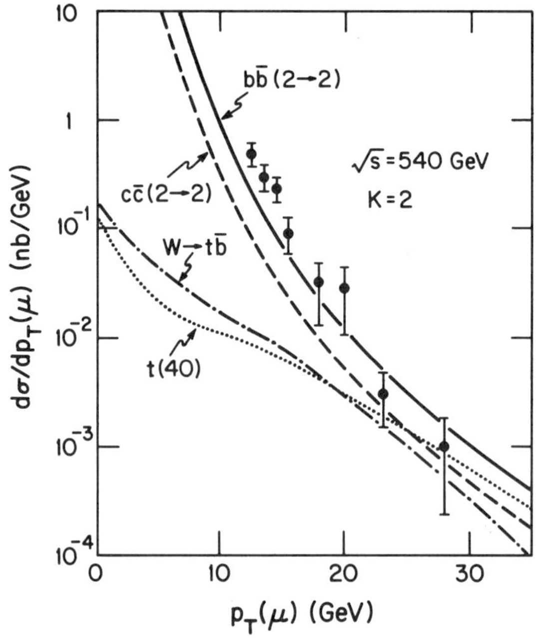

FIGURE 6. Comparison of $p_T(e)$ distribution from $O(\alpha_s^2)$ $b\bar{b}$ and $c\bar{c}$ fusion subprocesses with UA1 single charged lepton measurements.

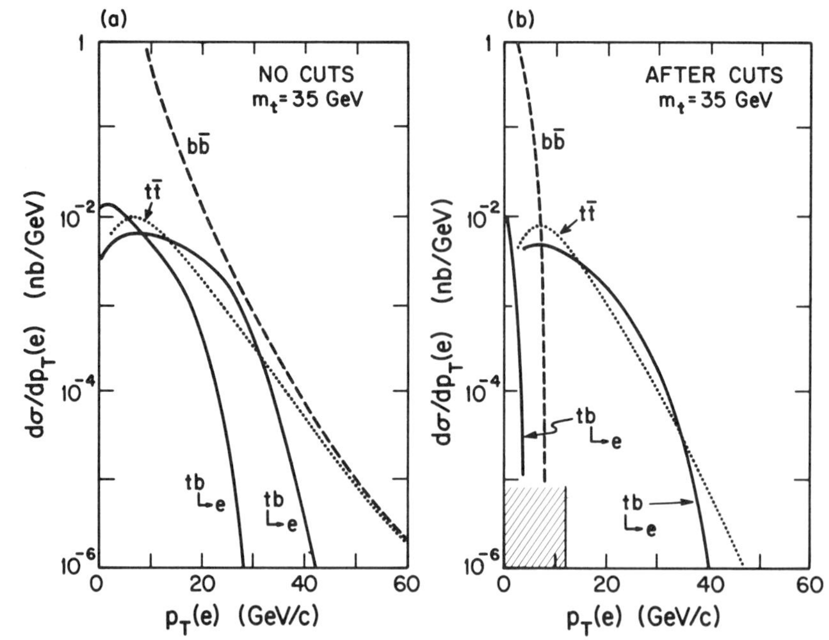

FIGURE 7. Expected $p_T(e)$ distribution in $e\not{p}_T j_1 j_2$ events from $q\bar{q}$, $gg \rightarrow \bar{b}b$, $t\bar{t}$ and $W \rightarrow t\bar{b}$ sources (a) without lepton isolation and (b) with lepton isolation.

hadronic debris. However, a 10 GeV electron from $b \to ce\nu$ decay necessarily has a fast b parent and must lie within 30° of the hadrons into which the c decay fragments. The only chance for high-p_T electrons from b or c decay to appear to be isolated is for the accompanying hadrons to be slow (which is kinematically possible to some extent). However, with $q\bar{q}$, $gg \to b\bar{b}$, $c\bar{c}$ production, isolated high-p_T leptons can only occur in events where one of the jets has little hadronic p_T; by requiring both lepton isolation and two or more high-p_T jets, contributions from this origin are suppressed.

The most dangerous remaining background comes from $b\bar{b}$ (or $c\bar{c}$) production with an extra light parton, via the $O(\alpha_s^3)$ QCD subprocesses,[32-37]

$$\begin{aligned} q\bar{q} &\to gb\bar{b} \\ qg &\to qb\bar{b} \\ gg &\to gb\bar{b}. \end{aligned} \tag{5.1}$$

In a small fraction of events, the electron from $b \to ce\nu$ decay carries most of the b energy and may escape the isolation veto, while the extra light parton gives the second jet.

In calculation of the $2 \to 3$ subprocesses, the divergence (from when a final light parton is collinear with one of the initial partons) must be cut off, but the choice of cut is not critical since the soft p_T region of phase-space is suppressed by the acceptance cuts. The $2 \to 3$ cross section is comparable to the $2 \to 2$ fusion cross section. The $p_T(e)$ distributions from the $2 \to 3$ processes are shown in FIGURE 8, with and without lepton isolation. The isolation cuts greatly suppress[32,33] this b, c background at high p_T. The small remaining background is found to have distinct, different distributions in other variables.[32] For two jets above the jet recognition threshold, the $2 \to 3$ backgrounds are less than 10% of the $W \to t\bar{b}$ signal. We conclude that the t signal can be separated from the b and c quark backgrounds by suitable experimental selection criteria.

FAKING THE TOP WITH SPARTICLES

A possible explanation of the UA1 monojet events,[38] which have a single narrow jet at high p_T and large $\not{p}_T$, is squark pair production[39-43] with $\tilde{q} \to \tilde{\gamma} q$ decays:

$$\tilde{q}\bar{\tilde{q}} \to \tilde{\gamma}\tilde{\gamma}q\bar{q} \to \not{p}_T + \text{jet}(s). \tag{6.1}$$

A squark mass of about 50 GeV and a higher gluino mass would be indicated along with a light photino.

It has been suggested[44] that the production and decays of supersymmetry particles might fake the observed $e\not{p}_T j_1 j_2$ signal for top if the wiggsino mass fell between the squark and lepton masses,

$$m_{\tilde{q}} > m_{\tilde{\omega}} > m_{\tilde{\ell}}. \tag{6.2}$$

Squark pair production and the decays,

$$\tilde{q} \to \tilde{\gamma} q, \quad \tilde{q} \to \tilde{\omega} q, \quad \tilde{\omega} \to e\tilde{\nu},\ \tilde{e}\nu \to e\nu\tilde{\gamma}, \tag{6.3}$$

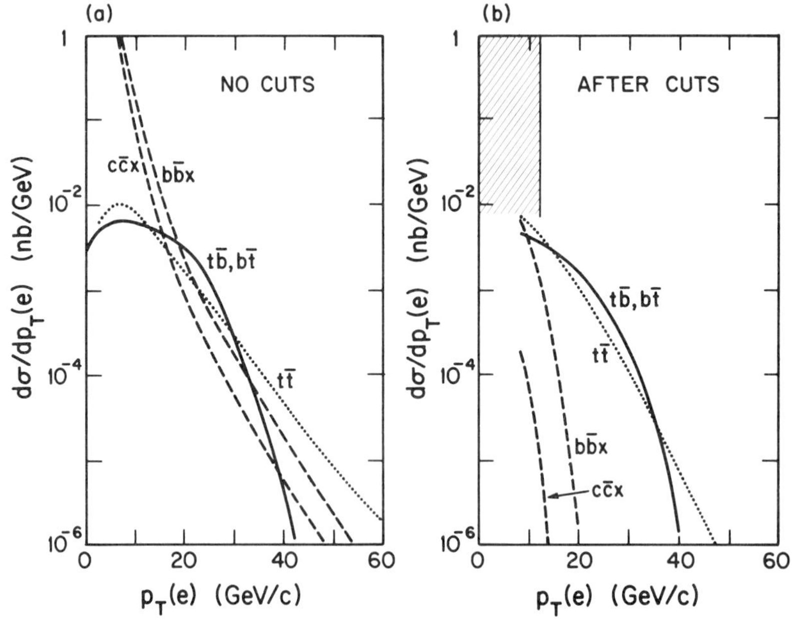

FIGURE 8. The $p_T(e)$ distribution in $e\not{p}_T j_1 j_2$ events from the $2 \rightarrow 3$ subprocesses, $b\bar{b}x$ and $c\bar{c}x$.

could then give t-faking final states:

$$\tilde{q}\bar{\tilde{q}} \rightarrow \tilde{\omega}\tilde{\gamma}q\bar{q} \rightarrow e\nu\tilde{\gamma}\tilde{\gamma}q\bar{q} \rightarrow e + \not{p}_T + \text{2-jets}. \qquad \textbf{(6.4)}$$

A substantial $m_{\tilde{q}}$, $m_{\tilde{\omega}}$ mass gap is needed in order to obtain a large two-jet cross section. The $\tilde{q} \rightarrow \tilde{\omega}q$ branching fraction can be as large as 20–30%, depending on the masses,[45] and the $\tilde{\omega} \rightarrow e\nu\tilde{\gamma}$ branching fraction is 33%. For $m_{\tilde{e}} \approx 23$ GeV and $m_{\tilde{\omega}} \approx 35$ GeV, the resulting acceptance corrected $e + \not{p}_T +$ two-jet cross sections from this SUSY source are potentially competitive with $W \rightarrow t\bar{b}$.

However, the calculated distributions[45] do not accurately fake $W \rightarrow t\bar{b}$ events, as shown by the comparisons in FIGURE 9. In particular, the SUSY results do not have the Jacobian peak structures in the M_T distributions predicted by the $W \rightarrow t\bar{b}$ source, which seem to be present in the data. Of course, SUSY $e\not{p}_T j_1 j_2$ contributions could be present in addition to $W \rightarrow t\bar{b}$ and could be possibly confused with $t\bar{t}$ contributions, in which case we would have the interesting situation of top as a background to other new physics. With the present low statistics it is not possible to give a definitive answer, but with more events it should be possible to discriminate between these two sources.

MULTILEPTONS FROM TOP

The cascade decays of heavy quarks lead to multiple charged lepton final states that provide further signatures for top.[14,15,18,39,46,47] For a common charged lepton acceptance cut, $p_T(\ell) > 6$ GeV, the predicted multilepton rates from top and bottom production and decays in $\bar{p}p$ collisions at $\sqrt{s} = 630$ GeV summed over $e^\pm$, $\mu^\pm$ are as follows:[47]

	$\sigma(t\bar{t}, t\bar{b})$	$\sigma(b\bar{b})$	
dileptons	160 pb	1000 pb	(7.1)
trileptons	6 pb	10 pb	

Although the $b\bar{b}$ rates dominate, azimuthal angle and mass correlations between leptons and lepton isolation make it possible to distinguish the top signal.

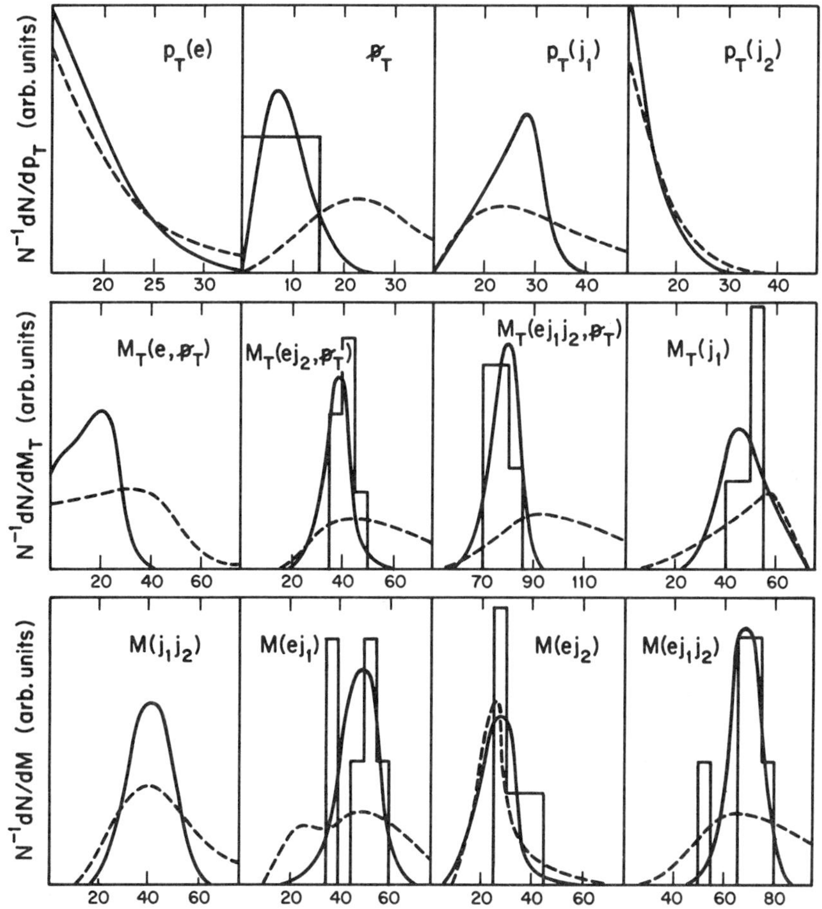

FIGURE 9. Comparison of a supersymmetry scenario for $e\not{p}_T j_1 j_2$ events with $W \to t\bar{b}$ predictions at $\sqrt{s} = 630$ GeV. Solid line indicates $W \to t(40)\bar{b}$; Dashed line indicates SUSY $\tilde{q}(50)$, $\tilde{\omega}(35)$, $\tilde{e}(23)$.

In dilepton events of $b\bar{b}$ origin, there is a peak at $\Delta\phi = 180°$ (where one lepton comes from b and the other from $\bar{b}$) and a peak at $\Delta\phi = 0°$ (where both leptons come from the same parent). The $2 \longrightarrow 3$ radiative processes fill in the azimuthal region between these peaks at a substantially lower level. In dilepton events from $t\bar{t}$ and $t\bar{b}$, the $\Delta\phi$ distributions are relatively flat. By making an isolation cut on one lepton only, charged leptons from the well-collimated b (or c) quarks can be rejected, with little

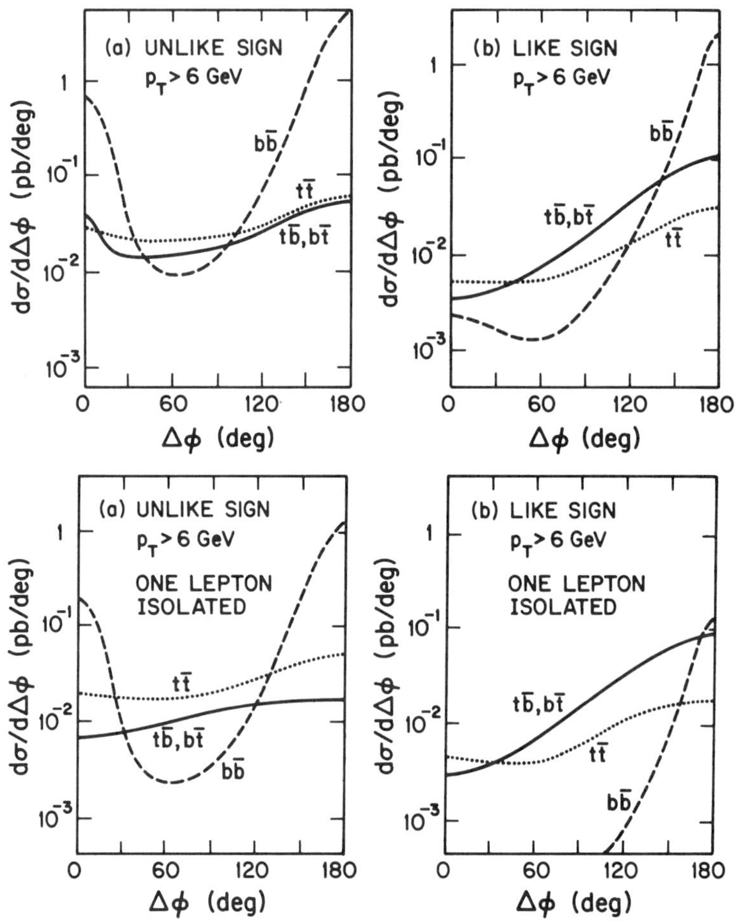

FIGURE 10. Dilepton cross sections at $\sqrt{s} = 540$ GeV with and without a loose isolation cut on one lepton.

cost to the top signals. FIGURE 10 shows typical results[46] obtained with using a rather loose single lepton isolation requirement. The t signal stands out over substantial ranges of $\Delta\phi$ and the jet topologies should further differentiate it from the $b\bar{b}$, $c\bar{c}$ backgrounds.

Trilepton events, although expected to be rare, would be valuable. Since at least two of the leptons must have a common parent, same-sign trileptons or events with opposite-sign pair masses greater than m_b must have top origins. Thus, the minimum pair invariant mass in such events would give a lower bound on the top mass.

OUTLOOK

The evidence for the top quark from $W \rightarrow t\bar{b}$ should soon be confirmed with higher statistics from the $\bar{p}p$ collider run that has just been completed. As statistics improve, it should soon become possible to identify dileptons from top decays.

Top production in $W \rightarrow t\bar{b}$ in conjunction with the $\bar{b}$ establishes its weak isospin doublet assignment. A more precise measure of the top mass is needed, though, because of the significance for e^+e^- physics. If a $Z \rightarrow t\bar{t}$ event were found, this would establish an upper bound, $m_t < \frac{1}{2}M_Z \approx 47$ GeV. The introduction of microvertex detectors in $\bar{p}p$ collider experiments will help clean up top signals by aiding in the identification of b and c jets through their quantum numbers. Further data will show whether diffractive $t\bar{t}$ production occurs in addition to fusion; this is an important question on which theory needs guidance from experiment.

Beyond the top, there may exist other new particles with masses comparable to top that are produced at the $\bar{p}p$ collider. Interesting possibilities include supersymmetry particles and the $Q = -\frac{1}{3}$ member of a fourth generation doublet (a, v). Therefore, the top signal that is the focus of present interest may then be a background to such undiscovered particles.

ACKNOWLEDGMENTS

I wish to thank Roger Phillips for comments and for providing some of the cross section calculations in this report. I thank H. Baer, W-Y. Keung, K. Hagiwara, A. D. Martin, and R. J. N. Phillips for collaborations that led to the results presented here.

REFERENCES

1. Arnison, G. *et al.* 1983. Phys. Lett. **122B:** 103.
2. Arnison, G. *et al.* 1983. Phys. Lett. **126B:** 398.
3. Arnison, G. *et al.* 1984. Phys. Lett. **134B:** 469.
4. Banner, M. *et al.* 1983. Phys. Lett. **122B:** 476.
5. Bagnaia, P. *et al.* 1983. Phys. Lett. **129B:** 130.
6. Arnison, G. *et al.* 1984. Phys. Lett. **147B:** 493.
7. Barger, V., A. D. Martin & R. J. N. Phillips. 1984. Phys. Lett. **125B:** 339, 343.
8. Barger, V., H. Baer, A. D. Martin & R. J. N. Phillips. 1984. Phys. Rev. **D29:** 887.
9. Adler, S. L. 1968. Phys. Rev. **177:** 2426.
10. Ball, J. S. & R. Jackiw. 1969. Nuovo Cimento **51:** 47.
11. Barger, V. & S. Pakvasa. 1979. Phys. Lett. **82B:** 445.
12. Kane, G. & M. Peskin. 1982. Nucl. Phys. **B195:** 29.
13. Avery, P. *et al.* 1984. Phys. Rev. Lett. **53:** 1309.
14. Abud, M., R. Gatto & C. A. Savoy. 1978. Phys. Lett. **79B:** 435.
15. Pakvasa, S., M. Dechantsreiter, F. Halzen & D. M. Scott. 1979. Phys. Rev. **D20:** 2862.

16. HORGAN, R. & M. JACOB. 1981. Phys. Lett. **107B:** 395.
17. CABIBBO, N. & L. MAIANI. 1979. Phys. Lett. **87B:** 366.
18. CHAU, L. L., W-Y. KEUNG & S. C. C. TING. 1981. Phys. Rev. **D24:** 2862.
19. GODBOLE, R. M., S. PAKVASA & D. P. ROY. 1983. Phys. Rev. Lett. **50:** 1539.
20. HAGIWARA, K. & W. F. LONG. 1983. Phys. Lett. **132B:** 202.
21. SEHGAL, L. M. & P. ZERWAS. 1984. Nucl. Phys. **B234:** 61.
22. HORGAN, R. & M. JACOB. 1984. Nucl. Phys. **B238:** 221.
23. BARGER, V., A. D. MARTIN & R. J. N. PHILLIPS. 1983. Phys. Rev. **28:** 145.
24. ROY, D. P. 1984. Z. Phys. **C21:** 333.
25. GODBOLE, R. M. & D. P. ROY. 1984. Dortmund preprint DO-TH 84/25.
26. DESAI, B. R. & J. LINDFORS. 1983. Phys. Lett. **131B:** 217.
27. VAN NEERVEN, W. L., J. A. M. VERMASEREN & K. J. F. GAEMERS. 1982. NIKHEF preprint H/82-20.
28. BARGER, V., A. D. MARTIN & R. J. N. PHILLIPS. 1983. Z. Phys. **C21:** 99.
29. ALTARELLI, G., R. K. ELLIS, M. GRECO & G. MARTINELLI. 1984. Nucl. Phys. **B246:** 12.
30. PETERSON, C., D. SCHLATTER, I. SCHMITT & P. M. ZERWAS. 1983. Phys. Rev. **D27:** 105.
31. BARGER, V., A. D. MARTIN & R. J. N. PHILLIPS. 1985. Phys. Lett. **151B:** 463.
32. BARGER, V., H. BAER, K. HAGIWARA, A. D. MARTIN & R. J. N. PHILLIPS. 1984. Phys. Rev. **D29:** 1923.
33. SCHMITT, I., L. M. SEHGAL, H. THOLL & P. M. ZERWAS. 1984. Phys. Lett. **139B:** 99.
34. KUNSZT, Z., E. PIETARINEN & E. REYA. 1980. Phys. Rev. **D21:** 733.
35. PAIGE, F. 1983. Unpublished.
36. BALLOCCHI, G. & R. ODORICO. 1984. Phys. Lett. **136B:** 126.
37. HALZEN, F. & P. HOYER. 1985. Phys. Lett. **154B:** 324.
38. ARNISON, G. *et al.* 1984. Phys. Lett. **139B:** 115.
39. BARGER, V., K. HAGIWARA & W-Y. KEUNG. 1984. Phys. Lett. **145B:** 147.
40. ELLIS, J. & H. KOWALSKI. 1984. Nucl. Phys. **B246:** 189.
41. ALLAN, A. R., E. W. N. GLOVER & A. D. MARTIN. 1984. Phys. Lett. **146B:** 247.
42. HABER, H. E. & G. L. KANE. 1984. Phys. Lett. **142B:** 212.
43. BARGER, V., K. HAGIWARA, W-Y. KEUNG & J. WOODSIDE. 1985. Phys. Rev. **D31:** 528.
44. HALL, L. & S. RABY. 1985. Phys. Lett. **153B:** 433.
45. BARGER, V., W-Y. KEUNG & R. J. N. PHILLIPS. 1985. Phys. Rev. Lett. **55:** 166; Phys. Rev. **D32:** 320, 1236.
46. BARGER, V. & R. J. N. PHILLIPS. 1984. Phys. Lett. **143B:** 259.
47. BARGER, V. & R. J. N. PHILLIPS. 1984. Phys. Rev. **D30:** 1890.

$W^{\pm}$ and Z^0 Production in the UA1 Experiment at the CERN Proton-Antiproton Collider

Presented by V. VUILLEMIN
UA1 Collaboration[a]
CERN CH-1211
Geneva 23, Switzerland

INTRODUCTION

In this presentation, I will summarize new physics results concerning the production and decay of the $W^{\pm}$ and Z^0 intermediate bosons obtained during the last year by the UA1 experiment at CERN. The data recorded in the spring of 1983 are from proton-antiproton collisions at a total center-of-mass energy of 540 GeV, which corresponds to an integrated luminosity of 136 nb^{-1}. Preliminary results presented in this paper are extracted from the data obtained during the autumn of 1984 at a center-of-mass energy of 630 GeV for a total recorded integrated luminosity of 270 nb^{-1}.

This paper deals with three main subjects. The first is the search for the decay of the Intermediate Vector Bosons (IVBs) into two hadronic jets. The second concerns the possible anomalies in the production or the decay of the IVBs. Such anomalies could uncover possible physics phenomena beyond the Standard Model of the electroweak interactions. They are: (a) the production of hard hadronic jets in association with the IVB and (b) the presence of the radiative Z^0's. The third subject is the observation of a new leptonic decay channel of the $W^{\pm}$ in the τ heavy lepton.

SEARCH FOR THE DECAY OF THE IVB INTO TWO HADRONIC JETS

The data sample used is based on an "inclusive jet" trigger defined by summing transverse energies over calorimeter cells in a sliding geometrical window. The window covered a region of $\Delta\eta = 1$, $\Delta\psi = 180°$ in the range, $|\eta| \leq 1.5$, and of $\Delta\eta = 1.5$, $\Delta\psi = 90°$ for $1.5 < |\eta| \leq 3$. The analysis is based on three data samples with different thresholds on the nominal E_T deposition in these windows, which are in detail:

$E_T \geq 15$ GeV (for an integrated luminosity of 6.1 events per nanobarn),
$E_T \geq 20$ GeV (40.1 events per nanobarn),
$E_T \geq 25$ GeV (64.6 events per nanobarn).

[a]This UA1 Collaboration includes Aachen, Amsterdam (NIKHEF), Annecy (LAPP), Birmingham, CERN, Harvard, Helsinki, Kiel, London (Queen Mary College), Padua, Paris (Coll. de France), Riverside, Rome, Rutherford Appleton Laboratory, Saclay (CEN), Vienna, and Wisconsin.

During reconstruction of events, detailed corrections for each calorimeter cell are applied to take into account both measured attenuation tables and an average response factor for hadrons in electromagnetic calorimeters. The corrected energies in the window are typically 10–15% higher than the nominal ones.

The algorithm used for defining jets in UA1 calorimeters has been described before.[1] It relies on local maxima, initiator cells with a threshold transverse energy (1.5 GeV), around which cells are added in order of decreasing transverse energy to form a cluster. The dependence of jets on the initiator threshold has been shown to be negligible for jet transverse energies of 15 GeV or more. All cells are considered provided they are geometrically close enough in a metric variable, ΔR, defined by

$$\Delta R = \sqrt{(\Delta\eta^2 + \Delta\psi^2)}.$$

Charged track information is not used for finding jets or for defining their energy. An average response factor of 1.13 was used for depositions in electromagnetic calorimeters to account for the hadron/π^0 mixture expected in jets.

The use of a clustering algorithm for defining a jet direction and energy entails a number of corrections:

(i) The retained jets invariably will not contain any jet debris that has been emitted further from the jet axis than the cut parameter, ΔR. From e^+e^- experiments[2] and currently accepted fragmentation models,[3] this is known to have a measurable effect, in particular on the total jet energy.

(ii) Any spectator remains, that is, parton debris not connected with the hard interaction, will be included in the jet by the algorithm if they happen to fall in the angular region defined for the jet.

(iii) The small uninstrumented zones in UA1's calorimeters cause some losses in jet energy as a function of the jet direction, whose average value can only be determined by Monte Carlo techniques using the detailed knowledge of the detector.

Corrections for these effects have been determined using extensive Monte Carlo calculations. Quantum chromodynamic (QCD) two-jet events were generated with varying E_T thresholds using the QCD event generator, ISAJET.[4] The final-state particles resulting from fragmentation and spectator particles were followed through the detector using, in particular, a suitable approximation for calorimeter deposition.[5] The jet algorithm was then applied, and reconstructed jets were compared with the emitted partons in energy and direction. If the angular difference was too large ($\Delta R > 1$), the event was rejected. Average values for corrections and their r.m.s. errors were evaluated as a function of transverse energy, E_T, rapidity, η, and azimuthal angle, ψ.

In FIGURE 1, we show the invariant mass of the jet-jet system. Jet energies have been corrected as described above. No structure is observed either in the IVB mass region or in the higher mass range.

The jet transverse energy resolution has been calculated by Monte Carlo and can be approximated by the following formula:

$$\Delta E_T/E_T = 1.5/\sqrt{E_T} \qquad [E_T \text{ in GeV}].$$

This translates to a resolution in jet-jet mass of about 20% in the IVB mass range.

JET ACTIVITY ASSOCIATED WITH $W^{\pm}$ AND Z^0 PRODUCTION

The majority of the IVBs produced at the CERN $p\bar{p}$ Collider are expected to come from the Drell-Yan mechanism,[6] which is the annihilation of a quark from the proton with an antiquark from the antiproton. The incoming annihilating quark and antiquark will radiate gluons. These initial-state bremsstrahlung gluons tend to go in the direction of the incoming beams and tend to have a low transverse momentum. However, occasionally one of these radiated gluons will have a sufficiently high transverse momentum to produce an observable hadronic jet in the event. In this QCD

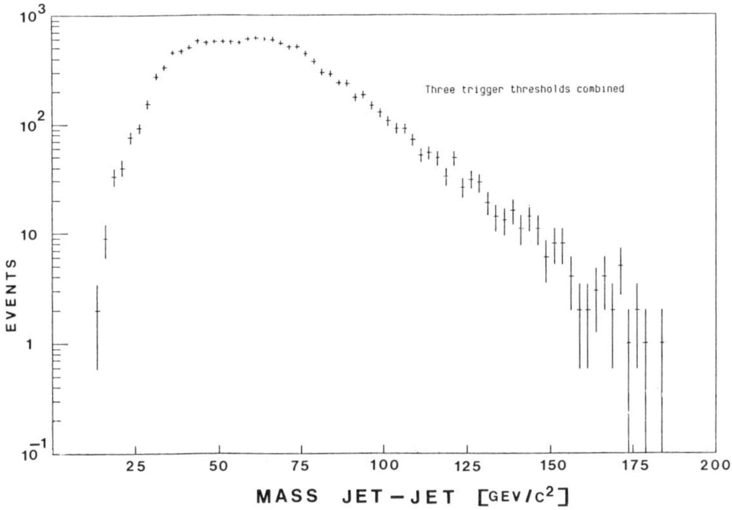

FIGURE 1. Observed two-jet mass spectrum. No significant structure is observed.

picture of IVB production, we expect:

(i) the occasional observation of a hadronic jet produced in association with W or Z^0;

(ii) that these jets should recoil against the IVB in the transverse plane, and their transverse momentum should tend to balance the IVB transverse momentum;

(iii) that this jet activity should be associated with higher transverse momentum IVBs.

A deviation from the QCD expectations would indicate the existence of an unexpected production mechanism creating the IVBs. For example, too many jets at large angle with respect to the beam direction might indicate the production of a massive object, X, which subsequently decays into an IVB plus one or more jets. On the other hand, too many jets at small angle with respect to the beam direction could indicate the production of W and Z^0 bosons by partons having an unusual color structure, etc.

In order to look for such phenomena, the standard UA1 jet algorithm[1] has been used to reconstruct the jet activity in our samples of

(i) 68 $W^{\pm} \rightarrow e^{\pm}\nu_e$ events[7] and
(ii) 9 $Z^0 \rightarrow L^+L^-$ events,[8]

which come from the 1982 and 1983 data-taking runs of the UA1 Collaboration. A detailed analysis of these early statistics has been published elsewhere.[9] The same analysis has also been pursued with the addition of the $W^{\pm}$ and Z^0 candidate events coming from the recent 1984 data-taking run. The overall statistics containing events and candidates (some of these events, though, have not been certified as showing the correct process at the Megatek scanning level) are the following:

(i) 151 $W^{\pm} \rightarrow e^{\pm}\nu_e$ events + candidates,
(ii) 42 $W^{\pm} \rightarrow \mu^{\pm}\nu_{\mu}$ events + candidates,
(iii) 31 $Z^0 \rightarrow L^+L^-$ events + candidates.

The 1984 data, though, are still highly preliminary and subject to small changes after the complete analysis of $W^{\pm}$ and Z^0 events has been performed. However, the main results will not be modified.

As for jet activity, it is indeed observed, and examples of it are shown in FIGURES 2 and 3, which display a graphic representation of UA1 events. FIGURE 2 shows a $W^{\pm}$ decaying into the $(e\nu_e)$ channel accompanied with the emission of a hard central jet of 23 GeV/c, and the muonic decay of a W^+ with the highest transverse momentum, p_T^W = 38 GeV/c. Both W's are recoiling back-to-back with the jets in the transverse plane. The same feature is also observed in Z^0 events containing jets as shown in FIGURE 3.

Using the entire statistics available on $W^{\pm}$ events, we verify that the jet activity reconstructed in $W^{\pm}$ events has the gross features that we expect for jets arising from initial-state gluon bremsstrahlung, namely:

(i) the jets tend to recoil against the IVB in the transverse plane, and their transverse momenta tend to balance the IVB transverse momenta;
(ii) the jet activity is associated with the higher transverse momentum IVBs (see FIGURES 4a, 4b, and 5).

In FIGURES 4a and 4b, the UA1 events are compared with the QCD model of Altarelli *et al.*[10] It can be seen that the number of events having W's of high momentum and containing a hard hadronic jet are quite consistent with what is expected from the tail of the QCD prediction. For comparison, also shown on these plots are the three published events from the UA2 experiment having the highest p_T of the W's.[11]

The angular distribution of jets reconstructed in W events is strongly peaked in the beam directions. This can be seen in FIGURE 6, where the distribution of $|\cos \theta^*|$ is shown, with θ^* being the angle between the jet and the average beam direction in the rest frame of the W and the jet. The shape of this distribution is in agreement with our expectation for initial-state gluon bremsstrahlung, namely $(1 - |\cos \theta^*|)^{-1}$.

Finally, for comparison with the jet activity in the IVB event samples, we also used the UA1 jet algorithm to reconstruct jets in a sample of 387 hadronic multijet events that satisfied the following requirements:

(i) There had to be at least two hadronic jets in the event;

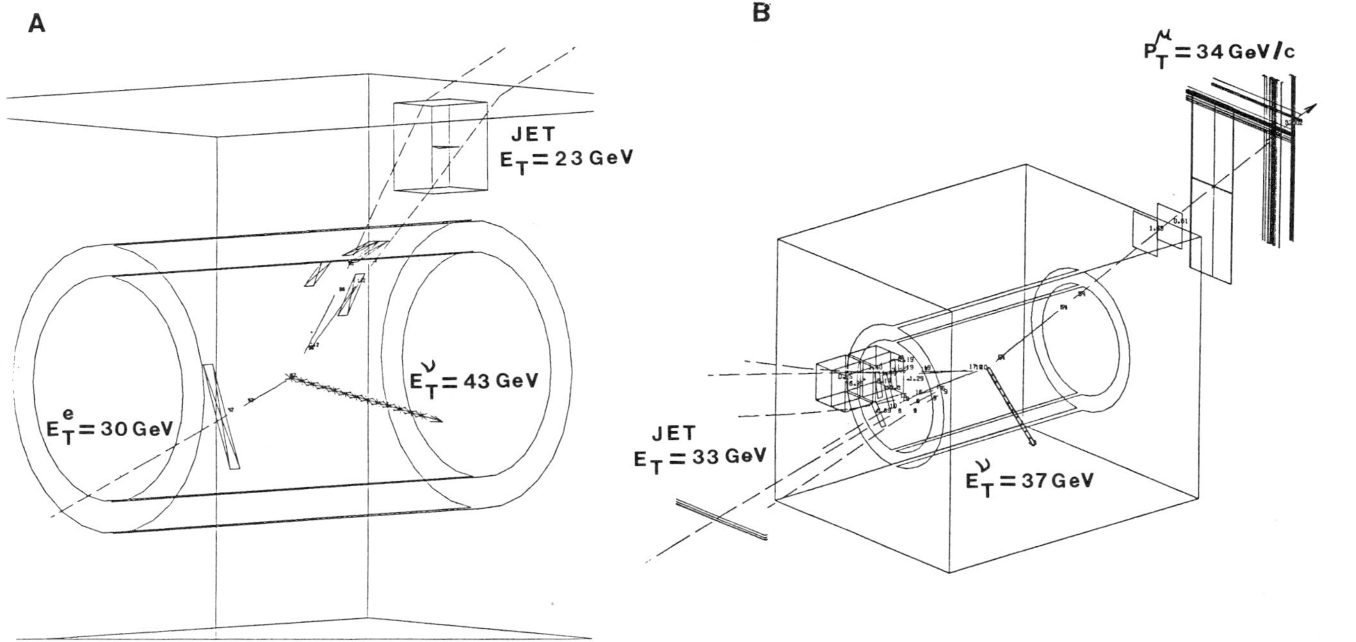

FIGURE 2. Graphic representation of: (A) $W \rightarrow e\nu_e$ event; (B) $W \rightarrow \mu\nu_\mu$ event having the W with the largest transverse momentum ($p_T^W = 38$ GeV/c). Both W's are recoiling against a hadronic jet with a reconstructed transverse energy that just balances the W transverse motion.

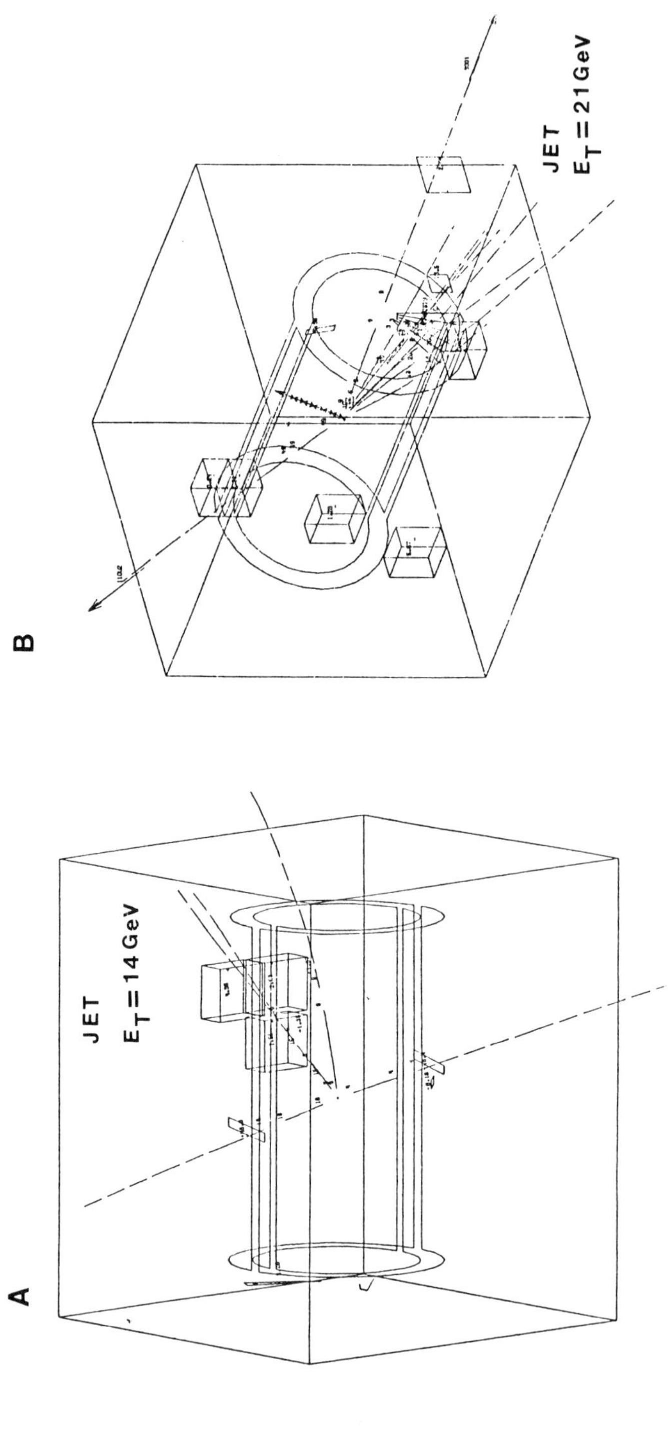

FIGURE 3. Graphic representation of: (A) $Z^0 \longrightarrow e^+e^-$ event; (B) $Z^0 \longrightarrow \mu^+\mu^-$ event. Both events contain, in addition, a hadronic jet recoiling against the Z^0.

(ii) The two highest transverse energy jets reconstructed by the algorithm had to have a two-jet invariant mass close to the W mass (70–90 GeV/c^2);

(iii) The angle, θ, between the direction of the two highest transverse energy jets and the average beam direction in the rest frame of the jet pair was required to satisfy $|\cos\theta| < 0.2$.

By selecting angles close to 90°, we maximize the separation between initial-state gluon bremsstrahlung, which will tend to go along the beam directions, and final-state

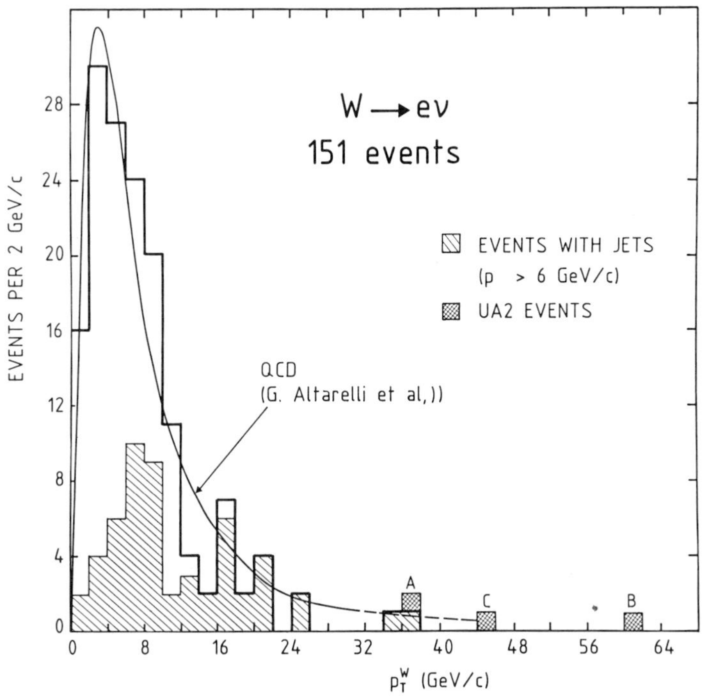

FIGURE 4a. W transverse momentum distribution for the decays in $e\nu$. The events containing hadronic jet activity dominate the tail of the distribution. The QCD curve is from reference 12. UA2 events are from reference 11.

gluon bremsstrahlung, which will tend to go along the jet-jet direction and be absorbed by the jet algorithm into the two highest transverse energy jets.

From our sample of 387 events satisfying these criteria, 163 events have "additional" jet activity; that is, jet activity in addition to the two highest transverse energy jets. The properties of these additional jets from hadronic multijet events compare well with the corresponding properties of the jets reconstructed in the IVB events and our QCD expectations for these properties.

Defining R_W to be the fraction of W events containing jet activity, and R_J to be the corresponding fraction of multijet events containing "additional" jet activity, we find,

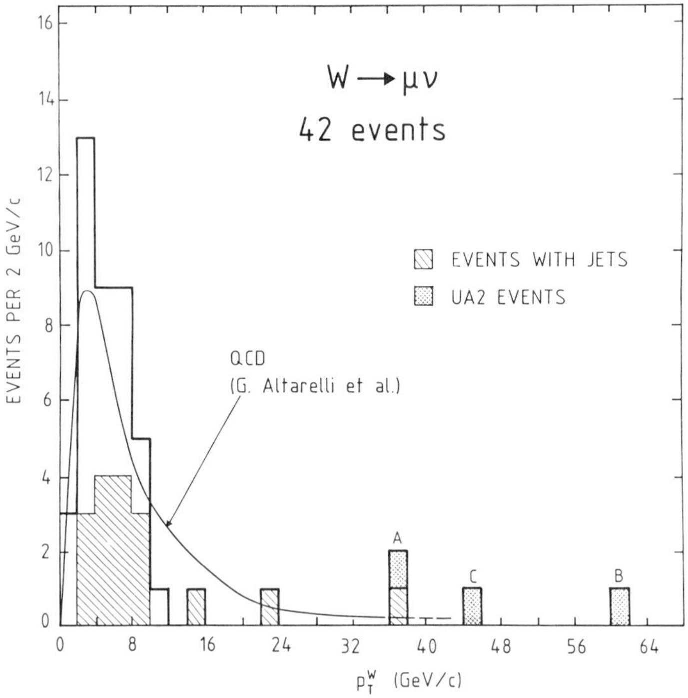

FIGURE 4b. *W* transverse momentum distribution for the decays in $\mu\nu$. The events containing hadronic jet activity dominate the tail of the distribution. The QCD curve is from reference 12. UA2 events are from reference 11.

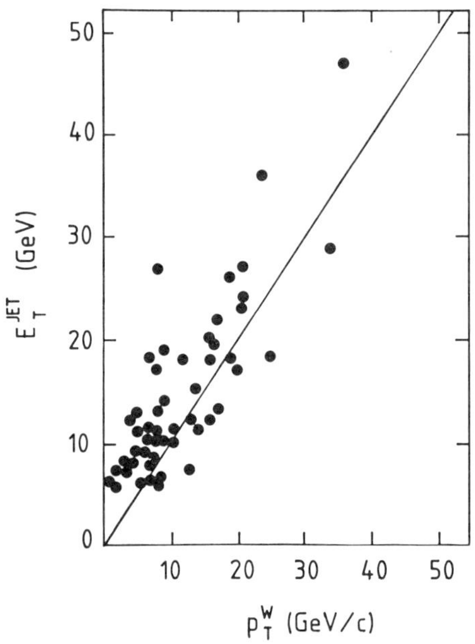

FIGURE 5. Correlation between *W* transverse momentum and jet transverse energy. For *W* events in which more than one jet has been reconstructed, only the highest E_T jet has been plotted.

after making a small correction for differences in the experimental acceptance, that $R_W/R_J = 0.7 \pm 0.2$. This is in excellent agreement with the lowest-order QCD expectation from reference 12 for this ratio, namely 0.69 to 0.73 (depending on the structure function used). The ratio, R_W/R_J, is expected to be insensitive to experimental and theoretical uncertainties common to both W and multijet events. Individually, however, R_W and R_J will depend on details such as the experimental efficiency for reconstructing low transverse energy jets and on the transverse energy threshold used to define a jet.

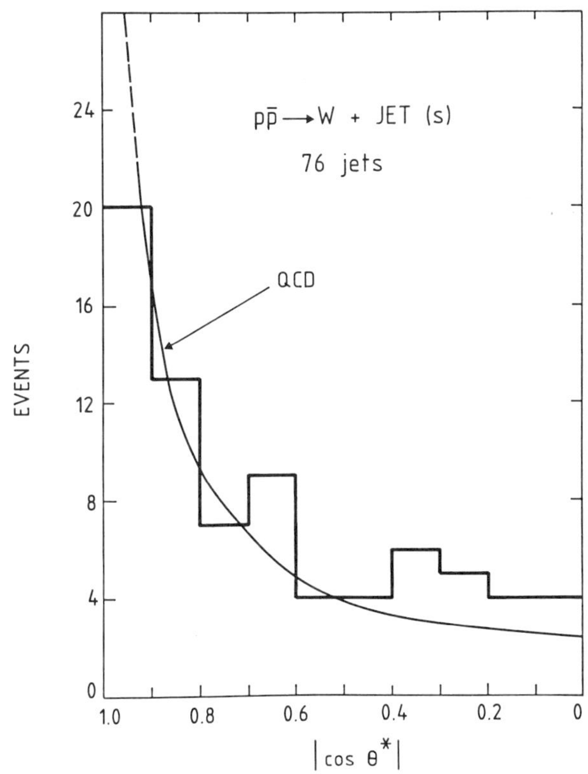

FIGURE 6. Angular distribution of jets in W events compared with the QCD distribution from reference 12.

The rates of "jetty" W and Z^0 events are also in good agreement with the lowest-order QCD expectation[12] mentioned above, as seen in FIGURES 7 and 8.

Before we conclude that the jet activity observed in W and Z^0 events is well understood in the framework of QCD, we should at least look at the (W + jet)-mass distribution (FIGURE 9) to see if there is any evidence for the production of a state, X, which subsequently decays into a W and a hadronic jet. There is no clear evidence for

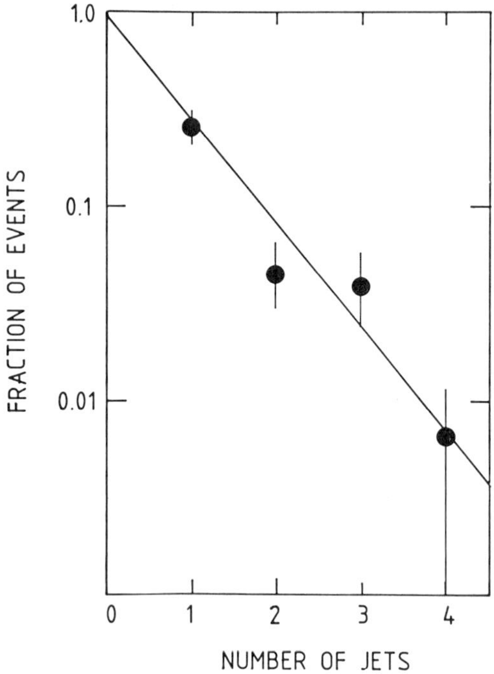

FIGURE 7. Distribution of the number of jets contained in $W \rightarrow e\nu$ events. The QCD curve is from reference 12. Symbols: ● = $W \rightarrow e\nu$; — = QCD (Stirling *et al.*).

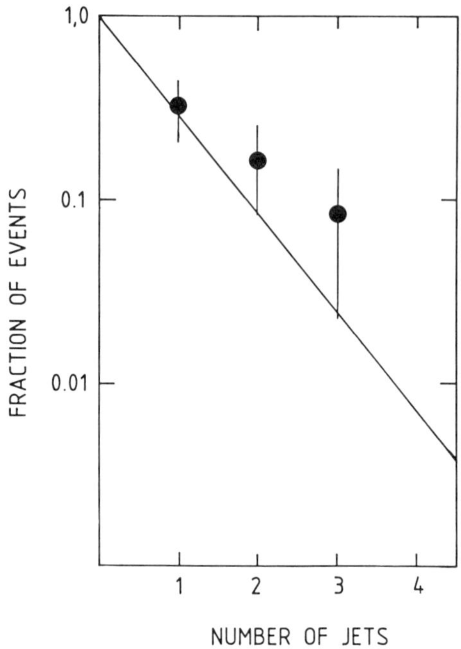

FIGURE 8. Distribution of the number of jets contained in $Z^0 \rightarrow e^+e^-$ and $Z^0 \rightarrow \mu^+\mu^-$ events. The QCD curve is from reference 12. Symbols: ● = $Z^\circ \rightarrow L^+L^-$; — = QCD (Stirling *et al.*).

such a process, and indeed the shape of the (W + jet)-mass distribution is well described by the lowest-order QCD calculation of reference 12, which also describes the shape of the (Z^0 + jet)-mass distribution.

To summarize, the jet activity observed in W and Z events is in agreement with our expectations for QCD corrections to the bare parton-model picture of Drell-Yan annihilation.

RADIATIVE Z^0's

The two events interpreted as $Z^0 \rightarrow L^+L^-\gamma$ decay may be a possible indication for new physics. A summary of the main parameters of these events is given in TABLE 1.

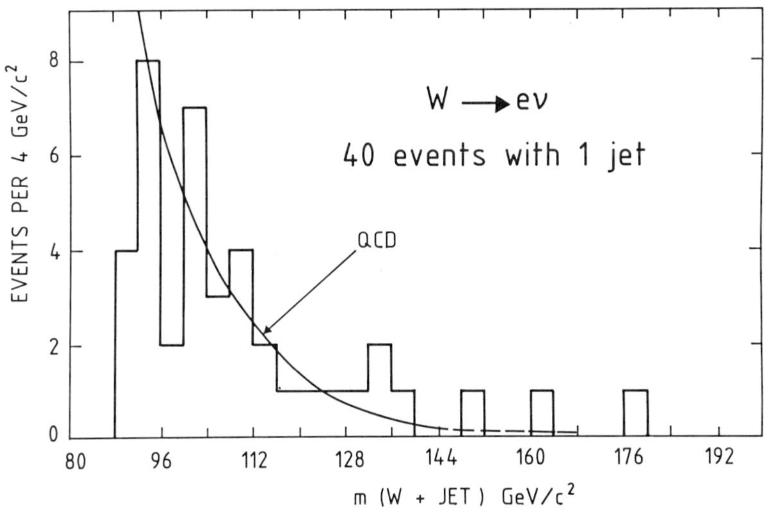

FIGURE 9. (W + jet)-invariant-mass distribution. The QCD curve is from reference 12.

The dominant feature is the high photon energy and the large opening angle between the charged lepton and the photon. The possibility of a bremsstrahlung with such an energy and angle observed is very small.

Out of a subset of the total statistics taken during the 1984 run and analyzed on-line, no event of identical type to the above has been found. Therefore, so far, these effects have not been confirmed. However, it is certainly too early to conclude before a full analysis has been carefully performed on all 1984 events.

EXPERIMENTAL EVIDENCE FOR THE HEAVY LEPTON DECAY, $W \rightarrow \tau\nu_\tau$

With the observation of the $W \rightarrow \tau\nu$ decay, the "program" on the leptonic decay channels of the IVB is complete.

In the case of a $W \rightarrow \tau\nu$ event where the τ decays in the hadronic mode, what we measure is a jet that includes charged tracks and the corresponding energy deposition in some calorimeter cells (both hadronic and electromagnetic). The measured jet represents the charged and neutral π's of the real τ. The total missing transverse energy reconstructed in the event (E_T^{miss}) is due to both the ν from the W decay and that from the τ decay. Therefore, events with missing transverse energy and one trigger jet were selected in the data recorded during the 1982 and 1983 runs (corresponding to an integrated luminosity of 136 nb^{-1}).

As a consequence of the e, μ, and τ universality of the weak current, the $W \rightarrow \tau\nu$ rate is expected to be relatively abundant. Almost half of the τ's decay into one charged π (10% without neutrals and 38% with) and a neutrino.[13] This suggests a clear signature with a reasonable rate: an isolated high-p_T track of a hadronic type and some missing transverse energy. In this sense, this search is rather similar to the one for the $W \rightarrow e\nu$, except that now we look for a hadronic track instead of an electromagnetic (e.m.) one and the pion will have a continuous energy spectrum.

TABLE 1. Properties of the $L^+L^- \gamma$ Events

	$e^+e^-\gamma$	$\mu^+\mu^-\gamma$
E_ν (GeV)	38.8 ± 1.5	28.3 ± 3
E_L^+ (GeV)	61.0 ± 1.2	$50.6\ ^{+5.8}_{-4.7}$
E_L^- (GeV)	9 ± 1	$42.2\ ^{+44.0}_{-14.3}$
$\Delta\alpha$ (L^+, γ) (·)[a]	132.0 ± 4.0	7.9
$\Delta\alpha$ (L^-, γ) (·)[a]	14.4 ± 4.0	99.0
$m(L^+L^-)$ (GeV/c^2)	42.7 ± 2.4	$70.9\ ^{+37.2}_{-12.4}$
$m(L^+L^-\gamma)$ (GeV/c^2)	98.7 ± 5.0	$88.4\ ^{+46.1}_{-15.2}$
$m(L^+\gamma)$ (GeV/c^2)	88.8 ± 2.5	5.0 ± 0.4
$m(L^-\gamma)$ (GeV/c^2)	4.6 ± 1.0	$52.5\ ^{+27.5}_{-9.3}$

[a] $\Delta\alpha$ is the angular difference in space.

The data are selected from the total statistics of the 1982 and 1983 $p\bar{p}$ runs, processed through an off-line filter characterized by a cut on the total missing transverse energy of 15 GeV as already described.[14] The following selection criteria are applied to the initial sample of 29,962 events:

(i) *A validation of the missing transverse energy* defined by $E_T^{miss} > 2\sigma$, where $\sigma = 0.7\ \sqrt{\Sigma E_T}$ (ΣE_T is the scalar sum, in GeV, of observed transverse energy contributions from all calorimeter cells). Due to the reduced efficiency of the calorimetry in the region ±15° around the vertical axis, we exclude events that have a jet axis in this region and events for which the sum of the p_T of all the central detector (CD) tracks in this region exceeds 15% of the total transverse energy.

(ii) *A validation of the trigger jet* defined as the calorimeter jet with highest transverse energy in the pseudorapidity range, $|\eta| < 1.5$. A matching is required between the calorimeter jet and the CD jet.

(iii) *No jet back-to-back with the trigger jet:* the backward region is defined by the angle, $\psi^{back} = 180° \pm 30°$ around the jet axis. It is required that there be no jet inside this backward region, with $E_T^{jet} > 7$ GeV or $p_T^{jet} > 5$ GeV for a jet rapidity of $\eta^{jet} \leq 2.5$. We are left with 550 events at this stage.

(iv) *Events with activity back-to-back to the trigger jet, but with no jet reconstructed by the jet algorithm* are rejected. We require that the total sum of the p_T of the tracks included in the backward zone ($\Sigma_{back}\, p_T$) be less than 4 GeV and the sum of the transverse energy in the calorimeter cells included in this backward zone ($\Sigma_{back}\, E_T$) be less than 6 GeV.

(v) We require that *the trigger jet has a leading track with a minimum p_T of 4 GeV/c.*

We finally remove all events recognized already as $W \to e\nu$ and scan the remaining sample. We end up with 16 events, which are subdivided into three topological categories:

(i) Five events with a single high-p_T charged track, which we call the "one-branch sample." Two of them have an associated jet with an $E_T^{jet} > 10$ GeV not back-to-back with the trigger jet.

(ii) Four events with two or three charged tracks, which we call the "three-branch sample."

(iii) Seven events with more than three charged tracks or E_T^{jet} larger than 40 GeV, which we call the "multi-branch sample."

This paper is concerned only with the one-branch and three-branch categories, which are regarded as τ candidates. The main properties of these events are summarized in TABLES 2 and 3. A view of these events is shown in FIGURE 10.

We now discuss some kinematical and physical properties of the τ candidates in order to test their consistency with the $W \to \tau\nu$ hypothesis. We compare some results for the process, $W \to \tau\nu$, with the corresponding ones in the now well-known case of $W \to e\nu$[7] and with the expectations from a Monte Carlo simulation of $W \to \tau\nu$ events. The Monte Carlo takes into account polarization effects[4,15] and includes the effects of trigger and off-line selections.

Hadronic Quality of the One-Branch Sample

To define this property, we use the experience gained during our previous searches for $W \to e\nu$[7] and $W \to t\bar{b}$,[16] where we studied in detail how to identify isolated high-p_T electrons.

Tracks with $p_T > 10$ GeV are classed as hadronic if the deposition in the hadron calorimeter cell behind two adjacent e.m. cells is at least 500 MeV or 1% of the total energy (in e.m. and hadronic calorimeters). In addition, a good e.m. shape in the energy depositions of the four e.m. segments is characterized by its χ_R^2;[17] we, thus, require this quality parameter to be larger than 35.

If the charged track has a $p_T < 10$ GeV, it is difficult to recognize the shape of the shower development. On the other hand, a charged hadron can be identified by a clear

TABLE 2. Main Characteristics of the One-Branch τ Candidates

	Track Characteristics					Missing Energy		"τ-Jet"						
Run/Event No.	p_T^{CD} (GeV)	η	ϕ (°)	Length (m)	$\Delta p/p$ (%)	ΣE_T (GeV)	E_T^{miss} (GeV)	E_T (GeV)	η	ϕ (°)	Acoplanarity (°)	Asymmetry ($\cos\theta^{\tau\pm}$)	$M_T(\nu,\tau)$ (GeV/c^2)	Second-Jet Characteristics
5181/126	4.5	−0.72	26.7	1.2	2.8	88.5	20.0	26.2	−0.74	23.2	45.8	−0.65	35.4	—no—
6024/1356	7.1	0.63	−62	1.0	4	48.8	16.7	15.4	0.77	−60	18.8	−0.62	29.5	—no—
6569/1064	6.4	0.23	106.2	0.8	4	80.7	18.6	16	0.31	108	22.2	0.28	33.1	—no—
7814/86	14.9	0.71	9.9	0.89	17	87.7	28.6	32.1	0.79	11	17.1	0.64	57.7	E_T^c: 15.8 GeV $\eta = -1.36$ $\phi = 125°$ p_T^{CD}: 12.4 GeV $\eta = -1.38$ $\phi = 115°$
8575/409	11.1	0.97	−66.1	1.5	7	74.9	15.6	19.2	1.02	−69	21.8	0.75	31.6	E_T^c: 13.6 GeV $\eta = 0.2$ $\phi = 171°$ p_T^{CD}: 2.6 GeV $\eta = 0.1$ $\phi = 137°$

TABLE 3. Main Characteristics of the Three-Branch τ Candidates

	Track Content of the "τ-Jet" Candidate					Missing Energy		"τ-Jet"						
Run/Event No.	p_T^{CD} (GeV)	η	ϕ (°)	Length (m)	$\Delta p/p$ (%)	ΣE_T (GeV)	E_T^{miss} (GeV)	p_T^{CD} E_T (GeV)	η^{CD} η^{calo}	ϕ^{CD} ϕ^{calo} (°)	Acoplanarity (°)	Asymmetry ($\cos\theta^{\tau\pm}$)	$M_T(\nu,\tau)$ (GeV/c^2)	Second Jet (non b/b)
3303/1289	+1.14 +1.30 −5.19	−0.12 0.53 0.29	−50.6 −62.9 −60.9	0.8 0.9 0.9	14 10 29	69.3	21.6	7.6 25.6	0.28 0.31	−60 −49	8.0	0.29	35.2	—no—
6378/213	+5.9 −1.0 +0.35	−0.72 −0.82 −0.12	215.1 209.9 229.9	1.1 1.0 1.9	69 23 –	77.0	15.5	9.9 19.2	−0.71 −0.72	−145 −158	0.4	−0.60	30.1	—no—
7255/103	+1.69 −4.16 −0.20	−1.22 −0.86 −1.44	72.1 58.8 82.6	1.6 1.3 1	6 15 —	60.6	17.6	6.0 18.9	−1.0 −1.16	63 72	18.3	+0.79	33.0	—no—
8032/1158	−4.78 −7.60	−0.25 −0.33	38.6 41.4	0.7 0.7	64 97	90.3	36.0	14.8 38.3	−0.20 −0.33	44 42	5.6	0.28	66.2	—no—

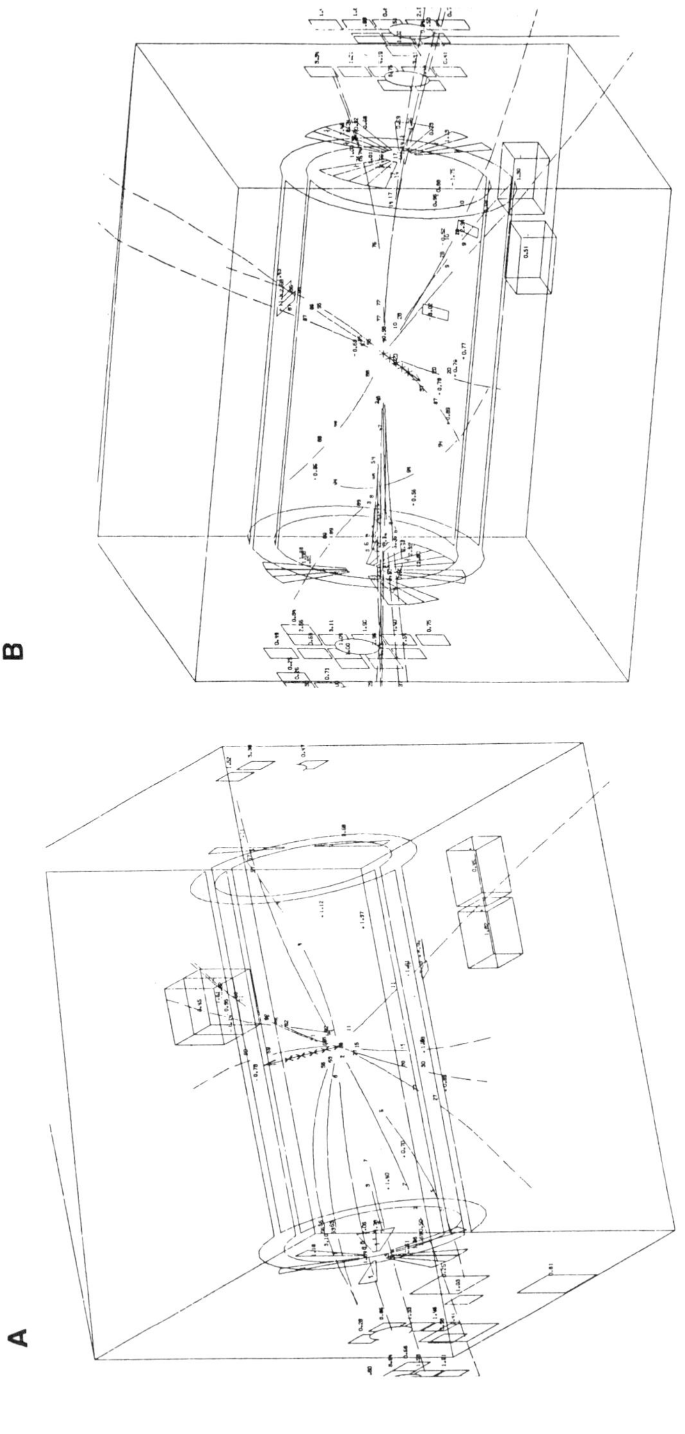

FIGURE 10. Megatek views of (A) a one-branch and (B) a three-branch event.

mismatch in space and/or energy when comparing the CD and the calorimeter information. At least one of the following three factors will be outside the e.m. bounds:

$$[(1/p - 1/E)/\sigma(1/p - 1/E)] > 3, \Delta x/\sigma_x > 3, \Delta\psi/\sigma_\psi > 3.$$

The first factor measures the matching between p, the momentum measurement in the CD, and E, the calorimeter energy deposition; $\Delta x/\sigma_x$ evaluates the matching along the beam direction; and $\Delta\psi/\sigma_\psi$ gives the matching in azimuth of the CD information compared to the calorimeter information.

In FIGURE 11, we give the hadronic characteristics of our five one-branch events. We compare them with the $W \rightarrow e\nu$ case and the expected results from the τ-generated events. We find that there is a significant disagreement with the $W \rightarrow e\nu$ events and a good agreement with the Monte Carlo expectation. We conclude that all five one-branch events pass the hadron quality test.

In addition, as we require the sum of the p_T of all tracks included in a 20° cone around the track candidate to be ≤ 1.5 GeV and no track with a $p_T > 1$ GeV in a cone of 30°, we conclude also that those five events are characterized by a single isolated hadron associated with missing transverse energy.

Kinematical Properties of the Selected Events

The display of E_T^{jet} versus E_T^{miss} for the one-branch and three-branch events is compared with the measured $W \rightarrow e\nu$ sample and the Monte Carlo expectations (FIGURE 12). Notice that, as expected, the τ candidates are systematically lower in the measured transverse energy and in the missing transverse energy than the $W \rightarrow e\nu$ events.

To calculate the invariant mass of the system, $\pi^{\pm}(n\pi^0)$ or $\pi^{\pm}\pi^{\mp}\pi^{\pm}(n\pi^0)$, we use the information given by the e.m. central calorimeter and the CD.[18] The energy deposited by π^0 ('s) in the electromagnetic calorimeters is (are) obtained by subtracting from the overall electromagnetic calorimeter deposition, the sum of charged track energies as measured in the CD diminished by the amount of energy seen in the hadronic cell(s) behind. An average π^0 impact point in the electromagnetic calorimeter can be computed, thereby allowing the effective mass of the charged and neutral pion(s) to be calculated, and thus checking the consistency with a τ decay hypothesis. The transverse mass of the (τ, ν) system is defined as the transverse mass of the system by E_T^{miss} and the measured jet. The masses are plotted in FIGURE 13 and show a consistency with the τ hypothesis for the one- and three-branch samples.

Background from QCD Bijets

The main source of background is due to standard bijets, where one or both jets are mismeasured because of either leakage or fluctuations in the calorimeter response. To estimate this background, we use an experimental sample of bijets that we fluctuate in energy according to the resolution of our apparatus. We have also generated QCD bijets using the ISAJET Monte Carlo program in different bins in E_T of the jets. The two samples are in good agreement with each other, yielding a rejection factor of 0(10) for jets with $E_T^{jet} < 25$ GeV and 0(5) for jets with $E_T^{jet} > 25$ GeV.

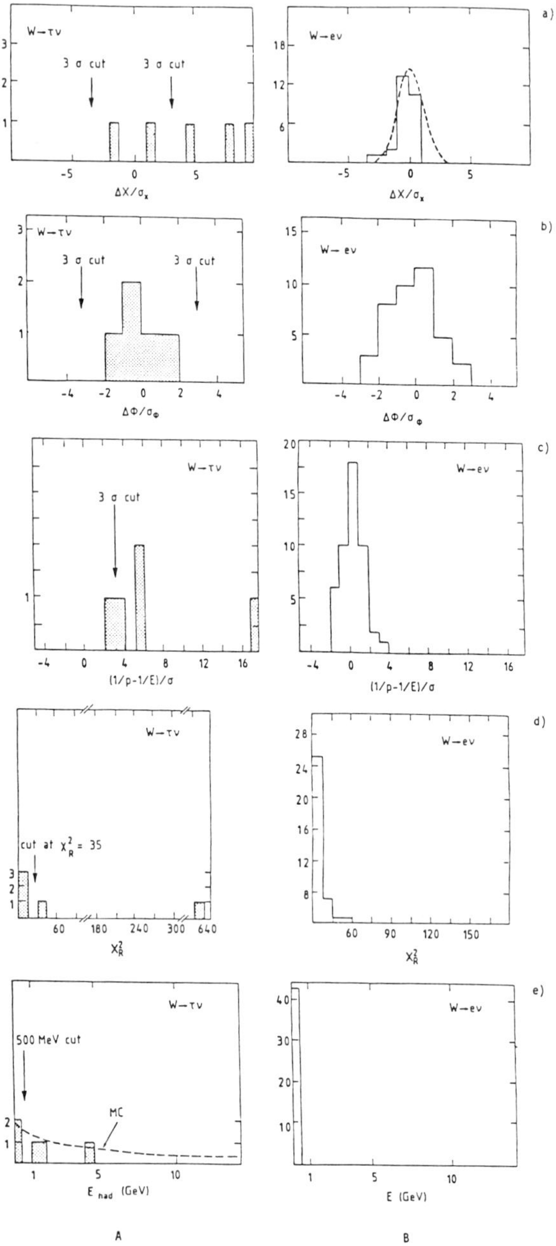

FIGURE 11. Hadronic properties of the one-branch events: the hadron quality of the one-branch sample (A, shaded) is compared with the electron quality of the control sample, $W \rightarrow e\nu$ (B). The quality variables shown are: (a) the matching along the beam direction: $\Delta X/\sigma_x$; (b) the matching in azimuth: $\Delta\psi/\sigma_\psi$; (c) the matching between the momentum measurement and the calorimeter energy deposition; (d) an overall quality parameter, χ^2_R, measuring the e.m. shape of the longitudinal shower profile and the pulse-sharing between the different calorimeter samples; (e) the energy deposited in the hadron calorimeter behind the e.m. shower.

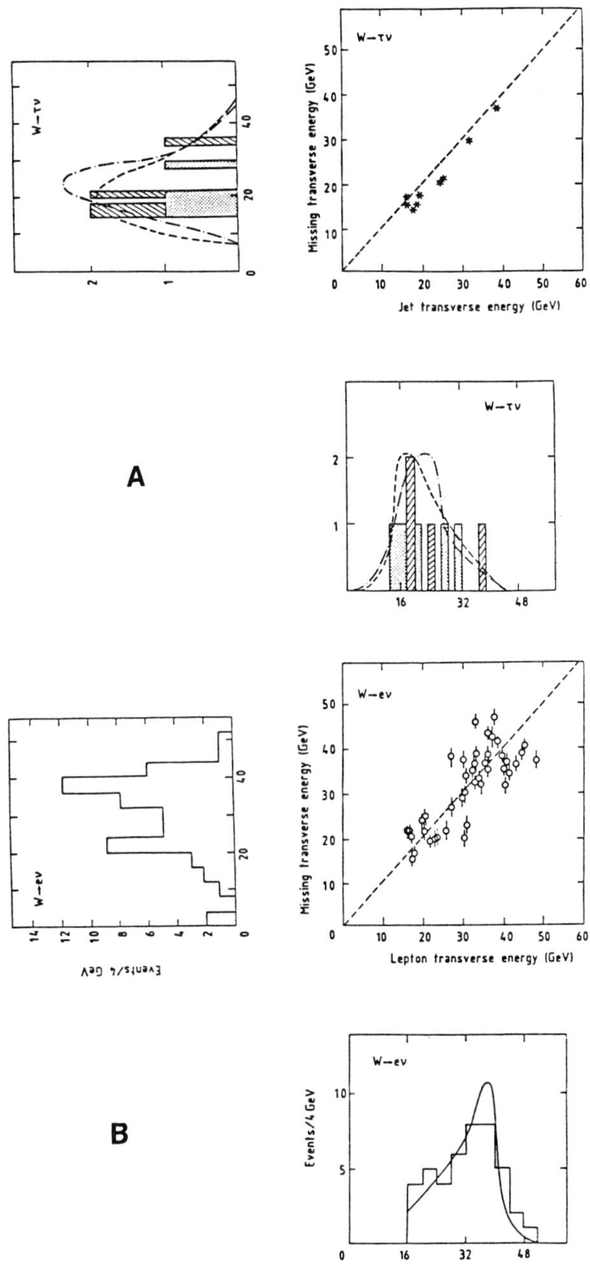

FIGURE 12. Scatterplot of the total missing transverse energy versus the lepton transverse energy of the sample (A) of the one-branch and three-branch events (candidates $W \rightarrow \tau\nu$), as compared to the same plot for the sample, $W \rightarrow e\nu$ (B). The missing transverse energy and the lepton transverse energy distributions for the one-branch and the three-branch samples (shaded) are compared with ISAJET Monte Carlo curves for one-branch (dashed curve) and three-branch τ's (dash-dotted curve). By lepton transverse energy, we mean, in the case of τ candidates, the transverse energy of the measured jet, where the jet includes one or three charged pions and possible additional neutral pions.

So far, we have not constrained the multiplicity of the trigger jet. To define the multiplicity, we only count, in a cone with $\Delta R = (\Delta\eta^2 + \Delta\psi^2)^{1/2} < 0.5$ (where ψ is the azimuthal angle measured in radians and η is the pseudorapidity), those tracks with $p_T \geq 0.5$ GeV. The multiplicities for experimental jets between 15 and 25 GeV are plotted in FIGURE 14. This shows that for one-charged track events, an additional reduction factor of 10 for jets with 15 GeV $\leq E_T^{jet} \leq$ 25 GeV and 20 for jets with $E_T^{jet} >$

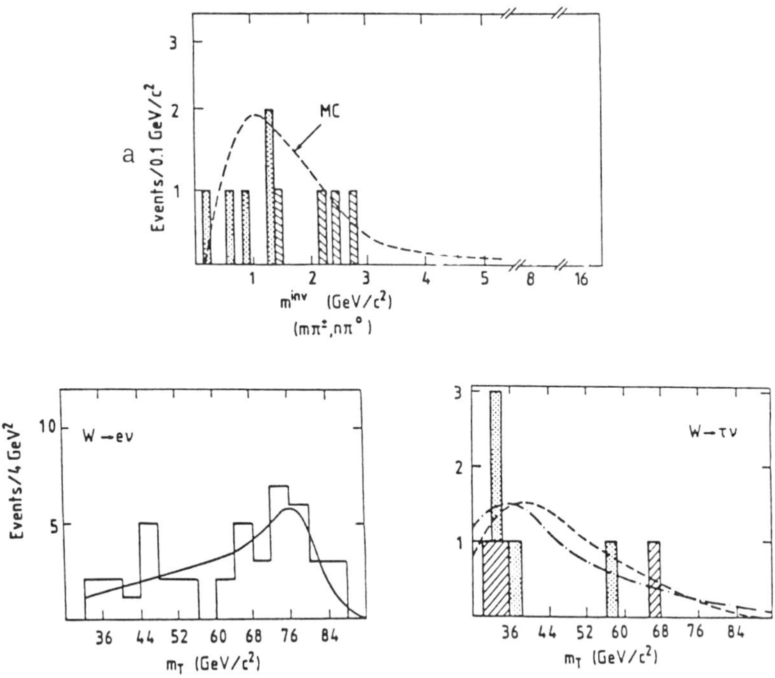

FIGURE 13. τ properties of the one-branch and three-branch events (shaded). It includes parameters that ensure the τ decay (such as the invariant mass) or parameters that ensure a $W \rightarrow \tau\nu$ decay (such as the transverse mass). (a) The invariant mass of the systems ($\pi^{\pm}$, $n\pi^0$) and ($\pi^{\pm}$, $\pi^{\mp}$, $\pi^{\pm}$, $n\pi^0$) is compared with the Monte Carlo predictions (dashed curve). This includes full apparatus simulation. (b) The transverse mass of the system ("τ jet", "ν") for the total sample of one-branch and three-branch events (shaded) is compared with the reference sample of $W \rightarrow e\nu$ events (solid line) and the one-branch (dash-dotted curve) and three-branch τ's (dashed line) generated with the ISAJET Monte Carlo. This includes full detector simulation.

25 GeV are obtained. For a multiplicity of two or three, the reduction factor is only two. Hence, we have an overall rejection factor using the CD information of 0(100) for the one-branch sample and 0(20) for the three-branch sample.

The background due to QCD jet fluctuation is 0.5 ± 0.1 (stat.) for the five one-branch events sample. Thus, the signal $W \rightarrow \tau\nu$ events, where the $\tau^{\pm} \rightarrow \pi^{\pm}(n\pi^0)\nu$, rises clearly above the QCD background. For the four events in the three-branch

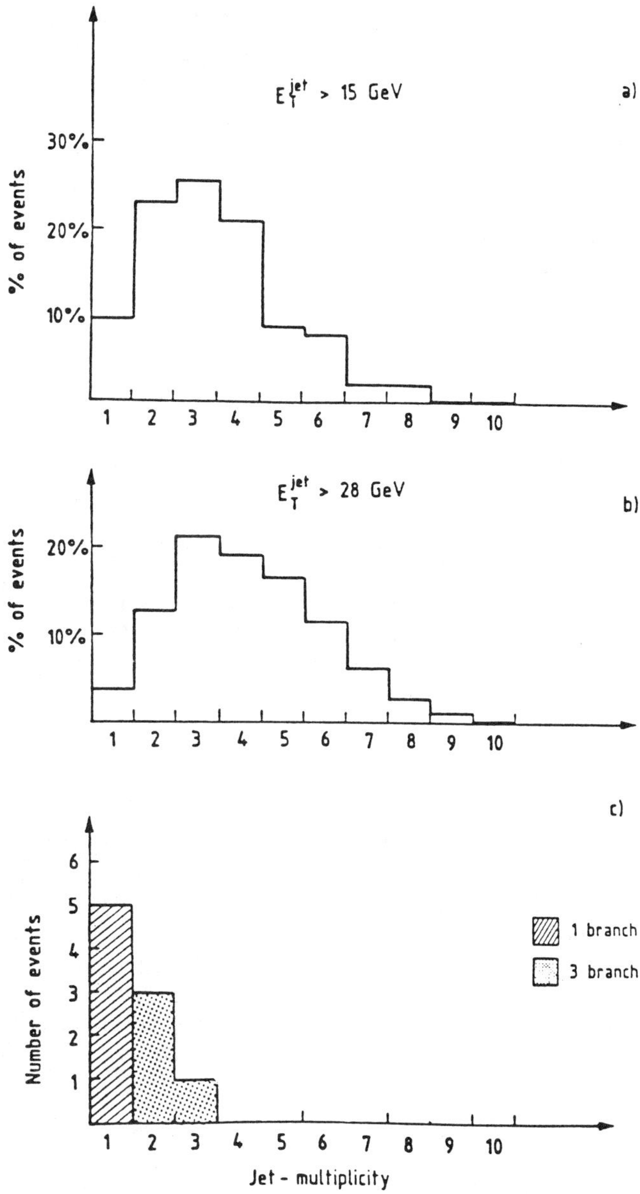

FIGURE 14. Multiplicity distributions: the multiplicity is defined in a cone in $\Delta R = (\Delta\eta^2 + \Delta\psi^2)^{1/2} < 0.5$ and counts only the tracks with $p_T > 0.5$ GeV in this cone. (a) Multiplicity spectrum for jets that have $E_T^{jet} > 15$ GeV and satisfy the electron trigger. (b) Multiplicity spectrum for jets that have $E_T^{jet} > 25$ GeV and satisfy the electron trigger. (c) Multiplicity of the one- and three-branch events.

TABLE 4. $W \rightarrow \tau$ Candidates from 130 nb^{-1} (1984 Data)

Run/Event No.	Charged Tracks p_T^{CD} (GeV)		Missing Energy E_T^{miss} (GeV)	"τ-Jet" E_T^J (GeV)	$M_T(\nu, \tau)$ (GeV/c^2)
10486/358	−6.3		36	31	66
10614/803	+13.1		23	26	49
11622/129	+19.7				
	−13.0	m = 1.76 ± 0.17	27	38	55
	+6.2				
11109/247	+23.4				
	−6.6	m = 2.7 ± 1.02	22	22	44
	−3.1				
11673/499	+12.2				
	−8.7	m = 0.99 ± 0.03	19	21	39
	−7.7				

sample, the background is 2 ± 1 (stat.). If we require $E_T^{miss} > 3\sigma$, we are left with three events, but now the background is reduced to 0.6 ± 0.2 (stat.). In these conditions, the signal of τ in three charged π's begins to be significant. A more detailed study of the background can be found elsewhere.[19]

The overall detection efficiency for the τ is 15%. We can then apply this factor to determine the branching ratio, R, of the process, $W \rightarrow \tau\nu$, with respect to the electron mode. From this search, we obtain five events compatible with $W^{\pm} \rightarrow \tau^{\pm}\nu$ where $\tau^{\pm} \rightarrow \pi^{\pm}(n\pi^0)\nu$. With similar selection criteria, we have obtained 47 $W \rightarrow e\nu$ events, which correspond to an overall detection efficiency of 65%. If we apply the factor of 48%, which is the decay rate of τ into one charged-pion and neutrals, we obtain

$$R = (W \rightarrow \tau\nu)/(W \rightarrow e\nu) = 0.96 \pm 0.45 \text{ (stat.)}.$$

The main results obtained from this study of the 1982 and 1983 data are: a clear τ signal has been evidenced for the first time in the channel, $W \rightarrow \tau\nu$ ($\tau \rightarrow$ hadronic model), and the branching ratio with respect to the electron mode is $W \rightarrow \tau\nu/W \rightarrow e\nu = 0.96 \pm 0.45$, which is in agreement with the universality of the weak current. The corresponding main background coming from QCD bijets is strongly suppressed by using topological CD information about jets.

About the same amount of data (around 130 nb^{-1}) has been analyzed from the 1984 sample with the only requirement being the need to have a validation of the missing transverse energy defined by $E_T^{miss} > 4\sigma$. This very strong requirement does not pretend to isolate all the $W \rightarrow \tau\nu$ candidates from 1984 data, but, on the other hand,

TABLE 5. $Z^0 \rightarrow \tau^+\tau^-$ Candidates from 130 nb^{-1} (1984 Data)

	Charged Tracks					
	τ^+		τ^-			
Run/Event No.	p_T^{CD} (GeV)	p^{CD} (GeV)	p_T^{CD} (GeV)	p^{CD} (GeV)	E_T^{miss} (GeV)	M (GeV/c^2)
(1) 11441/438	24.9	24.9 ± 7.1	13.2	22.3 ± 1.1	21	42 ± 6 ($M_{\mu^+e^-}$)
(2) 11246/146	22.5	25 ± 6	8.2	8.6 ± 0.5	24	29 ± 3 ($M_{e^+\pi^-}$)
(3) 11130/100	5.5	5.7 ± 0.3	2.6	−7.4 ± 0.04	25	35 ($M_{jet,\pi/e}$)

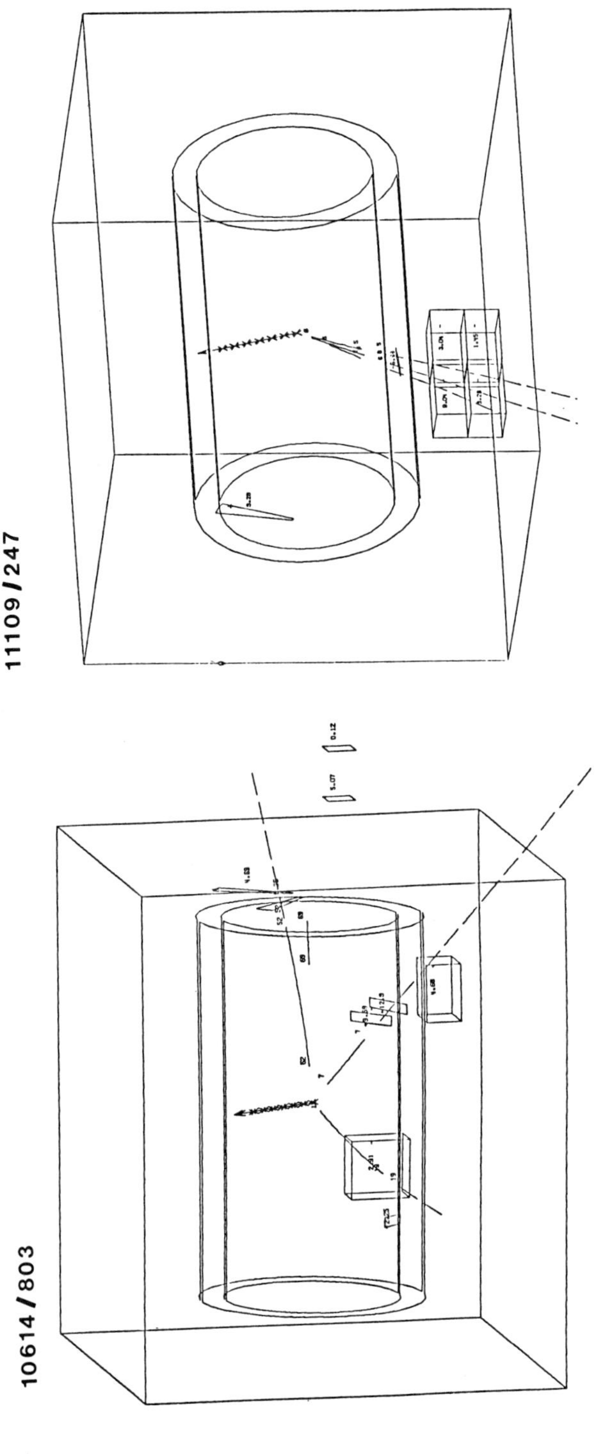

FIGURE 15. Megatek views of a one-branch and a three-branch event selected from 1984 data with the only requirement being a validation that the missing transverse energy is bigger than 4σ. A 1 GeV cut is applied on the transverse energy of the calorimeter cells, as well as a 1 GeV/c momentum cut on the charged tracks.

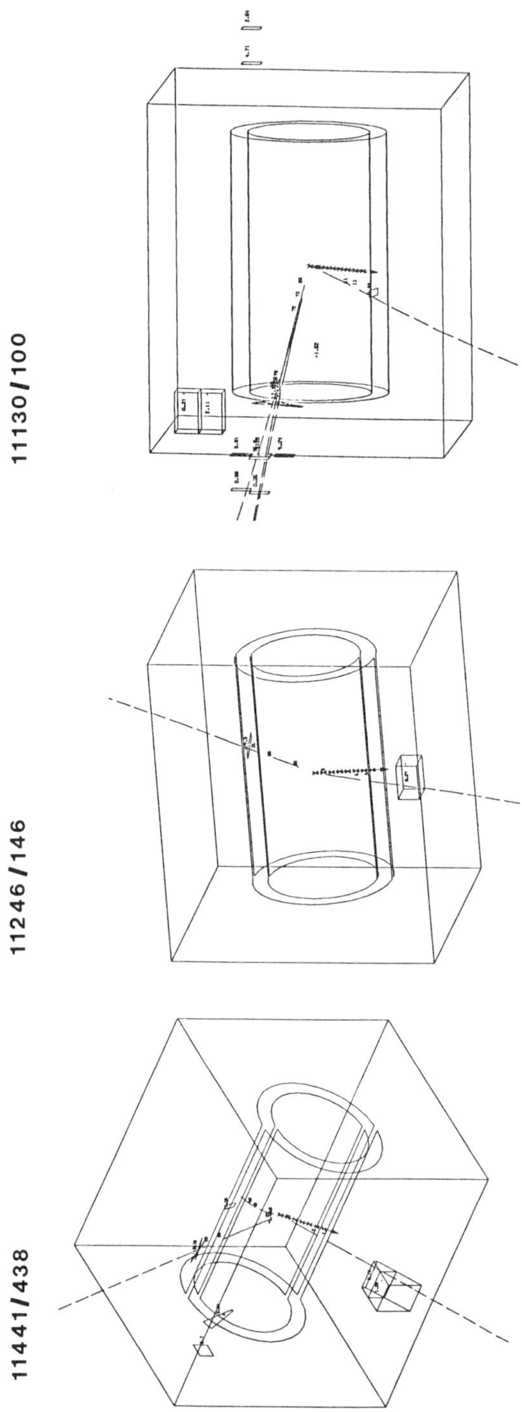

FIGURE 16. Megatek views of the three $Z^0 \longrightarrow \tau^+\tau^-$ candidates selected from 1984 data with the only requirement being a validation that the missing transverse energy is bigger than 4σ. A 1 GeV cut is applied on the transverse energy of the calorimeter cells, as well as a 1 GeV/c momentum cut on the charged tracks. The event topologies are described in the text.

the $W \rightarrow \tau\nu$ candidates extracted by this method have a very small background. Two one-branch and three three-branch candidates have been found. The characteristics of those events are listed in TABLE 4 and two of them are shown in FIGURE 14. These very clean events appear to be a nice confirmation of the 1983 data; the whole statistics of the 1984 run, however, still remain to be fully analyzed.

$Z^0 \rightarrow \tau^+\tau^-$ CANDIDATES

Out of the same selection criteria described above, three events show topologies and missing energies compatible with the $Z^0 \rightarrow \tau^+\tau^-$ hypothesis. They have the following decays:

$$Z^0 \rightarrow \tau^+(\tau^+ \rightarrow \mu^+\nu\nu)\,\tau^-(\tau^- \rightarrow e^-\nu\nu) \quad (1)$$

$$Z^0 \rightarrow \tau^+(\tau^+ \rightarrow e^+\nu\nu)\,\tau^-[\tau^- \rightarrow \pi^-(n\pi^0)\nu] \quad (2)$$

$$Z^0 \rightarrow \tau^+\,(\tau^+ \rightarrow e^+/\pi^+ \ldots)\,\tau^-\,[\tau^- \rightarrow \pi^-\pi^+\pi^-(n\pi^0)\nu] \quad (3)$$

The characteristics of these three events are listed in TABLE 5 and pictures of them are shown in FIGURES 15 and 16. Again, the complete analysis of the 1984 run and the searching for such processes should clarify the situation.

CONCLUSIONS

(i) No peak structure is observed in the jet-jet mass distribution.

(ii) The properties of the jet activity produced in association with charged IVBs are in agreement with our expectations for jets arising predominantly from initial-state gluon bremsstrahlung off the incoming annihilating quark and/or antiquark. This is in agreement with our expectations for QCD corrections to the bare parton-model picture of Drell-Yan annihilation.

(iii) Five events with a single hadronic high-p_T charged track and missing energy are observed in the data of the 1982 and 1983 runs. This signal rises clearly above the background due to the fluctuations of QCD bijets. We interpret these events as evidence of $W \rightarrow \tau\nu$, where $\tau^{\pm} \rightarrow \pi^{\pm}(n\pi^0)\nu$. The branching ratio with respect to the electron mode is 0.96 ± 0.45 (stat.).

(iv) Possible candidates for the decay, $Z^0 \rightarrow \tau^+\tau^-$, are observed in the 1984 data.

NOTES AND REFERENCES

1. ARNISON, G. *et al.* UA1 COLLABORATION. 1983. Phys. Lett. **123B:** 115; ARNISON, G. *et al.* UA1 COLLABORATION. 1983. Phys. Lett. **132B:** 214.
2. See, for example: WU, S. L. 1984. e^+e^- physics at PETRA. The first five years. DESY 84-028.
3. See, for example: GOTTSCHALK, T. D. 1983. QCD-cluster models. Proc. 14th Int. Symposium on Multiparticle Dynamics, Granlibakken, Lake Tahoe, California, p. 394. World Scientific. Singapore.
4. PAIGE, F. E. & S. D. PROTOPOPESCU. 1981. ISAJET program BNL 29777.
5. DELLA NEGRA, M. & PH. GHEZ. 1984. Internal Report CERN UA1-TN/84-15.

6. DRELL, S. & T. M. YAN. 1970. Phys. Rev. **25:** 316; 1971. Ann. Phys. **66:** 578.
7. ARNISON, G. *et al.* UA1 COLLABORATION. 1983. Phys. Lett. **122B:** 103; ARNISON, G. *et al.* UA1 COLLABORATION. 1983. Phys. Lett. **129B:** 273.
8. ARNISON, G. *et al.* UA1 COLLABORATION. 1983. Phys. Lett. **126B:** 398; ARNISON, G. *et al.* UA1 COLLABORATION. 1984. Phys. Lett. **147B:** 241.
9. GEER, S. 1984. Hadronic jet activity associated with intermediate vector boson production at the SPS Collider. Talk given at the 19th Rencontre de Moriond, La Plagne; UA1 COLLABORATION, presented by GEER, S. 1984. UA1 results on jet activity associated with $W^{\pm}$ and Z^0 production at the CERN proton-antiproton collider. Talk given at the DPF Annual meeting, Santa Fe, New Mexico, October 31–November 3, 1984 and preprint CERN-EP/84-160.
10. ALTARELLI, G., R. K. ELLIS, M. GRECO & G. MARTINELLI. 1984. Preprint CERN-TH.3851.
11. BAGNAIA, P. *et al.* UA2 COLLABORATION. 1984. Phys. Lett. **139B:** 105.
12. GEER, S. & W. J. STIRLING. 1984. Preprint CERN-TH.4063.
13. AGUILAR-BENITEZ, M. *et al.* 1984. Particle data group. Particle properties data booklet.
14. ARNISON, G. *et al.* UA1 COLLABORATION. 1984. Phys. Lett. **139B:** 115.
15. Introduction in ISAJET of the matrix elements taking into account the polarization for the process, $W \rightarrow \tau\nu$, is due to M. Mohammadi.
16. ARNISON, G. *et al.* UA1 COLLABORATION. 1984. Phys. Lett. **147B:** 493.
17. EISENHANDLER, E. *et al.* 1984. Internal Report CERN UA1-TN/84-64. χ^2_R measures the goodness of fit to the electron hypothesis. This parameter has been developed from a study of electron test-beam data and makes full use of the longitudinal profile of the shower measured in the four samplings of the e.m. calorimeter. It takes into account the electron energy and angle of incidence.
18. COLAS, P. *et al.* 1984. Internal Report CERN UA1-TN/84-74.
19. ARNISON, G. *et al.* UA1 COLLABORATION. 1984. Experimental evidence for the heavy lepton decay, $W \rightarrow \tau\nu_\tau$, of the charged intermediate vector boson at the CERN $p\bar{p}$ Collider. To be published.

Production of High Mass $e\nu$ and e^+e^- Pairs in the UA2 Experiment at the CERN $\bar{p}p$ Collider

Presented by JOHN RENNER HANSEN

UA2 Collaboration[a]

CERN CH-1211

Geneva 23, Switzerland

INTRODUCTION

We report here the results from a search for electrons with $p_T > 15$ GeV/c produced at the CERN $\bar{p}p$ collider ($\sqrt{s} = 540$ GeV) during its 1982 and 1983 periods of operation.

Following a general discussion of the topology of the events containing an electron candidate, we shall compare the data with expectations in the framework of the electroweak standard model[1] for the reactions,

$$\bar{p} + p \rightarrow W^{\pm} + \text{anything}$$
$$\rightarrow e^{\pm} + \nu\,(\bar{\nu}) \qquad (1)$$

$$\bar{p} + p \rightarrow Z^0 + \text{anything}$$
$$\rightarrow e^+ + e^- \text{ or } e^+ + e^- + \gamma, \qquad (2)$$

where $W^{\pm}$ and Z^0 are the postulated charged and neutral Intermediate Vector Bosons (IVBs), respectively.

According to the amount of data collected, we are now in the position to study some details of the IVB production, e.g., the influence of emission of gluon radiation on the distribution of the W transverse momentum (FIGURE 1).

Preliminary results from the study reported here have already been presented elsewhere[2] and a more complete discussion can be found in a recent publication.[3]

THE DETECTOR

The experimental apparatus, shown in FIGURE 2, has been described in detail elsewhere.[4] At the center of the apparatus, a system of cylindrical chambers (the vertex detector)[5] measures charged particle trajectories in a region without magnetic field. The vertex detector consists of: (a) four multiwire proportional chambers (C1 to C4) having cathode strips with pulse-height readouts at $\pm 45°$ to the wires and (b) two drift chambers with measurement of the charge division on a total of 12 wires per track. The

[a]This UA2 Collaboration includes Bern, CERN, Copenhagen, Orsay, Pavia, and Saclay.

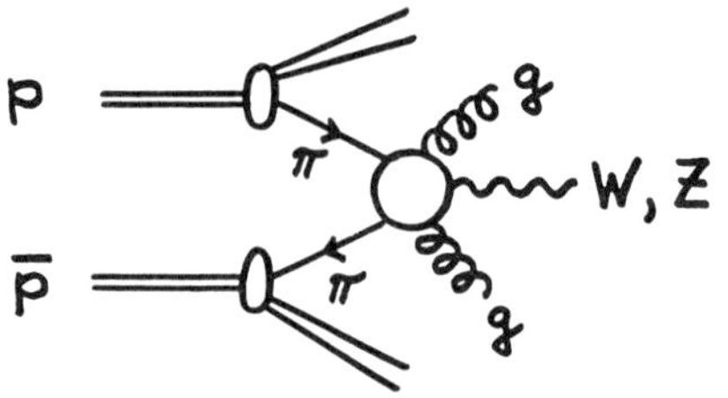

FIGURE 1. Typical diagram for W and Z production that takes into account the emission of gluon (g) radiation (π: parton in p or $\bar{\text{p}}$).

drift chambers are used to obtain both tracking information and to evaluate the most likely ionization, I_0, associated with each track. From the reconstructed tracks, the position of the event vertex is determined with a precision of ± 1 mm in all directions.

The vertex detector is surrounded by an electromagnetic and hadronic calorimeter (central calorimeter),[6] which covers the full azimuth and a polar angle interval, $40° < \theta < 140°$. The calorimeter is segmented into 240 independent cells, each covering 10° in θ and 15° in ϕ, and built in a tower structure pointing to the center of the interaction region. The cells are segmented longitudinally into a 17 radiation lengths thick electromagnetic compartment (lead-scintillator) followed by two hadronic compartments (iron-scintillator) of ≈ 2 absorption lengths each.

In the angular region covered by the central calorimeter, a cylindrical tungsten converter (1.5 radiation lengths thick), which is followed by a cylindrical proportional chamber (C5), is located just after the vertex detector. This device localizes electromagnetic showers initiated in the tungsten with a precision of ± 3 mm, as verified using test-beam electrons.

For the first 15 nb^{-1} of integrated luminosity, collected during the autumn of 1982, the azimuthal coverage of the central calorimeter was only 300°. The remaining

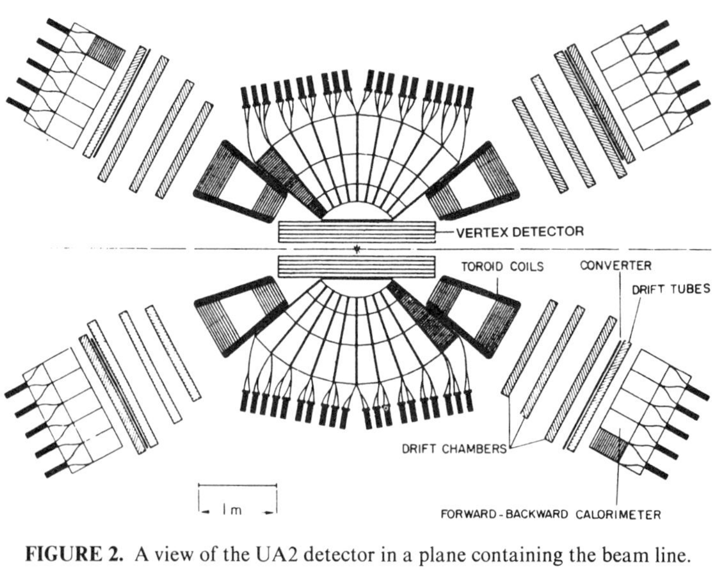

FIGURE 2. A view of the UA2 detector in a plane containing the beam line.

interval ($\pm 30°$ around the horizontal plane) was covered by a magnetic spectrometer that included a lead-glass array to measure charged and neutral particle production.[7]

The two forward regions ($20° < \theta < 37.5°$ and $142.5° < \theta < 160°$) are each equipped with twelve toroidal magnet sectors with an average bending power of 0.38 Tm. Each sector is instrumented with:

(a) Three drift chambers[8] located after the magnetic field region. Each chamber contains three planes with wires at $-7°$, $0°$, and $+7°$ with respect to the magnetic field direction;
(b) A 1.4 radiation lengths thick lead-iron converter, followed by a preshower counter that consists of two pairs of layers of 20 mm diameter proportional tubes (MTPC), staggered by a tube radius and equipped with pulse-height measurement.[9] This device localizes electromagnetic showers initiated in the converter with a precision of ± 6 mm;
(c) An electromagnetic calorimeter consisting of lead-scintillator counters assembled in ten independent cells, each covering 15° in ϕ and 3.5° in θ. Each cell is subdivided into two independent longitudinal sections, 24 and 6 radiation lengths thick, respectively, with the latter providing rejection against hadrons.

The systematic uncertainty in the energy calibration of the electromagnetic calorimeters for the data presented here amounts to an average value of $\pm 1.5\%$. The cell-to-cell calibration has a distribution with a r.m.s. of 2.2%. The energy resolution for electrons is measured to be $\sigma_E/E = 0.14/\sqrt{E}$ [6] in the central calorimeter and $0.17/\sqrt{E}$ in the forward ones (E in GeV).

DATA TAKING AND DATA REDUCTION

In order to implement a trigger sensitive to electrons of high transverse momentum, the photomultiplier gains in all calorimeters were adjusted so that their signals were proportional to the transverse energy.

Because of the cell dimensions, electromagnetic showers initiated by electrons may be shared among adjacent cells. Trigger thresholds were applied, therefore, to linear sums of signals from matrices of 2×2 cells rather than to individual cells. In the central calorimeter, all possible 2×2 matrices were considered; in the two forward ones, we included only those consisting of cells belonging to the same sector.

A W trigger signal (for Z^0 trigger, see reference 12) was generated whenever the linear sum from at least one such matrix exceeded a threshold that was typically set at 8 GeV. To suppress background from sources other than $\bar{p}p$ collisions, we required a coincidence with two signals obtained from scintillator hodoscopes covering the polar angle interval, 0.47°–2.84°, with respect to the beams on both sides of the collision region. These hodoscopes, which were part of an experiment to measure the $\bar{p}p$ total cross section,[10] gave a coincidence signal in more than 98% of all nondiffractive $\bar{p}p$ collisions.

Approximately 7×10^5 W triggers were recorded during the 1982 and 1983 runs, corresponding to an integrated luminosity of $\mathcal{L} = 131\ \text{nb}^{-1}$.

A first data reduction is made by requiring the presence of an energy cluster with a transverse energy greater than 15 GeV. In the central calorimeter, clusters are obtained by joining all electromagnetic cells that share a side and contain at least 0.5 GeV. A halo contribution from the cells having at least one side in common with a cluster is also added. The forward calorimeter clusters consist of, at most, two adjacent cells having the same azimuth (the cells are far from the interaction point and are much larger than the lateral extension of an electromagnetic shower; the dead region between cells at different azimuths does not allow clustering across it).

In the surviving events, a search is made for configurations consistent with the presence of a high-p_T electron among the collision products. An electron is identified from the observation of:

(a) the presence of a cluster of energy deposition in the first compartment (electromagnetic) of the calorimeters with a small lateral size and only a small energy leakage in the hadronic compartment;
(b) the presence of a reconstructed charged particle track that points to the energy cluster. The pattern of energy deposition must agree with that expected from an isolated electron incident along the track direction;
(c) the presence of a hit in the preshower counter with an associated pulse height larger than that of a minimum ionizing particle (m.i.p.). The distance of the hit from the track must be consistent with the space resolution of the counter itself.

A set of cuts has been defined according to these requirements. A detailed description of these cuts can be found in reference 3. The efficiencies of the simultaneous application of these cuts are 76% and 80% in the central and forward regions, respectively.

TOPOLOGY OF EVENTS CONTAINING AN ELECTRON CANDIDATE

After application of the electron cuts, the sample is reduced to 225 events, which contains genuine electrons and fake electrons resulting from misidentification of hadrons. The p_T distribution is shown in FIGURE 3.

The background of fake electrons can be shown to fall mainly into two categories. In approximately 70% of the cases, we are dealing with "overlaps," i.e., jets fragmenting into a hard π^0 with a charged pion nearby in angle. The rest of the background results from π^0's undergoing Dalitz decays or conversions in the beam-pipe.

From studies of hadron jets,[11] we expect the fake electrons to be accompanied by other high-p_T jets at approximately opposite azimuth. We shall, therefore, search for high-p_T jets by grouping together adjacent cells with energy into clusters using an algorithm described elsewhere.[11] Clusters with more than 3 GeV of transverse energy are retained and called jets. For minimum bias triggers, such clusters occur in only 15% of the events.

We find that 45 events contain no jet, with the electron candidate being the only high-p_T particle observed. Their p_T distribution is shown in FIGURE 4a. Such events contain either a neutrino, as in the case of $W \rightarrow e\nu$ decay, or other high-p_T particles having escaped detection. In the latter case, we would expect a rapidly falling p_T-spectrum typical of the jet p_T distribution.[11]

The remaining 180 events contain at least one jet (according to our definition), and FIGURE 5 shows the azimuthal separation of the highest p_T jet from the electron candidate. The events in the peak at $\Delta\phi = 180°$ are likely to be background. Hence, we sum all transverse momenta of clusters having $\Delta\phi > 120°$ and define

$$\rho_{opp} = -\vec{p}_T^{\,e} \cdot \Sigma\vec{p}_{T,jet} / |\vec{p}_T^{\,e}|^2, \tag{3}$$

where $\vec{p}_T^{\,e}$ and $\vec{p}_{T,jet}$ are two-dimensional vectors obtained by projecting $\vec{p}_e$ and $\vec{p}_{jet}$ on a plane perpendicular to the beams.

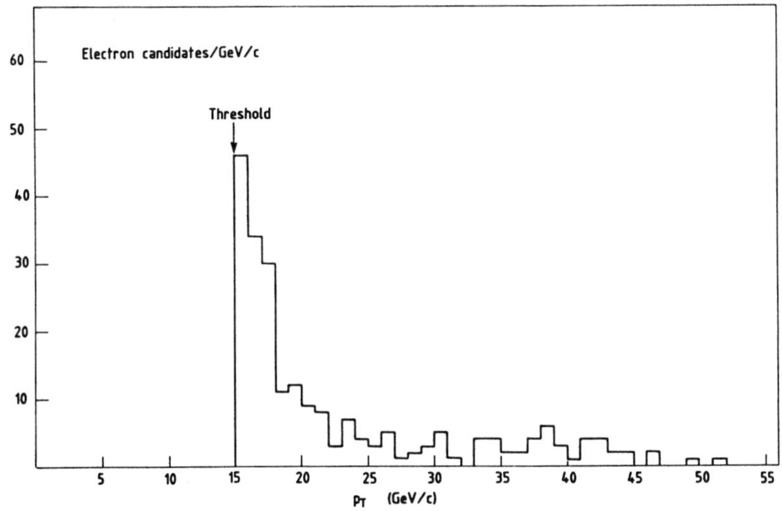

FIGURE 3. Transverse momentum distribution of the 225 electron candidates satisfying the electron cuts.

The distribution of ρ_{opp} is shown in FIGURE 6. We reject the 156 events having $\rho_{opp} > 0.2$. Their p_T^e distribution is shown in FIGURE 4b. This sample contains the eight $Z^0 \rightarrow e^+e^-$ (or $e^+e^-\gamma$) events recently reported.[12]

The p_T^e distribution of the remaining 24 events is shown in FIGURE 4c. These events do contain, in addition to the electron candidate, jets that do not carry a significant transverse momentum in the direction opposite to the electron ($\rho_{opp} < 0.2$).

BACKGROUND TO THE ELECTRON SPECTRA

For the present, we shall regard the 148 events containing a jet back-to-back with the electron candidate as a background sample of fake electrons (after removal of the

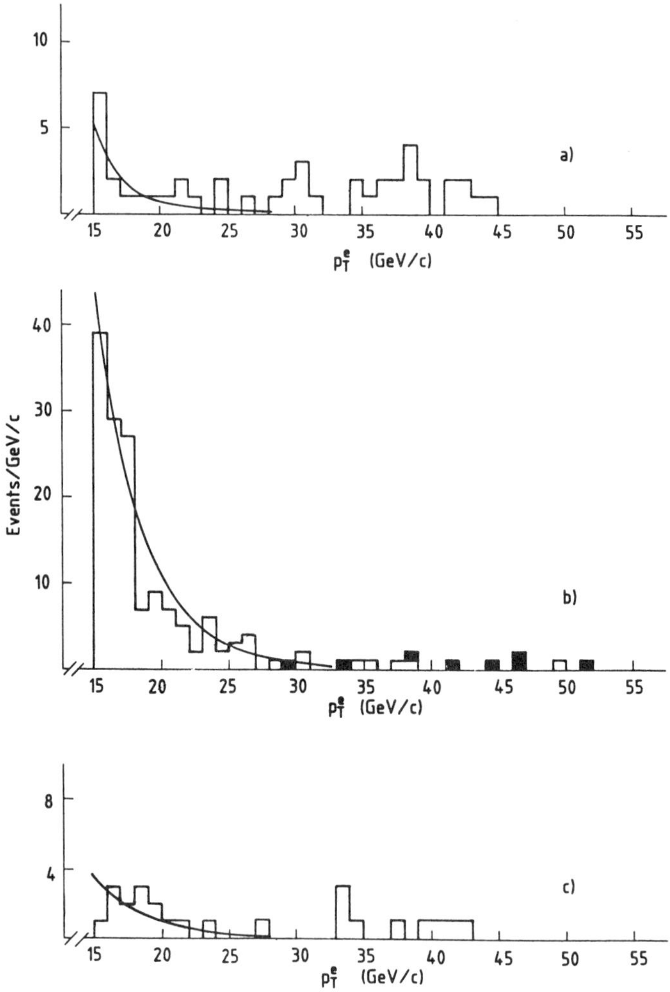

FIGURE 4. (a) Transverse momentum distribution of the electron candidates in events with no additional jet. (b) Transverse momentum distribution of the electron candidates in events having $\rho_{opp} > 0.20$. Dark points correspond to electrons from Z^0 decays. (c) Transverse momentum distribution of electron candidates in events with additional jets having $\rho_{opp} < 0.2$.

eight Z^0 events). These events occur at a rate lower than that of jet production by a factor of 3.6×10^4 for the central detector and 2.6×10^5 for the forward ones.

The backgrounds to the two event samples of FIGURES 4a and 4c are now estimated as follows. We consider a sample of events containing a manifestly fake electron not surviving the cuts. These events are now divided into three classes of different topologies, called samples A, B, and C, which correspond to the three classes of electron candidates of FIGURE 4.

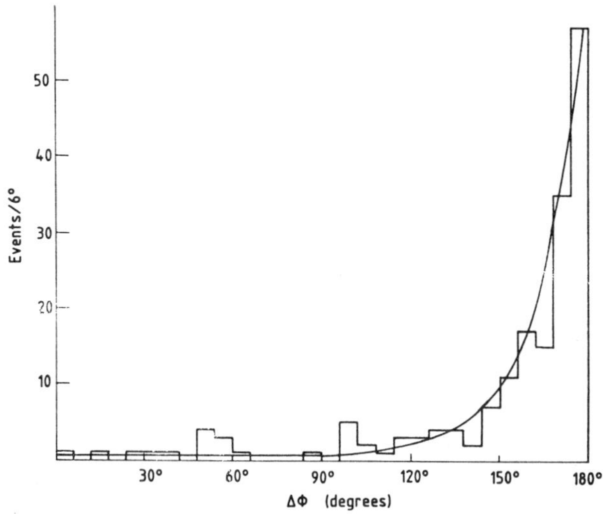

FIGURE 5. Distribution of the azimuthal separation, $\Delta\phi$, between the electron candidate and the associated jet having the highest p_T. Curve: estimated background.

We assume that the p_T distributions of the samples A, B, and C are similar to those of the background events contained in the corresponding electron samples. We can check this assumption for the 148 electron candidates with $\rho_{opp} > 0.2$, which, as mentioned above, are taken as a pure background sample, and we do indeed find that they have the same p_T distribution as the fake electrons of sample B. The background contributions to the 69 electron candidates in the other two topologies (45 with no additional jets and 24 with jets having $\rho_{opp} < 0.2$) are then estimated directly from the

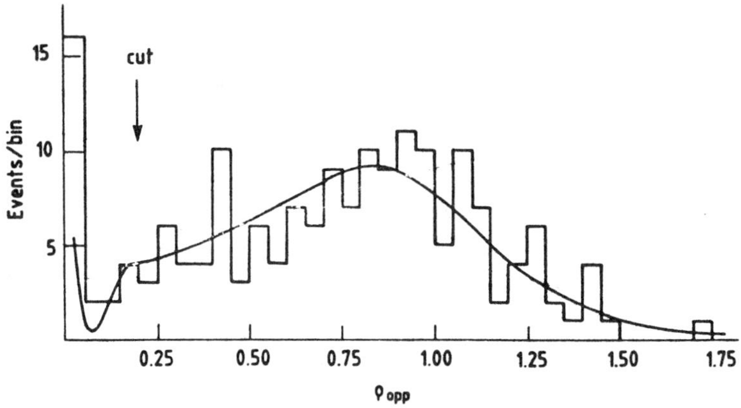

FIGURE 6. Distribution of ρ_{opp} for events containing an electron candidate and at least one associated jet. Curve: estimated background.

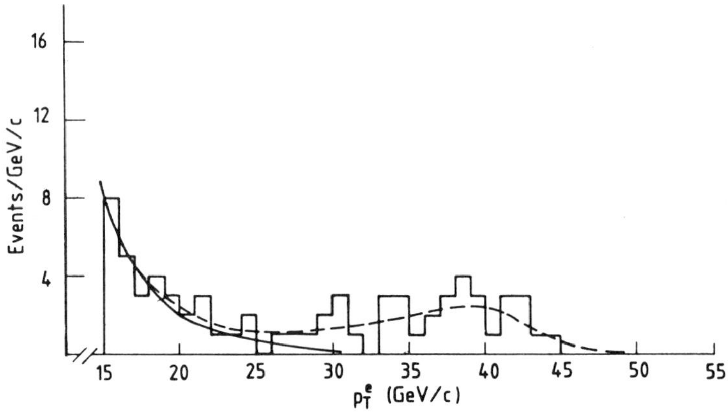

FIGURE 7. Transverse momentum distribution of the electron candidates in the $W \rightarrow e\nu$ event sample. Full curve: background estimate. Dashed curve: sum of all the contributions as discussed in DETERMINATION OF THE W MASS.

p_T distributions of samples A and C, multiplied by a factor equal to the ratio of the total number of electron candidates with $\rho_{opp} > 0.2$ (148 events) to the total number of events in sample B. These background estimates are shown as smooth curves in FIGURES 4a and 4c.

THE $W \rightarrow e\nu$ EVENT SAMPLE

The combined p_T^e distribution of the 69 events from FIGURES 4a and 4c is shown in FIGURE 7 together with the background estimate. The presence of a significant signal above the background is taken as evidence that most of the electron candidates shown here are indeed electrons and are associated with a high-p_T neutrino ($\vec{p}_T^{\nu} = \vec{p}_T^W - \vec{p}_T^e$). There is a clear accumulation of events near $p_T^e = 40$ GeV/c, which is distinctive of the Jacobian peak as expected for $W \rightarrow e\nu$ decay. For $p_T^e > 25$ GeV/c, the distribution contains 37 events with an estimated background of 1.5 ± 0.1 events.

We estimate the transverse momentum, $\vec{p}_T^W$, of the W by

$$\vec{p}_T^W = -(\Sigma\,\vec{p}_{T,jet} + \xi\,\vec{p}_T^{sp}), \tag{4}$$

where the sum extends over all observed jets and $\vec{p}_T^{sp}$ is the total p_T of all observed particles not belonging to jets. In an ideal detector, the correction factor, ξ, would be one. For an incomplete coverage, some fraction of the particles in the event is lost (typically among the low transverse momentum particles). We estimate, using the eight Z^0 events, that ξ should be 2.2 ± 0.5 in order to satisfy equation (**4**) on the average.

The distribution of p_T^W is shown in FIGURES 8a and 8b, using $\xi = 1$ and $\xi = 2.2$, respectively. The mean value (for $\xi = 2.2$) is $\langle p_T^W \rangle = 6.9 \pm 1.0$ GeV/c. QCD predictions,[13] illustrated by the curve in FIGURE 8, are consistent with the observed distribution. The event with the highest value of p_T^W, 29.6 GeV/c, is interpreted as a W

recoiling against a high-p_T jet. Among the W candidates with $p_T^e < 25$ GeV/c, we find two events with even higher values for p_T^W, which are clearly separated from the rest of the events. These events are discussed elsewhere.[14]

We note that the level of activity in the W events is about twice as high as that in minimum bias events. This feature is reflected in the average observed sum of the transverse energies of all particles other than the electron and jets (if any) with

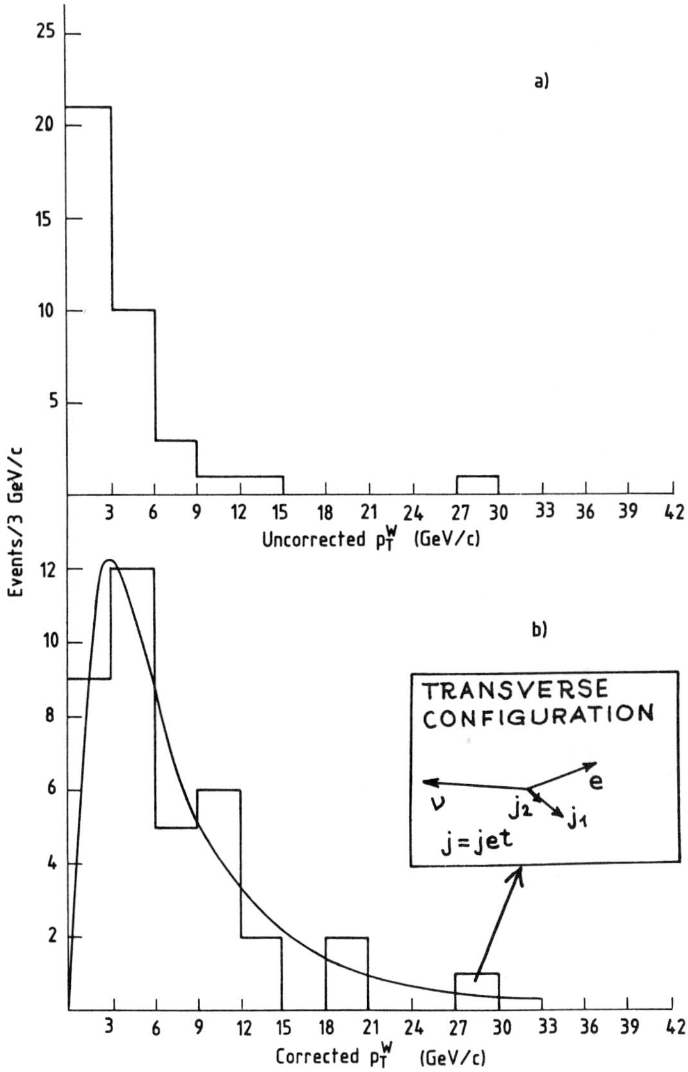

FIGURE 8. (a) Uncorrected distribution of the W transverse momentum [Eq. (**4**): $\xi = 1$]. (b) Corrected distribution [Eq. (**4**): $\xi = 2.2$]. Curve: QCD prediction.[13]

$\langle \Sigma E_T \rangle = 14.1 \pm 1.7$ GeV and the average value of the charged multiplicity, $\langle n_{ch} \rangle$, equal to 19.3 ± 1.9 in the rest of the event. The corresponding values for minimum bias events are $\langle \Sigma E_T \rangle = 8.8$ GeV and $\langle n_{ch} \rangle = 14.7$.

DETERMINATION OF THE *W* MASS

The mass, M_W, of the W is determined from the $W \rightarrow e\nu$ candidates with $p_T^e > 25$ GeV/c by performing a maximum likelihood fit to their two-dimensional distribution of $d^2n/dp_T^e d\theta_e$, where θ_e is the measured electron polar angle. In the Monte Carlo program that is used to generate the distribution, $d^2n/dp_T^e d\theta_e$, for different values of M_W, we make the following assumptions:

(a) the W longitudinal momentum distribution is obtained using the quark structure functions of the proton (antiproton) with scaling violation, as given by Glück *et al.*;[15]
(b) the p_T^W distribution is taken from reference 13, which allows for variations of $\langle p_T^W \rangle$ in order to take into account uncertainties of QCD predictions;
(c) the $W \rightarrow e\nu$ decay's angular distribution is described by the standard V–A coupling;
(d) the M_W distribution is generated according to a Breit-Wigner curve with a fixed value of the W width, $\Gamma_W = 2.7$ GeV/c^2.

The detector response is taken into account as well.

The sample of 37 measured events with $p_T^e > 25$ GeV/c is contaminated by three background sources:

(a) misidentified electrons from two-jet background: 1.5 ± 0.1 events;
(b) electrons from $Z^0 \rightarrow e^+e^-$ decay with one electron undetected: 2.5 ± 0.9 events. This contribution is evaluated by Monte Carlo technique to normalize the result to the total number of $Z^0 \rightarrow e^+e^-$ detected in this experiment (eight events, reference 12);
(c) electrons from $W \rightarrow \tau\nu$, $\tau \rightarrow e\bar{\nu}_e\nu_\tau$ decay chain: 0.9 ± 0.1. The detector acceptance for electrons from the $W \rightarrow \tau \rightarrow e$ decay chain has been evaluated by Monte Carlo technique and we assume a branching ratio, B_τ, equal to 0.17 for the decay, $\tau \rightarrow e\bar{\nu}_e\nu_\tau$.[16]

The sum of all contributions is shown as a dashed curve in FIGURE 7. After subtraction of the background events, we are left with a sample of 32.1 ± 6.0 $W \rightarrow e\nu$ decays. The best fit to the experimental $d^2n/dp_T^e d\theta_e$ distribution including the mentioned background contributions gives

$$M_W = 83.1 \pm 1.9 \text{ (stat.)} \pm 1.3 \text{ (syst.) GeV}/c^2.$$

An uncertainty of ± 1 GeV/c^2, which results from the effect of varying $\langle p_T^W \rangle$ between 4 and 10 GeV/c in the fit, is added in quadrature to the statistical error. The systematic error reflects the uncertainty in the overall mass scale arising from the absolute calibration of the calorimeter ($\pm 1.5\%$) and from small differences in the relative calibration of various cells of the calorimeter.

Within the quoted errors, this value agrees with the result of the UA1 experiment,[17] $M_W = 80.9 \pm 1.5$ (stat.) ± 2.4 (syst.) GeV/c^2.

CROSS SECTION FOR INCLUSIVE W PRODUCTION

The cross section, σ_W^e, for the inclusive process, $\bar{p} + p \longrightarrow W^{\pm}$ + anything, followed by the decay, $W \longrightarrow e\nu$, at $\sqrt{s} = 540$ GeV, is obtained from the relation,

$$N_W^e = \mathcal{L}\, \sigma_W^e\, \epsilon\, \eta,$$

where $N_W^e = 32.1 \pm 6.0$ is the number of $W \longrightarrow e\nu$ decays obtained by subtracting from the electron sample the background events as described in the previous section, $\mathcal{L} = 131\ \text{nb}^{-1}$ is the integrated luminosity, $\epsilon = 0.60 \pm 0.01$ is the detector acceptance, which includes the effect of the p_T^e threshold, and $\eta = 0.77 \pm 0.05$ is the overall efficiency of the electron identification criteria averaged over the central and forward detectors. The effect of the cut, $\rho_{opp} < 0.2$, previously described, is taken into account in the acceptance evaluation. We finally obtain

$$\sigma_W^e = 0.53 \pm 0.10\ \text{(stat.)} \pm 0.10\ \text{(syst.) nb},$$

where the systematic error reflects a $\pm 20\%$ uncertainty in the knowledge of $\mathcal{L}$. This value is in agreement with QCD predictions[13,18] and with the results of the UA1 experiment.[17]

CHARGE ASYMMETRY

It is known that, as a consequence of the V $-$ A coupling, the W is always produced with full polarization along the direction of the incident $\bar{p}$ beam and a distinctive charge asymmetry can be observed in the decay, $W \longrightarrow e\nu$. In the W rest frame, the angular distribution has the form, $(1 + \cos\theta^*)^2$, for electrons and $(1 - \cos\theta^*)^2$ for positrons, where θ^* is the angle between the momentum of the charged lepton and the direction of the incident protons. However, precisely the same configuration would result from V+A coupling because in this case all helicities change sign. In order to maintain full generality and allow for different amounts of V and A couplings, we write the angular distribution in the form,

$$dn/d(\cos\theta^*) \propto (1 - q\cos\theta^*)^2 + 2\, q\, \alpha \cos\theta^*, \tag{5}$$

where q is -1 for electrons and $+1$ for positrons and α depends on the ratio, x, between the A and V couplings (time reversal invariance requires x to be real). Under the assumption that x is the same for both $Wq\bar{q}$ and $We\nu$ couplings, α is given by

$$\alpha = [(1 - x^2)/(1 + x^2)]^2, \tag{6}$$

which gives $\alpha = 0$ for $|x| = 1$. (In the UA2 detector, a determination of the charge sign is only possible in the forward detectors where a magnetic field is present.)

We consider the eight events having an electron with $p_T^e > 20$ GeV/c in the forward detectors. The estimated background is 0.2 events. A comparison between the electron momentum, p, and energy, E, is made in FIGURE 9, which shows the position of these events in the plane, $(\hat{p}^{-1}, E^{-1})$, where $\hat{p}$ is the electron momentum with the sign of the product, $q \cdot \cos\theta_e$ (θ_e: laboratory angle of the electron momentum with respect to the proton direction). The horizontal error bars in FIGURE 9 represent the uncertainty on p^{-1}, which is 0.01 $(\text{GeV}/c)^{-1}$.

A clear asymmetry is visible in FIGURE 9: all events lie on one side of the plot. In order to extract a value of x from these data, we compute two-dimensional distributions, $f^{\pm}(p_T^e, \theta_e)$, for positrons and electrons separately, using equations (**5**) and (**6**) and taking into account the W longitudinal motion. To each event, we assign a likelihood, $Q_i = f^+\eta^+ + f^-\eta^-$, where $\eta^+(\eta^-)$ is the probability density that the observed particle was a positron (an electron) of momentum, p. The functions, $f^{\pm}$, are calculated at the observed values of p_T^e and θ_e. The probability densities, $\eta^{\pm}$, reflect the uncertainty in the determination of the charge sign resulting from the error in the momentum measurement. Maximizing the likelihood, $\Pi_i Q_i$, we obtain $|x| = 1.0^{+0.5}_{-0.3}$ for the ratio between the strengths of the A and V couplings.

We remark that our event sample consists of seven positrons detected in one

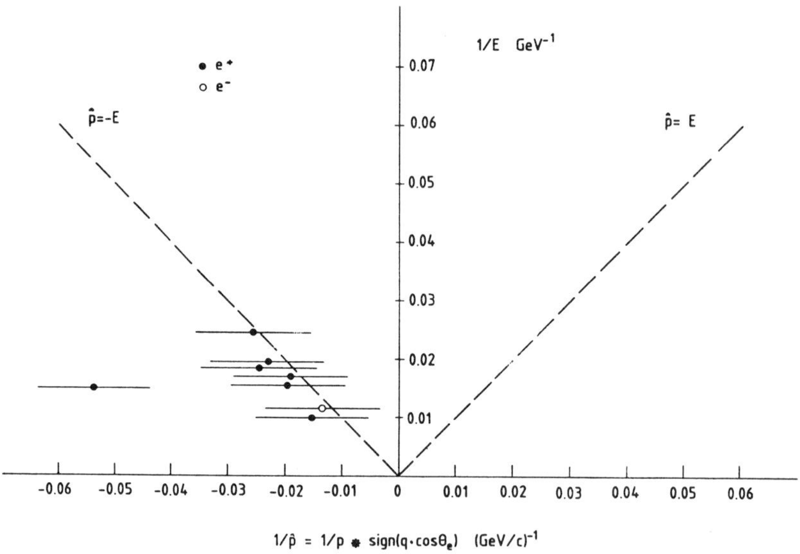

FIGURE 9. Plot of $1/E$ versus $1/\hat{p}$ for the eight $W \to e\nu$ candidates with $p_T^e > 20$ GeV/c detected in the forward regions. The quantity, $\hat{p}$, is the electron momentum with the sign of the product, $q \cdot \cos\theta$ [$q = +1$ (-1) for e^+ (e^-)].

hemisphere and one electron detected in the opposite one. The probability to observe no more than one electron in one of the two hemispheres is $\approx 7\%$.

THE DECAY, $Z^0 \to e^+e^-$

The observation in this experiment of seven $Z^0 \to e^+e^-$ decays and one $Z^0 \to e^+e^-\gamma$ decay has already been reported.[12] Following a recent recalibration of the calorimeters, the invariant mass values of these events and their errors have been slightly modified. The updated value of the Z^0 mass based on a weighted average[3] is

$$M_Z = 92.7 \pm 1.7 \text{ (stat.)} \pm 1.4 \text{ (syst.) GeV}/c^2. \qquad (7)$$

We recall that this value is obtained using only the four events for which the energy of both electrons (and that of the photon in the $e^+e^-\gamma$ event) is unambiguously determined. Within errors, this result agrees with the Z^0 mass value determined in the UA1 experiment.[19]

In order to extract an estimate of the Z^0 width, Γ_Z, from these four events, we first note that the r.m.s. deviation of the four mass values from the value of M_Z given by equation (7) is 2.0 GeV/c^2, which is almost the same as the weighted average of the errors, $\sigma = 2.1$ GeV/c^2. To obtain an upper limit to Γ_Z, we use a Monte Carlo program that generates a large number of event samples, each consisting of four $Z^0 \to e^+e^-$ decays, according to a Breit-Wigner shape and taking into account the energy resolution of the detector. As an estimate of the upper limit to Γ_Z at the 90% confidence level, we use the value that gives an r.m.s. of less than 2.0 GeV/c^2 in 10% of the event samples. This value is $\Gamma_Z < 6.5$ GeV/c^2 at the 90% confidence level.

Within the standard model, this upper limit can be related to the number of additional light neutrinos, ΔN_ν. We find $\Delta N_\nu < 22$ at the 90% confidence level, assuming $\sin^2\theta_W = 0.22$ and a value of the t-quark mass, $m_t > M_Z/2$.

An independent estimate of Γ_Z can be obtained within the standard model by measuring the ratio, $R = \sigma_Z^e/\sigma_W^e$, where σ_Z^e is the cross section for inclusive Z^0 production followed by the decay, $Z^0 \to e^+e^-$.[20] We obtain σ_Z^e after corrections that take into account the detector acceptance and the efficiency of the electron identification criteria. In this case, we use all eight events for which at least one electron passes all cuts and we find

$$\sigma_Z^e = 0.11 \pm 0.04 \text{ (stat.)} \pm 0.02 \text{ (syst.) nb}, \tag{8}$$

which is approximately twice as large as the value predicted by QCD.[13,21] For comparison, we quote the result found by the UA1 experiment:

$$\sigma_Z^e = 0.050 \pm 0.020 \text{ (stat.)} \pm 0.009 \text{ (syst.) nb.}^{19}$$

The error on R is dominated by statistics because the value of the total integrated luminosity cancels out. We find $R = 0.21 \pm 0.08$ and $R > 0.116$ at the 90% confidence level. QCD estimates of the ratio between Z^0 and W production cross sections[20] provide a relation between R and the ratio, Γ_W/Γ_Z:

$$\Gamma_W/\Gamma_Z = (9.3 \pm 0.9)\,R, \tag{9}$$

where the error reflects the uncertainty of the QCD calculations. Using the standard model value of $\Gamma_W = 2.77$ GeV/c^2 (which corresponds to a t-quark mass, $m_t = M_Z/2$), we find $\Gamma_Z < 2.6$ GeV/c^2 at the 90% confidence level.

As before, we can extract upper limits to the number of additional light neutrinos, ΔN_ν. We find $\Delta N_\nu \leq 0$ at the 90% confidence level. (For more details, see reference 3.)

THE DECAY, $Z^0 \to e^+e^-\gamma$

We have reported[12] a $Z^0 \to e^+e^-\gamma$ event containing a photon with an energy, k = 24 GeV, and an 11 GeV electron separated by an angle, ω_{lab} = 31°, excluding, therefore, external bremsstrahlung. In reference 12, we estimated the probability to be

$\approx 5 \times 10^{-3}$ per event, that in a $Z^0 \rightarrow e^+e^-$ decay, a photon at least as hard as the observed one is emitted as a result of radiative corrections[22] and the e^+e^- opening angle is equal to or smaller than the measured one. This calculation, which was performed in the Z^0 rest frame, should not be considered as an estimate of the probability of such a $Z^0 \rightarrow e^+e^-\gamma$ decay because it does not take into consideration all configurations that are less likely than the observed one.

There are several possible ways to define the relative likelihood of $Z^0 \rightarrow e^+e^-\gamma$ configurations. For cases of noncollinear $e\gamma$ pairs, the event distribution in the Z^0 rest frame is given by the differential cross section,

$$\frac{d^2\sigma}{dx_1 dx_2} = \sigma_0 \frac{\alpha}{2\pi} \frac{x_1^2 + x_2^2}{(1-x_1)(1-x_2)}, \tag{10}$$

where σ_0 is the total cross section for $Z^0 \rightarrow e^+e^-$ without radiative corrections and $x_i = 2E_i/M_Z$, with E_i being the electron energies. We say that a configuration is less likely than the observed one if its differential cross section is smaller than that calculated at the point corresponding to the observed $Z^0 \rightarrow e^+e^-\gamma$ event. Integrating equation (**10**) over all configurations that are less likely than the observed one, we find a probability of 1.4% per event or 11% to observe at least one such event in a sample of eight. It should be noted that detectable $Z^0 \rightarrow e^+e^-\gamma$ decays can be divided into two classes of configurations, the first consisting of three clearly resolved energy clusters and the second containing unresolved $e\gamma$ pairs that result in energy clusters inconsistent with an isolated electron. The corresponding probabilities are $\approx$1.0% and $\approx$0.1% per event, respectively. The remaining $Z^0 \rightarrow e^+e^-\gamma$ decays correspond to configurations that are not detectable in the UA2 apparatus.

Further possibilities to calculate a probability for observing such an event are given in reference 3.

SEARCH FOR THE DECAY, $W \rightarrow e\nu\gamma$

Given the interest in unexpected decay modes of the Z^0, we have also looked for events compatible with the decay, $W \rightarrow e\nu\gamma$, in the full data sample. In the central detector, we search for events containing an electron candidate with $p_T^e > 8$ GeV/c, which passes the electron cuts, and an additional photon with a momentum, k, in excess of 8 GeV/c. A photon is defined as an energy cluster, which satisfies the same criteria on size and hadronic leakage as an electron, but has no charged particle track pointing to it. If the photon cluster is seen also in the central detector, we require an angular separation of $\omega > 30°$ between the cluster centroids in order to resolve the $e\gamma$ pair. Six events satisfy these conditions. However, none of them survives the additional requirement of a transverse momentum imbalance compatible with the presence of a neutrino having $p_T^\nu > 10$ GeV/c. A Monte Carlo estimate of the number of $W \rightarrow e\nu\gamma$ decays that are expected to satisfy all of these requirements as a result of radiative corrections gives 0.1 events.

In the forward detectors, we can identify $e\gamma$ pairs with very small opening angles by releasing the condition that the electron momentum, p, and the energy, E, agree within the measuring errors. In this case, however, we limit our search to electron transverse momenta in excess of 20 GeV/c (measured in the calorimeter), for which we expect a background contribution of 0.2 events. We find one event that contains a cluster of transverse energy, $E_T = 41$ GeV ($E = 86.2$ GeV), and a track of 3.6 GeV/c momentum pointing to the cluster and satisfying all other electron identification criteria. Large missing p_T is detected in this event as expected in the case of an associated neutrino. The estimated background for $p_T^e > 35$ GeV/c is 0.01 events. Since the opening angle of this $e\gamma$ pair is compatible with zero, we also consider the effect of external bremsstrahlung and we obtain a probability of 0.5% per $W \rightarrow e\nu$ decay or 4.5% to observe one such event in the sample of nine $W \rightarrow e\nu$ and $e\nu\gamma$ candidates with $p_T^e > 20$ GeV/c detected in the forward detectors.

COMPARISON WITH THE $SU(2) \otimes U(1)$ MODEL

The IVB mass values predicted in the framework of the standard model, taking into account radiative corrections, are $M_W = 83.0^{+2.9}_{-2.7}$ GeV/c^2 and $M_Z = 93.8^{+2.4}_{-2.2}$ GeV/c^2.[23] These values are in excellent agreement with our experimental results.

We can extract a value of $\sin^2\theta_W$ from the definition of $\sin^2\theta_W = 1 - M_W^2/M_Z^2$, where the systematic errors on the mass scale resulting from the uncertainty in the calorimeter calibration cancel out. We find

$$\sin^2\theta_W = 0.196 \pm 0.047 \qquad \textbf{(11)}$$

in good agreement with the world average result of deep-inelastic neutrino experiments (including radiative corrections), which gives $\sin^2\theta_W = 0.217 \pm 0.014$.[24]

We can also extract a more precise value of $\sin^2\theta_W$ from the relation, $M_W = A/\sin\theta_W$, where the numerical value of $A = 38.65 \pm 0.04$ GeV/c^2 is obtained by taking into account radiative corrections.[23] In this case, we find

$$\sin^2\theta_W = 0.216 \pm 0.010\ \text{(stat.)} \pm 0.007\ \text{(syst.)}. \qquad \textbf{(12)}$$

A test of the standard model is provided by the relationship of $\rho = M_W^2/[M_Z^2(1 - A^2/M_W^2)]$, which should be equal to one for the minimal Higgs structure. We find

$$\rho = 1.02 \pm 0.06. \qquad \textbf{(13)}$$

CONCLUSIONS

We have studied the production of electrons with very high transverse momentum at the CERN $\bar{p}p$ collider. From a sample of events containing an electron candidate with $p_T > 15$ GeV/c, we have extracted a clear signal resulting from the production of the charged intermediate vector boson, $W^\pm$, which subsequently decays into an electron and a neutrino.

We have also given new and more refined results on the production and decay of the neutral vector boson, Z^0.

Our experimental results show good agreement with the predictions of the standard model of the unified electroweak theory. Furthermore, the production cross section and the distributions of the transverse momenta of the intermediate vector bosons are within the predictions of QCD calculations.

During the recent $\bar{p}p$ run, ending at Christmas of 1984, we collected 310 nb^{-1} of data. In a very fast and preliminary analysis, we found eight electron pairs with an invariant mass greater than 80 GeV/c^2 and 61 electrons with p_T greater than 25 GeV/c and $\rho_{opp} < 0.2$. No new $Z^0 \rightarrow e^+e^-\gamma$ was found, making the explanation, that the 1983 $e^+e^-\gamma$ event is a QED radiative decay, the most likely.

REFERENCES

1. ELLIS, J. *et al.* 1982. Ann. Rev. Nucl. Part. Sci. **32:** 443.
2. SAUVAGE, G. & J. SCHACHER. UA2 COLLABORATION. 1983. Latest results from the UA2 experiment at the CERN $\bar{p}p$ Collider. *In* Procedings of the International Europhysics Conference on High Energy Physics, Brighton, July 1983, p. 472; CLARK, A. G. 1983. Results from the UA2 experiment at the CERN $\bar{p}p$ Collider. *In* Proceedings of the 1983 International Symposium on Lepton and Photon Interactions at High Energies, Cornell, August 1983, p. 53.
3. BAGNAIA, P. *et al.* 1984. A study of high transverse momentum electrons produced in $\bar{p}p$ collisions at 540 GeV. Z. Phys. C. Part. Fields. **1241:** 1.
4. MANSOULIE, B. 1983. The UA2 apparatus at the CERN $\bar{p}p$ Collider. *In* Proceedings of the Third Moriond Workshop on $\bar{p}p$ Physics, p. 609. Editions Frontières. France.
5. DIALINAS, M. *et al.* 1983. The vertex detector of the UA2 experiment. LAL-RT/83-14.
6. BEER, A. *et al.* 1984. The central calorimeter of the UA2 experiment at the CERN $\bar{p}p$ Collider. Nucl. Instrum. Methods **224:** 360.
7. BANNER, M. *et al.* 1982. Phys. Lett. **115B:** 59; BANNER, M. *et al.* 1983. Phys. Lett. **122B:** 322.
8. CONTA, C. *et al.* 1984. The system of forward-backward drift chambers in the UA2 detector. Nucl. Instrum. Methods **224:** 65.
9. BORER, K. *et al.* 1984. Multitube proportional chambers for the localization of electromagnetic showers in the UA2 detector. Nucl. Instrum. Methods **227:** 29.
10. BATTISTON, P. *et al.* 1982. Phys. Lett. **117B:** 126.
11. BAGNAIA, P. *et al.* 1983. Z. Phys. **C20:** 117; BAGNAIA, P. *et al.* 1984. Phys. Lett. **138B:** 430.
12. BAGNAIA, P. *et al.* 1983. Phys. Lett. **129B:** 130.
13. ALTARELLI, G., R. K. ELLIS, M. GRECO & G. MARTINELLI. 1984. Vector boson production at colliders: a theoretical reappraisal. Ref. TH. 3851-CERN; HALZEN, F. & W. SCOTT. 1978. Phys. Lett. **78B:** 318.
14. BAGNAIA, P. *et al.* 1984. Phys. Lett. **139:** 105; HANSEN, J. R. Exotic events from UA2. Aspen Winter Physics Conference. Ann. N.Y. Acad. Sci. **461:** 314–324.
15. GLÜCK, M. *et al.* 1982. Z. Phys. **C13:** 119.
16. REVIEW OF PARTICLE PROPERTIES. 1982. Phys. Lett. **111B:** 1.
17. ARNISON, G. *et al.* 1983. Phys. Lett. **129B:** 273.
18. MINKOWSKI, P. 1984. Remarks on *W* production in $\bar{p}p$ collisions at $\sqrt{s}$ = 540 GeV. University of Bern preprint BUTP-83/22. To be published in Phys. Lett. B.
19. ARNISON, G. *et al.* 1983. Phys. Lett. **126B:** 398.
20. C. JARLSKOG, C. & F. J. YNDURAIN. 1983. Phys. Lett. **102B:** 361; HALZEN, F. & K. MURSULA. 1983. Limits to the number of neutrinos—a comment on Z^0 discovery. University of Helsinki preprint HU-TFT 83-36; HIKASA, K. November 1983. Counting neutrino species at high-energy proton-antiproton collisions. University of Wisconsin preprint MAC/PH/144.
21. HUMPERT, B. & W. L. VAN NEERVEN. 1980. Phys. Lett. **93B:** 456.
22. ALBERT, D., W. J. MARCIANO, D. WYLER & Z. PARSA. 1980. Nucl. Phys. **B166:** 460; BERENDS, F. A. & R. KLEISS. November 1983. Hard photon effects in $W^{\pm}$ and Z^0 decay. University of Leiden, The Netherlands.
23. MARCIANO, W. J. 1983. Electroweak interactions. *In* Proceedings of the 1983 International Symposium on Lepton and Photon Interactions at High Energies, Cornell, August 1983, p. 80.
24. SIRLIN, A. & W. J. MARCIANO. 1981. Nucl. Phys. **B189:** 442; SMITH, C. LLEWELLYN & J. WHEATER. 1981. Phys. Lett. **105B:** 486.

Heavy Quark Production at the CERN Proton-Antiproton Collider

Presented by ANNE KERNAN
UA1 Collaboration[a]
University of California
Riverside, California 92521

INTRODUCTION

Heavy quark production in strong interactions has been intensively studied in the c.m. energy range of $20 < \sqrt{s} < 63$ GeV.[1] In $p\bar{p}$ interactions at 540 GeV c.m. energy, we can study both strong and weak production of heavy flavors. We report here:

(1) A search for the decays, $W \rightarrow c\bar{s}$, $Z \rightarrow c\bar{c}$, $b\bar{b}$, and a simultaneous search for strong central production of $c\bar{c}$, $b\bar{b}$;[2]
(2) Observation of events that are kinematically consistent with production of a top quark of mass, $30 < m_t < 50$ GeV$/c^2$, in the process, $W \rightarrow t\bar{b}$, $t \rightarrow b\ell\nu$ ($\ell \equiv e, \mu$);[3]
(3) Production of the charmed D^* meson in high p_T jet fragmentation.[4,5]

All the results reported here come from the 118 nb^{-1} data recorded during the 1983 collider run.

SEARCH FOR $W \rightarrow c\bar{s}$, $Z \rightarrow b\bar{b}$, $c\bar{c}$

The standard electroweak model predicts that more than 70% of W's and Z's decay into quark-antiquark pairs. These two-jet modes are difficult to observe because of the large background from hard scattering of gluons and, to a lesser extent, of quarks. FIGURE 1 from Horgan and Jacob shows the predicted yield of jet pairs versus p_T from both strong and weak production processes.[6] We note that if at least one of the jets is required to come from a heavy quark (c, b, t), the strong production background is reduced to a level comparable to that of W and Z decay. In the UA1 experiment, c- and b-jets are identifiable by the presence of a muon in the jet, which is found in the (approximately) 10% of cases in which semileptonic muonic decay occurs.

We report here on a search for the processes, $W \rightarrow c\bar{s}$, $Z \rightarrow b\bar{b}$, $c\bar{c}$, in two-jet events in which one jet contains a muon. The search for $W \rightarrow t\bar{b}$, followed by semileptonic decay of the top quark, is described in the section entitled THE SEARCH FOR TOP.

Event Selection

The UA1 detector and the identification of jets and muons have been described previously.[7,8] The following selection criteria were used to find W or Z decaying to c or

[a]This UA1 Collaboration includes Aachen, Amsterdam (NIKHEF), Annecy (LAPP), Birmingham, CERN, Harvard, Helsinki, Kiel, London (Queen Mary College), Padua, Paris (Coll. de France), Riverside, Rome, Rutherford Appleton Laboratory, Saclay (CEN), Vienna, and Wisconsin.

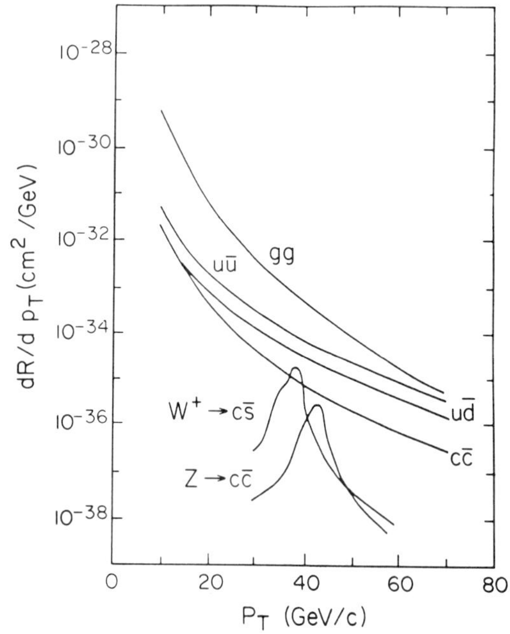

FIGURE 1. Quark decays of Z^0 and $W^{\pm}$ compared with QCD quark and gluon jet production.[6]

b with subsequent semileptonic decay:

- a muon with $p_T > 8$ GeV/c;
- a nearby jet ($\phi_{\mu\text{-jet}} < 40°$) with $p_T > 6$ GeV/c and $|\eta| < 2$;
- a second jet with $p_T > 10$ GeV/c and $|\eta| < 2$;
- no additional jets with $p_T > 6$ GeV/c and $|\eta| < 2$;
- a neutrino (i.e., missing transverse energy) of $p_T > 5$ GeV/c with $\phi_{\nu-\mu} < 90°$; a Monte Carlo calculation shows that $\phi_{\nu-\mu}$ may be as large as 90° due to the large measurement error (typically 3.5 GeV/c for each transverse component) on missing transverse energy.

To avoid a bias towards high transverse energy events, we used only events triggered by a muon. In 108 nb^{-1} of 1983 data, a total of 22 events satisfied all criteria.

Jet finding in the calorimeters is done with a standard algorithm that associates electromagnetic and hadronic calorimeter modules in a cone in (η, ϕ) space of a volume of $(\Delta\eta^2 + \Delta\phi^2) < 1$ around the jet axis. The resulting jet energy is underestimated because some jet fragments fall outside this cone. The energy outside the cone has been calculated as a function of E, ϕ, and η of the jet;[9] it varies from roughly 20% of the total E_T at 15 GeV to 10% of E_T at 60 GeV. All jet energies in the 22 events discussed below have been corrected by the appropriate factor.

Results

FIGURE 2a shows $M(jj\mu)$, the invariant mass of the muon-jet-jet system, for these 22 events. As indicated in FIGURE 1, this plot should contain $c\bar{c}$ and $b\bar{b}$ pairs produced by hard parton-parton scattering in addition to the W and Z decay component.

A complete neutrino momentum vector can be reconstructed on the assumption that the neutrino has the same rapidity as the associated muon-jet system. This allows an estimate of $M(jj\mu\nu)$ as shown in FIGURE 2b.

The dashed line in FIGURE 2 is a leading log QCD calculation of the distribution expected for $c\bar{c}$ and $b\bar{b}$ production.[10] The calculation incorporates scale violating structure functions and the coupling constant, along with the Peterson *et al.* heavy quark fragmentation function.[11] The resulting curve is normalized to the total number of events in the plot; this normalization requires a reduction by a factor of three of the $c\bar{c}$, $b\bar{b}$ cross sections of FIGURE 1; this reduction is compatible with the expected uncertainties of this QCD calculation.[12]

An excess of about five events around 100 GeV/c^2 is seen above the curve in FIGURE 2. A Monte Carlo calculation based upon the measured cross sections for W and Z production predicts three such events in the proportions, $Z(b\bar{b})$: $Z(c\bar{c})$: $W(c\bar{s})$, of 3:1:2.[10] The dominance of $Z \rightarrow b\bar{b}$ is due primarily to the hardness of b-quark fragmentation ($\langle z \rangle_b \simeq 0.8$) relative to that of the c-quark ($\langle z \rangle_c \simeq 0.6$).[13] The Monte Carlo also predicts that the experimentally observed W/Z peak will be shifted to 99 GeV/c^2 due to measurement errors. FIGURE 3 shows the track and calorimeter information for an event with $M(jj\mu\nu)$ of 103.5 GeV/c^2.

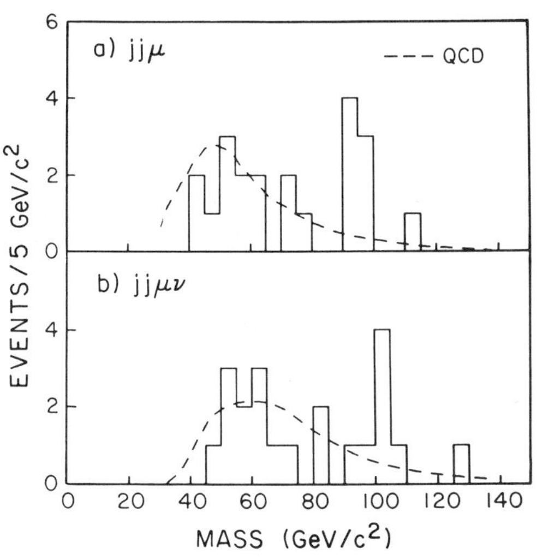

FIGURE 2. (a)$M(jj\mu)$ and (b) $M(jj\mu\nu)$ for two-jet events with a muon in one jet. (There is an additional event off the scale around 190 GeV/c^2.) The curve, which is normalized to the total number of events, is a perturbative QCD calculation of $c\bar{c}$ and $b\bar{b}$ production with semileptonic decay of one quark.[10]

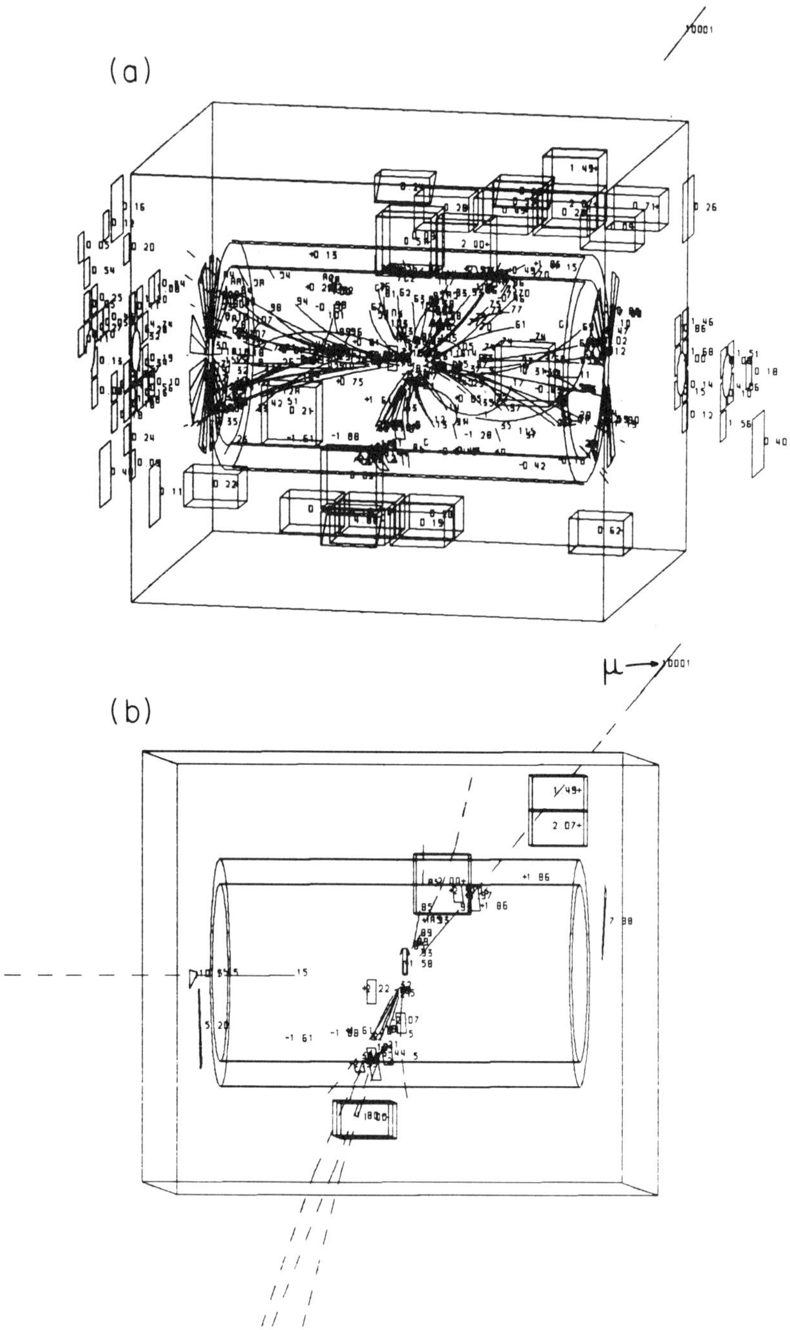

FIGURE 3a, b. Display of an event (run 7296, event 369) with $M(jj\mu\nu)$ of 103.5 GeV/c^2: (a) complete event; (b) with $p_T > 1$ GeV/c thresholds for tracks and $E_T > 1$ GeV for calorimeter modules.

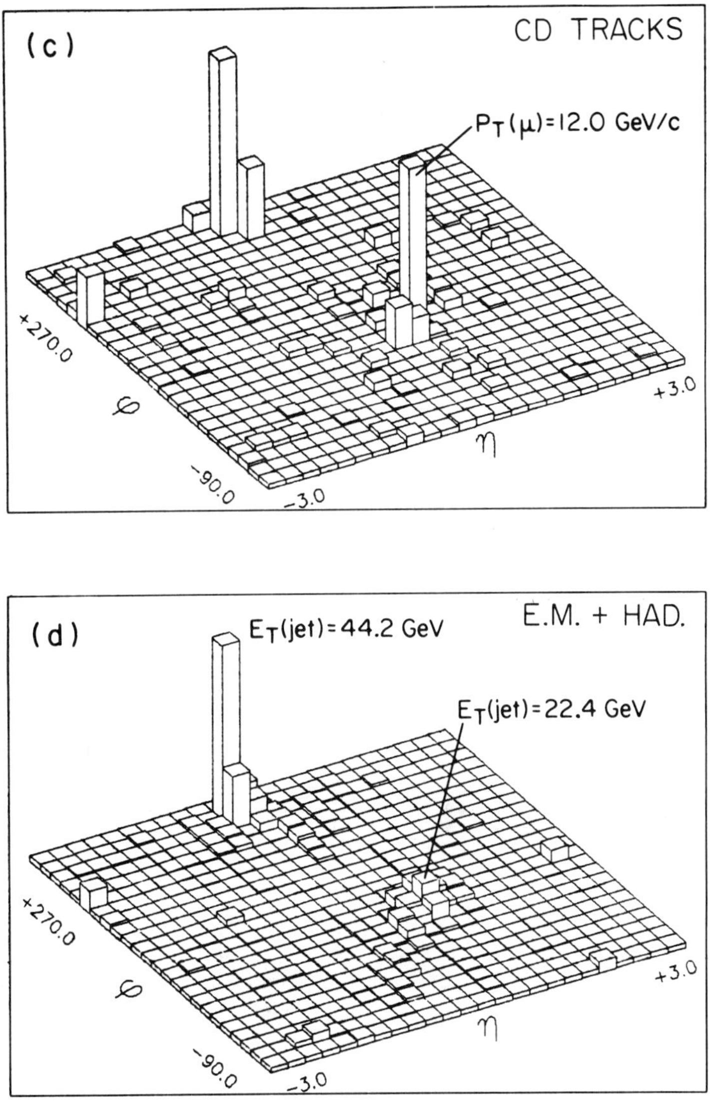

FIGURE 3c, d. Display of an event (run 7296, event 369) with $M(jj\mu\nu)$ of 103.5 GeV/c^2: (c) energy in charged particle tracks as a function of η and ϕ; (d) calorimetric energy deposition as a function of η and ϕ.

Background

The decays in flight, $\pi/K \rightarrow \mu\nu$, can simulate prompt leptons. In particular, the kaon decay may simulate a muon of higher momentum than the parent kaon.[8] Using the approach described in the next section (see DECAY IN FLIGHT MUONS), we have verified that background from this source is negligible. A second source of background arises from the possibility of associating a relatively low p_T muon in the outer muon drift chambers with the wrong (and higher p_T) track in the same jet in the central detector. A study of this background is still in progress. However, we note that both backgrounds lead to an overestimate of the muon energy; this in turn simulates a "neutrino" directly opposite the muon ($\phi_{\nu-\mu} \simeq 180°$). Hence, these events would be removed by the $\phi_{\mu-\nu} < 90°$ cut.

Summary

The structure around 100 GeV/c^2 in $M(jj\mu\nu)$ is consistent with expectations for W/Z decays to c- and b-quarks. The continuation of this study with additional data will test Standard Model predictions for W/Z decays to quarks and should also provide a calibration for multiquark spectroscopy, thereby refining the determination of the top quark mass (see THE $W \rightarrow t + \bar{b}$ HYPOTHESIS).

THE SEARCH FOR TOP

The search for the top quark in e^+e^- colliding beams has established a mass limit, $m_t \geq 22$ GeV/c^2.[14] At the Sp$\bar{\text{p}}$S collider, the decay channel, $W \rightarrow t\bar{b}$, makes it possible to extend this search to masses of around 65 GeV/c^2. The signal of 68 decays of $W^{\pm} \rightarrow e^{\pm}\nu$ observed in the UA1 experiment[15] implies a total of (181 ± 20) $W \rightarrow t\bar{b}$ decays (for $m_t = 40$ GeV/c^2) in the same data sample.

We recently reported[3] the observation of six events that are kinematically consistent with the process:

$$W^+ \rightarrow t\bar{b},\ t \rightarrow \ell^+\nu b\ (\ell \equiv e, \mu).$$

This process has the topology: charged lepton + two jets + missing energy. The isolation of the electron or muon, due to the large Q value in top decay, provides a distinctive signature. For a semileptonic branching ratio of 1/9 and the following requirements on transverse energy [$E_T(\bar{b}) > 8$ GeV; $E_T(b) > 7$ GeV; $p_T(\ell) > 12$ GeV/c], we expect (4 ± 0.3) events for each leptonic channel. (Additional requirements are discussed below.)

Electron and muon identification must be considerably improved with respect to the search for purely leptonic decays of $W \rightarrow \ell\nu_\ell$ since now the signal is only $\sim 1/20$ of the $W \rightarrow \ell\nu_\ell$ rate. In addition, the average lepton energy is a factor of about three smaller, greatly enhancing the probability of a hadron simulating the charged lepton. Furthermore, the presence of two jets in the event increases, relative to $W \rightarrow \ell\nu_\ell$ decay, backgrounds due to hard parton-parton scattering.

The Electron Sample

Electrons are identified by comparing the momentum measured by magnetic curvature in the central tracking chamber with the energy deposited in the surrounding electromagnetic calorimeter. The 27 X_0 lead/scintillator calorimeter hodoscope is segmented four times in depth (3 X_0, 7 X_0, 10 X_0, 7 X_0) and is followed by a hadron calorimeter in which only a small (<600 MeV) residual energy, E_{had}, is expected for electrons. Each of the four segments of the lead/scintillator calorimeter cells is read out by four independent photomultipliers in a way that permits the determination, by pulse division, of the centroid of the energy deposition in two orthogonal directions.

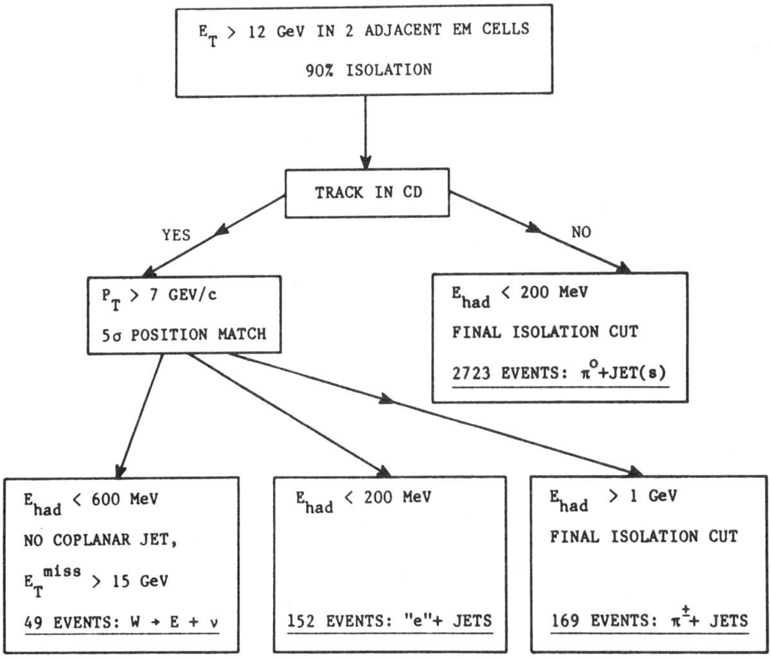

FIGURE 4. The selection process for events with an isolated electron of E_T above 12 GeV.

FIGURE 4 diagrams the initial selection process for an isolated electron with transverse energy exceeding 12 GeV/c. Isolation is enforced by specifying that a maximum of 10% of the electron energy is allowed in the cone ΔR surrounding the electron axis; $\Delta R \equiv (\Delta\eta^2 + \Delta\phi^2)^{1/2}$, where ϕ is the azimuthal angle measured in radians and η is the pseudorapidity. The tight requirement of less than 200 MeV leakage to the hadron calorimeter leaves us with a sample of 152 events with an "electron" and one or more jets. Simultaneously, we obtain a sample of 49 $W \rightarrow e\nu_e$ events, 2723 [π^0 + jet(s)] events, and 169 [π^+ + jet(s)] events; these jet events are used to estimate backgrounds due to strong production of jets (see section on π^0 BACKGROUND).

In the sample of 152 [electron + jet(s)] events, a total of 43 electrons were identified as arising from photon conversions in the beam tube and in the walls of the central drift chamber. In FIGURE 5, the electron parameters of the remaining 109 events are compared with those of the 49 $W \rightarrow e\nu$ events found in the same search. In FIGURE 5a, the excess of events with $p < E$ indicates the presence of background due to

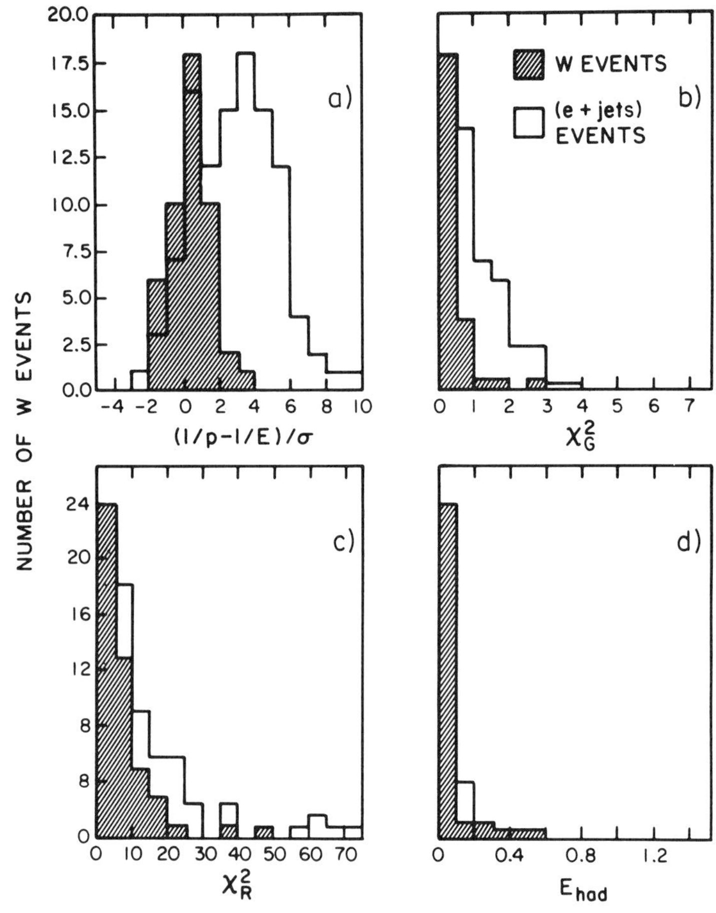

FIGURE 5. The 109 isolated "electrons" (with arbitrary normalization) are compared with the 49 electrons in $W \rightarrow e\nu$ decay found in the same search. The comparison is made in terms of (a) the difference between the momentum, p, measured by magnetic curvature and the energy, E, measured in the electromagnetic calorimeter; (b) the parameter, χ^2_G, which is a measure of the agreement between the direction of the electron measured in the central tracking chamber and in the electromagnetic calorimeter; (c) the parameter, χ^2_R, which indicates if the pattern of energy deposition in the four calorimeter segments is electromagnetic; and (d) the energy deposited in the hadron calorimeter behind the electromagnetic shower.

neutral pions overlapping with a single charged hadron. Similarly, the distributions in χ_G^2 (FIGURE 5b), which measures the spatial matching between the electron track in the central detector and the energy deposition in the electromagnetic calorimeter, suggest the presence of the same background.

In order to eliminate this background, the electron transverse energy threshold is raised to 15 GeV and the following conditions are applied:

(i) a good match between the momentum measurement and the energy deposition in the calorimeters, $|(1/\mathrm{p}) - (1/E)| < 3\sigma$;
(ii) a good electromagnetic shape characterized by $\chi_R^2 < 20$ for the energy depositions in the four e.m. calorimeter segments;
(iii) a stricter isolation requirement for the electron track, namely that $\Sigma \mathrm{p}_T$ and ΣE_T be each less than 1 GeV in a cone of $\Delta R \leq 0.4$ around the track; $\Sigma \mathrm{p}_T$ and ΣE_T are respectively summed over the tracks and the calorimetric energy within the cone.

These cuts result in a final sample of 12 events: 7 one-jet events, 3 two-jet events, and 2 three-jet events. In FIGURE 6, the electron parameters of the 5 multijet events are compared with those of the 49 $W \rightarrow e\nu$ events and seem to be in good agreement.

The Muon Sample

The "isolated muon" events must have at least one jet with E_T greater than 8 GeV and a muon satisfying the following criteria:

— transverse momentum in the central detector above 12 GeV/c with at least 30 digitizings and projected length in the bending plane $\geq$ 40 cm;
— good geometrical matching between the track in the muon chambers and in the central detector;
— isolation, namely $\Sigma \mathrm{p}_T < 0.1\ \mathrm{p}_T(\mu)$, $\Sigma E_T < 0.2\ \mathrm{p}_T(\mu)$ for $\Delta R < 0.4$. The 40 events satisfying these criteria were scanned for $K \rightarrow \mu\nu$ decays in flight in the central detector, cosmic rays, and other identifiable backgrounds. Finally, an additional isolation requirement that no jet be within $\Delta R = 1$ of the muon was imposed. This left 7 one-jet, 4 two-jet, and 1 three-jet events.

Background Studies

π^0 Background

The dominant background is expected to come from three-jet events in which one jet consists of a high p_T charged hadron and one or more overlapping neutral pions ("pileup"). This background has been calculated from the 169 isolated charged hadron events (FIGURE 4) found in the course of the isolated electron search. This sample contained 68 events with two or more jets in addition to the $(h^\pm + n\pi^0)$ system. The background calculation used these 68 events as follows:

(i) from the measured p_h^+, E_{had}, and $E_{e.m.}$ for a given event, that π^0 energy, E_{π^0}, was obtained;

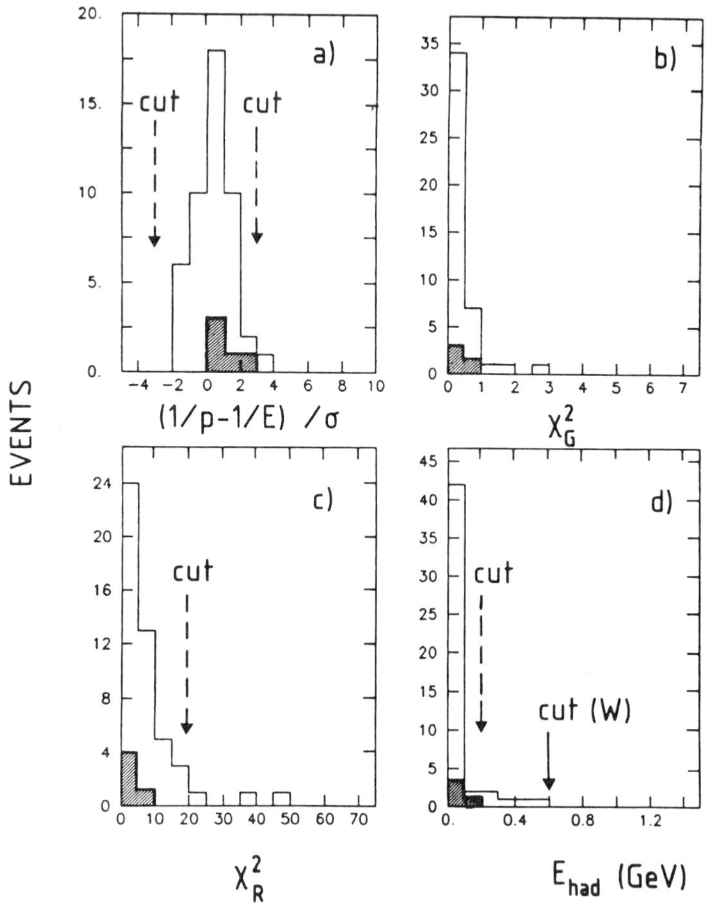

FIGURE 6. Comparison of electrons with the topology [isolated electron + ≥ two jets] with electrons from 49 W ⟶ $e\nu$ decays. The comparison is made in the same four parameters as FIGURE 5. □ = $W \rightarrow e\nu$ (49 events); ▨ = e + ≥2 jets (five events).

(ii) the probability, $P_e(E_{h^\pm}, E_{\pi^0})$, that the event will fluctuate to satisfy the criteria for an isolated electron was calculated using test-beam data ($\pi^\pm$ beams incident on UA1 calorimeter modules) and Monte Carlo simulations of π^0 showers. This procedure gave an estimate of 0.1 background events in the (electron + two-jet) sample.

Decay in Flight Muons

Here, the dominant background is expected to be three-jet events in which one jet fragments to a single charged pion or kaon that subsequently decays in flight ($\pi/K \rightarrow \mu\nu$). This background has been calculated using a 5 nb^{-1} data sample collected with a relatively low threshold jet trigger: $E_T > 15$ GeV. (This trigger is satisfied by the multijet isolated muon events.) From this data, the subset of events with two jets and an isolated charged hadron was used to estimate the background. The decay in flight of the isolated hadron was simulated assuming a charged hadron composition of 50% $\pi^{\pm}$ and 25% $K^{\pm}$. Simulated track digitizations were then passed through the standard track reconstruction program. This calculation gave an estimate of 0.17 background events in the (μ + two-jet) sample.

Charm and Beauty Production

The process,

$$g + g \rightarrow Q + \overline{Q} + g \; (Q \equiv c, b),$$

followed by semileptonic decay of the c/b quark can give rise to an (isolated lepton + two-jet) topology if the charged lepton carries most of the energy in the semileptonic decay. From the magnitude of the $Q\overline{Q}$ production signal observed in this experiment (see SEARCH FOR $W \rightarrow c\bar{s}$, $Z \rightarrow b\bar{b}$, $c\bar{c}$), we have verified that background from this source is negligible.

Kinematics of Background Events

Next, we compare the kinematics of the candidate top events with 933 events ($n\pi^0$ + $\geq$2 jets) found during the isolated electron search (FIGURE 4). The comparison is made in terms of the variables, cos θ^* and E_T^{out}. The variable, E_T^{out}, is the component of

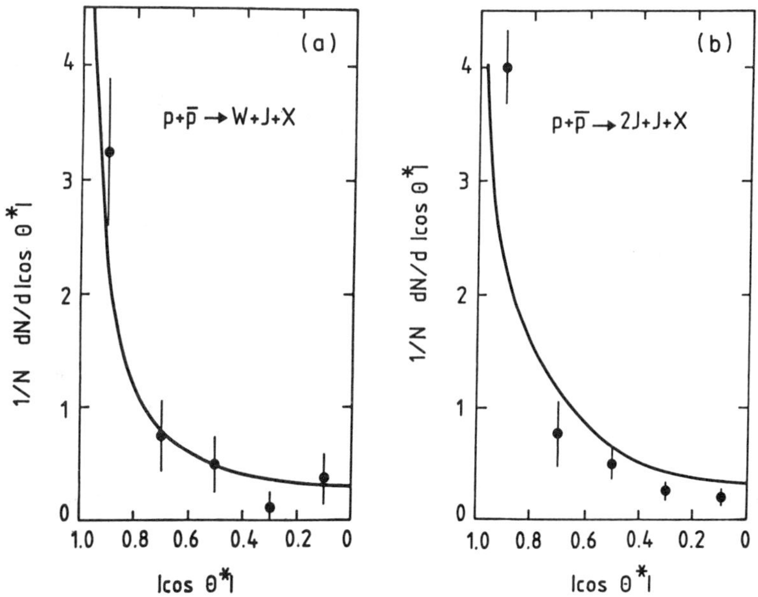

FIGURE 7. Distribution in the cos θ^* for (a) W + jet and (b) three-jet production. The curves are QCD predictions.[16]

the transverse energy of $\pi^0/\ell^\pm$ perpendicular to the direction of j_l, the highest E_T jet. The variable, θ^*, is defined in the rest frame of the ($\pi^0/\ell^\pm\ j_l j_2$) system where j_2 is the second highest E_T jet in the event; θ^* is the angle between the average $p\bar{p}$ beam axis and j_2. FIGURE 7 shows the $|\cos\theta^*|$ distribution for (a) the jet accompanying W production and (b) the lowest E_T jet in a three-jet final state.[16] The distribution is strongly peaked towards unity, reflecting the fact that the dominant source of these jets is bremsstrahlung of the incident partons.

FIGURE 8 shows cos θ^* versus E_T^{out} for (a) (π^0 + 2/3 jets) and (b) (isolated lepton + 2/3 jets) events. We note that the distribution of the isolated lepton events in FIGURE 8b differs considerably from the background distribution and is in good agreement with the Monte Carlo calculation for decay of top produced in $W \to t\bar{b}$.

The W $\to$ t + $\bar{\text{b}}$ *Hypothesis*

For the final sample of candidate top events, we make a cut, $|\cos\theta^*| < 0.8$, and retain only two-jet events. This leaves three electron and three muon events with the

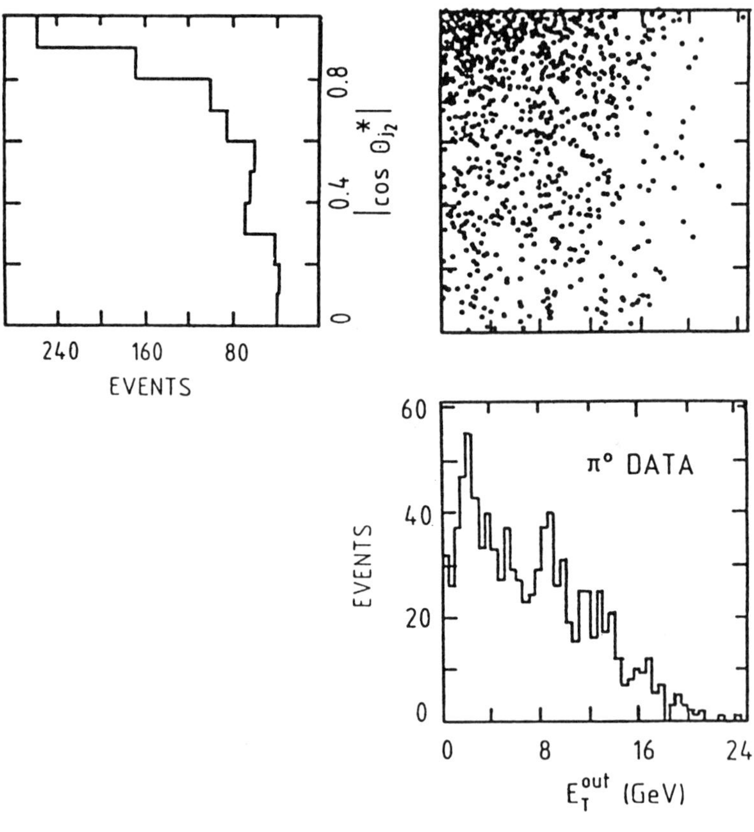

FIGURE 8a. Cos θ^* versus E_T^{out} for the topologies, (π^0 + $\geq$ two jets), 933 events. The curves are predictions for the process, $W \to t\bar{b}$, $t \to \ell b\nu$ ($\ell \equiv e, \mu$).

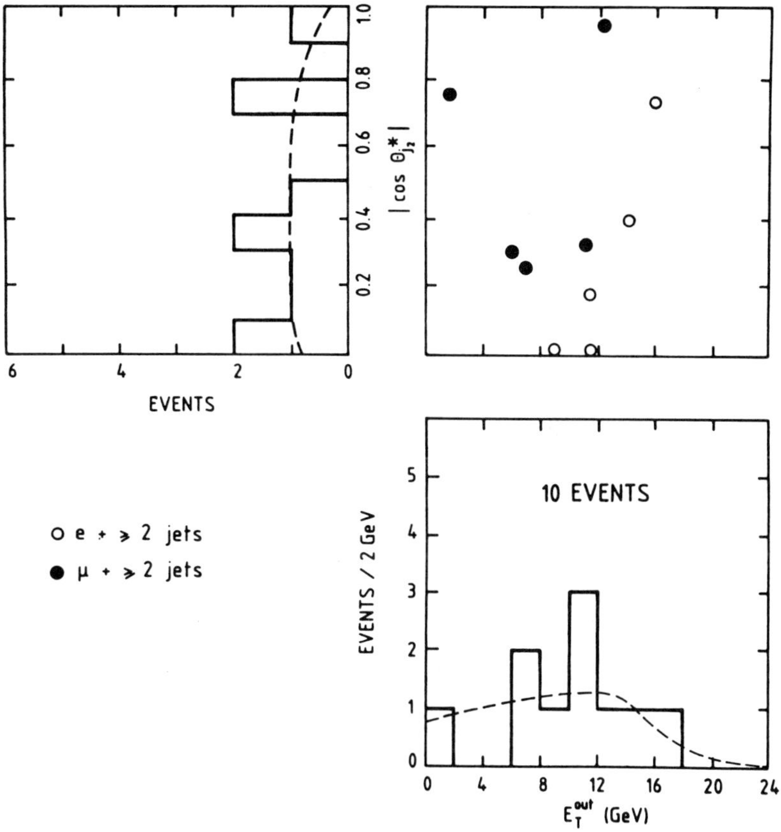

FIGURE 8b. Cos θ^* versus E_T^{out} for the topologies, (isolated lepton $\pm \geq$ two jets), 12 events. The curves are predictions for the process, $W \to t\bar{b}$, $t \to \ell b\nu$ ($\ell \equiv e, \mu$).

estimated background reduced to 0.05 and 0.10 events, respectively, following the cos θ^* cut. The parameters of these events are listed in reference 3. FIGURE 9 shows an example of one of these events.

FIGURE 10 shows the invariant mass of $m(\ell\nu_T j_2)$ versus $m(\ell\nu_T j_1 j_2)$ for the six candidate top events; the missing transverse energy is assigned to a neutrino. The jet energies in this plot have been corrected as detailed in the section, EVENT SELECTION. In FIGURE 11, the mass distributions for the six isolated lepton events are compared with the corresponding background distributions. FIGURE 12 shows $m(\ell\nu_T j_1)$ versus $m(\ell\nu_T j_2)$; for W decay with m_t around 40 GeV/c^2, one expects the top quark to be associated with the lower E_T jet in the majority of cases.

The shapes of the mass distributions in FIGURE 10 are consistent with the decay hypothesis of $W \to t\bar{b}$ for $m_t \simeq 40$ GeV/c^2. The observation of six top quarks from W decay in this experiment would imply a semileptonic leptonic branching ratio of 0.23 ± 0.09, or roughly double what is expected. However, the statistical error is large. Additionally, Godbole and Roy have pointed out that in $p\bar{p}$ interactions at $\sqrt{s}$ of 540 GeV, the weak and strong production mechanisms lead to very similar kinematic

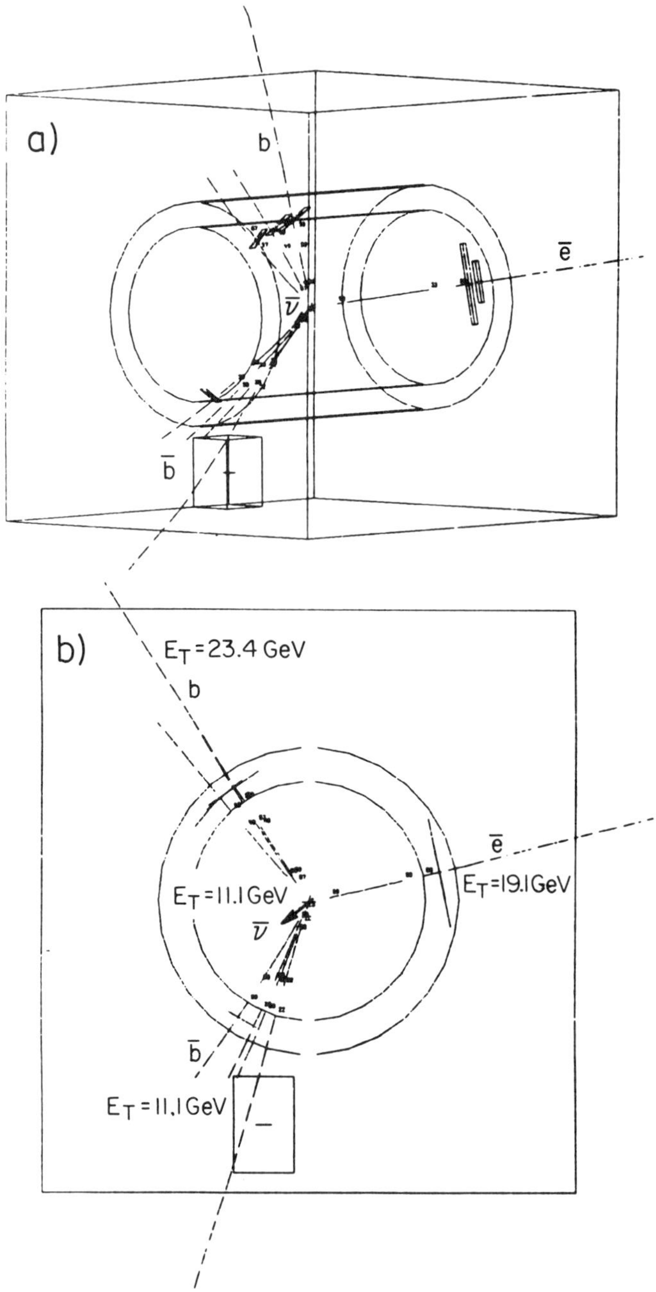

FIGURE 9. Graphic display of calorimeter cells ($E_T > 2$ GeV) and charged tracks ($p_T > 1.5$ GeV) observed in the UA1 detector for a $W^+ \rightarrow t\bar{b}$ candidate event (run 7443, event 509): (a) general view and (b) view looking along the beam direction.

distributions for (lepton + two-jet) events.[17] Indeed, if one assumes that the (lepton + three-jet) events reported here are due to strong $t\bar{t}$ production, it would follow that approximately half of the (lepton + two-jet) events arise from strong production of top.

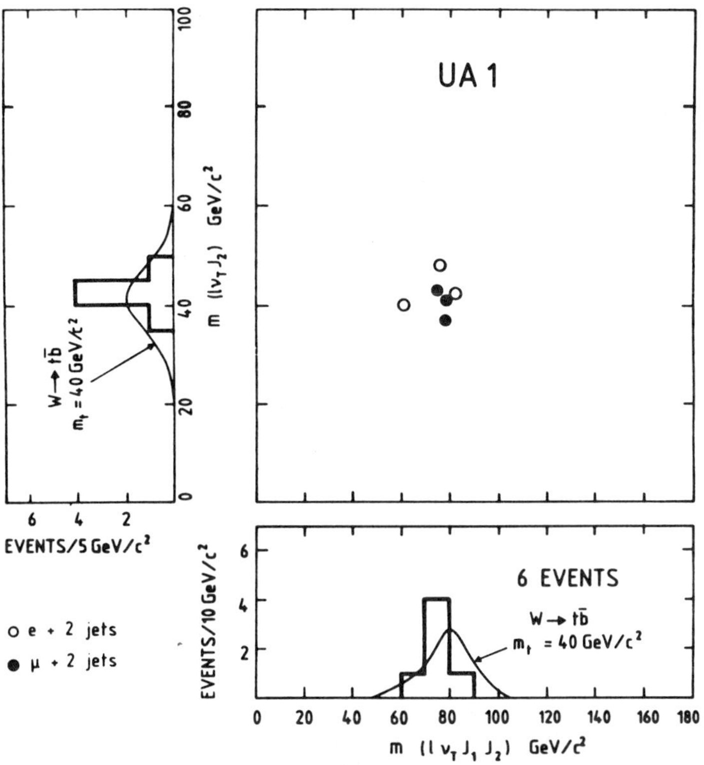

FIGURE 10. Four-body versus three-body mass distribution for the six $W \rightarrow t\bar{b}$ candidate events; $j_1 j_2$ are the higher and lower E_T jets, respectively, and ν_T represents the missing transverse energy vector. The curves show the expected distributions for $W \rightarrow t\bar{b}$, taking into account the experimental resolution. Allowance should be made for a systematic error arising from uncertainties in the jet reconstruction (± 10 GeV/c^2).

D^* PRODUCTION IN JETS

The UA1 collaboration has observed charged D^* mesons in jets.[4,5] We briefly summarize these observations here.

The result comes from the subset of the 1983 "electron trigger" data, which also satisfied a global E_T trigger requiring more than 60 GeV of total transverse energy in the barrel ($|\eta| < 1.5$) calorimeters. These events contain about 30% of all jets having E_T greater than 20 GeV.

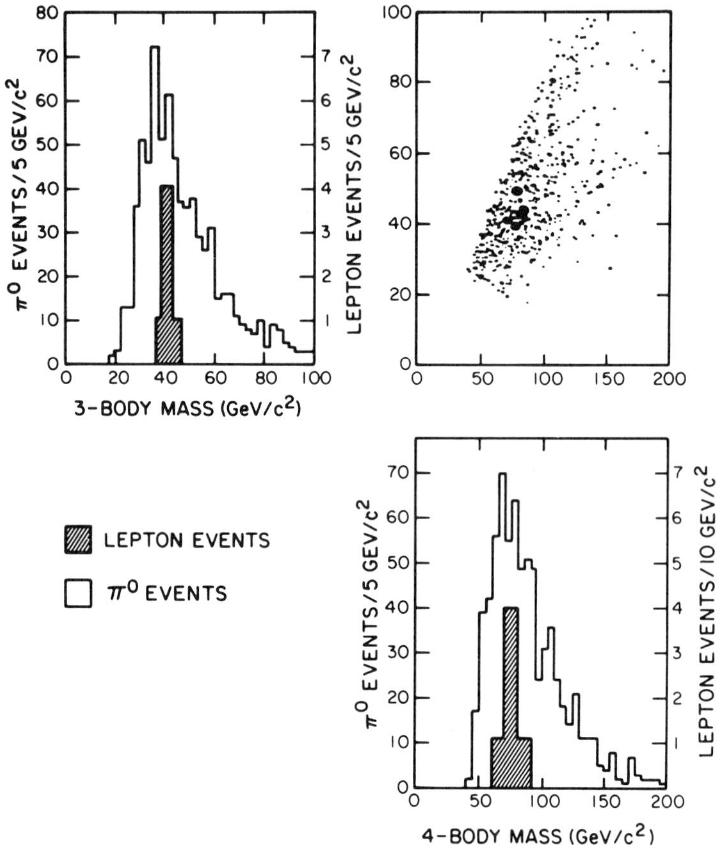

FIGURE 11. Comparison of $m(\ell\nu_T j_2)$ and $m(\ell\nu_T j_1 j_2)$ from FIGURE 10 with the corresponding distributions for the 933 background events in which e/μ is replaced by $n\pi^0$.

In jets containing a minimum of three charged particles, the $K^-\pi^+$ and $K^-\pi^+\pi^+$ mass combinations (and their charge conjugates) were formed; no particle identification was used to distinguish between K and π.

FIGURES 13a and 13b show the mass difference, $\Delta M = M(K^-\pi_1^+\pi_2^+) - M(K^-\pi_1^+)$, for (a) all events and (b) $1.83 < M(K^-\pi_1^+) < 1.92$ GeV/c^2. FIGURES 13c and 13d show the $K^-\pi_1^+$ invariant mass distribution for (c) all events and (d) $146 < \Delta M < 148$ MeV/c^2. A fit of a Gaussian distribution to the peak in ΔM (FIGURE 13b) gives a mean of 147.0 MeV/c^2 and an rms deviation of 0.6 MeV/c^2. The shift of 1.6 MeV/c^2 from the standard value of 145.4 $\pm$ 0.2 MeV/c^2 is associated with multiple scattering of the slow π^+ in the steel beam pipe. The fitted mean and width for the $K^-\pi^+$ peak (FIGURE 13d) are 1.870 GeV/c^2 and 23 MeV/c^2, which are compatible with the known D^0 mass and the estimated mass resolution.

The peaks in FIGURES 13b and 13d contain an estimated 20 $D^{*+} \rightarrow K^-\pi^+\pi^+$ events on a background of eight. FIGURE 14 shows the distribution in $z = \mathrm{p}_D^* \cdot \mathrm{p}_{jet}/(\mathrm{p}_{jet})^2$ for a subset of seven events in which there was excellent geometrical acceptance for both the charged particle jet and its corresponding calorimeter jet; p_D^* is the D^* momentum measured in the central drift chamber and p_{jet} is now the modulus of the total (charged plus neutral) jet momentum determined from the energy deposited in the calorimeter modules. The values of p_{jet} for these events range from 25 to 45 GeV/c. The sensitivity of $\langle z \rangle$ to various alternative definitions[13] for z is less than 10%.

The 20 observed $D^{*+} \rightarrow K^- \pi^+ \pi^+$ decays were found in 3.4×10^3 jets having at least three tracks. The corresponding total number of jets was estimated to be 4.0×10^3. The ratio, $N(D^{*\pm})/N(\text{jet})$, is then $0.65 \pm 0.19 \pm 0.33$ for $z > 0.1$. Although the systematic error is large (with roughly equal contributions from branching ratio and efficiency), this number of $D^{*\pm}$ per jet seems surprisingly high. Furthermore, the number of D^{*0} could presumably be equal to $D^{*\pm}$. However, we note that this measurement applies only to jets in events satisfying the combined electron and global E_T triggers and may well be different for an unbiased jet selection.

FIGURE 14 implies a charm fragmentation that is much softer than that determined in e^+e^- annihilation, where the fragmentation distributions were found to have mean

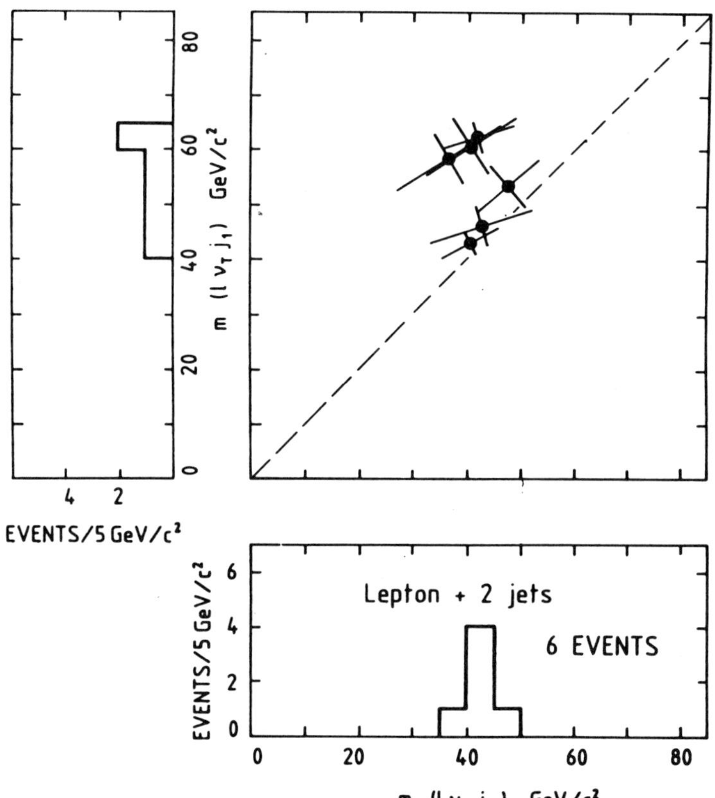

FIGURE 12. $m(\ell\nu_T j_1)$ versus $m(\ell\nu_T j_2)$ for the six events with an isolated e/μ and two jets.

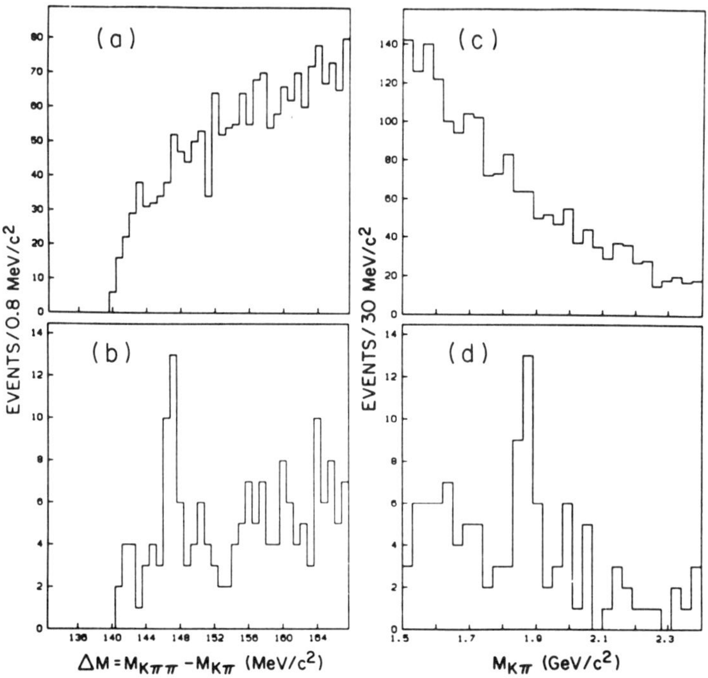

FIGURE 13. (a) ΔM distribution in the region of $M(D^{*+}) - M(D^0)$ with no $M_{K\pi}$ cut. (b) ΔM distribution for $1.83 < M_{K\pi} < 1.92$ GeV$/c^2$. (c) $M_{K\pi}$ distribution in the region of $M(D^0)$ with no ΔM cut. (d) $M_{K\pi}$ distribution for $146 < \Delta M < 148$ MeV$/c^2$.

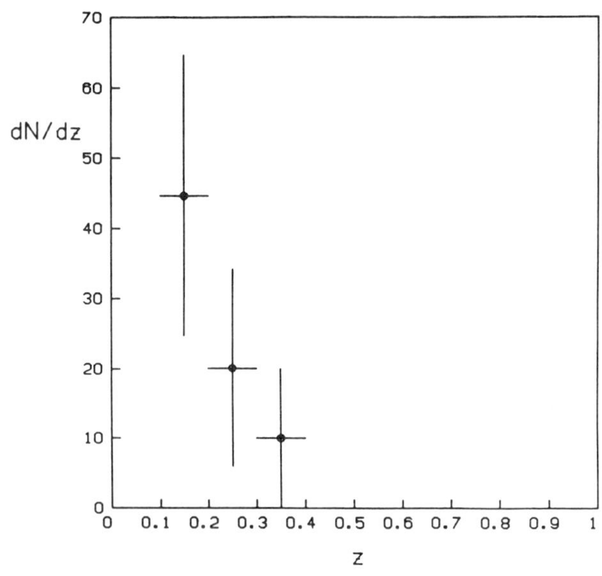

FIGURE 14. D^{*+} fragmentation distribution for $z > 0.1$. The data has been corrected for the track finding inefficiency for the π^+ from $D^{*+} \rightarrow D^0\pi^+$. The mean value of z for $z > 0.1$ is 0.2.

values near z of 0.6. This suggests that the charm production found here for $p\bar{p}$ collisions does not result from the fragmentation of jets initiated by a c-quark. In fact, the number of observed $D^{*\pm}$ is about a factor of 10^2 higher than predicted for c-quark jets.[6] Since the jets studied here are expected to be predominantly initiated by gluons, a likely conclusion is that the observed D^* production results from fragmentation of gluon jets into charm quark pairs.

Halzen and Hoyer have recently shown that the pertubative component of $gg \rightarrow gQ\bar{Q}$ exceeds the fusion process, $gg \rightarrow Q\bar{Q}$, by roughly an order of magnitude.[18] They point out that an explanation of the magnitude of D^* production in UA1 may call for an additional nonperturbative fragmentation mechanism.[19,20]

CONCLUSIONS

(1) The results of a search for $W \rightarrow c\bar{s}$, $Z \rightarrow b\bar{b}$, $c\bar{c}$, while of limited statistical significance, are in agreement with the expectations of the Standard Model.
(2) For the topology, $\ell + 2$ jets ($\ell \equiv e, \mu$), the requirement of an isolated well-identified charged lepton gives six events that are compatible with the cascade decay, $W \rightarrow t\bar{b}$, $t \rightarrow \ell b\nu$. These events suggest a mass of $30 < m_t < 50$ GeV/c^2 for the top quark.
(3) Copious D^* production is observed in jets of p_T above 20 GeV.

These results are from 118 nb^{-1} of data recorded in 1983 by the UA1 detector at the CERN proton-antiproton collider operated at 540 GeV c.m. energy.

REFERENCES

1. KERNAN, A. & G. VAN DALEN. 1984. Phys. Rep. **106:** 299.
2. RANSDELL, J. Search for $W \rightarrow c\bar{s}$ and $Z \rightarrow b\bar{b}$, $c\bar{c}$ at the CERN $p\bar{p}$ Collider. Proceedings of the 1984 Annual Meeting of the Division of Particles and Fields, AIP Conference Proceedings. In press.
3. ARNISON, G. *et al.* 1984. Phys. Lett. **147B:** 493.
4. ARNISON, G. *et al.* 1984. Phys. Lett. **147B:** 222.
5. FREY, R. 1984. Ph.D. thesis. University of California, Riverside, California.
6. HORGAN, R. & M. JACOB. 1981. Nucl. Phys. **179B:** 441.
7. ARNISON, G. *et al.* 1983. Phys. Lett. **132B:** 214.
8. ARNISON, G. *et al.* 1984. Phys. Lett. **134B:** 469.
9. DELLA NEGRA, M. *et al.* 1984. CERN UA1-TN/84-43; DELLA NEGRA, M. & P. GHEZ. 1984. CERN UA1-TN/84-15.
10. LINDFORS, J. Private communication. See LINDFORS, J. & D. P. ROY. 1984. Z. Phys. **C24:** 271.
11. PETERSON, C. *et al.* 1983. Phys. Rev. **D27:** 105.
12. JACOB, M. & J. LINDFORS. Private communications.
13. SÖDING, P. 1983. Proceedings of International Europhysics Conference on High Energy Physics, Brighton, July 1983, p. 567.
14. ALTHOFF, M. *et al.* TASSO COLLABORATION. 1984. Phys. Lett. **138B:** 441; YAMADA, S. 1983. Proc. 1983 Intern. Symp. on Lepton and Photon Interactions at High Energies (Cornell) and DESY report 83-100; ADEVA, B. *et al.* MARK-J COLLABORATION. 1984. Phys. Rep. **109:** 1; WU, S. 1984. Phys. Rep. **107:** 1.
15. ARNISON, G. *et al.* 1983. Phys. Lett. **122B:** 103; ARNISON, G. *et al.* 1983. Phys. Lett. **129B:** 273.

16. GEER, S. 1983. UA1 results on jet activity associated with $W^{\pm}$ and Z^0 production at the CERN proton-antiproton Collider. Proceedings of the 1983 Annual Meeting of the Division of Particles and Fields, AIP Conference Proceedings. In press; GEER, S. & W. STIRLING. 1984. Jet activity associated with W and Z^0 production. CERN preprint TH.4063.
17. GODBOLE, R. M. & D. P. ROY. 1984. Distinguishing the top production mechanism at the $p\bar{p}$ Collider. University of Dortmund preprint DO-TH 84/25, October 1984; BARGER, V. These proceedings.
18. HALZEN, F. & P. HOYER. 1984. University of Wisconsin report MAD/PH/224, December 1984.
19. HALZEN, F. & F. HERZOG. University of Wisconsin report MAD/PH/190 (Phys. Lett. B. to be published) and MAD/PH/185 (Phys. Rev. D. to be published).
20. BARGER, V. & R. J. N. PHILLIPS. 1984. University of Wisconsin report MAD/PH/202; HAGIWARA, K. & S. JACOBS. 1984. University of Wisconsin report MAD/PH/192.

Physics at the Tevatron Collider

G. BELLETTINI

Dipartimento di Fisica
Università di Pisa
Istituto Nazionale di Fisica Nucleare
Sezione di Pisa
Pisa, Italy

I accepted the title of this talk as it was proposed to me since it was so challenging. However, since the scope of the material was so immense, I selected a few items that are currently being discussed within CDF and elaborated on them at length.

TOTAL AND ELASTIC CROSS SECTIONS

Measuring σ_T and $d\sigma/dt$ at the Tevatron Collider will now be more exciting than it was anticipated one year ago since we are confronted with a new situation. Fits to the ISR data had produced a consistent picture of the proton-proton interaction, which was pictorially named geometrical scaling. In impact parameter space, the dominant imaginary scattering amplitude appeared to be energy independent at $b = 0$ (according to a constant, $\sigma_{el}/\sigma_T \sim 0.175$), while the growth of $\sigma_T \propto \ln^2 s$ and the shrinking of the forward elastic peak, $B \propto \ln^2 s$, were attributed to an increasing interaction radius, $R \propto \ln s$. This also implies that at all momentum transfers,

$$\frac{1}{\ln^2 s} \cdot \frac{d\sigma}{d(t \ln^2 s)} = \varphi(t \ln^2 s). \tag{1}$$

How well this was verified is shown in FIGURE 1, where data at the lowest and highest ISR energies were put together and scaled according to equation (**1**).[1] FIGURE 2 shows[2] the σ_{el}/σ_T data up to ISR energies, whose tendency towards a constant value at high energies was not contradicted by the preliminary UA4 data. FIGURE 2b shows the final UA4 result,[3]

$$\frac{\sigma_{el}}{\sigma_T} = 0.215 + 0.005,$$

which indicates a clear growth of this ratio. In addition, geometrical scaling fails to predict correctly the height of the shoulder observed at $\sqrt{s} = 540$ GeV in $d\sigma/dt$ at $t \sim 1$ GeV^2. FIGURE 3 shows[4] that the height of the $t \sim 1$ GeV^2 shoulder is underestimated by a factor of about ten.

The simpler way to represent the situation is in terms of the old multiple scattering model of Chou and Yang[5] and its more elaborate implementation,[6] in which the amplitude is written in terms of the eikonal $\chi\,(s, b)$,

$$f(s, b) = \frac{i}{2}(1 - e^{i\chi}),$$

with Im χ representing the opaqueness of the nucleon. If one lets Im χ depend on s simply as

$$\text{Im}\,\chi(s, b) = K(s)\,\rho(b),$$

one can represent the data with a suitable $K(s)$ and obtain a good fit as seen in FIGURE 3.[7] In the dynamical model of Cheng and Wu,[8] one predicts an asymptotic $K(s) \propto s^c$ (with c = positive constant), and

$$\text{Im}\,f(s, 0) \rightarrow \frac{1}{2}, \qquad \frac{\sigma_{el}}{\sigma_T} \rightarrow \frac{1}{2}, \qquad \sigma_T \propto \ln^2 s.$$

Whether this will happen and whether $\sigma_T \propto \ln^2 s$ will asymptotically saturate the Froissart bound, we do not really know. For the time being, we must accept that we are not in an asymptotic regime and, thus, we must rely again on experiment to explore the higher energy regime. For example, the observed energy dependence of $\sigma_T \propto \ln^2 s$—an apparent solid fact—may sound now rather as a transient accident, and one may seriously wonder why it should continue at the Fermilab Collider. This, therefore, shows the importance of accurate σ_T and $d\sigma/dt$ measurements at this accelerator.

This uncertainty is, of course, only one aspect of the main difficulty, which is we do

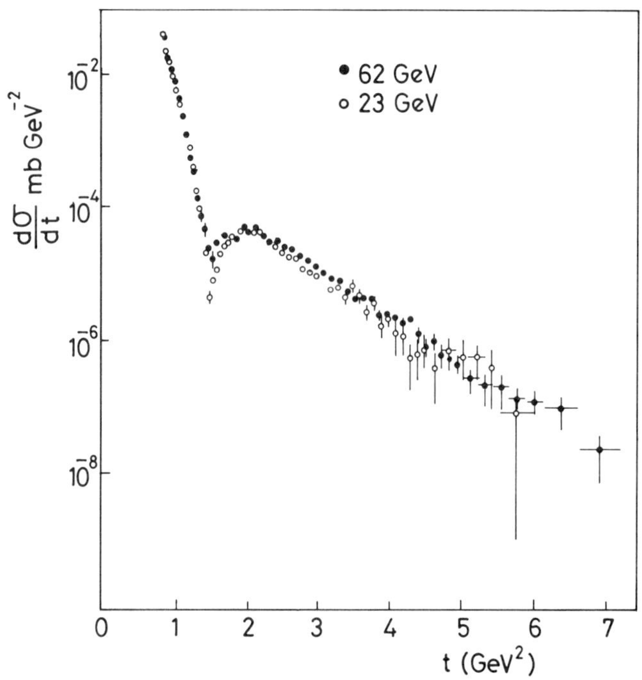

FIGURE 1. Proton-proton elastic scattering cross section at the lowest ($\sqrt{s}$ = 23 GeV) and the highest ($\sqrt{s}$ = 62 GeV) ISR energy, scaled according to the geometrical scaling formula (**1**) (reference 1).

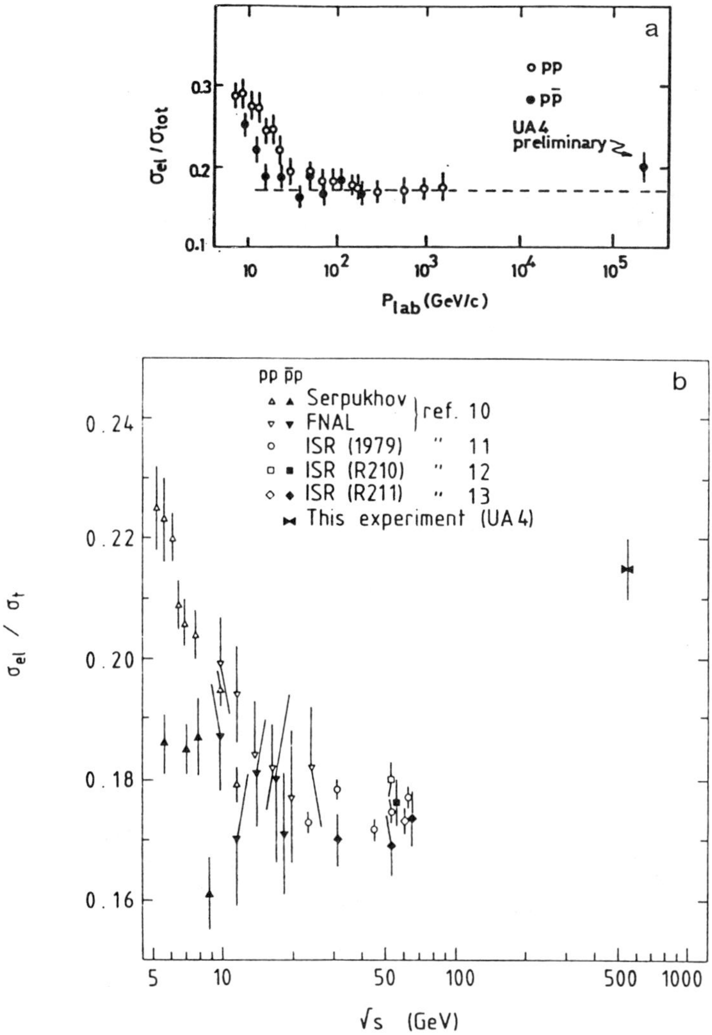

FIGURE 2. A summary of σ_{el}/σ_T data for pp and p̄p as quoted by M. Jacob in 1983 (FIGURE 2a, reference 2) and as presented by the UA4 Collaboration in 1984 (FIGURE 2b, reference 3).

not know why σ_T is growing. Even forgetting about such a fundamental question, we do not know for sure which inelastic channels contribute predominantly to the growth of σ_T. For example, Halzen and Herzog[9] observe that even the energy increase of the hard jet-jet cross section could be made responsible of the growth of σ_T (FIGURE 4). Truly enough, this would need to use the perturbative QCD cross section down to $p_T \sim 2$ GeV, and in addition, the numerical result would depend critically on the adopted lower p_T cut. However, the observation is still valid as a means to illustrate the uncertainty of the role that the onset of hard scattering events has in determining the

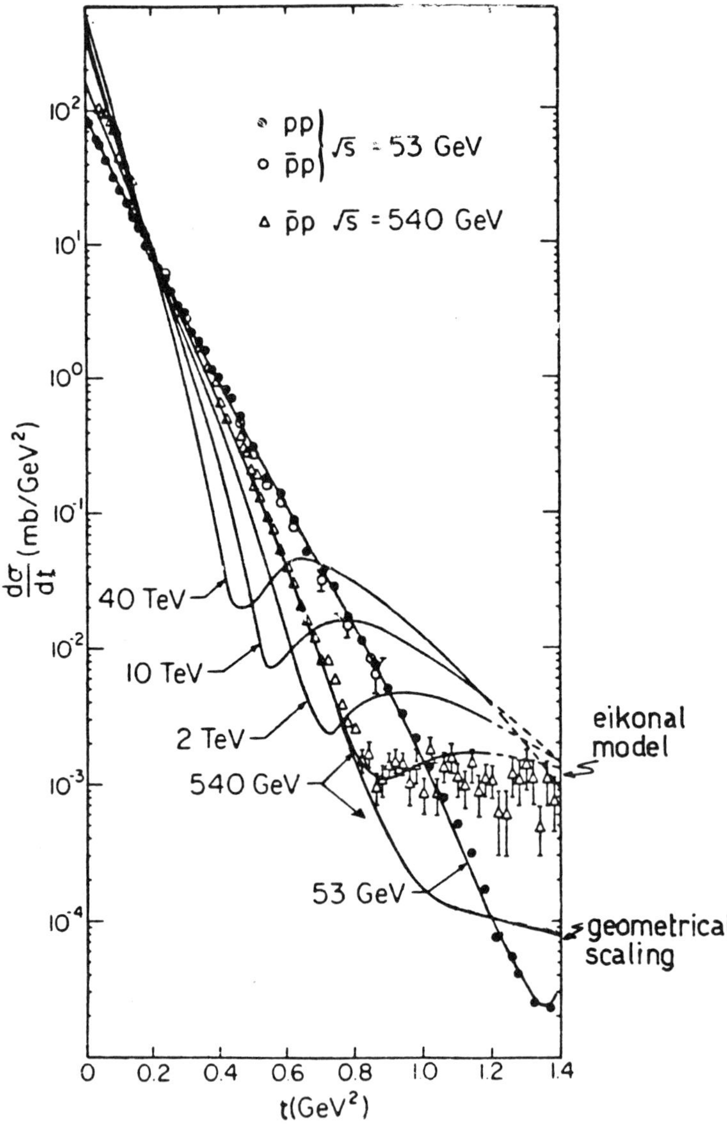

FIGURE 3. Eikonal model fits ($\sqrt{s}$ = 53 GeV and $\sqrt{s}$ = 540 GeV) and extrapolation ($\sqrt{s}$ = 2000 GeV) of the pp elastic scattering data (reference 6). The $\sqrt{s}$ = 540 GeV curve labeled "geometrical scaling" is from reference 4.

total interaction rate. Therefore, the opinion that this role might be important can be maintained.

K. Goulianos has an opposite opinion.[10] He studies the available information on single and double diffraction up to ISR energies and builds a model in which all the growth of σ_{in} is due to these channels. The hard-core cross section, though, is kept energy independent (FIGURE 5). Therefore, again, even if the limited quality of the data makes this interpretation not compulsory, this point can still be maintained.

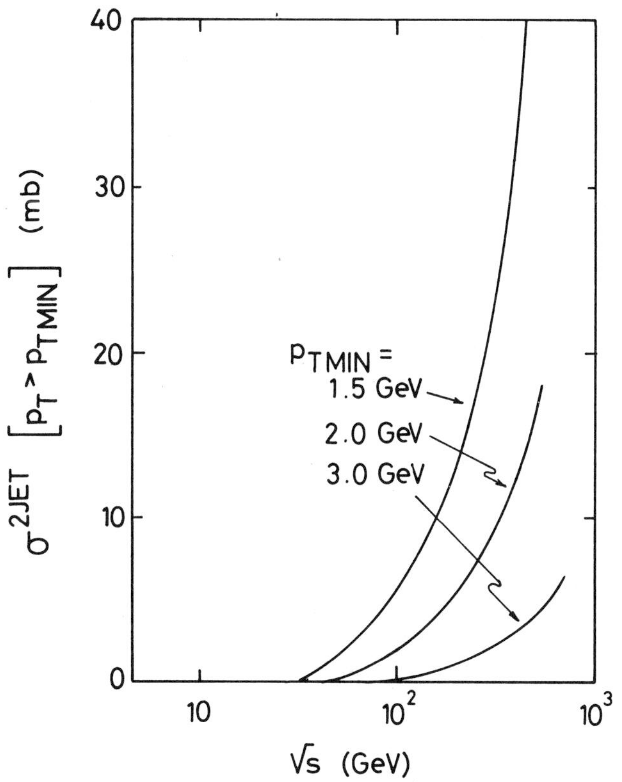

FIGURE 4. Total jet-jet cross section as a function of $\sqrt{s}$ in $\bar{p}p$ collisions. The cross section depends very strongly on a rather arbitrary low-p_T cut parameter, as emphasized in reference 9.

DIFFRACTION STUDIES

While it would be impossible to measure the total jet-jet cross section, it turns out that it will be possible to measure single diffraction very well at the Collider, both down to $M \sim m_p$ and up to $M \sim \sqrt{s}/2$. This is a unique opportunity and should be done in CDF by exploiting a system of silicon detectors inside the Collider vacuum pipe.[11]

FIGURE 6 shows the Monte Carlo predicted distribution of diffractive events as a function of the aperture of their prongs.[12] If at $\sqrt{s} = 2$ TeV the diffractive cross section

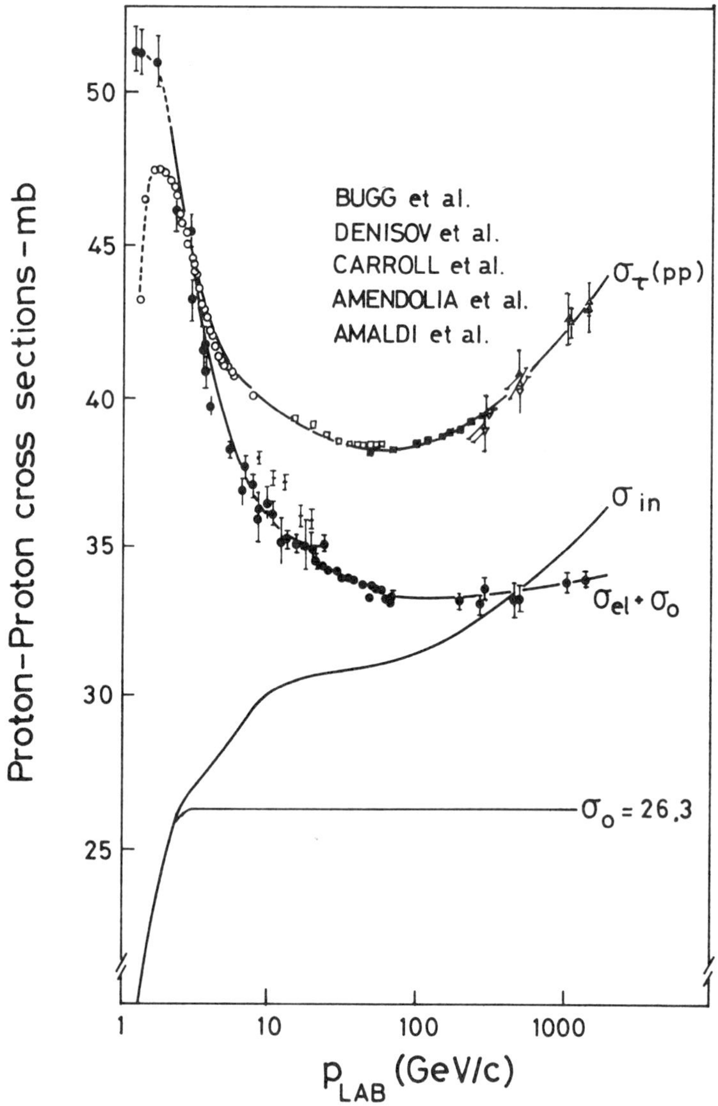

FIGURE 5. Goulianos fits to σ_T, σ_{in}, and σ_{el} for p-p collisions up to ISR energies.[10] All the growth of σ_{in} is attributed to single and double diffraction, while the hard-core cross section, σ_o, stays constant.

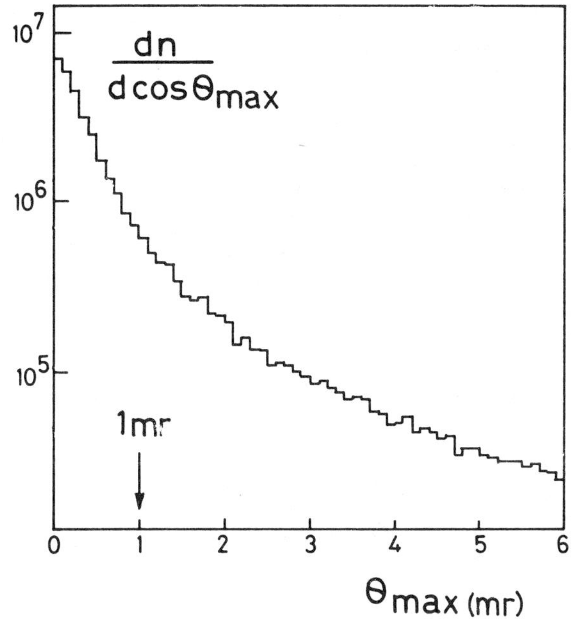

FIGURE 6. Event distribution as a function of the maximum angle of their decay prongs for very forward collimated events.[12]

at small masses is as large as at low energies, one predicts a large rate of events concentrated at $\theta \lesssim 1$ mr. Therefore, this is why the CDF hodoscopes extend down to ~0.2 mr, with complete azimuthal coverage.

FIGURE 7 shows the coverage in pseudorapidity of various CDF components. The small-angle inelastic prongs are detected by the "intermediate" and "forward" silicon insertions at $|\eta| \gtrsim 5.7$, thus covering the laboratory angles from 5 to ~0.2 mr. FIGURE 8 shows (not to scale) the location of the silicon inserts around the main CDF detector.

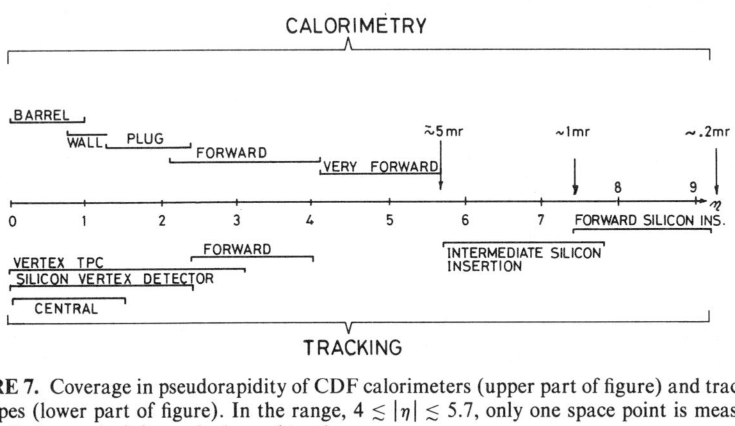

FIGURE 7. Coverage in pseudorapidity of CDF calorimeters (upper part of figure) and tracking telescopes (lower part of figure). In the range, $4 \lesssim |\eta| \lesssim 5.7$, only one space point is measured (inside the beam pipe) for each charged track.

The insertions contain both the high resolution detectors for measuring elastic and quasi-elastic scattering and the detectors with complete azimuthal coverage to tag inelastic prongs down to ~0.2 mr. On the left, insertions 3, 2, and 1 form a high resolution spectrometer served by an 18 m long string of machine bending magnets. On the right, similar detectors in insertions 6 and 7 measure the scattered proton momentum. In addition, insertions 3 and 6 contain the forward detectors, while insertions 4 and 5 contain the intermediate detectors whose coverage is shown in FIGURE 7.

FIGURE 9 illustrates the mechanics of the insertions 4 and 5. When the beams are stable, six separate azimuthal silicon "petals" are flipped from their parallel rest position to the transverse data-taking position. FIGURE 10 shows the polar and azimuthal splitting of the electrodes on the front and rear of the petals. FIGURE 11 shows the mechanics of the spectrometer insertions (1, 2, 3, 6, and 7). With stable beams, crystals are advanced to the minimum distance from the beam that is allowed by beam halo. FIGURE 12a shows the position of the forward diffraction telescopes ($1 \gtrsim$

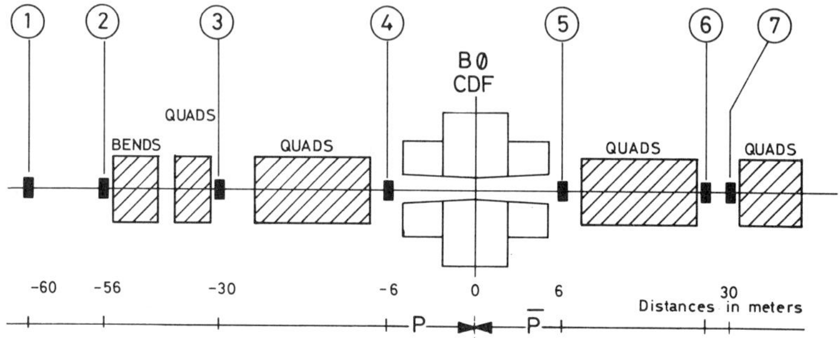

FIGURE 8. Sketch showing the locations of the seven silicon inserts around the main CDF detector.

$\theta \gtrsim 0.2$ mr) in high β optics and illustrates the geometry of the electrodes. In low β runs, these hodoscopes are still useful, but they will have to be "open-up" as illustrated in FIGURE 12b. FIGURE 13 shows the high precision x-y detectors of the antiproton (stations 3, 2, 1) and proton (stations 6, 7) spectrometer. (The position of these detectors relative to the diffraction hodoscopes in stations 3 and 6 is sketched in FIGURE 12b.) The x-y resolution of these detectors is ~50 μm.

TABLE 1 gives the number of electrodes in each detector. FIGURE 14 shows the first prototype telescope of petals that will be tested on the Collider ring in the near future. The momentum resolution of the antiproton spectrometer is $\lesssim$0.2%, as shown in FIGURE 15. Since the coverage of S_3 for inelastic antiprotons is less than for S_2 and S_1, this resolution was computed by neglecting the S_3 information and averaging over the transverse source size. The proton spectrometer has a resolution of the order of 1% at $\theta \gtrsim 2$ mr, which deteriorates quickly at smaller angles. If the vertex of the interaction is measured to ±5 cm, the resolution improves by ~0.5%. This shows that while an accurate measurement of proton diffraction can be made with the antiproton spectrometer, only a rather qualitative measurement will be possible for antiproton

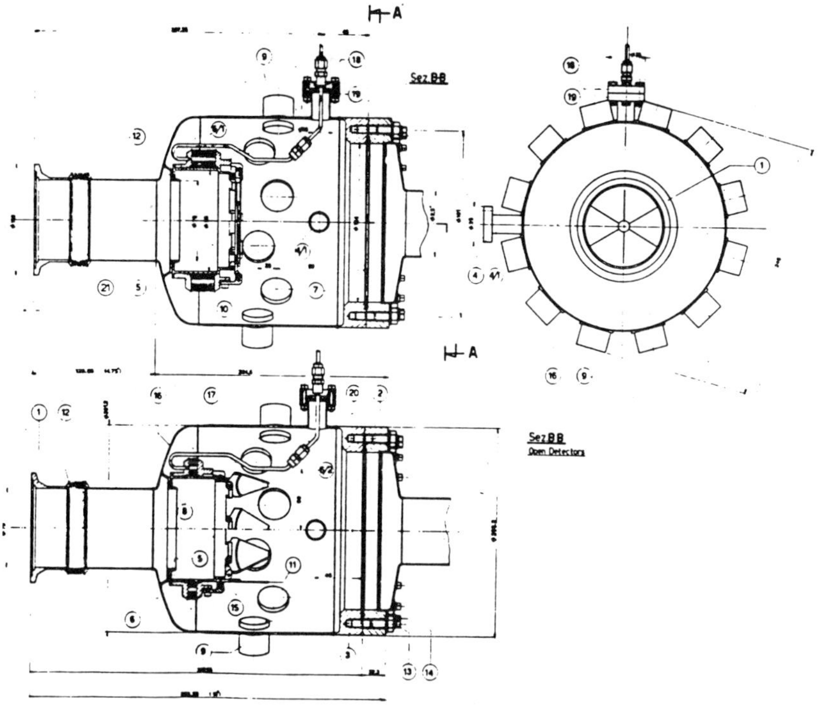

FIGURE 9. Engineering drawing of the vacuum inserts at locations nos. 4 and 5. With stable beams, the silicon "petals" that are usually parallel to the beam (lower left view) are flipped normal to it (upper left view).

diffraction using the proton spectrometer. FIGURE 16 shows the recoil mass resolution in the two spectrometers. In the antiproton arm, $\Delta M/M \sim 20\%$ at 100 GeV and $\Delta M/M \sim 10\%$ for $M \sim 200$ GeV. The high quality of this spectrometer offers unique opportunities in physics to CDF.

The acceptance of the antiproton spectrometer is also large. This is shown in

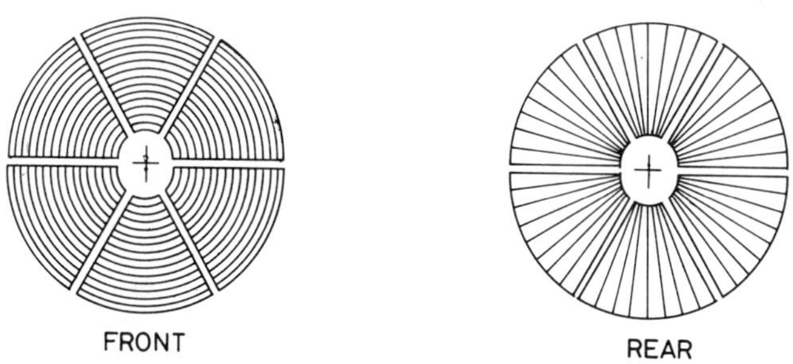

FIGURE 10. Illustration of the polar and azimuthal electrode structure of the silicons at stations 4 and 5 (S_4, S_5 "petals").

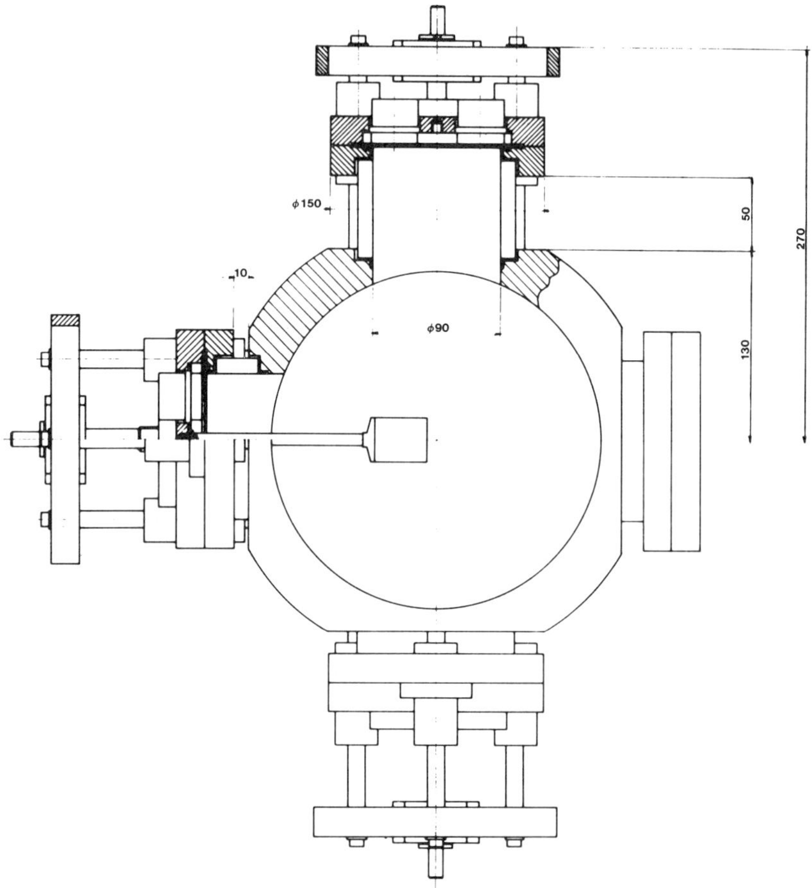

FIGURE 11. Schematic drawing of the mechanics of the silicon spectrometer insertions. Under remote control, the support moves the silicon towards or away from the beam. Positions are known and reproducible to a few μm.

FIGURE 17. The cross section of reference 12 was assumed and the partial cross section accepted was computed.[13] One sees that for $M \gtrsim 300$ GeV, one accepts ~0.6 mb. This number is large enough to make detailed studies of exclusive channels within large mass diffraction possible.

It is important to realize that in large mass diffraction, the secondaries invade the full solid angle. This is illustrated in FIGURE 18, which shows the excess of rate as a function of η in diffractive relative to central collisions.[14] For $M \sim 90$ GeV, the excess extends to $\eta \sim -4$, which is very near the inelastic antiproton detected in the spectrometer. One can easily guess that for large masses ($M \sim 500$ GeV), all of the CDF detector will be involved in the measurement. Such a wide distribution of diffraction prongs is likely to be obtained as a superposition of events of very different types. One possibility, which is suggested by the large forward production of charm observed at the ISR, is that heavy flavors (including possible new ones) may be copiously produced in large mass diffraction since the fragmentation of M is possibly flavor independent. This can be studied at the Collider and would provide a complement to studies of heavy flavor production in large p_T jets. In a forward jet, besides the mass, the momentum is also precisely measured in the spectrometer and is in the range, 900 to 1000 GeV. At large p_T, the momentum is measured with the calorimeters in the range, $p \sim p_T \lesssim 500$ GeV, with the limit being determined by the vanishing rate. Thus, the sample of largest momentum jets will be provided by forward diffraction. In large p_T jets, one can attempt to measure the mass only via the

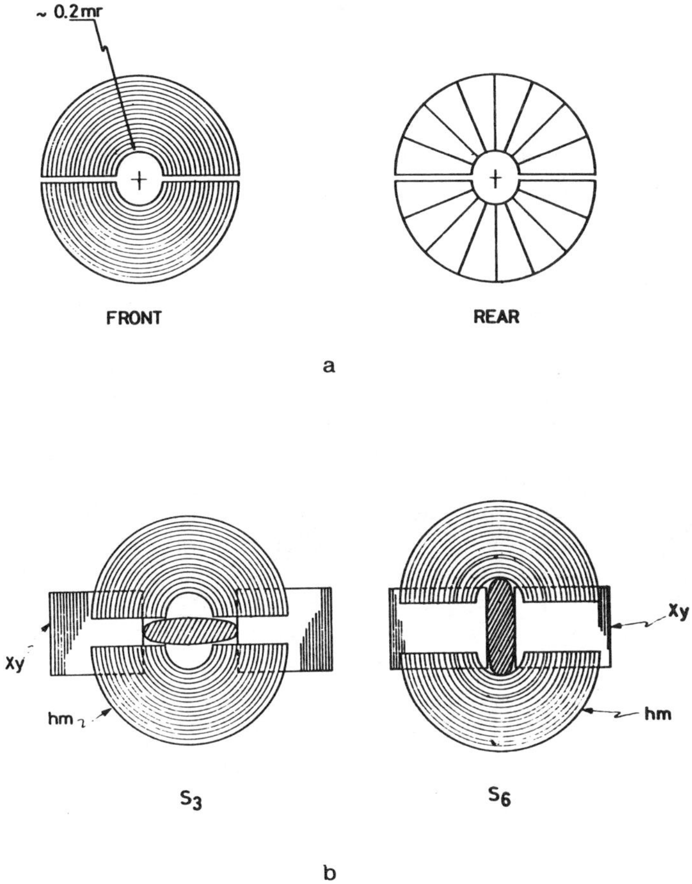

FIGURE 12. The position of the two half-moon crystals of stations 3 and 6 (S_3^{hm}, S_6^{hm}) in high β runs (a). The splitting into front θ-rings and rear φ-pads is also shown. The position of the crystals relative to the beam in low β runs is shown in part (b) of the figure (where the crystals of the high resolution spectrometer are also sketched).

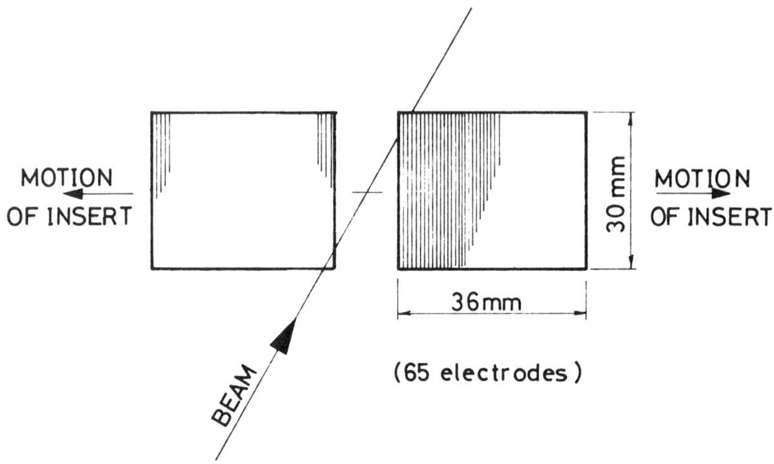

FIGURE 13. Sketch of the high resolution spectrometer crystals (front view, showing the 65 vertical electrodes). Identical crystals are located in insertions 3 (named S_3^{xy}), 2, 1, 6 and 7 (i.e., S_3^{xy}, S_2, S_1, S_6^{xy}, S_7).

transverse mass as measured in the calorimeters. However, in the fragmentation process, any jet would acquire an apparent mass that can be quite large. For example, for $p_T \sim 450$ GeV, the $\langle n_{TOT} \rangle \sim 60$, the $\langle p_{T,jet} \rangle \sim 0.7$ Gev, and the apparent mass is $M \sim 40$ GeV. The calorimeter measurement, moreover, introduces additional fluctuations. Therefore, while in the case of jets originated by quarks of very large mass one might expect to see a trace of it in the measured transverse mass, in no case can one reliably deduce M from M_T. Hence, we conclude that the dependence of jet properties on jet mass can only be studied in large mass diffraction.

In order to indicate how the knowledge of the jet mass might be useful, imagine that a new quark, v, exists with mass, $m = 50$ GeV, which can be "democratically" produced in diffraction. At $50 < M < 100$ GeV, $\bar{v}v$ pairs cannot be produced. At $150 < M < 250$ GeV, they can be produced. (Unfortunately, though, the cross section would probably be partly suppressed by the threshold effects.) At $M = 75$ GeV, the total multiplicity would be $\langle n \rangle = k\ln^2 M \sim 30$. At $M = 200$ GeV, $\langle n \rangle \sim 40$, while in the $M \rightarrow$

TABLE 1. Summary of Insert Channels[a]

Insert	Crystals	x − Bins per Crystal	y − Bins per Crystal	Θ − Bins per Crystal	Φ − Bins per Crystal
S_1	2	65	30		
S_2	2	65	30		
S_3^{xy}	2	65	30		
S_3^{hm}	2			16	8
S_4	6			32	8
S_5	6			32	8
S_6^{hm}	2			16	8
S_6^{xy}	2	65	30		
S_7	2	65	30		

[a] 1526 channels in total.

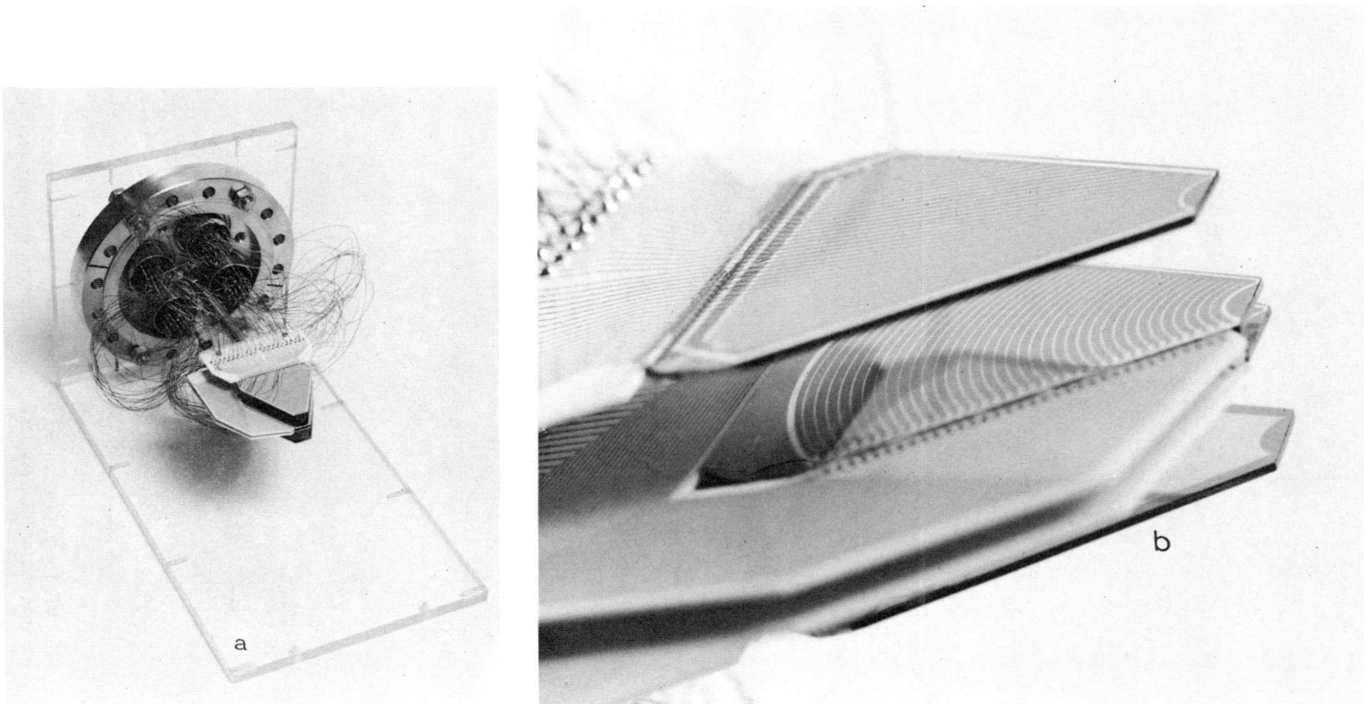

FIGURE 14. Photograph of the prototype telescope of silicon "petals" (insertions 4 and 5). In part (b) of the figure, one can notice the θ, φ structure of the electrodes.

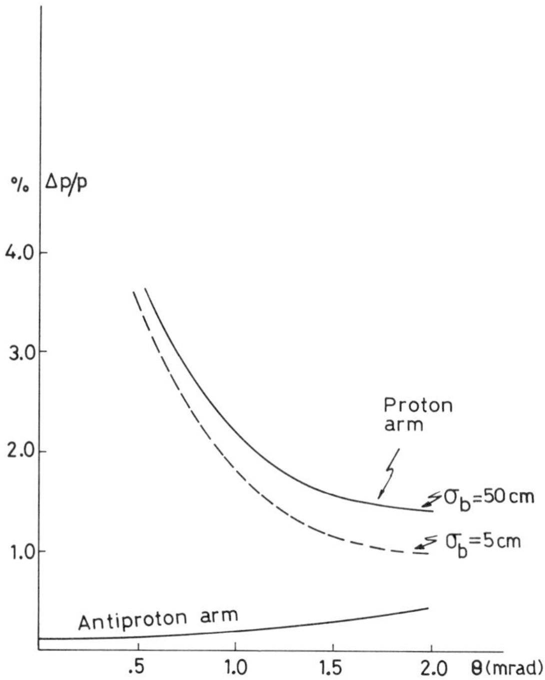

FIGURE 15. Momentum resolution in the antiproton and proton spectrometer. The resolution in the proton arm deteriorates at small angles because of the reduced bending power of the quadrupoles. The improvement obtainable if the longitudinal position of the vertex is measured to an accuracy of 5 cm is also shown.

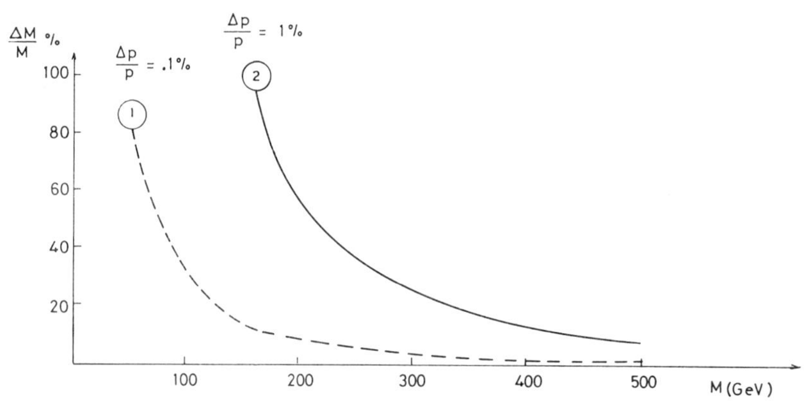

FIGURE 16. Resolution on diffractive mass corresponding to a momentum resolution in the spectrometer of 0.1% (typical of the antiproton spectrometer) and of 1% (typical of the proton spectrometer).

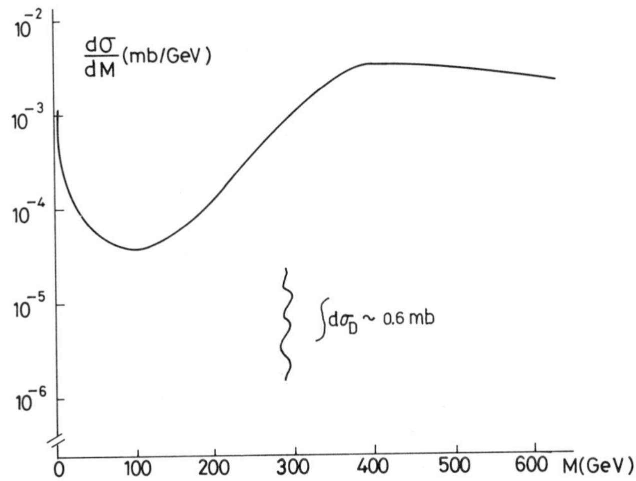

FIGURE 17. Accepted partial cross section of diffractive events.

$\bar{\nu}\nu X$ events, the total multiplicity would be $\langle n \rangle = 2\langle n \rangle_\nu + k\ln^2 M_x$. The tracking system and the calorimeters of CDF can be used to measure n and M_T. In this situation, if one represents the events as points in a M_T, n plane, one might imagine finding a picture such as the one illustrated in FIGURE 19. In the $M = 75$ GeV data, the events spread around $n \sim 30$ and $M_T \sim 75 \pm 20$ GeV. In the $M = 200$ GeV plot, in addition to the bulk of the events at $n \sim 40$ and $M_T \sim 200 \pm 20$ GeV, two more groups of events are born at

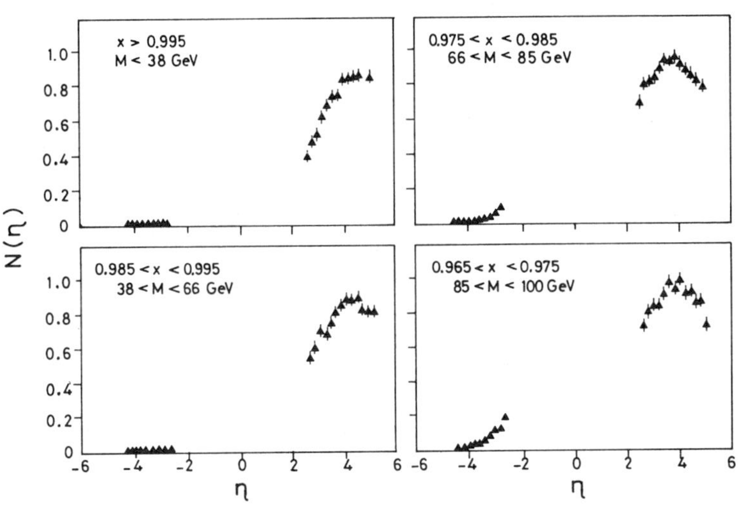

FIGURE 18. UA4 measured ratio, $N(\eta)$, between η-distribution of charged prongs of diffractive events and of inclusive events as accepted by the double-arm trigger. The data is shown either as a function of decreasing relative antiproton momentum, x, or increasing diffracted mass, M.[14]

lower multiplicity. The group labeled (2) would correspond to production of $\nu\bar{\nu}$ at low x, while group (1), at a lower apparent M_T, would correspond to events with a leading heavy-flavored baryon escaping detection through the calorimeter hole at $\theta \sim 0$. Of course, the above is only a pictorial and daring example aiming at illustrating the sort of games that one can play. Once M is known, threshold effects will be looked for in lepton production, jet and dijet production, and so on. If an indication like the one illustrated above for a new family of events is found, one would be able to study the

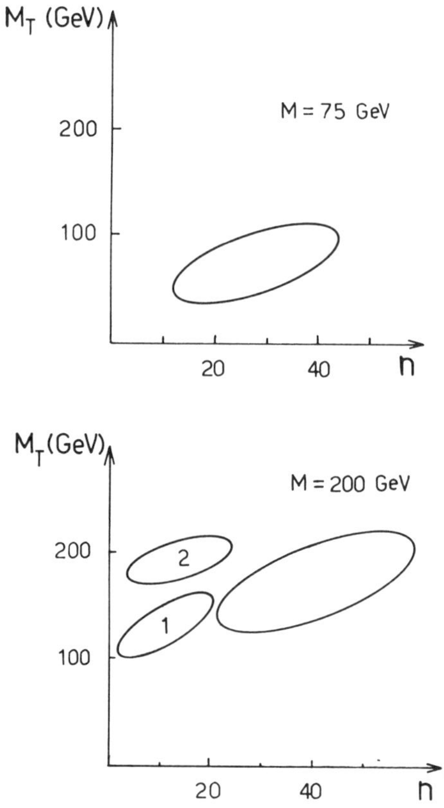

FIGURE 19. Hypothetical example of how large mass diffraction events might group in a transverse mass versus multiplicity plot. At $M = 75$ GeV, large mass $\bar{\nu}\nu$ pairs cannot be produced. At $M = 200$ GeV, $\bar{\nu}\nu$ events give rise to lower multiplicity events, which possibly split into those with two soft ν's (1) and those with a leading ν (2).

internal structure of these events—like minimum mass of a fraction of its tracks—in a sample of reduced background.

Talking of a possible important production of heavy flavors is very topical because of the recent observation by UA1 of abundant production of charm at low x in jets of moderate p_T.[15] One can derive from their measurement an estimate of $c\bar{c}$ production as a function of jet momentum, which would indicate about one $\bar{c}c$ pair at $p_j \sim 40$ GeV.[16] One might tentatively scale this number and say that on the average, one $\bar{b}b$ pair can be expected in jets of $p_j = 400$ GeV. For a quark of ten times the beauty mass, one might expect the rate to be down by 100 because of the ratio of its mass to the beauty mass and down by an additional large factor because of threshold effects. However, with ~0.6 mb accepted cross section at large masses (FIGURE 17), even if the cross section is further reduced by four orders of magnitude, one would still get comparable rates as for W, Z production.

As mentioned before, one may distinguish heavy flavor production in diffractive events at small x and production at large x (leading heavy-flavored baryons). The former will be at large angles and, thus, will be measurable in the main detector. Leading heavy-flavors, though, might be more difficult in the present CDF detector. On the one hand, the design of CDF leaves room for extensive upgrades at $\theta \lesssim 10°$. Leading heavy-flavored baryons would be peculiar of forward jets since they have no counterpart in large p_T jet production. On the other hand, though, heavy flavor production at small x in large mass diffraction should be measurable soon with CDF, which should allow an interesting comparison with the same production in large p_T jets.

FIGURE 20 shows the splitting in pseudorapidity and azimuthal angle of the CDF projective hadron calorimeter towers at angles, $\theta \gtrsim 2°$. Such a detailed splitting should be fully adequate for an accurate jet study. An interesting question is where will "important" jets go in polar angle. In hard collisions, the larger the mass of a produced particle, the more central its production. In addition, this means that its decay jets will tend to be more abundant at small pseudorapidity. On the other hand, the angular distribution of possible jets in large mass diffractive events will favor smaller angles. To illustrate this, we take an example from reference 17, where the tentative hypothesis is made that the Pomeron has a parton structure. Occasionally, hard jets might arise from hard scattering between one parton in the Pomeron and one parton in the proton. FIGURE 21 illustrates the process. In the Pomeron-proton center-of-mass system, there will be two back-to-back hard jets and two spectator jets. The center-of-mass energy squared will be $\hat{s} = s(1 - x_{\P}) = M^2$. In order to predict the angular and momentum distribution of these jets in their center of mass, one would have to rely on a model. However, kinematics is also instructive. Let us consider the possible situation of a jet of c.m. momentum, $p_j^* = 90$ GeV, produced in a diffracted mass at $M = 450$ GeV. FIGURE 22 shows the energy-angle relationship for this jet in the laboratory frame. Production at $\theta^* = 90°$ would correspond to $\theta \sim 25°$. For a forward-backward symmetric production in the center of mass, one half of the jets will have $\theta \lesssim 25°$ and momentum, $p_j \gtrsim 90$ GeV. At large angles, the associated jet would still be measured up to angles as large as ~90° ($p_j \sim 20$ GeV). This example illustrates the importance of accurate jet studies in the forward cones.

TAGGING HEAVY FLAVORS

The detection of heavy flavors at the Collider will be a very important goal for a number of reasons, ranging from the search for a fourth generation of quarks and for new heavy particles having large decay widths into heavy flavors (like the Higgs), to a study of CP violation effects. The use of a specific hadronic channel as done by UA1,[15] ($D^* \longrightarrow D\pi$), cannot serve this scope because of the low branching ratio and reconstruction efficiency. For $\overline{Q}Q$ production, the basic signature will be leptons and dileptons, both of unequal and equal sign, which have come from crossed leptonic decays as illustrated in FIGURE 23b. It is understood by an inspection of this figure that secondary vertices (both in the hadronic channels, FIGURE 23a, and in the semileptonic ones, FIGURE 23b) and the measurement of whether leptons come from primary or secondary vertices (FIGURE 23b) provide additional signatures for the event, which can serve to distinguish among the production of different heavy flavors.

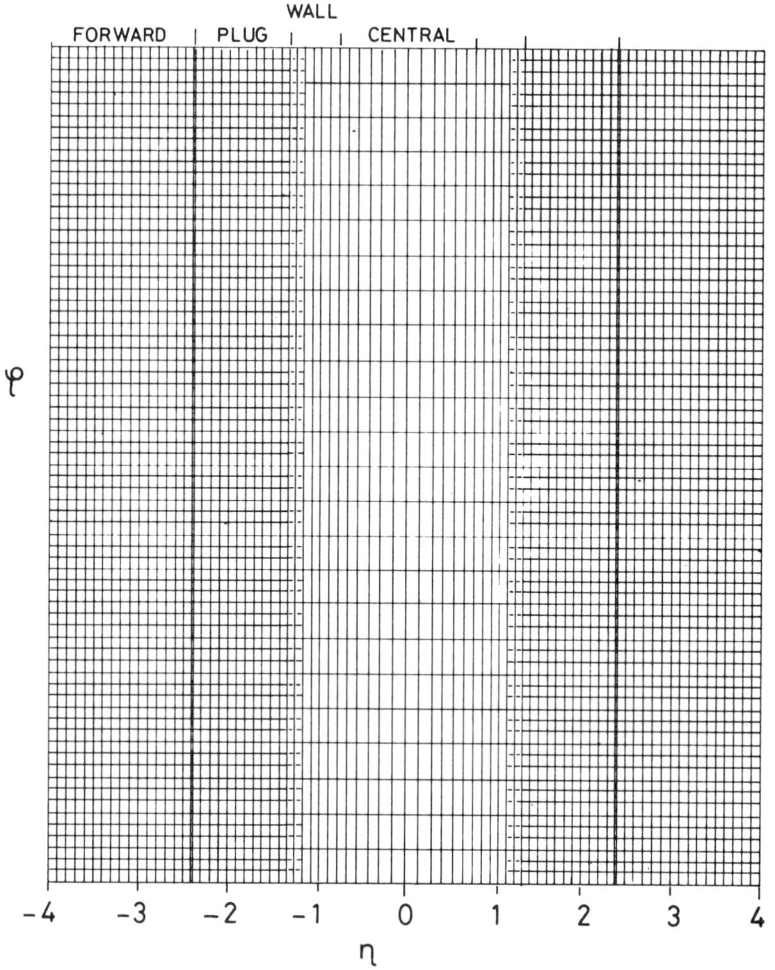

FIGURE 20. Illustration of the splitting of the CDF hadron calorimeters at $\theta \gtrsim 2°$ into bins of pseudorapidity and polar angle.

The detection of secondary vertices is attempted in CDF by accurate tracking around the interaction point using silicon vertex detectors. FIGURE 24 gives a sketch of the detector geometry. More information can be found in reference 18. There will be 540 rectangular crystals, 5 cm long and a few cm wide, arranged into four layers at a

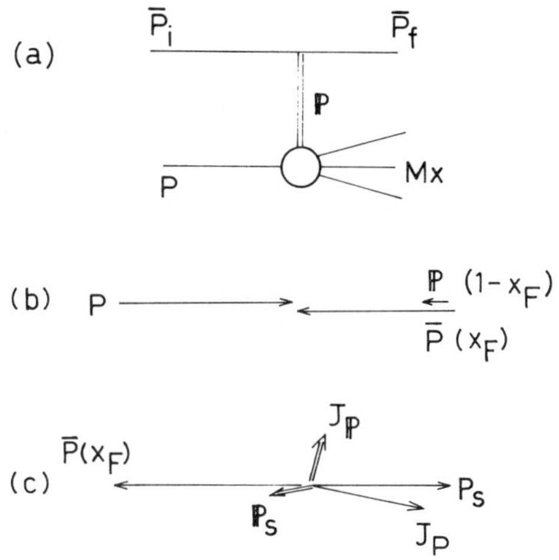

FIGURE 21. The Pomeron exchange diffractive process, $\bar{\mathrm{p}}\mathrm{p} \rightarrow \bar{\mathrm{p}}X$ (a), can be realized with a Pomeron of momentum, 1-x_F, accompanying the antiproton of momentum, x_F, hitting the proton (b). In the final state, the spectator antiproton continues with momentum, x_F. Occasionally, hard scattering can take place between a parton in the Pomeron and a parton in the proton to give rise to two large p_T jets, $J_{¶}$ and J_p (c). The rest of the Pomeron and the proton appear as spectator jets, $¶_s$ and p_s (reference 17).

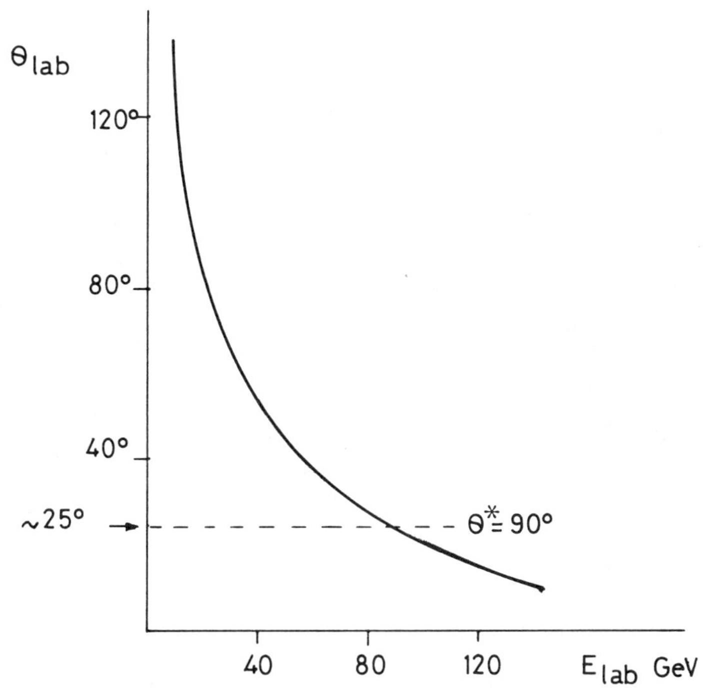

FIGURE 22. Example of angle-energy correlation for a jet generated by a parton in the Pomeron for diffracted masses accessible at the Fermilab Collider (M(¶ − p) = 450 GeV; $\sqrt{s}$ = 1900 GeV).

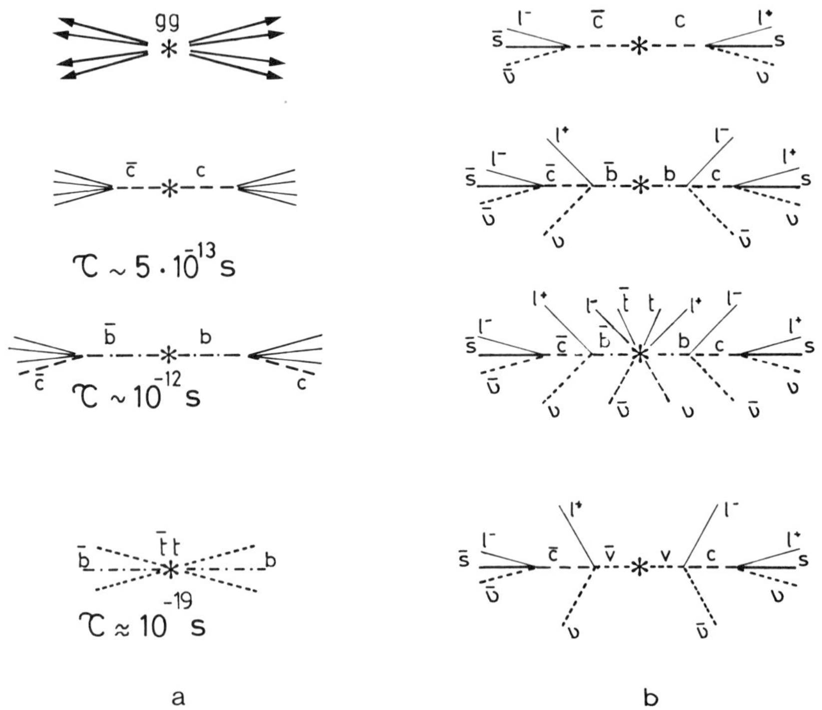

FIGURE 23. Illustration of the secondary vertex structure and multilepton signature in heavy flavor hadronic (a) and semileptonic (b) decay.

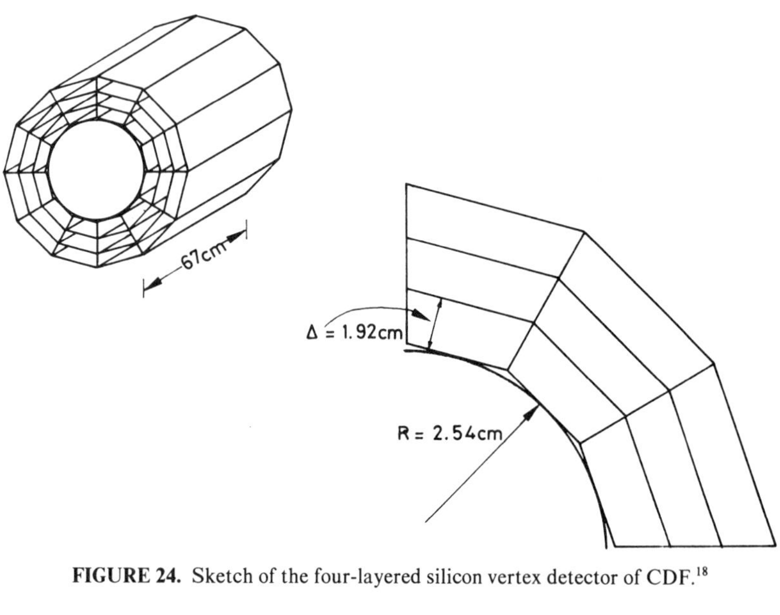

FIGURE 24. Sketch of the four-layered silicon vertex detector of CDF.[18]

distance between ~2.8 and ~8.5 cm from the beam, and containing 16,565 channels. The transverse R resolution will be approximately 25 μm. The performance of this detector was estimated using Isajet events and a realistic detector simulation including multiple scattering, photon conversion, and reconstruction inefficiencies. The search for secondary vertices in the transverse $R\varphi$ plane is done in a simpler and more efficient way than searching for points of minimum approach in two dimensions.[18] Each track is represented as a point in a conjugate plane, in which track originating from a common vertex must lie on a straight line. These points have an error that is determined by multiple scattering, which is known since the momentum is measured in the main CDF central tracking system.

FIGURE 25 shows how well the impact parameter of the production point can be measured for a single track, depending on its transverse momentum. The problem of

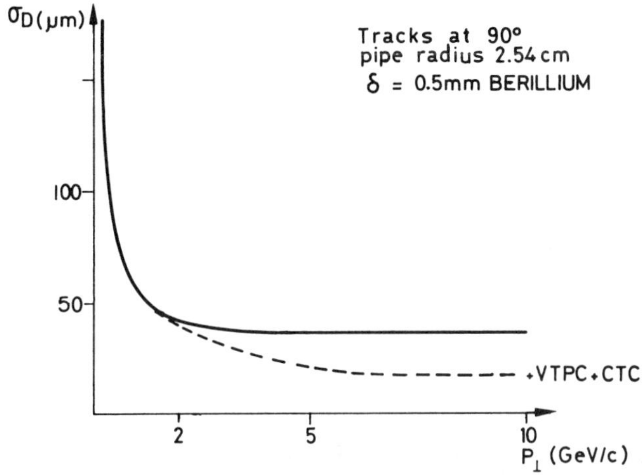

FIGURE 25. Resolution of the reconstructed impact parameter in CDF as a function of track transverse momentum.[18]

finding secondary vertices is reduced to a problem of pattern recognition of straight lines through points in the representative plane. FIGURE 26 shows a favorable example of a hard scattering process leading to a $\bar{b}b$ pair, in which two vertices were found and all tracks were correctly attributed to the proper vertex.

TABLE 2 shows the efficiency of tagging charm and beauty events, which is defined as the probability of finding at least one secondary vertex containing two or more tracks. The events were QCD hard scattering events, $\bar{\mathrm{p}}\mathrm{p} \rightarrow \bar{Q}QX$, generated with Isajet. While the efficiency for charm is very low, the numbers are very encouraging for beauty. The small increase of ϵ when the source length is reduced to zero shows that most losses are not due to the finite detector coverage. The improvement of ϵ from ~30% to ~70% with a perfect detector shows the importance of improving the detector resolution. The efficiency, $\epsilon \sim 70\%$, should be considered as an ultimate goal since when the $b \rightarrow c \rightarrow K, \pi$ decay chain has $n_{ch} \leq 1$, the vertex cannot be detected.

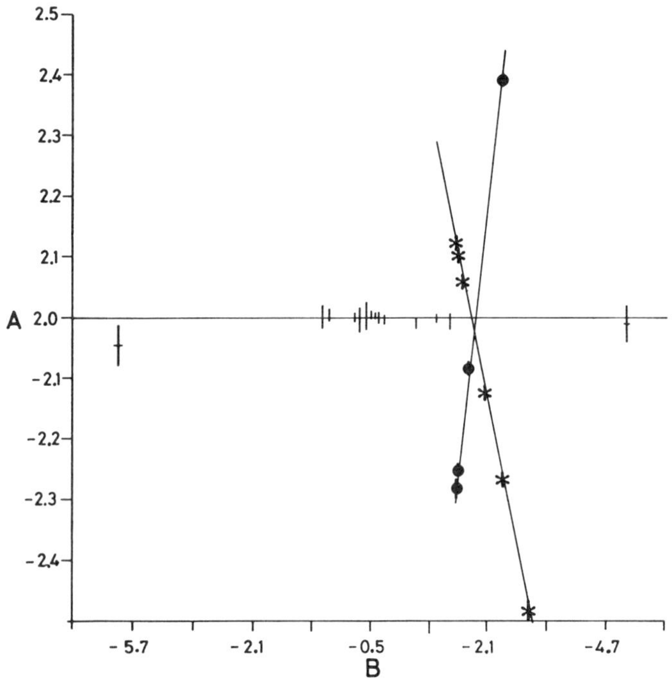

FIGURE 26. Example of a $\bar{p}p \rightarrow \bar{b}bX$ event in which two secondary vertices were found as two not-horizontal lines in the representative plane of the event.[18]

Based on this simulation, we[19] have performed the following gedanken experiment. We have addressed the general problem of how well the $W \rightarrow t\bar{b}$ decay can be separated from the remaining $W \rightarrow j\bar{j}$ decays and from the QCD background arising from jj production. We started with a basically unbiased event sample generated by Isajet, requesting two jets of $p_T > 15$ GeV and $|y| < 2$. FIGURE 27 shows the dijet mass distribution in this sample. At the peak, the Z^0, $W \rightarrow jj$ rate is 1/2000th of the inclusive QCD rate. The situation is much more favorable if one compares Z^0, W decays into b and t to $t\bar{t} + b\bar{b}$ production (signal/background ~30%). The efficiency of tagging the main competing processes is given in TABLE 3 for vertices containing three

TABLE 2. Tagging Efficiency for Charm and Beauty Production[a]

σ_z (cm)	$\Delta R\varphi$ (m)	Charm (%)	Beauty (%)
0	0	10	69
0	250	3.5	35
20	250	2	30
0	500	0.5	14

[a] σ_z = standard half-width of Gaussian beam–beam overlap; $\Delta R\varphi$ = electrode pitch in silicon detectors. The $R\varphi$ resolution is typically 10% of the pitch.

tracks (ϵ_3) and four tracks (ϵ_4). One sees that the tagging efficiency is the same for $W \rightarrow t\bar{b}$ and for $b\bar{b}$ production. $t\bar{t}$ was not simulated in this first study because of its lower rate relative to $b\bar{b}$. However, one also expects for this channel an efficiency similar to $b\bar{b}$.

Comparing ϵ_3 and ϵ_4, one sees how much one loses in efficiency, but improves in discrimination against lighter quarks when one imposes tighter constraints on the vertex. In particular, ϵ_4 is only 0.4% for c, u, d, s, g dijets, while still being ~15% for $W \rightarrow t\bar{b}$. The ratio between these efficiencies—discrimination factor—is about 40, which is an extremely interesting number. When applied to a specific process, similar to the one illustrated in FIGURE 27, this would indicate that a distinct peak would be seen in the m_{jj} peak at $\int Ldt \sim 10^{37}\ \text{cm}^{-2}$, while an integrated luminosity several times higher would be needed in order to notice the peak in the untagged sample. In general, the chance of discovering a new peak in the m_{jj} distribution will depend on both the signal/background ratio as well as on the integrated luminosity.

It should be noted that a more favorable discriminating factor will be obtained for processes when more than two t's or b's are produced. For example, consider an

TABLE 3. Tagging Efficiency for Vertices with Multiplicity ≥ 3 or ≥ 4 ($p_{jet} > 15$ Gev, $|y|_{jet} < 2$)

Process	ϵ_3	ϵ_4
$W \rightarrow t\bar{b}$	30.5%	15.5%
$W \rightarrow c\bar{s}$	2.0%	0^a
$b\bar{b}$ (QCD)	35.3%	22.4%
$c\bar{c}$ (QCD)	1.6%	0.4%[b]
u, d, s, g (QCD)	0.8%	0.4%[c]

[a]Out of 250 events.
[b]One event.
[c]Eight events.

associate Z^0H production for a Higgs mass in the region of 40–50 GeV or a double W production followed by decays into heavy flavors:

$$
\begin{aligned}
\bar{p}p \rightarrow\ & Z\ H\ X \\
& \quad\ \llcorner\!\rightarrow b\bar{b} \\
& \llcorner\!\rightarrow b\bar{b},\ t\bar{t}
\end{aligned}
$$

$$
\begin{aligned}
\bar{p}p \rightarrow\ & W^+ W^- X \\
& \quad\ \llcorner\!\rightarrow t\bar{b} \\
& \llcorner\!\rightarrow \bar{t}b.
\end{aligned}
$$

These processes are expected to have a $\sigma \cdot B$ of the order of picobarns at the Fermilab Collider and are, therefore, within reach. TABLE 4 gives the efficiency for finding 1, 2, etc., vertices for these processes as well as for the competing $\bar{b}b$ events. These efficiencies are estimated on the basis of those obtained in the previously mentioned simulation, using simple combinatorial arguments. The discrimination factor against $b\bar{b}$ grows from two, if simply one or more vertices are requested, to six, if specifically two vertices are requested. One thus sees that by using ϵ^2, an appreciable discrimina-

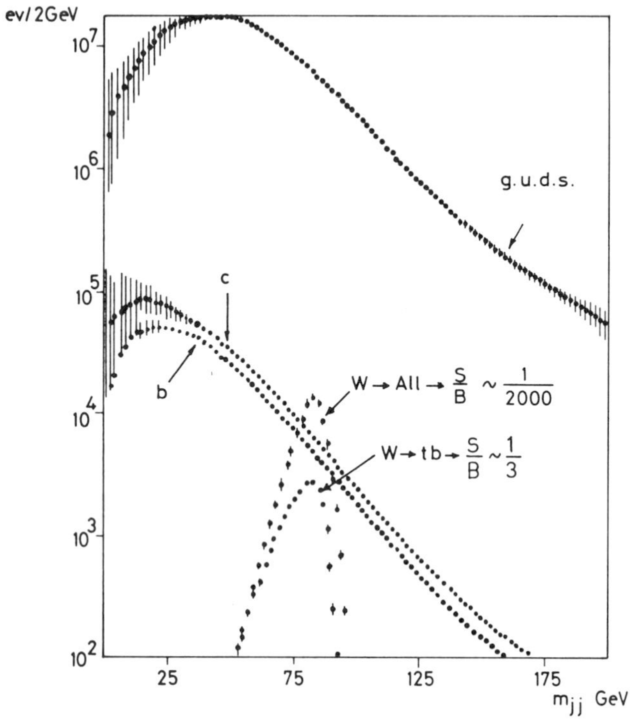

FIGURE 27. Simulated mass distribution of dijets in CDF for QCD jets and W, $Z^0 \rightarrow jj$ decays. The symbol "W" in the figure stands for W, Z^0 ($\sqrt{s} = 2000$ GeV, $\int L\,dt = 10^{37}$ cm^{-2}, $b_{T_j} > 15$ GeV, $|y| < 2$).

tion can be obtained against $b\bar{b}$ pairs while keeping an efficiency of ~13%. Of course, the discrimination against lighter quarks would be extremely strong. We conclude that the discovery of new phenomena of this type is indeed possible with CDF.

ACKNOWLEDGMENTS

The dedicated and intelligent work of Mrs. Clara Scopsi has been essential for the quick and accurate typing of this manuscript.

TABLE 4. Efficiency for Finding N-Vertices in Events Containing Beauty

No. of b's	ϵ^{any}	ϵ^1	ϵ^2	ϵ^3
0	5×10^{-4}	5×10^{-4}	2.5×10^{-7}	0
2	3×10^{-1}	2.7×10^{-1}	2.7×10^{-2}	0
4	5.1×10^{-1}	3.8×10^{-1}	1.1×10^{-1}	1.5×10^{-2}

NOTES AND REFERENCES

1. NAGY, E. *et al.* 1979. Nucl. Phys. **B150:** 221.
2. JACOB, M. 1983. Collider Physics—Present and Prospects. SLAC Topical Conference, July 27–29, 1983. TH. 3693 CERN.
3. BOZZO, M. *et al.* UA4 COLLABORATION. 1984. Measurement of the proton-antiproton total end elastic cross section at the CERN SPS Collider. Phys. Lett. **147B:** 392.
4. DIAS DE DEUS, J. & P. KROLL. 1983. J. Phys. **G9:** L81.
5. CHOU, T. T. & C. N. YANG. 1968. Phys. Rev. **170:** 1591.
6. HAYOT, F. & U. P. SUKHOTME. 1973. Phys. Rev. **D10:** 2183; CHENG, H., J. K. WALKER & T. T. WU. 1974. Phys. Lett. **44B:** 97; CHOU, T. T. & C. N. YANG. 1979. Phys. Rev. **D19:** 3268.
7. An accurate and up-to-date discussion of the implications of existing pp and $\bar{p}p$ data on elastic scattering and total cross section is given by CASTALDI, R. & G. SANGUINETTI. 1985. Elastic scattering and total cross sections at very high energies. CERN EP/85-24. To appear in Ann. Rev. Nucl. Part. Sci. **35.** FIGURE 3 is taken from CHIU, C. 1984. Phys. Lett. **142B:** 309.
8. CHENG, H. & T. T. WU. 1970. Phys. Rev. Lett. **24:** 1456.
9. HALZEN, F. & F. HERZOG. July 1984. MAD/PH/195. 15th Symposium on Multiparticle Dynamics, Lund, 1984.
10. GOULIANOS, K. December 1983. Diffractive interactions of hadrons at high energies. Physics Rep. **101:** n. 3.
11. BERTOLUCCI, S. *et al.* The small angle silicon detector system of CDF. CDF 285.
12. BELFORTE, S. & K. GOULIANOS. Monte Carlo study of single diffractive trigger and the luminosity monitor. CDF 257.
13. BEDESCHI, F. Semi-analytic calculation of acceptance corrected mass distributions for single diffraction and double Pomeron processes. CDF 270.
14. VANNINI, C. UA4 COLLABORATION. 1984. Measurement of proton-antiproton elastic scattering, total cross section, and diffractive dissociation at the CERN SPS Collider. Proceeding of the 19th Recontre de Moriond, La Plagne, March 4, 1984.
15. FREY, R. UA1 COLLABORATION. 1984. Proceeding of the 4th Topical Workshop on Proton Antiproton Collider Physics, Bern.
16. HALZEN, F. & F. HERZOG. June 1984. MAD/PH/185.
17. INGELMAN, G. & P. E. SCHLEIN. 1985. Jet structure in high mass diffractive scattering. Phys. Lett. **152D:** 256.
18. BEDESCHI, F. & G. CHIARELLI. Simulation of the silicon vertex detector. CDF 284.
19. I am reporting here of unpublished work, due primarily to F. BEDESCHI and G. CHIARELLI, performed within the Italian CDF group.

A Review of σ_{tot}, ρ, and B for p$\bar{\text{p}}$ and pp Elastic Scattering[a]

M.M. BLOCK

Department of Physics and Astronomy
Northwestern University
Evanston, Illinois 60201

INTRODUCTION

The advent of p$\bar{\text{p}}$ collider physics at the CERN ISR and SPS during the last four years has extended the maximum p$\bar{\text{p}}$ center-of-mass energy from $\sqrt{s} \sim 20$ GeV to $\sqrt{s} =$ 540 GeV. At the SPS, experimental groups have measured σ_{tot}, the total cross section, and B, the nuclear slope parameter, at $\sqrt{s} = 540$ GeV. In the energy range of 30 GeV $<$ $\sqrt{s} < 62$ GeV, groups at the ISR have made precision measurements of these quantities for both p$\bar{\text{p}}$ and pp, with the same apparatus used to compare the p$\bar{\text{p}}$ and the pp system. Moreover, new ISR measurements of elastic scattering in the Coulomb interference region have made possible accurate determinations of ρ, the ratio of the real to the imaginary portion of the forward nuclear scattering amplitude, for both p$\bar{\text{p}}$ and pp. The latter data, taken together with earlier results, now enable us to make a critical comparison of the p$\bar{\text{p}}$ and pp elastic scattering parameters in the high energy domain from $\sqrt{s} = 5$ GeV to $\sqrt{s} = 62$ GeV, and allow theoretical extrapolation to higher energies. The agreement between our results with the SPS results gives us some confidence in the larger extrapolations to $\sqrt{s} = 2$ TeV, the Tevatron Collider energy, and to $\sqrt{s} = 40$ TeV, the energy for the proposed SSC.

We will deal exclusively with p$\bar{\text{p}}$ and pp collisions and, in particular, we will concern ourselves with the analysis of elastic scattering in the low $|t|$ region, $-t < 0.02$ $(\text{GeV}/c)^2$, where t is the four-momentum transfer squared. The treatment follows an article by Robert Cahn and myself that appeared in *Rev. Mod. Phys.* (April 1985. **57:** 2).

Elastic scattering will be discussed in terms of an impact parameter representation. We will express, in impact parameter space, the nuclear slope parameter, B $(= d/dt[\log(d\sigma/dt)_{t=0}])$, and the curvature parameter, C $(= \frac{1}{2}\, d^2/dt^2[\log(d\sigma/dt)_{t=0}])$, where $d\sigma/dt$ is the invariant differential elastic nuclear scattering cross section. The gray disk and Chou-Yang[1] models will be discussed and it will be shown that the value of the curvature, $C = 0$, is a convenient criterion for the onset of "asymptopia," which is defined as the energy domain where the elastic scattering distribution approaches that of a sharp disk.

A model-free analysis will be made of the experimental quantities, σ_{tot}, ρ, and B.

[a]This work was supported in part by Department of Energy contract no. DE-AC02-76-ER-02289 Task B.

Traditionally, the requirements of analyticity have been compared with experimental data by means of dispersion relations. We will demonstrate how the same ends can be achieved more easily and transparently through direct use of simple analytic functions. The success of our fits is an experimental confirmation of the principle of analyticity. The comprehensive fit to p$\bar{\text{p}}$ and pp scattering reveals an unexpected regularity between the two systems over the full energy domain considered.

ANALYSIS OF $t = 0$ AMPLITUDES

Conventional Amplitudes

We define F as the analytic continuation of the forward scattering amplitude into the complex E plane, where E is the complex energy (E is the pp laboratory energy if E is real and $\geq m$, which is the nucleon mass). The F's are real analytic functions having cuts on the real axis from $+m$ to ∞ and from m to $-\infty$. We choose the normalizations such that, for fits with no odderons (unconventional odd amplitudes), the even continuation is

$$4\pi F_+ = -\sqrt{(m+E)(m-E)}$$
$$\left\{A + \beta \frac{\{[\log(2m(m+E)/s_0) + \log(2m(m-E)/s_0)]/2\}^2}{1 + a\{[\log(2m(m+E)/s_0) + \log(2m(m-E)/s_0)]/2\}^2}\right.$$
$$\left. + \frac{C}{2\sin(\pi\mu/2)}[[2m(m+E)]^{\mu-1} + [2m(m-E)]^{\mu-1}]\right\} \quad \textbf{(2.1a)}$$

and the odd continuation is

$$4\pi F_- = \sqrt{(m+E)(m-E)}$$
$$\times \frac{D}{2\cos(\pi\alpha/2)}[[2m(m+E)]^{\alpha-1} - [2m(m-E)]^{\alpha-1}], \quad \textbf{(2.1b)}$$

where A, β, s_0, a, C, μ, D, and α are real constants to be fitted by the data. Clearly, $F_+(E) = F_+(-E)$ and $F_-(E) = -F_-(-E)$. To find the scattering amplitudes for pp scattering, f_+ and f_-, we evaluate $F_+(E + i\epsilon)$ and $F_-(E + i\epsilon)$ in the limit of real E and $\epsilon \rightarrow 0$ (for the p$\bar{\text{p}}$ amplitudes, we evaluate the F's at $-E - i\epsilon$). We obtain

$$\left(\frac{4\pi}{\text{p}}\right)f_+ = i\left\{A + \frac{\beta[\log(2m\text{p}/s_0) - i\pi/2]^2}{1 + a[\log(2m\text{p}/s_0) - i\pi/2]^2}\right.$$
$$+ C\left[[(2m)(E-m)]^{\mu-1}e^{i\pi(1-\mu)/2}\right.$$
$$\left.\left. + \frac{[(2m)(E+m)]^{\mu-1} - [(2m)(E-m)]^{\mu-1}}{2\sin(\pi\mu/2)}\right]\right\} \quad \textbf{(2.2a)}$$

and

$$\left(\frac{4\pi}{\mathrm{p}}\right)f_- = -D\left\{[(2m)(E-m)]^{\alpha-1}e^{i\pi(1-\alpha)/2} + i\frac{[(2m)(E+m)]^{\alpha-1}-[(2m)(E-m)]^{\alpha-1}}{2\cos(\pi\alpha/2)}\right\}. \quad \textbf{(2.2b)}$$

The optical theorem relates the cross sections, σ^+ and σ^-, to the above by

$$\sigma^\pm = \left(\frac{4\pi}{\mathrm{p}}\right)\mathrm{Im}\, f_\pm, \quad \textbf{(2.3)}$$

where p is the laboratory momentum. Hence, the imaginery portions of **(2.2a)** and **(2.2b)** give the appropriate cross sections, from which we form

$$\sigma(\mathrm{p\bar{p}}) = \frac{(\sigma^+ + \sigma^-)}{2} \quad \textbf{(2.4a)}$$

and

$$\sigma(\mathrm{pp}) = \frac{(\sigma^+ - \sigma^-)}{2}. \quad \textbf{(2.4b)}$$

The formulae **(2.2a)** and **(2.2b)** simplify greatly in the limit of $E \gg m$, where s is given by $s \sim 2mE \sim 2m\mathrm{p}$. Using the notation $\langle f \rangle$ for the limit of f as $E \longrightarrow \infty$, we find

$$\left(\frac{4\pi}{\mathrm{p}}\right)\langle f\rangle_+ = i\left\{A + \beta\frac{[\log(s/s_0) - i\pi/2]^2}{1 + a[\log(s/s_0) - i\pi/2]^2} + Cs^{\mu-1}e^{i\pi(1-\mu)/2}\right\} \quad \textbf{(2.5a)}$$

and

$$\left(\frac{4\pi}{\mathrm{p}}\right)\langle f\rangle_- = -D\{s^{\alpha-1}e^{i\pi(1-\alpha/2)}\}. \quad \textbf{(2.5b)}$$

If we put $a = 0$ in **(2.5a),** we find by inspection of the real and imaginary parts of **(2.5a)** and **(2.5b)**, the very useful and simple formulae:

$$\sigma(\mathrm{pp}) = A + \beta\left[\log^2(s/s_0) - \frac{\pi^2}{4}\right] + C\sin\left(\frac{\pi\mu}{2}\right)s^{\mu-1} + D\cos\left(\frac{\pi\alpha}{2}\right)s^{\alpha-1}, \quad \textbf{(2.6a)}$$

$$\sigma(\mathrm{p\bar{p}}) = A + \beta\left[\log^2(s/s_0) - \frac{\pi^2}{4}\right] + C\sin\left(\frac{\pi\mu}{2}\right)s^{\mu-1} - D\cos\left(\frac{\pi\alpha}{2}\right)s^{\alpha-1}, \quad \textbf{(2.6b)}$$

$$\rho(\mathrm{p\bar{p}}) = [\beta\pi\log(s/s_0) - C\cos(\pi\mu/2)s^{\mu-1} + D\sin(\pi\alpha/2)s^{\alpha-1}]/\sigma(\mathrm{pp}), \quad \textbf{(2.6c)}$$

$$\rho(\mathrm{p\bar{p}}) = [\beta\pi\log(s/s_0) - C\cos(\pi\mu/2)s^{\mu-1} - D\sin(\pi\alpha/2)s^{\alpha-1}]/\sigma(\mathrm{pp}). \quad \textbf{(2.6d)}$$

We have essentially used the forms **(2.6a–d)** in our earlier work,[2] where we introduced only the coefficients A, β, s_0, D, α, and a. We interpret the even amplitude,

$Cs^{\mu-1}$, as an even Regge exchange term, with the odd amplitude, $Ds^{\alpha-1}$, as an odd Regge exchange term. The term in β gives the $\log^2 s/s_0$ rising cross section and A corresponds to a constant cross section. It will turn out that the coefficients using (**2.5a, b**) (i.e., using (**2.6a–d**)) are nearly identical to those using the kinematically correct equations, (**2.2a, b**). The only important difference is that (**2.2a, b**) give an improved χ^2 for the fit. This is because the low energy kinematics (the cut structure in E) is treated correctly in (**2.2a, b**) for $\sqrt{s} \sim 5$ GeV, where they are of importance. For $\sqrt{s} > 10$ GeV, the results using either (**2.2a, b**) or (**2.5a, b**) are numerically indistinguishable. The introduction of an even Regge trajectory is a departure from our earlier treatment.[2] We note that the $\log^2 s/s_0$ term in the even amplitude, for $s < s_0$, simulates this term in the cross section. We have fixed the power, μ, to be 0.5, since we expect it to be about the same as α, which turns out to be ~0.5.

TABLE 1. Results of Fits to Total Cross Sections and ρ Values[a]

Fit	#1	#2	#3	#4
A	41.74	41.30	41.70	41.11
	± 0.04	± 0.28	± 0.04	± 0.23
β	0.66	0.62	0.64	0.59
	± 0.01	± 0.03	± 0.02	± 0.02
s_0	338.5	293.6	332.7	275.1
	± 7.7	± 28	± 7.9	± 22
C		8.4		10.9
		± 5.1		± 4.1
μ		0.5		0.5
D	−39.37	−40.51	−39.20	−41.32
	± 1.6	± 1.8	± 1.5	± 1.9
α	0.48	0.47	0.48	0.46
	± 0.01	± 0.01	± 0.01	± 0.01
a			0.0056	0.0082
			± 0.003	± 0.003
$\chi^2/d.f.$	1.165	1.146	1.127	1.058
$d.f.$	76	75	75	74

[a] The parameters, A, β, and ϵ, are in mb, s_0 is in GeV^2, α, μ, and a are dimensionless, C is in mb GeV$^{2(1-\mu)}$, and D is in mb GeV$^{2(1-\alpha)}$.

Clearly, setting $a = 0$ in (**2.2a**) gives rise to a cross section that continues to rise indefinitely as $\log^2 s/s_0$. The introduction of a small positive value of a in (**2.2a**) gives us a functional form that will have the cross section rise locally as $\log^2 s/s_0$ (in the energy region of 5 GeV < 62 GeV). However, as $s \rightarrow \infty$, i.e., at very high energies, the cross section will flatten out and go to the constant cross section, $A + (\beta/a)$, for positive a. Thus, we model the case where the Froissart bound is not truly saturated (it rises as $\log^2 s$ only locally), and eventually the cross section rise stops, going to a constant cross section at ∞. We consider this extreme case a measure of the possible error due to extrapolation beyond the energy region of our fit.

The fits were made using seven different types of experimental quantities, $\sigma(pp)$, $\sigma(p\bar{p})$, $\rho(pp)$, $\rho(p\bar{p})$, $\Delta\sigma = \sigma(p\bar{p}) - \sigma(pp)$, $\Delta\rho = \rho(p\bar{p}) - \rho(pp)$, and $\rho_{av} = [\rho(p\bar{p}) + \rho(pp)]/2$. The χ^2 was minimized using the seven quantities and their quoted errors. No attempt was made to adjust any of the data systematically. The sources of the data are given in reference 2. Details of the fitted parameters are given in TABLE 1. The units are such that σ is in mb, if E, m, p, and $\sqrt{s}$ are in GeV.

For Fit #1 ($C = a = 0$), we get an excellent reproduction of the data using only five parameters, with a $\chi^2/d.f.$ (χ^2/degrees of freedom) of 1.17 for 76 degrees of freedom. Fit #2 ($a = 0$) introduces an even Reggeon amplitude and gives an equally satisfactory fit, with $\chi^2/d.f. = 1.15$ for 75 degrees of freedom. The introduction of $a \neq 0$ results in Fits #3 and #4. The $\chi^2/d.f.$ is not significantly changed and we find positive values for a, which are between 2 and 2.5 standard deviations from zero. For Fit #3, $\sigma \rightarrow 156$ mb, and for Fit #4, $\sigma \rightarrow 113$ mb. For $C \neq 0$, the fits are plotted in FIGURE 1a for σ and FIGURE 1b for ρ.

At the present time, the values of $\sigma(\mathrm{p\bar{p}})$ at the SPS are not sufficiently accurate to distinguish between the cases of $a = 0$ and $a \neq 0$. The only evidence bearing on whether the cross sections continue their rise indefinitely as $\log^2 s/s_0$ is preliminary cosmic ray data, where they[3] measure the p-air cross section and reduce it to a lower limit for the pp cross section at a mean energy of 10 TeV. This lower limit is shown in FIGURE 1a and is evidence in favor of the cross section continuing to rise logarithmically. It should be noted in this regard that the value of β, the coefficient of $\log^2 s/s_0$, is about 0.6 mb. This is to be contrasted with the Froissart-Martin bound, which says that σ must rise less rapidly than $(\pi/m_\pi^2)\log^2 s/s_0 \sim 60 \log^2 s/s_0$ mb, where m_π is the pion mass. Hence, it does not seem appropriate to claim that we have "saturated the Froissart Bound," in spite of the fact that the cross section seems to rise as $\log^2 s/s_0$.

Can a Log s *Rise Fit the Data?*

To verify the form of the rise, we have made a fit using an even amplitude that would cause the cross section to rise only as $\log s/s_0$, i.e., we modified the term in β that appeared in f_+ in (**2.2a**) to be

$$\beta\left[\log\left(\frac{2mp}{s_0}\right) - i\frac{\pi}{2}\right], \qquad \textbf{(2.7)}$$

and used the remainder of (**2.2a**), along with f_-, given by (**2.2b**), and repeated the fit. This was done as a test to see how well the data could be reproduced using only a $\log s/s_0$ term, as contrasted to a $\log^2 s/s_0$ term. The fit was very poor, giving a $\chi^2/d.f. =$ 7.2 for 76 degrees of freedom. The predicted value for the SPS was much too low. Fundamentally, one could not use only a $\log s/s_0$ term and simultaneously reproduce both the cross section and the ρ values in the energy domain of 5 GeV $< \sqrt{s} <$ 62 GeV. We conclude that fitting all the available data in this energy region requires the presence of a term varying essentially as $\log^2 s/s_0$, and that only a $\log s/s_0$, or indeed, any lower power, is ruled out.

Odderon Amplitudes

Up until now, we have only concerned ourselves with fitting the data using the odd amplitude given by (**2.2b**), which, in the limit of $s \rightarrow \infty$, has $\Delta\sigma, \Delta\rho \rightarrow 0$. The form of (**2.2b**) is suggested by an odd Reggeon exchange amplitude. However, the requirements

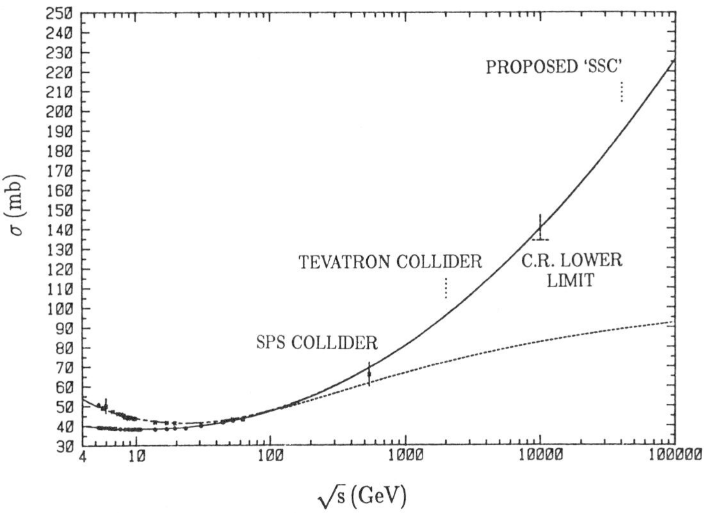

FIGURE 1a. The total cross sections, σ_{tot}, in mb, for $\bar{p}p$ and pp, as a function of $\sqrt{s}$, in GeV. Below $\sqrt{s} = 100$ GeV, the upper curve corresponds to pp and the lower to $\bar{p}p$, using Fit #2 ($C \neq 0, a = 0$). Above this energy, the difference in cross sections is too small to be visible. The upper curve above this energy corresponds to the Froissart bound form, with $a = 0$ ($C \neq 0$, Fit #2), while the lower curve is for $a \neq 0$ ($C \neq 0$, Fit #4). The experimental data used in the fit were in the energy interval, 5 GeV $< \sqrt{s} <$ 62 GeV. The experimental SPS $\bar{p}p$ cross section at $\sqrt{s} = 540$ GeV is appended for comparison, as is the cosmic ray lower limit for the pp cross section. To guide the reader, the energies of the Tevatron collider and the SSC are shown.

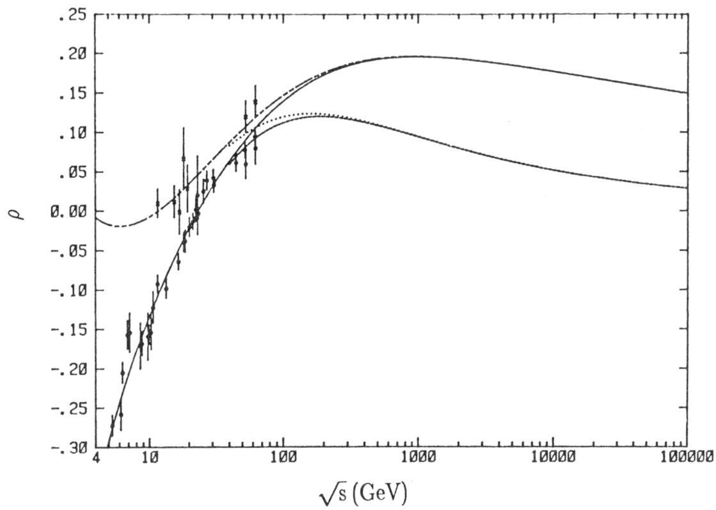

FIGURE 1b. The ρ values for $\bar{p}p$ and pp. Below $\sqrt{s} = 40$ GeV, the upper curve is for $\bar{p}p$, using Fit #3 ($a = 0, C \neq 0$). Above this energy, the curves split, with the lower of each pair corresponding to $a \neq 0$ ($C \neq 0$, from Fit #4) and the upper to $a = 0$ ($C \neq 0$, from Fit #2). At very high energies, the differences between $\bar{p}p$ and pp disappear and the two $a = 0$ curves (Fit #2) coalesce, as do the two $a \neq 0$ curves (Fit #4).

of analyticity are compatible with odd amplitudes—the "odderons"[4,5]—which give rise to $\Delta\sigma \to$ nonzero constant or, even, $\Delta\sigma \to \log s/s_0$, as $s \to \infty$. To test for the presence of these odderon terms, we introduce three types of odderon amplitudes, $f_-^{(j)}$, where $j = 0$, 1, or 2, via

$$4\pi f_-^{(0)} = -\epsilon^{(0)}E, \tag{2.8a}$$

$$4\pi f_-^{(1)} = -\epsilon^{(1)}E\left[\log\left(\frac{2mp}{s_0}\right) - i\frac{\pi}{2}\right], \tag{2.8b}$$

$$4\pi f_-^{(2)} = -\epsilon^{(2)}E\left[\log\left(\frac{2mp}{s_0}\right) - i\frac{\pi}{2}\right]^2, \tag{2.8c}$$

where the ϵ's are real constants. We form a new odd amplitude, $f_-^{new,(j)} = f_-^{old} + f_-^{(j)}$,

TABLE 2. Results of Fits to Total Cross Sections and ρ Values Including Odderons[a]

Fit	#5	#6	#7	#8	#9	#10
A	41.73	41.51	41.70	41.61	41.66	41.36
	± 0.04	± 0.30	± 0.04	± 0.29	± 0.05	± 0.26
β	0.68	0.65	0.67	0.66	0.65	0.62
	± 0.02	± 0.03	± 0.01	± 0.04	± 0.01	± 0.03
s_0	340.4	317.4	345.8	336.0	350.3	316.1
	± 0.5	± 31	± 8.1	± 33	± 9.6	± 30
C		4.3		1.7		6.0
		± 5.6		± 5.7		5.0
μ		0.5		0.5		0.5
D	−42.07	−42.31	−41.35	−41.47	−35.34	−36.70
	± 2.2	± 2.2	± 1.7	± 1.8	± 2.1	± 2.4
α	0.46	0.45	0.48	0.48	0.50	0.49
	± 0.02	± 0.02	± 0.01	± 0.01	± 0.02	± 0.02
ϵ	−0.29	−0.25	−0.12	−0.11	−0.04	−0.04
	± 0.12	± 0.13	± 0.04	± 0.04	± 0.02	± 0.02
$\chi^2/d.f.$	1.105	1.112	1.059	1.072	1.089	1.084
$d.f.$	75	74	75	74	75	74

[a]The parameters, A, β, and ϵ, are in mb, s_0 is in GeV^2, α, μ, and a are dimensionless, C is in mb GeV$^{2(1-\mu)}$, and D is in mb GeV$^{2(1-\alpha)}$. Fits #5 and #6 correspond to odderon 0, Fits #7 and #8 to odderon 1, and Fits #9 and #10 to odderon 2.

where the old odd amplitude is given by equation (**2.2b**). The results of these fits are summarized in TABLE 2. All odderon fits used $a = 0$. All of these fits gave satisfactory χ^2, with the $\chi^2/d.f.$ ranging from 1.08 to 1.11. The coefficients, $\epsilon^{(j)}$, are all small and negative, being around 2 to 2.5 standard deviations from zero.

It is very difficult to rule out the presence of odderon terms from the existing data. Although the odderon amplitudes are exceedingly small (in comparison with the constant amplitude, A, they are $<1\%$), they, of course, dominate the odd amplitude as $s \to \infty$. The data do not require adding an odderon amplitude, but they also do not rule out the presence of these terms at a two standard deviation level. If the odderons are as large as in our fits, comparisons of $\sigma(\mathrm{p\bar{p}})$ and $\sigma(\mathrm{pp})$ at $\sqrt{s} > 500$ GeV would easily confirm their presence (and one would need precise ρ measurements to rule out odderon 0).

SLOPE ANALYSIS OF NEARLY FORWARD ELASTIC SCATTERING DATA

The near-forward hadronic amplitude for $\mathrm{p\bar{p}}$ and pp elastic scattering is reflected in three experimentally accessible parameters—the total cross section, σ_{tot}, the ρ value, and the nuclear slope parameter, $B(s)$, defined as

$$B(s) = \frac{d}{dt}\left[\log\left(\frac{d\sigma}{dt}\right)\right]_{t=0}. \tag{3.1}$$

In the previous section, we analyzed $t = 0$ data for σ_{tot} and ρ in order to extract the forward hadronic amplitudes, f_+ and f_-. In this section, we will use the results of Fit #2 to obtain the s dependence of the slopes' B for pp and $\mathrm{p\bar{p}}$ elastic scattering, using experimental data in the near-forward direction (defined as the small $|t|$ region, $-t < 0.02$ $(\mathrm{GeV}/c)^2$). We write the invariant hadronic differential scattering cross section as

$$\frac{d\sigma}{dt} = \frac{\pi}{\mathrm{p}^2}\left| f_+ g_+(t, s) \pm f_- g_-(t, s)\right|^2, \tag{3.2}$$

where p is the laboratory momentum. We have assumed real, exponential "form factors" in equation (**3.2**), with $g_\pm(t, s) = \exp(B^\pm t/2)$. Since we are only concerned with very small $|t|$, the assumption of an exponential is the practical equivalent of replacing $e^{Bt/2}$ with $1 + Bt/2$. We rewrite equation (**3.2**) as

$$\frac{d\sigma}{dt} = \frac{\pi}{\mathrm{p}^2}\{[\mathrm{Re} f_+ \exp(B^+t/2) \pm \mathrm{Re} f_- \exp(B^-t/2)]^2 + [\mathrm{Im} f_+ \exp(B^+t/2) \pm \mathrm{Im} f_- \exp(Bt/2)]^2\} \tag{3.3}$$

with the plus sign for $\mathrm{p\bar{p}}$ and the minus sign for pp.

In our analysis, we will include slope data in the energy region of 5 GeV $< \sqrt{s} < 62$ GeV and only those data measured in the low $|t|$ region, $-t \sim 0.02$ $(\mathrm{GeV}/c)^2$, in order to reasonably approximate the definition of B made in equation (**3.1**). The data do not form a smooth set, unlike the situation for σ_{tot} and ρ. In an earlier paper,[6] we discussed the choice of experimental data that we have employed. Since the total cross section is assumed to rise as $\log^2 s/s_0$, it is important to parameterize the B^+ term, the ven slope, with a corresponding $\log^2 s$ term. We choose

$$B^+(s) = C^+ + D^+\log s + E^+\log^2 s \tag{3.4a}$$

and

$$B^-(s) = C^- + D^-\log s, \tag{3.4b}$$

where s is measured in $(\mathrm{GeV})^2$. No attempt is made to adjust the data for systematic errors. The results for Fit #2 are:

$$C^+ = 10.94 \pm 0.45,$$
$$D^+ = -0.09 \pm 0.15,$$
$$E^+ = 0.043 \pm 0.013,$$
$$C^- = 23.30 \pm 1.4,$$
$$D^- = 0.94 \pm 0.13,$$

with $\chi^2/d.f. = 1.51$ for 52 degrees of freedom.

FIGURE 2 is a plot of our predicted values of B versus $\sqrt{s}$, where the data used in the fit were in the energy interval, 5 GeV $< \sqrt{s} <$ 62 GeV. Also plotted in FIGURE 2 for comparison is the recent SPS measurement for $B(p\bar{p})$.[7] The agreement is satisfactory.

We define Σ_{el} as

$$\Sigma_{el} = \sigma_{tot}^2(1 + \rho^2)/16\pi B. \tag{3.5}$$

Thus, the ratio of the elastic cross section to the total cross section is given by

$$\Sigma_{el}/\sigma_{tot} = \sigma_{tot}/16\pi B. \tag{3.6}$$

We plot the ratio of Σ_{el}/σ_{tot} versus $\sqrt{s}$ for both the pp and the $p\bar{p}$ systems (using Fit #2) in FIGURE 3. The measured value at the SPS is 0.21 $\pm$ 0.01, which is in good

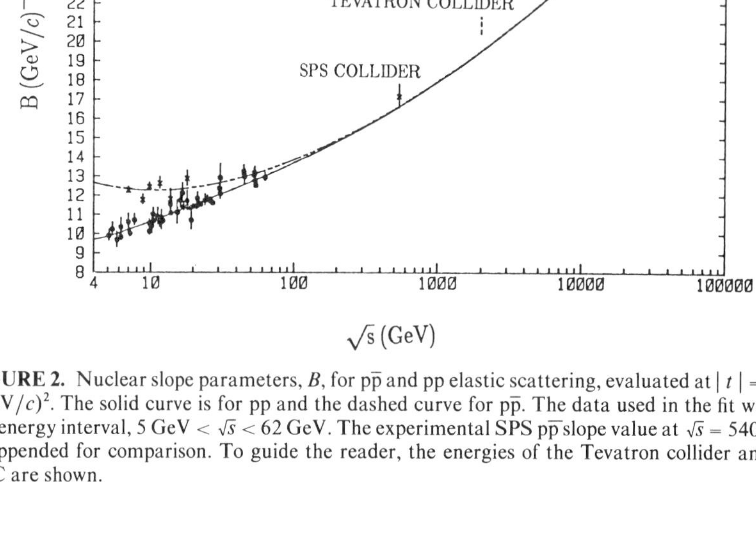

FIGURE 2. Nuclear slope parameters, B, for $p\bar{p}$ and pp elastic scattering, evaluated at $|t| = 0.02$ $(\text{GeV}/c)^2$. The solid curve is for pp and the dashed curve for $p\bar{p}$. The data used in the fit were in the energy interval, 5 GeV $< \sqrt{s} <$ 62 GeV. The experimental SPS $p\bar{p}$ slope value at $\sqrt{s} = 540$ GeV is appended for comparison. To guide the reader, the energies of the Tevatron collider and the SSC are shown.

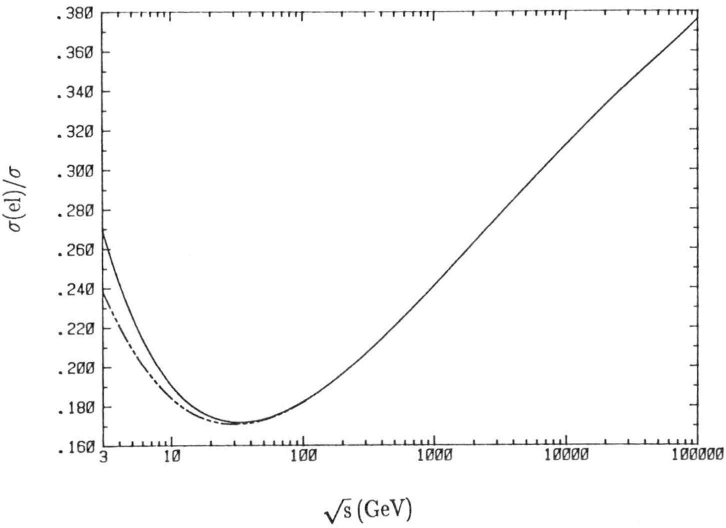

FIGURE 3. The ratio of Σ_{el}/σ_{tot}, for $p\bar{p}$ and pp, as a function of $\sqrt{s}$. The solid curve is for pp and the dashed curve for $p\bar{p}$. The curve was computed using Fit #2 ($C \neq 0$ and $a, \epsilon = 0$).

agreement with our prediction. It is easy to show that asymptotically, the ratio,

$$\Sigma_{el}/\sigma_{tot} \longrightarrow \beta/16\pi E^{+} \text{ as } s \longrightarrow \infty. \tag{3.7}$$

Thus, at infinite energy, the ratio, $\Sigma_{el}/\sigma_{tot} = 0.74 \pm 0.23$. Indeed, this result is also consistent with the black disk prediction of 0.5. We remark that even at as high an energy as $\sqrt{s} = 500$ TeV, the ratio has only grown to 0.44.

TABLE 3 gives a summary of our high energy predictions for σ_{tot}, ρ, and B, using Fits #1 and #2.

TABLE 3. Predictions for σ_{tot}, ρ, and B at High Energies[a]

Fit	$\sqrt{s}$ (TeV)	σ (mb)	ρ	B $(\text{GeV}/c)^{-2}$	Σ_{el}/σ
#1	0.540 (SPS)	70.37 ± 0.62	0.200 ± 0.002	16.65 ± 0.74	0.227
#2	0.540 (SPS)	69.32 ± 0.89	0.194 ± 0.005	16.66 ± 0.59	0.223
#1	2.00 (Tevatron)	98.30 ± 1.17	0.198 ± 0.002	19.53 ± 1.41	0.270
#2	2.00 (Tevatron)	95.93 ± 1.87	0.193 ± 0.004	19.57 ± 1.12	0.263
#1	20.0	169.46 ± 2.57	0.172 ± 0.001	26.03 ± 3.15	0.347
#2	20.0	163.39 ± 4.55	0.168 ± 0.002	26.13 ± 2.50	0.332
#1	40.0 (SSC)	196.38 ± 3.10	0.163 ± 0.001	28.34 ± 3.81	0.368
#2	40.0 (SSC)	188.84 ± 5.60	0.160 ± 0.002	28.47 ± 3.03	0.352
#1	100.0	235.87 ± 3.88	0.152 ± 0.001	31.64 ± 4.78	0.394
#2	100.0	226.15 ± 7.15	0.149 ± 0.002	31.81 ± 3.79	0.376
#1	500.0	316.00 ± 5.46	0.134 ± 0.000	38.14 ± 6.75	0.436
#2	500.0	301.75 ± 10.4	0.133 ± 0.001	38.39 ± 5.35	0.413

[a] The predictions are the same for pp and $p\bar{p}$ since no odderons are included. Both Fits #1 and #2 have cross sections growing as $\log^2 s$.

A REGULARITY OF THE $p\bar{p}$ AND pp SYSTEMS

We notice from FIGURE 3 that for all s, the ratios, Σ_{el}/σ_{tot}, are surprisingly close to being the same for both the $p\bar{p}$ and the pp systems, in spite of the fact that at the lower energies, the total cross sections, ρ values, and nuclear slope B values are very different. An even more surprising regularity is found when we examine the quantity, A, defined as

$$\begin{aligned} A &= \sigma_{tot}\sqrt{1+\rho^2}/16\pi B \\ &= \Sigma_{el}/(\sigma_{tot}\sqrt{1+\rho^2}), \end{aligned} \tag{4.1}$$

which is proportional to the magnitude of the impact parameter amplitude at zero

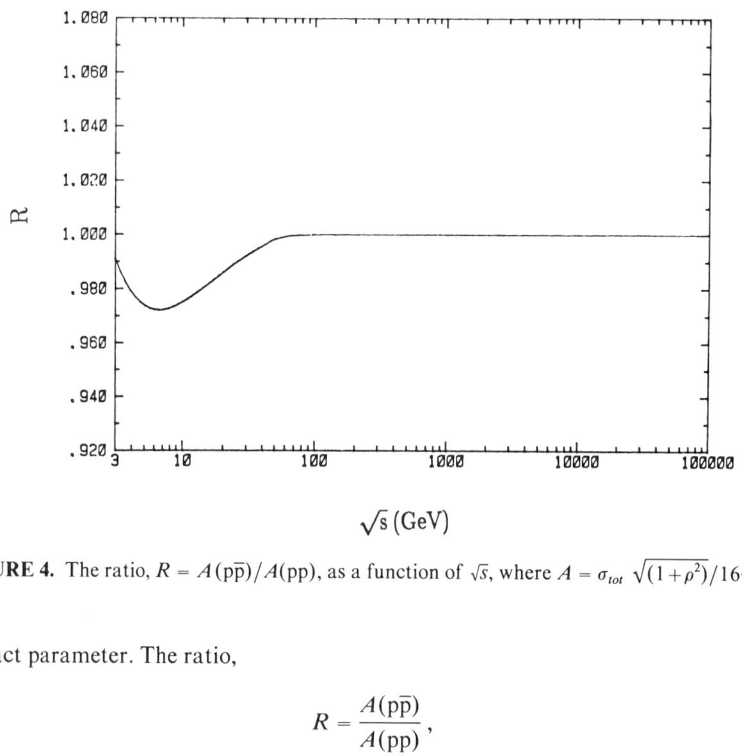

FIGURE 4. The ratio, $R = A(p\bar{p})/A(pp)$, as a function of $\sqrt{s}$, where $A = \sigma_{tot}\sqrt{(1+\rho^2)}/16\pi B$.

impact parameter. The ratio,

$$R = \frac{A(p\bar{p})}{A(pp)}, \tag{4.2}$$

is plotted versus $\sqrt{s}$ in FIGURE 4. We see that the ratio is compatible with unity, within the errors of our fitted values. It is, of course, not surprising that $R = 1$ for the high energy data, where $\Delta\sigma$, $\Delta\rho$, and $\Delta B \to 0$ as $s \to \infty$. What is strange and important is that at the lowest energies (3 GeV $< \sqrt{s} <$ 15 GeV), where the total cross sections and slopes are very different (up to ~35%), the ratio, R, is compatible with 1, within fitting errors of <2.5%. This implies that at all s, the magnitude of the impact parameter amplitude at zero impact parameter is the same for both the $p\bar{p}$ and the pp system. It remains a fundamental problem for any model of elastic $p\bar{p}$ and pp scattering at high energies to explain this newly observed regularity.

MODELS AND THE APPROACH TO "ASYMPTOPIA"

We will restrict our attention to two models—the gray sharp disk and the Chou-Yang[1] model. As is known, if the cross section is sufficiently large, the Chou-Yang model evolves into a black disk. Thus, these two models, which make vastly different predictions at ISR and SPS energies, will merge at very high energies. A sensitive indicator for this onset of "asymptopia" is the curvature parameter, C, defined as

$$C = \frac{1}{2}\frac{d^2}{dt^2}\left[\log\left(\frac{d\sigma}{dt}\right)\right]_{t=0}. \tag{5.1}$$

For small $-t$, this is the equivalent of parameterizing the elastic scattering cross section as

$$\frac{d\sigma}{dt} = \left(\frac{d\sigma}{dt}\right)_{t=0} \exp\,(Bt + Ct^2). \tag{5.2}$$

As will be shown, C is negative at all energies for a disk, whereas in the Chou-Yang model, C is positive at the ISR and at the SPS. (At the ISR, $C \sim 5\ (\mathrm{GeV}/c)^{-4}$.) Since the Chou-Yang model evolves with increasing energy into a disk, C must go through zero and eventually become negative.

We will work in the cms, where f is the scattering amplitude, k is the momentum, b is the impact parameter, and $a(b, s)$ is the impact parameter amplitude. We note that $t = -q^2$, where $q = 2k \sin(\theta/2)$, with θ as the scattering angle. In impact parameter space,

$$f(s, t) = \frac{k}{\pi}\int d^2b\, e^{i\vec{q}\cdot\vec{b}}\, a(b, s). \tag{5.3}$$

Therefore,

$$\frac{d\sigma}{dt} = \frac{\pi}{k^2}\left|f\right|^2 \text{ and } \sigma_{el} = \int_{-\infty}^{0}\left(\frac{d\sigma}{dt}\right)dt, \tag{5.4}$$

$$\sigma_{tot} = \frac{4\pi}{k}\,\mathrm{Im}\, f(s, 0) = 4\int d^2b\, \mathrm{Im}\, a(b, s), \tag{5.5}$$

and thus, from **(5.3)** and **(5.4)**,

$$\sigma_{el} = 4\int d^2b\, |\, a(b, s)\,|^2. \tag{5.6}$$

If we assume that the phase of $a(b, s)$ is independent of b, we can further write

$$B(s) = (N/2)\int d^2b\, b^2 a \tag{5.7}$$

and

$$C(s) = (N/32) \int d^2b\, b^4 a - (1/16)\, [N \int d^2b\, b^2 a]^2, \tag{5.8}$$

where we define

$$N = \frac{1}{\int d^2b\, a} \tag{5.9}$$

and $a = a(b, s)$.

For a gray sharp disk, with radius, R,

$$a = i\frac{A}{2}, \text{ if } 0 < b \leq R, \text{ and } a = 0, \text{ if } b > R, \tag{5.10}$$

with $0 < A \leq 2$. The black disk corresponds to $A = 1$. It can easily be shown from equations (**5.5–5.9**) that for a disk,

$$\sigma_{el}/\sigma_{tot} = A/2 = \sigma_{tot}/16\pi B, \tag{5.11}$$

$$B = R^2/4 \text{ and } C = -R^4/192. \tag{5.12}$$

Thus, for a black disk, $\sigma_{el}/\sigma_{tot} = 0.5$. In terms of the Chou-Yang parameters, X and Y, defined as

$$X = \sigma_{el}/\sigma_{tot} \text{ and } Y = \sigma_{tot}/16\pi B, \tag{5.13}$$

we find for the disk that $X = Y$.

We use for the impact parameter amplitude in (**5.3**) the purely imaginary value of

$$a(b, s) = \frac{i}{2}(1 - e^{-\Omega(b,s)}), \tag{5.14}$$

where we assume that Ω factors into $A(s)g(b)$, with $g(b)$ being a universal function. In the Chou-Yang model, $g(b)$ is taken proportional to the Fourier-Bessel transform of the square of $G(q^2)$, which is the electromagnetic form factor for e-p scattering. If $G(q^2)$ is assumed to be of the dipole form,

$$G(q^2) = \left(\frac{\Lambda^2}{q^2 + \Lambda^2}\right)^2, \tag{5.15}$$

where $\Lambda^2 = 0.71\ (\mathrm{GeV}/c)^2$, one obtains[8]

$$g(b) = \tfrac{1}{8}\, (\Lambda b)^3\, K_3(\Lambda b), \tag{5.16}$$

where K_3 is the modified Bessel function of the second kind. For small $x = \Lambda b$, $g = 1 - x^2/8$, and for large x, $g \longrightarrow cx^{5/2}e^{-x}$, where c is a constant. The model has only one adjustable parameter, $A(s)$, which is determined by fitting the experimental value of σ_{tot}. It is clear from (**5.14**) that if $A(s)$ is sufficiently large, $e^{-\Omega(0)} \sim 0$ and hence, $a(0) = i/2$. The amplitude, $a(b)$, will be essentially $i/2$ up to $b = R$, where R is such that for large $R\Lambda$, $\Omega \sim 1$. Since Ω falls exponentially with $R\Lambda$, this means $R\Lambda$ varies as $\log(A(s))$. Thus, if $\Lambda(s)$ increases as a power of s, then R grows as $\log s$, and $a(b) \sim i/2$

for $b \leq R$ and ~ 0 for $b > R$, where $R = R_0 \log(s/s_0)$, with s_0 being a scale factor. We see that the total cross section must grow as $\log^2 s/s_0$, which is what we assumed for our Fits #1 and #2. Indeed, this is a pseudoderivation of the Froissart bound. The amplitudes assumed in (**5.10**) and (**5.14**) were pure imaginary. In order to take into account the known filling of the diffraction minima due to the small, but finite real part of the amplitude, we modify the differential elastic cross section given in (**5.4**) by an heuristic correction term[9] and write

$$\frac{d\sigma}{dt} = \left(\frac{\pi}{k^2}\right)(|f|^2 + \rho^2 |d/dt(tf)|^2)\,. \qquad \textbf{(5.17)}$$

The amplitude, f, in (**5.17**) is taken to be pure imaginary and is given by (**5.10**) for

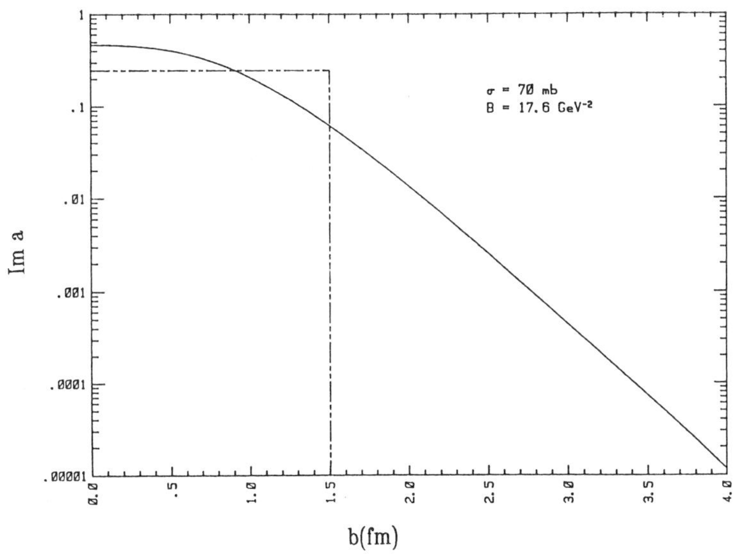

FIGURE 5a. A plot of the imaginary portion of the impact parameter amplitude, Im $a(b)$, versus b, the impact parameter, in fermis, for the Chou-Yang model (solid curve) and the sharp disk (dashed curve). The energy is $\sqrt{s} = 540$ GeV.

the disk and (**5.14**) for the Chou-Yang model. Thus, we find the simple result for the disk,

$$\frac{d\sigma}{dt} = \pi A^2 R^4 \left[(J_1(qR)/qR)^2 + \frac{\rho^2}{4}(J_0(qR))^2 \right]. \qquad \textbf{(5.18)}$$

We, therefore, have calculated the Chou-Yang coefficient, $A(s)$, for our predicted total cross sections, using (**5.5**). We, then, calculate the slope, B, from (**5.7**), the curvature, C, from (**5.8**), and the elastic scattering from (**5.6**), and obtain the parameters, X and Y. Finally, we require that the disk reproduce the same σ_{tot} and B

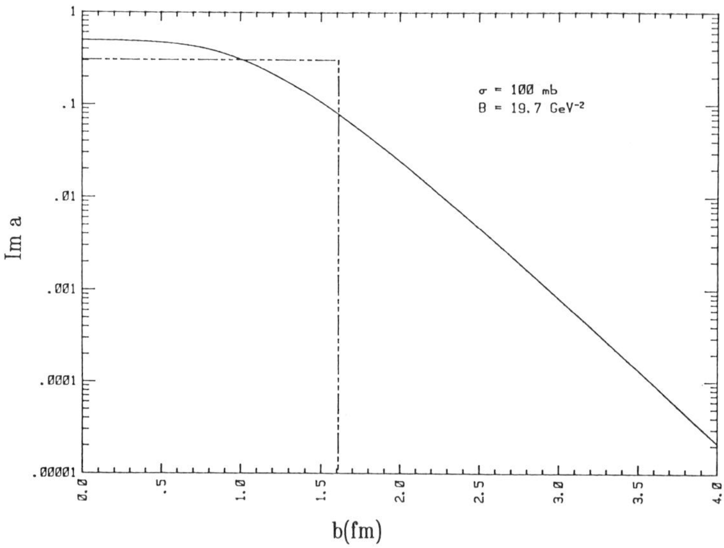

FIGURE 5b. See caption of FIGURE 5a. The energy, though, is $\sqrt{s} = 2$ TeV.

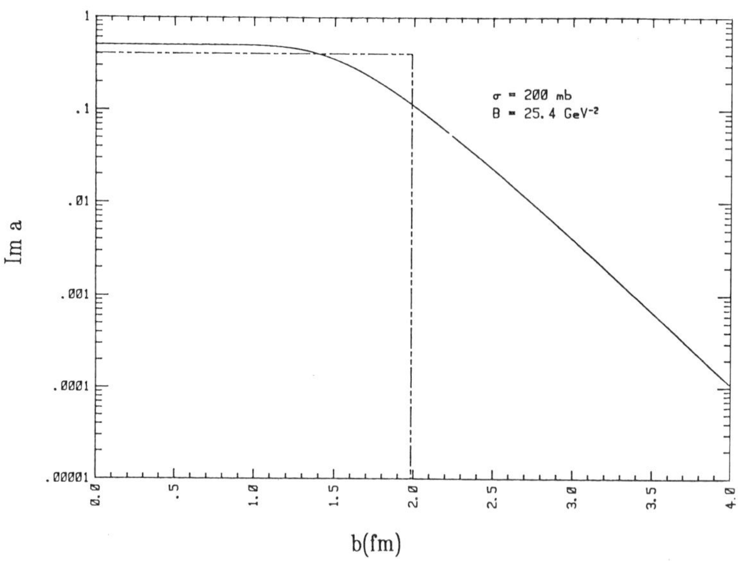

FIGURE 5c. See caption of FIGURE 5a. The energy, though, is $\sqrt{s} = 40$ TeV.

TABLE 4. Chou-Yang and Gray Disk Predictions[a]

Chou-Yang						Disk		
$\sqrt{s}$ (TeV)	$A(s)$	$B(1+\rho^2)$	C	X	Y	A	R	C
0.54	2.7	15.1	3.9	0.251	0.246	0.49	1.50	−18.0
2.0	4.9	17.4	−0.03	0.301	0.307	0.61	1.61	−22.1
40.0	24.5	26.1	−25.2	0.374	0.402	0.81	1.99	−53.5

[a] B is in $(\text{GeV}/c)^{-2}$, C is in $(\text{GeV}/c)^{-4}$, and R is in fermis.

values for each energy and thus fix R and A. Using the differential cross section of (**5.17**), it is simple to show that for $\rho \neq 0$, $B \rightarrow (1+\rho^2)B$ and $\sigma_{el} \rightarrow (1+\rho^2)\sigma_{el}$.

Semilog plots of $|a(b)|$, the impact parameter amplitude, versus b are shown in FIGURES 5a–c, for the energies, $\sqrt{s}$ = 540 GeV, 2 TeV, and 40 TeV, respectively, for both a gray disk and the Chou-Yang model. It is clear that the disk and the Chou-Yang model amplitudes are quite different at 540 GeV, with the Chou-Yang having a positive curvature, C, and the disk, of course, being negative, as indicated in TABLE 4. At 2 TeV, the Chou-Yang model has $C \sim 0$ and is just beginning to develop a "flattop." At 40 TeV, the Chou-Yang model has C strikingly negative, thus starting to resemble the disk distribution. These plots clearly show that $C = 0$ begins the onset of "asymptopia."

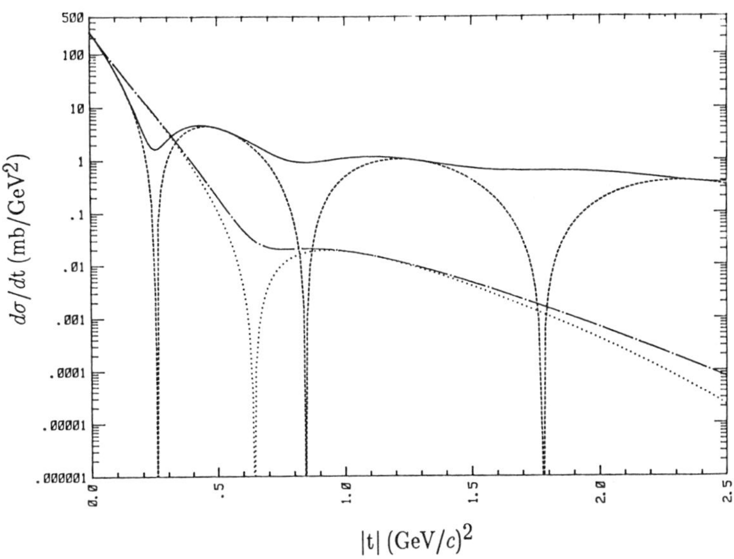

FIGURE 6a. A plot of $d\sigma/dt$, the elastic differential scattering cross section, for both the Chou-Yang model and the sharp disk, using the impact parameter amplitudes shown in FIGURE 5. The dotted curves (for $\rho = 0$) and the dash-dotted curves (for $\rho \neq 0$) are the Chou-Yang predictions, whereas the dashed curves (for $\rho = 0$) and the solid curves (for $\rho \neq 0$) are the predictions for the sharp disk. The energy is $\sqrt{s}$ = 540 GeV.

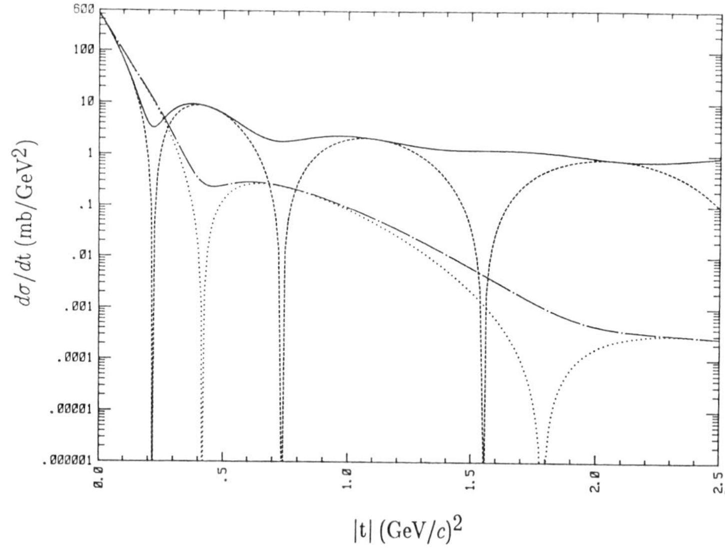

FIGURE 6b. See caption of FIGURE 6a. The energy, though, is $\sqrt{s} = 2$ TeV.

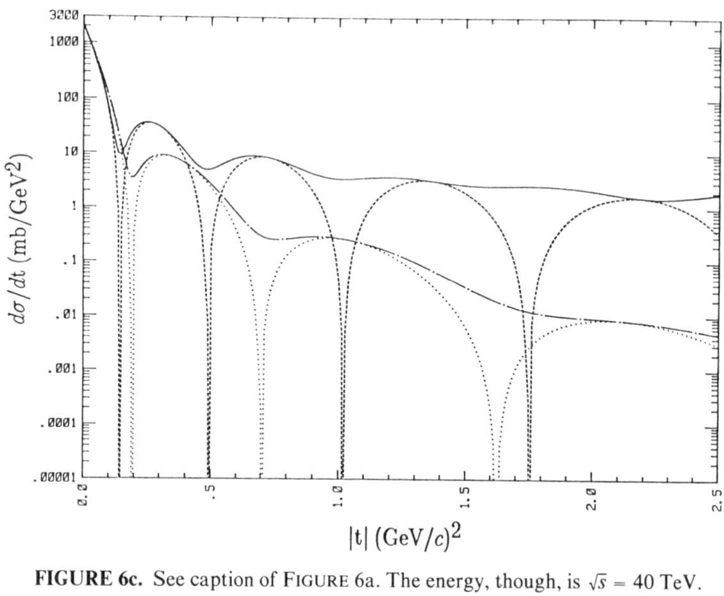

FIGURE 6c. See caption of FIGURE 6a. The energy, though, is $\sqrt{s} = 40$ TeV.

The elastic scattering differential cross sections versus t for the same three energies are shown in FIGURES 6a–c for both the model and the disk, which include the cases with $\rho = 0$ and $\rho \neq 0$. The ρ values used are from our Fit #2 to $t = 0$ data. The curves for $\rho \neq 0$ are expected to be a more realistic estimate of the true distributions since they fill in the diffraction minima significantly, thus resulting in a "shoulder" structure in the t distribution. As was already inferred from FIGURE 5a, we note that the two models predict widely different distributions at $\sqrt{s} = 540$ GeV. At 2 TeV, they are beginning to come together. At 40 TeV, the first minimum from the Chou-Yang model and the sharp disk are going to be almost impossible to distinguish experimentally, as seen in FIGURE 6c.

It is clear from the above that we expect the Chou-Yang model, at 40 TeV, to go essentially into a disk, i.e., we have entered "asymptopia" at an energy that will hopefully soon be realized experimentally.

ACKNOWLEDGMENTS

I would like to thank the Aspen Center for Physics for its hospitality during the carrying out of this research.

REFERENCES

1. CHOU, T. T. & C. N. YANG. 1968. Phys. Rev. **170:** 1591; 1983. Phys. Lett. **128B:** 457.
2. BLOCK, M. M. & R. N. CAHN. 1982. Phys. Lett. **120B:** 224.
3. YODH, G. B. 1983. Proceedings of the Third Moriond Workshop, p. 213. LaPlagne, France, March 13–19, 1983.
4. LUKASZUK, L. & B. NICOLESCU. 1973. Nuovo Cimento Lett. **8:** 405.
5. KANG, K. & B. NICOLESCU. 1975. Phys. Rev. **D11:** 2461.
6. BLOCK, M. M. & R. N. CAHN. 1982. Phys. Lett. **120B:** 229.
7. UA4 COLLABORATION. 1982. Phys. Lett. **115B:** 333; UA4 COLLABORATION. 1982. Phys. Lett. **117B:** 126; UA1 COLLABORATION. 1983. Phys. Lett. **128B:** 336.
8. DURAND, L. & R. LIPES. 1968. Phys. Rev. Lett. **20:** 637.
9. MARTIN, A. 1973. Nuovo Cimento Lett. **7:** 811; MARTIN, A. 1983. Proceedings of the Third Topical Workshop on p-$\bar{p}$ Physics, Rome; DIAS DE DEUS, J. & P. KROLL. 1978. Acta Phys. Pol. **B9:** 147; DIAS DE DEUS, J. & P. KROLL. 1982. CMFC preprint E21/82.

Results from the UA4 Experiment on Total Cross Section and Elastic Scattering at the CERN SPS Collider

G. MATTHIAE

Department of Physics of the University
Istituto Nazionale di Fisica Nucleare
Napoli, Italy

INTRODUCTION

The UA4 experimental apparatus at the SPS Collider was designed to measure total cross section, elastic scattering, and diffraction dissociation. It consists of a system of "Roman pots" and a two-arm vertex detector, which is symmetric with respect to the crossing point. The "Roman pots" are placed in the machine tunnel behind quadrupoles of the machine lattice. They detect scattered particles up to a maximum angle of 4.5 mrad, while the minimum accessible scattering angle depends on the machine optics. Momentum analysis is performed on the scattered particles with resolution of 0.6%. The vertex detector covers the angular region from 0.5° up to about 10°, which corresponds to the pseudorapidity range from 2.5 up to 5.6. The large angle region is covered by the central detector of experiment UA2. The performance and operation of the overall system are described in detail in reference 1.

TOTAL CROSS SECTION

The total cross section, σ_t, was obtained[2] from the measurement of the low-t elastic scattering and the inelastic interaction rate, making use of the optical theorem. This method does not require the knowledge of the machine luminosity and provides a measurement of the quantity, $(1 + \rho^2)\sigma_t$, where ρ is the ratio of the real to the imaginary part of the forward elastic amplitude. The measurement of the inelastic rate was performed by a vertex detector with almost complete solid angle coverage, using a fully inclusive trigger. Events having the special topology of diffractive dissociation were included in the trigger. Vertex reconstruction allowed clear identification of beam-beam interactions. The fraction of the inelastic events that are missed because all particles are confined inside the beam pipe was estimated to be not more than 1%.

The result of the measurement is $(1 + \rho^2)\sigma_t = 63.3 \pm 1.5$ mb. Using the value, $\rho = 0.15$, as suggested by the dispersion relation fit of reference 3 and by the more recent fit of reference 4, the result on the total cross section is $\sigma_t = 61.9 \pm 1.5$ mb. The uncertainty on the estimate of ρ reflects a change of the value of σ_t, which is well inside the quoted error. The Collider result is shown in FIGURE 1, together with $\bar{p}p$ and pp data at lower energies. Using the elastic scattering data and a determination of the machine luminosity (affected by a 10% systematic error), the result of $\sigma_t = 61 \pm 3$ mb was obtained, which agrees with the more accurate measurement from the luminosity independent method. The elastic cross section is $\sigma_{el} = 13.3 \pm 0.6$ mb, while the ratio, σ_{el}/σ_t, which is independent of any assumption on the real part, is equal to 0.215 ± 0.005. These results are shown in FIGURES 2 and 3, together with lower energy data.

A fit to the $\bar{p}p$ and pp total cross section and real part data using the logarithmic term, $(\log s)^{\gamma}$, in addition to the usual power terms needed to describe the low-energy

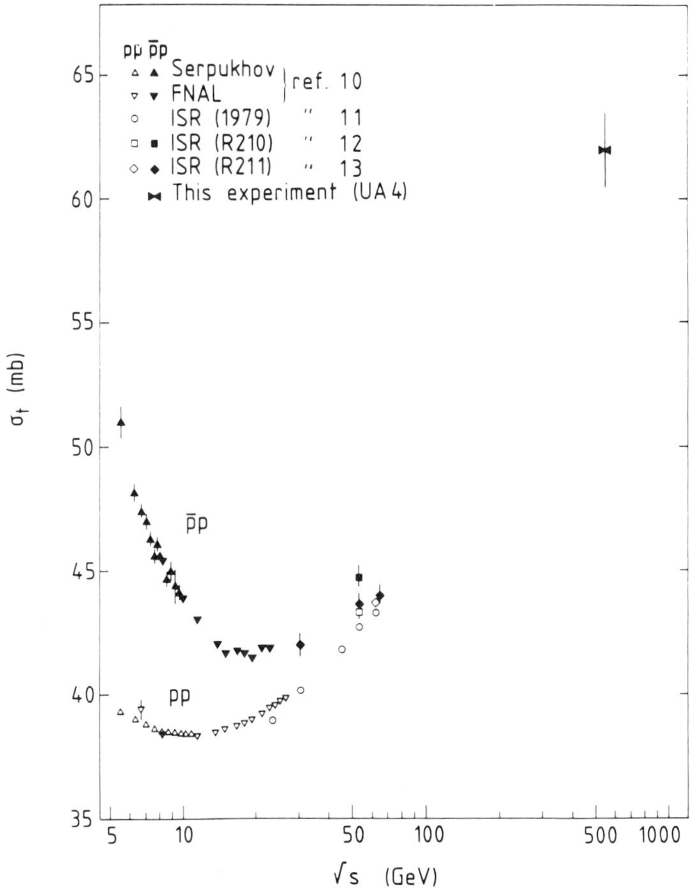

FIGURE 1. The Collider result on the $\bar{p}p$ total cross section at $\sqrt{s} = 546$ GeV is shown together with lower energy $\bar{p}p$ and pp data.

region, gave the result of $\gamma = 1.96 + 0.05$. This indicates that the total cross section rises in the presently explored energy range in a way that is already close to the $(\log s)^2$ behavior of the Froissart bound. The ratio of the elastic to the total cross section is definitely larger at the Collider than at the ISR, where it is nearly constant around 0.175.

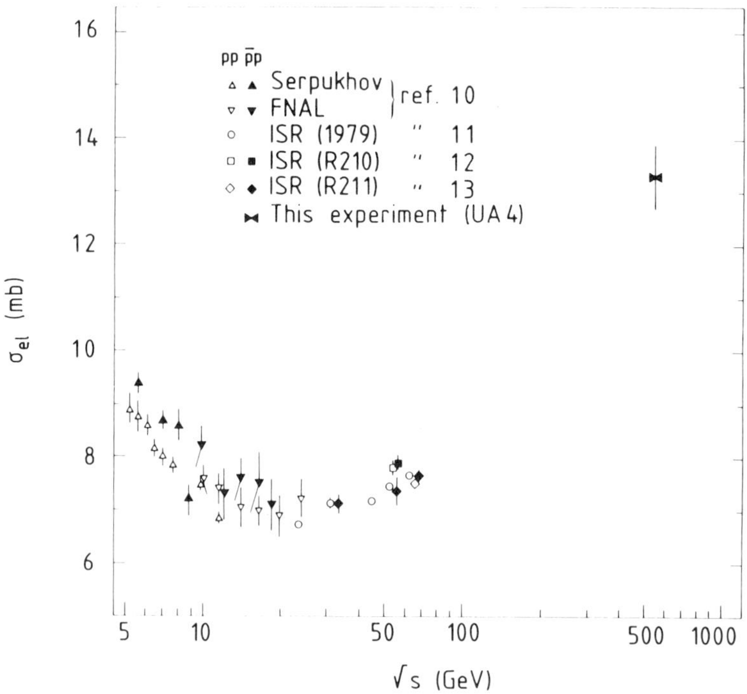

FIGURE 2. The $\bar{p}p$ and pp elastic cross section as a function of energy.

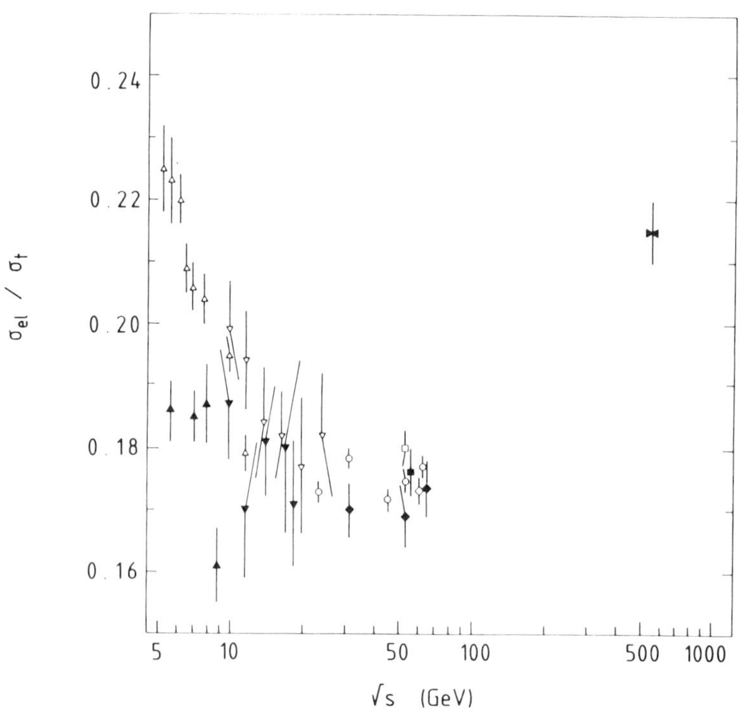

FIGURE 3. The ratio of the elastic to the total cross section for $\bar{p}p$ and pp scattering as a function of energy.

ELASTIC SCATTERING

Proton-antiproton elastic scattering was measured at $\sqrt{s} = 546$ GeV using a system of "Roman pots" and different beam optics.[5,6] For each set of beam optics, a different range of scattering angles was covered. In the high-β mode, the range was $0.03 \leq -t \leq 0.32$ GeV^2, in the medium-β mode, it was $0.21 \leq -t \leq 0.50$ GeV^2, and in the low-β mode, it was $0.45 \leq -t < 1.55$ GeV^2. The absolute normalization of the low-t data was obtained from the optical point. The other data samples were normalized by smoothly joining in the overlapping regions. The measured differential cross section for $-t \leq 0.5$ GeV^2 is shown in FIGURE 4.

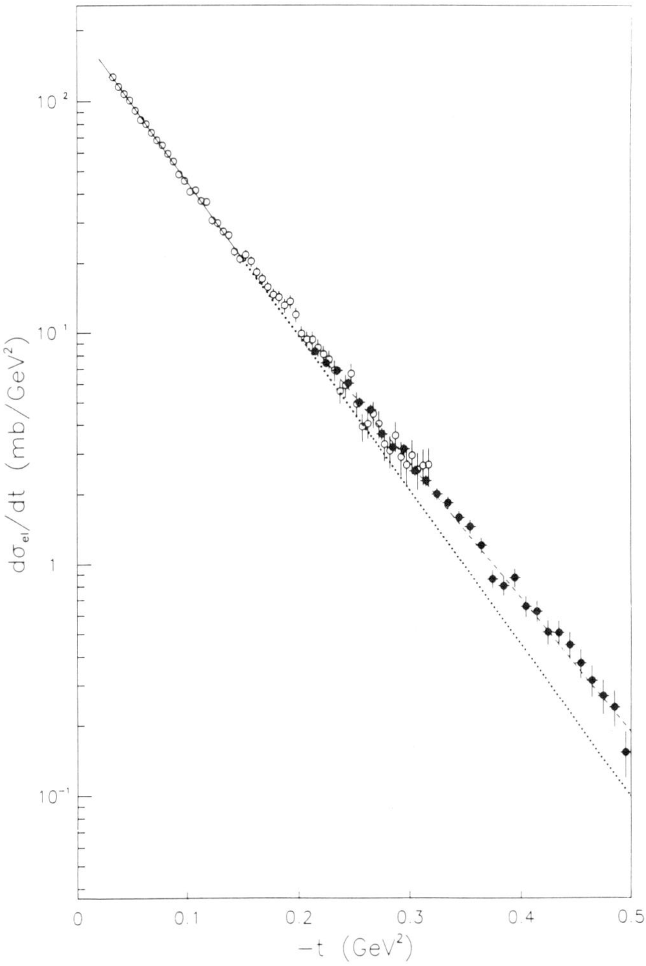

FIGURE 4. The proton-antiproton differential cross section of elastic scattering at a c.m.s. energy of 546 GeV in the low-t region ($-t \leq 0.5$ GeV^2).

From the study of the t-dependence, it was found that for $-t \leq 0.15$ GeV^2, the data can be well described by a single exponential, exp (bt). The same is true in the range, $-t \geq 0.2$ GeV^2. The slope parameter, b, has however quite different values in the two regions: $b = 15.3 \pm 0.3$ GeV^{-2} for $0.03 \leq -t \leq 0.10$ GeV^2 and $b = 13.4 \pm 0.3$ GeV^{-2} for the $0.21 \leq -t \leq 0.5$ GeV^2. A rapid change of the local slope parameter takes place between these two regions of momentum transfer. A fit of the data with two constant slopes, leaving also as a free parameter the t-value, t', where the sudden change of slope occurs, gives the result of $-t' = 0.14 \pm 0.02$ GeV^2. The exponential form with quadratic t-dependence, exp $(Bt + Ct^2)$, is not a good parametrization of the data in the range of $-t \leq 0.5$ GeV^2.

This behavior of the collider data is very similar to that observed at the ISR,[7] where a change of slope was found around $-t = 0.1$ GeV^2. As the energy increases from the ISR to the SPS Collider, the overall t-distribution of elastic scattering shows a very

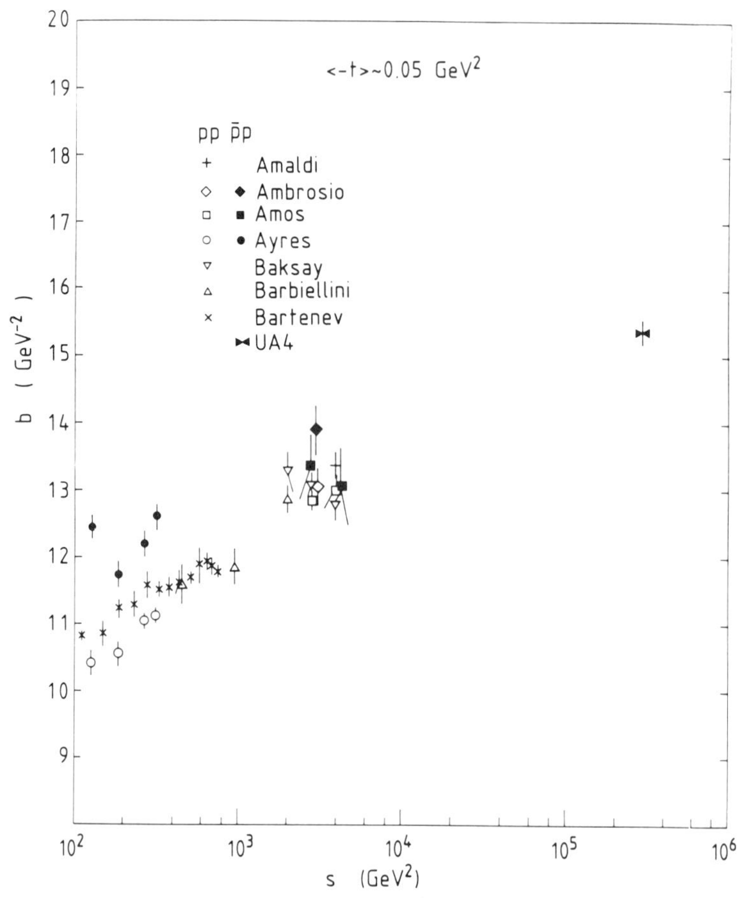

FIGURE 5. Energy dependence of the forward slope parameter ($-t \leq 0.1$ GeV^2) for $\bar{p}p$ and pp scattering as a function of energy.

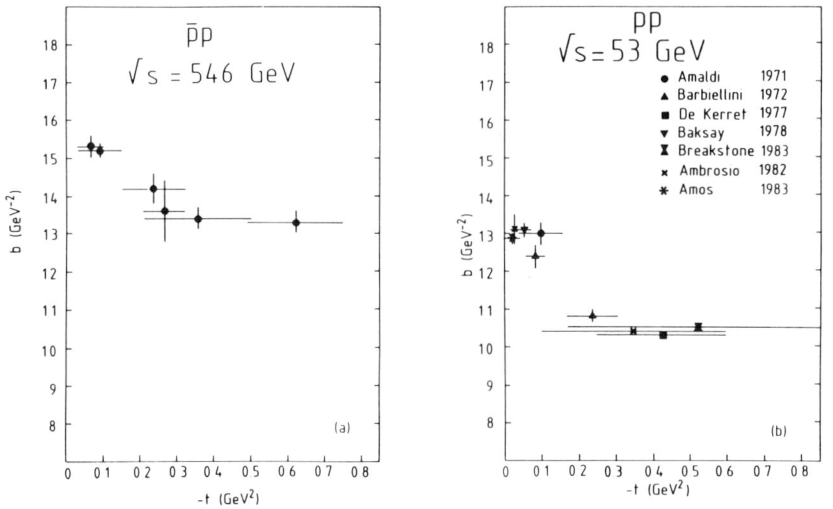

FIGURE 6. The dependence of the local slope parameter on t. The horizontal bar indicates the t-interval where the $\exp(bt)$ fit was performed. (a) The UA4 results at the Collider; (b) Compilation for pp scattering at $\sqrt{s} = 53$ GeV.

clear shrinkage. The shape of the t-distribution, however, keeps the feature of a rapid variation of the slope between 0.1 and 0.2 GeV^2. The energy dependence of the "forward" slope ($-t \leq 0.1$ GeV^2) is shown in FIGURE 5. Data on the local value of the slope, $b(t)$, as measured in various t-intervals at the Collider and in pp scattering at $\sqrt{s} = 53$ GeV, are presented in FIGURE 6.

The full t-distribution of elastic scattering at the Collider is shown in FIGURE 7. No dip is observed in the data, which show an exponential decrease with momentum transfer and a break at $-t \approx 0.9$ GeV^2 that is followed by a shoulder. Above 0.9 GeV^2, the differential cross section is of the order of 1 $\mu b/\text{GeV}^2$ and decreases slowly with t.

The proton-proton elastic scattering data at ISR energies[8] were fitted using a sum of two interfering amplitudes with exponential t-dependence,

$$d\sigma/dt \sim |e^{b(t-t_0)/2} + e^{d(t-t_0)/2 + i\phi}|^2.$$

Complete destructive interference would occur at $t = t_0$ if the relative phase, ϕ, was equal to π. Also, the present data were well represented by the previous expression. A fit in the range, $0.21 \leq -t \leq 1.55$ GeV^2, gave the following results[6]:

$$b = 12.9 \pm 0.1 \text{ GeV}^{-2}$$

$$d = 2.4 \pm 0.2 \text{ GeV}^{-2}$$

$$-t_0 = 0.827 \pm 0.004 \text{ GeV}^2$$

$$\phi = \pi - (0.82 \pm 0.04)$$

with $\chi^2 = 51$ for 56 degrees of freedom. The relatively large value of the phase difference, $\phi - \pi$, reflected the absence of a dip in the data.

The Collider results on $\bar{p}p$ elastic scattering at $\sqrt{s} = 546$ GeV are plotted in FIGURE 8, together with pp data measured at the ISR at $\sqrt{s} = 53$ GeV. The value of the differential cross section on the second maximum is more than one order of magnitude higher at the Collider than at the ISR energies. While the presence of a very narrow dip at the Collider cannot be ruled out, the t-resolution of the experiment ($\Delta t = 0.06\sqrt{|t|}$ for $-t \geq 0.5$ GeV^2) would be sufficient to observe such a prominent dip-bump structure as is seen at the ISR.

Recent $\bar{p}p$ results at $\sqrt{s} = 53$ GeV[9] in the range of $0.1 \leq -t \leq 3.5$ GeV^2 have shown that at this energy, $\bar{p}p$ elastic scattering is essentially indistinguishable from pp scattering except in the region of the dip that appears to be significantly less pronounced in $\bar{p}p$ scattering.

The dramatic increase with energy of the differential cross section at the second maximum, together with the increase of the ratio of the elastic to the total cross section,

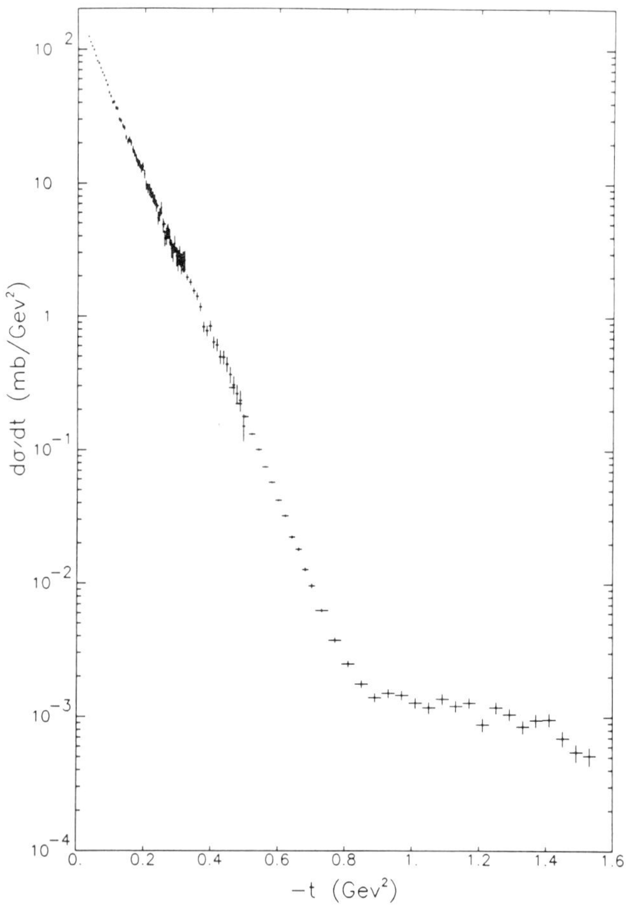

FIGURE 7. The proton-antiproton differential cross section of elastic scattering at the Collider.

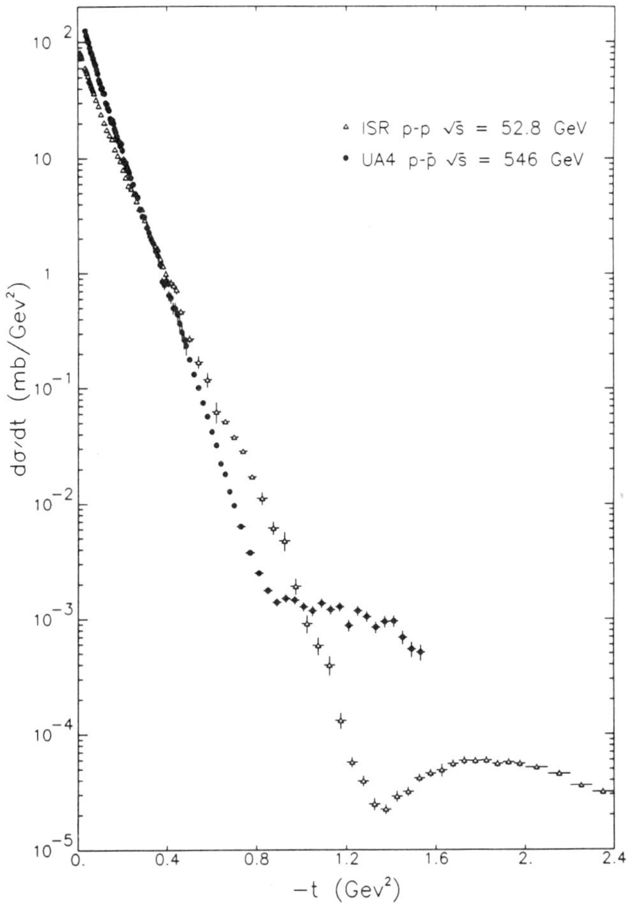

FIGURE 8. The elastic scattering data at the Collider are compared to pp data at $\sqrt{s} = 53$ GeV.

confirm at least qualitatively the prediction of the impact picture,[10] which is inspired by the study of general properties of quantum field theory at very high energy.[11]

The preliminary data of this experiment were discussed in several theoretical papers.[12] A model independent analysis[13] of these data, performed in terms of the impact parameter representation of the scattering amplitude, strongly suggests the following interpretation, in agreement with the predictions of references 10 and 11. As the energy goes up from the ISR to the Collider, the effective interaction radius of the proton increases. In addition, the proton itself becomes more absorbing in the central region at small values of the impact parameter.

REFERENCES

1. Battiston, R. *et al.* UA4 Collaboration. 1985. Nucl. Instrum. Methods **A238:** 35.
2. Bozzo, M. *et al.* UA4 Collaboration. 1984. Phys. Lett. **147B:** 392.

3. AMALDI, U. *et al.* 1977. Phys. Lett. **66B:** 390.
4. BLOCK, M. M. & R. N. CAHN. 1983. Phys. Lett. **120B:** 224.
5. BOZZO, M. *et al.* UA4 COLLABORATION. 1984. Phys. Lett. **147B:** 385.
6. BOZZO, M. *et al.* UA4 COLLABORATION. 1985. Phys. Lett. **155B:** 197.
7. BARBIELLINI, G. *et al.* 1972. Phys. Lett. **39B:** 663.
8. NAGY, E. *et al.* 1979. Nucl. Phys. **B150:** 221.
9. BREAKSTONE, A. *et al.* 1984. Nucl. Phys. **B248:** 253; ERHAN, S. *et al.* 1985. Phys. Lett. **152B:** 131; BREAKSTONE, A. *et al.* 1985. Phys. Rev. Lett. **54:** 2180.
10. CHENG, H., J. K. WALKER & T. T. WU. 1973. Phys. Lett. **44B:** 97; CHOU, T. T. & C. N. YANG. 1979. Phys. Rev. **D19:** 3268.
11. CHENG, H. & T. T. WU. 1970. Phys. Rev. Lett. **24:** 1456.
12. DONNACHIE, A. & P. V. LANDSHOFF. 1983. Nucl. Phys. **B231:** 189; BOURRELY, C., J. SOFFER & T. T. WU. 1984. Nucl. Phys. **B247:** 15; CHIU, C. 1984. Phys. Lett. **142B:** 309; GAURON, P. & B. NICOLESCU. 1984. Phys. Lett. **143B:** 253; GLAUBER, R. J. & J. VELASCO. 1984. Phys. Lett. **147B:** 380; ISLAM, M. M., T. FEARNLEY & J. P. GUILLAUD. 1984. Nuovo Cimento **81A:** 737.
13. HENZI, R. & P. VALIN.1983. Phys. Lett. **132B:** 443; HENZI, R. 1984. Proceedings of the Fourth Topical Workshop on Proton-Antiproton Collider Physics, Bern.

Multiplicities in High Energy Interactions

M. DERRICK

Argonne National Laboratory
Argonne, Illinois 60439

INTRODUCTION

Although multiparticle production had been studied for several decades using the cosmic radiation as the incident beam, there was a revival of interest in looking at the global properties of high energy inelastic interactions when precise inclusive data became available from the CERN Intersecting Storage Rings.[1] A number of unifying theoretical ideas, such as limiting fragmentation, and the multiperipheral model, with its prediction of short-range rapidity correlations, provided a framework in which to discuss the data.[2] The first accurate measurements of the charged particle multiplicity distributions that came from Fermilab[3] were crucial in limiting the range of theoretical options. These data also established the validity of scaling in the multiplicity distributions when expressed as a function of the mean, which became known as KNO scaling.[4]

Results from interactions such as $\bar{p}p$ annihilation and deep inelastic neutrino scattering showed narrower multiplicity distributions. This, then, lead to the suggestion that the diffractive scattering, which is absent in such processes, was responsible for the difference with pp interactions.[5] Comparisons with high energy e^+e^- data, show that this suggestion is an oversimplification.[6]

The first results from the CERN $\bar{p}p$ collider[7] at $\sqrt{s} = 540$ GeV showed a dramatic violation of the KNO scaling observed up to the ISR energy range. These highest energy data, which show many more events with multiplicities of three to four times the mean, stimulated a revival of interest in the subject. We are still far from a settled understanding.

In this paper, I review the results, point out the similarities and differences in the data from different reactions, and comment on the models that are currently extant.

MEAN CHARGED MULTIPLICITY

The energy variation of the mean charged multiplicity for e^+e^- annihilation[8] is shown in FIGURE 1. These data include charged particles from K^0 and Λ^0 decay, and at $\sqrt{s} \doteq 29$ GeV such particles contribute an average 1.4 units to $\langle n_{ch} \rangle$.[9] These charged particles are included in the e^+e^- data since logically, if they were to be removed, one should also remove the decay products of D and B mesons. However, these heavy particles carry a large fraction of the available energy, leaving a much lower effective energy for the hadronization of the remaining particles.

There are new quark thresholds in the 3 GeV and 9 GeV region, which, in principle, could introduce some discontinuities in the data. The systematic uncertainties in the data, which are typically ~ 0.5 units, mask any such effects.

The results of FIGURE 1 can be roughly represented by a power law, $\langle n_{ch} \rangle = as^{1/4}$, as originally suggested on statistical grounds by Fermi.[10] The full line shows this variation with $a = 3.0$.

The QCD motivated variation[11] of

$$\langle n_{ch} \rangle = a \left[\ln \frac{Q}{\Lambda} \right]^b \exp \left(c \ln \frac{Q}{\Lambda} \right)^{1/2} \tag{1}$$

also provides a reasonable representation of the data above the bottom quark threshold. Normalizing to the HRS result of $\langle n_{ch} \rangle = 13.1$ at $Q = \sqrt{s} = 29$ GeV gives $a = 0.12$, with $b = -4.92$, $c = 6.26$, and $\Lambda = 0.25$ GeV fixed by the QCD model.

A better representation of the data over the full energy range is provided by the power law expansion in $\ln s$:

$$\langle n_{ch} \rangle = a + b \ln s + c \ln^2 s, \tag{2}$$

with $a = 3.33 \pm 0.11$, $b = -0.4 \pm 0.08$, and $c = 0.26 \pm 0.01$. This fit is shown by the dashed line in FIGURE 1.

The pp data up to ISR energies,[3,12,13] shown in FIGURE 2, can also be fit by the $\ln s$

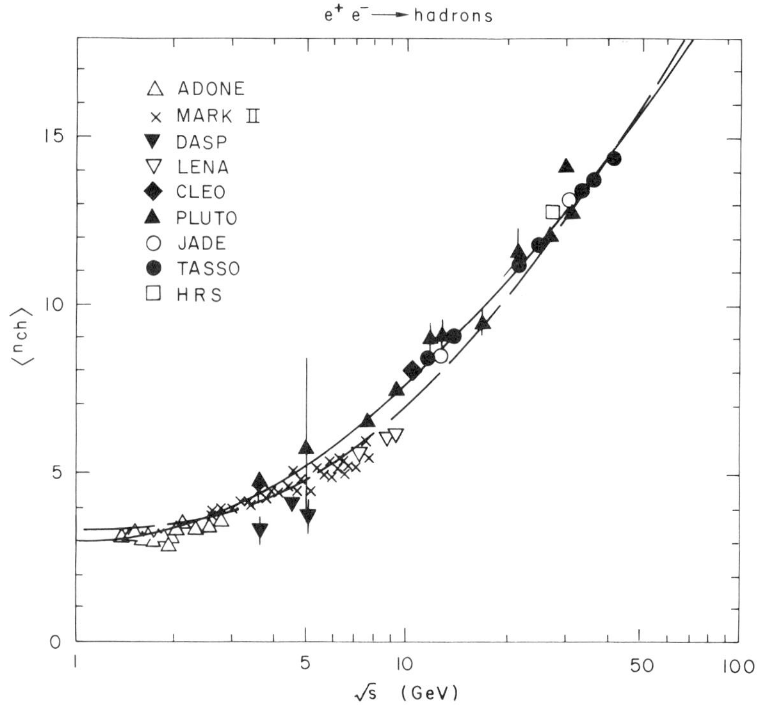

FIGURE 1. Center-of-mass energy variation of mean charged multiplicity, $\langle n_{ch} \rangle$, for e^+e^- annihilation. The full line shows the power law, $\langle n_{ch} \rangle = 3.0\, s^{1/4}$; the dashed line shows $a \ln s$ plus $\ln^2 s$ variation.

plus $\ln^2 s$ variation of equation (**2**), but with $a = 0.80 \pm 0.12$, $b = 0.47 \pm 0.05$, and $c = 0.114 \pm 0.005$ with $\sqrt{s}$ in GeV. This gives $\langle n_{ch} \rangle = 24.8$ for $\sqrt{s} = 540$ GeV, which is in good agreement with the value of 26.8 ± 2.1 for $\bar{\text{p}}\text{p}$ interactions reported by the UA5 collaboration.[14] The differences between pp and $\bar{\text{p}}\text{p}$ interactions at collider energies are not expected to be significant and so are ignored in the subsequent discussions.[15]

FIGURE 3 shows the two fits of equation (**2**) together with the QCD fit of equation (**1**) (dashed-dot) over a wide energy range. The full line, which fits the data for

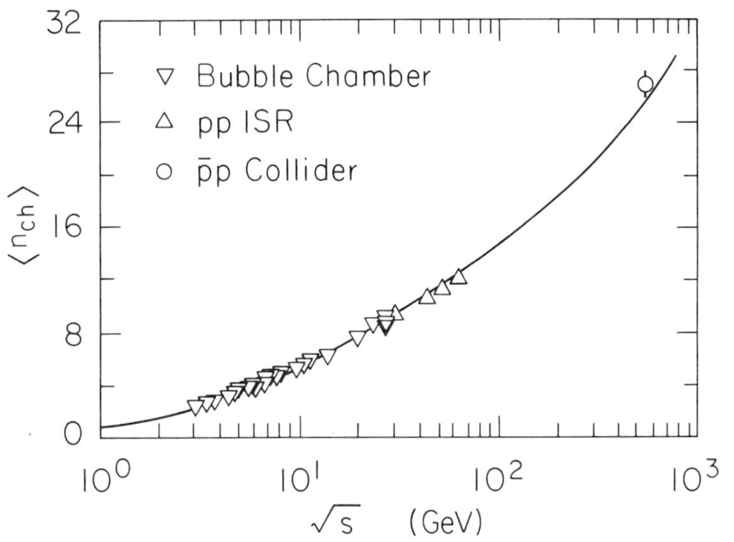

FIGURE 2. Center-of-mass energy variation of mean charged multiplicity, $\langle n_{ch} \rangle$, for pp interactions. The full line represents the fit, $\langle n_{ch} \rangle = 0.80 + 0.47 \ln s + 0.114 \ln^2 s$.

hadronic collisions when extrapolated to $\sqrt{s} = 40$ TeV, predicts $\langle n_{ch} \rangle = 62.0$. It is possible that this may be an underestimate since the fraction of hard collisions that populate the central rapidity region and that give much higher than average $\langle n_{ch} \rangle$ values is increasing rapidly with energy. The hard e^+e^- processes have a much stronger variation of $\langle n_{ch} \rangle$ with $\sqrt{s}$ than does the pp data.

In the energy range from 10–30 GeV in the c.m., the pp data, if plotted at $\sqrt{s}/2$, agree with the e^+e^- data consistent with the simple expectation that the available energy is a dominant variable[16] and that the leading particles take half of the center-of-mass energy on average. This result is shown in FIGURE 4. The fits to the pp and e^+e^- data of FIGURE 3 indicate that this simple scaling will probably not continue at the higher energies.

RAPIDITY DENSITY

The early expectation that the rapidity density would be independent of $\sqrt{s}$ and that the plateau would merely widen as the energy is increased leads to an $a + b \ln s$

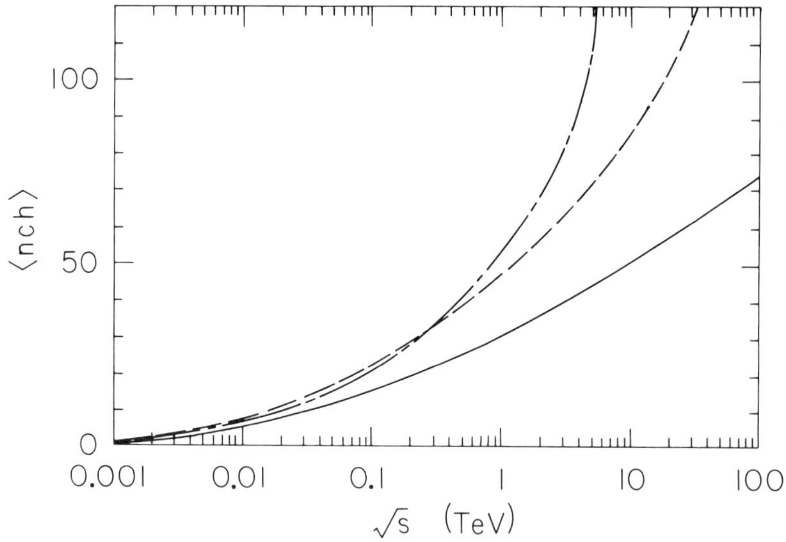

FIGURE 3. Extrapolation of $\langle n_{ch} \rangle$ for e^+e^- annihilation and pp interactions to SSC energies. The full line represents the pp data of FIGURE 2; the dashed line the e^+e^- data of FIGURE 1. The dashed-dotted line is the QCD fit to the e^+e^- data.

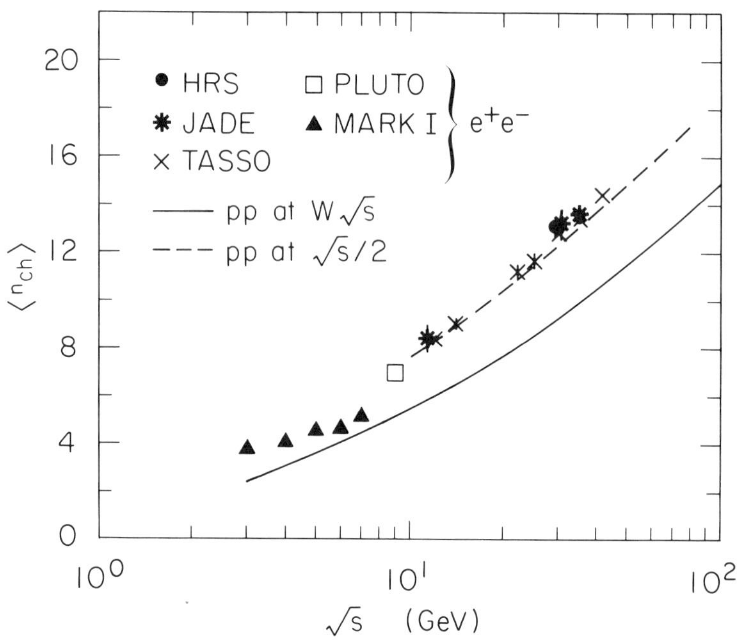

FIGURE 4. Comparison of $\langle n_{ch} \rangle$ for e^+e^- and pp data.

variation of $\langle n_{ch} \rangle$ that is clearly not in accord with data. The rapidity density at $Y = 0$ increases with energy for both e^+e^- data and pp data,[9] as shown in FIGURE 5. Separate, simple $a + b \ln s$ variations of $1/\sigma \, d\sigma/dY|_{Y=0}$, as shown by the lines, represent both sets of data well. The pp bubble chamber data[17] fit well with the collider results,[18] even though the latter measurements are made using the pseudorapidity variable, η ($\eta = \frac{1}{2} \ln \tan \theta/2$), and so lie typically 10–15% below the measurements made using the true rapidity variable, Y. The ISR point[19] is higher than the trend of the bubble chamber measurements, suggesting that perhaps the variation of the $\bar{p}p$ data is faster than is indicated in FIGURE 5. The dashed line connecting the hadronic data, if extrapolated to 40 TeV, predicts a value of $1/\sigma \, d\sigma/dY|_{Y=0}$ of 5.5 charged particles per unit of rapidity.

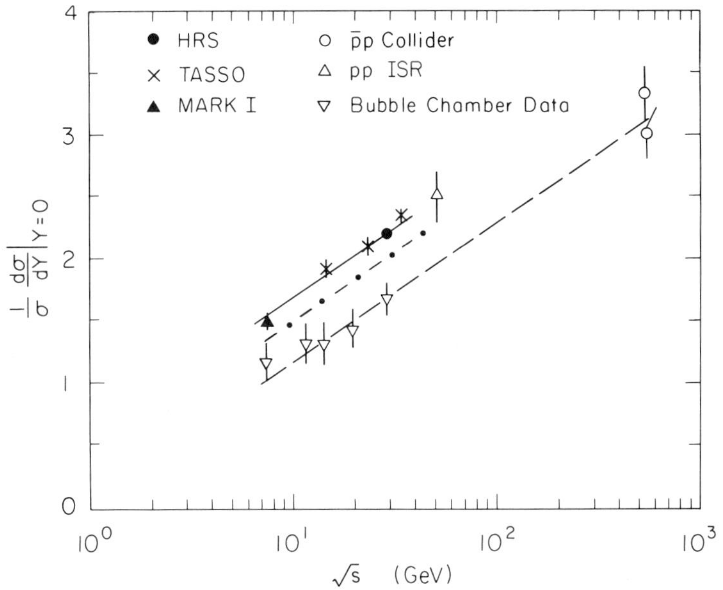

FIGURE 5. Energy variation of the rapidity density for e^+e^- and pp data.

The short dashed line shows the pp data plotted at $\sqrt{s}/2$. In this case, an agreement with the e^+e^- data is not obtained by such a simple scaling. This is perhaps connected with the fact that the shape of the pp and e^+e^- rapidity distributions are different.[9] As seen in FIGURE 6, the pp data peak at $Y = 0$, whereas the e^+e^- data have a shallow dip in the central region. Part of this difference clearly results from the breakup of the resonances containing the heavy quarks (c and b) that are responsible for almost half of the e^+e^- events and only a fraction of a percent of the hadronic interactions. The combination of the hard fragmentation and the restricted Q value of the heavy meson decays give a peak in the rapidity range, $1 \lesssim Y \lesssim 2$. This is illustrated by the DELCO data[20] on b quark fragmentation shown in FIGURE 7. The decay of the first rank mesons yields a distribution with a strong dip at $Y = 0$, which is only partially filled in by the higher rank fragmentation.

NEUTRAL PION PRODUCTION

The number of neutral pions produced in pp collisions, as measured by the number of observed photons, increases proportionally to the charged multiplicity. A variation,

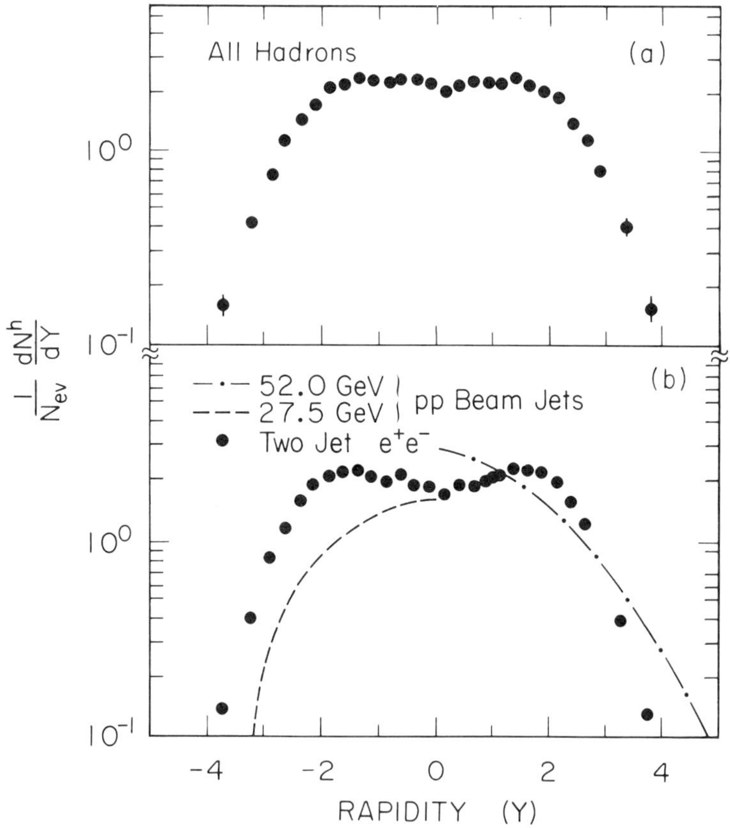

FIGURE 6. Rapidity distributions for e^+e^- and pp data. (a) HRS data at $\sqrt{s} = 29$ GeV; (b) comparison of HRS two-jet data with pp interactions at similar $\sqrt{s}$ values.

$$\langle n_{\pi^0} \rangle = \beta + \alpha n_-,$$

represents the data well and the strength of the correlation (α) increases with $\sqrt{s}$ from approximately zero in the 2- to 5-GeV range, up to about 3.5 for the collider energies as seen in FIGURE 8. The lower energy data[21] come from the inclusive pp bubble chamber experiments; the typical ISR results shown, continue the trend of α increasing with $\sqrt{s}$.

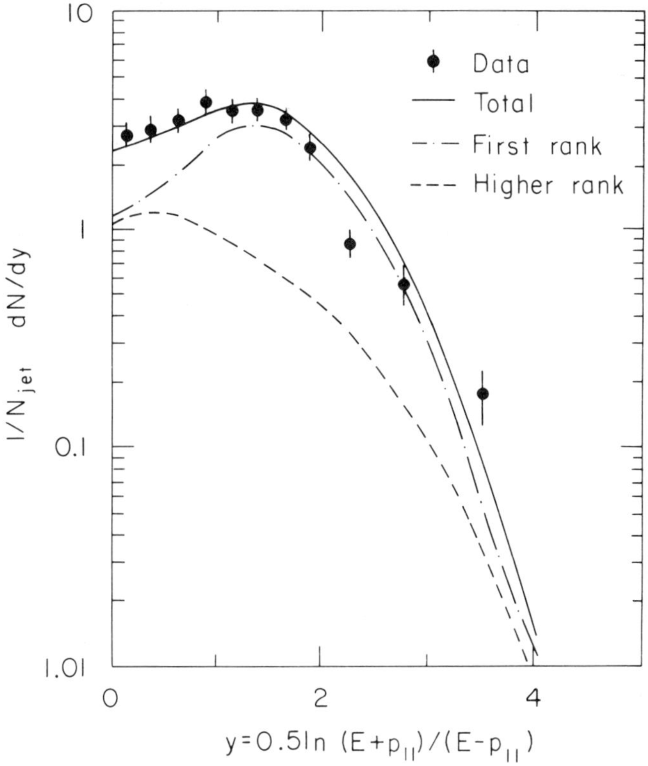

FIGURE 7. Rapidity distribution for $e^+e^- \rightarrow b\bar{b}$ events at $\sqrt{s}$ = 29 GeV, as measured by DELCO. The dashed-dotted line shows the Y distribution expected from the B meson decay products and the dashed line shows the remaining hadrons.

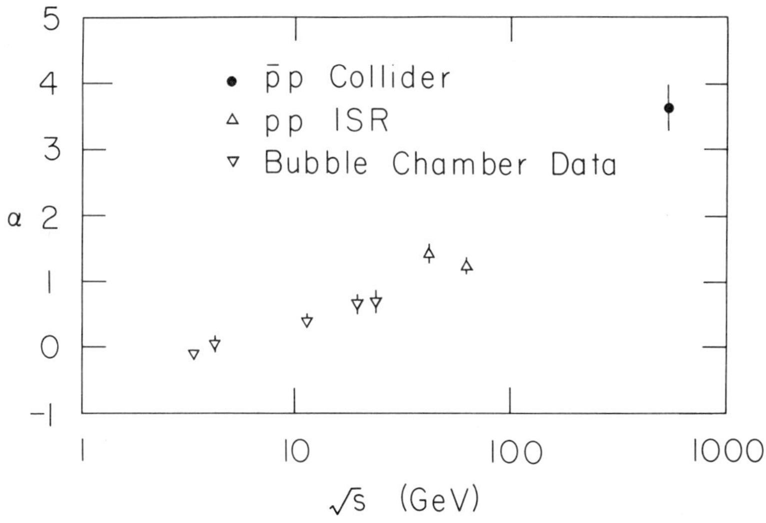

FIGURE 8. Energy variation of neutral pion: charged particle multiplicity correlations.

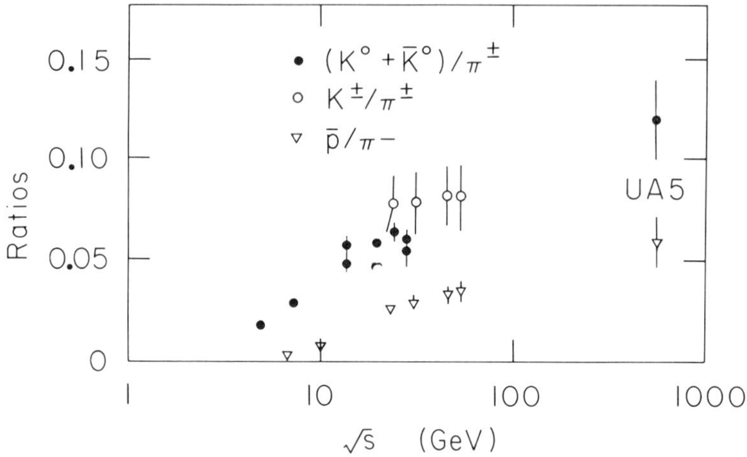

FIGURE 9. Energy variation of the heavy particle fractions for pp collisions.

The UA5 results can be represented by the expression, $\langle n_\gamma \rangle = 8 \pm 3 + (0.9 \pm 0.08)$ n_{ch}, which, assuming all γ rays come from π^0 decay, gives $\langle n_{\pi^0} \rangle = 16 \pm 6 + (3.6 \pm 0.32)\ n_-$.

The latter assumption may not be as valid as integrating up to $|\eta| = 5$ since the UA5 group measure $\langle n_\gamma \rangle = 34 \pm 2$ as compared to $\langle n_{ch} \rangle = 26.5 \pm 1$ in the same pseudorapidity range. If the charged particles are corrected for the heavy particle content, then $\langle n^\pi_{ch} \rangle = 22.5 \pm 1$, which, when compared to $\langle n_\gamma \rangle$, suggests strong η^0 production since, for pions, the relationship of $\langle n^\pi_{ch} \rangle = 2\ \langle n^\pi_0 \rangle$ is expected to hold. An η/π^0 ratio of ~ 0.32 is required in this interpretation.

HEAVY PARTICLE FRACTIONS

The yields of K mesons and baryons continue to increase up to the highest energies. FIGURE 9 shows the mean multiplicities as a function of $\sqrt{s}$ for pp collisions. The values are still increasing even at $\sqrt{s} = 540$ GeV. The average event[22] at this highest energy $\bar{p}p$ collision contains 2.5 $K^\pm$, 2.7 $(K^0,\bar{K}^0)$, 1.5 $p^\pm$, 0.25 $(\Sigma^\pm\Sigma^\pm)$, 0.5 $(\Lambda,\bar{\Lambda})$, and $\sim$ 0.2 Ξ^- particles, with a total of 26.5 charged particles. This may be contrasted to e^+e^- annihilation,[24] where at $\sqrt{s} \sim 34$ GeV, there are a total of 11.7 charged particles (removing the $K^0 \rightarrow \pi^+\pi^-$ and $\Lambda \rightarrow p\pi^-$ decays), including 10.3 $\pi^\pm$, 2.0 $K^\pm$, 1.6 $(K^0\bar{K}^0)$, 0.8 $p^\pm$, 0.28 $(\Lambda,\bar{\Lambda})$, and 0.026 Ξ^- particles. Clearly, the K meson fraction is higher for e^+e^- annihilation and the baryon fraction is about the same. There is, of course, no reason to make the comparison for these particular energies since, in hadronic collisions, there is a strong dependence of the particle fractions on $\sqrt{s}$.

MULTIPLICITY DISTRIBUTIONS

Comparison of e^+e^- *and pp Data*

The cross sections, as a function of prong number for e^+e^- annihilation at $\sqrt{s} = 29$ GeV[25] and pp scattering at 205 GeV/c[26] ($\sqrt{s}$= 19.6 GeV), are shown in FIGURE 10.

Even though $\langle n_{ch} \rangle = 7.68 \pm 0.07$ for the pp case, one event with 26 prongs was observed corresponding to a fractional cross section of $\sim 1.3 \times 10^{-4}$; a similar rate to that observed in the e^+e^- annihilation data of FIGURE 10, but where $\langle n_{ch} \rangle$ is much higher ($12.62 \pm 0.03 \pm 0.5$).[27] This difference remains if the comparison is made with the nondiffractive pp data for which $\langle n_{ch} \rangle = 8.56 \pm 0.12$. It is clear that when expressed as a fraction of the mean values, the e^+e^- data are significantly narrower than the hadronic data, even after accounting for the diffractive component in the latter.

The scaling in the mean first suggested by Koba, Nielsen, and Olesen[4] on the basis of Feynman scaling and a logarithmic multiplicity growth, both of which are now known to be violated, holds well for the different data sets. If $z = n_{ch}/\langle n_{ch} \rangle$, then KNO scaling says that

$$\psi(z) = \langle n_{ch} \rangle \frac{\sigma_n}{\Sigma \sigma_n}, \tag{3}$$

where $\psi(z)$ is a universal function for a given type of reaction and σ_n is the cross section for production of the final state containing n charged particles.

FIGURE 11 shows a compilation of e^+e^- results.[8,25] Similar data for a sample of

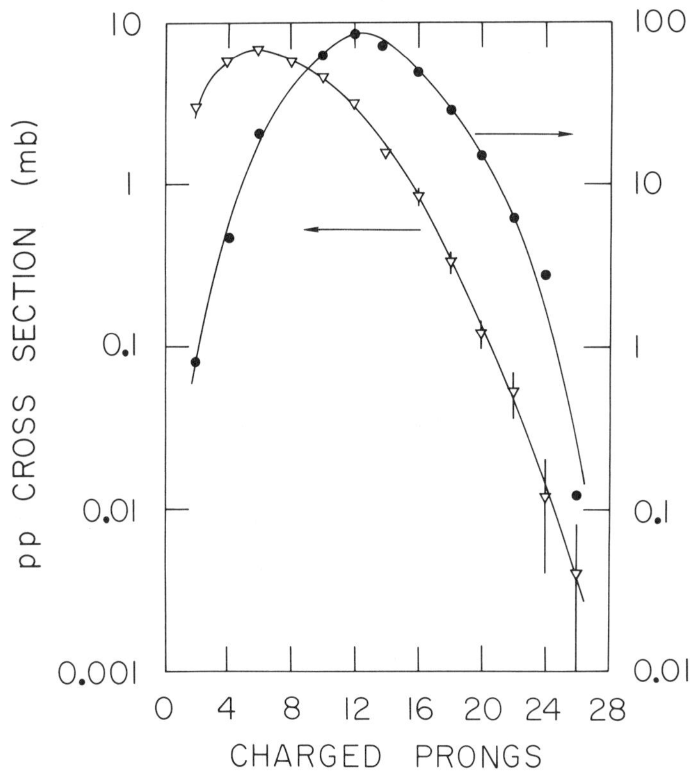

FIGURE 10. Cross sections for charged particle production in e^+e^- annihilation at $\sqrt{s} = 29$ GeV (● scale on right) and pp collisions at $\sqrt{s} = 19.6$ GeV (▽ scale on left).

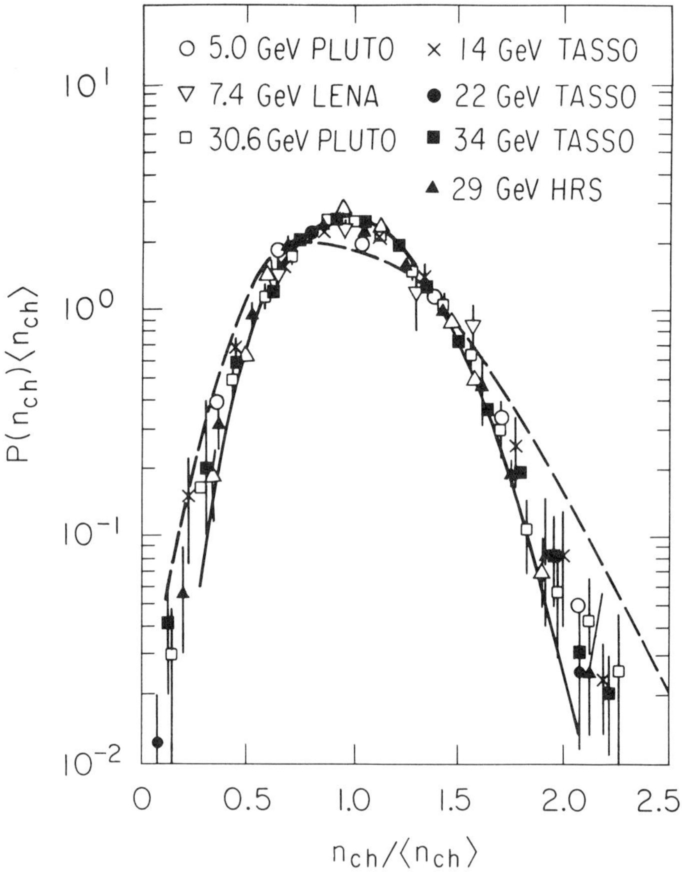

FIGURE 11. KNO scaling distribution for e^+e^- annihilation. The full line shows the prediction of the Webber model and the dashed line shows the fit to nondiffractive pp data.

nondiffractive pp interactions from Fermilab[26] and the ISR[13] are shown in FIGURE 12. Exact KNO scaling requires constant values for the moments of the distributions,

$$C_q = \frac{\langle n_{ch}^q \rangle}{\langle n_{ch} \rangle^q}, \tag{4}$$

and, in particular for $\langle n_{ch} \rangle / D$, where D is the dispersion of the distribution,

$$D^2 = \langle n_{ch}^2 \rangle - \langle n_{ch} \rangle^2. \tag{5}$$

As pointed out by Breakstone *et al.*,[13] it is only for the nondiffractive data that $\langle n_{ch} \rangle / D$ is constant, being fit by the expression,

$$D = (0.439 \pm 0.007)\, \langle n_{ch} \rangle - (0.02 \pm 0.06). \tag{6}$$

The nondiffractive pp data of FIGURE 12 are reasonably represented by the expression,[28]

$$\psi(z) = e^{-3.5z}(-2.72\,z + 75.9\,z^3 - 16.35\,z^5 + 0.89\,z^7), \tag{7}$$

which is shown by the line. The same function is also shown dashed on the e^+e^- data of FIGURE 11. It is clear that the nondiffractive pp data are wider than the e^+e^- results.

Another way to exhibit the differences between the hadronic and the e^+e^- data is in terms of the $\langle n_{ch}\rangle/D$ ratios. FIGURE 13 shows a compilation of such results: the mean

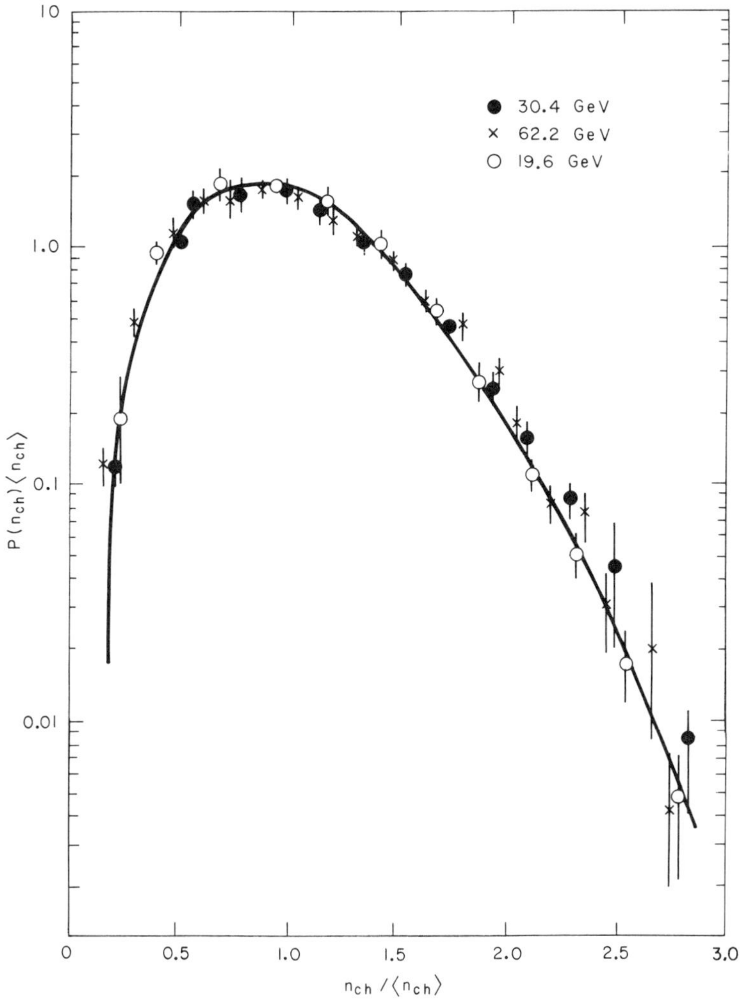

FIGURE 12. KNO scaling distribution for nondiffractive pp scattering observed in the Fermilab bubble chamber (19.6 GeV) and at the ISR (30.4 GeV and $\sqrt{s} = 62.2$ GeV).

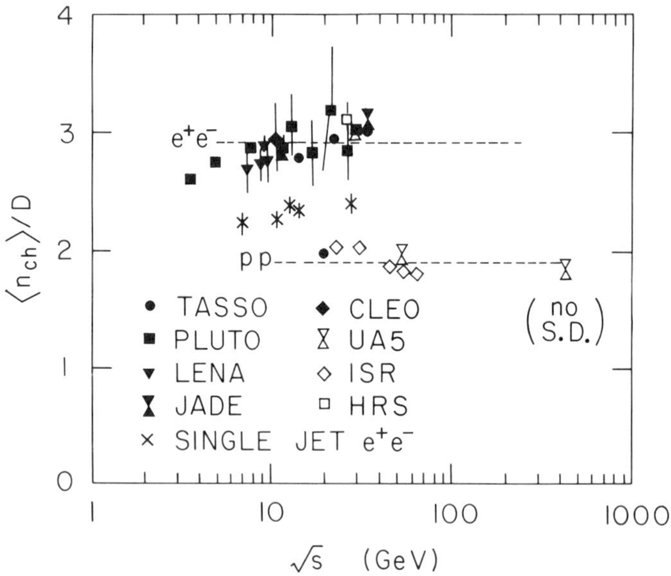

FIGURE 13. $\langle n_{ch} \rangle / D$ ratios for e^+e^- annihilation and pp interactions. The data points (×) between the two lines show the single jet e^+e^- $\langle n_{ch} \rangle / D$ ratios.

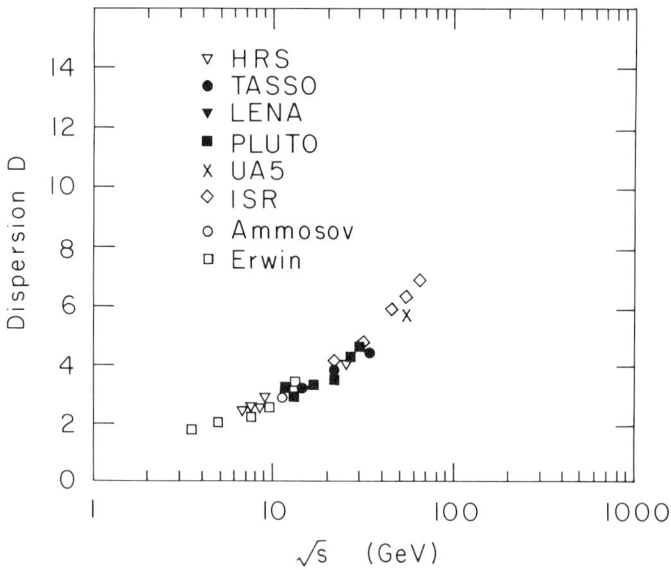

FIGURE 14. Energy variation of the dispersion, D, for e^+e^- annihilation and pp inelastic scattering.

e^+e^- values are near 3, whereas the hadronic data are somewhat below 2. The $\langle n_{ch} \rangle / D$ ratios for the single jet e^+e^- data, shown by the crosses, lie below the values for the whole event. Even though these ratios are different, the dispersion measurements themselves[8] make a continuous curve as shown in FIGURE 14. This remarkable result is presumably accidental.

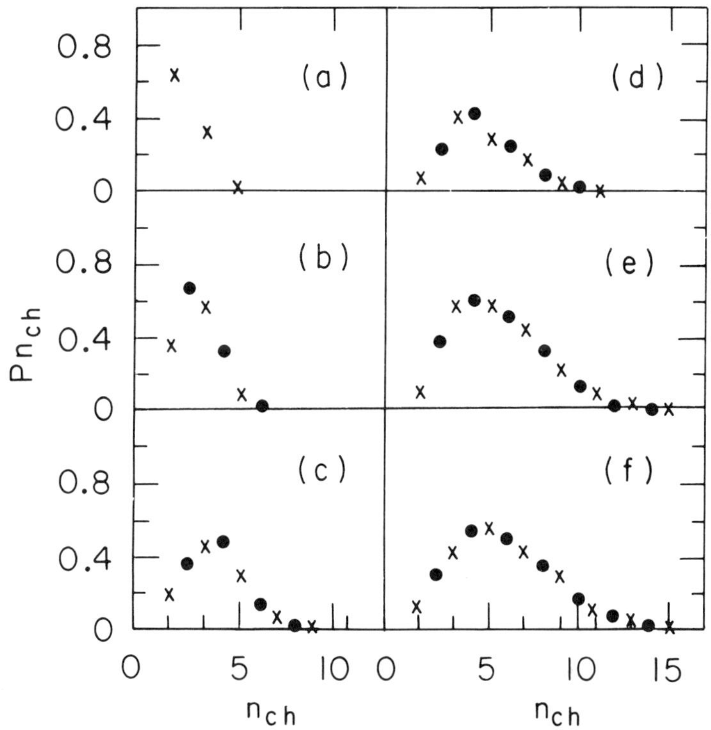

FIGURE 15. Multiplicity distribution for pp → pX (×) compared to pp → X (●) for different selections of the mass (M) of the system, X. (a) $0 < M^2 < 5$ GeV^2, (b) $5 < M^2 < 10$ GeV^2 and $p_{lab} = 5.5$ GeV/c, (c) $0 < M^2 < 25$ GeV^2 and $p_{lab} = 13$ GeV/c, (d) $25 < M^2 < 50$ GeV^2 and $p_{lab} =$ 26 GeV/c, (e) $50 < M^2 < 100$ GeV^2 and $p_{lab} = 50$ GeV/c, and (f) $100 < M^2 < 150$ GeV^2 and $p_{lab} =$ 69 GeV/c.

Diffractive Data

We have discussed the differences between the results from the nondiffractive hadronic collisions and the e^+e^- data. We now turn to a consideration of the diffractive data set itself. There are several difficulties in discussing diffractive multiplicities: the definition of a diffractive event is not uniquely specified so that different groups use different selection criteria, the multiplicities are low since the effective mass of the diffractive cluster is low, resonant N^* states (or, in the case of pion interactions, the A_1

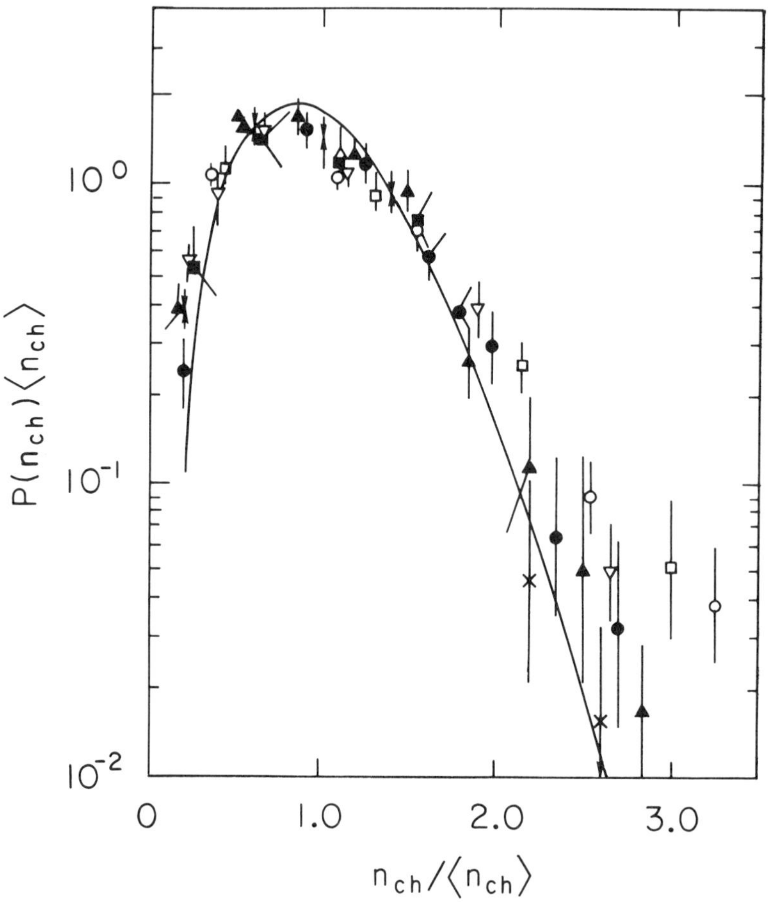

FIGURE 16. KNO scaling distribution for diffractive pp scattering observed in the Fermilab bubble chamber.

resonance) play a major role, and finally, the effective c.m. energy has to be measured from the events themselves.

One of the striking early results from Fermilab[29] was the observation, shown in FIGURE 15, that the multiplicity distributions in the reaction, pp → pX, at 205 GeV/c followed those of the inclusive pp → X reaction when compared at the same mass of the system, X. Of course, since the charge of X is different in the two cases, a comparison using n_- would not provide such good agreement. It is not clear which is the more fundamental variable.

A more detailed comparison is given in FIGURE 16. The data points show the KNO scaling distributions for diffractive events from 102, 205, and 405 GeV/c pp interactions.[30] Although the errors are large, the data show a general accord with a scaling expectation. The line that shows the fit to the nondiffractive data [equation (7)] indicates that the diffractive events, selected in the bubble chamber with a low effective mass at the target vertex and with a rapidity gap between this system and the beam proton,[31] have a somewhat broader and flatter KNO distribution than do the nondiffractive inclusive data.

An extensive series of more recent measurements have been made by the Rockefeller group[32] using incident $\pi^{\pm}$, $K^{\pm}$, and $p^{\pm}$ beams of 100 GeV/c and 200 GeV/c. The recoil protons in the range, $0.025 < |t| < 0.095$ (GeV/c)^2, were detected in a vertex spectrometer surrounding a gas target. The multiplicity of the forward system was measured by the pulse heights in two scintillation counters. The results, shown in FIGURE 17, indicate that the mean multiplicity values are independent of the projectile when compared at the same values of the available energy. When expressed in the KNO form, this diffractive data is more symmetric and, as seen in FIGURE 18, can be reasonably represented by the Gaussian function:

$$\frac{\sigma_n}{\Sigma\sigma_n} = \frac{2}{\sqrt{2\pi}D} e^{\frac{-(n-n_0)^2}{2D^2}} \quad (8)$$

with $n_0 = 2M^{1/2}$ and $n_0/D = 2$ or

$$n_0 \frac{\sigma_n}{\Sigma\sigma_n} = \frac{\sqrt{8}}{\pi} e^{2(1-n/n_0)^2}. \quad (9)$$

These results, then, differ from the earlier bubble chamber data. The source of this difference could lie in the very different techniques used to define a diffractive event and to measure the multiplicity.

Goulianos[33] further shows that the KNO distribution for the inclusive pp system

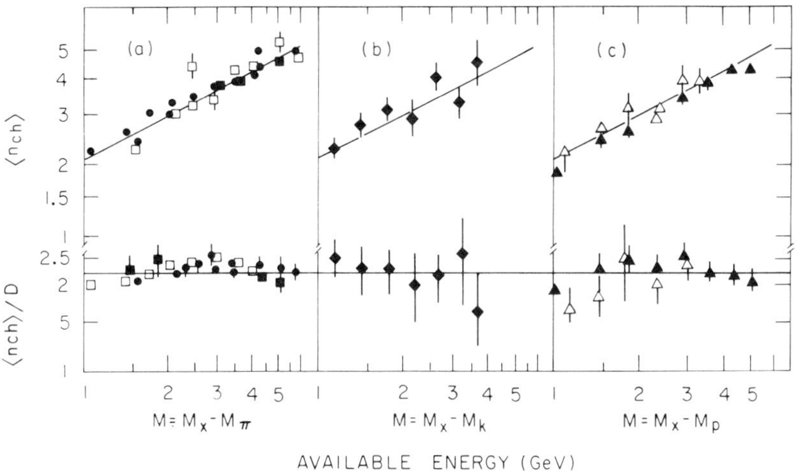

FIGURE 17. Variation of $\langle n_{ch}\rangle$ and $\langle n_{ch}\rangle/D$ with available energy for (a) πp, (b) Kp, and (c) pp collisions.

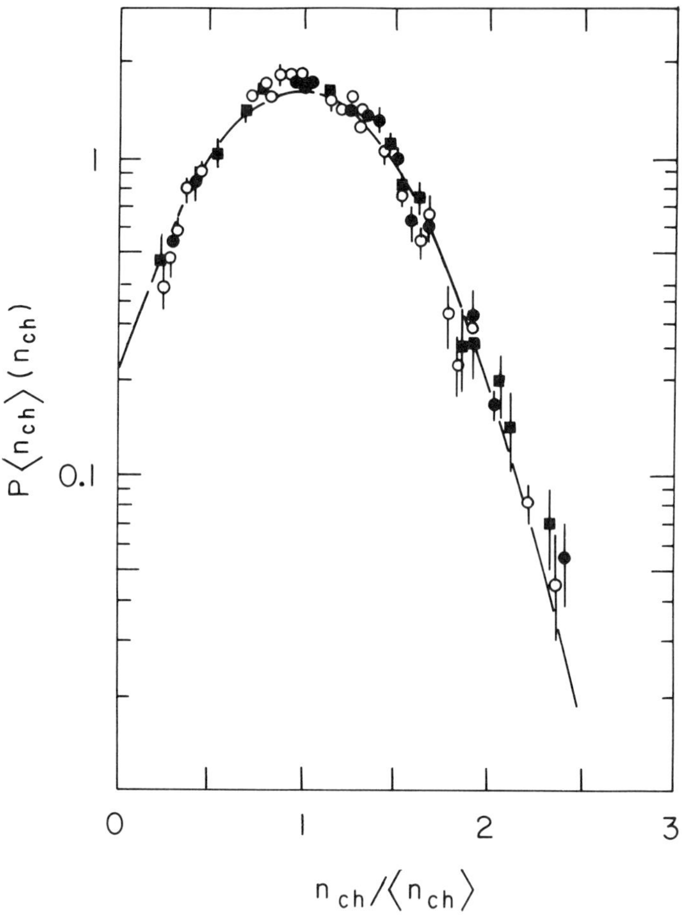

FIGURE 18. Comparison of diffractive KNO distribution with Gaussian curve.

can be represented by a sum of the nondiffractive and diffractive distributions, each of which is represented by a Gaussian distribution.

Neutrino Data

Multiplicity measurements in lepton-nucleon scattering have been primarily done for neutrino scattering observed in the large bubble chambers. The events are very clean, but the experiments suffer from low statistics and a low available energy, and also have some difficulty in assigning the correct energy to each event. Nevertheless, several groups have reported results from both $\bar{\nu}p$ and νp scattering. The mean charged hadron multiplicity from three $\bar{\nu}$p experiments[34] is shown in FIGURE 19. The curve represents the fit done by Wroblewski[28] to nondiffractive π^-p interactions.

The KNO distributions for the νp and $\bar{\nu}$p data from the high statistics BEBC experiment[35] are shown in FIGURE 20. Since the experiments are done with a wideband beam, the scaling can be exhibited by data from a single experiment. The $\bar{\nu}$p distribution is broader than measured for νp. This difference has been interpreted by Zieminska[36] to mean that one should only use the produced particles in such comparison. Since the νp ($\bar{\nu}$p) experiments lead to a final hadronic state of charge 2(0) and the total number of particles is small, this effect is strong for the neutrino data. Higher energy data, which would settle this issue, will not now become available. The curves shown in FIGURE 20 represent the modified Poisson distribution,[37]

$$\psi(z) = \frac{C^{(CZ+1)}}{\Gamma(CZ+1)} e^{-C}, \tag{10}$$

with $C = 8.64 \pm 0.13$ for νp and $C = 6.25 \pm 0.18$ for $\bar{\nu}$p interactions. These low energy neutrino distributions are sharper and narrower than either the hadronic data or the e^+e^- results of FIGURE 11.

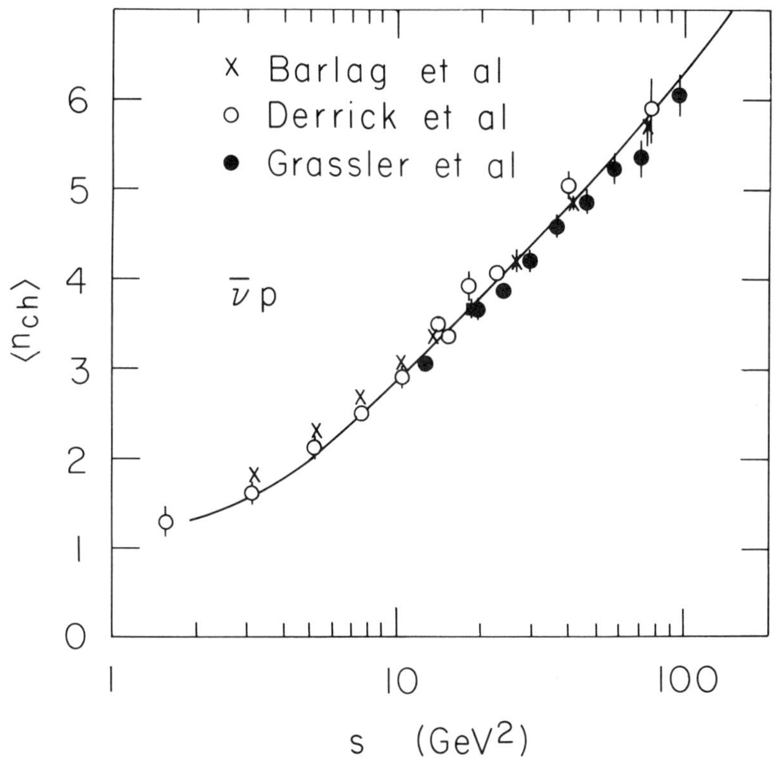

FIGURE 19. Energy variation of the mean charged multiplicity for $\bar{\nu}$p scattering.

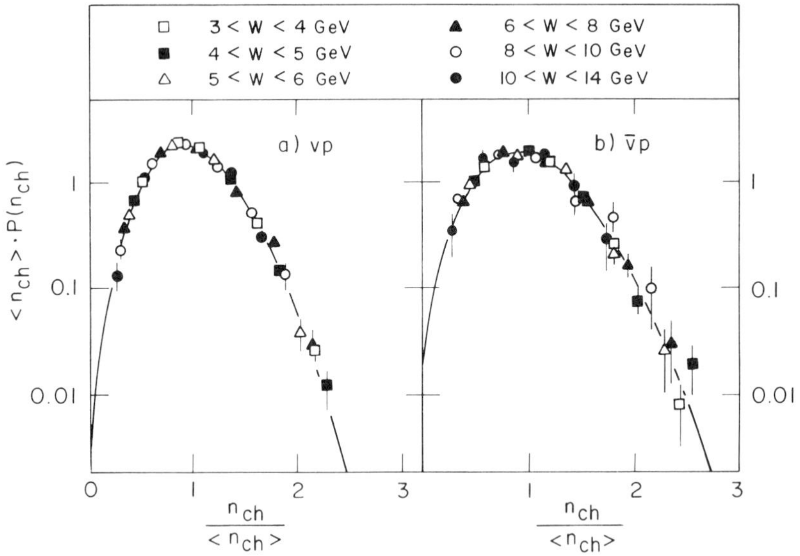

FIGURE 20. KNO scaling distributions for νp and $\bar{\nu}$p scattering as measured in BEBC.

Collider $\bar{p}p$ *Data*

As we have seen, although there are some differences in the shape of the KNO distributions for the different reactions, within a given reaction type, the data scale well. The first results from $\bar{p}p$ scattering at $\sqrt{s} = 540$ GeV, which showed a dramatic violation of the scaling that holds up to 62.5 GeV, the highest energy of the ISR, was unexpected. This comparison is shown in the compilation of the moments of the multiplicity distribution of FIGURE 21. The higher moments, C_4 and C_5, increase by about a factor of two between 60 GeV and 540 GeV.

In addition, the multiplicity distributions for selected rapidity ranges differ with the central region, $|\eta| < 0.5$, showing almost an exponential falloff with z.[38]

FORWARD-BACKWARD (F:B) MULTIPLICITY CORRELATIONS

Another aspect of the data in which hadronic collisions and e^+e^- annihilation show different behavior is in the long-range correlations that give rise to correlated multiplicities in the two hemispheres.

FIGURE 22, which shows e^+e^- data at $\sqrt{s} = 29$ GeV,[39] indicates that strong F:B multiplicity correlations are absent in such interactions. This may be contrasted with the strong correlation observed at the ISR and particularly at the CERN collider.[40] The later can be expressed as:

$$n_F = a + b\,\langle n_B \rangle. \tag{11}$$

The energy variation of b is given in FIGURE 23.

These data have been interpreted in terms of the independent emission of clusters,[40] which is consistent with the known short-range rapidity correlations. However, e^+e^- data show very similar rapidity correlations, yet no F:B multiplicity correlations. The hadronic data also have been interpreted in terms of the dual mode.[41]

INTERPRETATIONS

What do all these results mean and, in particular, is there any simple underlying physics? The first comment is that a scaling distribution results from a tree branching process, such as is indicated in FIGURE 24, and which represents the underlying QCD parton level process at least for e^+e^- annihilation.[42] Such a branching process is implemented in the Monte Carlo simulation of e^+e^- annihilation written by Webber.[43] Although such a calculation yields the distribution of partons, it is still necessary to account for the final hadronization by some reasonable prescription. There is no

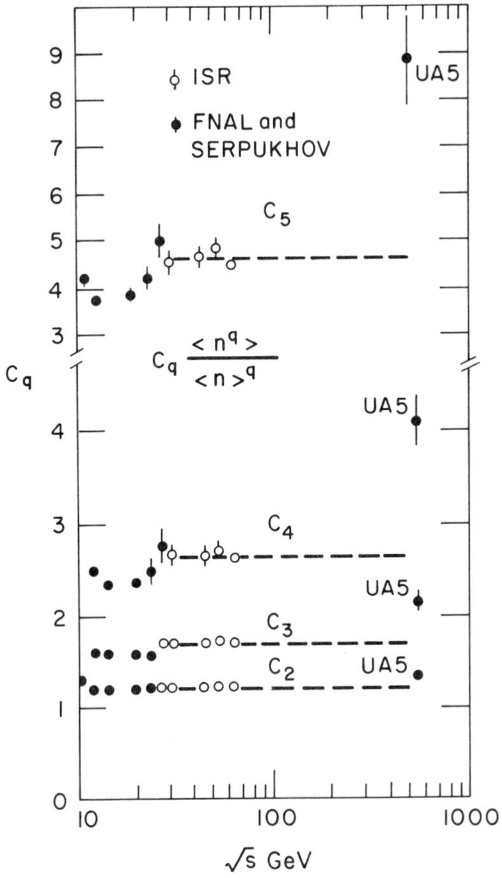

FIGURE 21. Energy variation of the moments of the KNO distribution for pp scattering.

fundamental understanding of how to handle this nonperturbative phase of the parton-hadron transition. With this caveat, the prediction of the KNO distribution for e^+e^- annihilation, predicted by this model, is shown by the full line in FIGURE 11. It is a little narrower than the data, but agrees well with the overall shape of the results. Since the final hadronization in this model is done via a simple two-body decay of the color neutral clusters, it would obviously be possible to adjust the parametrization to improve the agreement. The calculation also agrees quite well with the energy variation of the yields of the stable particles, π, K, and p, as seen in FIGURE 25.

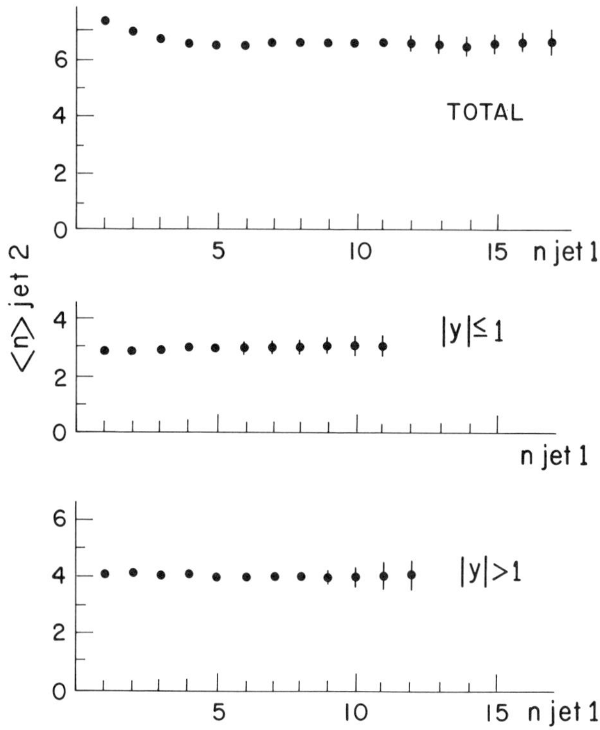

FIGURE 22. Forward (F): Backward (B) multiplicity correlations for e^+e^- annihilation at $\sqrt{s}$ = 29 GeV.

Since we do not yet have a simulation of hadronic interactions,[44] there is no agreed explanation for the broadening of the multiplicity distribution observed at the collider. It seems that a new process may have set in and several possibilities have been discussed[45]: The dual parton model[46] provides one explanation. As the energy is increased, diagrams with additional chains, as shown in FIGURE 26, start to contribute a higher multiplicity component. The results of one explicit calculation[47] are given in FIGURE 27.

Geometrical models have also been discussed. The basic idea is that the central

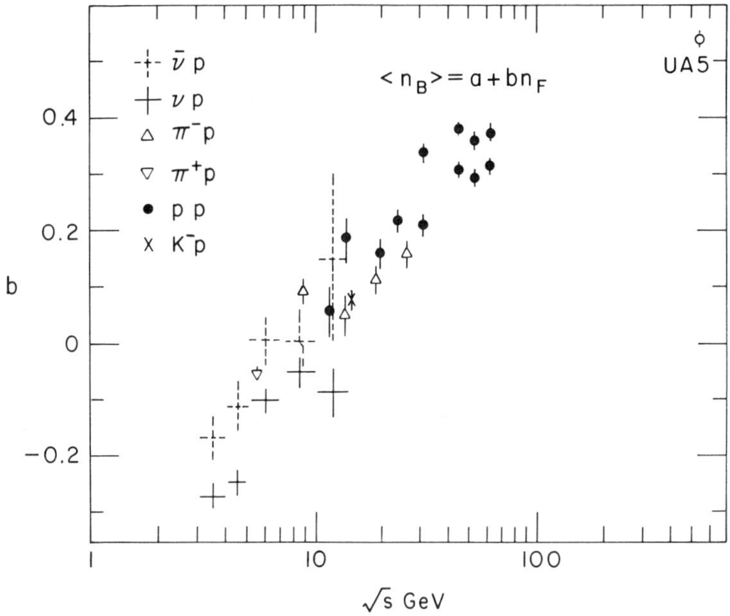

FIGURE 23. Energy variation of F:B multiplicity correlations in pp scattering.

collisions have a different dependence of $\langle n_{ch} \rangle$ on energy than do the peripheral collisions. The physical picture is attractive and, indeed, the idea of the violent collisions proceeding in a two-step process—first, the production of one or more blobs of excited hadronic matter called fireballs, followed by the fireball decay—goes back to the earliest days of cosmic ray studies. Such ideas, though, are difficult to quantify and cast into a theory with predictive power. However, it is possible to explain the data *a posteriori*.[48] The broader KNO distribution observed for the central rapidity region is consistent with these ideas.

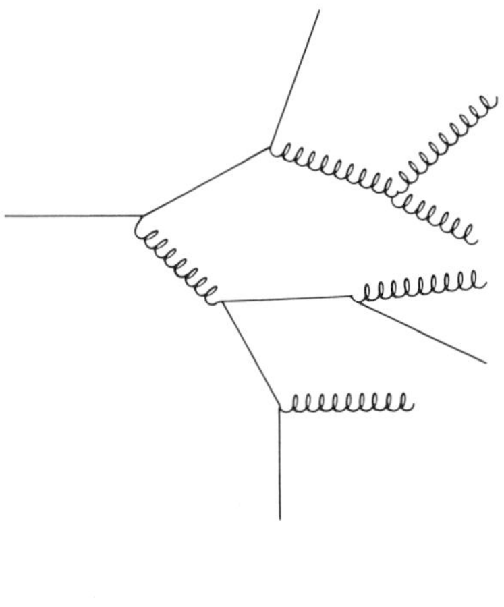

FIGURE 24. Tree branching process.

A somewhat related idea arises from the observation that the multijet events that result from hard parton-parton collisions constitute a rapidly increasing fraction of the total cross section as the energy is increased. The events with the highest multiplicities receive contributions from the high p_T jets, as well as the beam fragmentations.[49] The striking observation that $\langle p_T \rangle$ increases with multiplicity is in accord with this idea.

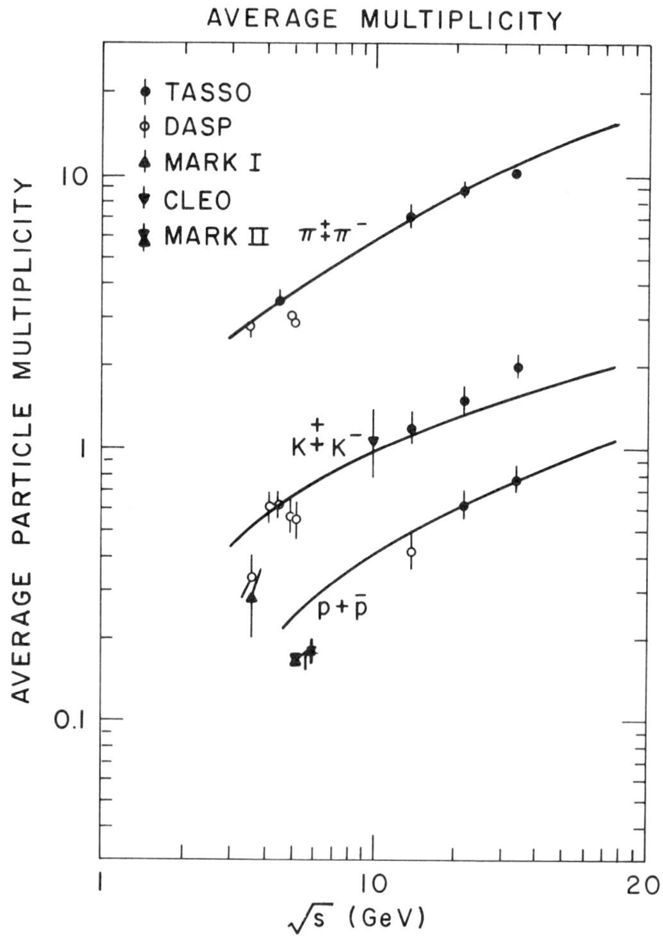

FIGURE 25. Energy variation of particle yields in e^+e^- annihilation as fit by the Webber model.

Another possibility that has been discussed is that the two components are, respectively, quark and gluon initiated, with the gluon multiplicity being 9/4 of the quark and the gluon/quark fraction growing with energy.[50] Carruthers and collaborators[51] have discussed a different approach. They point out that KNO scaling is a rather general result in physics, ranging from photoelectron emission in quantum optics to the

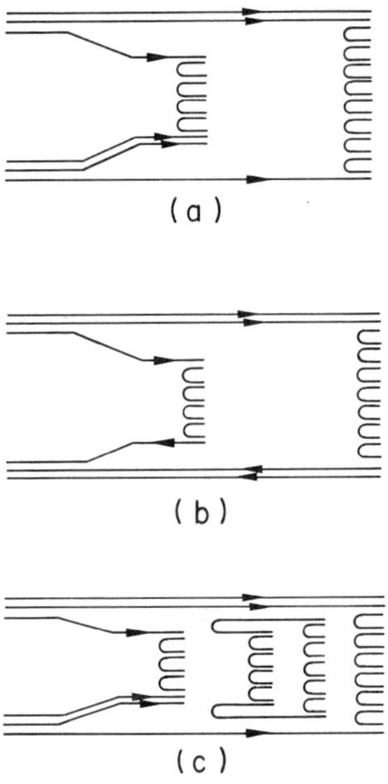

FIGURE 26. Dual model diagrams.

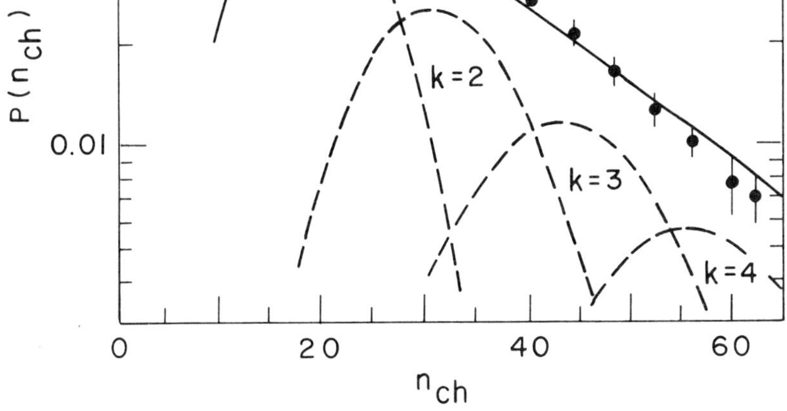

FIGURE 27. Dual model calculation compared to KNO distribution at $\sqrt{s} = 540$ GeV.

distribution of the number of galaxies in clusters. They speculate that the differences in the KNO distributions may reflect different underlying mechanisms. A generalization of the Bose-Einstein distribution results from emission by k identical sources:

$$\mathrm{P}_n^k = \frac{(n+k-1)!}{n!\,(k-1)!} \cdot \frac{\left(\dfrac{\langle n\rangle}{k}\right)^n}{\left(1+\dfrac{\langle n\rangle}{k}\right)^{n+k}}. \tag{12}$$

For n and $\langle n\rangle$, large compared to k, this reduces to

$$\psi(z) = \frac{k^k}{(k-1)!}\, z^{k-1}\, e^{-kz}, \tag{13}$$

which fits the hadronic KNO distributions well if $k \sim$ 3–4. However, to obtain a narrower distribution, k must be increased, which is difficult to understand intuitively. However, the single jet e^+e^- distributions are broader than those measured for the complete event, as seen from the $\langle n_{ch}\rangle/D$ ratios plotted in FIGURE 13. The whole e^+e^- event is then a superposition of two independent, but equal, quark emitters.

CONCLUSIONS

Although an abundance of beautiful data exist on multiplicities and related subjects, no one has illuminated the real underlying physics. The subject is very active and more detailed information will become available both at current energies, as well as from higher energy e^+e^- and $\mathrm{p\bar{p}}$ collisions.

ACKNOWLEDGMENTS

I have benefited from multiple discussions with colleagues over the past several years. Particular thanks are due to P. Carruthers, B. Durand, S. Ellis, D. Goulianos, and A. Wroblewski.

NOTES AND REFERENCES

1. FOA, L. 1975. Phys. Rep. **22:** 1; GIACOMELLI, G. & M. JACOB. 1979. Phys. Rep. **55:** 1.
2. FRAZER, W. R. *et al.* 1972. Rev. Mod. Phys. **44:** 284.
3. WHITMORE, J. 1974. Phys. Rep. **10C:** 273; 1976. Phys. Rep. **27C:** 187.
4. KOBA, Z., H. B. NIELSEN & P. OLESEN. 1972. Nucl. Phys. **B40:** 317.
5. DERRICK, M. *et al.* 1978. Phys. Rev. **D17:** 1.
6. BERGER, CH. *et al.* 1980. Phys. Lett. **95B:** 313.
7. ALPGARD, K. *et al.* 1983. Phys. Lett. **121B:** 209.
8. ALTHOFF, M. *et al.* 1984. Z. Phys. **C22:** 307.
9. BENDER, D. *et al.* 1985. Phys. Rev. **D31:** 1.
10. FERMI, E. 1950. Prog. Theor. Phys. **5:** 570.
11. BASSETTO, A., M. CIAFALONI & G. MARCHESINI. 1978. Phys. Lett. **83B:** 207.

12. THOME, W. *et al.* 1977. Nucl. Phys. **B129:** 365.
13. BREAKSTONE, A. *et al.* 1984. Phys. Rev. **D30:** 528.
14. ALNER, G. J. *et al.* 1984. Phys. Lett. **138B:** 305.
15. Fits to ISR data (AMBROSIO, M. *et al.* 1983. Proceedings of the Third Topical Workshop on Proton-Antiproton Collider Physics, Rome) give $\sigma(\bar{p}p) - \sigma(pp) \simeq 60\ s^{-0.56}$ mb, which corresponds to 0.05 mb at $\sqrt{s} = 540$ GeV.
16. BASILE, M. *et al.* 1981. Nuovo Cimento **65A:** 414.
17. SIMS, W. H. *et al.* 1972. Nucl. Phys. **B41:** 317; BUMAZSHNOV, V. A. *et al.* 1974. Phys. Lett. **50B:** 283; BROMBERG, C. *et al.* 1976. Nucl. Phys. **B107:** 82; KAFKA, T. *et al.* 1977. Phys. Rev. **D16:** 1261.
18. ALPGARD, K. *et al.* 1981. Phys. Lett. **107B:** 310; **107B:** 315; ARNISON, G. *et al.* 1983. Phys. Lett. **123B:** 108.
19. BREAKSTONE, A. *et al.* 1983. Phys. Lett. **132B:** 458.
20. SAKUDA, M. *et al.* 1984. Phys. Lett. **152B:** 399.
21. DAO, F. T. & J. WHITMORE. 1973. Phys. Lett. **46B:** 252.
22. RUSHBROOKE, J. G. 1984. *In* Proceedings of the Workshop on $\bar{p}p$ Options for the SSC, p. 176.
23. ALPGARD, K. *et al.* 1982. Phys. Lett. **115B:** 71.
24. WU, S. L. 1984. Phys. Rep. **107:** 60.
25. DERRICK, M. (HRS COLLABORATION). 1984. Proceedings of the DPF Meeting at Sante Fe.
26. BARISH, S. J. *et al.* 1974. Phys. Rev. **D9:** 2689.
27. This value represents an analysis of a larger data set as compared to $\langle n_{ch} \rangle = 13.1 \pm 0.05 \pm 0.6$ given in reference 9.
28. WROBLEWSKI, A. 1983. *In* Proceedings of the XIV Symposium on Multiparticle Dynamics, p. 573.
29. WHITMORE, J. & M. DERRICK. 1974. Phys. Lett. **50B:** 280.
30. BARSHAY, S. *et al.* 1974. Phys. Rev. Lett. **32:** 1390.
31. BARISH, S. *et al.* 1974. Phy. Rev. **D9:** 1171; DERRICK, M. *et al.* 1974. Phys. Rev. **D9:** 1215.
32. COOL, R. L. *et al.* 1982. Phys. Rev. Lett. **48:** 1451.
33. GOULIANOS, K. 1983. Phys. Rep. **101:** 169. A reasonable fit to the e^+e^- data using the Gaussian form of equations (8) and (9) requires $\langle n \rangle / D \sim 2\sqrt{2}$. A Monte Carlo study shows that the non-Gaussian shape of the Breakstone data (reference 13) may be due to an incomplete removal of events coming from the single and double diffraction (GOULIANOS, K. Private communication).
34. BARLAG, S. *et al.* 1982. Z. Phys. **C11:** 283; DERRICK, M. *et al.* 1982. Phys. Rev. **D25:** 624; GRASSLER, M. *et al.* 1983. Nucl. Phys. **B223:** 269.
35. SCHMITZ, N. 1983. Proceedings of the XIV Symposium on Multiparticle Dynamics, p. 68.
36. ZIEMINSKA, D. 1983. Phys. Rev. **D27:** 502.
37. LEVY, D. 1973. Nucl. Phys. **B59:** 583.
38. ALNER, G. J. *et al.* 1985. Phys. Lett. **160B:** 193.
39. VALDATA NAPPI, M. 1983. Proceedings of the XIV Symposium on Multiparticle Dynamics, p. 75.
40. UHLIG, S. *et al.* 1978. Nucl. Phys. **B132:** 15; ALPGARD, K. *et al.* 1983. Phys. Lett. **123B:** 361.
41. FIALKOWSKI, K. & A. KOTANSKI. 1982. Phys. Lett. **115B:** 425.
42. ELLIS, S. 1983. Proceedings of the 11th SLAC Summer Institute, SLAC report #267; DURAND, B. and S. D. ELLIS. 1984. Proceedings of the 1984 Summer Study on the Design and Utilization of the Superconducting Super Collider, Snowmass, Colorado.
43. WEBBER, B. R. 1984. Nucl. Phys. **B238:** 492; MARCHESINI, G. & B. R. WEBBER. 1984. Nucl. Phys. **B238:** 1.
44. A review of such attempts is given by DERRICK, M. & T. GOTTSCHALK. 1984. Report of the Snowmass hadronization group. *In* Proceedings of the 1984 Summer Study on the Design and Utilization of the Superconducting Super Collider, Snowmass, Colorado, p. 49.
45. PANCHIERI, G. 1984. Proceedings of the XV Symposium on Multiparticle Dynamics, Lund, p. 108.

46. CAPELLA, A. & J. TRAN THAN VAN. 1982. Phys. Lett. **114B:** 450; KAIDALEV, A. B. & K. A. TER-MARTIROSYAN. 1982. Phys. Lett. **117B:** 247.
47. KANKI, T. 1984. Nucl. Phys. **B243:** 44.
48. LIAN-SOU, L. & M. TA-CHUNG. 1983. Phys. Rev. **D27:** 2640.
49. PANCHIERI, G. in a talk given at the Annual Meeting of the Division of Particles and Fields of the American Physical Society, 1984, Sante Fe, New Mexico, showed that the UA1 events containing high p_T jets have a narrow KNO distribution similar to that observed in e^+e^- annihilation.
50. PANCHIERI, G., Y. SRIVASTAVA & M. PALLOTTA. 1984. Harvard preprint HVTP 84/A081.
51. CARRUTHERS, P. & C. C. SHIH. 1983. Phys. Lett. **127B:** 242; 1984. Phys. Lett. **137B:** 425.

Total Cross Sections at Ultrahigh Energy from Cosmic Ray Experiments[a]

G. B. YODH

Department of Physics and Astronomy
University of Maryland
College Park, Maryland 20742

INTRODUCTION

Cosmic rays incident upon the earth provide us with a beam of high energy particles extending over twelve decades of energy, from 10 GeV to 10^{10} GeV. The energy dependence of interactions of hadrons over a wide energy range can be investigated using this beam.

Over ten years ago, rise of the proton-proton total cross section from 20 GeV to 540 GeV in the center of mass was deduced from an analysis of cosmic ray data on the propagation of cosmic rays in the atmosphere.[1-3] The predicted increase was subsequently observed for proton-proton interactions at the ISR[4,5] in the 20 to 63 GeV range of energies in the center of mass. In the last few years, the continuation of this violation of scaling has been measured in antiproton-proton interaction at the SPS-$\bar{p}p$ collider[6,7] ($\sqrt{s}$ = 540 GeV) and in ISR-$\bar{p}p$ experiments at 20 to 63 GeV.[8,9]

At energies beyond 10^6 GeV, the smallness of the cosmic ray flux necessitates their observation using air shower methods rather than by observing them directly in a detector to study individual interactions. Hence, it is important to summarize what is known about the mix of primary cosmic rays up to the highest energies.

The cosmic ray beam consists of nuclei of various elemental species. The major components are protons, alpha particles, carbon-nitrogen-oxygen (CNO) groups of elements, the elements in the Mg–Si range (medium heavies, MH), and the iron group (heavies, H). The integral all-particle spectrum of cosmic rays, in units of particles per m^2, sterad second, as a function of energy per nucleus of the primary, is shown in FIGURE 1. The fluxes of the various components up to the highest energies where they have been directly measured are shown in FIGURE 2. In this figure, the differential spectra have been multiplied by the energy per nucleus raised to the 2.5 power for convenience of display. Also shown in this figure are two different models for the cosmic ray energy spectra that lead to different mixes of elemental species in the 10^7–10^8 GeV region, while adequately fitting the direct measurements.[10,11] We note that the interpretation of air shower data to derive the behavior of high energy interactions must be done with care if one wants to obtain information about proton-air cross sections. The air shower data must be selected in a manner that preferentially studies showers initiated by proton primaries. In addition, careful Monte Carlo studies must be made to estimate the effects of contamination of this proton enhanced sample

[a]This work has been supported in part by National Science Foundation grant no. PHY 82-07425.

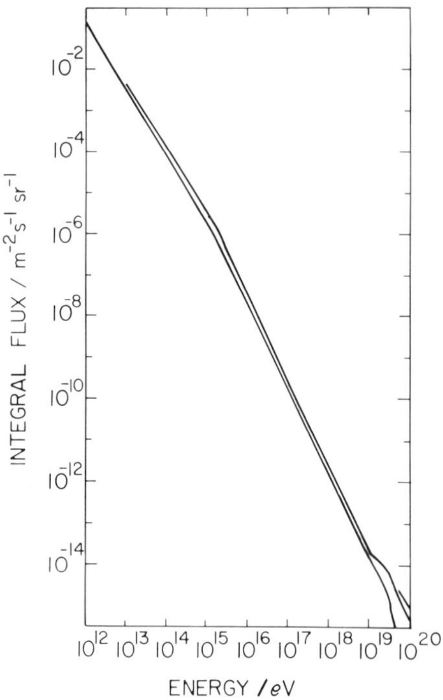

FIGURE 1. Composite integral spectrum of all particles of primary cosmic rays. The curve beyond 10^6 GeV is derived from air shower observations. The abscissa is energy per nucleus.

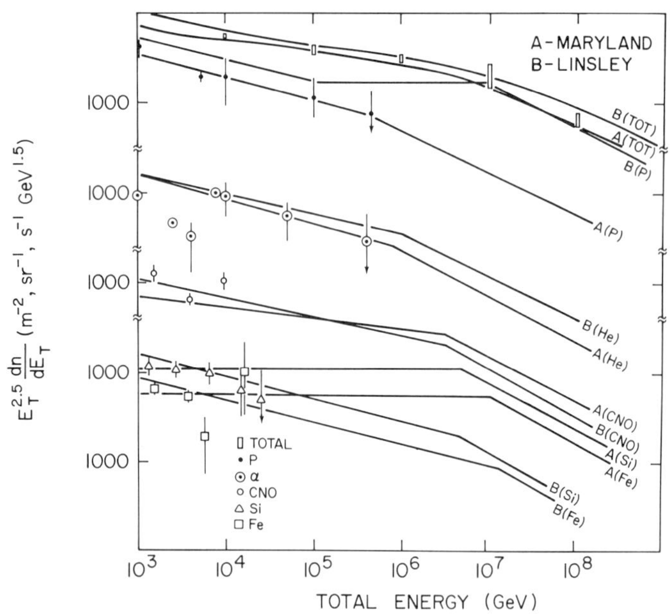

FIGURE 2. Two models of primary spectra compared with data obtained by direct measurements at the top of the atmosphere. These models lead to a quite different mix of elemental species beyond the bend in the spectra around 10^6 GeV per nucleus.

by heavier species in the cosmic ray beam. Having warned the reader of some of the inherent problems in dealing with cosmic ray data on air showers, we outline the principal methods used to analyze air shower data to derive proton-air inelastic cross sections at the highest energies.

EXPERIMENTAL METHODS AND RESULTS

General Information

An air shower is the nuclear-hadronic-electromagnetic cascade generated by a cosmic ray in the atmospheric "calorimeter" of vertical thickness of 1030 gm/cm^2 at sea level. This calorimeter has a depth of about ten interaction lengths (measured in terms of the proton-air inelastic cross section, $\sigma^{inel}_{\text{p-}air}$ (E), and given by $\lambda_{\text{p-}air}$ (E) = $24{,}070/\sigma^{inel}_{\text{p-}air}$ (E) in gm/cm^2, where σ is in mb) and some 25 radiation lengths. This cascade can be described by three functions defining the longitudinal development of the electron-photon, N_e, the hadron, N_h, and the muon, N_μ, components and three function families giving the lateral structure of each component at different longitudinal depths. We shall be mainly concerned with two functions describing the longitudinal development of the electron-photon and muon components, given by

$$N_e(E_o, x), N_\mu\,(\text{E}_o, E_\mu, x),$$

where E_o is the energy of the primary, x is the longitudinal depth coordinate of the shower expressed in gm/cm^2, and E_μ is the energy of a muon crossing the depth, x. The hadronic component, of course, is the source of both the electron-photon and muon cascades; however, it is difficult to sample in air shower experiments and very few experiments at these high energies actually measure it. Thus, we do not emphasize it.

Although the complex cascade of the air shower obscures the study of individual interactions, it is possible to extract the interaction cross section of the particle that initiates the shower by the study of fluctuations in the location of the point of first interaction. This was first pointed out over twenty years ago by a Japanese air shower group.[12]

Zenith Angle Distribution of Air Showers with Fixed Muon and Electron Sizes

Discussion of the Method

FIGURE 3(a) shows the longitudinal development functions, $N_e\,(x)$ and $N_\mu(>E_\mu, x)$, for showers generated by protons of energy, E_o, with two different points of first interaction, x_I. Monte Carlo calculations show that the essential difference in the shapes of N_e and N_μ curves is that beyond the maximum, the number of muons attenuate rather slowly, while the number of electrons attenuate significantly. This means that at observation levels deep in the atmosphere, such as x_{obs} shown in FIGURE 3(a), measurement of N_μ can give a good estimate of the energy of the initiating primary as indicated in FIGURE 3(b). The rapid attenuation of $N_e(E_o, x)$, however, makes it possible to use a measurement of shower size as an indicator of the location of the point of first interaction for those showers whose muon size is fixed. The

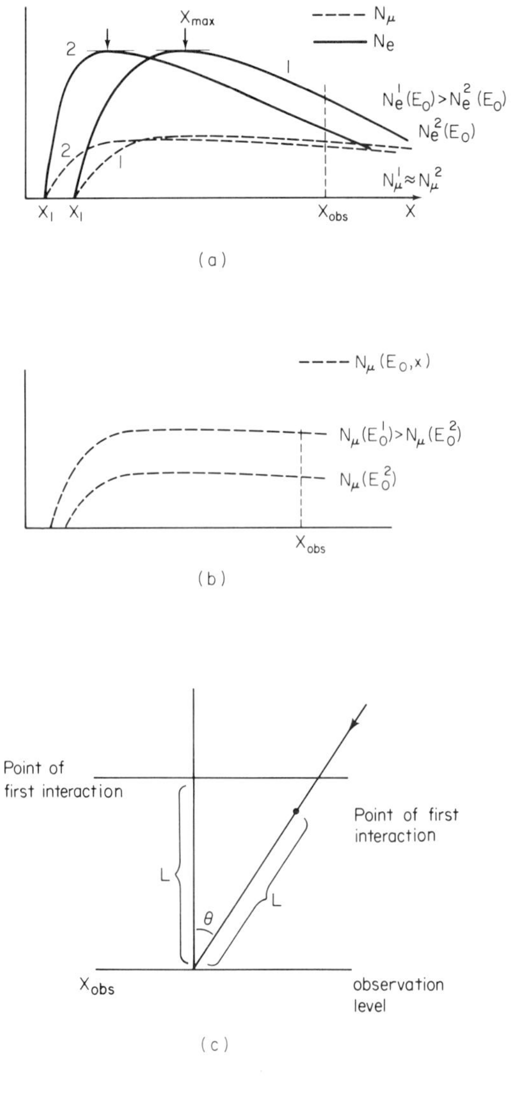

FIGURE 3. (a) Comparison of longitudinal development of electrons and muons for proton showers of fixed energy, but with different starting points. (b) A graph illustrating the use of N_μ as an indicator of primary energy. (c) The principle of the Akeno method for obtaining $\sigma_{p\text{-}air}$. To get the same N_μ and N_e at the observation level from two different zenith angles, the inclined shower must penetrate deeper. Here, L is in gm/cm^2.

experimental technique is to fix the muon size, N_μ, and the electron size, N_e, and to measure the event rate as a function of the zenith angle of the arrival direction of the shower. To get the same shower size from zenith angle θ as from $\theta = 0$, the primary has to penetrate deeper before having the first interaction, which causes the "distance" between x_{obs} and x_l to be the same for the inclined shower as for the vertical shower, as shown in FIGURE 3(c). It is simple to show that

$$N(\theta) = N(0) \exp\left[-x_o (\sec\theta - 1)/\lambda_{obs}\right].$$

If there were no fluctuations in the shower development other than those of the point of first interaction and if there were no nuclei in the cosmic ray beam other than protons, λ_{obs} would give the proton-air interaction length. To estimate the effect of fluctuations in the cascade development after the first interaction, Monte Carlo calculations must be done that relate λ_{obs} to λ_{true}. In general, $\lambda_{obs} > \lambda_{true}$ since fluctuations tend to make for deeper penetration of showers. The effect of contamination due to heavier nuclei is minimized experimentally by selecting showers with $N_e > N_e$ (threshold), independent of zenith angle. This is so because for a fixed N_μ, air showers initiated by primaries of atomic number larger than protons have generally a smaller shower at deep observation depths and, therefore, selecting on large N_e antiselects heavier nuclei because of the steep energy dependence of cosmic ray spectra. This method has been used recently

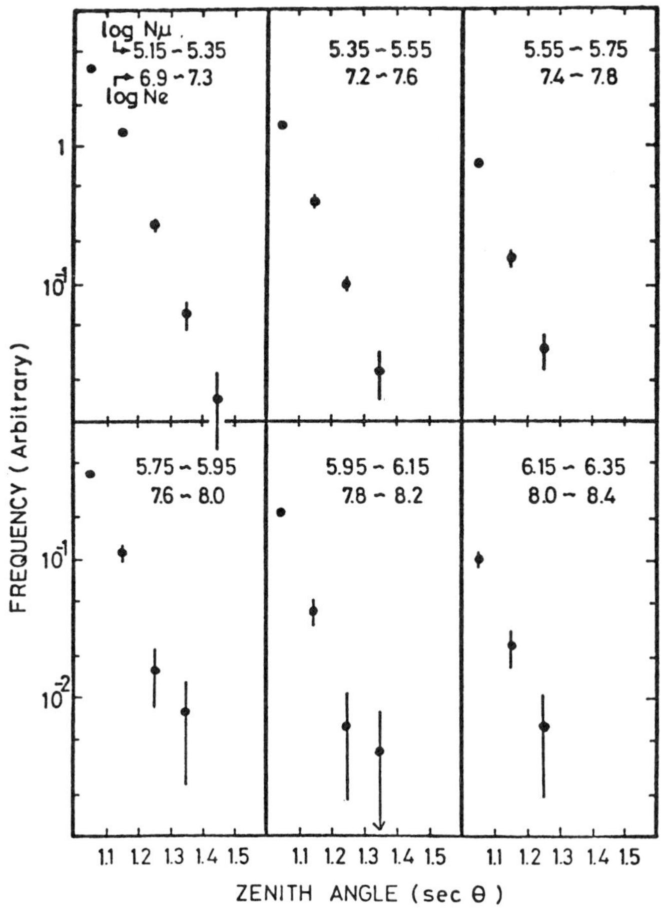

FIGURE 4. Akeno data on zenith angle distributions for different $N_\mu - N_e$ bins.

by the Akeno air shower group with good success in deriving $\sigma^{inel}_{\text{p-air}}$ at $\sqrt{s} \sim 20$ TeV.[13] Their results are described next.

Experimental Results from the Akeno Air Shower Experiment

The Akeno air shower array is situated 920 m above sea level. It has over 150 scintillator counters distributed over 1 km^2 to measure N_e and θ, and nine muon detectors of 25 m^2 each. The threshold energy of muons required is 1.0 sec θ GeV. They determine θ to $\pm$ 4°, and percent errors in size determinations are 5 percent for electrons and 15 percent for muons.[14] The experimental results on the zenith angle distributions are shown in FIGURE 4 for different $N_\mu - N_e$ ranges. These curves were fitted to obtain λ_{obs}. The observed values of the interaction length were then converted to a λ_{true} and subsequently to $\sigma^{inel}_{\text{p-air}}$ by doing Monte Carlo simulations of the air cascades. In TABLE 1, we give the results on proton-air inelastic cross sections of the Akeno measurements, as given in their paper.[13] Their selection of $N_e > N_e$ (threshold) in order to define the sample is very important because it reduces the contamination from particles other than protons significantly. In fact, the authors of this work assert from detailed Monte Carlo studies that "so long as fraction of protons is higher than 10%, our conclusion is not affected." We note that the experiment "measures" proton-air inelastic cross section and is insensitive to elastic scattering, diffractive excitation of target nucleons, and quasi-elastic scattering. The value of $\sigma^{inel}_{\text{p-air}}$ has doubled in going from 20 GeV to 20 TeV as shown in FIGURE 5. Next to be examined are other methods to obtain information on total cross sections.

TABLE 1. Results of Proton-Air Inelastic Cross Sections

Estimated Energy GeV	$\sqrt{s}$ TeV	$\sigma^{inel}_{\text{p-air}}$ mb	Ref.
$(1.78–3.16) \times 10^7$	7.03	539 ± 56	13
$(3.55–6.31) \times 10^7$	9.93	562 ± 12	13
$(7.08–12.6) \times 10^7$	14.0	667 ± 45	13
$(1.41–2.51) \times 10^8$	19.8	601 ± 74	13
$(2.82–5.01) \times 10^8$	28.0	674 ± 89	13
$(5.62–10.0) \times 10^8$	39.5	603 ± 60	13
5×10^8	~30	530 ± 66	18

Fluctuations in the Depth of Shower Maxima

The Method

An examination of the longitudinal profiles of showers in FIGURE 3(a) shows that the location of the maximum of a shower is coupled to the point of first interaction. A

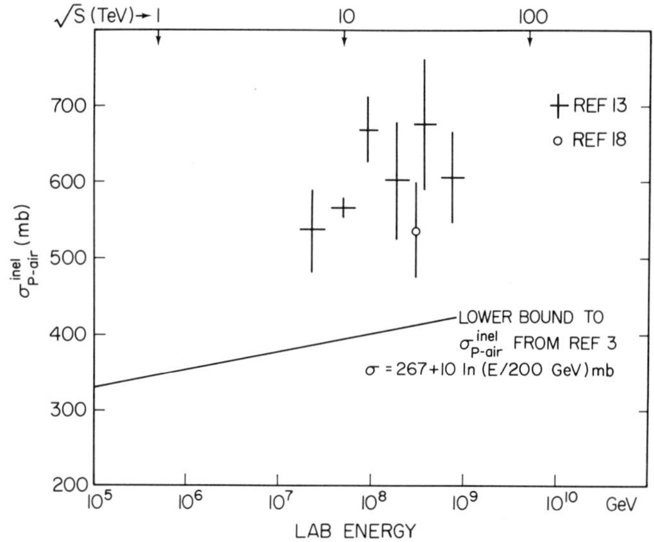

FIGURE 5. The proton-air inelastic cross section deduced from air shower experiments of the Akeno and the Fly's Eye groups. The lower bound was previously derived through the study of the propagation of unaccompanied hadrons in the atmosphere.

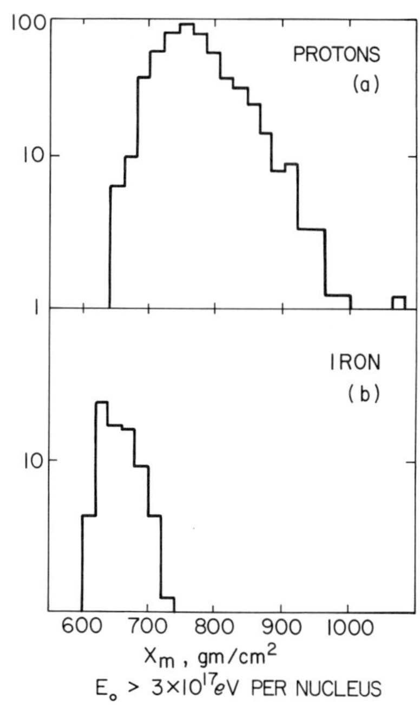

FIGURE 6. The distribution of depths of shower maxima for simulated showers generated by protons (a) and iron nuclei (b). Note the shift in the peaks of the curves and the narrow width of the iron curve.

study of the fluctuations in x_{max} should reflect the fluctuations in the point of first interaction. It is, therefore, of interest to study the shape of the distribution of depths of shower maximum.[15] The study of longitudinal development of air showers has recently become feasible through the study of time structure of Cherenkov light from air showers[16,17] and through the observations of scintillation light from nitrogen fluorescence in the atmosphere, which is induced by the traversal of large air showers.[18] The last technique being used by the Fly's Eye detector in Utah is capable of measuring shower profiles of individual events from the start of the shower to its demise. We will call experiments that can measure the full profile of the shower, class A experiments. The fluctuations in x_m can also be inferred from measurements of shower parameters at a fixed observational level, which are sensitive to longitudinal developments such as a pulse rise time, the muon to electron ratio, and the lateral distribution function.[19,20] These techniques are rather more indirect and will be referred to as class B experiments. In order to estimate proton-air cross sections, it is important to restrict the data sample from such experiments in order to minimize the contamination due to primaries other than protons. This is difficult to do unless one can estimate the primary energy (for instance, as done by the Akeno group using N_μ) and then select on some other parameter to enhance protons ($N_e > N_e$ (threshold), in the case of the previously mentioned Akeno experiment). The class A experiments provide an opportunity to enrich the sample used in analysis in protons; for class B experiments, though, one must use the full sample.

Detailed Monte Carlo simulations of air showers produced by various primary species have been carried out.[15] Events were generated from an isotropic distribution at the top of the atmosphere and on a E^{-3} differential energy spectrum. FIGURES 6(a) and (b) show the distribution of the depths of shower maxima for showers from proton and iron primaries with $E \geq 3 \times 10^{17}$ eV. The Monte Carlo simulation used a nuclear fragmentation model based upon emulsion data[21] and a particle production model consistent with ISR and SPS-$\bar{p}p$ collider results.[22,23] The sensitivity of the analysis to particle interaction models with substantial scale violations has been studied[15] and will be discussed below.

The qualitative features of the distributions in FIGURES 6(a) and (b) are:

(1) Iron showers generally have an x_m higher in the atmosphere than proton showers by about 100 gm/cm^2.
(2) The width of the x_m distribution, σ_{x_m}, is much narrower for iron showers than for proton showers: $\sigma \sim 35$ gm/m^2 for Fe and $\sigma \sim 60$ gm/m^2 for p.
(3) The large depth tails ($x_m > 760$ gm/cm^2) of these distributions can be fitted by exp $(-x_m/\Lambda_m)$, where the effective attenuation length, Λ_m, depends on the atomic mass of the primary, with Λ_m being considerably larger for protons.
(4) The large depth part of the distribution is dominated by protons (unless they are totally absent in the primary flux).

A little reflection shows that if one had a sample of events whose energy is known (class A) and whose depths of maxima are required to lie at depths beyond about 760 gm/cm^2, then one has a sample enhanced in protons. A value of Λ_m can be obtained from this data. However, how does Λ_m depend on $\sigma^{inel}_{p\text{-}air}$? How sensitive is this relation

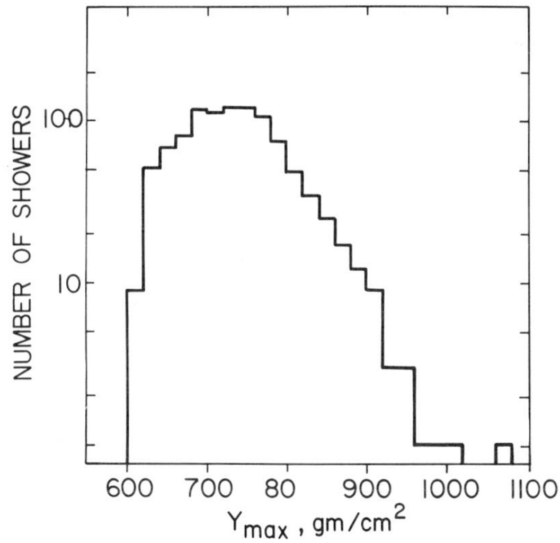

FIGURE 7a. The distribution of depths of shower maxima for simulated showers generated for two different mixes of primary cosmic rays: A "light" mix, *L*.

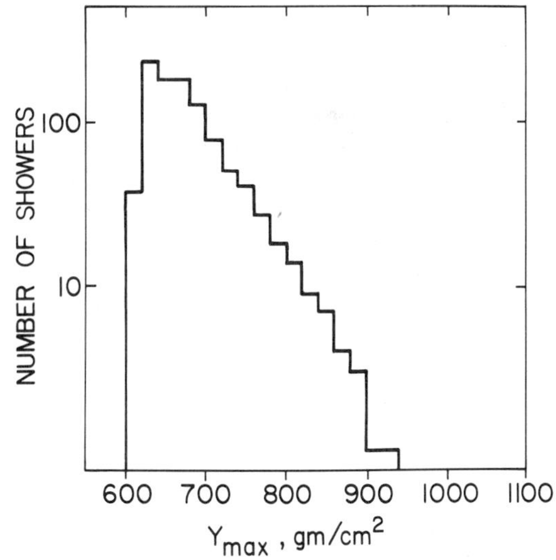

FIGURE 7b. The distribution of depths of shower maxima for simulated showers generated for two different mixes of primary cosmic rays: A "heavy" mix, *H*.

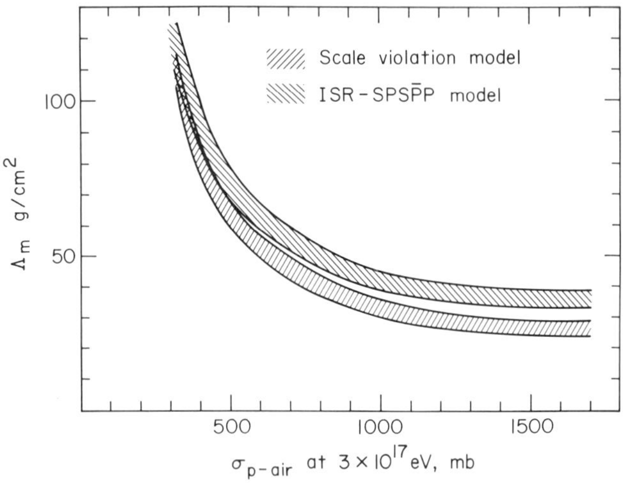

FIGURE 8. Sensitivity of the effective attenuation length, Λ_m, of the depth of the shower maximum distribution to input parameters of the energy variation of cross section and the energy variation of particle production models.

to the details of the model of high energy interactions? Also relevant is the question, how much is Λ_m influenced by different mixes of primary nuclei?

We address the last question first. We have Monte Carlo simulated the combined distribution of x_m for two different assumed primary compositions:[15] a composition dominant in light nuclei, like that at very low energies, called *L*, and a composition enhanced in MH and H nuclei called *H*. These are shown in FIGURES 7(a) and (b). The value of Λ_m is decreased by 13% for the *H* composition and only by 7% for the *L* composition, relative to the case for pure protons. A further examination of the figures also shows that the region with $x_m < 750$ gm/cm^2 is particularly sensitive to the abundance of heavies.

The sensitivity of Λ_m to $\sigma_{\text{p-air}}^{inel}$ is shown in FIGURE 8 for two different high energy physics models.[22,23] The curves are for proton showers chosen from a power-law energy spectrum. The bands represent the statistical uncertainty of the simulation. We note the following two points: (1) For $\sigma_{\text{p-air}}^{inel} \geq 600$ mb, the sensitivity of Λ_m to $\sigma_{\text{p-air}}$ is reduced because for larger cross sections, the interaction length becomes comparable to the radiation length in air ($\sim$37 gm/cm^2) and (2) that for a specifically measured Λ_m, strong scale violating[23] models would give a somewhat smaller $\sigma_{\text{p-air}}$ than for models with a less drastic scale violation.[22] The difference in calculated values of Λ_m between models of references 22 and 23 are no more than 10 gm/cm^2.

Similar analysis can be done for the width, σ_{x_m}, of the x_m distribution. It is, however, not easy to enhance the proton content of the data sample. FIGURE 9 shows both the dependence of σ_{x_m} on different models of the energy variation of proton-proton cross sections and on different composition in the energy range of 4×10^{17} eV to 1.6×10^{18} eV. The upper side of each square represents a set of calculations with *L* composition and the lower boundary represents *H* composition and scale violations. To differentiate between models of cross sections and composition, one must measure σ_{x_m} to about three percent, a difficult task as we shall see below.

Experimental Results from the Fly's Experiment (Class A Experiment)

The Fly's Eye detectors of the University of Utah are located at Dugway, at an altitude corresponding to an overburden of 830 gm/cm^2 of air. It detects and measures the longitudinal profiles of air showers due to ultrahigh-energy cosmic rays by observing the nitrogen fluorescence generated from the ionization produced by the shower particles in their passage through the atmosphere.[24]

The detector consists of 67 mirrors, of 1.6 m diameter and 1.6 m focal length each, oriented so as to cover 2π steradians of the sky. The viewing is done with 14 photomultipliers (p.m.) located at the focus of each mirror, with a cone angular coverage of 10 mr per p.m. tube. Ionization produces about five photons per meter of path traversed by a relativistically charged particle, independent of altitude. By recording the arrival times and pulse integrals of each photomultiplier viewing this fluorescent light, longitudinal development of air showers can be reconstructed on an event-by-event basis. The specific quantities that can be derived for each shower are: (1) its trajectory and arrival direction and (2) its shower size as a function of slant depth, $N_e(x)$, as indicated in FIGURE 10. Therefore, one determines primary energy, E_o, which is proportional to the integral of $N_e(x)$ over the trajectory, location of

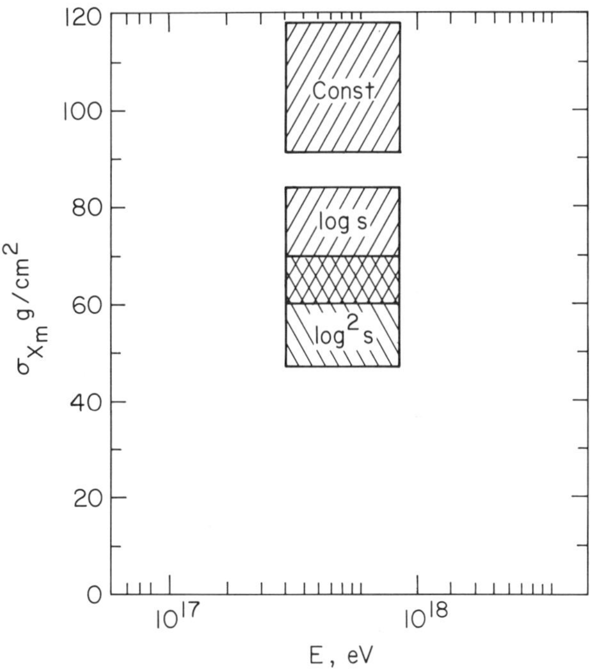

FIGURE 9. σ_{x_m} calculated for showers of 4×10^{17} eV $< E_o < 1.6 \times 10^{18}$ eV and different models for the energy variation of $\sigma_{p\text{-}air}$. See text.

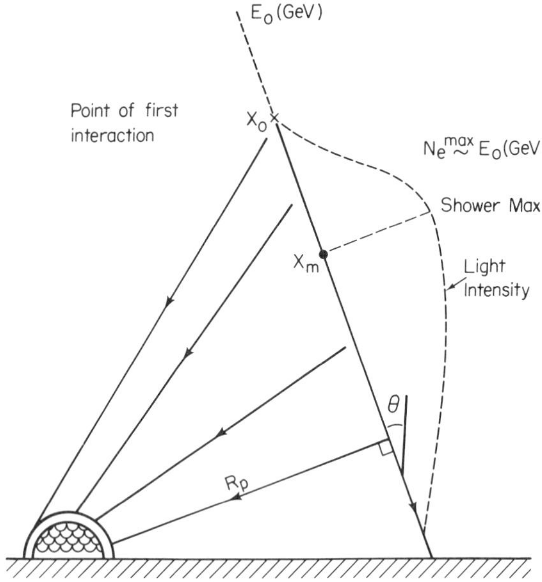

FIGURE 10. A schematic illustration to explain the principles of the Fly's Eye technique.

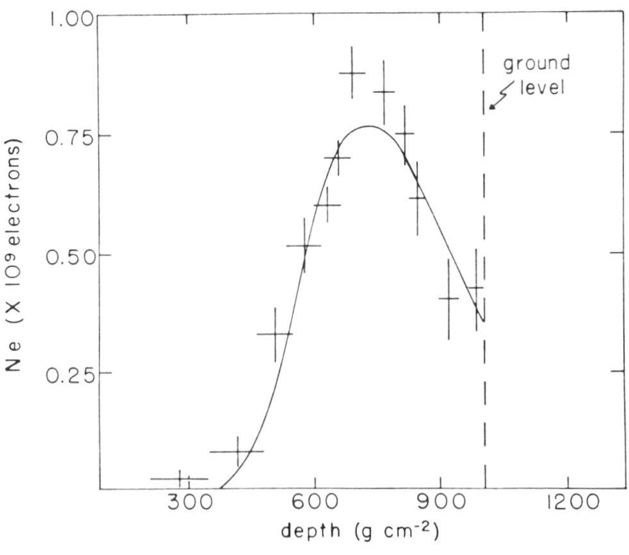

FIGURE 11. The longitudinal profile for a well-reconstructed shower.

shower maximum, and shower width.[25] A special feature of this detector is that since acceptance increases with E_0, one can view showers 20 km away for $E > 10^{10}$ GeV.

The group reported results from a study of the distribution of x_m that was based upon a sample of 1366 events, taken during a run time of 439 hours, that has been reconstructed geometrically well enough to make a determination of the longitudinal profile for each event.

A typical reconstructed profile of a high energy shower is shown in FIGURE 11. Its energy is 1.3×10^{18} eV and the location of the shower maximum is at 727 ± 33 gm/cm^2. The distribution of depths of maxima for events whose fitting errors on x_m are estimated to be smaller than 125 gm/cm^2 is shown in FIGURE 12. A best fit to the shape of the tail of this curve for $x_{max} > 830$ gm/cm^2 (cut made to enhance the proton content of the sample) with the form, $\exp(-x_m/\Lambda_m)$, gave $\Lambda_m = 73 \pm 9$ gm/cm^2. From the upper band in FIGURE 8, one calculates $\sigma^{inel}_{\text{p-air}} = 530 \pm 66$ mb at an average energy of

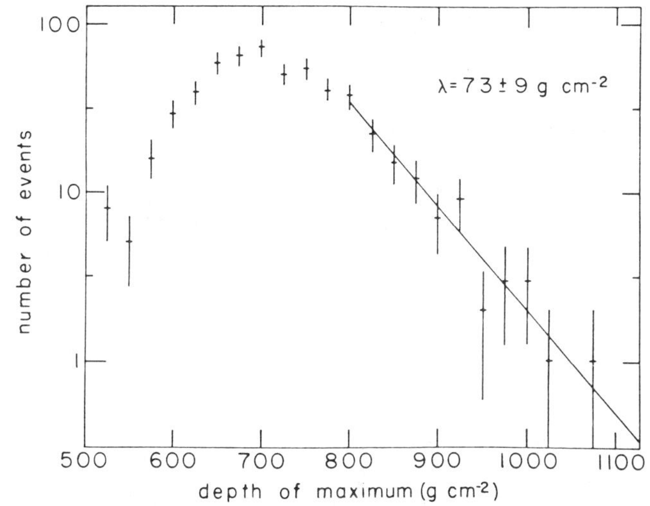

FIGURE 12. The distribution of the depth of shower maxima as measured by Fly's Eye.

5×10^{17} eV. The width, σ_{x_m}, of the curve is seen to be of the order of 80 gm/cm^2. Another point to be noted is the presence of a substantial fraction of events with $x_m <$ 650 gm/cm^2, indicating that according to the curves of FIGURES 6(a) and (b), the width, σ_{x_m}, is larger than what would be expected for pure protons because of the presence of heavier species that have x_{max} below 650 gm/cm^2. The Fly's Eye experiment is continuing to collect data and should be able to measure the energy variation of $\sigma^{inel}_{\text{p-air}}$ in the future. The value of proton-air inelastic cross section is in good agreement with the Akeno result presented in the previous section.

Experimental Results on σ_{x_m} (Class B Experiments)

As pointed out earlier,[19,20] fluctuations in the depth of shower maximum, σ_{x_m}, can also be inferred from measurements of shower parameters that are sensitive to the longitudinal development. This kind of analysis does not measure the shape of the individual shower curves, but obtains a measure of the height of x_m on an event-by-event basis. For instance, the full width at half maximum (FWHM) of the pulse of Cerenkov light from air showers has been shown to be a monotonic function of x_m.[19] Hence, a measurement of FWHM gives an estimate of x_m. In addition, by measuring either the shower size, the muon size, or the Cherenkov size, an estimate of the primary energy may be made. This estimate has, in general, greater fluctuations on an event-by-event basis than one gets by the Fly's Eye technique of measuring the track length integral.

The procedure adopted by several groups to obtain a measure of σ_{x_m} from air shower data is to study the variation of some shower parameter, P, with x_m for fixed energy of the primary, under the assumption that nonlinear dependence on the increment, $(x - \bar{x}_m) \equiv \Delta x_m$, can be neglected.[20,26] (The fixed energy cut is nontrivial to achieve and it is most important to simulate the systematics, which is a task not adequately covered by the authors of these papers.) An examination of the distribution of x_m obtained by the Fly's Eye detector, as displayed in FIGURE 12, shows that the distribution is clearly asymmetric and is broader than what would be expected for a pure component like those in FIGURES 6(a) and (b). The values of σ_{x_m} obtained from such analysis should be treated with caution and it is not easy to "enhance" the protons in the sample. FIGURE 13 shows a summary of values of σ_{x_m} obtained by various experimenters.[19,20,26,27] One notices the enormous spread in measured values, and in some cases, measured values differ from each other by five or six standard deviations. It is clear, however, that σ_{x_m} is generally larger than what one would get from a pure

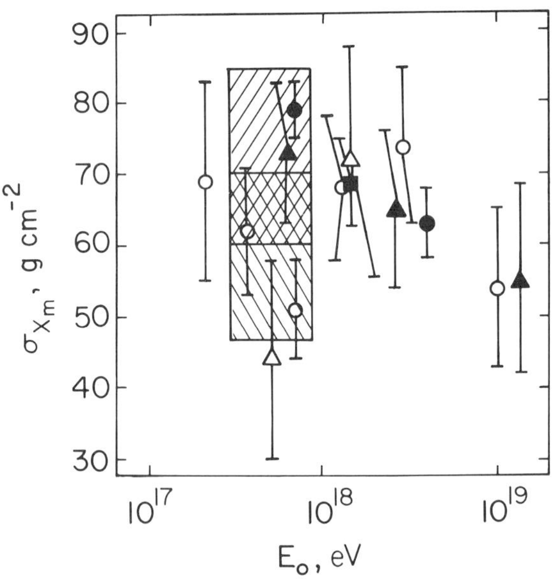

FIGURE 13. Energy dependence (if any) of the estimations of the fluctuations in the depths of shower maxima. Pure species have small values of σ_{x_m}. The points indicate a mixed composition.

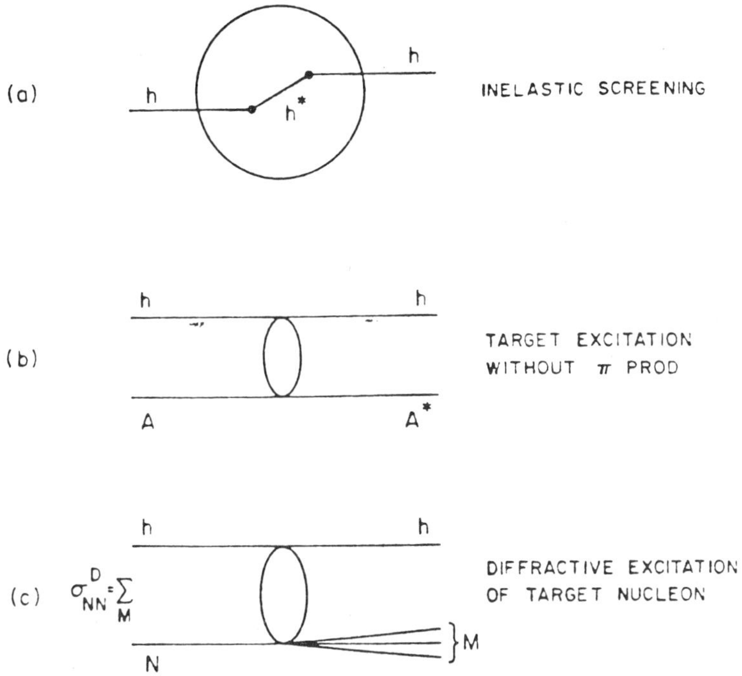

FIGURE 14. Schematic representations of the Glauber corrections for: (a) inelastic screening, (b) quasi-elastic scattering, and (c) diffractive dissociation of target nucleon (σ_{hA} (inelastic) = $\sigma_{hA}^{total} - \sigma_{hA}^{el} - \sigma_{hA}^{*} - \sigma_{hA}^{D} - \Delta\sigma_{inel}^{Abs}$).

species. In addition, in the same figure are graphed the predictions from simulations for different assumptions about the energy variation of the cross section and the composition models. One sees that σ_{x_m} data does not have the discriminating power to determine the proton-air inelastic cross section.

ASYMPTOTIC BEHAVIOR OF σ_{pp}^{tot}: RELATING $\sigma_{p\text{-}air}^{inelastic}$ TO σ_{pp}^{tot}

If the elastic scattering amplitude in hN interaction is known, then one can calculate the total elastic and inelastic cross sections for $h - A$ interactions using Glauber multiple scattering formalism.[28] In particular, as input, one must know σ_{tot} $(h - N)$, the slope parameter, B_{hN}, and the cross section for diffractive processes. In addition, one must assume a form for the nuclear wave function. Furthermore, the Glauber formalism must be extended to include shadowing effects due to inelastic screening.[29–32]

Most experiments that measure inelastic cross sections in hadron nuclear collisions are insensitive to detection of some or all of the processes in which the incident hadron

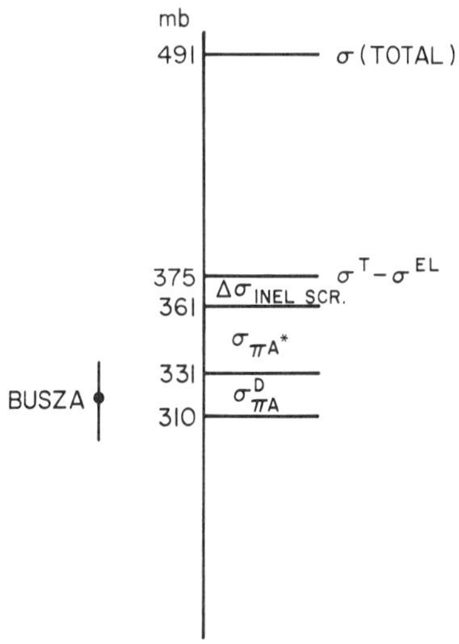

FIGURE 15. An illustration of the importance of the various terms in calculating $\sigma_{hA}^{inelastic}$ from *h-h* parameters for the case of 50 GeV protons on aluminum. (Pion-aluminum cross section, 175 GeV.)

retains nearly all of its energy.[33] Two of these processes are:

(a) Quasi-elastic scattering, in which the target nucleus is excited or broken up without pion production, called σ_{hA}^{qe}.
(b) Diffractive dissociation of a target nucleon, called σ_{hA}^{D}.

Thus, in order to compare theory with experiment, full absorptive hadron-nuclear cross section, $\sigma_{hA}^{Abs} = \sigma_{hA}^{tot} - \sigma_{hA}^{el}$, must be corrected for inelastic screening, $\Delta\sigma_{hA}^{Abs}$ (inel.sc.), σ_{hA}^{qe}, and some fraction of σ_{hA}^{D}. Thus, we can write:

$$\sigma_{hA}^{inel} = \sigma_{hA}^{Abs} - \Delta\sigma_{hA}^{Abs}\,(\text{inel. sc.}) - \sigma_{hA}^{qe} - \sigma_{hA}^{D}. \tag{1}$$

The effects discussed above are shown schematically in FIGURE 14. Some typical values of the cross sections and corrections for protons on aluminum at 50 GeV are: $\sigma_{hA}^{Abs} = 488$ mb, $\Delta\sigma$(inel.sc.) $= 7$ mb, $\sigma_{hAl}^{qe} = 38$ mb, and $\sigma_{hAl}^{D} = 21$ mb. (Calculations were done using a Gaussian nuclear density with an r.m.s. radius of 3.11 fermis.) The importance of making all these corrections before confronting theory with experimental data is shown in FIGURE 15, where calculations are compared with measurements done at Fermilab[34] for 175 GeV pions on aluminum. We have done such calculations for various projectiles with laboratory momenta ranging from 50 GeV to 300 GeV incident upon various nuclei with atomic masses ranging from 9 to 208 and compared them with experimental data[34–36] in FIGURE 16. One observes that the Glauber calculations agree with data to better than ± 5 percent over a wide range of projectile momenta and nuclear masses. This gives us some confidence in applying the same technique to extract σ_{pp}^{tot} from $\sigma_{\text{p-}air}^{inel}$ at cosmic ray energies.

In order to relate $\sigma_{\text{p-}air}^{inel}$ to σ_{pp}^{tot} according to the technique discussed above, we have to specify the values of σ_{pp}^{tot}, the elastic slope parameter, B_{pp}, at $t = 0$, the ratio of the single diffractive cross section to the inelastic cross section, and the dependence of

$(d^2\sigma/dtdM^2)_{t=0}$ on M^2 for proton-proton interactions, as a function of energy. We must also assume a nuclear shape for "air" nucleus. (A Gaussian profile has been used with a r.m.s. radius of 2.6 fermis for "air.")

To calculate $\sigma^{Abs}_{p\text{-}air} = \sigma^{tot}_{p\text{-}air} - \sigma^{el}_{p\text{-}air}$, we need to know $\sigma^{tot}_{pp}(E)$ and $B_{pp}(E)$. Four different energy variations have been used: (1) Block-Cahn fit number 1,[37] (2) Block-Cahn fit number 2,[37] (3) $\log^2 s$ model with geometric scaling (GS2),[15] and (4) log s model with geometric scaling (GS1).[15] The Block-Cahn fit 2 corresponds to a $\log^2 s$ energy dependence of σ^{tot}, while for fit 4, σ^{tot} saturates at high energies at about 90 mb. In models (3) and (4), $B_{pp} = 0.296\ \text{GeV}^{-2}\ \sigma^{tot}$ (mb) is used. The single diffractive cross section is assumed to be asymptotically reaching a fraction of 0.13 of σ^{inel}. If, however, the fit to the single diffractive cross section according to Goulianos[38] is used, the fraction would become somewhat larger. For the inelastic screening correction, we follow the procedure of Murthy *et al.*[31] and use their fit for $(d\sigma/dtdM^2)_{t=0}$. The quasi-elastic scattering correction is largest for Block-Cahn fit 2, where it increases with energy from 32 mb to 50 mb for $\sqrt{s} \sim 20$ GeV to $\sqrt{s} \sim 20$ TeV, respectively.

The curves relating $\sigma^{inel}_{p\text{-}air}$ to σ^{tot}_{pp} for the above four models are shown in FIGURE 17 for $\sigma^{tot}_{pp} \gtrsim 40$ mb. We note that the fractional increase in $\sigma^{inel}_{p\text{-}air}$ is only about 50% as large as the fractional increase in σ^{tot}_{pp} at 140 mb. This was pointed out earlier in reference 2.

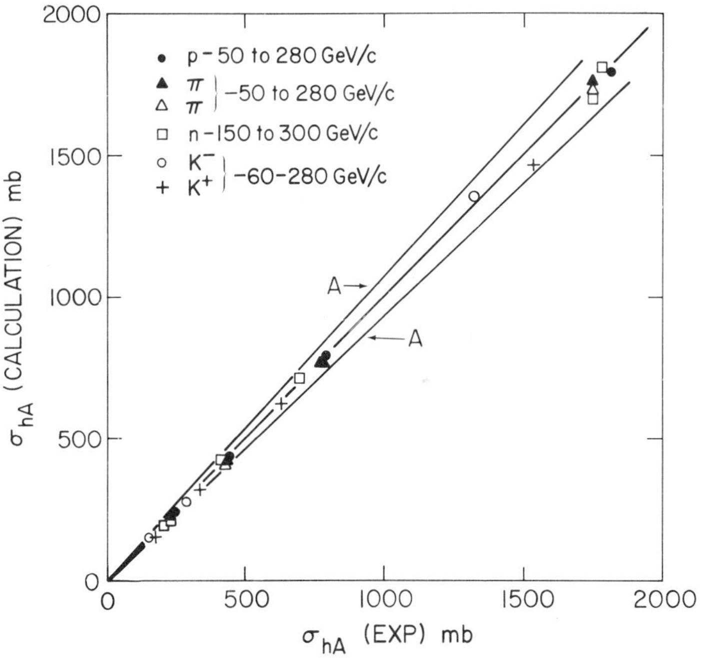

FIGURE 16. Comparison of experimental inelastic cross sections and theoretical calculations for different projectiles and various target nuclei, covering an energy range from 20 to 300 GeV and mass numbers from 9 to 208.

Finally, in FIGURE 18, we compare the prediction of the four models for $\sigma_{p\text{-}air}^{inel}$ with data discussed in EXPERIMENTAL METHODS AND RESULTS. It is seen that even with the large errors, the data strongly favor a $\log^2 s$ increase of σ_{pp}^{tot}.

CONCLUSIONS AND DISCUSSION

We have reviewed the experimental methods used to obtain a measure of proton-air inelastic cross sections from cosmic ray air shower data above 10^7 GeV. The results of the Akeno and Fly's Eye experiments show an increase of $\sigma_{p\text{-}air}^{inel}$ by a factor of two between Fermilab energies, $\sqrt{s} \sim 20$ GeV, and air shower energies, $\sqrt{s} \gtrsim 15$ TeV (FIGURE 5).

We have outlined a method based upon Glauber theory to relate $\sigma_{p\text{-}air}^{inel}$ to σ_{pp}^{tot}. The proton-air absorptive cross section has to be corrected for effects of quasi-elastic excitation of the target nucleus, for diffractive dissociation of target nucleons, and for screening due to diffractive inelastic processes in order to calculate what the air shower experiments measure. For four different models of energy variation of proton-proton elastic parameters, the relation between $\sigma_{p\text{-}air}^{inel}$ and σ_{pp}^{total} was shown in FIGURE 17. A comparison of the air shower data with the predictions of the models (FIGURE 18) shows that the data favor a $\log^2 s$ variation of σ_{pp}^{total} with energy. The experimental cross sections cannot distinguish between models with or without geometrical scaling.

It has been suggested[39] that the Glauber methods used in this work may not be valid at very high energies because the rise in total cross sections may be entirely due to diffractive phenomena. The critical Pomeron Regge field theory predicts that all cross sections (hadron-hadron and hadron-nucleus) increase with the same power of log s at high energies. This power, called the critical exponent, has been calculated to be 0.26 ±

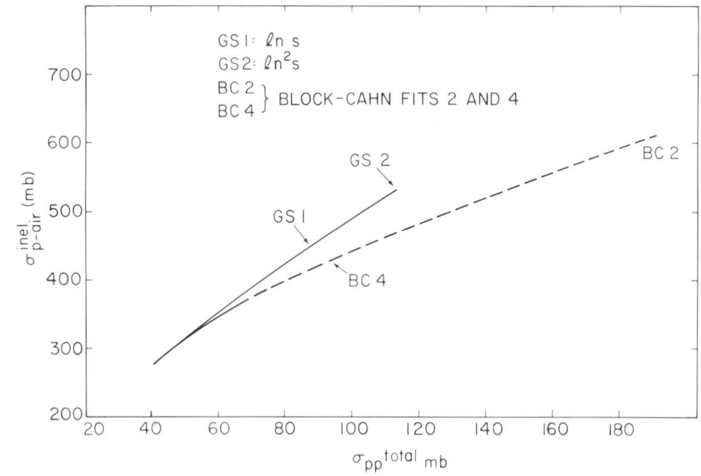

FIGURE 17. Relationships between $\sigma_{p\text{-}air}^{inel}$ and σ_{pp}^{total} for the four models of pp interaction discussed in the text: Block-Cahn fits 2 and 4, $\log^2 s$, and $\log^2 s$ variation with geometric scaling.

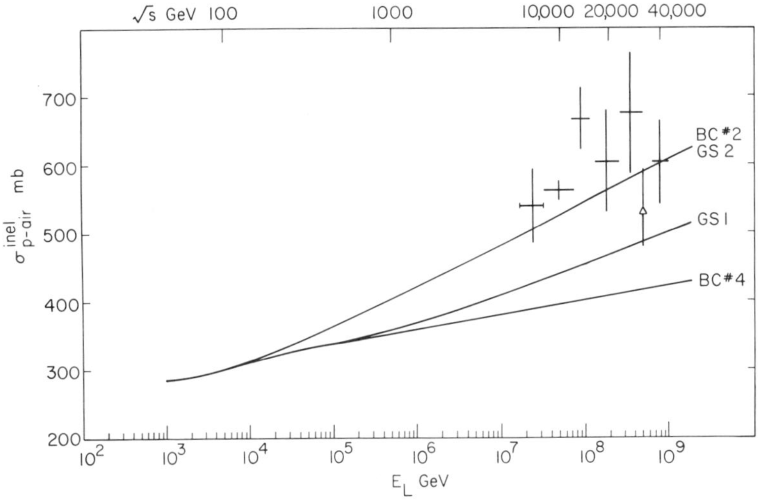

FIGURE 18. Energy variation of $\sigma^{inel}_{\text{p-}air}$ for the four models (see text) compared with air shower data. The experimental cross sections favor $\log^2 s$ variation of σ^{total}_{pp} with energy.

0.02.[39] If we assume this is correct, then the ratio, $\sigma^{inel}_{\text{p-}air}$ $(E = 10^8\ \text{GeV})/\sigma^{inel}_{\text{p-}air}$ $(E = 10^3$ GeV), should be given by $(8/3)^{0.26}$, which predicts a cross section of 365 mb at 10^8 GeV. This value does not agree with the values of the proton-air cross sections obtained from the air shower experiments given in TABLE 1. It is more than 3σ away from the air shower average value. It may be that "asymptopia," where the critical Pomeron Regge field theory represents the full picture, has not been reached at $\sqrt{s} \sim 10$ TeV.

ACKNOWLEDGMENTS

This work has been done in collaboration with T. K. Gaisser and C. Noble. I want to acknowledge valuable discussions I have had with my colleagues, R. W. Ellsworth, S. C. Tonwar, and T. Stanev.

REFERENCES

1. YODH, G. B., Y. PAL & J. S. TREFIL. 1972. Evidence for rapidly rising p-p total cross sections from cosmic ray data. Phys. Rev. Lett. **28:** 1005–1008; YODH, G. B., Y. PAL & J. S. TREFIL. 1973. Phys. Rev. **D8:** 3233–3236.
2. BARGER, V., F. HALZEN, T. K. GAISSER, C. J. NOBLE & G. B. YODH. 1974. Proton-proton total cross section above 10^4 GeV: can cosmic rays give the answer? Phys. Rev. Lett. **33:** 1051–1054; GAISSER, T. K. & G. B. YODH. 1974. Energy dependence of $\sigma_{\text{p-}air}$ up to 20 TeV. Nucl. Phys. **B76:** 182–188.
3. YODH, G. B., R. W. ELLSWORTH, T. STANEV & T. K. GAISSER. 1981. Cosmic rays in the atmosphere and characteristics of high energy interaction. Nucl. Phys. **B183:** 12–28; YODH, G. B., S. C. TONWAR, T. K. GAISSER & R. W. ELLSWORTH. 1983. Proton-proton cross section from 1 to 100 TeV. Phys. Rev. **D27:** 1183–1186.

4. AMENDOLIA, S. R. *et al.* 1973. Phys. Lett. **44B:** 119; AMALDI, U. *et al.* 1973. Phys. Lett. **44B:** 112.
5. AMALDI, U. *et al.* 1977. Phys. Lett. **66B:** 390.
6. BATTISTON, R. *et al.* 1982. Measurement of the proton-antiproton elastic and total cross section at a center of mass energy of 540 GeV. Phys. Lett. **115B:** 333 and **117B:** 126. (These papers give the first results of the UA4 experiment.)
7. ARINSON, G. *et al.* 1982. Phys. Lett. **121B:** 77. (This gives the first results of the UA1 experiment.)
8. AMBROSIO, M. *et al.* 1983. Measurement of total cross section in proton-proton and proton-antiproton collisions at ISR energies. *In* Proceedings of the Third Moriond Workshop on Antiproton-Proton Physics and the *W* Discovery. J. Tran Thanh Van, Ed.: 49–56. Edition Frontiers. France.
9. AMOS, N. *et al.* 1983. Comparison of p̄p and pp elastic scattering at $\sqrt{s} = 62.5$. *In* Proceedings of the Third Moriond Workshop on Antiproton-Proton Physics and the *W* Discovery. J. Tran Thanh Van, Ed.: 57–63. Edition Frontiers. France.
10. YODH, G. B., J. A. GOODMAN, S. C. TONWAR & R. W. ELLSWORTH. 1984. Composition of primary cosmic rays at energies 10^{15} eV from data on high-energy muons in extensive air showers. Phys. Rev. **D29:** 892–901.
11. LINSLEY, J. 1983. Spectra, anisotropy, and composition of cosmic rays above 1000 GeV. *In* Proceedings of the 18th International Cosmic Ray Conference, Bangalore, India. N. Durgaprasad, S. Ramadurai, P. V. Ramana Murthy, M. V. S. Rao & K. Sivaprasad, Eds.: **12:** 135–192. Indian National Scientific Documentation Center. New Delhi, India.
12. FUKUI, S., H. HASEGAWA, T. MATANO, I. MIURA, M. ODA, K. SUGA, G. TANAHASHI & Y. TANAKA. 1960. Prog. Theor. Phys. Suppl. **16:** 1.
13. HARA, T. *et al.* 1983. Inelastic p-air cross section at energies between 10^{16} and 10^{18} eV estimated from air showers. Phys. Rev. Lett. **50:** 2058.
14. TAN, Y. H., M. NAGANO & S. KAWAGUCHI. 1982. Institute of Cosmic Ray Research, University of Tokyo, ICR report-99-82-2 (unpublished); KAMATA, K. 1979. Akeno Air Shower Project, Cosmic Rays, and Particle Physics, 1978. T. K. Gaisser, Ed.: 443–450. AIP Conference Proceedings No. 49, Particle and Field Subseries, No. 16. New York.
15. ELLSWORTH, R. W., T. K. GAISSER, T. STANEV & G. B. YODH. 1982. Ultrahigh-energy cross sections from study of longitudinal development of air showers. Phys. Rev. **D26:** 336–339; STANEV, T. & T. K. GAISSER. 1982. Model dependence of the determination of cross section from large air shower experiments. *In* Proceedings of the Workshop on Very High Energy Cosmic Ray Interactions. M. L. Cherry, K. Lande & R. I. Steinberg, Eds.: 125–135. University of Pennsylvania Press. Philadelphia.
16. KHRISTIANSEN, G. B. 1979. Proceedings of the 17th International Cosmic Ray Conference, Kyoto, Japan. Conference Papers, Institute of Cosmic Ray Research, University of Tokyo, Tokyo, Japan. **14:** 360–381.
17. OXFORD, K. & K. E. TURVER. 1976. Nature (London) **264:** 727–729; ANDAM, A. A. *et al.* 1982. Phys. Rev. **D26:** 23–26.
18. CASSIDAY, G. L. *et al.* 1982. *In* Proceedings of the Workshop on Very High Energy Interactions of Cosmic Rays. M. Cherry, K. Lande & R. I. Steinberg, Eds.: 72–82. University of Pennsylvania Press. Philadelphia; BATRUSAITIS, R. M. *et al.* 1984. Phys. Rev. Lett. **52:** 1380–1383.
19. MCCOMB, T. J. L. & K. E. TURVER. 1982. The average depth of cascade maximum in large cosmic-ray showers from particle measurements. J. Phys. G: Nucl. Phys. **8:** 1119–1129.
20. COY, R. N. *et al.* 1982. Fluctuations of depth of maximum deduced from measurements of the lateral distribution of showers at 3×10^{17} eV. *In* Proceedings of the Workshop on Very High Energy Interactions of Cosmic Rays. M. Cherry, K. Lande & R. I. Steinberg, Eds.: 120–124. University of Pennsylvania Press. Philadelphia; DYAKONOV, M. N. *et al.* 1981. *In* Proceedings of the 17th International Conference on Cosmic Rays, Paris (Centre d'Etudes Nucleares de Saclay, Gif-sur-Yvette France) **6:** 43–46.
21. GAISSER, T. K., T. STANEV, P. FREIER & C. J. WADDINGTON. 1982. Phys. Rev. **D25:** 2341–2350.
22. ELLSWORTH, R. W. *et al.* 1981. Phys. Rev. **D23:** 764–770.
23. WDOWCZYK, J. & A. W. WOLFENDALE. 1979. Nuovo Cimento **54A:** 443–449.

24. CASSIDAY, G. L. *et al.* 1979. *In* Report on the Utah Fly's Eye Cosmic Rays and Particle Physics. T. K. Gaisser, Ed.: 417–441. AIP Conference Proceedings No. 49, Particle and Fields No. 16.
25. BALTRUSAITIS, R. M. *et al.* 1984. The total proton-proton cross section at $\sqrt{s} = 30$ TeV. Phys. Rev. Lett. **52:** 1380–1383.
26. DYAKONOV, M. N. *et al.* 1983. Flux density fluctuations of the EAS atmospheric Cherenkov light with primary energy above 10^{17} eV. *In* Proceedings of the 18th International Cosmic Ray Conference, Bangalore, India (see reference 11 for details) **6:** 110–113.
27. WALKER, R. & A. A. WATSON. 1982. J. Phys. G: Nucl. Phys. **8:** 1131–1140; WATSON, A. A. & J. G. WILSON. 1974. J. Phys. A: Math. Nucl. Gen. **7:** 1119; CHANTLER, M. P. *et al.* 1983. J. Phys. G: Nucl. Phys. **9:** L27.
28. GLAUBER, R. J. & G. MATTHIAE. 1970. Nucl. Phys. **B21:** 135.
29. KARMANOV, V. A. & L. A. KNODRATYUK. 1973. Zh. Eksp. Teor. Fiz. Pis'ma Red. **18:** 451. (JETP Lett. **18:** 266).
30. JONES, L. *et al.* 1974. Phys. Rev. Lett. **33:** 1440.
31. MURTHY, P. V. R. *et al.* 1975. Neutron total cross sections on nuclei at Fermilab Energies. Nucl. Phys. **B92:** 269–308.
32. BIEL, J. *et al.* 1976. Phys. Rev. Lett. **36:** 1004.
33. GAISSER, T. K. *et al.* 1975. On the relation between proton-proton and proton-nucleus cross sections at very high energies. *In* Proceedings of the 14th International Cosmic Ray Conference, Munich, vol. 7, p. 2161–2166.
34. BUSZA, W. *et al.* 1975. Phys. Rev. Lett. **34:** 836.
35. DENISOV, P. *et al.* 1973. Nucl. Phys. **B61:** 62.
36. CARROLL, A. S. *et al.* 1979. Absorption cross sections of $\pi^{\pm}$, $K^{\pm}$, p, and $\bar{p}$ nuclei between 60 and 280 GeV/*c*. Phys. Lett. **80B:** 319–322; CARROLL, A. S. *et al.* 1979. Total cross sections of $\pi^{\pm}$, $K^{\pm}$, p, and $\bar{p}$ on proton and deuterons between 200 and 370 GeV/*c*. Phys. Lett. **80B:** 423–427.
37. BLOCK, M. M. 1985. A review of σ_{tot}, ρ, and B for $\bar{p}$p and pp elastic scattering. Aspen Winter Physics Conference. Ann. N.Y. Acad. Sci. **461:** 186; BLOCK, M. M. & R. N. CAHN. 1984. Lawrence Berkeley Lab. Report No. LBL-17522.
38. GOULIANOS, K. 1983. Diffractive interactions of hadrons at high energies. Phys. Rep. **101:** 170–219.
39. WHITE, A. R. 1981. Diffraction in acceleration, colliders, and Q.C.D. *In* Proceedings of the Proton-Antiproton Collider Physics, 1981. AIP Conference Proceedings No. 85, Particles and Fields Subseries No. 26; WHITE, A. R. 1982. XIII International Symposium on Multiparticle Dynamics, Volendam, Netherlands. W. Kittel, W. Metzger & A. Stergiou, Eds.: 799–814. World Scientific Publications.

Cross Sections in the Multi-TeV Range[a]

T. K. GAISSER[b] AND F. HALZEN

Department of Physics
University of Wisconsin
Madison, Wisconsin 53706

Interpretation of air shower experiments to extract the cross section in the 40 TeV range depends on the model of hadronic interactions used to simulate the experiments. We argue that if the observed scaling violations in hadronic collisions are associated with the hard scattering of constituent partons, then the scaling violations will be confined largely to the central region and will have little effect on the cross section determination.

The only way at present to study hadronic cross sections at $\sqrt{s} \sim 50$ TeV is with cosmic rays. The flux at this energy is of the order, one per km^2 per week, so direct observation of the primaries is out of the question and one must resort to ground-based air shower arrays with large areas exposed for long times. Since the primary particles and the initial interactions are not then directly observed (but only their products deep in the atmosphere), it is necessary to assume a model of hadronic interactions to interpret the data. Moreover, the chemical composition of the primary nuclei is not directly observed either. Thus, cross section and composition of the beam must be determined from the same data.

Both Fly's Eye[1] and Akeno[2] experiments have recently reported measurements of the proton cross section around 10^{18} eV ($\sqrt{s} \simeq 40$ TeV), as shown in FIGURE 1. Both analyses circumvent the problem of an unknown mixed composition by using only the most deeply penetrating showers for analysis. The assumption is that heavier nuclei with larger cross sections will have been attenuated. If the primary beam does not contain at least 10% protons, the method clearly cannot work.

For purposes of illustration, we briefly review the method used in reference 1 to obtain a measurement of the cross section and we discuss its sensitivity to the model of hadronic interactions. We also discuss how the data may in principle be used to determine some aspects of the primary composition, a problem of great astrophysical interest. Though the details differ, similar remarks will apply to other experiments, such as that of reference 2. We conclude with a discussion of the properties of hadronic interactions involving hard scattering of constituents, which leads us to argue that approximate hadronic scaling holds at high energies, thus reducing the ambiguity in interpretation of these experiments.

By observing the atmospheric scintillation light that is emitted as an air shower goes across the sky, the Fly's Eye experiment is able to reconstruct each shower trajectory with a degree of accuracy that depends on the orientation of the shower

[a]This research was supported in part by the University of Wisconsin Research Committee, with funds granted by the Wisconsin Alumni Research Foundation, and in part by the Department of Energy under contract no. DE-AC02-76ER00881.

[b]On leave from Bartol Research Foundation, University of Delaware, Newark, Delaware 19716.

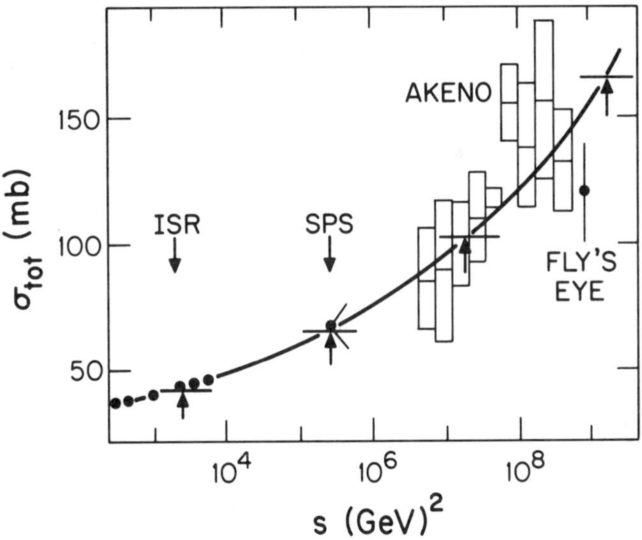

FIGURE 1. The increasing cross section associated with jets of moderate transverse momentum (see TABLE 1) is superimposed on a constant total cross section of 38 mb. The result is compared to data for different center-of-mass energies.

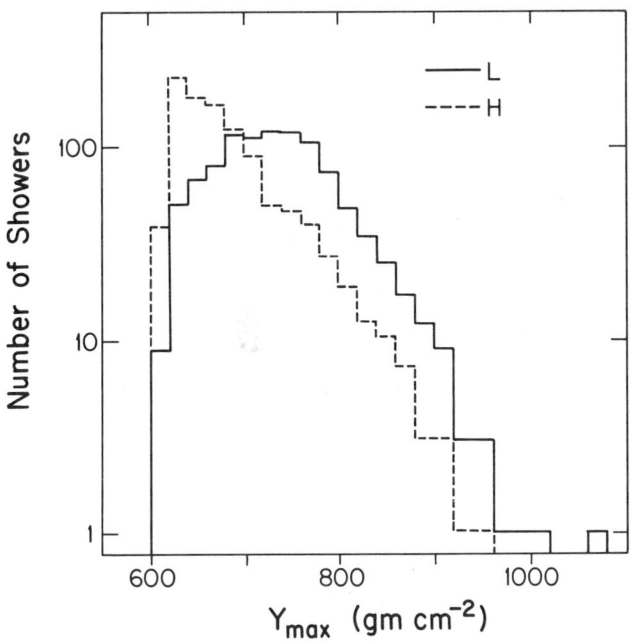

FIGURE 2a. Distribution of depth of maximum for showers of energy $\geq 3 \times 10^{17}$ eV per nucleus for two compositions: p : α : CNO + Mg : Fe = 0.55 : 0.21 : 0.16 : 0.08 as at low energy (denoted L); and p : α : CNO + Mg : Fe = 0.2 : 0.08 : 0.07 : 0.65, denoted H. (From reference 3.)

relative to the detector. From the pulse heights in the individual detector elements, it is possible then to reconstruct for each shower the number of ionizing particles as a function of slant depth measured along the shower axis. The slant depth at which each shower reaches maximum development is taken as a measure of the depth of shower penetration. Fluctuations in the depth of maximum result from fluctuations in starting point and from fluctuations in shower development after interaction. Starting point fluctuations reflect both composition and cross section per species. Development fluctuations, in addition, reflect the rate at which energy is transferred from the hadronic to the electromagnetic component of the shower, which in turn reflects properties of inclusive hadronic interactions at high energy.

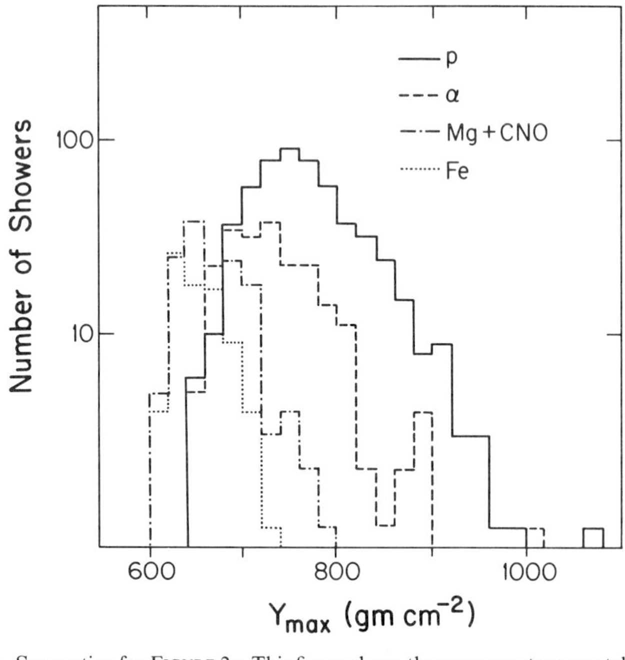

FIGURE 2b. See caption for FIGURE 2a. This figure shows the components separately for the low energy composition. (From reference 3.)

FIGURE 2 shows the simulated depth of maximum distribution for two different primary compositions.[3] The tail of this distribution is characterized by an exponential with characteristic length, Λ_m.[4] There are two points to be noted from FIGURE 2:

(1) The fraction of early developing showers is sensitive to composition because heavy primaries dissipate energy more quickly than a proton of the same total energy per nucleus. (An air shower is a calorimetric measurement, so it is natural to compare different nuclei at the same energy per nucleus rather than at the same energy per nucleon.)

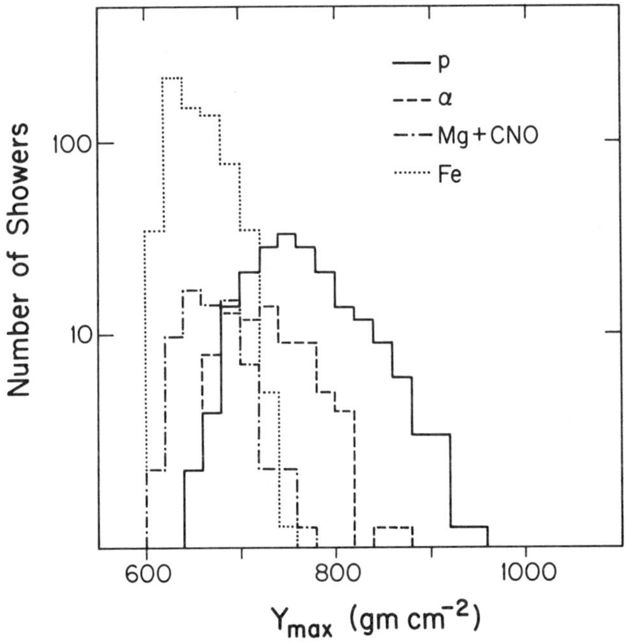

FIGURE 2c. See caption for FIGURE 2a. This figure shows the components separately for the heavy composition. (From reference 3.)

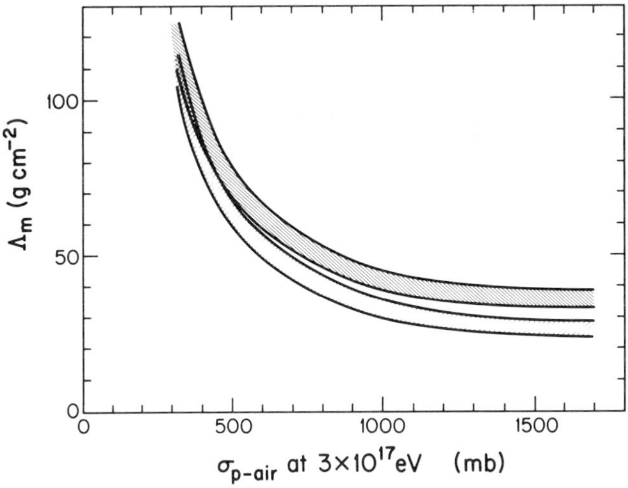

FIGURE 3. Λ_m versus $\sigma_{p\text{-}air}$ at $\sqrt{s} \sim 40$ TeV for proton showers chosen from a power law energy spectrum (integral index = 1) with $E_o > 3 \times 10^7$ eV. The upper band represents the scaling model and the lower one represents the scaling violation with $\alpha = 0.09$. Band width indicates statistical uncertainty of the calculation. (From reference 5.)

TABLE 1. Properties of Jet Cross Sections with $p_T > p_{Tmin}$ at Different Energies

E (TeV)	$\sqrt{s}$ (GeV)	p_{Tmin} (GeV)	σ_{jet} (mb)	$\langle x \rangle_{jet}$	$\langle y \rangle_{jet}$
1	43.0	1.25	4	0.17	1.8
150	540.0	2	26	0.07	2.9
10^4	4330.0	3.2	63	0.05	4.2
10^6	43,300.0	6	127	0.05	5.9

(2) The shape of the tail of the distribution is insensitive to the composition, reflecting the deeply penetrating proton (and alpha particle) primaries. If $N_p > N_\alpha$, then Λ_m should be sensitive primarily to development of proton-initiated showers.

In FIGURE 3, we show Λ_m versus $\sigma_{p\text{-}air}$ for two rather different models of hadronic interactions. These are simulations[5] based on a model with hadronic scaling, in which

$$\frac{E}{\sigma}\frac{d\sigma}{dp^3} \rightarrow f(x, p_T), \tag{1a}$$

and on a model with violation of scaling[6] such that

$$\frac{E}{\sigma}\frac{d\sigma}{dp^3} \rightarrow \left(\frac{s}{s_o}\right)^\alpha f\left(x\left(\frac{s}{s_o}\right)^\alpha, p_T\right). \tag{1b}$$

Here $x = E/E_o$, where E_o and E are the energies of incident and produced particles, respectively, in the hadron-nucleon c.m. system. In reference 6, α and s_o are parameters of the model. In the example shown in FIGURE 3, we took $\alpha = 0.09$ and $s_o = m_p\,2$ TeV. If α is larger, the effect is bigger.

In reference 1, the scaling curve from FIGURE 3 (as calculated in reference 3) was compared with a measured value of $\Lambda_m = 73 \pm 9$ g/cm^2 to obtain a value of $\sigma_{p\text{-}air} = 530 \pm 66$ mb at $E_o \sim 10^{18}$ eV.[7] If the scaling violation curve was used instead, one would obtain $\sigma_{p\text{-}air} \cong 450 \pm 50$. Thus, the uncertainty due to the model-dependence of the hadronic inclusive cross section is comparable to that arising from the experimental limitations on the accuracy of Λ_m.

This model dependence will also interfere with determination of the composition,[8] so it is clearly desirable to reduce this theoretical source of uncertainty as much as possible. Of primary importance for development of atmospheric cascades is the behavior of hadronic inclusive cross sections in the forward fragmentation region because this governs the rate at which energy is fed into the electromagnetic component, which produces the bulk of ionizing particles in air showers. Experimental information on this behavior is difficult to obtain at $\bar{p}p$ colliders because the forward particles go down the beam pipes.

To circumvent this problem, we have made[9] a connection[10] between the hard scattering component and the increasing pp cross section, which allows an extrapolation of the inclusive cross section to high energies. The basic idea is to integrate the perturbative QCD expressions for hard scattering of constituents down to an energy-dependent $p_{T\,min}$ defined by

$$\sigma_{tot} = \sigma_o + \sigma_{jet}(p_{T\,min}), \tag{2}$$

where (schematically) $\sigma_o = 38$ mb and

$$\sigma_{jet}(p_{T\,min}) = \int\int\int_{p_{T\,min}} F(x_1)F(x_2)\sigma(\theta). \tag{3}$$

$F(x_i)$ are the probabilities for finding constituents with fractional momenta, x_i, in the incident hadrons and $\sigma(\theta)$ is the elementary cross section for scattering of constituents to angle θ defined in the parton-parton c.m. system. The factorization displayed in equation (3) is an approximate result suitable for practical calculations.[11]

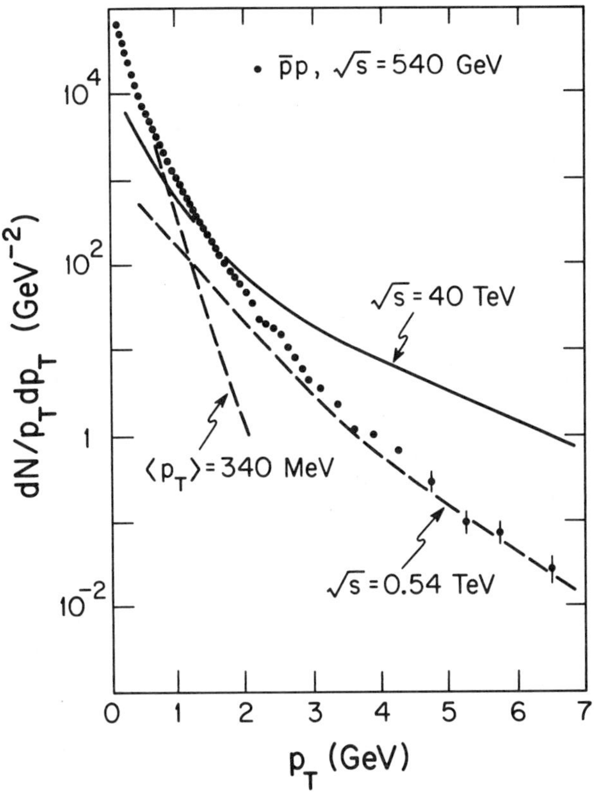

FIGURE 4. Two-component description of the charged particle transverse momentum distribution.[15] The low-p_T component is described by an exponential falloff with $\langle p_T \rangle = 340$ MeV. The "high"-p_T component corresponds to hadrons that are the fragments of parton jets with $p_T > p_{T\,min}$ chosen as in TABLE 1. The jets have been fragmented according to the distribution of reference 14. Data from the UA1 experiment.[16]

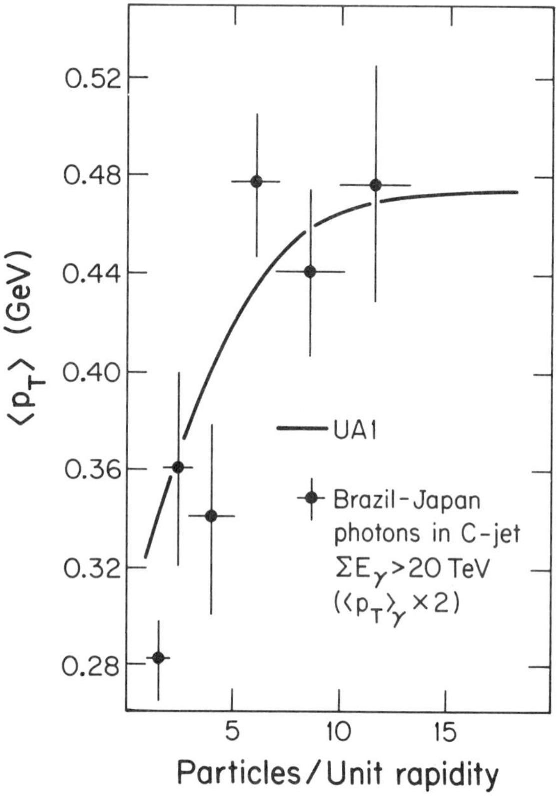

FIGURE 5. Increase of $\langle p_T \rangle$ associated with increasing multiplicity. The data[16] from the Brazil-Japan collaboration are compared with an interpolation of the UA1 data at $\sqrt{s} = 540$ GeV.

The formalism represented by equation (**3**) has been verified at the $\bar{p}p$ collider for jets with large p_T.[12] At small p_T, perturbative QCD eventually breaks down. We circumvent this problem by the procedure outlined above, which forces $\sigma_{jet} < \sigma_{tot}$. (Otherwise, as s increases for fixed $p_{T\,min}$, the integrated hard scattering cross section will eventually exceed the total cross section.[13]) TABLE 1 shows the values of σ_{jet} and $p_{T\,min}$, along with the mean values of Feynman x and the rapidity of the scattered jet obtained from the integrals using the structure functions of reference 14. The values of $p_{T\,min}$ have been chosen so that equation (**2**) reproduces the total cross sections shown in FIGURE 1. It has been shown[15] that this value of σ_{jet} reproduces the transverse momentum distribution observed at the $p\bar{p}$ collider (see FIGURE 4). Moreover, the observed[16] correlation between multiplicity and mean p_T per event (FIGURE 5) is in qualitative agreement with this picture since gluon radiation from jets produces high multiplicity events.

Because of the large cross sections involved, such semihard processes can be expected to produce dramatic effects at high energy. FIGURE 6 illustrates the expected energy dependence of the rapidity distribution at various energies of interest. We have assumed that beam fragments in hard collisions produce rapidity distributions of secondary pions as in soft collisions, with two charged particles per unit of rapidity. The scattered jets are assumed to produce rapidity distributions with the same shape as beam fragments at the kinematic limits and also with an asymptotic density of two charged particles per unit of rapidity. In the absence of significant scaling violation effects in the fragmentation functions, we would predict an asymptotic rapidity density of four charged particles.

By energy conservation, one expects a violation of scaling in the fragmentation region as the central component grows with energy.[17] The magnitude of this effect can be estimated from the values of $\langle x_{jet} \rangle$ tabulated above. At high energies, where semihard scattering events represent a significant fraction of the cross section, $\langle x_{jet} \rangle \simeq$ 0.05, so we expect fragmentation region scaling violation only at the level of 5%. In contrast, even with the small value of the scaling violation parameter illustrated above ($\alpha = 0.09$), one would expect a large fraction of the hadronic energy to be shifted to very small x. For example, at $\sqrt{s} \sim 40$ TeV and $\alpha = 0.09$, produced hadrons with $x >$ 0.05 would carry only ~15% of the incident energy as compared to ~30% at $s = s_o$. (We assume the fragment nucleon carries an energy-independent average 50% fraction.)

We, therefore, conclude that if the association between the hard scattering and the increasing cross section described here is qualitatively correct, then scaling violations should have little effect on the behavior of the electromagnetic component of the cosmic ray cascades because these depend primarily on the fragmentation region.[18] In particular, use of scaling-type models to interpret the Fly's Eye experiment should be essentially correct. A nontrivial result of the calculation that increases our confidence in the correctness of the association between growing cross section and hard scattering is represented by the last column of TABLE 1. The jet rapidities that come out of the

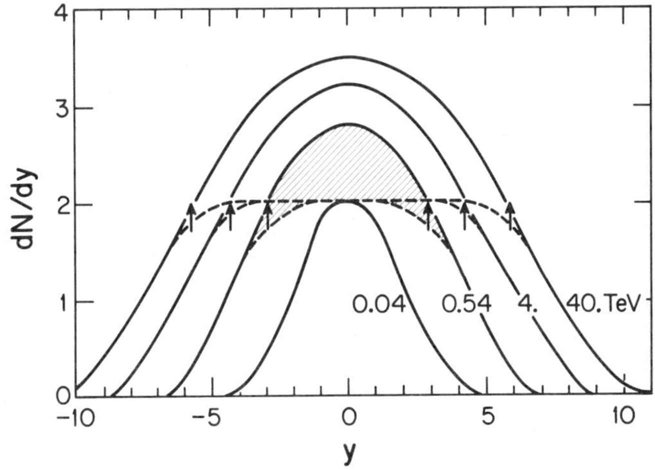

FIGURE 6. Sketch of the two-component rapidity distribution corresponding to the two-component total cross section shown in FIGURE 1. At each energy, the rapidity distribution consists of a plateau on which is superimposed the excess of hadrons near $y = 0$ due to fragmentation of the scattered jets. Rapidities of the scattered jets are indicated by arrows. The excess is crosshatched for $\sqrt{s} = 0.54$ TeV.

QCD calculation are such that the excess rapidity density from the scattered jets joins smoothly onto the underlying rapidity plateau from beam fragments. The Gaussian-type rapidity distribution observed at $\sqrt{s} = 540$ GeV is, thus, naturally reproduced.

ACKNOWLEDGMENTS

We are grateful to Todor Stanev for helpful discussions.

NOTES AND REFERENCES

1. BALTRUSAITAS, R. *et al.* 1984. Phys. Rev. Lett. **52:** 1380.
2. HARA, T. *et al.* 1983. Phys. Rev. Lett. **50:** 2058.
3. ELLSWORTH, R. W., T. K. GAISSER, T. STANEV & G. B. YODH. 1982. Phys. Rev. **D26:** 336.
4. Note that if the distribution were for shower starting points, then the corresponding Λ would be the interaction length. Unfortunately, shower starting points are invisible to any air shower experiment.
5. STANEV, T. & T. K. GAISSER. 1982. Proceedings of the Workshop on Very High Energy Cosmic Ray Interactions. M. L. Cherry, K. Lande & R. I. Steinberg, Eds.: 125. University of Pennsylvania Press. Philadelphia.
6. WDOWCZYK, J. & A. W. WOLFENDALE. 1984. J. Phys. **G10:** 257 and references therein.
7. The cross section determined in an air shower experiment is the proton-air inelastic cross section (excluding quasi-elastic processes). The problem of determining σ_{pp}^{tot} from $\sigma_{p\text{-}air}$ using the Glauber multiple scattering theory (GLAUBER, R. J. & G. MATTHIAE. 1970. Nucl. Phys. **B21:** 135) has been discussed by BARGER, V. *et al.* 1974. Phys. Rev. Lett. **33:** 1051 and GAISSER, T. K., G. B. YODH, V. BARGER & F. HALZEN. 1975. Proceedings of 14th International Cosmic Ray Conference, Munich. **7:** 2161.
8. A method for comparing simulations to the Fly's Eye data to obtain information on primary composition around 10^{18} eV is discussed by GAISSER, T. K. 1984. *In* Proceedings of 4th Moriond Astrophysics Meeting, La Plagne, France.
9. GAISSER, T. K. & F. HALZEN. 1984. University of Wisconsin preprint PH/225.
10. The possible connection between the rise of σ_{tot} and the increased production of large-p_T secondaries with energy has been previously noted: CLINE, D., F. HALZEN & J. LUTHE. 1973. Phys. Rev. Lett. **31:** 491; HALZEN, F. 1974. Nucl. Phys. **B92:** 404; HALZEN, F. & F. HERZOG. 1985. Phys. Rev. **D30:** 2326. For a related discussion on differences between particle and antiparticle-proton cross sections, see ANTONIOUS, N. G. *et al.* 1980. Phys. Lett. **93B:** 472 and KATSOUFIS, E. C. *et al.* 1981. Phys. Lett. **106B:** 231.
11. HALZEN, F. & P. HOYER. 1983. Phys. Lett. **B130:** 326; COMBRIDGE, B. & MAXWELL, C. 1984. Nucl. Phys. **B239:** 429; COHEN-TANNOUDJI, G. *et al.* 1983. Phys. Rev. **D28:** 1626.
12. UA1 COLLABORATION: ARNISON, G. *et al.* 1984. Phys. Lett. **136B:** 294; UA2 COLLABORATION: BAGNAIA, P. *et al.* 1984. Phys. Lett. **144B:** 283.
13. DURAND, L. 1984. Proceedings of the 1984 Snowmass Workshop on SSC Physics.
14. BAIER, R., J. ENGELS & B. PETTERSON. 1980. Z. Phys. **C6:** 309.
15. DING, L. K., K. KASAHARA, S. TORII & T. YUDA. 1983. ICR preprint 107-83-1 (Tokyo).
16. LATTES, C. M. G., Y. FUJIMOTO & S. HASEGAWA. 1980. Phys. Rep. **65:** 151; UA1 COLLABORATION: ARNISON, G. *et al.* 1982. Phys. Lett. **118B:** 167; 1983. Phys. Lett. **123B:** 115; 1983. CERN preprint EP/83-198; UA2 COLLABORATION: BANNER, M. *et al.* 1982. Phys. Lett. **118B:** 203. 1983. Z. Phys. **C20:** 117.
17. GAISSER, T. K. 1981. Phys. Lett. **100B:** 425; SLAVATINSKY, S. A. 1984. Proceedings of the International Symposium on Cosmic Rays and Particle Physics. Tokyo, Japan; HALZEN, F. 1984. Proceedings of the International Symposium on Cosmic Rays and Particle Physics. Tokyo, Japan.
18. The low-energy muon component should be more significantly affected, but this should not interfere with determination of the cross section.

Comments on Monojets and Other Anomalous Events[a]

G. L. KANE

Randall Physics Laboratory
University of Michigan
Ann Arbor, Michigan 48109

INTRODUCTION

The Standard Model (SM) is increasingly well tested and it continues to describe all experiments to the present limits of accuracy. Most particle physicists hope that phenomena will be found that do not fit the SM predictions, not because it will contradict the SM (which is now so well tested that it must be a good description of nature on the 100 GeV scale), but because it will provide clues to physics beyond the SM. There are still many questions whose answers we would like to have, such as why the SM takes the form it does, what is the origin of mass and flavor, can all forces be unified, is supersymmetry (which is an allowed symmetry in a quantum field theory) realized in nature, and more. Even most theorists feel that experimental clues may be needed to untangle these questions or at least will greatly speed up the process.

In recent years, a number of possible effects beyond the SM have been reported, such as Higgs boson candidates with properties inconsistent with a single neutral SM Higgs boson, too large a cross section for production of like-sign dileptons in ν reactions, several categories of events in the UA1 and UA2 detectors at the CERN collider, and others. Unfortunately, most of them have not stood up to subsequent experimental examination. Even the long-standing like-sign dileptons are not appearing in recently improved experiments.[1] As we have heard[2,3] at this meeting, of all the candidates at the CERN collider, only the UA1 monojet events (and perhaps some associated dijet events with missing momentum) have become a stronger signal as the data improved statistically and systematically.

On the assumption that others kinds of anomalous events might occur, perhaps at new machines later this decade, I will divide this talk into two parts. First, I will consider some of the kinds of constraints that must apply to any attempts to understand the origins of anomalous events, with an emphasis on high energy colliders. The first requirement on a theory is obviously that some events of the "observed" kind are produced, i.e., that a specific mechanism exists. Equally important is that the rate of production of such events is reasonable and that other associated (but not observed) events are not produced. Most published models do not do well in this regard. It is very instructive to see the kinds of things that go wrong; today's unobserved prediction may be tomorrow's signal.

Then I will concentrate on the predictions of supersymmetry[4] and whether it can describe the monojet events. As first emphasized in references 5 and 6 and by F. Paige

[a]This research was supported in part by the U.S. Department of Energy.

(unpublished), extensive further analysis is necessary to compare theoretical predictions with data under actual experimental conditions; this is described fully in the following. The calculations I present are from an extensive collaboration[7] with M. Barnett and H. Haber. Supersymmetry is a suprisingly well constrained theory in terms of comparison with experiment. Only the masses of the superpartners are unknown; once they are specified, all cross sections and distributions are fixed. Some kinds of events must occur (such as one-sided events with one or two jets) and others must not occur (e.g., any event with no missing momentum or any event with a hard γ). It is by no means trivial to find events of the right kind also occurring at the right rate and with the right distributions in jet momentum and in missing momentum. As we will see, at the present time there is (at least) one choice of masses that appears to be consistent with the monojet data and with other information concerning anomalous events.

It is very interesting that of the many types of anomalous events for which there were candidates during 1984, only the events with missing momentum (that arise naturally if supersymmetric partners are being produced) still seem to exist in the 1985 data.

POSSIBLE SIGNATURES OF ANOMALOUS EVENTS

There are a number of categories of events where anomalous effects might stand out. It may be useful to list some of these. Normally, if one kind of event occurs in a model, so do others.

(A) Signatures with missing transverse momentum ($\not{p}_T$):

Topology	SM Background
$j\not{p}_T$	$W \to \tau\nu_\tau$ ($\tau \to$ hadrons), $g + Z$ ($Z \to \nu\bar{\nu}$), QCD
$jj\not{p}_T$	$g + W$ ($W \to \nu\bar{\nu}$)
$\ell^\pm\not{p}_T$	$W \to \ell\nu$
$\ell^\pm j\not{p}_T$	$g + W$ ($W \to \ell\nu$)
$\ell^+\ell^-\not{p}_T$	$Z \to \tau^+\tau^-$

In all cases, the background can be measured experimentally when sufficient data is available and it can be quite accurately calculated as well. All of the above signatures are predicted to occur, in varying amounts, in supersymmetric theories.

(B) Signatures involving peaks:

Topology	SM Background
$\ell^+\ell^-$	toponium, η_t
jj	η_t
jjj	$W + j$, $Z + j$ + cuts and phase-space
jW	$W + j$, $Z + j$ + cuts and phase-space
jZ	$W + j$, $Z + j$ + cuts and phase-space

(C) Signatures involving hard γ's:

Topology	SM Background
$\gamma\gamma$	η_t, QCD
$\ell^+\ell^-\gamma$	bremsstrahlung
$\gamma \not{p}_T$	γZ
γj	QCD
$\gamma\nu\bar{\nu}$	γZ
$\ell^\pm\gamma\not{p}_T$	
$\gamma\gamma\not{p}_T$	
γZ	QCD
γW	QCD

(D) Other constraints that are easily gotten wrong can be listed:

(1) The *number of events* has to be right, i.e., $\sigma \times$ BR, where σ is the production cross section for the mechanism in question and BR is the branching ratio for the decays that lead to the signature in question. Many proposals give wrong rates; we will return to this below.

For explanations involving production of a single particle, there are "unitarity" constraints that exclude some (proposed) interpretations; they are not rigorous "unitarity" constraints since they require adding a physical condition that necessitates that the production partial width is smaller than the available energy region, but they give important constraints in practice.[8,9]

The basis is that for production of a single particle, X, of arbitrary spin and color from constituents, a/b, one has

$$\left.\frac{d\sigma}{dy}\right)_{y=0}(\bar{\mathrm{p}}\mathrm{p} \to X + \text{----}) = \frac{2J_X + 1}{(2S_a + 1)\,(2S_b + 1)}\,\frac{N_c^X}{N_c^a N_c^b}\,\frac{\Gamma(X \to ab)}{M_X^3}\,\Lambda,$$

where N_c is the color multiplicity (1, 3, 6, 8, ---) and Λ depends on the structure functions,

$$\Lambda = 16\pi^2\,[F_{a/\mathrm{p}}(x)F'_{\bar{b}/\bar{\mathrm{p}}}(x) + (\mathrm{p} \leftrightarrow \bar{\mathrm{p}})].$$

Clearly, if $\Gamma(X \to ab)$ is limited, either by outside information or by the observation that (say) all events are in a region of 40 GeV (or whatever value is appropriate), the possible rate is limited. At $M_X \simeq 100$ GeV, the constraint is important and for $M_X \geq 150$ GeV, it is very serious.

Barger *et al.*[8] integrate over y and define

$$\sigma(\mathrm{p}\bar{\mathrm{p}} \to X + \text{---}) = (2J_X + 1)N_c^X \sum_{ab} \Gamma(X \to ab)\, R_{ab},$$

thus giving useful graphs of R_{ab} versus M_X at various energies. Dusedau *et al.*[9] study the constraints in detail for scalar particles.

Note that what is actually observed is

$$\sigma \times \mathrm{BR}(X \to f) = \Lambda' \frac{\Gamma}{M_X^3}\,\mathrm{BR}(X \to ab)\;\mathrm{BR}\,(X \to f),$$

where f is the final state of interest. Taking $\Gamma(X \rightarrow ab) \leq 50$ GeV (assuming the events show up in a limited energy region) and

$$\text{BR}\,(X \rightarrow ab)\ \text{BR}\,(X \rightarrow f) \leq 1/4,$$

there is an absolute constraint. Often, of course, the product of branching ratios is much less than 1/4.

In general, it is possible to conclude that to explain anomalous events with single particle production, one must have $2J_X + 1$ and/or N_c^X greater than unity and that major branching ratios must be involved.

(2) If a Z^0 is involved in the mechanism, a *number of constraints* operate:

- (i) if a large branching ratio is involved, it can increase Γ_Z;
- (ii) probably, rates are large enough to see[10] any anomalous Z^0 decay at PETRA, or PEP as well, via $e^+e^- \rightarrow Z \rightarrow ?$
- (iii) if a Z^0 mode is the origin of anomalous decay, other Z^0 modes must appear; e.g., if $\ell^+\ell^-\gamma$ events were supposed to be $Z^0\gamma$ with $Z^0 \rightarrow \ell^+\ell^-$, then $\gamma\nu\bar{\nu}$ events must appear with $\sigma(\gamma\nu\bar{\nu})/\sigma(\gamma e^+e^-) = 6$, and γjj events must appear with a known,[11] large branching ratio before corrections for soft and coalesced jets.

(3) A *number of distributions* are characteristic of specific mechanisms. One can plot p_T^j, $\not{p}_T$, $\theta(\ell^\pm\gamma)$, $M_T(j\not{p}_T)$, etc. We will see below that the distributions can be as powerful as one might expect in distinguishing competing mechanisms.

(4) The *characteristics of jets* are very important. They are also very sensitive to cuts. Typical jet charged particle multiplicities in the momentum range in question are large, with $n_{ch} \sim 10$–20. However, most of these particles are soft. If a cut is made to take only particles with $p_T > 1$ GeV, then n_{ch} drops greatly for quark jets, with 50–90% of the particles being softer than 1 GeV. If, as sometimes happens, a jet is a combination of two jets coalesced, the charge multiplicity will be double. Similar effects happen to jet effective masses.

The UA1 monojets have rather low mass and charged multiplicity. It seems to be very unlikely that they could be pairs of particles such as $q\bar{q}$, qg, gg. Whether a single q or g can fragment into that few particles ($n_{ch} \leq 4$ for most monojets) is not yet clear, but it seems reasonable. Paige[12] has examined this question with the ISAJET Monte Carlo, which gives a satisfactory description of jet multiplicities and masses at e^+e^- colliders. When he produces events with hard quark jets and UA1 cuts, he finds the mean n_{ch} is 4.2, with 39% having $n_{ch} \leq 3$. However, when he examines a jet where $\tilde{g} \rightarrow q\bar{q}\tilde{\gamma}$ and $q\bar{q}$ coalesce, he finds, as expected, about double the multiplicity.

Probably, it is fair to conclude at present that the "narrow" jets observed can originate with single hard quarks and perhaps single gluons; however, any mechanism where they originate with coalesced jets must provide an explanation of the jet properties in order to be taken seriously. Further, if the jets originate as single quarks, some less narrow ones must appear, so the mean $n_{ch} \simeq 4$.

Many examples could be given of models that can provide one type of anomalous event, but that all too often provide too many of another. A few illustrations are:

(i) Any model based on heavy quarkonium states (formed from known or hypothetical new quarks) that give decays with $\ell^+\ell^-\gamma$ or decays with missing momentum produces[13] a large number of $\ell^+\ell^-$ events at the quarkonium mass. This occurs since the direct annihilation occurs: $Q\bar{Q} \rightarrow \gamma$ or $Z^0 \rightarrow \ell^+\ell^-$. No energetic $\ell^+\ell^-$ pairs are observed, excluding all such models.

(ii) Most models that produce a Z^0 in the decay of a heavy object also produce decays with the Z^0 replaced by a $W^\pm$ and by a γ. The absence of many candidates for events with hard γ's is a severe constraint on models.

(iii) There is an event reported by the CELLO detector at DESY, at 44 GeV near the top machine energy, of $jj\mu^+\mu^-$. The four particles are in one plane, which is very unlikely unless the event is due to a pair of heavy particles produced essentially at rest. However, if the particles are leptoquark states, $LQ \rightarrow j\mu$, the cross section to produce them at CERN is large (they would be like 20 GeV t-quarks) and their signature is very clear, so this is excluded. On the other hand, though, they could be heavy neutrinos, with $\nu' \rightarrow \mu q\bar{q}$, but then the jet would be a coalesced $(q\bar{q})$ one of large effective mass and multiplicity, while the experimental one is a typical light quark jet. Therefore, no interpretation is consistent.

We will see more examples of constraints below.

CATEGORIES OF INTERPRETATIONS

The interpretations of anomalous events can be put into four main categories:

(1) New Z^0 decay modes;
(2) Decays of a singly produced new particle;
(3) Associated production of new particles;
(4) Profound.

The last category, of which examples are given in references 15 and 16, argues that a new interaction may be showing up, but not in a simple way, via new low-spin particles. Supersymmetry either combines categories (2) plus (3) or just uses category (3), depending on the masses involved.

APPROACHES TO MONOJET EVENTS

Since currently the monojet events are the main candidate for a signal for anomalous events, I will comment on various approaches to their interpretation. First, I will describe three approaches that—if the data is taken at face value—do not easily give an acceptable interpretation, but involve interesting and instructive physics. Then I will turn to the supersymmetric approach and describe how some choice of masses may successfully describe the present monojet data.

On the basis of the comments in the previous pages and from the reported monojet (and dijet) data, I will require any explanation to have as necessary conditions:

(i) about the right number of events, including effects of experimental cuts, detection efficiencies, etc.;
(ii) some one-sided dijets (even though the experimental situation and the comparison of experiment and theory is not straightforward here—see below);
(iii) most jets must be due to a single quark and probably not to coalesced jets;
(iv) the monojet distributions, especially the $M_T(j\not{p}_T)$ distribution, must look like the data;
(v) no other wrong predictions.

CAN Z^0 DECAYS EXPLAIN THE MONOJETS?

Two interesting approaches to monojets via Z^0 decays are (a) that of Krauss,[17] followed by references 18 and 19, where $Z^0 \rightarrow \nu'\nu''$, and where the objects, ν' and ν'', are new or familiar neutrino-like particles, which may either escape the detector providing $\not{p}_T$, or decay, providing narrow hadronic jets; and (b) that of Glashow and Manohar,[10] Moffat,[20] and King,[20] where $Z^0 \rightarrow H_l^0 H_2^0$, with H_l^0 and H_2^0 being two light neutral scalar or pseudoscalar (perhaps Higgs) bosons that arise in models with two doublets or with extra gauge groups. One H_l^0 is constrained to be effectively invisible, either by requiring it to be very light so only $\gamma\gamma$ and e^+e^- modes are allowed or by largely decoupling it. The other H_l^0 is light enough to appear as a hadronic jet when it decays.

These approaches do not seem to me to give likely explanations of the data, for reasons I list below, but they have some nice points and should be kept in mind as the data improves. Among their virtues are that they can easily give the right number of events since the associated Z^o branching ratios are in the 1–3% range, which is about what is needed, and that the events would be centered near M_Z in the M_T distribution. As the data increases, the latter point seems to be less of a virtue because the data does not peak in M_T and it spreads out over 80 GeV $\leq M_T \leq$ 120 GeV.

Their problems are:

(1) The experimental M_T distribution is too broad and not sharply peaked;
(2) For H_l^0 giving the monojet, it seems very hard to get $n_{ch} \leq 4$ for most events. The H_l^0 will probably have to decay to $c\bar{c}$, which will give $n_{ch} \geq 5$ for most events;
(3) For ν' giving the monojet, one always needs to have a charged lepton in the jet (since $\nu' \rightarrow \ell \pm q\bar{q}$), but almost none of the monojets are reported to contain a hard charged lepton;
(4) No primary dijet events are generated by this mechanism, so the dijets must come from the primary production of gZ^0 and are suppressed in rate.

CAN TECHNICOLOR EXPLAIN THE MONOJETS?

Technicolor was and is a good idea,[21] that the new physics required by spontaneous symmetry breaking is due to the dynamics of some new fermions. Some bound states also play the role of Higgs bosons. Then, normally, other new particles arise and give a number of interesting signatures. Some of the particles, their masses, and their decays are listed in TABLE 1 (N is a group label, with $N \geq 4$ expected). Although the simplest masses, with $N = 4$, are not appropriate for the current CERN data, perhaps N is larger or maybe a different symmetry gives lower masses; or perhaps such signatures will appear at a future collider. From the table,[21] we can immediately see both some interesting signatures, such as $t\bar{t}$, busy Z's or busy $W^{\pm}$ (Z's or W's accompanied by jets), and $g\gamma$ events. The $g\gamma$ events would be very dramatic, appearing as $j\gamma$, and would have a large cross section. Monojet events would appear as gZ^0 with $Z \rightarrow \nu\bar{\nu}$ and would then require events with $j\ell^+\ell^-$, $j\gamma$, and jjj, all of which are not seen. Pair production of leptoquarks (LQ) would be important if $M(LQ)/M(\eta_T \text{ or } \pi_T) \simeq 2/3$ was maintained, again giving interesting signatures. Altogether, it is clear that technicolor signatures are not what is being observed, but it will be interesting to look again in the future.[22]

TABLE 1. Possible Anomalous Events from Technicolor

Particle	Mass (GeV)	Color	Decays
		Singly Produced	
η_T	$\sim 240\sqrt{4/N}$	8	$t\bar{t}$ 90% gg 5% gZ^0 10%
$\pi_T^{\pm}$	$\sim 240\sqrt{4/N}$	8	$t\bar{b}$ 75% $gW^{\pm}$ 20%
π_T^0	$\sim 240\sqrt{4/N}$	8	gZ^0 30% $g\gamma$ 70%
		Pair Produced	
LQ	$\sim 160\sqrt{4/N}$	3	$t\tau^+$, $t\bar{\nu}$, $b\tau^+$, $b\bar{\nu}$, $j\mu$, je, $j\nu$

CAN *WW* "BOUND STATES" GIVE MONOJETS?

Although proponents of complicated dynamics[15,16] have not proposed specific models for monojets, one can examine a few alternatives to see if any simple interpretation arises. Although none has so far, this does not necessarily imply that the approach is wrong; only that more subtle thinking would be needed if such approaches were relevant. The following remarks are my own approach to such questions and do not reflect the views of other authors.

Two general problems with such approaches will be (i) that the decay $q\not{p}_T$ of a color-singlet WW bound state is not possible, so it will be hard to get narrow, low multiplicity jets plus missing momentum, and (ii) one might expect a single, rather narrow state so far below threshold, which would give problems with the M_T distribution.

One way to get narrow jets would be a vector dominance mechanism, with one $W \to \pi, \rho$. Then monojets could arise from a $W^{\pm}Z^0$ state with

$$\begin{array}{ll} W^{\pm} & Z^0 \\ \quad \to \pi, \rho & \quad \to \nu\bar{\nu} \end{array}$$

or a Z^0Z^0 state with

$$\begin{array}{ll} Z^0 & Z^0 \\ \quad \to \pi, \rho, \omega, \phi, \psi. & \quad \to \nu\bar{\nu} \end{array}$$

The main problems with this mechanism are rate/branching ratio and other modes. In particular, if this occurred, one would also expect

$$\begin{array}{ll} W^{\pm} & Z^0 \\ \quad \to \pi, \rho & \quad \to \ell^+\ell^-, jj \end{array}$$

and

$$\begin{array}{ll} W^{\pm} & W^{\pm} \\ \quad \to \pi, \rho & \quad \to \ell^{\pm}\nu \end{array}$$

giving large numbers of $\ell^{\pm} j \not{p}_T$ events, which may not be observed. (However, note the UA2 events.)

CAN SUPERSYMMETRY EXPLAIN MONOJET EVENTS?

We have seen that no proposed approach we considered above seems to give a clean explanation of the monojet events. It has been known for a long time[23] that events with jets and missing momentum would occur if supersymmetric partners were being produced. Now, we will consider whether the detailed characteristics of supersymmetric particle production are consistent with the monojet data. All the work I describe in this section is based on a collaboration[7] with M. Barnett and H. Haber, plus the work of other authors. We will see that there is indeed a satisfactory description, perhaps even competing ones, as far as the monojet data is concerned. Unbalanced dijet events are also predicted and some are found. The detailed properties of the predicted ones and the observed ones are not quite the same, however, and whether this is a problem is not yet clear.

The minimal spectrum of new particles expected if supersymmetry were a symmetry of nature as realized on the weak scale is that each of the quarks, leptons, and gauge and Higgs bosons should have a partner whose properties are identical to the particle except that it differs in spin by half a unit. A useful notation and nomenclature is indicated in TABLE 2. The names indicate both the origin of the particle and, for the partners of gauge bosons and Higgs bosons, how the partner interacts; because of weak interaction eigenstates mixing to form mass eigenstates, the interaction properties can be subtle, but they are important to determine observability. Left-(L) and right-handed(R) fermions are separated since they have different weak interactions.

If supersymmetry were an unbroken symmetry, the partners would have the same mass, color, charge, etc., as the particles, and they would have been easily detected. Since they have not been, we consider them to be heavier because of symmetry breaking. If supersymmetry is relevant to explaining the weak scale, presumably it makes sense to expect the partners to have mass, $\sim M_W$. If partners are not discovered

TABLE 2. The Minimal Particle Spectrum in a Supersymmetric Theory

Normal Particles	Superpartners	Names
g	$\tilde{g}$	gluino
u_L	$\tilde{u}_L$	up-squark
u_R	$\tilde{u}_R$	·
·	·	·
·	·	·
e_L	$\tilde{e}_L$	selectron
·	·	·
·	·	·
$W^{\pm}$	$\tilde{W}^{\pm}$	wino } charginos
		whiggsino } charginos
$H^{\pm}_{1,2}$	$\tilde{h}^{\pm}_{1,2}$	higgsino } charginos
γ	$\tilde{\gamma}$	photino } neutralinos
		zino } neutralinos
Z	$\tilde{Z}$	higgsino } neutralinos
$H^0_{1,2}$	$\tilde{h}^0_{1,2}$	

on the weak scale, eventually the limits on their masses will be pushed up to well beyond the weak scale and they will become irrelevant to the understanding of physics on the weak scale.

Since the way in which the supersymmetry is broken is not understood, the masses of the superpartners are unknown. It is necessary to proceed by assuming a set of masses (in practice that means masses for mainly $\tilde{g}$, $\tilde{q}$, $\tilde{w}$, $\tilde{\ell}^{\pm}$, $\tilde{\nu}$, and $\tilde{\gamma}$), calculating predictions, and testing them.

Fortunately, the couplings of the superpartners are known because they are mainly gauge couplings. In fact, they are the actual measured gauge couplings, so no uncertainty arises at all. The rule that emerges from the theory is that every vertex in the standard theory has added other vertices obtained by replacing particles by their partners in pairs, while keeping the interaction strength the same. Thus, the $\bar{q}qg$ coupling gives $\tilde{q}\tilde{g}q$ and $\bar{\tilde{q}}\tilde{q}g$ couplings, all of strength, g_3 ($g_3^2/4\pi = \alpha_s$), the $We\nu$ coupling gives $\tilde{w}\tilde{e}\nu$, $\tilde{w}e\tilde{\nu}$, and $W\tilde{e}\tilde{\nu}$ couplings, all of strength, g_2, etc. With this rule, one can draw all the Feynman graphs that are relevant to whatever process is being considered and estimate any cross section or decay.

Since supersymmetric partners occur in pairs, they can either be pair produced or a single one can be produced via $\tilde{q}$ or $\tilde{g}$ in a proton.[24,25] (The other supersymmetric particle goes down the beam pipe with other soft partners.) We will consider both mechanisms; it turns out that if gluinos are lighter than about 25 GeV, the gluino content of a proton is significant.

Once a superpartner is produced, it will decay into normal particles plus a lighter superpartner; the lightest superpartner will be stable (one can consider modified cases where the lightest superpartner is unstable and carry through the analysis in a way analogous to the following, but little changes and we will not pursue this alternative here). The lightest superpartner can be a partner of a gauge boson or a Higgs boson, or it can be a sneutrino. For our purposes, it will not matter much what is the lightest superpartner, just as long as its essential property is that it normally escapes collider detectors. This is so because to interact, it must excite a superpartner in the detector, and since the partners of quarks and leptons are heavy, the interaction cross section is at most of order, $\alpha^2/\tilde{m}^2$, with $\tilde{m} \gtrsim 20$ GeV, which gives too small a cross section to see.

Thus, the basic signature of the production of supersymmetric partners is missing momentum, accompanied by jets or charged leptons in characteristic patterns. If a squark is produced, it will decay into $q\tilde{g}$ and $q\tilde{\gamma}$ if both are kinematically allowed, in a ratio (ignoring phase-space) of

$$\frac{\Gamma(\tilde{q} \longrightarrow q\tilde{\gamma})}{\Gamma(\tilde{q} \longrightarrow q\tilde{g})} = \frac{3e_q^2\alpha}{4\alpha_s}.$$

For $\tilde{u}$, this is $\alpha/3\alpha_s \simeq 0.02$. If $\tilde{m}_q \gg \tilde{m}_w, M_q$, then

$$\frac{\Gamma(\tilde{q} \longrightarrow q\tilde{w})}{\Gamma(\tilde{q} \longrightarrow \tilde{q}g)} \simeq \frac{1}{2}\frac{3\alpha_2}{8\alpha_s} \simeq 0.05,$$

where the 1/2 appears because only $\tilde{q}_L$ has a coupling to $\tilde{w}$. Similarly, if $\tilde{\ell}^\pm$ appear, they will decay as $\tilde{\ell}^\pm \longrightarrow \ell^\pm\tilde{\gamma}$. Gluinos will mainly decay via a virtual $\tilde{q}$, $\tilde{g} \longrightarrow \bar{q}q\tilde{\gamma}$, with a branching ratio (that depends on other masses) of perhaps 1% to $g\tilde{\gamma}$.

If sleptons or sneutrinos are lighter than winos or zinos, the latter will have useful signatures,

$$\tilde{w} \longrightarrow \tilde{\ell}^\pm\nu \longrightarrow \ell^\pm \not{p}_T,$$

$$\tilde{w} \longrightarrow \ell^\pm\tilde{\nu},$$

or

$$\tilde{z} \longrightarrow \tilde{\ell}^+\ell^- \longrightarrow \ell^+\ell^- \not{p}_T.$$

If sleptons and sneutrinos and also $\tilde{g}$ are heavier than winos and zinos, the latter still may have a useful signature,

$$\tilde{w} \longrightarrow \bar{q}q\tilde{\gamma} \longrightarrow jj\not{p}_T,$$

$$\tilde{z} \longrightarrow \bar{q}q\tilde{\gamma} \longrightarrow jj\not{p}_T.$$

If only g are lighter than $\tilde{w}$, $\tilde{z}$, the latter will decay into $jjjj\tilde{\gamma}$, becoming essentially unobservable. In the approach we favor, with $\tilde{M}_g \simeq 10$ GeV, this is perhaps the most likely situation, though it could well happen that sleptons or sneutrinos are lighter than $\tilde{w}$ or $\tilde{z}$.[26]

PRODUCTION OF SUPERPARTNERS

There are two kinds of ways to produce superpartners at a hadron collider.

(A) Starting with quarks and gluons in the proton, we have[27]

$$\sigma(p\bar{p} \rightarrow \tilde{a}\tilde{b}X) = \int dx_1 dx_2 F_c(x_1) F_{\bar{d}}(x_2) \hat{\sigma}(c\bar{d} \rightarrow \tilde{a}\tilde{b}),$$

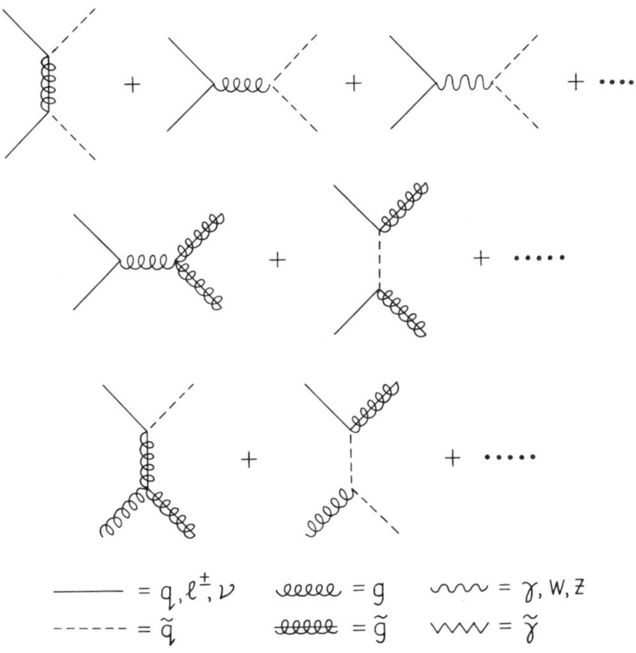

FIGURE 1. Some ways to produce superpartners. At $\tilde{q}\tilde{q}$ vertices, $\tilde{g}$ can be replaced by $\tilde{w}$, $\tilde{z}$, or $\tilde{\gamma}$, giving production of $\tilde{w}\tilde{w}$, $\tilde{w}\tilde{g}$, $\tilde{g}\tilde{\gamma}$, $\tilde{z}\tilde{g}$, etc.

where F_c is the structure function for constituent c in a proton, F_d is the structure function for constituent d in a proton, and $\hat{\sigma}$ is the constituent cross section. FIGURE 1 shows a number of constituent processes to produce superpartners; which ones dominate will depend on masses, kinematic regions, cuts, etc. In all cases, we include all subprocesses in our calculations. We have calculated all cross sections to produce $\tilde{q}\tilde{q}$, $\tilde{g}\tilde{g}$, $\tilde{g}\tilde{q}$, $\tilde{w}\tilde{g}$, $\tilde{g}\tilde{\gamma}$, etc., and comparted with earlier results in reference 27. We use the EHLQ structure functions[28] for calculations, with $\Lambda = 0.29$ GeV. Then, events are generated by a Monte Carlo integration technique[29] and final state particles are decayed, using a procedure due to two of us.[29] (Note that we have not included any effects of QCD gluon radiation or "K-factors" in the calculations.)

Ultimately, an important test of whether supersymmetric partners are being produced will be the determination of the spins of the new particles. While it is clearly too early to be doing that, the spins do affect the distributions and rates we predict, so their effects are being included. At the present level of calculation, no polarization effects are expected; transverse polarizations are absent since we are assuming parity conservation. Supersymmetric processes in general do not conserve parity since masses of $\tilde{q}_L$ and $\tilde{q}_R$ are not in general equal. By including such mass differences, longitudinal

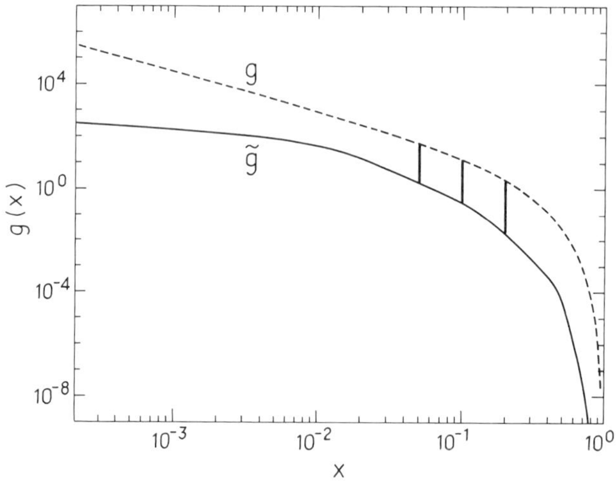

FIGURE 2. Gluino distribution function, from reference 7.

polarizations would be generated, but at the present time such effects are highly model dependent and we will neglect them.

(B) If colored superpartners were light, they would occur[24,25] in hadrons as well as normal colored quarks and gluons. Since squarks are known to be heavier than about 20 GeV, they will not appear in significant quantities in a proton. The probability of a gluon splitting into a gluino pair is six times larger than the probability of it splitting into a quark pair of the same mass, so even if $\tilde{M}_g$ is 20–25 GeV, there is a significant[7] gluino content to the proton.

The question of gluinos in a proton has been raised in reference 24, and Herrano *et al.*[25] have specifically suggested that single squark production would occur via this mechanism. We have calculated the gluino structure functions, including mass effects,[7] by using the technique proposed by Gluck,

Hoffman, and Reya (based in turn on the Altarelli, Parisi equations) and by checking against the EHLQ calculations of the heavy quark content of a proton. Since $\tilde{m}_g^2/Q^2 \lesssim 1/16$ and even though there are corrections of the order, $\tilde{M}_g^2/Q^2$ and $(\tilde{M}_g^2/Q^2)^2$, involved, the lowest order ones, which are the most likely to be correct, are dominant. FIGURE 2 shows the gluino structure functions obtained. The curve for $d\sigma/dx$ peaks around $x \simeq 0.08$, where the calculation is expected to be reliable, with x neither large nor small when compared to $\tilde{M}_g^2/Q^2$ for the values of $\tilde{M}_g^2$ we find to be relevant.

EXPERIMENTAL CUTS

In a real experimental environment, it is necessary to apply cuts to the data[30] in order to exclude conventional sources of anomalous events. The most important cut is on the minimum $\not{p}_T$ required to count the event as a monojet event. Data has been presented[2] in various ways, with fixed cuts such as $\not{p}_T > 40$ GeV and with a "4σ" cut. We will discuss how we implement these[7] in the theory in the following section; it is subtle to make the "4σ" cut in a way where theory is being correctly compared to experiment. Other groups[5,6,31–35] have already done similar analyses. It is a difficult procedure and not all analyses agree, though convergence is occurring.

In addition, an anomalous event must have one jet with $p_T^j > 25$ GeV. Since $\not{p}_T$ and p_T^j are both large, clearly this biases all other activity in the event to be softer. A jet is defined[2,30] as all activity in a bin, $\Delta\eta^2 + \Delta\phi^2 \leq 1$, where η is the rapidity and ϕ is the azimuthal angle.

Some geometrical cuts are applied, as described in the UA1 paper,[30] and we have included these in our analysis.

THEORETICAL CUTS

In discussing the data, it is useful[7] to consider three different cuts on $\not{p}_T$: (a) $\not{p}_T > 40$ GeV, (b) $\not{p}_T > 32$ GeV, and (c) $\not{p}_T > 2.8\sqrt{E_T}$ (i.e., the 4σ cut). The first two are straightforward. The recently reported data is most complete for the $\not{p}_T > 40$ GeV cut. The $\not{p}_T > 32$ GeV results are presented as a comparison and as a prediction of how the number of monojet events changes as the cut is varied.

For the cuts on jet p_T, we use the same values as the experimental analysis; $p_T^j > 12$ GeV for others. This is somewhat of a subtle point, as it is not clear whether a 25 GeV parton corresponds to a 25 GeV jet in the detector. Future analyses can improve on this.

The most subtle aspect of the analysis is defining the 4σ cut. It is not necessary to consider this cut for the main part[7] of our analysis (where we examine the $\not{p}_T > 40$ and $\not{p}_T > 32$ distributions for all $\tilde{q}$ and $\tilde{g}$ masses), but it is necessary to compare with previous analyses, particularly of pair production of lighter gluinos, and it will be necessary for a full analysis in the future of all $\not{p}_T$ events due to both Standard Model and anomalous sources since they overlap near the 4σ cut. Therefore, we go into some detail.[7]

The experimental missing energy fluctuations are characterized by a $\sigma = 0.7\ \sqrt{E_T}$, i.e., the amount of energy missing in an event goes as $\sqrt{E_T}$. To be safe in assigning an event to be new physics, it is required that $\not{p}_T$ be greater than 4σ. The problem is how to relate what happens in a theory event to E_T. The so-called "minimum bias" collider events (those with no large p_T activity) have a distribution in E_T with a mean of $\langle E_T \rangle \simeq$ 20–25 GeV. Further, normal QCD jet-jet events are observed to have the property that when the two jets are removed, the remaining event has $\langle E_T \rangle \simeq$ 40–50 GeV.

Now suppose the monojets are due to the production of supersymmetric partners. Then presumably the total E_T is given by

$$E_T^{Tot} = E_T^0 + E_T^{q,g},$$

where E_T^0 is related to what is observed to QCD jet-jet events after the two jets are removed and where $E_T^{q,g}$ is the total energy associated with all of the partners that arise from the decay of supersymmetric particles, whether they pass the jet cuts or not. We take E_T^0 to have mean of 40 GeV (using the lower side of the range because there is one less jet to radiate) and we use a distribution about E_T^0 to make the simulation as realistic as possible.[7]

This procedure gives a larger $E_T = E_T^{Tot}$ with which to calculate σ, so fewer events pass the cuts, and we do not find too many monojets arising from light gluinos that are pair produced. As a consequence, we do not find that any $\tilde{M}_g$ is clearly excluded by predicting significantly more monojet events than there are candidates for. In this connection, it should be noted that as the gluino mass decreases, it becomes less probable that a given $\tilde{g}\tilde{g}$ event passes the cuts, while the cross section for making the $\tilde{g}\tilde{g}$ pair grows. Since the probability of passing the cuts gets quite small ($\lesssim 10^{-3}$ when $\tilde{M}_g \simeq$ 5 GeV), it is necessary to generate a large number of Monte Carlo events to have reliable results. While we have been careful to do this, we think further consideration of the low mass gluino pairs may be worthwhile.

MONOJET EVENTS

We have concluded that the regions of the $\tilde{M}_g$-$\tilde{M}_q$ plane that could describe the data are those shown in FIGURE 3. In the next sections, we study the allowed regions further. For monojets,[b] one can get about the same conclusions from the distributions in $M_T(j, \not{p}_T)$, p_T^j or $\not{p}_T$, so we mainly use M_T, which is easy to interpret physically. We find that, taking the data at face value, there seems to be a preference for the region with $\tilde{M}_q \simeq 110$ GeV and $\tilde{M}_g \simeq$ 10–20 GeV where most of the monojets arise from single $\tilde{q}$ production via gluinos in the proton. While it is too soon to conclude that this is the only allowed region or that it is indeed the origin of the monojets, at present it may give a better description of the data than the other alternatives.

It is important to note that the reported monojet data has not been corrected for p_T^j mismeasurement.[36] Indications from other jet information[36] are that p_T^j, M_T, etc., will

[b] Note that for dijet events, it is not so simple because of the extra jet. Our dijet predictions give a good description of the reported M_T distribution, but not of the $\not{p}_T$ values.

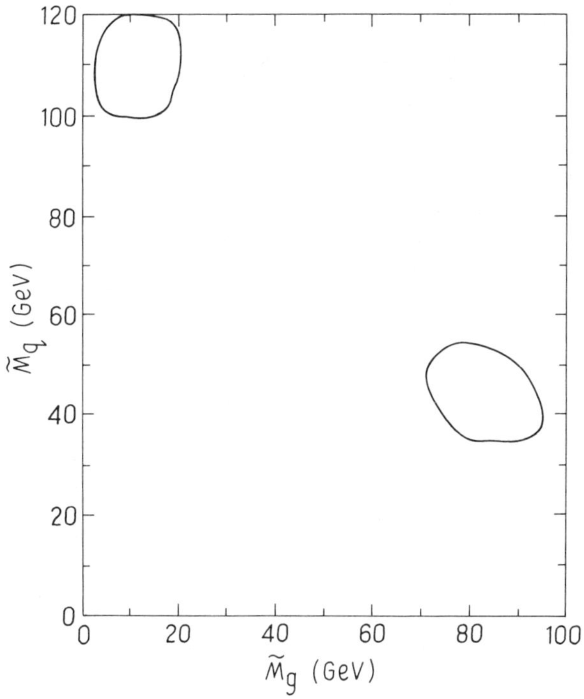

FIGURE 3. Allowed regions in the $\tilde{M}_g$-$\tilde{M}_q$ plane for production of supersymmetric partners to explain monojet events, according to the criteria of reference 7.

be corrected up by 5–10% eventually. We do not show such corrections, but they should be kept in mind.

SINGLE SQUARK PRODUCTION

We have explained above how the single squark is produced via the $\tilde{g}$ content of the proton.[7] There are three sources of actual monojet events:

(1) $q\tilde{g} \to \tilde{q} \to q\tilde{g}$
$\qquad\qquad\qquad\quad |\!\!\to q\bar{q}\tilde{\gamma}$

This provides the majority of monojet events. The q from $\tilde{q}$ decay is energetic and passes the jet cuts, and 5–10% of the time, the photino from $\tilde{g}$ decay passes the $\not{p}_T$ cut, giving an event that passes the UA1 cuts. There is, though, some residual energy from the two soft q, $\bar{q}$ from the $\tilde{g}$ decay.

(2) $q\tilde{g} \to \tilde{q} \to q\tilde{\gamma}$

This is about a 2% branching ratio, but it essentially always passes the cuts. It provides 10–15% of the monojet events.

(3) Since gluinos are 10–20 GeV, their pair production cross section is large. As we discussed above, of the order 10^{-3} of the candidates pass the cuts. We find about 25–35% of the monojet events arising this way. Once $\tilde{M}_g^2 \ll Q^2 \simeq (100\ \text{GeV})^2$, the number of monojets arising from this mechanism is basically independent of $\tilde{M}_g$, as might be expected physically.

We note from the M_T distribution with various cuts (FIGURE 4) that if the cut is placed at $\not{p}_T > 32$ GeV, almost as many new monojet events should occur as above $\not{p}_T > 40$ GeV. While not quite that many candidates are in the reported data, several candidates do exist, so at present this prediction is not inconsistent with the data. However, to accept the heavy squark interpretation, it will be crucial to see the additional events as the $\not{p}_T$ cut is reduced.

Note that there is some sensitivity to $\tilde{q}$ and $\tilde{g}$ masses, but not too much flexibility. The fraction of the events from $\tilde{q} \rightarrow q\tilde{\gamma}$ does shift as $\tilde{M}_q$ shifts, but because the other sources of events depend on the $\tilde{g}$ decay, they are not as sensitive. When high statistical data is available, an accurate determination of masses will be possible. If the M_T distribution is shifted up 5–10% from jet energy calibration, as expected, one can shift masses to compensate, but it is not a simple situation because cross sections will then decrease. If one takes the present numbers very seriously, though, $\tilde{M}_g \simeq 10$–15 GeV probably, but as the data improves this could change.

SQUARK PAIR PRODUCTION

The M_T distribution for squark pair production[37] is shown in FIGURE 5. If only the $\not{p}_T > 40$ distribution were considered, this mechanism does as well as the single heavy

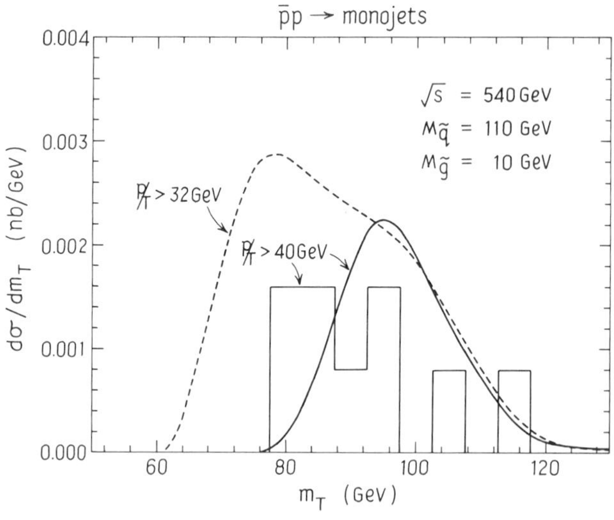

FIGURE 4. M_T distribution for $\tilde{M}_q = 110$ GeV, $\tilde{M}_g = 10$ GeV, from reference 7. Data from UA1, reference 3.

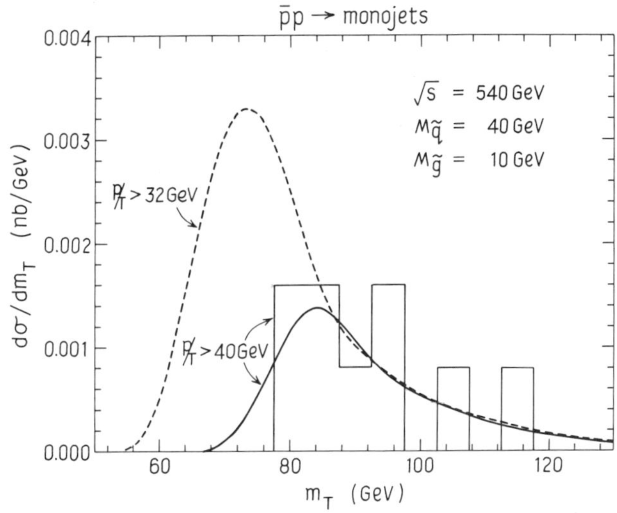

FIGURE 5. M_T distribution for $\tilde{M}_q = 40$ GeV, $\tilde{M}_g = 80$ GeV, from reference 7. Data from UA1, reference 3.

squark. However, the full M_T distribution for $\tilde{q}\tilde{q}$ peaks strongly at $M_T \simeq 70$–75 GeV, while the data is flat or a little peaked nearer to 90–100 GeV. If the $\not{p}_T$ cut is relaxed from 40 to 32 GeV, at least 1.5 times as many events should appear as were already found. However, the reported data contains only about two-thirds as many candidate events in that region as above $\not{p}_T = 40$ GeV. If one tries to shift the distribution up by increasing $\tilde{M}_q$, the cross section decreases. Therefore, if the data continues to be very flat or peak above 85–90 GeV, it will not be possible to obtain a consistent description of the data with $\tilde{q}$ pair production. The situation will get worse if the entire M_T distribution is shifted up by 5–10 GeV because of jet energy calibration. This is the basis of our preference for the single squark mechanism.

HEAVY SQUARK AND VERY LIGHT GLUINOS

Barger *et al.* have argued[33,35] that an alternative mechanism, with a heavy $\tilde{q}$ ($\tilde{M}_g \simeq$ 3 GeV), gives a description of the data. While we find this is perhaps correct as to rate and perhaps as to the M_T distribution, for two reasons we are concerned[7] about the validity of this interpretation. Since the gluino in this scenario is either long-lived or fragments softly so that the resulting photino and jets essentially never pass the cuts, all of the monojet events were assumed[35] to arise from

$$qg \to \tilde{q}\tilde{g}$$
$$\quad\quad \lfloor\!\to q\tilde{\gamma}.$$

This gives a relatively narrow distribution, but one that is spread out by the recoil $\tilde{g}$. It is also necessary to include the $\tilde{g}$ in the proton when it is so light; this gives additional events from

$$q\tilde{g} \rightarrow \tilde{q} \rightarrow q\tilde{\gamma}$$

and there is a possible problem from getting too many monojet events from the two mechanisms. Further, for every monojet generated by this latter mechanism, there are 50–60 jet-jet pairs with an effective mass of $\tilde{M}_q$; the $\tilde{q}$ width is 8–10 GeV, comparable with the experimental resolution, so this may be an observable peak. Our calculations indicate[7] about 500 jj events from this source, with M_{jj} about 100 GeV. It is not clear whether such a signal would be observed.

COMMENTS

In this discussion of the anomalous events, I have first given a general analysis of how to approach the study of new (possible) experimental phenomena that might take us beyond the Standard Model and I have discussed the constraints on ideas. Then I gave a short review of the kinds of anomalous events expected if supersymmetric partners were being produced. It is very interesting that at the present time the only candidates for anomalous events are the kind that are expected if partners of quarks and gluons are produced. Indeed, other possible kinds of events, which could not have been explained by supersymmetry, have tended to disappear as the data have improved. I have not considered at all whether the anomalous events constitute a signal of phenomena not interpretable within the Standard Model; that is up to the experimenters, who have now proposed such status for the monojets.

Given that situation, it is next necessary to examine whether supersymmetry can quantitatively interpret the data; in this case, that means mainly the UA1 monojet data (the situation with dijets plus missing momentum is somewhat unclear experimentally and also unclear theoretically because of the possibility of QCD radiation). It would be of extraordinary significance if new phenomena were discovered at all and of even greater significance if the new phenomena were related to supersymmetry which has such potential theoretical value. It does indeed seem to be the case that at present, there is at least one valid supersymmetric interpretation of the data. Finally, I have discussed choosing among different supersymmetric interpretations (i.e., choices of superpartner masses), which is of importance particularly to workers within the field and to model builders. Further discussion along these lines is available in references 7, 25, 26, and 31–35 and in several papers that are surely in preparation by various authors.

ACKNOWLEDGMENTS

I appreciate a number of useful conversations with V. Barger, S. Glashow, L. Hall, I. Hinchliffe, A. Kernan, M. Mohammadi, F. Paige, and V. Vuillemin. I would like to thank M. Block and his family and Ruth Goldhor for their hospitality and their very effective efforts to make this meeting successful.

NOTES AND REFERENCES

1. For other reviews, see HALL, L. J., R. L. JAFFE & J. ROSNER. 1984. Proceedings of the Workshop on Electroweak Symmetry Breaking, June 1984. LBL-18571; BARGER, V. 1984. Invited talk at Santa Fe DPF Meeting, October 1984, MAD/Ph/229.
2. FNAL COLLABORATION. BROOKHAVEN, COLUMBIA. To be published; BALTAY, C. Private communication.
3. See the talks of MOHAMMADI, M. & J. ROHLF, UA1, at this conference; see the talk of HANSEN, J., UA2, at this conference.
4. For a review, see HABER, H. E. & G. L. KANE. 1985. Phys. Rep. **117:** 75.
5. ELLIS, J. & H. KOWALSKI. 1984. Phys. Lett. **142B:** 441.
6. REYA, E. & D. P. ROY. 1984. Phys. Rev. Lett. **53:** 881.
7. BARNETT, R. M., H. E. HABER & G. L. KANE. 1985. Implications of a systematic study of the CERN monojets for supersymmetry (LBL-18900/SLAC-PUB-3551). Phys. Rev. Lett. **54**; Preprint LBL-20102, Aug. 1985.
8. BARGER, V., H. BAER & K. HAGIWARA. 1984. Phys. Rev. Lett. **146B:** 257.
9. DUSEDAU, D., D. LUST & D. ZEPPERFELD. 1984. Phys. Lett. **148B:** 234.
10. This has been emphasized by GLASHOW, S. L. & A. MANOHAR. 1985. Phys. Rev. Lett. **54:** 526.
11. For general constraints from any $SU(2) \times U(1)$ current, see DUNCAN, M. J. & M. VELTMAN. 1984. Phys. Lett. **139B:** 310.
12. PAIGE, F. Private communication.
13. KANE, G. L. & L. MAIANI. 1984. Phys. Lett. **147B:** 469.
14. BEHREND, H. J. *et al.* CELLO COLLABORATION. 1984. Phys. Lett. **141B:** 145.
15. VELTMAN, M. 1984. Phys. Lett. **139B:** 139.
16. HIKASA, K-I. 1984. Phys. Lett. **145B:** 139 (UT-433, May 1984).
17. KRAUSS, L. M. 1984. Phys. Lett. **143B:** 248.
18. GRONAU, M. & J. ROSNER. 1984. Phys. Lett. **147B:** 217.
19. HOLDOM, R. 1984. Toronto preprint.
20. MOFFAT, J. Toronto preprint; KING, S. F. 1985. Phys. Rev. Lett. **54:** 528.
21. WEINBERG, S. 1976. Phys. Rev. **D13:** 974; SUSSKIND, L. 1979. Phys. Rev. **D20:** 2619; DIMOPOULOS, S. & L. SUSSKIND. 1979. Nucl. Phys. **B155:** 237; EICHTEN, E. & K. LANE. 1980. Phys. Lett. **90B:** 125. All numbers are from DIMOPOULOS, S., S. RABY & G. L. KANE. 1981. Nucl. Phys. **B182:** 77.
22. For related signatures, see BERGER, E. L. & M. JACOB. 1984. Phys. Lett. **147B:** 197.
23. KANE, G. L. & J. P. LEVEILLE. 1982. Phys. Lett. **112B:** 227.
24. CAMPBELL, B. A., J. ELLIS & S. RUDAZ. 1982. Nucl. Phys. **B198:** 1; ANTONIADIS, I., C. KOUNNAS & R. LACAZE. 1983. Nucl. Phys. **B211:** 216; KOUNNAS, C. & D. ROSS. 1983. Nucl. Phys. **B124:** 317; JONES, S. K. & C. H. LLEWELLYN SMITH. 1983. Nucl. Phys. **B217:** 145; HERRERO, M. J., C. LOPEZ & F. J. YNDURAIN. 1984. Nucl. Phys. **B244:** 207.
25. HERRERO, M. J., L. E. IBANEZ, C. LOPEZ & F. J. YNDURAIN. 1983. Phys. Lett. **132B:** 199; 1984. Phys. Lett. **145B:** 430.
26. For further discussion of the possible implications of supersymmetry for other experimental signatures, see HALL, L. J. & J. POLCHINSKI. Implications of supersymmetric origins for monojets. HUTP-84/AO85; ELLIS, J. & M. SHER. 1984. Phys. Lett. **148B:** 309; BAER, H., J. ELLIS, D. V. NANOPOULOS & X. TATA. 1984. CERN-TH4059/84; HALL, L. J. Talk at DPF Santa Fe Meeting; and reference 7.
27. KANE, G. L. & J. P. LEVEILLE. 1982. Phys. Lett. **112B:** 227; HARRISON, P. R. & C. H. LLEWELLYN SMITH. 1983. Nucl. Phys. **B213:** 223 (1983. E. **B223:** 542); ANTONIADIS, I., L. BAULIEU & F. DELDUC. 1984. Z. Phys. **C23:** 119; DAWSON, S., E. EICHTEN & C. QUIGG. 1984. FERMILAB-PUB-83/82-THY.
28. EICHTEN, E., I. HINCHLIFFE, K. LANE & C. QUIGG. 1984. Rev. Mod. Phys. **56:** 579.
29. See BARNETT, R. M. & H. E. HABER. 1984. Simulating supersymmetry at the SSC. To be published in the Proceedings of the 1984 DPF Summer Study on the Design and Utilization of the SSC, Snowmass, Colorado.

30. ARNISON, G. *et al.* 1983. Phys. Lett. **132B:** 214.
31. ELLIS, J. & H. KOWALSKI. 1984. Nucl. Phys. **B246:** 189.
32. REYA, E. & D. P. ROY. 1984. Dortmund preprint DO-TH 84/19.
33. BARGER, V., K. HAGIWARA, W. Y. KUENG & J. WOODSIDE. 1984. Wisconsin preprint MAD/PH/197.
34. ALLAN, A. R., E. W. N. GLOVER & A. D. MARTIN. 1984. Phys. Lett. **146B:** 247; ALLAN, A. R., E. W. N. GLOVER & S. L. GRAYSON. 1984. Durham preprint DTP/84/28.
35. BARGER, V., K. HAGIWARA & J. WOODSIDE. 1984. Phys. Rev. Lett. **53:** 641; BARGER, V., K. HAGIWARA & W. Y. KEUNG. Phys. Lett. **145B:** 147.
36. I appreciate a discussion with A. Kernan on this point.
37. Previous analyses are in references 5, 31, 33, and 34. We have carried out our own calculations as described in reference 7, and FIGURE 5 is from reference 7.

Experimental Observation of Events with Large Missing Transverse Energy[a]

Presented by MOHAMMAD MOHAMMADI

UA1 Collaboration[b]
CERN CH-1211
Geneva 23, Switzerland

INTRODUCTION

This presentation will summarize the results of the missing energy analysis from the UA1 experiment at CERN. The talk is divided into two parts. The first part will review the 1983 data taken at a total center-of-mass energy of 540 GeV. These data correspond to an integrated luminosity of 113 nb^{-1}. The second part is a summary of the new data from the first 130 nb^{-1} of the 1984 run taken at a total center-of-mass energy of 630 GeV.

REVIEW OF THE 1983 DATA

Missing Transverse Energy

The UA1 detector has 4π calorimetry and detectors covering the full angular range down to 0.2°, with a minimal fraction of insensitive areas. For standard events, such as events with no noninteracting particles, the energy balance is expected to be determined by overall calorimeter resolutions. The longitudinal component of energy balance cannot, of course, be measured, owing to the loss of energetic particles down the beam pipe. However, the transverse components are fairly well measured. FIGURE 1 shows the observed distribution of the vertical component of transverse energy, E_T, for minimum bias events.[1] The transverse components, ΔE_y and ΔE_z, of E_T seem to be neatly centered around zero and have a Gaussian shape with an r.m.s. width that can be parametrized as $0.5\ \sqrt{|E_T|}$, where $|E_T|$ is the scalar sum of transverse energies observed in all calorimeter cells. The modulus of the resultant missing E_T vector, $|\Delta E_M| = (\Delta E_y^2 + \Delta E_z^2)^{1/2}$, is distributed exponentially in the variable, $|\Delta E_M|^2$. The same parametrization of the missing E_T resolution also holds for events that have high transverse momentum jets. This is demonstrated in FIGURE 2, where the missing E_T observed for a sample of two-jet events is shown. The curve shows the result of a Monte Carlo calculation in which our parametrization of the resolution for each transverse

[a]This work was supported by U.S. Department of Energy contract no. DE-AC02-76ER00881.

[b]This UA1 Collaboration includes Aachen, Amsterdam (NIKHEF), Annecy (LAPP), Birmingham, CERN, Harvard, Helsinki, Kiel, London (Queen Mary College), Paris (Coll. de France), Riverside, Rome, Rutherford Appleton Laboratories, Saclay (CEN), Vienna, and Wisconsin.

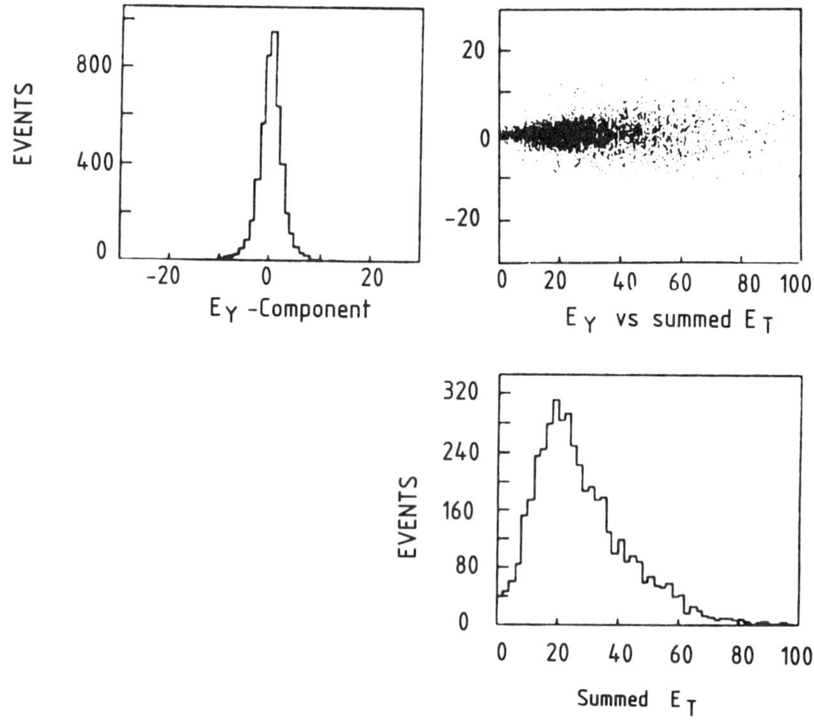

FIGURE 1. Scatter plot of the transverse component (y component) of the missing transverse energy versus the total transverse energy observed in all calorimeter cells.

component was used to fluctuate the jets and the underlying event, using a sample of balanced events containing a minimum of two jets. The missing E_T distribution of these fluctuated events seems to be in good agreement with data.

The apparent missing E_T measured in an event can be associated with the emission of one or more neutral noninteracting particles, such as neutrinos or photinos of supersymmetric theories. As a practical example, the missing E_T distribution of our 1983 $W \rightarrow e\bar{\nu}$ events is shown in FIGURE 3. A sharp Jacobian peak at about one-half of the W mass is clearly demonstrated.

Event Selection

The selection of events with large missing E_T is described in detail elsewhere.[2] Here, it is emphasized that since there was no dedicated trigger provided by the missing E_T alone, an additional requirement was necessary to trigger the event. This was (a) a jet of $E_T > 25$ GeV, (b) an electromagnetic cluster of $E_T > 10$ GeV, (c) a muon of $p_T > 5$ GeV/c, or (d) a scalar transverse energy, $|E_T| > 60$ GeV, in the region, $|\eta| < 1.5$. After requiring $|\Delta E_M| > 4\sigma$, where $\sigma = 0.7\ \sqrt{|E_T|}$, and that the ΔE_M vector did not point to within $\pm 20°$ of the vertical, the selected events were all scanned by physicists. A total of 77 events were found to pass the final selection and the ΔE_M were declared genuine. FIGURE 4a shows a plot of $|\Delta E_M|^2$ versus $|E_T|$ for these events. Out of the 77 events, 50 were found to be known $W \to e\bar{\nu}$ decays and 27 were new. These events are shown in FIGURE 4b. The events are classified as having a single electromagnetic cluster, a single jet ($E_T > 12$ GeV), two jets, or three or more jets.

Isolated Electromagnetic Clusters

Events G and H of FIGURE 4b have a single isolated electromagnetic cluster without an associated hard track. These are called the "photon" events. An inclusive "photon" search without a missing energy requirement was performed. The results are

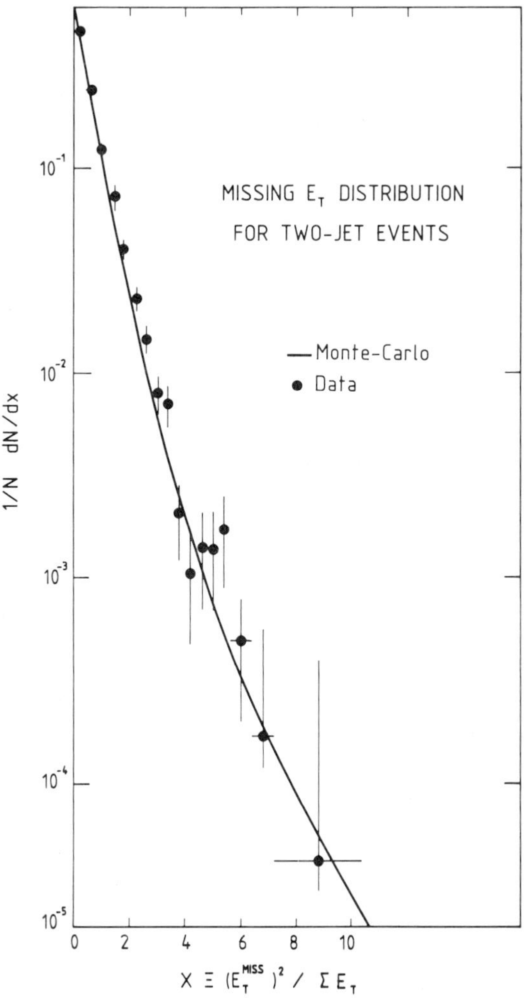

FIGURE 2. Missing transverse energy squared normalized to the total scalar transverse energy observed for a sample of two-jet events. The solid curve is the result of a Monte Carlo simulation (see text). The number of standard deviations is approximately equal to $\sqrt{2X}$.

shown in FIGURE 5, where the "photon" transverse energy is plotted against the missing transverse energy squared. As can be seen, the low missing E_T events are consistent with jet-jet events where one of the jets has one or more leading π^0's, while the other fluctuates down and appears to be very soft. However, events G and H have missing E_T far larger than our expected ΔE_M resolution and they exhibit quite larger values of missing energy than do the other events.

Events G and H have clear interaction vertices and their electromagnetic showers point to the vertex. One potential background comes from $W \rightarrow e\bar{\nu}$ decays, where the electron track is not recorded in the central tracking chambers. FIGURE 6 shows the polar versus azimuthal angle of all observed $W \rightarrow e\bar{\nu}$ events. The crosshatched area corresponds to the insensitive volume of the central detector (CD), which is due to the

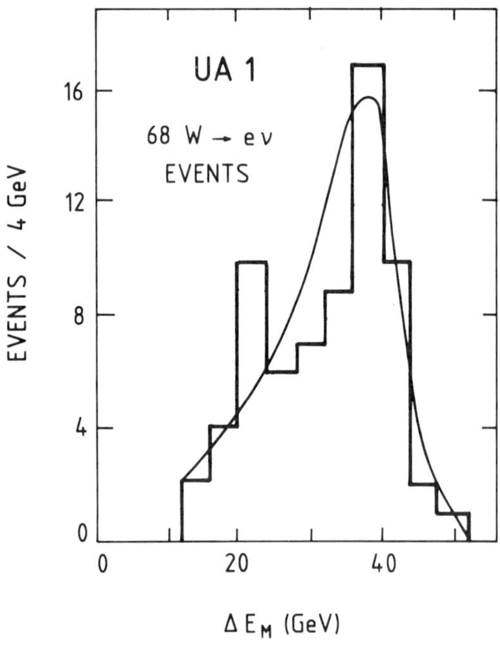

FIGURE 3. Missing transverse energy distribution for $W \rightarrow e\bar{\nu}$ events. The curve is the expectation for a W mass of 81 GeV/c^2, including our expected resolution.

presence of a foam septum. Although event G is inside this "blind" area, event H is clearly in an active region of the chamber.[2] However, there are no hits found pointing to the shower.

Event H is shown in FIGURE 7. The electromagnetic shower has an E_T of 54 GeV. The transverse mass of the system of the "photon" and ΔE_M is 93 ± 5 GeV/c^2.

The following is a summary of possible backgrounds considered for the "photon" events:[3]

(i) A $W \rightarrow e\bar{\nu}$ event with the electron track not recorded in the central detector. This is possible for event G. For event H, there is the possibility that a hard π^0 hit the same shower counter as did an electron from W decay, thus shifting the

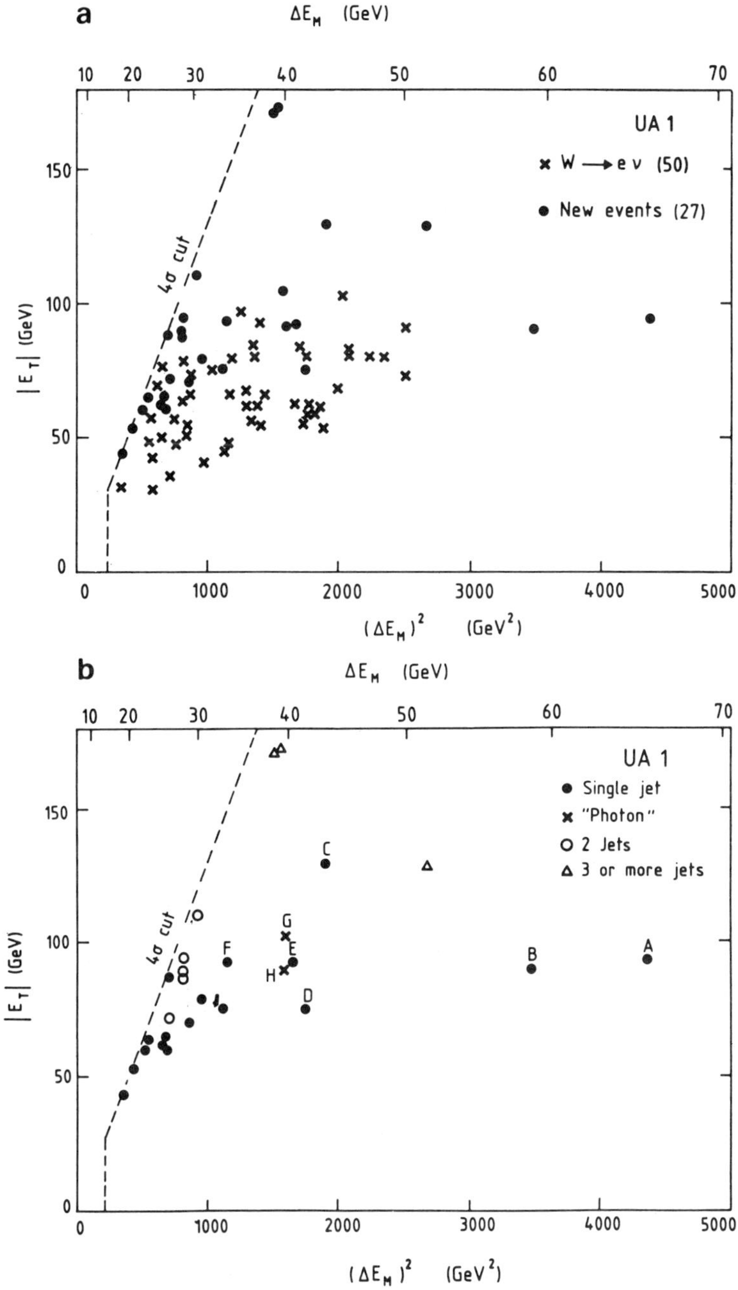

FIGURE 4. (a) Scatter plot of missing transverse energy squared versus the total scalar transverse energy for selected events that have missing E_T more than four standard deviations from zero. The $W \rightarrow e\bar{\nu}$ decays are labeled with a cross. (b) Same as (a), but with $W \rightarrow e\bar{\nu}$ events removed. The events are labeled according to their topology.

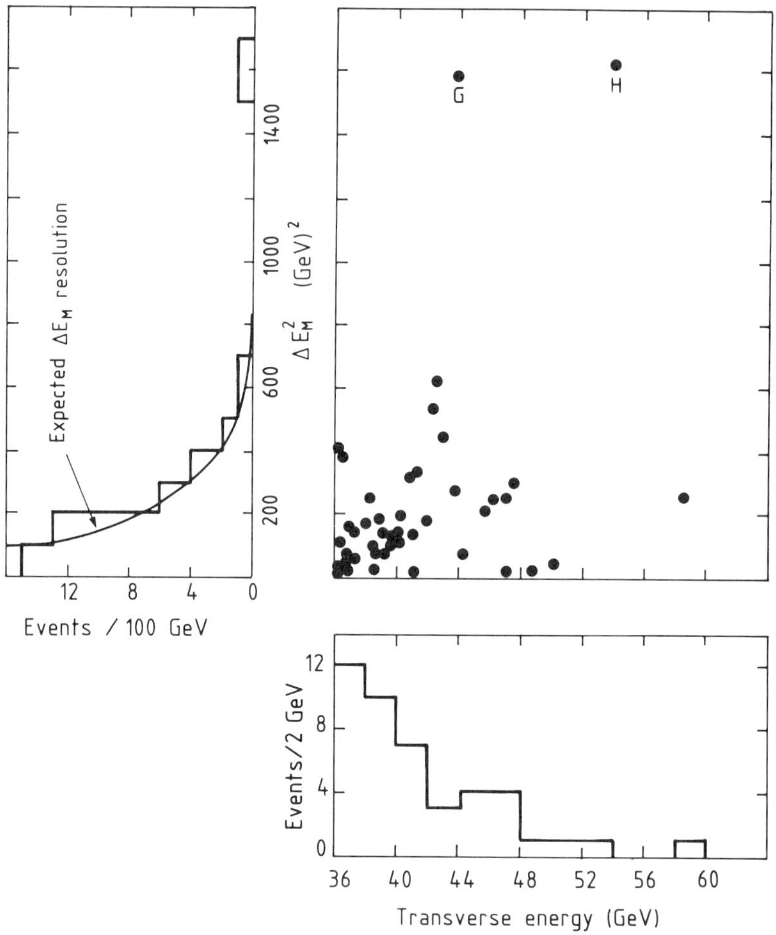

FIGURE 5. Scatter plot of the transverse energy versus the missing E_T squared for all selected "photon" events without a missing energy requirement.

shower position away from the blind region of the CD. This background is measured, by looking at the π^0 rate in W events, to be about 0.002 event.

(ii) Fake photons, i.e., neutrals striking the same electromagnetic counter up and down and faking a single shower with missing E_T. We expect less than 0.007 event.

(iii) Cosmic rays. A normal collision in coincidence with a cosmic ray may produce an electromagnetic shower profile. This background is about 0.001 event.

(iv) Jet-jet fluctuations. As discussed before, for events G and H, this background is very small, being less than 0.001 event.

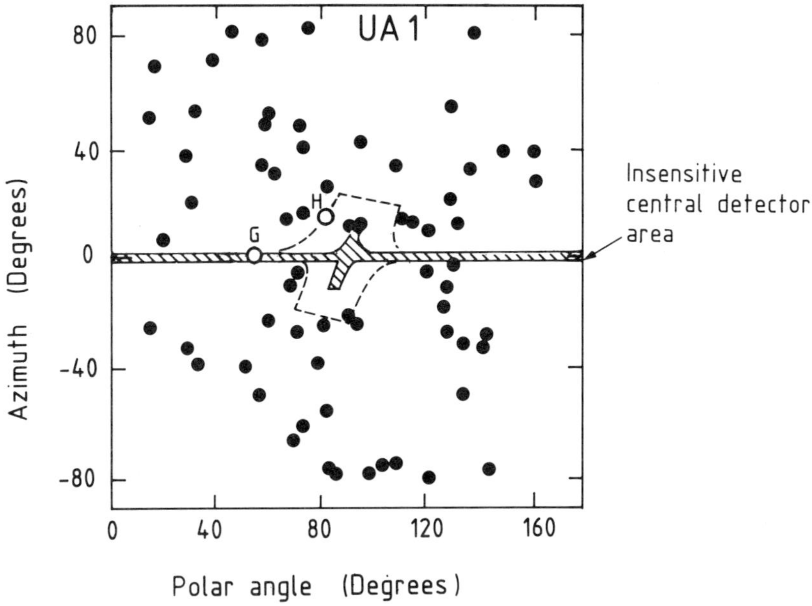

FIGURE 6. Electron polar angle (with respect to the beam axis) versus azimuthal angle for the 1983 sample of $W \to e\bar{\nu}$ events. The photon candidates are indicated by open circles. The dashed contour shows the area beyond which at least 20 hits are expected in the central detector. (● denotes W events; ○ denotes "Photons" with no charged track.)

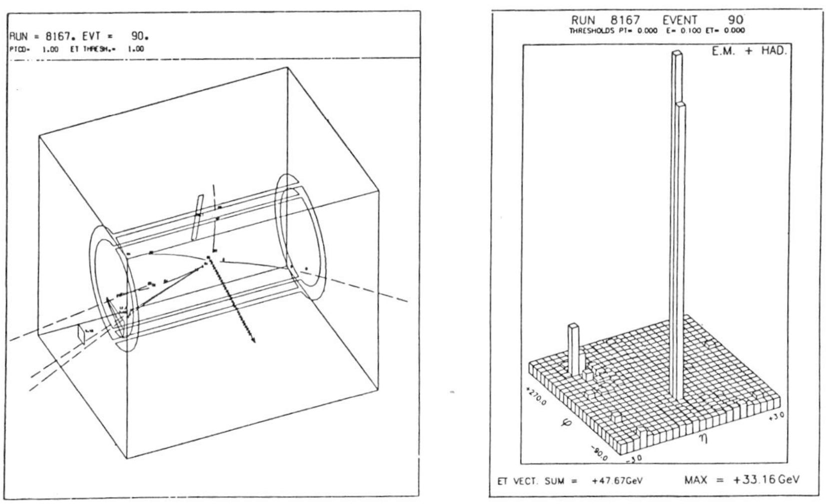

FIGURE 7. Event display and transverse energy flow in azimuthal angle (ϕ) and pseudorapidity (η) space for event H.

In summary, event H seems to be inconsistent with the backgrounds considered, while for event G, the $W \rightarrow e\bar{\nu}$ background cannot be excluded.

Single-Jet Events

There are 17 events with a single isolated jet. To estimate the backgrounds for these events, the ΔE_M cut was relaxed from 4σ to 2σ or 15 GeV, whichever was greater. In addition, we required some isolation cuts on the ΔE_M vector[2] and no vertical jets in the event. At this point, we were left with 117 events. The main background is from quantum chromodynamics (QCD) jets, in which, owing to calorimeter fluctuations, all jets, but one, become soft. This background was further rejected by cutting on $\Delta\phi$, the angle in the transverse plane that the rest of the event (jet subtracted and E_T recalculated from the rest of the calorimetry) makes with respect to the jet. The distribution of $(\Delta E_M)^2$ for the remaining events is shown in FIGURE 8. The solid curve shows the background expected from jet-jet fluctuations as calculated from our Monte Carlo. It is evident that the events with $\Delta E_M < 30$ GeV are consistent with background.

Other contributions to the monojet events are as follows:

(i) $W \rightarrow \tau\bar{\nu}$ events, where τ decays hadronically. The missing E_T distribution is shown in FIGURE 8. Nine events are expected in total, with about two events

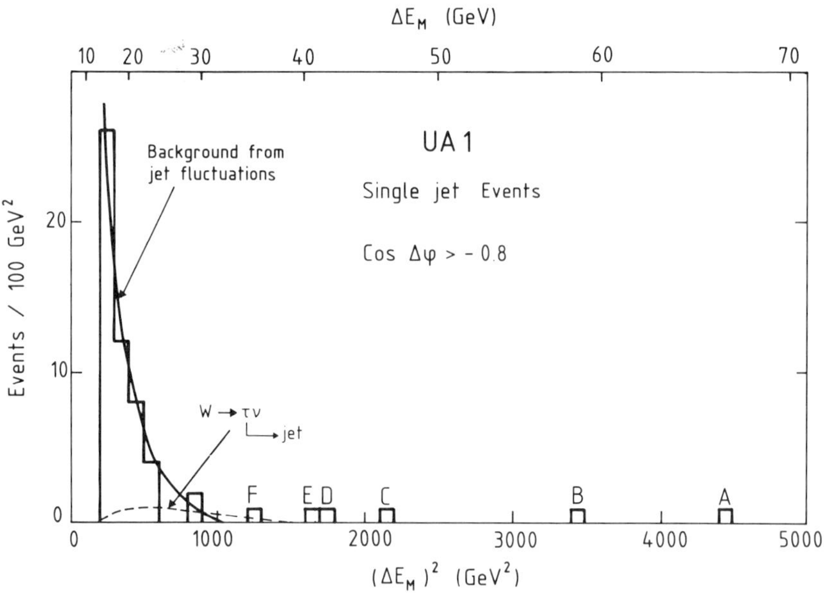

FIGURE 8. Distribution of missing E_T squared for single-jet events that pass the 2σ cut and $\cos\Delta\phi > -0.8$. The solid curve is the expected background from jet-jet fluctuations. The dashed curve is the expected contribution from $W \rightarrow \tau\bar{\nu}$ decays.

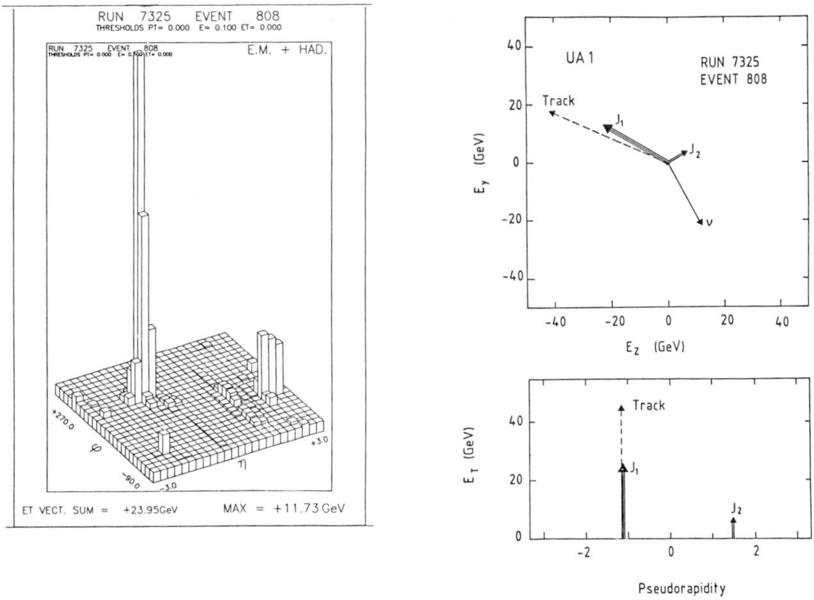

FIGURE 9. Transverse energy flow (ϕ versus η) seen in the calorimeters for monojet event A. Also shown are the transverse view and the pseudorapidity distribution of the jets.

with $\Delta E_M > 30$ GeV. Event F, because of its low jet mass, lower jet E_T, and three-prong multiplicity, is a $W \rightarrow \tau\bar{\nu}$ candidate. A hypothetical new sequential lepton of mass >20 GeV/c^2 can also contribute at the level of a few events with a softer spectrum.

(ii) $Z^0 + g$. A QCD jet can be produced in association with a Z^0, which in turn decays into a pair of neutrinos (the branching ratio is expected to be about 19%). This can give a monojet signature. Some W + jet or Z^0 + jet events have been observed, but in the 1983 data sample, there are no events with a jet of $E_T > 25$ GeV. Thus, this background is small for our values of E_T.

(iii) Heavy quark decays. Heavy quark pairs ($c\bar{c}$, $b\bar{b}$, $t\bar{t}$), with one of them decaying into a leading neutrino, can produce a monojet event. Less than 0.1 event is expected for $\Delta E_M > 40$ GeV. Also, a direct search for lepton + jet events with the lepton $E_T > 40$ GeV gives a negative result.

Therefore, there are five events that are inconsistent with the backgrounds considered above. FIGURE 9 shows the transverse energy flow and transverse view for event A. The unusual feature of this event is the presence of an energetic muon within the jet. The momentum of the muon is measured to be 80^{+20}_{-13} GeV/c in the CD and 105^{+59}_{-27} GeV/c as measured independently in the muon system. Except for the muon track, there are no other charged tracks in the jet. The missing E_T, including the muon, is 66 GeV.

The transverse energy flow and transverse view for event B is shown in FIGURE 10.

The jet has three charged tracks and a low invariant mass (0.8 GeV/c^2). Its E_T is 48 GeV and the missing E_T is 59 GeV. FIGURE 11 shows the transverse view for three events: C, D, and E. For more information on these events, see reference 2.

Multijet Events

There are eight such events. All but one are compatible with QCD background and heavy-flavor decays. The remaining event is a three-jet event. FIGURE 12 shows the transverse energy flow along with the transverse view for this event. The jet transverse energies are 51 GeV, 20 GeV, and 14 GeV. The event has a missing E_T of 55 GeV.

The 1983 results are basically summarized in FIGURE 13. In (A), the scatter plot of

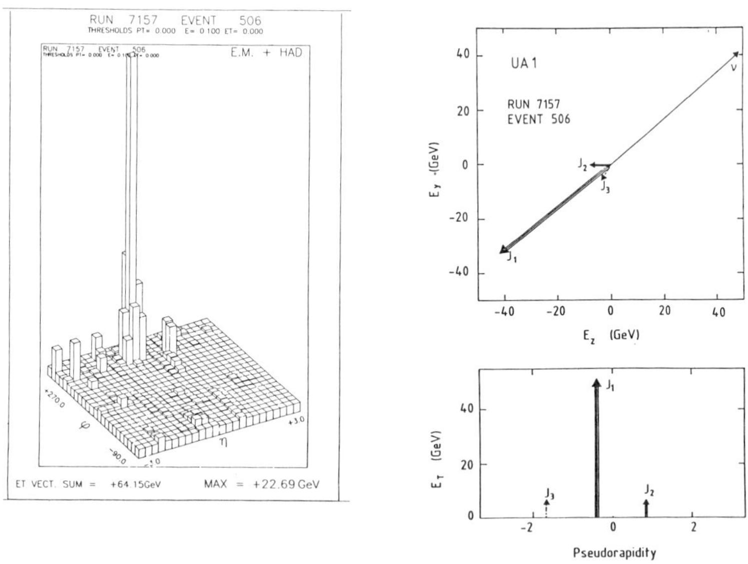

FIGURE 10. Same as FIGURE 9 for monojet event B.

ΔE_M versus E_T of the jet is shown for events A to H; it is evident that ΔE_M is mainly due to a single unbalanced jet. In (B), the transverse mass distribution of the jet and ΔE_M for events A to H is compared with that of a sample of W events with $\Delta E_M > 30$ GeV, and one can see that $W \rightarrow e\bar{\nu}$ decays have relatively smaller transverse masses.

NEW DATA FROM 1984

1984 Run Conditions

As mentioned in the INTRODUCTION, the 1984 run was taken at a total center-of-mass energy of 630 GeV. Another change from the 1983 run was the addition of a

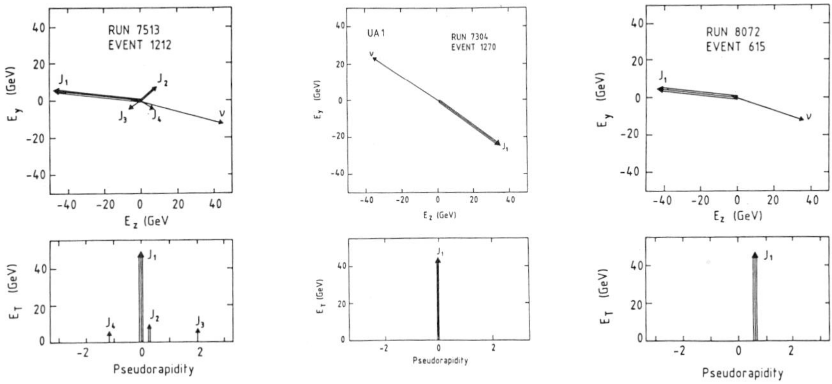

FIGURE 11. Same as FIGURE 9 for monojet events C, D, and E (from left to right).

new dedicated trigger to record events with a large amount of missing transverse energy. This trigger worked in the following way. The scalar sum of transverse energy was calculated once from all calorimeter cells in the right half of the detector ($|E_T|_R$) and a second time from all cells in the left half of the detector ($|E_T|_L$). If the difference between $|E_T|_R$ and $|E_T|_L$ was greater than 17 GeV *and* there was also a jet with $E_T > 15$ GeV, the event was recorded. This "E_T imbalance" trigger was active throughout the run, as well as other standard triggers. Those relevant to the missing E_T analysis were: (i) electromagnetic cluster of $E_T > 10$ GeV, (ii) jet of $E_T > 25$ GeV, and (iii) $|E_T|$ ($|\eta| < 1.5$) > 80 GeV.

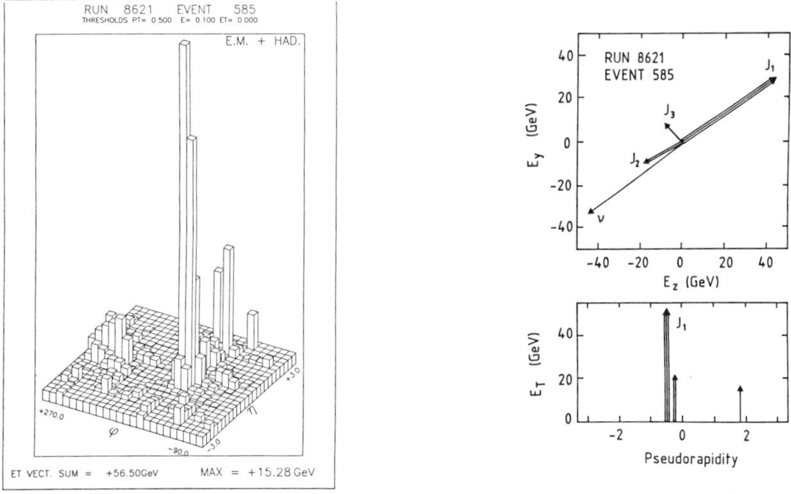

FIGURE 12. Same as FIGURE 9 for the three-jet event.

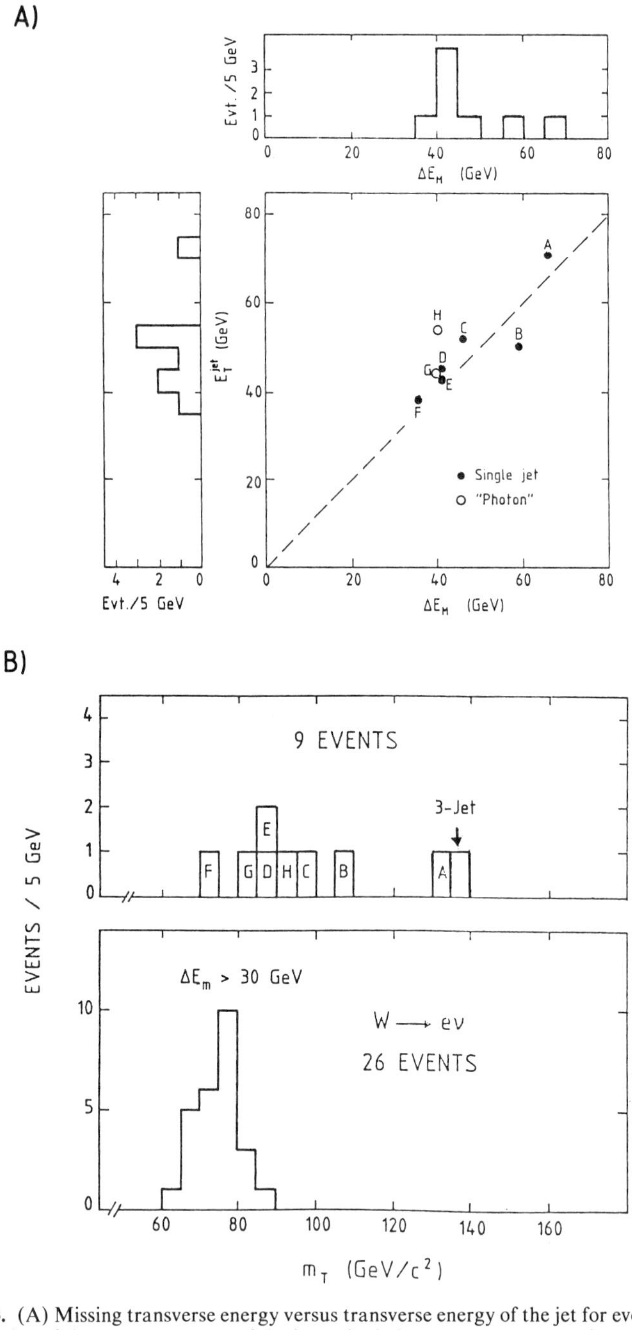

FIGURE 13. (A) Missing transverse energy versus transverse energy of the jet for events A to H. (B) Distribution of the transverse mass of the jet and missing transverse energy for events A to H and the three-jet event. Also shown is the M_T distribution for a sample of $W \rightarrow e\bar{\nu}$ events.

Event Selection

For 1984, all events with ΔE_M larger than 3.5σ ($\sigma = 0.7\ \sqrt{|E_T|}$) were written onto special "Express Line" tapes for rapid processing. A simple $\Delta E_M > 4\sigma$ cut was then applied to these events. A total of 1796 events were selected from the first 130 nb^{-1}. This sample included $W \rightarrow e, \nu, \tau$ decays and some $Z^0 \rightarrow \tau^+\tau^-$ candidates, as well as some junk events such as beam halo and cosmic and vertical jets. After removing $W \rightarrow e, \mu$ decays and rejecting all junk events, the remaining events were scanned by physicists on the Megatek facility. A number of single-jet and two-jet candidates were found, which will be described next. However, it should be emphasized that to date the results are very preliminary and the events presented have to be carefully studied in terms of various potential backgrounds.

Single-Jet Events

FIGURE 14 shows the scatter plot of the missing E_T squared versus $|E_T|$ for all single-jet and two-jet events. There are fifteen single-jet events in total. Eight of them are classified as monojet candidates, whereas the rest are considered to be $W \rightarrow \tau\bar{\nu}$ candidates. The separation of monojet and $W \rightarrow \tau\bar{\nu}$ events is based on two observations: (i) the magnitude of the missing transverse energy ($\Delta E_M > 40$ GeV for monojets) and (ii) the characteristics of the jet in terms of size and multiplicity (low mass and ≤ 3 prongs for the τ-jet). For example, an event with $\Delta E_M < 40$ GeV, but a jet mass or multiplicity inconsistent with τ, would be considered to be a monojet. Nevertheless, this classification is by no means definitive and the results should not be taken as final.

The $W \rightarrow \tau\bar{\nu}$ candidates will be discussed later. As an example of monojet candidates, FIGURE 15 shows the transverse energy flow and transverse view for event 9501/822. This event has a missing transverse energy of 46 GeV ($\sigma = 7.4$) and the jet E_T is 44 GeV (28 GeV in electromagnetic and 16 GeV in hadronic energy). The jet has six charged tracks, each with transverse momentum, p_T, over 1 GeV/c, including one with p_T as high as 14 GeV/c. Its charged mass is 2.7 ± 0.5 GeV/c^2 and the transverse mass of the ΔE_M + jet is 90 GeV/c^2.

As mentioned in the subsection, SINGLE-JET EVENTS, there are other contributions to the monojet events, namely: $Z^0 + g$,[4] heavy quark decays,[5] and $W \rightarrow \tau\bar{\nu}$ decays. A preliminary estimate of these backgrounds (per 100 nb^{-1}) is demonstrated in FIGURE 16. Another background, not shown in the plot, comes from W + jet events, where W decays into an electron or a τ that happens to overlap with the jet, faking a monojet signature. This background, still to be calculated, could be as large as 0.5 event per 100 nb^{-1} (from naive phase-space arguments) for $\Delta E_M > 40$ GeV.

Two-Jet Events

In our selection of events with large missing energy, there was no requirement on the number of jets in the event. After events were selected, we simply counted the number of jets with $E_T > 12$ GeV. In this manner, we found eight two-jet events. FIGURE 14 shows $(\Delta E_M)^2$ versus $|E_T|$ for these events.

There are basically two topologies observed in two-jet events:

(i) Viewed in the transverse plane, the jets are on the same side and the missing transverse energy is balancing them on the other side. FIGURE 17 shows an example of this configuration. The missing transverse energy of this event is measured to be 65 GeV ($\sigma = 8.3$) and the jet E_T's are 49 and 16 GeV with E_T^{em}/E_T^{had} equal to 41/8 and 9/7, respectively. There are six charged tracks with $p_T > 1$ GeV/c in the big jet and three positively charged tracks in the small

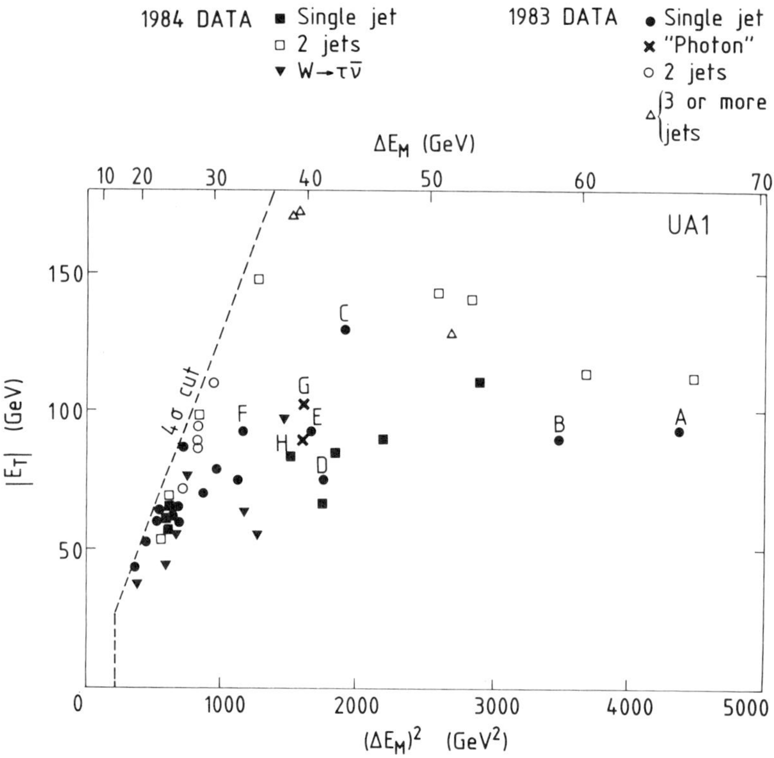

FIGURE 14. Scatter plot of missing transverse energy squared versus the total scalar transverse energy for events with $\Delta E_M > 4\sigma$ from 130 nb^{-1} of 1984 data. The results of 4σ selection from 1983 are also shown. (1983 Data: ● = Single jet, × = "Photon," ○ = Two jets, △ = Three or more jets; 1984 Data: ■ = Single jet, □ = Two jets, ▼ = $W \rightarrow \tau\bar{\nu}$.)

one. The transverse mass of the ΔE_M + jets is 130 GeV. There are two events of this type in the sample.

(ii) The second topology is when the two jets are back-to-back in the transverse plane and the missing E_T is parallel to one of the jets, clearly the smaller one. Event 9986/574, shown in FIGURE 18, is an example of this topology. The missing E_T is 50 GeV and the jet transverse energies are 74 and 27 GeV.

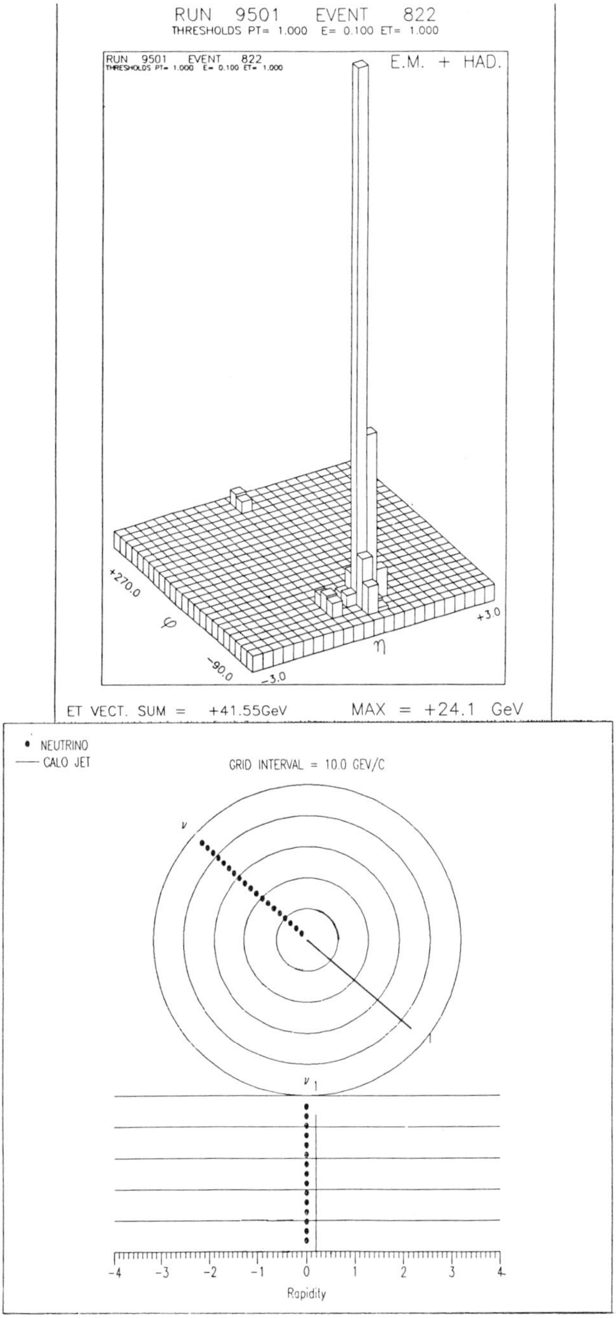

FIGURE 15. Transverse energy flow observed in the calorimeters and the transverse view for the monojet event 9501/822. Distribution of pseudorapidity of the jets is also shown.

Events of this type are much harder to evaluate in terms of backgrounds such as jet-jet fluctuations and QCD production of two-jet events with the energy of one jet incorrectly measured, owing to the leakage or the presence of energetic muons not detected in the muon chambers. Therefore, for the present, they are excluded from the sample.

W $\rightarrow \tau\bar{\nu}$, τ *Decaying into a Jet*

There are seven such events. Their ΔE_M distribution is shown in FIGURE 14. Out of seven candidates, three are three-prong and four are one-prong. FIGURES 19 and 20

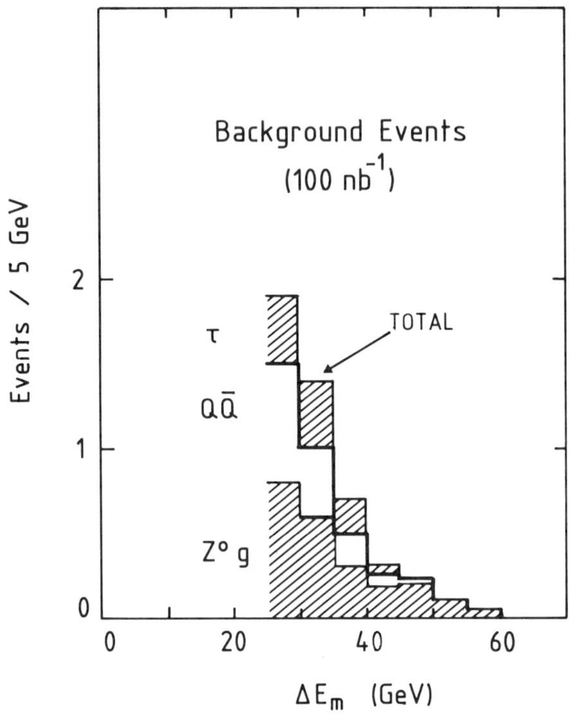

FIGURE 16. Distribution of the missing transverse energy for $Z^0 + g$, $Q\bar{Q}$, and $W \rightarrow \tau\bar{\nu}$ decays.

show the transverse energy flow and the transverse view for a three-prong and a one-prong event, respectively. The three-prong event (11,673/499) has a missing E_T of 19 GeV ($\sigma = 4.5$) and the jet E_T is 21 GeV ($E_T^{em}/E_T^{had} = 7/14$). The missing E_T for the one-prong event (10,486/358) is 36 GeV ($\sigma = 7.1$) and the E_T of the jet is 31 GeV ($E_T^{em}/E_T^{had} = 25/6$). The transverse masses of these events are 39 and 66 GeV, respectively. FIGURES 21 and 22 show the CD view for the above events. It can be seen that the events have clean vertices and are relatively quiet, as expected from W decays.

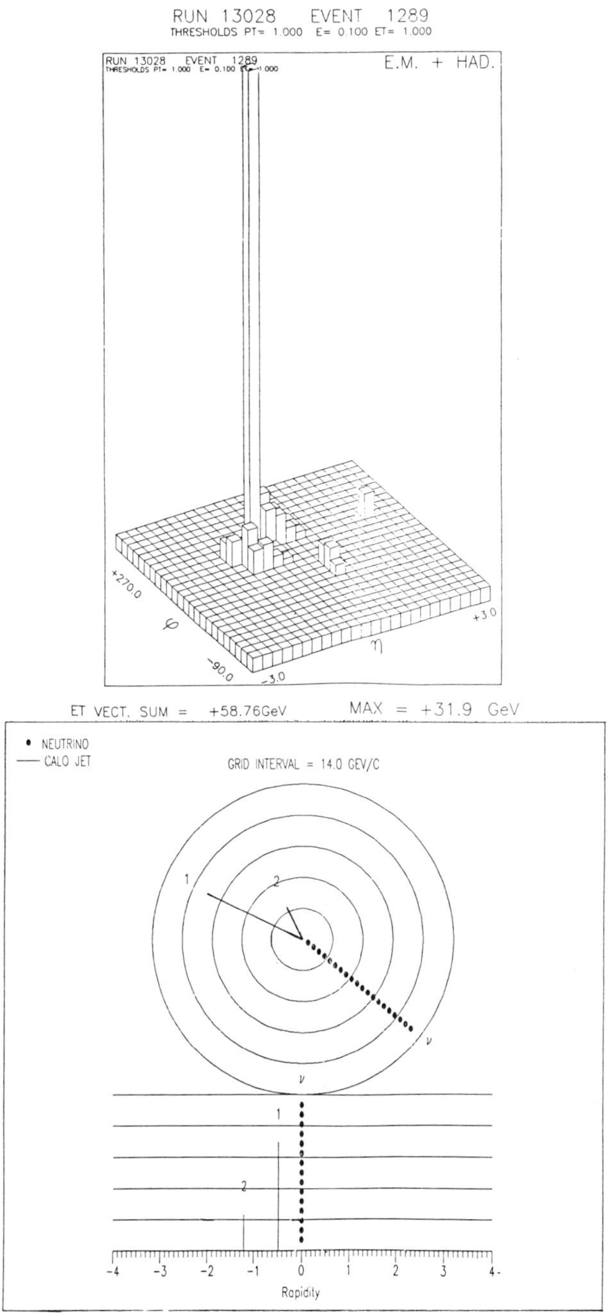

FIGURE 17. Same as FIGURE 15 for two-jet event 13,028/1289.

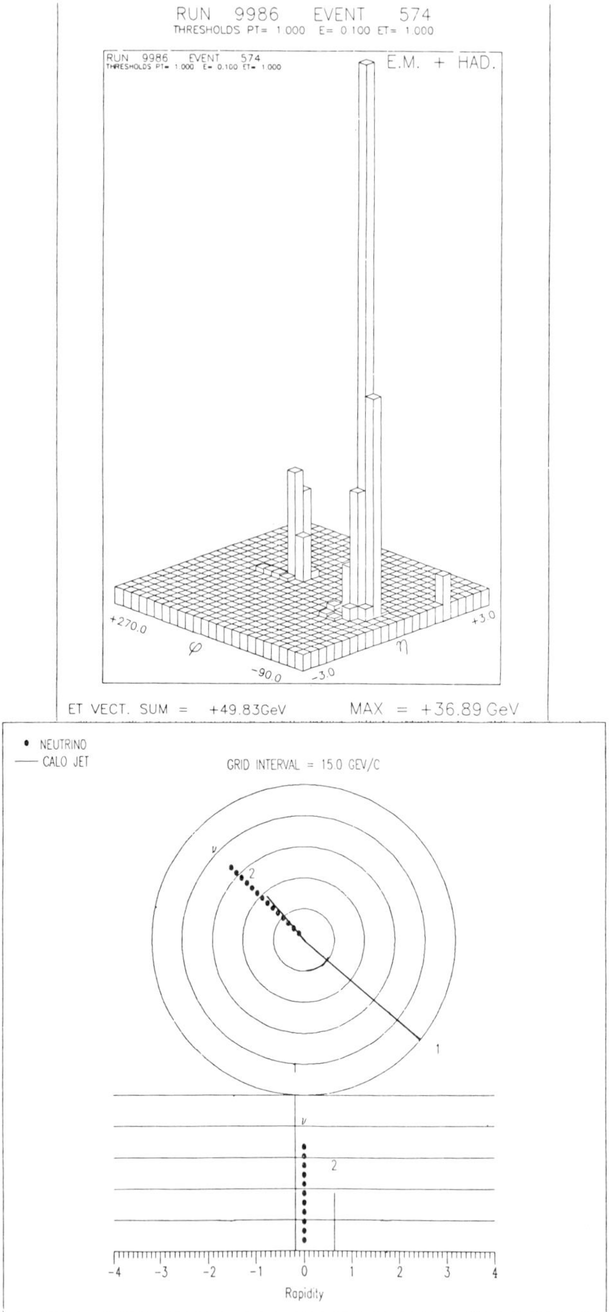

FIGURE 18. Same as FIGURE 15 for two-jet event 9986/574.

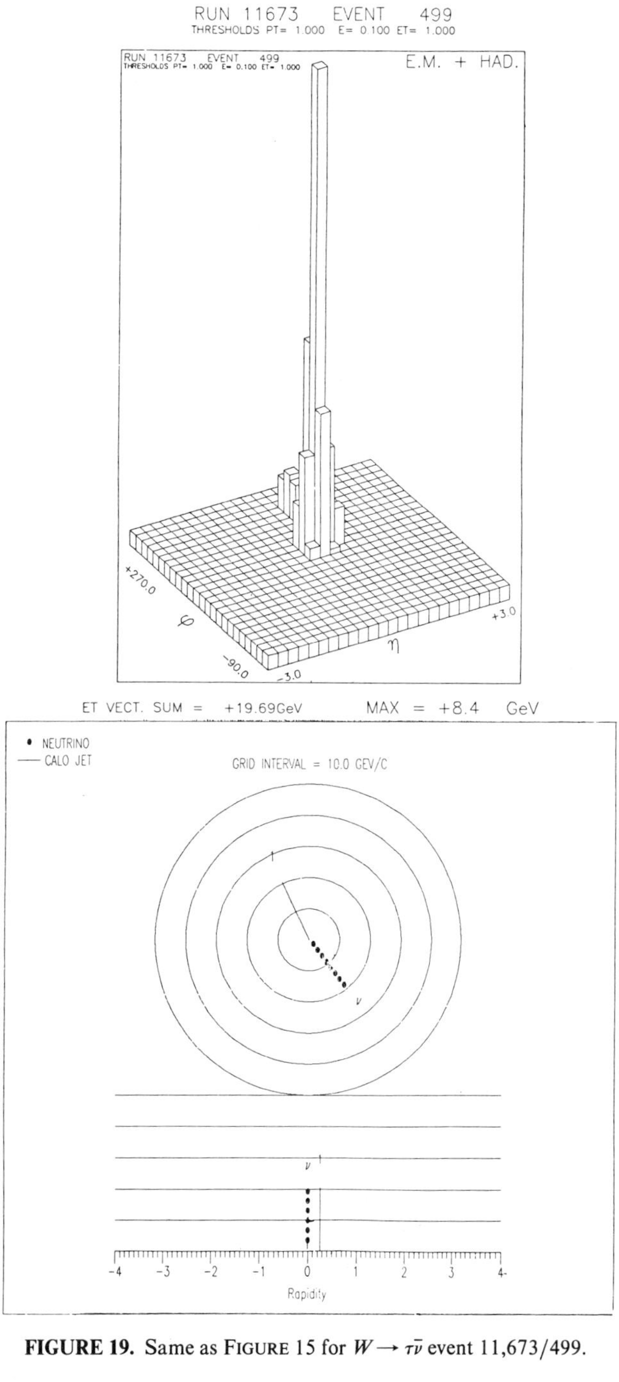

FIGURE 19. Same as FIGURE 15 for $W \rightarrow \tau\bar{\nu}$ event 11,673/499.

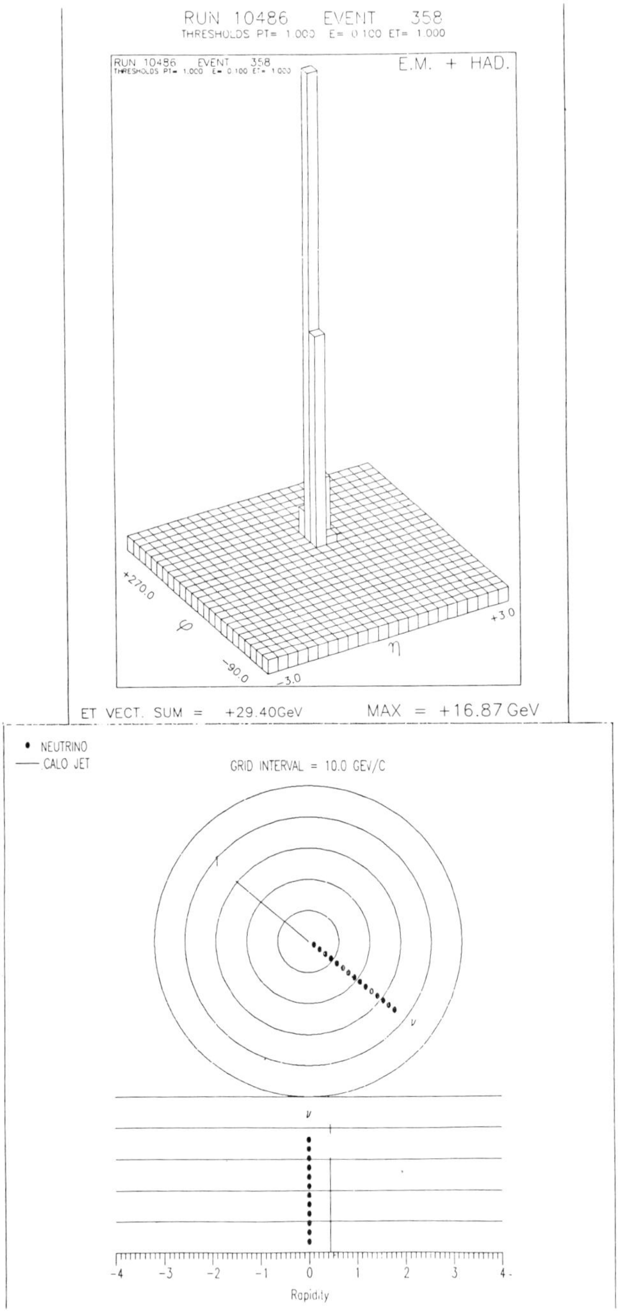

FIGURE 20. Same as FIGURE 15 for $W \rightarrow \tau\bar{\nu}$ event 10,486/358.

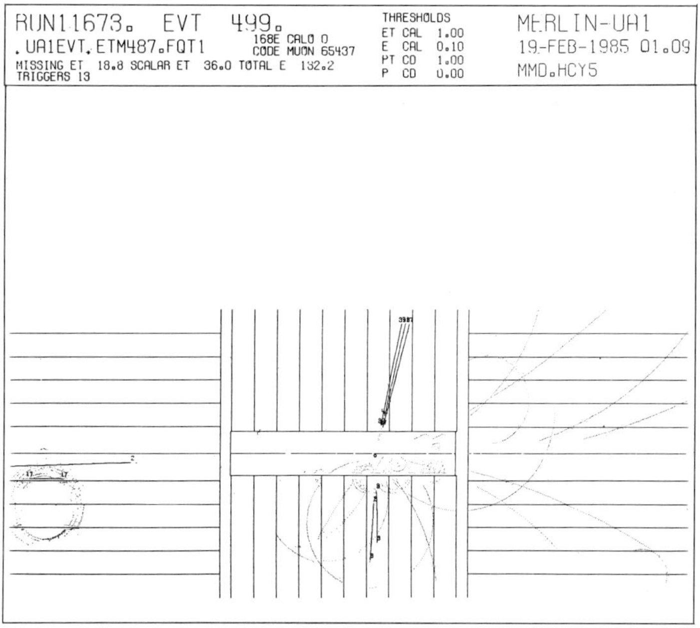

FIGURE 21. Central detector display of digitization of tracks in event 11,673/499.

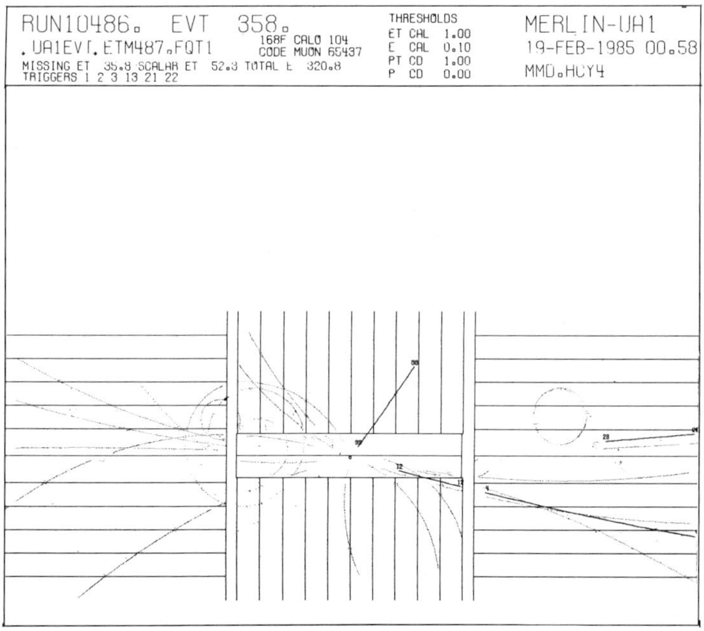

FIGURE 22. Same as FIGURE 21 for event 10,486/358.

The jet in event 11,673/499 has three charged tracks ($Q \cdot p_T = +12.2, -8.7, -7.7$ GeV/c) rather collimated. Its charged mass is 1.0 ± 0.03 GeV/c^2. It is interesting to note that the recording of this event was only due to the E_T imbalance trigger. The one-prong event has $Q \cdot p_T = -6.3$ GeV/c and is a clear candidate for τ decay into $\pi^- + n\pi^0$.

In order to understand better our analysis of events with missing energy and also to be able to separate our $W \rightarrow \tau\bar{\nu}$ and monojet candidates effectively, it is very important that we establish a signal in the $W \rightarrow \tau\bar{\nu}$ channel and realize the backgrounds involved.

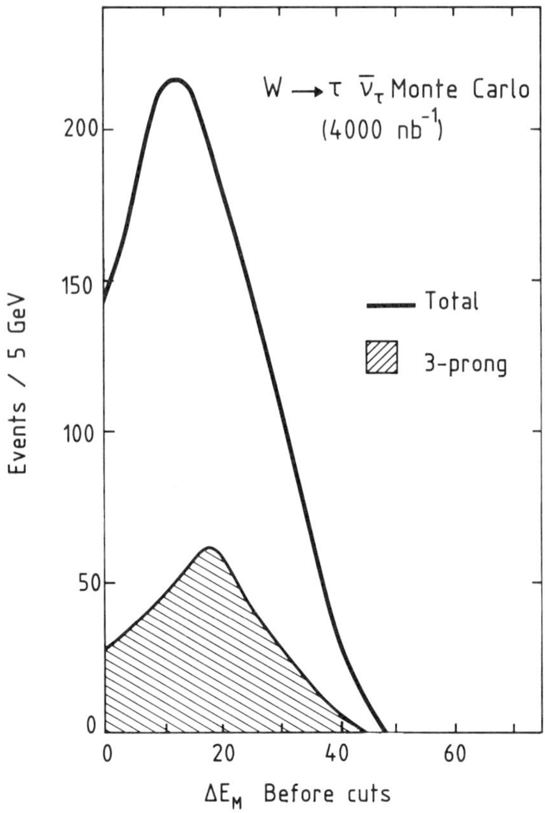

FIGURE 23. Distribution of the missing transverse energy of $W \rightarrow \tau\bar{\nu}$ Monte Carlo events. No selection cuts applied; however, detector smearing effects are included.

To this end, a series of Monte Carlo calculations were carried out using the ISAJET program with certain additions and modifications.[6] Various hadronic decays of τ were calculated with proper branching ratios. After a full detector simulation, the results are shown in FIGURE 23, where the ΔE_M of these events is plotted before applying any selection cuts. The Monte Carlo sample corresponds to a total integrated luminosity of 4000 nb^{-1}. The dashed area shows the contribution from $\tau \rightarrow$ three-prong decays. Notice that a large fraction of events have $\Delta E_M < 15$ GeV. FIGURE 24 shows the

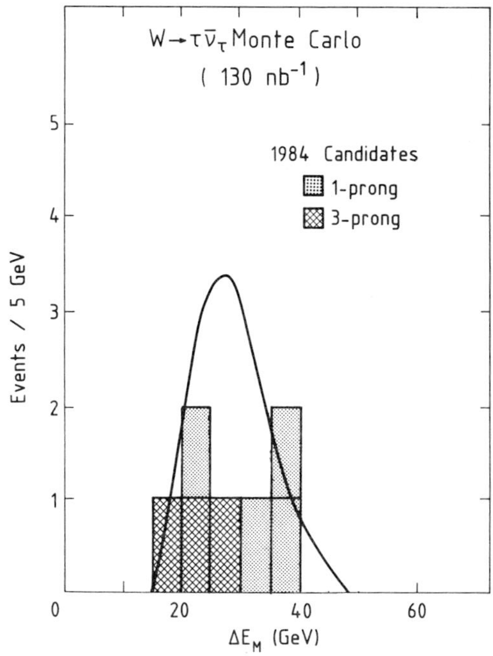

FIGURE 24. Same as FIGURE 23 with trigger and selection cuts included. The 1984 three-prong and one-prong candidates are also shown.

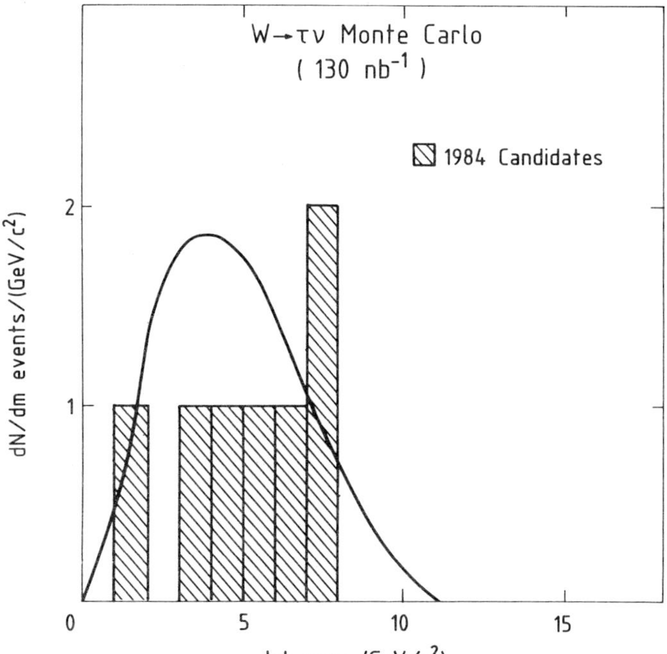

FIGURE 25. Distribution of the calorimetric jet mass (see text) for $W \rightarrow \tau\bar{\nu}$ Monte Carlo events. The dashed area shows the seven 1984 $W \rightarrow \tau\bar{\nu}$ candidates.

distribution of ΔE_M after including trigger efficiencies and applying our selection cuts of $\Delta E_M > 4\sigma$ and not vertical. This plot is normalized to 130 nb^{-1} total integrated luminosity. A total of eleven events are expected. Our seven candidates seem to agree with the Monte Carlo predictions rather well.

To get an idea of how the τ-jets compare with the jets in our monojet events or with the QCD jets, a calorimetric jet mass is defined. This is basically the invariant mass of all energy cells in calorimeters associated with the jet by the UA1 jet algorithm.[7] This is clearly not a valid measurement of the mass; however, it is useful for the purpose of comparison. FIGURES 25 and 26 show the distributions of the calorimetric jet mass for the Monte Carlo events, the monojet candidates, and the QCD jets. It is clear that the

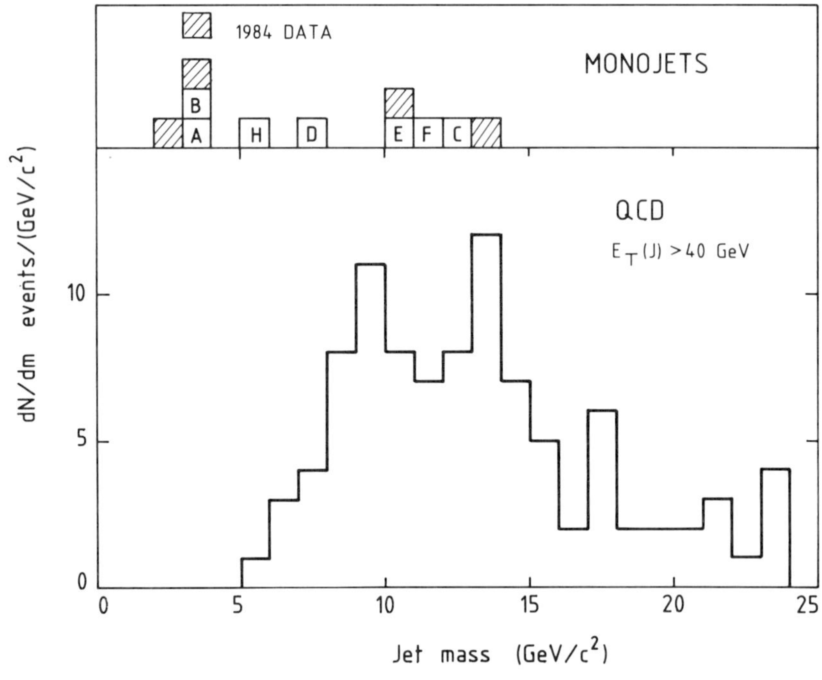

FIGURE 26. Same as FIGURE 25 for the monojet events and the QCD jets with $E_T > 40$ GeV.

τ-jets have lower masses than the QCD ones, but some jets of the monojet events seem to have relatively large masses.

CONCLUSIONS

In the first 130 nb^{-1} of the 1984 data, we have observed events in which a large missing transverse energy is associated with a single jet. This observation confirms the existence of the monojet events seen in the 1983 data. In addition, we have also observed events in which the missing E_T is associated with two jets.

Decays of the W boson into a τ lepton have been observed and they seem to be in good agreement with our expectations in terms of missing transverse energy and characteristics of the τ-jets; namely, mass and multiplicity.

Any statement about the interpretation of the monojets and the two-jet events would be premature at this time. The remaining data should be analyzed and the backgrounds further studied in order to determine the physics behind these events.

ACKNOWLEDGMENTS

The UA1 experiment is a collaboration of many institutes. We are all grateful to the management and staff of CERN for the support given to this experiment.

I would like to thank Professor Carlo Rubbia and Professor David Cline for providing me with the opportunity to work on the UA1 experiment and present these results, which have been arrived at by the hard work of many collaborators.

I am grateful to the organizers of this conference for a pleasant visit to Aspen.

NOTES AND REFERENCES

1. Minimum bias events are events in which one or more charged particles pass through each of two large hodoscopes around the beam pipe. See ASTBURY, A. *et al.* 1978. CERN/SPSC/78-06 for a description of the detector.
2. ARNISON, G. *et al.* 1984. Phys. Lett. **139B:** 115.
3. By "photon," we mean any number of collinear photons of low invariant mass (a few GeV/c^2).
4. ALTARELLI, G. *et al.* 1984. Fermilab-Pub-84/107-T. [Assuming the number of $(Z^0 \rightarrow \nu\bar{\nu}) = 6 \times (Z^0 \rightarrow e^+e^-)$. No selection cuts applied.]
5. ARNISON, G. *et al.* 1984. Phys. Lett. **147B:** 493. [Using $(d\sigma/dp_T)$ $(p\bar{p} \rightarrow \mu^{\pm} + X)$ and replacing μ with ν.]
6. PAIGE, F. E. & S. D. PROTOPOPESCU. 1981. ISAJET program, BNL 29777. The program was modified by the author (M.M.) to include the polarization effects in τ decays. Also, A1 resonance was added along with proper mass widths for A1 and ρ.
7. A description of the jet algorithm can be found in ARNISON, G. *et al.* 1984. Phys. Lett. **132B:** 214.

Exotic Events in UA2

Presented by JOHN RENNER HANSEN
UA2 Collaboration[a]
CERN CH-1211
Geneva 23, Switzerland

INTRODUCTION

In the paper entitled, "Production of High Mass $e\nu$ and e^+e^- Pairs in the UA2 Experiment at the CERN $\bar{p}p$ Collider" (which is on page 125 of this volume), and in reference 1, we have studied a sample of events recorded by UA2[2] at the Sp$\bar{p}$S collider[3] that contains an electron candidate in the final state with a transverse momentum, $p_\perp(e)$, in excess of 15 GeV/c. This sample includes events having a genuine electron in the final state, as well as a contamination of background events for which the electron identification is erroneous. In the former category, we expect that from lepton number conservation, the electron will be associated with another electron of opposite charge or with a neutrino. We have reported elsewhere[4] on the eight events containing an electron pair in the final state, all of which were associated with the production and decay of the electroweak boson, Z^0. We have also reported[1] on events containing an electron-neutrino pair in the final state and we have given a detailed analysis of these events in terms of the production and decay of the electroweak boson, $W^\pm$. This latter analysis discarded events in which a significant amount of transverse energy was detected at the opposite azimuth to the electron candidate. This procedure efficiently rejected two-jet events in which one of the jets was misidentified as an electron. Other events, however, which contained a genuine electron in the final state, may have also been rejected by this cut.

The purpose of the present paper is to report on a systematic search for events containing an electron-"neutrino" pair in the final state, no matter what the transverse energy at the opposite azimuth to the electron may be. The loss of rejection power against background, resulting from having relaxed this constraint, is compensated by the requirement that both the electron and the "neutrino" have large transverse momenta.

Our detector does not distinguish between a neutrino and other possible noninteracting particles, such as the photino postulated by supersymmetry theories.[5] Therefore, for the remainder of this paper, the word, "neutrino," must be understood in a broad sense.

APPARATUS

The UA2 detector has been described in detail elsewhere.[2] We briefly recall its main features.

[a]This UA2 Collaboration includes Bern, CERN, Copenhagen (NBI), Orsay (LAL), Pavia, and Saclay (CEN) Collaborations.

Apart from two narrow cones along the beams, the detector provides full azimuthal coverage in three distinct regions of polar angles: $40° < \theta < 140°$, which is the central region, and $20° < \theta < 40°$ and $140° < \theta < 160°$, which are the forward regions. In the center of the detector, a set of coaxial cylindrical drift and proportional chambers detect the charged particles produced in the collision and measure the position of the event vertex.

An array of 480 calorimeter cells, each cell covering a similar domain of longitudinal phase-space (15° of azimuth and $\simeq$ 0.2 units of rapidity), measures electron energies, $E(e)$, to a good accuracy. Each cell is segmented longitudinally to provide electron-hadron separation. However, while hadron showers are usually contained in the 4.5 absorption lengths of the central calorimeter, thus providing a measurement of jet energies, they only deposit a fraction of their energy in the forward calorimeters, which are $\simeq$ 1.0 absorption length thick. These forward regions are equipped with magnetic spectrometers that provide additional rejection power against both hadrons and converted photons when searching for electrons. In addition, they measure the momenta of charged jet fragments, with the energy of π^0's being measured in the calorimeter cells.

In both the central and forward regions, electron identification is significantly improved by preshower counters that accurately measure the match between the observed incident track and the developing shower.

DATA SAMPLE

The present analysis deals with the sample of 225 events containing an electron candidate having $p_\perp(e) > 15$ GeV/c and obeying the selection criteria described in table 1 of reference 1. This event sample corresponds to an integrated luminosity of 131 nb^{-1}. We discard from this sample the events collected in the 1982 period because during this time the azimuthal coverage of the UA2 detector was incomplete.[2] We also discard ten events in which the electron candidate is observed near the interface between the central region and one of the forward regions and is associated with nearby calorimeter energy in each of the two regions. In such cases, when the electron candidate is in fact a misidentified jet, a significant fraction of the jet energy is likely to be lost in the magnet coils at the interface of the two regions and may simulate a neutrino. Together, the initial sample is reduced to 190 events and the corresponding integrated luminosity to 116 nb^{-1}. Also to be noted is that the global detection efficiency of the electron selection criteria is estimated[1] to be 80% in the forward regions and 76% in the central region.

In each event, we reduce the final state to a set of transverse energy clusters according to simple algorithms that have been described elsewhere.[1,2] One of the clusters is associated with the electron candidate. Each of the other clusters is usually associated with a multiparticle jet, under the provision that it carries large transverse energy. For this reason, in what follows, clusters having a transverse energy in excess of 3 GeV shall be called "jets."

For each event, we calculate the momentum, $\bar{p}$ (transverse component, $\bar{p}_\perp$, and longitudinal component, $p_\parallel$), and the energy, E, of the electron candidate (e) and of each individual jet (j) assumed to be massless. We also evaluate the momentum, $\bar{p}$, energy, E, and mass, m, of various sets of particles, such as the system of all jets (J) or the system of electrons and jets, (Je). The quantity, $p_\perp(Je)$, measures the missing transverse momentum in the set of all the detected particles resulting in clusters having $E_\perp > 3$ GeV. In a typical event, the softer particles carry together only a small transverse momentum and, if no large transverse momentum particle has escaped detection, the observation of a large $p_\perp(Je)$ reveals the presence of a neutrino (ν) with $p_\perp(\nu) \simeq p_\perp(Je)$.

The distribution of $p_\perp(Je)$ is shown in FIGURE 1a. The background evaluation follows the method described in reference 1 and measures the probability that a multijet event contains both a misidentified electron and an undetected jet (or jets) escaping the UA2 acceptance. Its distribution is shown as a curve superimposed on the data of FIGURE 1a. The background contribution amounts to 10.0 ± 0.9 events for $p_\perp(Je) > 20$ GeV/c, 3.4 ± 0.3 events for $p_\perp(Je) > 25$ GeV/c, and 1.2 ± 0.1 events for $p_\perp(Je) >$ GeV/c.

In the remainder of this paper, we restrict the analysis to the sample of 35 events that have $p_\perp(Je) > 25$ GeV/c and are therefore candidates for containing a large transverse momentum neutrino in the final state.

DATA ANALYSIS AND BACKGROUND EVALUATION

From the results of the analysis presented in reference 1, we expect our sample to contain a number of events in which the observed electron carries most of the detected transverse energy and is therefore accompanied by no jet or by jets having small transverse energies. The distribution of the sum of the jet transverse energies, $E_\perp(J)$, confirms this expectation (FIGURE 1b). The 31 events having $E_\perp(J) < 30$ GeV all belong to the samples of figures 4a and 4c in reference 1, where they were interpreted in terms of $W \rightarrow e\nu$ decays. We do not comment further on these events. Instead, we consider the four events having $E_\perp(J) > 30$ GeV. The distribution of the events in the $p_\perp(Je)$, $E_\perp(J)$ plane is shown in FIGURE 2 for each of the signal and background samples. The background contamination expected in the region, $p_\perp(Je) > 25$ GeV/c, $E_\perp(J) < 30$ GeV, is $\simeq 0.45 \pm 0.04$ events. In each of the four signal events, the electron candidate is detected in the central region of the UA2 detector. Their transverse momentum configurations are illustrated in FIGURE 3.

For each event, in which $p_\perp(Je)$ and $E_\perp(J)$ take the values p_0 and E_0, respectively, we evaluate the expected background contaminations, B_1 and B_2, corresponding to the following configurations:

$$B_1 : p_\perp(Je) > p_0,\ E_\perp(J) > 30 \text{ GeV},$$

$$B_2 : p_\perp(Je) > 25 \text{ GeV}/c,\ E_\perp(J) > E_0.$$

In each of the four events, either B_1 or B_2 is always less than 0.02. It is remarkable that there is no background event in the region, $p_\perp(Je) > 50$ GeV/c, $E_\perp(J) > 30$ GeV, which contains events A to C. We infer from this a background contamination of at most 0.02 events (90% confidence level) in this region.

As a check on the quality of the electron identification, we compare in FIGURE 4 the values taken by a few characteristic parameters in the present sample and in the

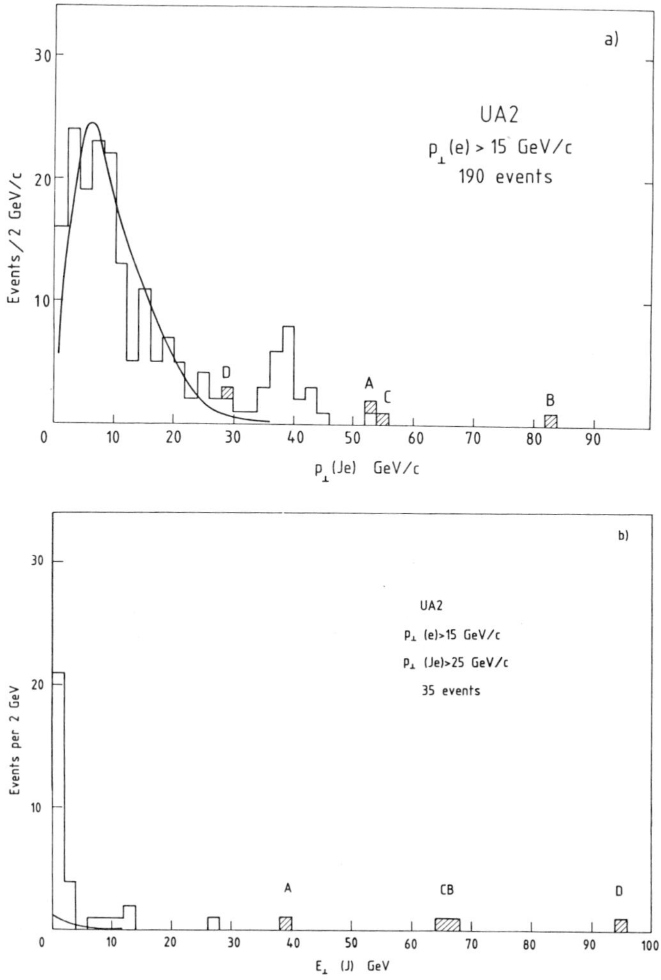

FIGURE 1. (a) Transverse momentum distribution of the system of electrons and jets in the initial sample of 190 events. (b) Transverse energy distribution of the system of jets in the sample of 35 events having $p_\perp(Je) >$ GeV/c. The event at $E_\perp(J) \simeq 26$ GeV is the largest transverse momentum W in figure 8 of reference 1. The lines correspond to the calculated background contaminations. The four events having $p_\perp(Je) > 25$ GeV/c and $E_\perp(J) > 30$ GeV are crosshatched.

sample of $W \rightarrow e\nu$ candidates having $p_\perp(e) > 25$ GeV/c.[1] Each of the four events has been examined in detail with the help of a high resolution graphics display facility. The track multiplicity is, of course, higher here than in most of the $W \rightarrow e\nu$ events studied in reference 1 and we cannot exclude that the increased complexity of the signal pattern in the central vertex detector could result in some deterioration of the electron identification power.

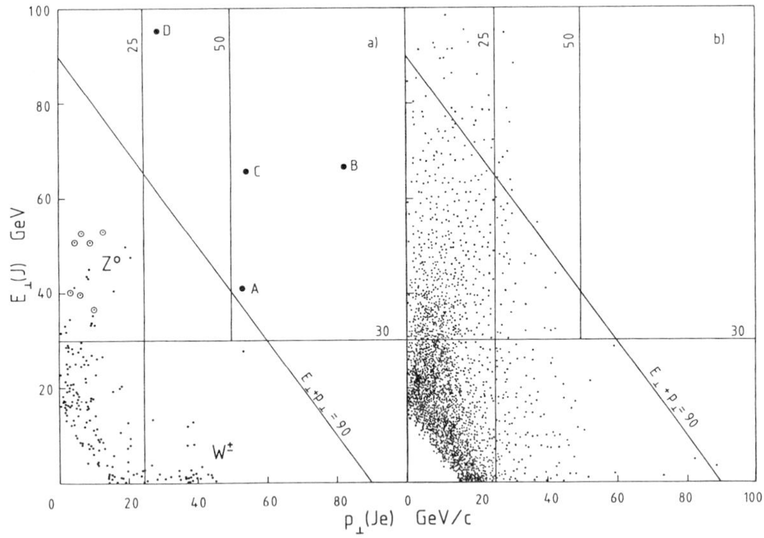

FIGURE 2. (a) Distribution of the 190 events of the initial sample in the $p_\perp(Je)$, $E_\perp(J)$ plane. Seven Z^0 events are circled (the eighth one was collected during the 1982 period, see reference 4). The W region is indicated. (b) Distribution of the background sample in the $p_\perp(Je)$, $E_\perp(J)$ plane. A reduction factor of 141 must be applied to infer from this sample the background contamination to the sample of FIGURE 2a. Lines of constant $p_\perp(Je) = 25$ to 50 GeV/c, $E_\perp(J) = 30$ GeV, and $p_\perp(Je) + E_\perp(J) = 90$ GeV are indicated.

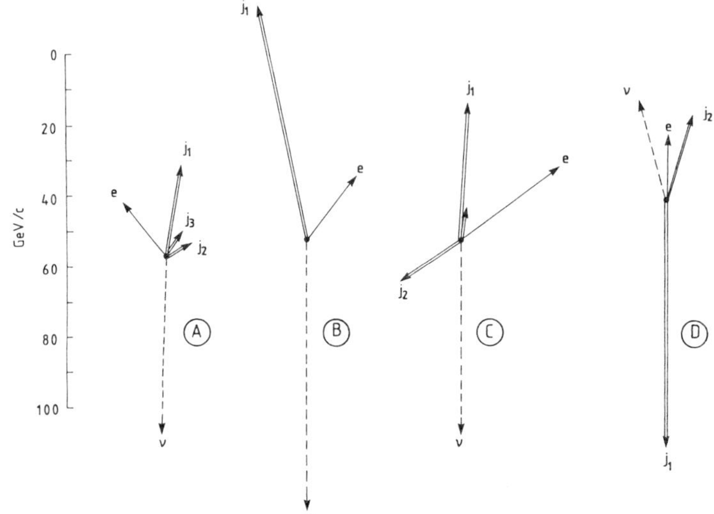

FIGURE 3. Transverse momentum configuration of the four events having $p_\perp(Je) > 25$ GeV/c and $E_\perp(J) > 30$ GeV. Their relative orientation is arbitrary.

As far as neutrino identification is concerned, we have checked that in each of the four events, there is no sign of large transverse momentum particles having hit passive parts of the UA2 detector, such as the magnet coils, in the azimuthal region where the neutrino is expected. However, in the case of event D, which has an azimuthal configuration similar to that of a two-jet event, such as interpretation cannot be completely excluded.

Large transverse momentum particles may have escaped detection because they were produced at small angles to the beam line. We considered for each event the

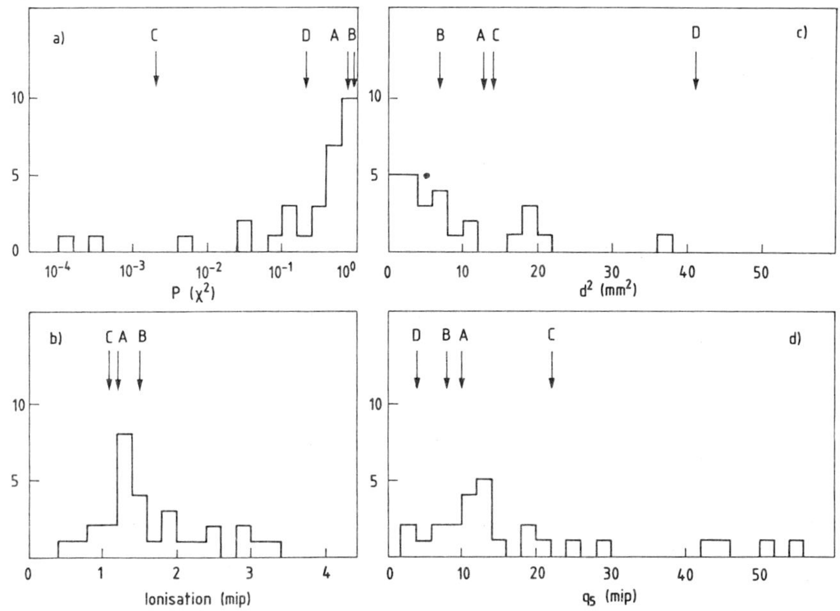

FIGURE 4. Electron quality parameters (see reference 1): (a) quality of the track-cluster match; (b) ionization measured in minimum ionizing particle (mip) equivalent. (This information is not available for event D.); (c) distance squared from the track to the preshower counter cluster centroid (measured in mm^2); (d) charge of the preshower counter cluster (measured in mip). The histograms are for the $W \to e\nu$ candidates presented in reference 1 and having $p_\perp(e) > 25$ GeV/c. The values taken by the quality parameters of the four events in FIGURE 3 are indicated by arrows.

possibility that a jet having the same transverse momentum as the neutrino candidate be produced at $\theta = 15°$ (or 165°). Under such an assumption, we can calculate a lower limit for the invariant mass of the system of all large transverse momentum particles produced in the event, including the small angle undetected jet. In the case of events A to C, they correspond to impossible or very unlikely kinematical configurations.

More generally an event appearing at (p_0, E_0) in the $p_\perp(Je)$, $E_\perp(J)$ plane (FIGURE 2) may be interpreted as containing an undetected jet, which would account, at least in part, for the measured value of p_0. If this jet had been detected, the same event would have appeared at (p_0', E_0') with $p_0' < p_0$ and $p_0' + E_0' \geq p_0 + E_0$. However, the only events having $p_\perp(Je) + E_\perp(J) > 90$ GeV are precisely the four events of FIGURE 3. The absence of other events in this region, even with low values of $p_\perp(Je)$, excludes such an interpretation.

A muon can in principle simulate a neutrino in the UA2 detector. Although we know of no mechanism that could produce a very massive electron-muon pair at a detectable rate, we looked, at each of the four events, for a track in the vertex detector near the neutrino azimuth and associated with calorimeter energy consistent with the response to a minimum ionizing particle. We found none.

In each of the four events in FIGURE 3, the sharing of the jet energies between the various calorimeter compartments[2] is consistent with expectation. Moreover, each jet contains several tracks having their origin at the event vertex. These observations exclude interpretations in terms of a cosmic ray or beam-gas background.

EVENT INTERPRETATION

In this section, we study possible sources for events A to D, under the assumption that they contain a genuine $e\nu$ pair.

While in three events (A to C) the $e\nu$ pair has a large azimuthal opening ($\Delta\phi > 120°$), event D contains a narrow $e\nu$ pair ($\Delta\phi \simeq 17°$). It consists of a large transverse momentum jet [$p_\perp(j_1) \simeq 70$ GeV/c] emitted at the opposite azimuth to the electron-neutrino pair and a smaller transverse momentum jet [$p_\perp(j_2) \simeq 25$ GeV/c]. The invariant mass of the ($e\nu j_2$) system depends upon the unknown value of $p_\parallel(\nu)$. It takes its minimum value, 23 ± 6 GeV/c^2, when the neutrino has the same rapidity as the (ej_2) system. In this case, $m(e\nu j_1 j_2) = 145 \pm 15$ GeV/c^2. The configuration of this event suggests an interpretation in terms of a quark-antiquark pair, one member of which decays semileptonically. However, because of the absence of other events with similar topologies and because of the similarity of its configuration with that of a two-jet event, we prefer to defer such an interpretation until other events of the same kind have been observed.

In the three other events (A to C), the transverse mass of the $e\nu$ pair (TABLE 1) ranges between 56 GeV/c^2 and 82 GeV/c^2, thus suggesting an interpretation in terms of a $W \rightarrow e\nu$ decay since the W boson is the only known particle with a large enough mass. It is indeed possible to adjust the unknown value of $p_\parallel(\nu)$ to obtain $m(e\nu) = m(W)$ and to describe the events in terms of associated W-jet(s) production. In general, there are two solutions to this problem, both associated with different values of both $p_\parallel(WJ) \equiv x_F(WJ)\sqrt{s}/2$ and of $m(WJ)$, which are listed in TABLE 1. The distribution of $m(WJ)$ for the solution giving the smaller value of $|x_F(WJ)|$ is shown in FIGURE 5 for all events having $p_\perp(Je) > 25$ GeV/c. While the 31 events having $E_\perp(J) < 30$ GeV cluster in the neighborhood of $m(WJ) = m(W)$, the three events of TABLE 1 populate the region, $160 \leq m(WJ) \leq 180$ GeV/c^2, which might suggest an interpretation in terms of a heavy object decaying into a W boson and a system, J, of other particles. However, the significance of this observation is weakened by the fact that the background events have a similar $m(WJ)$ distribution (FIGURE 5), thus implying that the clustering in mass might simply result from kinematics. In addition, such an interpretation should account for the fact that in events A and B, J consists essentially of a single jet, while in event C, it consists of a large mass pair of jets.

We remark that a reasonable upper limit to the rate of occurrence of events containing a $W \rightarrow e\nu$ produced via known processes, in association with a system, J, of large transverse momentum jets, can be obtained from the relation,

$$\sigma(p\bar{p} \rightarrow W + J + \ldots)/\sigma(p\bar{p} \rightarrow W + \ldots)$$

$$\simeq \sigma(p\bar{p} \rightarrow j_1 j_2 + J + \ldots)/\sigma(p\bar{p} \rightarrow j_1 j_2 + \ldots),$$

TABLE 1. Event Parameters

		Events			
		A	B	C	Units
Electron	$p_T(e)$	18.3 ± 0.8	22.0 ± 0.9	34.4 ± 3.2	GeV/c
	$\eta(e)$[a]	0.02	−0.23	0.24	
	$P(\chi^2)$[b]	0.97	0.76	0.002	
	d^2[b]	13	7	14	mm^2
	q_5[b]	10	8	22	mip
Jets[f]	$p_T(j_1)$	27 ± 3	67 ± 7	38 ± 5	GeV/c
	$\eta(j_1)$[a]	−0.59	−0.26	0.07	
	$\Delta\phi(j_1)$[c]	50	310	311	degrees
	$p_T(j_2)$	6 ± 1		21 ± 3	GeV/c
	$\eta(j_2)$[a]	0.50		0.12	
	$\Delta\phi(j_2)$[c]	93		183	degrees
	$p_T(j_3)$	5 ± 1		7 ± 1	GeV/c
	$\eta(j_3)$[a]	−1.09		−1.38	
	$\Delta\phi(j_3)$[c]	70		316	degrees
	$E_T(J)$	39 ± 4	67 ± 7	66 ± 6	GeV
	$m(J)$			63 ± 5	GeV/c^2
	$p_T(\nu)$	51 ± 4	86 ± 6	57 ± 5	GeV/c
	$\Delta\phi(\nu)$[c]	220	141	141	degrees
	$m_T(e\nu)$	56 ± 2	81 ± 3	82 ± 4	GeV/c^2
1	$m(WJ)$	179 ± 9	176 ± 9	162 ± 8	GeV/c^2
	$x_F(WJ)$	0.45 ± 0.04	−0.01 ± 0.06	−0.04 ± 0.05	
2	$m(WJ)$	141	166	164	GeV/c^2
	$x_F(WJ)$	−0.55	−0.31	0.13	
	B_1[d]	<0.016(90% C.L.)	<0.016(90% C.L.)	<0.016(90% C.L.)	
	B_2[d]	0.33 ± 0.03	0.06 ± 0.01	0.06 ± 0.01	
	N_{QCD}[e]	0.4	$1.2 \cdot 10^{-2}$	$7 \cdot 10^{-3}$	

[a] The pseudorapidity, η, is positive in the proton direction.
[b] Electron quality parameters are defined in FIGURE 4 and reference 1.
[c] $\Delta\phi$ is the azimuth difference with respect to the electron (in degrees).
[d] Background evaluations, B_1 and B_2, are described in the text (see section entitled DATA ANALYSIS AND BACKGROUND EVALUATION).
[e] Number of W-jets events expected from known production processes (see text).
[f] Jet energies are expected to be smaller than the parent parton energies (see text). This has not been corrected for and affects all parameters depending upon jet energies.

where $j_1 j_2$ is a pair of jets having the same configuration as the $e\nu$ pair ascribed to a W decay. This relation holds approximately for processes in which $j_1 j_2$ couples via a gluon to a quark-antiquark pair (although the W-current, contrary to the gluon-current, is flavor changing). We expect the contribution of other QCD subprocesses and interferences among them to alter only slightly the right-hand side of the above relation. We evaluate such upper limits for events A to C from the sample of jet events described in reference 6. We take as J any jet (jet pair) having a transverse momentum (invariant mass) at least as large as that of the corresponding jet (jet pair) in events A and B (C).

The results, listed in TABLE 1, include the different acceptances of the UA2 apparatus to $e\nu$ pairs and to jet pairs. They indicate that events B and C, if they indeed contain a genuine $W \to e\nu$, are difficult to understand in terms of associated W-jet(s) production via known processes. Event A, on the other hand, has an unlikely high value of $x_F(WJ)$, which corresponds to a W longitudinal momentum of nearly 150 GeV/c.

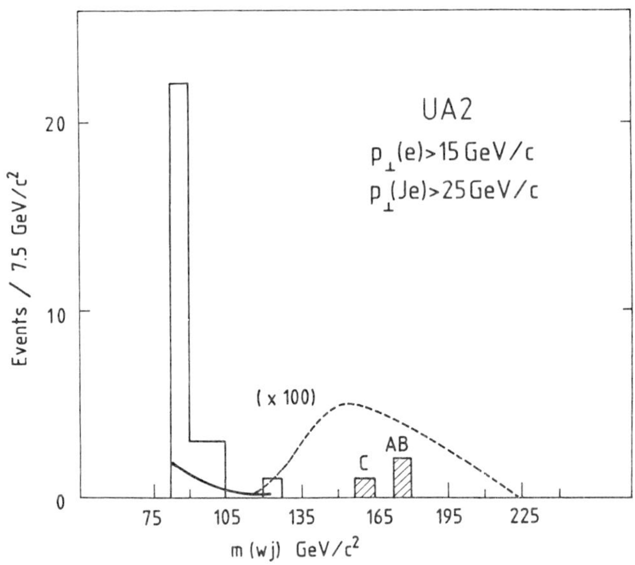

FIGURE 5. Distribution of $m(WJ)$ for the 35 events having $p_\perp(Je) > 25$ GeV/c. Events also having $E_\perp(J) > 30$ GeV are crosshatched. The event at $m(WJ) \simeq 120$ GeV/c^2 is the event having $E_\perp(J) \simeq 26$ GeV (see FIGURE 1b and reference 1). Event D is omitted from the figure because its narrow $e\nu$ pair does not suggest an interpretation in terms of $W \to e\nu$. Such an interpretation would imply a very large value of $p_\perp(W)$ and $m(WJ) \simeq 230$ GeV/c^2. The smooth line corresponds to the calculated background. The dotted line is the background distribution (multiplied by 100) for events having $E_\perp(J) > 30$ GeV.

More generally, we note that if the $e\nu$ pair in events A to C is coupled to the initial state via a real or virtual W, as expected in standard theories, other events in which the $e\nu$ pair is replaced by a jet pair should occur as well and at a rate at least six times as high (for the two lighter quark families). However, this factor is reduced by acceptance effects and a search for such events is beyond the scope of the present paper.

The uncertainties quoted in TABLE 1 account for measurement errors on energies and angles. The neutrino transverse momentum may be evaluated from the large transverse momentum particles alone, $\overline{p}_\perp(\nu) = -\overline{p}_\perp(Je)$, or by including all other softer particles that do not result in transverse energy clusters exceeding 3 GeV. The differences between the two evaluations are small and have been accounted for in TABLE 1.

In addition to the above instrumental uncertainties, the measured jet energies are expected to be somewhat smaller than that of the parent partons (we neglect the jet mass and we do not include jet fragments that do not contribute to the transverse energy cluster). These effects are not corrected for in TABLE 1. Using the ISAJET program[7] to simulate *W* decays into a pair of light quarks, we find that the measured two-cluster mass is expected to be $\simeq$ 15 GeV/c^2 lower than $m(W)$. Therefore, the value, $m(J) = 63 \pm 5$ GeV/c^2, obtained for event C is not inconsistent with the hypothesis that *J* results from a *W* decay.

Finally, we note that the interpretation of the $e\nu$ pair in terms of a *W* decay in events A to C is by no means mandatory because the missing transverse momentum does not need to be ascribed to a single neutrino, but might be shared among several undetected particles.

CONCLUSION

A search for events containing an electron-neutrino pair having $p_{\perp}(e) > 15$ GeV/c and $p_{\perp}(\nu) > 25$ GeV/c has resulted in a sample of 35 events, the majority of which have been previously studied[1] and interpreted in terms of $W^{\pm}$ production, with characteristic properties in agreement with QCD predictions.

Four events have been found in which the $e\nu$ pair is produced in association with a jet, or a system of jets, having a very large transverse energy. We have given a detailed description of these events and we have considered possible sources of background contaminations.

Three of these events contain a large transverse mass $e\nu$ pair and have been interpreted in terms of *W*-jet(s) associated production. However, their configurations are such that their production via known processes is very unlikely for at least two of them. In each of the three events, the invariant mass of the *W*-jet(s) system is measured to be in the vicinity of 170 GeV/c^2, but the significance of this observation is weakened by the fact that this mass region is kinematically favored by the selection criteria.

POSTSCRIPTUM

UA2 has collected new data in 1984 equivalent to 310 nb^{-1}. A very preliminary analysis of these data does not add more events of the type A–C to the sample of electrons with missing p_T. It makes it more likely that events A–C come from *W* production in association with a hard QCD subprocess, but the result is too preliminary to base any final conclusion on.

REFERENCES

1. BAGNAIA, P. *et al.* UA2 COLLABORATION. 1984. A study of high transverse momentum electrons produced in $p\bar{p}$ collisions at 540 GeV. Z. Phys. C. Particles and Fields **24:** 1.
2. MANSOULIÉ, B. 1983. The UA2 apparatus at the CERN $p\bar{p}$ Collider. *In* Proceedings of the

Third Moriond Workshop on p$\bar{p}$ Physics, p. 609. Editions Frontières. France; DIALINAS, M. *et al.* 1983. The vertex detector of the UA2 experiment. LAL-RT/83-14, ORSAY; CONTA, C. *et al.* 1983. The system of forward-backward drift chambers in the UA2 detector. CERN-EP/83-176. Submitted to Nucl. Instrum. Methods; BORER, K. *et al.* 1983. Multitube proportional chambers for the localization of electromagnetic showers in the UA2 detector. CERN-EP/83-177. Submitted to Nucl. Instrum. Methods; BEER, A. *et al.* 1983. The central calorimeter of the UA2 experiment at the CERN p$\bar{p}$ Collider. CERN-EP/83-175. Submitted to Nucl. Instrum. Methods.
3. THE STAFF OF THE CERN p$\bar{p}$ PROJECT. 1981. Phys. Lett. **107B:** 231.
4. BAGNAIA, P. *et al.* UA2 COLLABORATION. 1983. Phys. Lett. **129B:** 130.
5. FAYET, P. & S. FERRARA. 1977. Phys. Rep. **32C:** 249 and references therein.
6. BAGNAIA, P. *et al.* UA2 COLLABORATION. 1984. Measurement of very large transverse momentum jet production at the CERN p$\bar{p}$ Collider. Phys. Lett. **138B:** 430.
7. PAIGE, F. & S. PROTOPOPESCU. 1981. ISAJET, BNL report 31987.

Supergravity and Superstrings

MURRAY GELL-MANN
Physics Department
California Institute of Technology
Pasadena, California 91125

Over the last decade, theorists have been searching hard for a unified quantum field theory of physical interactions with or without gravitation. The "standard" $SU_3 \times SU_2 \times U_1$ theory has three independent coupling constants and numerous dimensionless parameters determining the mass ratios, the weak coupling matrix, etc. The number of sets of Higgs bosons and the representation to which they belong appear arbitrary. The theory cries out to be embedded in a bigger and more nearly unified theory. The fact that the coupling constants appear to come together somewhere between 10^{15} GeV and the Planck mass of $\sim 2 \times 10^{19}$ GeV suggests a unified Yang-Mills theory, for example with SU_5 and three fermion families (fermions in $\underset{\sim}{5} + \overline{\underset{\sim}{10}}$). However, such a theory, with one coupling constant, still has numerous dimensionless parameters and arbitrary Higgs boson behavior.

Renormalizability is assured, but the constants, although they absorb all the infinities, are not calculable and the unified Yang-Mills theory is still obviously incomplete. In particular, the very small and rather similar ratios of Λ_{QCD}/M_{UNIF}, various $M_{FERMION}/M_{UNIF}$, and $M_{INT.BOSONS}/M_{UNIF}$ must be explained. Of course, Λ_{QCD}/M_{UNIF} is related to exp $- const/\alpha_{UNIF}$, but then the value of the unified coupling constant near the unification mass needs explanation, which is like a modern version of the old problem of calculating $^1\!/_{137}$.

Although there is as yet no direct evidence, an attractive modification involves switching to N = 1 super-Yang-Mills theory, in which spin 1 bosons are accompanied by new spin ½ gluinos, photinos, etc., spin ½ fermions are accompanied by new spin zero squarks and sleptons, and spin zero higgsons are accompanied by new spin ½ shiggsons! The super-Yang-Mills theory is more easily compatible with experiments on proton decay, and besides, it may provide the only way to keep some Higgs boson(s) light.

We may note here that there is a set of theories finite at least to two loops, including just one of phenomenological interest, with SU_5 as the gauge group, three quark-squark, lepton-slepton families of $\underset{\sim}{5} + \overline{\underset{\sim}{10}}$, and higgsons-shiggsons as follows: one $\underset{\sim}{24}$, four $\underset{\sim}{5}$'s, four $\overline{\underset{\sim}{5}}$'s. The theory has a number of parameters and can fit all experimental evidence to date. (The extra higgsons, mostly heavy, might be useful for producing enough baryons in the early universe.)

While N = 1 supergravity, generalizing Einstein's gravity theory, is not necessarily very divergent itself (first divergences setting in perhaps at three loops), it is terribly divergent when coupled to external N = 1 supermatter, such as N = 1 super-Yang-Mills theory with N = 1 supermultiplets of spin ½ and spin zero.

Such N = 1 supermatter coupled to N = 1 supergravity with a cutoff has been used to show how important modifications are introduced by supergravity coupling into such quantities as the proton decay rate and branching ratios; however, such a divergent theory really does not make sense. We still require a finite, parameter-free theory and one that unifies all of physics, including Einsteinian gravitation.

Three paths are being explored in the search for the ultimate unified theory of physics. The first path involves N > 1 supergravity in four dimensions, without external supermatter, particularly the largest such theory, N = 8 supergravity, where there is no room for external supermatter. The N = 8 supergravity supermultiplet itself contains all the haplons (fundamental fields of the theory).

N = 8 Supergravity
Haplons:

		J_z
1	graviton	+2
8	gravitinos	+3/2
28	vector bosons	+1
56	spin 1/2 fermions	+1/2
70	spin 0 bosons	0
56	spin 1/2 fermions	−1/2
28	vector bosons	−1
8	gravitinos	−3/2
1	graviton	−2

Here, divergences probably do not set in until we get to three or perhaps seven loops, but there is no known reason to exclude them. N = 8 supergravity has Newton's constant, κ^2, and may also have a dimensionless self-coupling parameter, e, for the gauge group, SO_8, with 28 generators. Unfortunately, SO_8 does not contain $SU_3 \times SU_2 \times U_1$. Moreover, the list of spin 1/2 particles, while long enough, does not quite match the list of quarks and leptons we know. If N = 8 supergravity is to match experience, then the elementary particles we know cannot be the haplons of the theory, but must be composites and/or solitons.

The second and third paths involve higher-dimensional theories, with ten or eleven dimensions, where six or seven, respectively, are supposed to roll themselves up spontaneously into a small structure that is not observable directly, but which is detected only through its effects on the spectrum and symmetries of the elementary particles and on very early cosmology.

Higher-dimensional theories that have been studied are either "superstring theories" in ten dimensions or else N = 1 or N = 2 supergravity theories in eleven and ten dimensions, respectively. The latter are much more recent and, I believe, much less interesting, but they have received much more attention.

If we perform trivial dimensional reductions (just circles for all the curled-up coordinates), then we obtain the relations:

N = 1, D = 11
supergravity
↓
"N = 2A," D = 10 supergravity (non-chiral) ↘

"N = 2B," D = 10 supergravity (chiral) ↙

N = 8, D = 4 supergravity
supergravity
(non-chiral)

Ten-dimensional supergravity theories (N = 1, N = 2A, and N = 2B) are all very singular, even for one loop. Eleven-dimensional supergravity theory is trivially nondivergent in one loop because it is odd-dimensional, but in two loops it also must be highly divergent. Nevertheless, many brilliant theorists have spent a great deal of time on these theories, concentrating especially on generalizing the old Kaluza-Klein idea of getting electromagnetism from a fifth dimension.

Here, they use six or seven extra dimensions (index, i), our usual four dimensions (index, μ), and components, $g_{\mu i}$, of the metric tensor to construct gauge vector potentials, where the gauge symmetries are isometries of the six- or seven-dimensional space.

While a trivial dimensional reduction (just a product of seven circles) turns N = 1 supergravity in eleven dimensions into N = 8 supergravity in four dimensions (with 28 vector bosons that can gauge SO_8), we notice that SO_8 does not include $SU_3 \times SU_2 \times U_1$ and remark that a different, nontrivial dimensional reduction could lead, for example, to Kaluza-Klein bosons gauging $SU_3 \times SU_2 \times U_1$.

However, this Kaluza-Klein program of accounting for gauge bosons is flawed. For one thing, it is very difficult, and perhaps impossible, to obtain a chiral theory with massless gauge bosons after dimensional reduction. For another, the dimensional reduction over a space with nontrivial isometries tends to give a huge cosmological constant at the classical level. In any case, the higher-dimensional supergravity theories are no doubt hopelessly divergent.

The cosmological constant is proportional to the vacuum expected value of the energy density. Supersymmetry makes it vanish in the simplest theories, where there is no spontaneous violation. In realistic theories, where supersymmetry is gauged and spontaneously violated, it is by no means obvious how to get rid of the cosmological constant. Yet the effective cosmological constant in astronomy is less than 10^{-118} times the natural unit involving the Planck mass. If the cosmological constant gets a large value at the classical level, then (1) we can try to cancel that against quantum corrections, hoping to do it convincingly to avoid introducing the biggest fudge factor in history, or (2) we can try to exploit (with Hawking) the possibility of quantum gravity giving a foaminess to space-time, thus disconnecting the theoretical cosmological constant from the astronomical one.

Another promising strategy, however, is to have the cosmological constant vanish classically and then try to make the corrections vanish too. The former is possible if we abandon Kaluza-Klein, but the latter may be harder.

The third path, as mentioned above, involves superstrings. A superstring theory has an infinite number of fundamental boson and fermion states, of all spins, initially (as coupling starts from zero) stable and lying on an infinite number of straight Regge trajectories, so that the states with spins increasing by two at a time are spaced equally in mass squared. The spacing is related to the "string tension." The coupling introduces level widths and modifies the trajectories.

String theories may be thought of as embodying the "bootstrap principle;" in a sense, the infinitely many particles of the theory are bound to one another by means of forces coming from the exchange of the same particles, which then form bound states that are again the same particles.

Although string theories can be written in terms of quantum fields that are functions of a one-dimensional string rather than a point, they are still local theories because when an interaction causes a string to divide into two strings, for example, the new ones just correspond to dividing the old one at a point.

Superstring theory was originally invented in 1971 by Neveu and Schwarz in an effort to describe hadrons. Unlike earlier string models, it incorporated not only a boson spectrum, but also a fermion spectrum proposed by Ramond. The theory seems to require ten dimensions; of course, one hopes that six dimensions curl up spontaneously. In 1971, the theory possessed a kind of supersymmetry, which was noted by Gervais and Sakita. In 1976, Gliozzi, Scherk, and Olive noticed that the theory could be improved by the consistent omission of certain pieces, after which one could show that there were no ghosts or tachyons. The theory could then be shown to possess regular N = 1 supersymmetry in ten dimensions and to contain N = 1 supergravity in ten dimensions as a sector of the theory. This was after the invention of space-time supersymmetry and supergravity, which, if they had not been invented in other ways, could have been discovered in superstrings.

Meanwhile, QCD had given a presumably correct description of hadrons, with strings only an approximation. Thus, around 1974, Scherk and Schwarz suggested that the string theory be used as a unified theory of all physics, generalizing Einsteinian gravitation, with the trajectory slopes changed from $\sim 1/(\mathrm{GeV})^2$ to $\sim 10^{-38}/(\mathrm{GeV})^2$.

Original (Type I) superstring theory has closed strings, generalizing supergravity, and open strings, generalizing super-Yang-Mills. It is known how to construct a formal superstring theory when the gauge group is classical (i.e., unitary, orthogonal, or symplectic); simple arguments then eliminate the unitary groups. At first, the choice of orthogonal or symplectic group seems arbitrary, but recently that has been found to be untrue.

Among the three constants, Newton's κ^2, string tension, T, and super-Yang-Mills coupling constant, γ, there is one obvious relation and now another has been shown to exist. Thus, the theory is parameter-free apart from M_{Planck}. As in any parameter-free theory, all the small quantities we know about must be calculable and must somehow come out small, in a way that depends on the nature of the spontaneous dimensional reduction and of the low-mass sector that results.

The initially massless states of the open strings in Type I superstring theory belong to the adjoint representation of the gauge group. The states of the first excited level (initially around the Planck mass) belong to a different representation. The states continue to alternate between the two representations (i.e., for the orthogonal and symplectic groups, which are known to give formal superstring theories). Type I superstring theory reduces, on truncation to the initially massless sector, to N = 1 supergravity in ten dimensions, coupled to N = 1 super-Yang-Mills in ten dimensions with the gauge group of the superstring theory, plus some interesting extra terms discussed later. The truncated theory is terribly divergent, of course, but shows the character of the superstring theory.

Extraction of results of superstring theory is crucially dependent on the nature of the spontaneous dimensional reduction. Trivial reduction of Type I theory (ten dimensions reduced to four dimensions and the product of six circles) gives N = 4 supergravity coupled to N = 4 super-Yang-Mills as the initially massless sector.

N = 4 Supergravity
Four Dimensions:

		J_z
1	graviton	$+2$
4	gravitinos	$+3/2$

	6	vector bosons	+1
	4	spin ½ fermions	+½
1+	1	spin 0 bosons	0
4		spin ½ fermions	−½
6		vector bosons	−1
4		gravitinos	−3/2
1		graviton	−2

<u>N = 4 Super-Yang-Mills</u>
<u>Four Dimensions:</u>

		J_z	
1	vector boson	+1	All in adjoint representation of the gauge group
4	spin ½ fermions	+½	
6	spin 0 bosons	0	
4	spin ½ fermions	−½	
1	vector boson	−1	

Breaking this last down to N = 1 supersymmetry, we obtain:

1 Super-Yang-Mills	+1, +½, −½, −1	× Adjoint
3 N = 1 Matter Supermultiplets	+½, 0 0, −½	× Adjoint

A more complicated dimensional reduction will give different results and can break down the gauge group.

During the last couple of years, Michael Green and John Schwarz have found that there are two more ten-dimensional superstring theories, IIA and IIB, with only closed strings. They reduce, on truncation to the initially massless sector, to N = 2A and N = 2B supergravity, respectively, in ten dimensions. However, the superstring theories are finite to one loop instead of divergent like the corresponding supergravities. In addition, we see that IIA and IIB superstrings, when truncated to the initially massless sector and trivially reduced to four dimensions, yield N = 8 supergravity.

All three superstring theories, although they have the traditional description as "S-matrix" theories on the mass shell, can also be written as field theories (with fields as functionals of strings, instead of functions of points) with local couplings. So far, the field description is not covariant. In the present description, there are just polynomial couplings like:

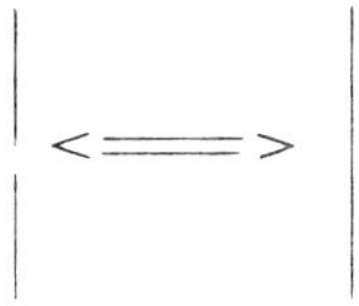

for open strings

(Basic super-Yang-Mills coupling)

or

for closed strings.

(Basic self-coupling of supergravity)

Higher couplings like fourth degree terms in Yang-Mills theory come from summing over exchanges of heavy intermediate excited states.

Type I and Type IIB superstring theories and N = 2B supergravity in ten dimensions appear at first to be threatened by the possible existence of "anomalies" (gravitational, Yang-Mills, and mixed) that would destroy gauge invariance and unitarity, as well as finiteness or renormalizability. (The last applies to strings since ten-dimensional supergravity is divergent anyway.) IIA superstring theory, being non-chiral, has all anomalies vanishing trivially and is finite to one loop. However, it is less interesting than IIB for comparison with observation, just because it is non-chiral. Last year, Witten and Alvarez-Gaumé were surprised to discover that all anomalies mysteriously cancel for N = IIB ten-dimensional supergravity. That same result could be demonstrated for the IIB superstring. That made IIB theory, which is also finite to one loop, an attractive candidate for the universal theory. Type I, with its arbitrary gauge group, seemed less attractive.

In the last few months, however, Green and Schwarz have shown that all of the anomalies of Type I string theory with a classical gauge group vanish to one loop if and only if the gauge group is $SO(32)$ and also that the theory is finite rather than renormalizable to one loop if and only if the gauge group is $SO(32)$. Here, therefore, we have a very impressive form of unification with (super)-gravitational closed strings and (super)-Yang-Mills open strings implying each other and controlling each other's behavior.

The string theories, IIA and IIB, as well as Type I with $SO(32)$ as gauge group, are all finite to one loop and all anomaly-free. They seem increasingly desirable in this order since Type IIA is non-chiral, while Type IIB lacks gauge bosons other than Kaluza-Klein ones, which give problems with chirality on reduction to four dimensions.

There is one more gauge group, the nonclassical (exceptional) group, $E_8 \times E_8$, that can yield finiteness and freedom from anomalies, provided a superstring theory can be built on it. I shall go into more detail below about that exciting prospect.

The finiteness of string theories in one loop is of a kind that strongly suggests the possibility of finiteness persisting to all orders, unlike the finiteness in one loop of N = 1, D = 11 supergravity, which appears to be merely a trivial consequence of D being odd and unlikely to persist in higher order. Also, the demonstration of the persistent absence of anomalies to all orders in string theory would really require this finiteness.

Where would the gauge bosons come in according to the different kinds of proposed unified theory?

N = 8, D = 4 supergravity: At least some would be composites or solitons.
N = 1, D = 11 supergravity: Kaluza-Klein.
Type I superstring, D = 10: Haplons that survive nontrivial dimensional reduction.
Type IIB superstring, D = 10: Again, composites or solitons.

(This last according to Gell-Mann and Zwiebach, who showed how in this theory two dimensions seem to want to roll up into a "teardrop" with a U_1 isometry, where the Kaluza-Klein boson acquires a mass and is useless as a gauge boson.)

It is presumably possible to integrate out all the initially massive states of a superstring theory and obtain a very complicated effective theory of the initially massless sector, which is not divergent if superstring theory finite. Expanding that effective theory at low energies for Type I superstring with $SO(32)$, we obtain first the truncated theory with N = 1 supergravity and N = 1 super-Yang-Mills theory in ten dimensions, with a correction term that breaks supersymmetry, but cancels anomalies, and then other terms that restore supersymmetry and so forth. The sum of these terms should give back the exact superstring theory with initially massive states integrated out. It may be that this kind of summation always leads to a superstring. Witten also has shown that freedom from anomalies in ten dimensions for Type I superstrings implies freedom from anomalies in four dimensions.

Kaluza-Klein aficionados want many isometries for their extra six- or seven-dimensional space in order to get massless gauge bosons. Type I superstring fans want as few as possible: The field equations may require $R_{ij} = 0$ (for the extra dimensions), where R_{ij} is the Ricci-Einstein tensor. (There can still be, though, some Riemann-Christoffel curvature, $R_{ikjl} \neq 0$.) The condition of $R_{ij} = 0$ (extra dimensions) will give no cosmological constant at the classical level and also no nontrivial isometries leading to extra massless gauge bosons. Even trivial U_1 isometries leading to massless Kaluza-Klein vector bosons may be eliminated by the requirement of surviving "massless" spin ½ fermions. Then there would be no Kaluza-Klein bosons at all. Note that in the dimensional reduction of a Type I superstring, if $\langle R_{ikjl} \rangle \neq 0$, then the background Yang-Mills field strength, $\langle G_{ij}^{x} \rangle \neq 0$ (as in the presence of a generalized monopole). Spatial and Yang-Mills symmetries are violated together by related amounts.

Now there are many unanswered questions about superstring theories and their dimensional reduction; I shall briefly mention just a few.

If the field equations allow many possible dimensional reductions, which one wins? Of course, to discuss quantum cosmology, we must adopt the many-worlds approach to quantum mechanics, which is a good idea anyway. Then, do we, following Hawking, transform to a Euclidean space, with $e^{iS} \longrightarrow e^{-S}$, and minimize the quantum-corrected action, S, so that by an inverse probability argument we have the most probable universe? Maybe the mathematics saves us the trouble and allows only one dimensional reduction.

What about sizes of extra dimensions? If we are using Kaluza-Klein ideas, then we are limited by the fact that the coupling parameters, e, are of order $R_{Planck}/R_{extra\ dimensions}$ for the Kaluza-Klein gauge bosons and these e's would be too tiny if the $R_{extra\ dimensions}$ were much larger than R_{Planck}. With Type I superstrings, we do not have this limitation. What are then the restrictions on $R_{extra\ dimensions}$ from observation? Could we even have some of them very much larger than the Planck length?

In superstring theories (Type I or II), the closed strings generalize supergravity, which generalizes Einsteinian gravitation, and the superstrings must obey a generalization of general relativity. Here, therefore, we have the reverse of Einstein's situation: theory first, principle afterward. What is the principle? It must be one of great importance.

Now let us return to the possible gauge group, $E_8 \times E_8$. If there is a Type I string theory based on $E_8 \times E_8$ (which has dimension 496 and rank 16 like $SO(32)$), it also would give freedom from anomalies and finiteness to one loop. Theorists have been investigating whether there is some mathematical trick that makes possible a Type I $E_8 \times E_8$ superstring. (We recall that existing methods permit only the classical orthogonal and symplectic groups, of which only $SO(32)$ is allowed by anomaly considerations.)

It is clear that $E_8 \times E_8$ is highly desirable as the gauge group of a superstring:

(1) With all initially massless spin 1, ½, and 0 particles in the adjoint representation of the gauge group, it is very suitable to have a group with only one pair of low-dimensional representations, $(\underset{\sim}{1}, \underset{\sim}{248})$ and $(\underset{\sim}{248}, \underset{\sim}{1})$, that serve, in a sense, as adjoint, fundamental, and spinor all at the same time.

(2) The adjoint representation, $(\underset{\sim}{1}, \underset{\sim}{248})$ and $(\underset{\sim}{248}, \underset{\sim}{1})$, contains a natural color, SU_3 (that of the non-complex-number part of the underlying octonions, as pointed out by Gürsey), and the right kind of flavor variables (if there is a suitable reduction of symmetry) to describe the spin 1, ½, and 0 particles we know and love.

Very recently, a new type of superstring theory in ten dimensions has been found by Gross, Harvey, Martinec, and Rohm. They call it a "heterotic" superstring theory, but I shall refer to it as Type III/II. It can be thought of as a hybrid of the Type I superstring in ten dimensions and the old Veneziano string in 26 dimensions; alternatively, the extra 16 coordinates can be treated as fermionic or else we can introduce internal degrees of freedom attached to a ten-dimensional superstring. Although it is a theory of closed strings only, it does involve an internal super-Yang-Mills gauge group in ten dimensions. That gauge group can only be $SO(32)$ or $E_8 \times E_8$.

The initially zero-mass sector of the Type III/II theory is the same as that of the Type I theory (for $SO(32)$, and also for $E_8 \times E_8$ if a Type I theory exists for that group), but the representations to which the higher-mass states belong are completely different. For $SO(32)$, the next states above the super-Yang-Mills particles in the adjoint representation $\underset{\sim}{496}$ belong to the representations $\underset{\sim}{1}$ and $\underset{\sim}{527}$ in Type I theory and the higher states oscillate between these and the adjoint. In Type III/II theory, what lies above the adjoint 496 is a 69752-dimensional reducible representation, where 69752 = 1 + 527 + 496 + 35960 + 32768. Here, 35960 is the four-index antisymmetric tensor in 32 dimensions and 32768 is a spinor. The higher-mass states belong to larger and larger reducible representations.

For $E_8 \times E_8$, the picture in Type III/II theory is similar and even the dimensions of the reducible representations are the same as for $SO(32)$. In this case, we have 496 = (1, 248) + (248, 1); 69752 = (1, 1) + (1, 1) + (1, 248) + (248, 1) + (248, 248) + (3875, 1) + (1, 3875); and so forth.

In each case, the sequence of representations 496, 69752, etc., together with a fictitious singlet, form an irreducible representation of the Kac-Moody algebra, $[\alpha_{mi}, \alpha_{nj}] = if_{ijk}\, \alpha_{m+n,k} + m\delta_{ij}\, \delta_{m+n,0}$, where m and n run over all the integers (positive, negative, and zero) and i and j run over the 496 generators of the group, $SO(32)$, or $E_8 \times E_8$.

The Type III/II superstring, regarded as a ten-dimensional theory, is no exception to the statement made earlier that Kaluza-Klein bosons are unnecessary and probably unwanted; however, if we do introduce sixteen extra commuting dimensions and then shrink them to zero in order to describe the $E_8 \times E_8$ gauge symmetry, then a kind of formal Kaluza-Klein procedure is involved at that stage. That may well be the only form in which the Kaluza-Klein idea survives.

Some work has already been done on the possible reduction of gauge symmetry in an $E_8 \times E_8$ superstring theory (Type III/II or, if it exists, Type I) as ten dimensions are reduced to four. Candelas, Horowitz, Strominger, and Witten have studied such reductions over a certain kind of Ricci-flat ($R_{ij} = 0$) six-dimensional space. Several of these "Calabi-Yau" spaces are known and more are being discovered. Candelas *et al.* have found at least one of considerable phenomenological interest. For the solution found in that space, the background field strength is like an SU_3 "monopole" and breaks down one of the E_8's to E_6. (We note that E_8 includes $E_6 \times SU_3$ and that for this embedding, E_6 is the group of transformations commuting with SU_3.) The occurrence of SU_3 in this connection seems to be the necessary and sufficient condition for an N = 1 supersymmetry to survive the dimensional reduction.

In this example, what survives below the Planck mass is a unified super-Yang-Mills theory with gauge group, E_6, and four families of spin ½ and spin 0 particles, each in a 27 of E_6. Of course, E_6 is an elegant generalization of SU_5, and the 27 is a generalization of the 5 + $\overline{10}$ of SU_5. There is also a super-Higgs phenomenon involving a nonlinear representation of E_6 such that the Higgs bosons are effectively in the adjoint representation 78. The symmetry is thus broken to something like $SU_3 \times SU_2 \times U_1 \times U_1 \times U_1$.

The remaining E_8 is unbroken and describes a kind of shadow world of particles coupled to known particles only through interactions of gravitational strength (including some processes involving virtual excitations of states above the Planck mass). We must check carefully what would be the consequences for cosmology.

It looks as if the truly unified quantum field theory of all particles and interactions may be at hand, but a great deal of work is needed on dimensional reduction and extraction of low-energy results and cosmological implications before a real comparison with experience is possible.

DISCUSSION OF THE PAPER

M. BLOCK (*Northwestern University, Evanston, IL*): What is the relation of this theory to proton decay?

M. GELL-MANN (*California Institute of Technology, Pasadena, CA*): The E_6 super-Yang-Mills theory presumably gives a rate of proton decay in the same range as that given by other supertheories. As usual, the coupling to supergravity introduces some uncertainty in practice because we do not yet know how to compute that coupling without a cutoff.

V. VUILLEMIN (*CERN, Geneva, Switzerland*): From the superstring theory, do you expect more generations of quarks and leptons?

GELL-MANN: I can answer with a firm "Maybe."

A. WHITE (*Argonne National Laboratory, Chicago, IL*): On the issue of the underlying principle, why should a theory of the universe be formulated on the light-cone? Alternatively, why should there not be a fundamental theory of surfaces rather than strings?

GELL-MANN: Whether the superstring theories, written as quantum field theories of strings, can be formulated in a manifestly covariant manner is a purely technical matter as long as the theory is really covariant, which it is.

As to surfaces or other still more complicated structures, no one has ever shown that they can compete with superstrings in furnishing a consistent, perturbatively finite, perfectly bootstrapped theory of the world that contains supergravity and super-Yang-Mills theory.

G. L. KANE (*University of Michigan, Ann Arbor, MI*): Do you think the fermions that come out of this theory at the Planck scale will be the same as those we see (quarks and leptons) at low energies?

GELL-MANN: The simplest situation would be for the dimensional reduction to lead to a unified super-Yang-Mills theory with families of spin ½ and spin 0 particles, all initially at zero mass, with a mechanism for spontaneous mass generation that leaves some of the particles very light; those would be the ones we can detect at accelerator energies.

KANE: It seems one should be leery of forcing the spectrum of elementary fermions to look like the familiar quarks and leptons. Quarks and leptons may be composite after all.

GELL-MANN: An alternative possibility is certainly for the fundamental theory to lead to a system of relatively low-mass entities that are not known particles, but rather constituents out of which the known particles are built as composites or solitons.

S. MESHKOV (*National Bureau of Standards, Washington, D.C.*): What is the SU_3 contained in the breakdown, $E_8 \supset SU_3 \times E_6$?

GELL-MANN: Physical SU_3 groups, including SU_3 of color, would have to come out of E_6. The SU_3 group about which you are asking is just the one that occurs in $E_8 \supset SU_3 \times E_6$ in such a way that the adjoint $\underset{\sim}{248}$ of E_8 becomes $(\underset{\sim}{8}, \underset{\sim}{1}) + (\underset{\sim}{1}, \underset{\sim}{78}) + (\underset{\sim}{3}, \underset{\sim}{27}) + (\overline{\underset{\sim}{3}}, \overline{\underset{\sim}{27}})$.

In the theory, the background E_8 field strength transforms like an SU_3 "monopole" and that is the SU_3 we are discussing.

KANE: If the particles we "know" (quarks, leptons, gauge bosons, and their supersymmetric partners) go in a $\underset{\sim}{248}$, what other particles can we expect since ours do not fill up $\underset{\sim}{248}$?

GELL-MANN: The resulting super-Yang-Mills theory with spin ½ and 0 superfamilies need not fit into a $\underset{\sim}{248}$. As we mentioned, one reduction gives E_6 for the former and four $\underset{\sim}{27}$'s for the latter. Note that dimensional and symmetry reduction take place together. Of course, the adjoint and 27 of E_6 do involve new particles.

D. SCHRAMM (*University of Chicago, Chicago, IL*): Is the shadow E_8 identical in structure to our E_8? Such a symmetry would cause a problem cosmologically since it would result in too many low-mass particles.

GELL-MANN: No. It appears that the shadow E_8 may be unbroken, which would produce even more low-mass (or zero-mass) particles. If that is really the case, then we must make sure that the fact that these particles are coupled to ours only through interactions of gravitational strength (or weaker) is sufficient to prevent a clash with experience.

Some Comments on SSC Physics[a]

IAN HINCHLIFFE

Lawrence Berkeley Laboratory
University of California
Berkeley, California 94720

INTRODUCTION

In this paper, I shall discuss some topics relevant to the physics of the SSC. I shall not give a review of the subject. Exhaustive reviews are provided by EHLQ,[1] by the Proceedings of the Lausanne Workshop,[2] and by the forthcoming Proceedings of the 1984 DPF Summer Study held at Snowmass.[3] I will first discuss the reliability of predictions for event rates at the SSC and then I will discuss the status of proposals to search for the minimal Weinberg-Salam model Higgs boson.

UNCERTAINTIES IN RATE ESTIMATES

The QCD parton model underlies almost all of the estimates of signals and backgrounds relevant to the SSC and other hadron colliders. Fundamental interactions that are probed in the search for new physics involve the collisions of quarks and gluons and, in some specialized cases, other particles, such as W's, which can appear as constituents of the proton. The rate of some new particle production is given by the following parton model formula,

$$\sigma(s) = \sum_{i,j} \int f_i(x_1,Q^2) f_i(x_2,Q^2) \sigma_{i,j}(sx_1x_2) dx_1 dx_2.$$

Here, $\sqrt{\hat{s}} = \sqrt{sx_1x_2}$ is the invariant mass of the parton-parton system and $\sqrt{s}$ is that of the proton-proton system. In this formula, $f_i(x, Q^2)$ is the probability of a parton of type i being inside the proton with fraction x of the proton's momentum. $\sigma_{i,j}$ is the cross section for the interaction of the two partons, i and j. The scale, Q, appearing in the distribution functions is characteristic of the momentum transfers in the partonic process. In the case of jet production, its value is somewhat ambiguous, but it is of the order of the jet transverse momentum. In the case of new particle production, the mass of the new particle is probably the appropriate value.

The distributions, $[f_i(x, Q^2)]$, are extracted from deep inelastic scattering data at low Q^2 and then extrapolated to the higher Q^2 values appropriate to the SSC using the Altarelli-Parisi[4] equations of QCD. The kinematical range of the x_1 and x_2 integrals is $x_1x_2 \geq \hat{s}_{\min}/s$, where $\sqrt{\hat{s}_{\min}}$ is the minimum kinematically allowed value of $\sqrt{\hat{s}}$. The distribution functions are rapidly falling fuctions of x, so the dominant region of the

[a]This work was supported by the Director, Office of Energy Research, Office of High Energy and Nuclear Physics, Division of High Energy Physics of the U.S. Department of Energy under contract no. DE-AC03-76SF00098.

integral comes from $x_1, x_2 \approx \sqrt{\hat{s}_{\min}}/s$. At $\sqrt{s} = 40$ TeV, the production of a W boson is dominated by $x \approx 0.002$ and x as small as 4×10^{-6} is possible. These small values of x are not probed by current experiments, which are confined to $x > 0.05$ and $Q^2 \leq 200$ GeV^2, so two problems arise.

First, does the lack of data imply a very poor knowledge of the structure functions in this small x region? As Q^2 increases, the uncertainties caused by the lack of data at small x wash out. This is illustrated in FIGURE 1, which shows the behavior of the gluon

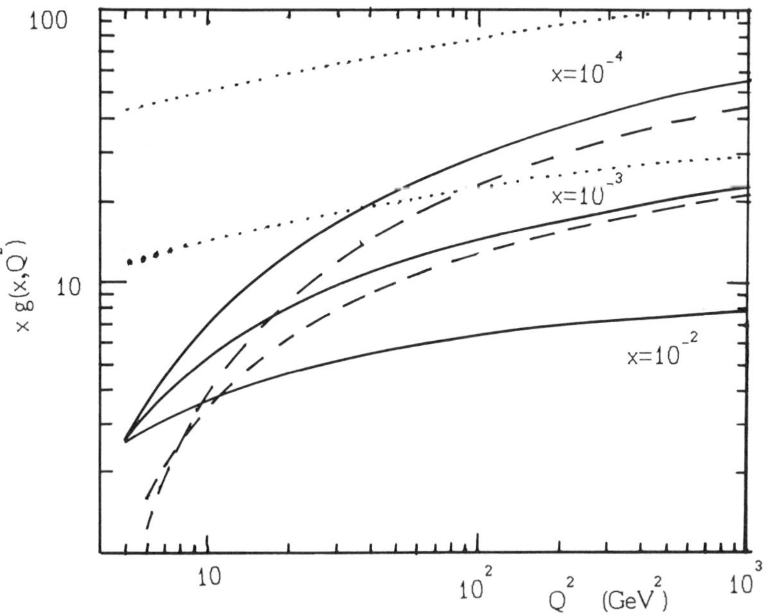

FIGURE 1. The evolution of $xg(x, Q^2)$ as a function of Q^2 for various values of x. The solid lines correspond to the starting distribution, (a), the dashed lines to (b), and the dotted lines to (c). (See text.)

distributions at small x as a function of Q^2. Three different gluon distributions were assumed at $Q^2 = 5$ GeV:[1]

(a) $x\, g(x, Q^2 = 5 \text{ GeV}^2) = (2.12 + 9.17x)(1 - x)^{5.9}$
(b) $x\, g(x, Q^2 = 5 \text{ GeV}^2) = $ (a) for $x > 0.01 = 25.56\, x^{1/2}$ for $x < 0.01$
(c) $x\, g(x, Q^2 = 5 \text{ GeV}^2) = $ (a) for $x > 0.01 = 0.44\, x^{-1/2} - 1.886$ for $x < 0.01$

The value of $g(x_0, Q^2)$ is controlled by that at $x > x_0$ at lower Q^2. The starting distributions were chosen to agree at $x > 0.01$, where data exist, so that eventually they will agree at smaller x. It is surprising and encouraging that the differences wash out so rapidly.

Secondly, does QCD perturbation theory in the form of the Altarelli-Parisi equations continue to apply at these small values of x? This vital question occupied one of the physics groups at Snowmass and the report by John Collins[5] should be consulted for details.

The QCD perturbation theory does indeed break down at small x, but the appropriate values of x are extremely small. An easy way to see that something must go wrong is to compute the cross section for, say, single W production as a function of s. In the region of small x, the structure function, $f_i(x, Q^2)$, is proportional to x^{-a} with a of order, 1.3.[1] The cross section for W production will then behave as $s^{(a-1)} \log s$. This behavior is unacceptable asymptotically and is an indication that the evolution of $f_i(x, Q^2)$ must change ultimately to cut off the growth at small x. This cutoff happens when the parton fraction, D, approaches one, where D is defined by[6]

$$D(x, Q^2) = xf_i(x, Q^2)m_x^2/Q^2.$$

There is a region in the (x, Q^2) plane where D is of the order of one or larger. In this region, partons start to overlap and cease to act individually, collective effects become important, and the simple parton model picture collapses. In his report, John Collins[5] has investigated this breakdown region. It is important not only that structure functions be used only when the fraction is small, but also that in the evolution of the Altarelli-Parisi equations from small Q^2, there are no sizeable contributions from regions of x and Q^2 where the equations are not valid. His conclusions are that the structure functions that are claimed to work for $x > 10^{-4}$ and $5\ \text{GeV}^2 \leq Q^2 \leq 10^8\ \text{GeV}^2$ are indeed valid over that range, so that the results obtained using them are reasonably reliable.

One final word of caution about distribution functions. Some of the radically different results for rates at the SSC that have appeared in the literature are due to misuse of distribution functions and do not represent some kind of extreme values showing the size of uncertainties. Care should be exercised when using a set of parameterizations of structure functions because these parameterizations usually have strictly limited ranges of applicability in x and Q^2 and can produce absurd results if used outside these ranges. Also, it is not legitimate to adjust the value of Λ_{QCD} in these parameterizations since they may not agree then with low energy data: Λ_{QCD} is usually strongly correlated with the gluon distribution and these correlations tend to reduce differences at higher Q^2.

Another problem is the heavy flavor content of the proton at high Q^2. EHLQ include bottom and top quarks in their distributions. Unfortunately, there is some ambiguity in the method for dealing with thresholds. The simplest method is to ignore heavy quarks (of mass m_Q) below threshold, i.e., when $Q^2 \leq 4m_Q^2$, and then to allow them to evolve in the same way as massless quarks once the threshold has been crossed. EHLQ have not done this, but rather have included some mass effects in the heavy quark evolution[7] that tend to slow their rate of growth above threshold. The ambiguity cannot be resolved completely since the different prescriptions are all equivalent up to the higher order QCD corrections that were neglected.

The reliability at small x has an interesting consequence. The fraction of events at the SSC that can be understood in terms of perturbative QCD will be much greater than the Sp$\bar{\text{p}}$S collider. The cross section for events in which there are two jets of transverse momentum greater than 10 GeV, each of rapidity, $|y| \leq 2.5$, is shown as a function of $\sqrt{s}$ in FIGURE 2. Estimates for the total cross section vary between 100 and 200 mb SSC energies, with the value at $\sqrt{s} = 540$ GeV being about 60 mb.[9] Of course, as s gets very large, the probability of a double parton scatter will increase and the jet mutiplicity will start to rise. The rate shown in this figure is very uncertain since it depends upon small x and Q^2.

Once we are certain that our predictions are free of small x disasters, the only uncertainty remaining concerns the accuracy of the evolved structure functions and the size of the higher order QCD corrections to the parton-parton scattering cross section, $\sigma_{i,j}$. These corrections are often ignored in making estimates, but they can be significant and they give some indication about the size of the error on these estimates. For example, there are expected to be corrections of the order of 30 percent, which increase the cross section for single W production.[10] The inclusion of these corrections at the Sp$\bar{p}$S collider improves the agreement between the predictions and the data.[11] For most processes, these corrections are not known; it would be valuable to have more

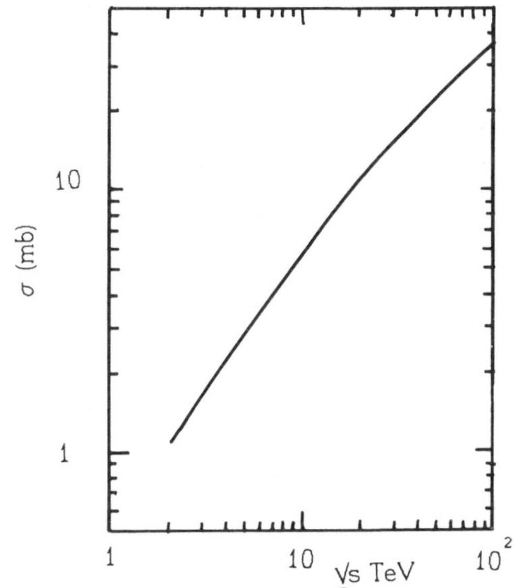

FIGURE 2. The cross section for producing a pair of jets each with transverse momentum greater than 10 GeV and with rapidity, $|y| < 2.5$.

calculations. In the case of jet production, the corrections are known only for a small fraction of the $2 \longrightarrow 2$ processes.[12] The knowledge of these corrections helps to resolve the ambiguities in the choice of the scale, Q^2, that appears in the distribution functions.

These corrections are sometimes incorporated into so-called K factors. It has become fashionable to multiply the lowest order QCD predictions by a K factor of the order of two before comparing with data. Except for the few processes where the higher order corrections are known, there is absolutely no justification for this procedure. In any case, the corrections are very unlikely to be a simple multiplicative factor; for example, in the case of jet production, the size of the correction will surely depend on the transverse momentum and the rapidity of the produced jet.

A very important issue concerns the mechanism for the production of new strongly

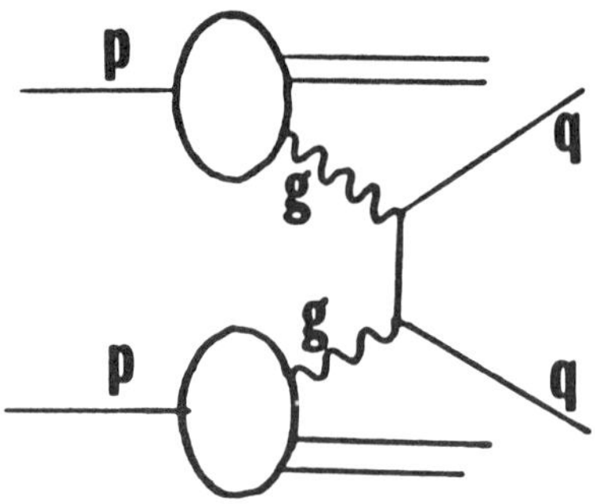

FIGURE 3. The gluon-gluon fusion mechanism for new particle production.

interacting particles. In the estimates most widely available, the pair production of such particles proceeds via the annihilation of a pair of gluons (or a quark and an anti-quark), one from each of the incoming beams (see FIGURE 3). This production is centered around zero rapidity. There has been much discussion in the literature[13] of the so-called intrinsic mechanism for such production. This mechanism exploits the possibility of exciting the pairs of heavy strongly interacting particles that are present in the proton's wave function (see FIGURE 4). If this latter mechanism is dominant, then the production of new, strongly interacting particles will be diffractive. Therefore, these particles will emerge at small angles to the beam (large rapidity).

Since the intrinsic mechanism is nonperturbative, a clean QCD prediction is not possible. The cross section has the following approximate form:

$$\sigma \approx f(M/\sqrt{s})/M^b.$$

It is important to understand the parameter, b, and hence the dependence of the intrinsic rate upon the heavy particle's mass (M). This will then allow an extrapolation from the existing upper bound on the intrinsic charm component at ISR energies to the intrinsic components of the larger mass objects at the SSC. The fusion process has a mass dependence that corresponds to $b \approx 2$. Brodsky *et al.*[14] have concluded that the mass dependence of the intrinsic mechanism is $b \approx 4$. Since the charm production rate at the ISR from the intrinsic mechanism cannot be much greater than that from the fusion mechanism, this result would imply that the intrinsic mechanism will not be important for the production at the SSC of particles heavier than about 30 GeV.

FIGURE 4. The intrinsic mechanism for new particle production.

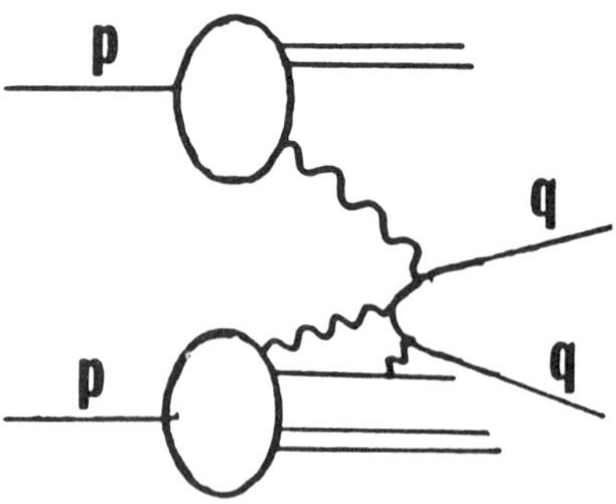

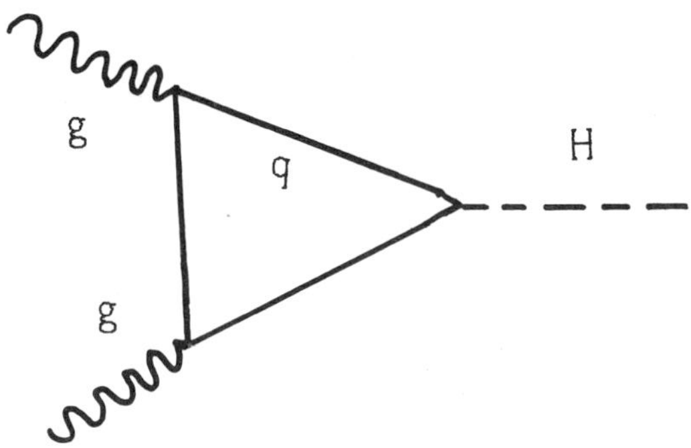

FIGURE 5. The process, gluon-gluon → Higgs.

Given the above discussion, it seems fair to conclude that the estimates given for production rates of new particles at the SSC should be good to a factor of two or so, at least for particles with masses greater than 100 GeV.

THE HIGGS BOSON

Several mechanisms have been proposed for the production of the Higgs boson at hadron colliders. First, it can be produced singly in gluon-gluon fusion via the diagram shown in FIGURE 5. The rate depends on the number and masses of quark flavors that can appear inside the loop.[15] In the mass range, $m_i \ll m_H$, the rate is roughly proportional to m_i^2, where m_i is the mass of the quark inside the loop. Thus, the existence of extra generations will greatly enhance the rate. The process shown in FIGURE 6 is less model dependent since it involves the coupling of the Higgs to the W and Z boson.[16] This coupling grows as the Higgs mass increases, so this mechanism will eventually dominate over the gluon-gluon fusion process. Although the final state consists of a Higgs and a quark pair, the quarks move almost parallel to the incoming hadrons. Therefore, this mechanism can also be considered as nonassociated production of a Higgs.

Two associated mechanisms have also been considered. The production in association with a W or Z boson[17] will have a smaller rate, but the presence of a W or Z in the final state could be used as a tag, thus improving the signal to background ratio.[18] The

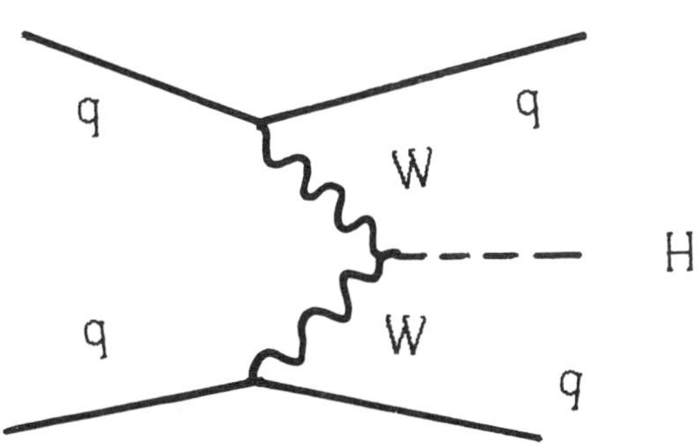

FIGURE 6. The process, $qq \to qqH$, via the gauge boson fusion process.

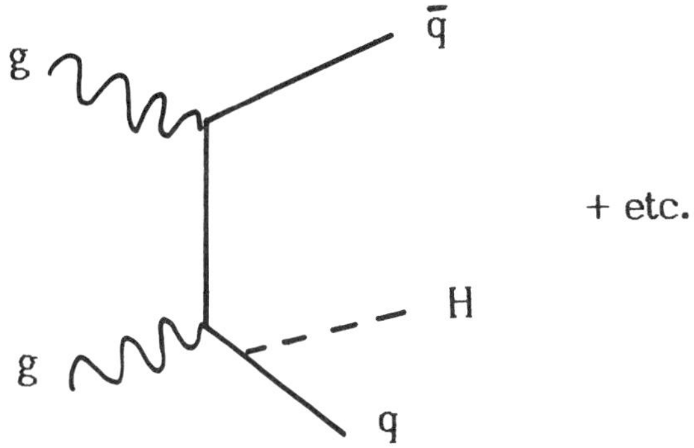

FIGURE 7. The associated production of a Higgs boson and a $t\bar{t}$ quark pair.

production in association with a top quark pair is shown in FIGURE 7. Again the rate is small, but the presence of a heavy quark pair could be useful.

FIGURE 8 compares the production rate from all four mechanisms. The single production rates are always dominant. Notice that for $m_H \geq 400$ GeV, the gauge boson fusion mechanism is dominant, so the rate is fairly model independent.

If the Higgs is lighter than about 70 GeV, it is likely to be found at the next

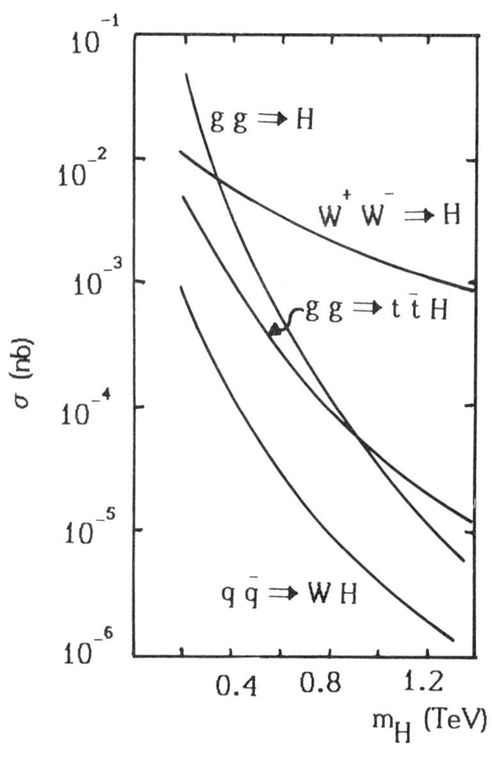

FIGURE 8. A comparison of the rate of Higgs boson production from the various mechanisms. (From reference 19.)

generation of e^+e^- machines, either in the process, $Z \rightarrow H\, e^+e^-$,[20] or via $e^+e^- \rightarrow Z + H$.[17] Consequently, I will only discuss Higgs masses larger than this. It is convenient to divide the discussion into three ranges of mass: (a) $m_H < 2\ m_w$; (b) $2\ m_w < m_H \leqslant 0(1)$TeV; (c)$m_H \geq 0(1)$ TeV. I will discuss the second region first.

If the Higgs mass is greater than twice the W or Z mass, then it will decay dominantly into W or Z pairs. The background in these channels from W or Z pair production via qq annihilation has been shown not to be a problem.[1,2] Even if the Z pair mode is detectable only via the leptonic decays of the Z's, it should still be possible to detect a Higgs boson if experiments are possible with an integrated luminosity of 10^{40}cm^{-2}. (See FIGURE 9.) FIGURE 9 shows the signal and background for $H \rightarrow W + W$. Here, both the W's are required to satisfy a rapidity cut of $|y_w| < 2.5$. The background has been estimated from $d\sigma/dM$ for the production of a pair of W's of invariant mass, M, multipled by ΔM, with ΔM being the greater of 10 GeV or the Higgs width.

The observability of W and Z final states in their hadronic decay modes will greatly ease the detection of such a Higgs boson. If the W or Z bosons decay hadronically, then the background will arise from final states with QCD jets in them. In the case of the WW final state where one W decays leptonically and the other decays hadronically, the background arises from the state of W + jet.

The report of Fernandez *et al.*[22] discusses in great detail how the final state WW can be distinguished from that of W + jet(s). In the case where the W momentum is small, the two jets for $W \rightarrow q\bar{q}$ are widely separated, so their invariant mass can be well measured. The UA2 group[23] working at the Sp$\bar{\text{p}}$S collider have attempted to detect the W in this manner. Their resolution is at present too poor and the statistics too low for the W to show up. Nevertheless, the situation is not hopeless. The better resolution for a proposed SSC detector should improve the situation and it should be possible to detect a slow W.[24] In her talk, Anne Kernan[25] discussed the possibility of detecting $W \rightarrow c\bar{b}$ via the semileptonic decay of the charm quark. The results give grounds for optimism.

If the W momentum is very large, then the jets from its decay will coalesce. One now has to measure the invariant mass of a single jet. A typical QCD jet of the same energy will tend to have a larger invariant mass and a higher multiplicity. Roughly, the probability of a jet of energy, E, having an invariant mass of M is[26]

$$\frac{dN}{d\theta} = 0.25e^{-4\theta},$$

where $\theta = M/E$. At the SSC, however, it is unlikely that this latter, high-momentum regime can be reached.

Fernandez *et al.*[22] considered the more difficult case where the W pair have an invariant mass of 1 TeV, which is a region of critical interest. They require that one W should decay leptonically and that the other decay hadronically. The background now arises from the final state of W + jet and is approximately 200 times the signal,[27] so a very strong background rejection is required. They find that such a rejection can be obtained by a series of cuts on the multiplicity, jet mass, distribution of particles within the jet, etc. A realistic detector simulation was used. FIGURE 10, reproduced from their article, shows the reconstructed jet mass from the W, which was assumed to have a mass of 80 GeV, and a QCD jet of the same energy. A calorimeter segmentation in

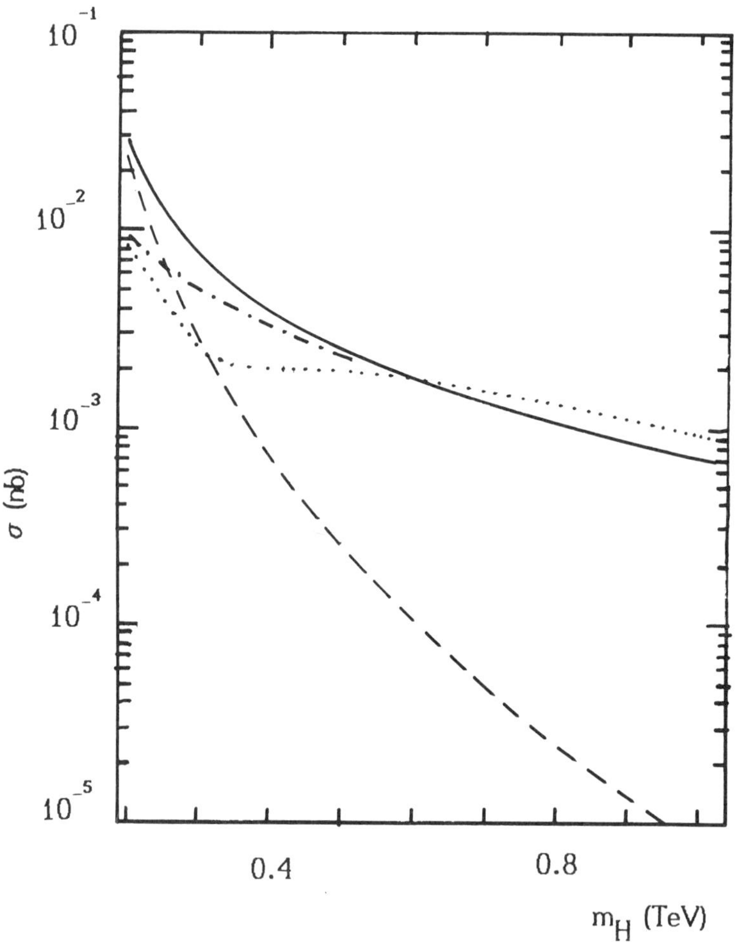

FIGURE 9. Cross section for the reaction, pp $\rightarrow$ $(H + W^+W^-)$ + anything, with $m_t = 30$ GeV/c^2, for $\sqrt{s} = 40$ TeV. The intermediate bosons must satisfy $|y_w| < 2.5$. The contributions of gluon fusion (dashed line) are shown separately. Also shown (dotted line) is $\Gamma_H d\sigma(\text{pp} \rightarrow W^+W^- + X)/dM$ with $|y| < 1.5$, $M = M_H$, and Γ_H = larger of the Higgs width and 10 GeV. (From reference 1.)

rapidity of 0.03 units and in azimuth of 0.03 radians is used. The segmentation of the calorimeter is vital. It is not possible to produce the required rejection if the calorimeter segmentation is 0.1 in both rapidity and azimuth. Segmentations smaller than 0.03 do not significantly improve the situation.

A rejection factor of 200 corresponds to an efficiency of about 25 percent for the $W \rightarrow q\bar{q}$ mode.[22] I have superimposed on FIGURE 10 a peak corresponding to the decay of a 90 GeV object. The separation between this peak and the one at 80 GeV gives one some idea as to how well a W could be distinguished from a Z if both decay hadronically. There is clearly some potential for separation.

In view of the large rejection factor obtained, it is reasonable to ask what are the uncertainties in their analysis. Undoubtedly, the largest source of uncertainty is in the ISAJET Monte Carlo that was used to simulate the background. After very strong cuts have been applied, one can become very sensitive to parts of the Monte Carlo that are not normally tested. The background final state is really W + two jets and the approximation used to generate these multijet states is not perfect.[28] In view of the importance of the result, a reanalysis using another Monte Carlo is probably required.

I will now discuss the ramifications of this very important result. If we require that in the W pair mode 100 detected events are needed in an run of integrated luminosity of 10^{40} cm^{-2}, then we can only observe processes with a cross section of more than 0.16 pb. For the production of a W pair via the decay of minimal Weinberg-Salam Higgs boson, the cross section is larger than 0.3 pb for all values of the Higgs mass greater than $2M_w$ and less than 1 TeV. This rate corresponds to that where the W's are produced centrally with $|y| \leq 1.5$. If this is relaxed to $|y| \leq 2.5$, the cross section is always larger than 0.8 pb.[1]

If the Higgs mass becomes very large, then its couplings to itself and to longitudinally polarized W's and Z's become (case (c)) very large. The width of the

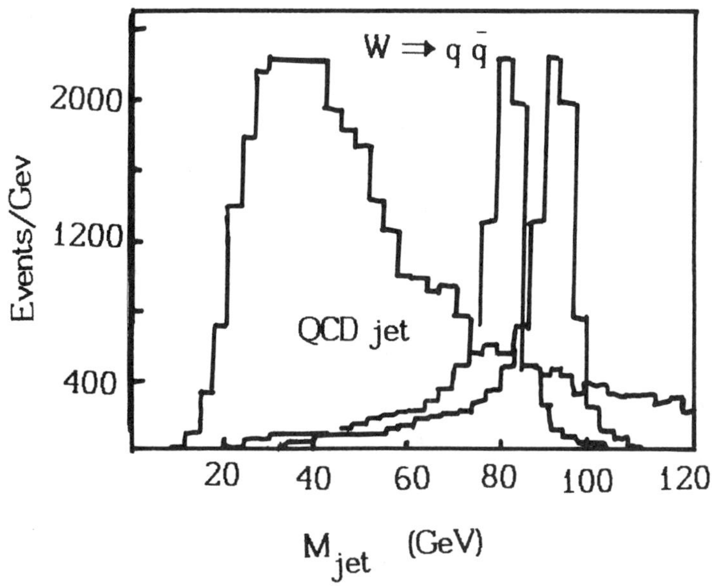

FIGURE 10. Reconstructed masses of $W \rightarrow q\bar{q}$ and QCD jets produced for W-jet pairs of invariant mass, 1 TeV. The W mass is assumed to be 80 GeV and a calorimeter segmentation of $\Delta y = \Delta\phi = 0.03$ is used.[22] The figure also shows the W peak displaced and centered at 90 GeV in order to give an indication of the W/Z separation that may be expected.

Higgs calculated perturbatively,

$$\Gamma_H \approx 58 \left(\frac{M_H}{500\ \text{GeV}}\right)^3 \text{GeV},$$

becomes very large and it is difficult to speak of a resonance. In this case, one is dealing with a strongly coupled system consisting of W's and Z's. The problems of predicting its behavior are similar to those of calculating hadron masses in QCD.

Several methods have been tried in order to make estimates of the consequences of such strong coupling. One method predicts the existence of a scalar bound state of mass below 1 TeV.[29] Such a bound state will behave in a similar way to the elementary Higgs.

There are several qualitative features that should occur.[30–32] Structure should appear in multi-W final states when the invariant mass of the multi-W system is of the order of 1 TeV. Examples of this structure include:

(i) A larger ratio of $3W$ and $4W$ final states to $2W$ final states than is predicted on the basis of quark, antiquark annihilation.
(ii) A ratio of Z pair to W pair cross sections that is larger than that expected from quark, antiquark annihilation. Furthermore, this ratio should be larger at large pair masses.
(iii) Variation of the W and Z pair cross sections with the invariant mass of the pair, which is different from what is expected from production via quark, antiquark annihilation.

In the case of two and three gauge boson final states, some estimates of the rates are possible by considering the production and decay of a 1 TeV Higgs boson produced either alone, which will generate a two gauge boson final state, or in association with a W or a Z, which will produce a three gauge boson final state. The ratio of cross sections with three and two gauge bosons obtained in this manner is similar to that obtained by using "soft W theorems,"[31] which are analogous to the soft pion theorems of QCD. Estimates[32,33] for the three boson final state yield about 200 fb in the interesting invariant mass region. The rates for final states with four gauge bosons will be less. Even if the W sector is not strongly interacting, there will be final states of three and four gauge bosons produced from $q\bar{q}$ annihilation. The rate will fall rapidly as the invariant mass of the multi-W system rises. This background, though, is not known; however, it may not be relevant since the observability of these small rates is in doubt.

The efficiency for detecting a $3W$ final state is difficult to assess. A detailed Monte Carlo study is really required, but I will try to make an educated guess. I will assume that one W decays leptonically into either an electron or a muon and that the others decay hadronically. The background now arises from final states with $W + W +$ jet(s) and $W +$ jets. The background rejection factor required is unknown. I will assume that the efficiency in the lepton channel is 1.0 and that in the hadronic channel is 0.25. The fraction of WWW events detected is then 0.017. The efficiency for the other final states with three bosons should not be radically different. This implies that in order to observe 20 events, 1200 must be produced. If the estimates presented earlier for the production rates are valid, this will require an integrated luminosity of order 10^{41} cm^{-2}, which is a seemingly impossible task. Using the same method, I can estimate the efficiency for a

$4W$ final state where one W decays leptonically to be 0.004. The same depressing conclusion concerning observability seems unavoidable.

The situation in case (a) is much more problematic. In this region, the Higgs will decay dominantly into a $t\bar{t}$ pair. I will assume a t quark mass of the order of 45 GeV in what follows. There is a large background from the production of $t\bar{t}$ pairs via gluon fusion. This background totally overwhelms the signal, thus making the observation exceedingly difficult.[1,33]

The associated production of a Higgs and a tt pair offers no more hope.[19] Here the

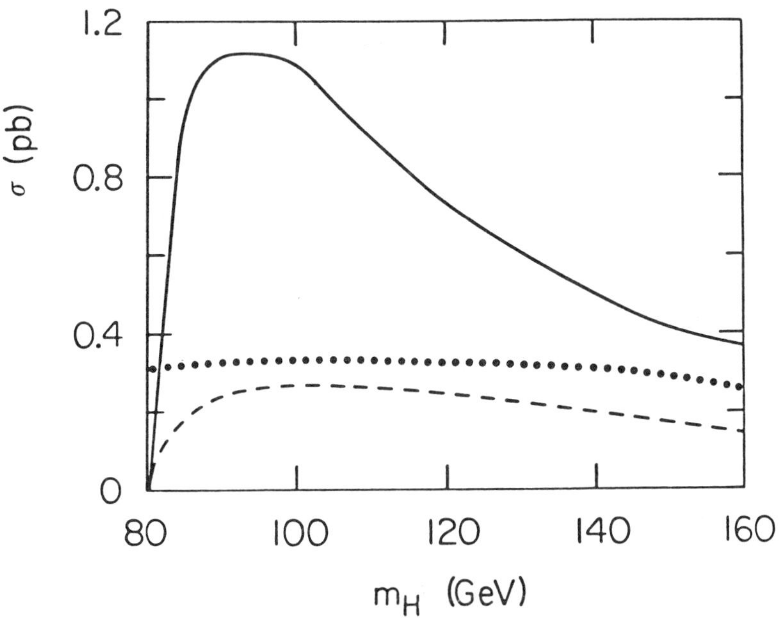

FIGURE 11. The rate, pp → $W + H$, at $\sqrt{s} = 40$ TeV (solid). Also shown is the background from pp → $Wt\bar{t}$ (dashed) and $Wt\bar{b}$ (dotted). The backgrounds are estimated from $\Delta M d\sigma/dM$, where M is the mass of quark fair and $\Delta M^2 = 0.1M^2$. The latter background has been divided by 100.

background arises from $gg \rightarrow t\bar{t}\ t\bar{t}$. The presence of extra t quarks in the final state enhances the signal, but the background still dominates.[19]

The best chance for detection seems to be by observing the final state, HW, where the W is tagged from its leptonic decay. The background now comes from final states of $W + t\bar{t}$, which is produced only by $q\bar{q}$ annihilation and is consequently smaller than backgrounds initiated by gluon fusion. Gunion *et al.*[18] have demonstrated that given a resolution of the order of 0.1 on $\Delta M^2_{t\bar{t}}/M^2_{t\bar{t}}$, the signal exceeds the background. Unfortunately, the cross section is rather small. If both the W and H are produced centrally, having $|y| \leq 2$, and the W has $p_T > 40$ GeV (both of these cuts enhance the signal to background ratio), then it is of the order of 120 fb for Higgs masses around 130 GeV if the W is required to decay either to $e\nu$ or to $\mu\nu$.

There is a large background from the process with a final state of $Wt\bar{b}$, which is produced by gluon-gluon fusion and which will be serious if good t/b rejection is not available. With the same cuts and resolution on ΔM^2, this rate is approximately 50 times the signal.[18] Good b quark rejection is therefore vital if this process is to be exploited; FIGURE 11 shows the signal and background as a function of m_H. In order to decide whether this method is feasible, a detailed Monte Carlo analysis is required.

F. Gilman and B. Cox[34] and G. Abrams and B. Cox[35] have carried out such an analysis for this case. The t and b quarks should be distinguishable by looking at the invariant mass of the jet formed by their hadronization. FIGURE 12 shows the reconstructed b and t quark masses from the ISAJET Monte Carlo.[26] The long tail on the t quark distribution is caused by missing energy carried off by neutrinos. The t quarks from the decay of a 130 GeV Higgs boson are moving rather slowly, so the jet produced by their fragmentation is not very clear. In contrast, the background from the final state, Wtb, has a clearer jet from the b quark. After cuts, a signal to background ratio of one-sixth is obtained.[34,35] Having identified the tt pair, the Higgs is searched for by looking for a peak in the tt mass spectrum.

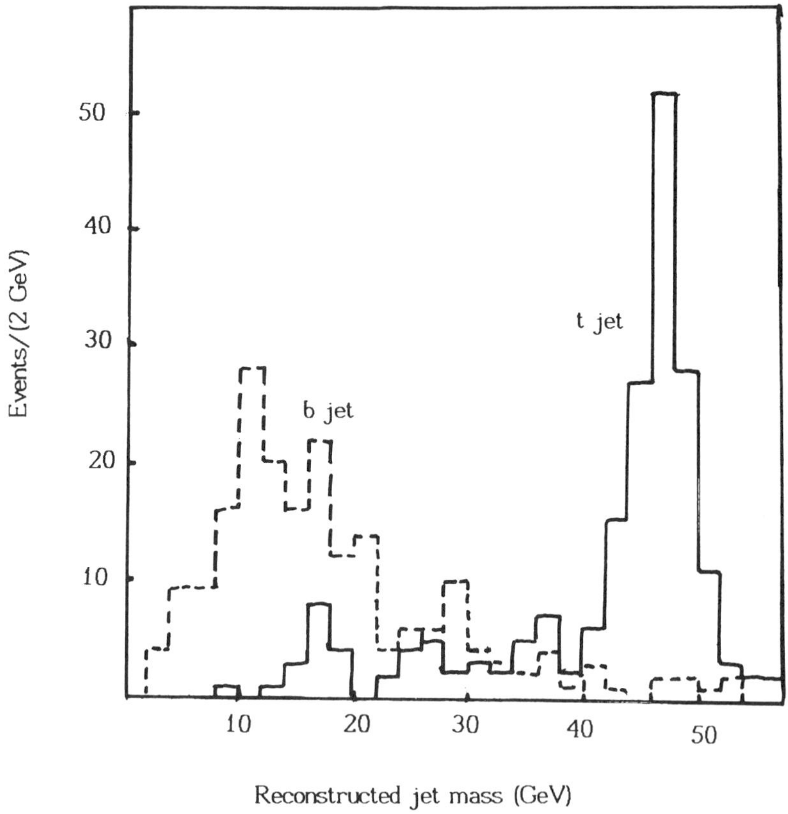

FIGURE 12. The reconstructed jet mass for a sample of b and t quark jets arising from the decay of a 130 GeV Higgs boson. (From reference 34.)

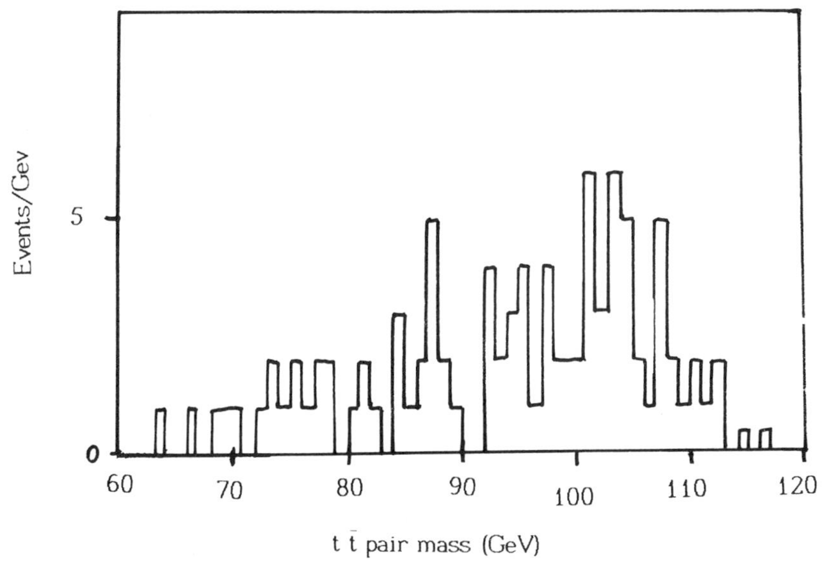

FIGURE 13. The $t\bar{t}$ mass spectrum resulting from the decay, $H^0 \rightarrow t\bar{t}$, with $m_H = 120$ GeV. Fragments with an energy less than 1 GeV are excluded. (From reference 34.)

Since the top quarks are rather soft, the resolution on the tt pair mass depends critically on the ability to detect and to correctly assign low transverse momentum hadrons. If the hadrons with $p_T < 1$ GeV are excluded, the peak in the tt mass spectrum disappears. (See FIGURE 13.) The peak also vanishes if either detector resolution is included or if fragments from the t jet, which are moving backwards, are rejected. The rather negative conclusion of their studies is disappointing since it means that we have no method with which we can confidently expect to find a minimal Higgs boson of mass less than $2M_W$.

The situation could be improved in nonminimal models of electroweak symmetry breaking with more than one Higgs doublet, such as, for example, models based on supersymmetry. In these models (see reference 36 for a recent review), there are several physical Higgs bosons, some of which can have enhanced couplings to quarks and leptons. The production rates could therefore be enhanced significantly over those discussed here.

ACKNOWLEDGMENTS

I am grateful to Martin Block for arranging such an enjoyable and useful conference. Much of the material in this talk was based on the outcome of the DPF Summer Study held at Snowmass, Colorado. I am very grateful to all the participants for their hard work and stimulation.

NOTES AND REFERENCES

1. EICHTEN, E. J., I. HINCHLIFFE, K. D. LANE & C. QUIGG. 1984. Rev. Mod. Phys. **56:** 579. Referred to as EHLQ.
2. PROCEEDINGS OF THE CERN/ICFA WORKSHOP. 1984. Large hadron collider in the LEP tunnel. CERN 84-10.
3. PROCEEDINGS OF THE DPF SUMMER STUDY ON THE DESIGN AND UTILIZATION OF THE SSC. 1985. J. Morfin & R. Donaldson, Eds. FNAL.
4. ALTARELLI, G. & G. PARISI. 1977. Nucl. Phys. **B126:** 298.
5. COLLINS, J. 1985. Theory problems at small x. (See reference 3.)
6. GRIBOV, L. V., E. M. LEVIN & M. G. RYSKIN. 1983. Phys. Rep. **100C:** 1.
7. GLUCK, M., E. HOFFMANN & E. REYA. 1982. Z. Phys. **C13:** 119.
8. BLOCK, M. & R. N. CAHN. 1984. Rev. Mod. Phys. **57:** 563.
9. BOZZO, M. *et al.* UA4 COLLABORATION. 1984. CERN/DP 84-91.
10. ALTARELLI, G., R. K. ELLIS & G. MARTINELLI. 1978. Nucl. Phys. **B143:** 521; **B146:** 544(E).
11. ELLIS, R. K. 1984. The QCD parton model at collider energies. *In* Proceedings of the Workshop on Electroweak Symmetry Breaking. T. Applequist, M. K. Gaillard & I. Hinchliffe, Eds. LBL June 1984, LBL-18571.
12. ELLIS, R. K. *et al.* 1980. Nucl. Phys. **B173:** 397.
13. BRODSKY, S. J. *et al.* 1980. Phys. Lett. **93B:** 451; BRODSKY, S. J., C. PETERSON & N. SAKAI. 1981. Phys. Rev. **D23:** 2745.
14. BRODSKY, S. J. *et al.* 1985. Intrinsic chevrolets at the SCC. (See reference 3.)
15. GEORGI, H. *et al.* 1978. Phys. Rev. Lett. **40:** 692.
16. DAWSON, S. & R. N. CAHN. 1984. Phys. Lett. **136B:** 196.
17. IOFFE, B. L. & V. A. KHOZE. 1976. Leningrad Report LINP-274.
18. GUNION, J. F. *et al.* 1984. SLAC Pub 3403.
19. KUNSZT, Z. 1984. Nucl. Phys. **B247:** 339.
20. BJORKEN, J. D. 1976. *In* Proceedings of the 1976 SLAC Summer Institute. M. Zipf, Ed.
21. LANE, K. 1984. *In* Proceedings of the Workshop on Electroweak Symmetry Breaking. T. Applequist, M. K. Gaillard & I. Hinchliffe, Eds. LBL-18571.
22. FERNANDEZ, E. *et al.* 1985. Identification of W Pairs at the SSC. (See reference 3.)
23. HANSEN, J. 1986. *In* Aspen Winter Physics Conference, vol. 461. M. Block, Ed. N.Y. Academy of Sciences. N.Y.
24. GORDON, H. A. & F. E. PAIGE. Internal Brookhaven Report. Unpublished.
25. KERNAN, A. 1986. Heavy quark production in UA1. *In* Aspen Winter Physics Conference, vol. 461. M. Block, Ed. N.Y. Academy of Sciences. N.Y.
26. PAIGE, F. & S. PROTOPOPESCU. 1980. Brookhaven Report BNL29777.
27. HINCHLIFFE, I. LBL-18742. To appear in reference 3.
28. ELLIS, R. K. & J. OWENS. 1985. See reference 3.
29. EINHORN, M. B. 1984. University of Michigan preprint UM-HE-84-3.
30. LEE, B. W., C. QUIGG & H. B. THACKER. 1977. Phys. Rev. **D16:** 1517.
31. CHANOWITZ, M. C. & M. K. GAILLARD. 1984. Phys. Lett. **142B:** 85.
32. GAILLARD, M. K. 1984. Strongly interacting W's and Z's. *In* Proceedings of the Workshop on Electroweak Symmetry Breaking. T. Applequist, M. K. Gaillard & I. Hinchliffe, Eds. LBL-18571.
33. EICHTEN, E. *et al.* 1985. See reference 3.
34. COX, B. & F. J. GILMAN. 1984. Standard electroweak interactions and Higgs bosons. (See reference 2.)
35. ABRAMS, G. & B. COX. Search for $H^0 \rightarrow tt$: Quark jet identification problems. (See reference 3.)
36. LANGACKER, P. 1984. *In* Proceedings of the Workshop Electroweak Symmetry Breaking. T. Applequist, M. K. Gaillard & I. Hinchliffe, Eds. LBL-18571.

Detector Problems at the SSC[a]

STANLEY G. WOJCICKI

Physics Department
and
Stanford Linear Accelerator Center
Stanford University
Stanford, California 94305

INTRODUCTION

During the last couple of years, there has been considerable concern expressed among the U.S. high energy community as to whether detector limitations would prevent one from being able to fully exploit a luminosity of 10^{33} cm^{-2} sec^{-1} at a hadron-hadron high energy collider. As a result of these concerns, a considerable amount of work has been done recently in trying to understand the nature of potential difficulties and the required R&D that needs to be performed. A lot of this work has been summarized in the 1984 DPF Summer Study at Snowmass.[1] This paper attempts to review some of these results.

This work is limited to the discussion of detector problems associated with the study of high energy hadron-hadron collisions. Even though there have been considerations of other possible future options for the SSC, like e^-p and fixed target capability, the detector questions for these options are quite different and will not be considered here.

We shall start with the discussion of the desirable features of the detectors and of the SSC environment in which they will have to work. After a brief discussion of the model 4π detectors, we shall discuss specific detector aspects: lepton identification, tracking, calorimetry, and computing and triggering. We shall end with some remarks about the possible future course of events.

DESIRABLE FEATURES

There are a number of characteristics that we would like to see in the SSC detectors. We enumerate them briefly below:

(a) Diversity. The SSC, hopefully, will be the first machine in a totally new energy regime. Thus, since it will be exploring a "*terra incognita,*" we want to be prepared for any potential surprise. This, of necessity, will require certain diversity in detectors since one cannot predict a priori the necessary features. More specifically, it is generally agreed that both magnetic and nonmagnetic

[a]This work was supported by the Department of Energy, contract nos. DE-AC03-76SF00515 and DE-AC03-ER40050, and The National Science Foundation.

4π detectors should be built. Furthermore, specialized experiments will require special purpose experimental setups.

(b) Capability to withstand high rate environment. The 10^8 interactions per second will severely tax the detectors. One should mention here several specific requirements necessary to cope with those rates: ability to withstand potential radiation damage, good time resolution, and ability to make quick trigger decisions.

(c) Hermeticity. The experience of the CERN $\bar{p}p$ collider clearly demonstrates that a lot of new physics may be signaled by missing transverse energy.[2] Theoretical considerations for the SSC energy regime reinforce this view.[3] This need, in turn, imposes two requirements on the detectors: minimization of cracks and identification and energy measurement of all the muons.

(d) Lepton identification. Here again we are guided by the CERN experience[4] and the theoretical forecasts for the SSC.[3]

(e) Good granularity. At these energies, the jets will be composed of very closely spaced tracks. Thus, identification of leptons within or in the neighborhood of jets demands good granularity. Good jet energy measurement also requires good granularity.

(f) Reasonable cost. The detector architecture and parameters will be, in the end, a compromise between the available funds and the desires of their proponents. It is important that these trade-offs be well thought-out.

(g) Integrated approach. The problems we are dealing with are sufficiently complex and interrelated, so we have to consider all of the detector aspects together in the design. More specifically, the detector hardware, the detector software, and the machine parameters affect each other quite closely.

MACHINE ENVIRONMENT

There is no doubt that the SSC will present quite a hostile environment to its detectors. For a bunch spacing of 10 m (33 nsec separation in time), we can expect for each crossing about three to four interactions that will give tracks in the detectors. This is based on the assumption of a total inelastic cross section of 90–120 mb[5] and the supposition that the detectors will not see any of the elastic scattering events.

The bunch spacing is a machine parameter that can be adjusted within some limits. The possible trade-offs[6] are indicated in FIGURE 1. Larger total beam current increases costs (synchrotron radiation, safety considerations) and thus must be kept in bounds. It is clear that a factor of two or so leeway exists, but not much more than that.

The multiplicity in typical events will be high; the present estimate[7] gives $dn/dy|_{y=o} \approx 6$. In addition, the jets will have their tracks close together. These factors will make track reconstruction difficult unless good granularity and sufficient redundancy are provided.

Finally, the high energy of the tracks imposes stringent requirements on the resolution if curvature measurements are to be used for track momentum determination.

4π DETECTORS

We have already mentioned that both magnetic and nonmagnetic 4π detectors have some special advantages of their own. It might be worthwhile to enumerate them briefly.

Advantages of nonmagnetic detectors:

(a) Tracking is considerably simpler.
(b) As a result of (a), the calorimeter can be brought considerably closer and hence becomes cheaper.
(c) The absence of coil makes the apparatus simpler and removes the "inert" coil material.
(d) Components of jets stay together. Thus, calorimetric energy measurement is easier.

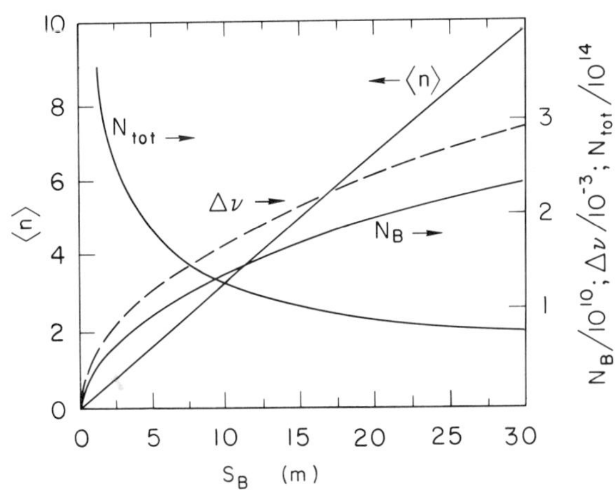

FIGURE 1. Plot of the average number of interactions per bunch crossing ($\langle n \rangle$), number of protons per bunch (N_B), linear tune shift due to single bunch collisions ($\Delta\nu$), and total number of protons in each ring (N_{tot}) as a function of the bunch spacing (S_B). The data used correspond to the Reference Design A parameters.

On the other hand, in favor of magnetic detectors one can mention:

(a) Lepton identification is improved since its momentum measurement from curvature can be cross-checked with other information.
(b) The sign of the electron's (and all hadrons') charge can be measured.
(c) Invariant mass measurements are possible.
(d) Momentum of all individual tracks can be measured.
(e) Magnetic field spreads out the jets and the additional separation may be helpful in track separation.

The characteristics of possible 4π detectors were studied in detail at the 1984 Snowmass meeting. Paper designs of three possible detectors were generated and

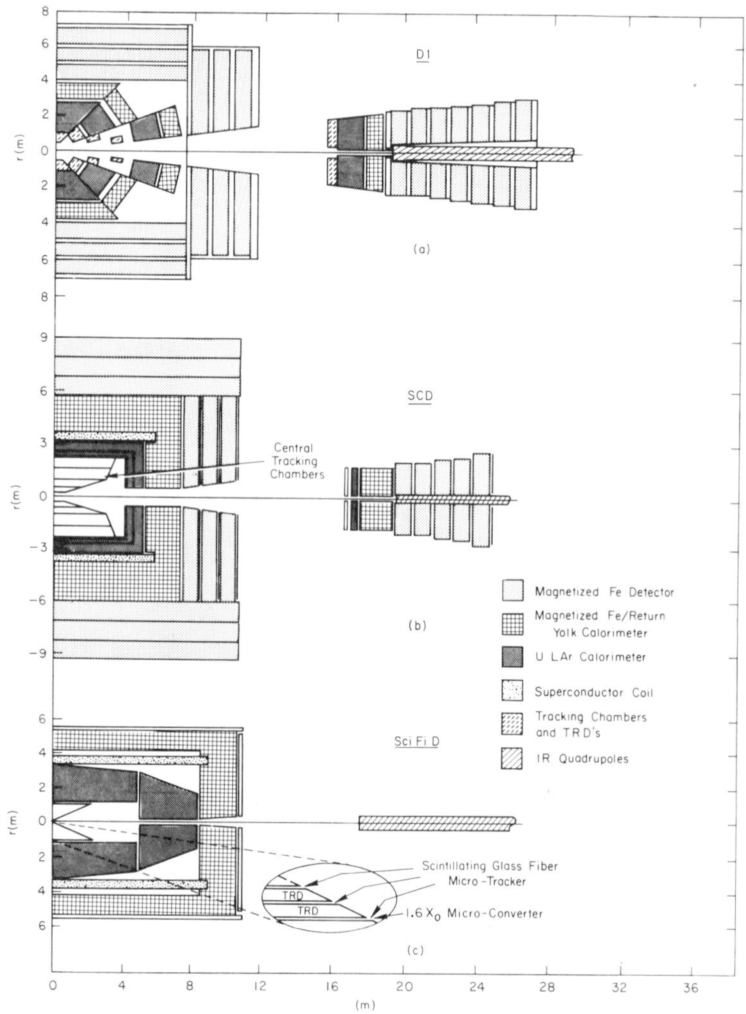

FIGURE 2. Schematic drawing of (a) the nonmagnetic detector, referred to as D1, (b) the "conventional" magnetic detector, called SCD, and (c) the scintillating fiber detector (SciFiD) (from reference 8).

costed.[8] The first two, one magnetic, the other nonmagnetic, are rather conventional insofar that they do not introduce any new technology. The third one, relying on the scintillating fibers as the main tracking medium, employs as yet unproven technology and hence is more speculative. The three detectors are displayed in FIGURE 2, reproduced from the Snowmass report. The relevant parameters of those three detectors (also from the Snowmass report) are shown in TABLES 1, 2, and 3, which give the parameters of their tracking systems, the parameters of their calorimeters, and their costs, respectively.

LEPTON IDENTIFICATION

We start out by making a case for the necessity of having the capability to identify the leptons. Some of the arguments are:

(a) New physics is very likely associated with leptons.
(b) Since the muon energy is not included in the total calorimetric energy measurement, the requirement of total jet energy measurement, as well as that of hermeticity, can be achieved only if the muons are identified and measured.
(c) Heavy quark jets can be tagged frequently by the presence of energetic leptons.
(d) Presence of energetic leptons can be a very useful element of level 1 trigger.

We proceed next to discuss specific details and problems of electron and muon identification. Beginning with the muons,[9] the sources of backgrounds are hadronic punch-throughs and μ's from π and K decay. Those are illustrated in FIGURE 3, which shows the rate of punch-throughs at $L = 10^{33}$ as a function of p_T and the rate of the decay muons (assuming 1 m of flight path before the absorber) as a function of amount of iron filter material (and, thus, effectively of minimum p_μ). The typical detector probably will have in excess of 3 m of iron equivalent in the absorber; thus, the muon background rate will be below 10^5. As we shall see later, this is probably low enough to be used directly in the level 1 trigger.

The resolution for muon momentum measurement is illustrated in FIGURE 4. The assumption made is that the spatial resolution in the three gaps between the iron layers

TABLE 1. Parameters of the Tracking Systems

	D1	SCD	SciFiD
Type	Wire chambers	Wire chambers[a]	Scintillating glass optical fibers
Principal coordinate	z	ϕ	ϕ
Secondary coordinate measurement technique	Charge division	Small angle stereo and cathode strips	Small angle stereo
Radial extent ($z = 0$)	50 to 100 cm	25 to 235 cm	10 to 100 cm
Rapidity range	0 to 4.7	0 to 1.5 (central) 1.5 to 5.0 (endcap)	0 to 1.5
Cell spacing	5 mm	1 to 4 mm	25 mm
Number of hits per track	20	100 (wires) 6 (strips)	130
Number of dE/dx layers	20	10	none
Number of channels	184.5 k	100 k wires[b] 32 k strips	150 M fibers 38 M cells[c] 920 CCDs
Resolutions:[d]			
principal coordinate	200 μm	200 μm	10 μm
secondary coordinate	1 cm	1 mm	70 μm/su per layer
track pair	5 mm	1–4 mm	50 μm
momentum	none	3×10^{-4}p	2.5×10^{-4}p $\oplus$ 0.031

[a] Inner silicon strip detectors are omitted from this table.
[b] There are an additional 30 k channels in the endcap tracking system.
[c] Cells are defined by the image intensifier resolution.
[d] Spatial resolutions are per measurement unless stated otherwise.

TABLE 2. Parameters of the Calorimeters

	D1	SCD	SciFiD
Type			
electromagnetic	U-Cu/liquid Argon	U-Cu/liquid Argon	U-Cu/liquid Argon
hadronic	U-Cu/liquid Argon	U-Cu/liquid Argon[a]	U-Cu/liquid Argon
catcher/flux return	Fe/wire chambers	Fe/wire chambers	Fe/wire chambers
Thickness			
electromagnetic	30 X_0(1.4 λ)	40 X_0(1.2 λ)	40 X_0(1.5 λ)
hadronic	5.6 λ	4.0 λ	8.5 to 13 λ
catcher/flux return	5.0 λ	9.0 λ	5 to 8 λ
Rapidity range	0 to 5.5	0 to 5.0	0 to 5.0
Sampling thickness			
electromagnetic	0.6 X_0	0.5 X_0	0.5 X_0
hadronic	0.03 λ	0.03 λ	0.03 λ
catcher/flux return	0.3 λ	0.3 λ	0.3 λ
Longitudinal segments	5 (3 in EM)	5 (3 in EM)	5 (3 in EM)
Tower size ($\Delta\eta = \Delta\phi$)			
electromagnetic	0.04	0.02 to 0.1	0.03
hadronic and catcher	0.04	0.04 to 0.2	0.06
Number of towers			
electromagnetic	31.5 k	45 k	48 k
hadronic and catcher	31.5 k	11 k	12 k
Number of channels	157.5 k	166 k	180 k
Mass			
U-liquid Ar	1330 T	2440 T	2400 T
Fe-gas	1780 T	6900 T	6300 T

[a]The forward hadronic calorimeter is constructed from iron-gas tube sandwiches.

TABLE 3. Parameters of the Costs

	D1		SCD		SciFiD	
Tracking system		37.0		28.8		26
mechanical costs	18.5		11.6		10	
electronics costs	18.5		17.2		16	
Transition radiation detectors		25.6		—		15
mechanical costs	9.2		—		2	
electronics costs	16.4		—		13	
U-liquid argon calorimeters		39.5		81.0		68
mechanical costs	33.2		73.7		60	
electronics costs	6.3		7.3		8	
Iron-gas calorimeters		17.0		21.4		14
mechanical costs	15.4		20.4		13	
electronics costs	1.6		1.0		1	
Superconducting coil		—		19.0		31
Muon detection system		35.3		49.5		16
mechanical costs	28.8		41.5		8	
electronics costs	6.5		8.0		8	
Trigger and data acquisition		9.9		10.0		10
Totals (before contingency)		164.3		209.7		180
mechanical total	105.1		166.2		124	
electronics total	59.2		43.5		56	
Contingency (25%)		41.1		52.4		45
Total[a]		205.4		262.1		225

[a]Off-line computing costs have not been included in this table. The Computing Working Group estimated that these costs would be about $40 M for a magnetic detector. They should be considerably less for a nonmagnetic detector.

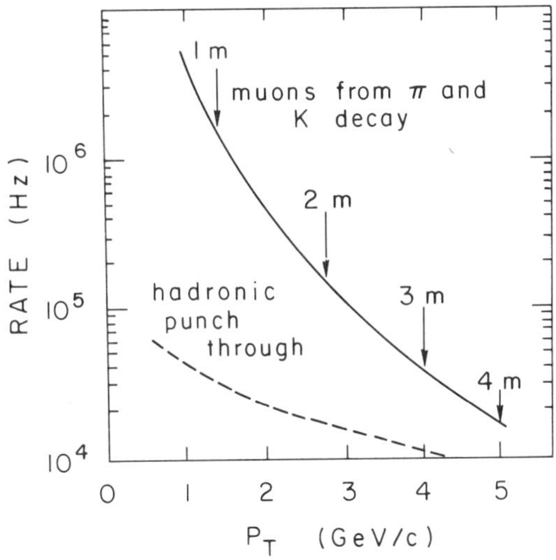

FIGURE 3. Integrated rate of μ's from π and K decay as a function of hadron transverse momentum. A decay path of 1 m and a luminosity of 10^{33} are assumed. The arrows indicate the muon range in iron. The dashed curve is the rate of pions that punch through ten interaction lengths of iron (from reference 9).

is 300 μ. The length indicated (2 m and 4 m) is the total length of the magnetized iron. For comparison, the advertised momentum resolution of both the L3 detector[10] at LEP and the SciFiD SSC detector (the third 4π detector discussed above) are also shown.

The problems with the electron identification and measurement[11] are quite different. The energy can be measured with a very high degree of accuracy in the electromagnetic calorimeter. The calculations indicate that a 1000:1 hadron rejection can be obtained at the SSC energies if the calorimeter is subdivided into at least three

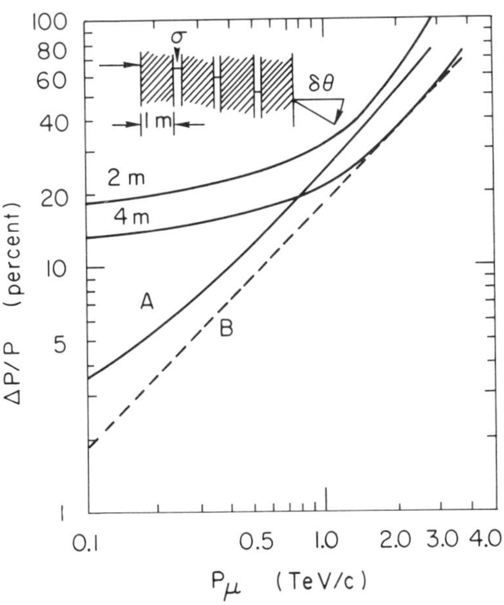

FIGURE 4. Comparison of four different momentum resolution functions for muons. The top two curves (marked 2 m and 4 m) correspond to tracking muons through 1 m thick magnetized iron slabs ($B = 1.8$ T). Input and output slopes are assumed known to $\delta\theta = \pm 0.15$ mrad and spatial resolution is taken to be 300 μ. The curve marked A is the resolution calculated for the scintillating fiber detector discussed above (with $B = 1.5$ T). The curve marked B is the expected resolution from the L3 detector at LEP (from reference 9).

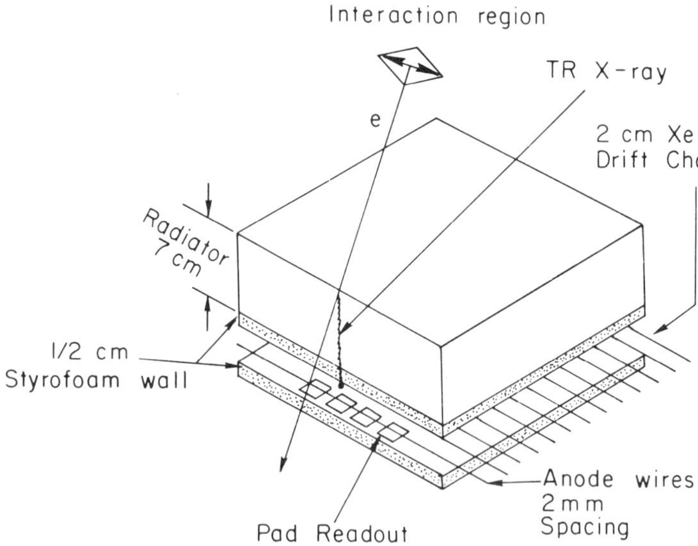

FIGURE 5. Sketch of a possible TRD module (from reference 11).

longitudinal segments. Some of the obvious backgrounds are the π^0 (or η) $\rightarrow 2\gamma$ decays, with one of the γ's either converting internally or in the beam pipe and with a γ-hadron overlap in the same calorimeter cell. The first background represents real electrons. Thus, it can be eliminated only by either detecting the other electron (in a magnetic detector) or by measuring dE/dx and thus discriminating between one or two tracks (in a nonmagnetic detector).

The other background can be reduced by a P versus E measurement comparison in a magnetic detector or by a transition radiation detector (TRD) in a nonmagnetic detector. The latter will work at electron energies below 200 GeV. One of its drawbacks is the fact that to obtain a high enough conversion efficiency for the transition radiation X rays, a layer of gas about 2 cm thick is required. This, in turn, will necessitate a long enough gate-time so that the problem with accidentals can become quite severe. This can, in turn, be alleviated by providing and reading out cathode pads near the anode wires; the cost of this solution, however, is many more

TABLE 4. Synchrotron Radiation Parameters[a]

E(GeV)	E_c(MeV)	ΔE(MeV)	$x^{\min} = \dfrac{5\text{MeV}}{\epsilon_c}$	$N(\epsilon > 5\text{ MeV})$
50	5	14.6	1.0	0.41
100	20	58	0.25	3.4
200	81	229	0.062	7.8
500	506	1430	0.010	12.3
1000	2025	5715	0.0025	16.5

[a] $Bl = 3$ Tesla meters; $B = 1.5$ Tesla.

channels of required electronics. A possible schematic of a TRD is illustrated in FIGURE 5.

Finally, one should mention the possibility of detecting electrons of very high energies by synchrotron radiation. In a strong field ($B \gtrsim 1$ Tesla), a significant number of photons above e^+e^- threshold will be generated. Their conversion early in the calorimeter will give a very large pulse height in the initial layers of the calorimeter. Whether this technique will work in practice needs to be verified experimentally; the relevant numbers are illustrated in TABLE 4.

TRACKING

We should consider here both the standard central tracking chambers as well as the vertex detectors.[12] The limitations on these systems at the SSC will be imposed by:

(a) Radiation damage considerations.
(b) Instantaneous rate considerations (current drawn, space charge).
(c) Occupancy rate.
(d) Reconstruction possibility.

We shall consider each one of these factors in turn. To review the radiation damage question, we must recall that 1 rad corresponds approximately to 3.5×10^7 mip/cm^2. The tolerance of different components to high radiation levels is known reasonably well and is displayed in FIGURE 6 as a function of the minimum radius at which the components could survive at the SSC.

The performance of the central tracker can be considered by taking a specific example. We assume that we want to cover $|y| \leq 1.5$, with a detector that starts at $r =$ 20 cm and has 600 cells with 2 mm spacing and 1 m long wires. With our previously stated assumptions about cross sections and multiplicities, we obtain a comparison between the expected performance and our present ideas of the values that can be tolerated (TABLE 5).

The Table assumes a rather low gain, yielding 10^6 electrons/particle, and the only considered mechanism for efficiency loss is due to double hits in a single cell (we shall count the "right" hit 50% of the time in such cases). Even though these parameters push the state of the art, there is no obvious reason why such a system should not work.

The question of tracking has not been sufficiently investigated. Some preliminary calculations by H. H. Williams on $t\bar{t}$ jets of 500 GeV/c indicate that the losses due to finite double-hit resolution are not prohibitive. However, more detailed work is needed on interference from other interactions in the same crossing.

For completeness, we reproduce in FIGURE 7 the three central tracking systems discussed at the last Snowmass meeting. One should note that the wires in the D1 detector (no magnetic field) run transversely to the beam axis.

The possibilities for the vertex detector are: drift chamber, scintillating fibers, and Si strips. Because of the futuristic aspects of the scintillating fiber technology, most of the work so far has been done with the other two techniques. Because of the radiation damage, however, most of the discussed designs are limited to the luminosity of 10^{32} cm^{-2} sec^{-1}. One such design for that luminosity is displayed in FIGURE 8.

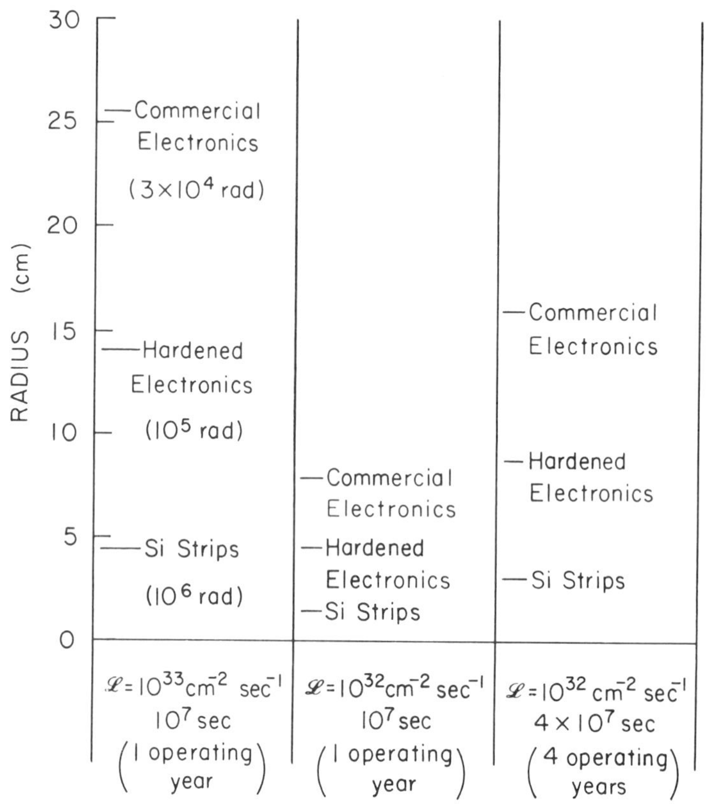

FIGURE 6. Minimum component radii for radiation damage at luminosities of 10^{32} and 10^{33} cm^{-2} sec^{-1} (from reference 12).

CALORIMETRY

Some of the obvious motivations for the importance of calorimetry[13] in the SSC detectors are:

(a) It provides a natural way to measure jet energy;
(b) The resolution improves the energy;
(c) The required depth grows only logarithmically with energy;

TABLE 5. Performance of a Central Tracker

	Estimated	Allowed Limit
Maximum drift time	20(gas) + 7(wire) ns	33 ns
Occupancy	15%	?
Efficiency	92%	?
Current	$0.5\mu a$	$1\mu a$
Particles/mm/sec	3×10^3	10^4
Electrons/mm/yr.	3×10^{16}	10^{18}

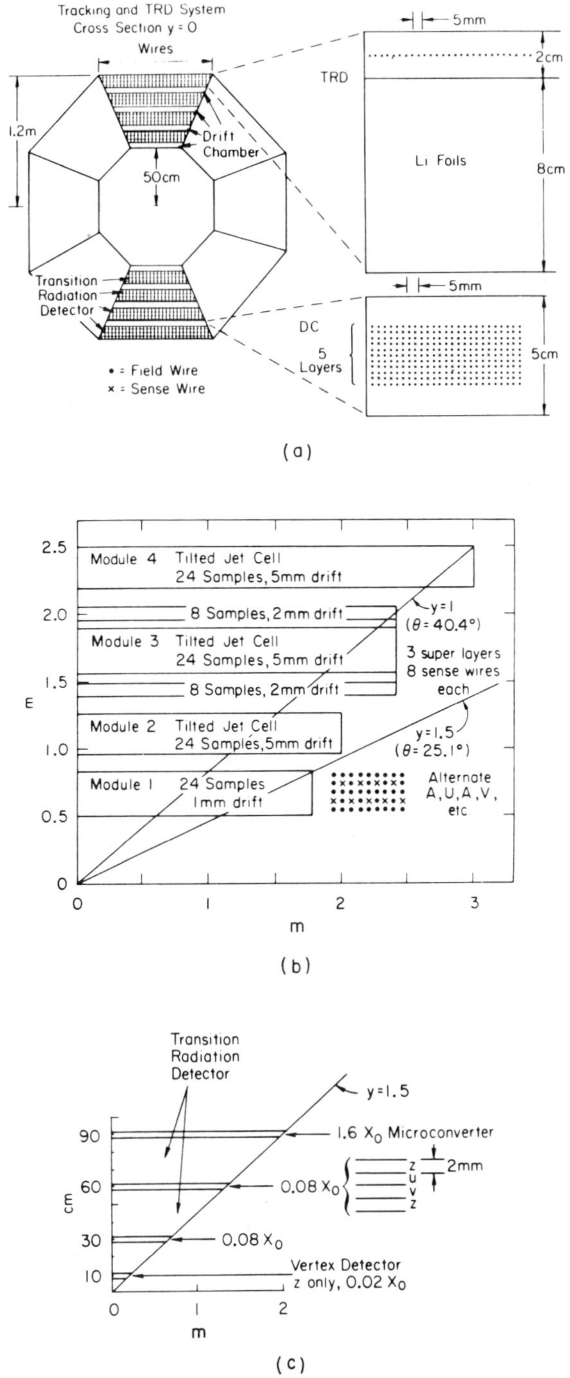

FIGURE 7. The central tracking systems discussed at Snowmass for (a) the D1 detector, (b) the SCD detector, and (c) the SciFiD detector (from references 8 and 12).

(d) It can provide a level 1 trigger capability;
(e) It automatically becomes the front part of the muon detection system.

The obvious questions that need to be addressed in designing the calorimeter are:

(a) Materials to be used. Some considerations that are relevant here are compactness, cost, integration time, ease of calibration, and similarity in response to the hadronic and electromagnetic components.
(b) Granularity. This has to be well matched to the rates expected and the nature of the physics to be investigated.
(c) Hermeticity. It is important to cover as large an area as possible with a minimum number of cracks. The relevant question is how much of a departure from a perfect system can we tolerate.

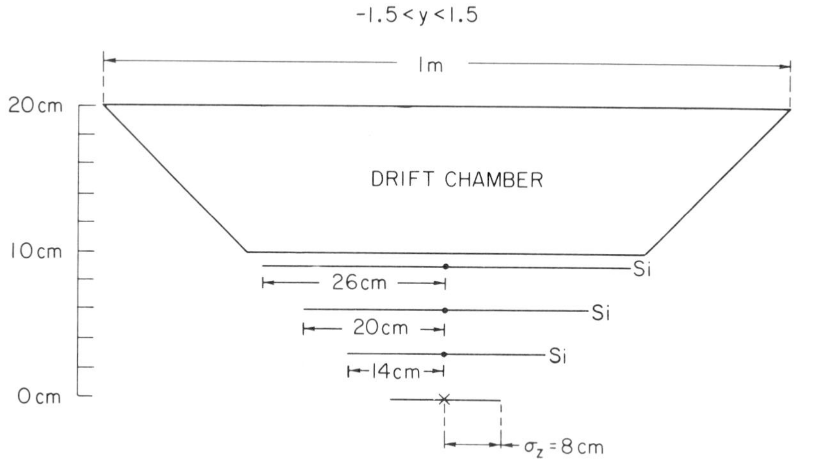

FIGURE 8. A possible vertex detector system for use at a luminosity of 10^{32} cm^{-2} sec^{-1} (from reference 12).

Regarding the material of the calorimeter, there is at present a strong preference in the community for the uranium-liquid argon mix. Some of the advantages here are: equality of hadronic and electromagnetic response, no radiation damage, high density and hence good compactness, and relative ease of calibration.

One of the difficulties of this system is its relatively long charge collection time. For a 2 mm gap, a 150 ns collection time is required, resulting in a high fraction of cells having some remnant of the previous shower. Thus, time measurement of the leading edge of the pulse will be necessary to discriminate against unrelated energy. This system would perform best with a minimum bunch spacing.

Other possible calorimetric media are warm liquids, iron and gas, heavy glass, silicon, and barium fluoride. They all appear to have serious disadvantages and/or require still intensive development.

The granularity choice has to be considered both in light of the nature of the showers and the nature of the physics studied. For electrons, the width of the shower is characterized by the Moliere radius, r_m, given by

$$r_m = \frac{21\text{MeV}}{\epsilon_c} X_0,$$

where ϵ_c is the critical energy and X_0 is the radiation length. Experimentally, a sandwich of 2 mm U and 2 mm Ar will contain 95% of the shower in 2 cm. The hadronic shower width is proportional to the interaction length, λ_0, with typically 95% of the shower energy being contained within a cylinder of that radius.

Regarding physics considerations, the Snowmass study looked at two specific cases. The resolution of the W mass, if W decays via $W \longrightarrow q\bar{q}$, improves down to a granularity of $\Delta\eta \simeq \Delta\varphi \approx 0.03$. For t quark decay by the electronic mode, 80% of the time the electron is isolated from the rest of the jet if $\Delta\eta = \Delta\varphi = 0.02$. Since a typical calorimeter will start about one to two meters away from the interaction point, these physics considerations lead to parameters that are comparable to those imposed by the requirements of the electromagnetic shower criteria.

HERMETICITY

There are three obvious experimental sources that can generate spurious missing transverse energy. These are the aperture provided by the beam pipe, the cracks in the calorimeter associated with boundaries of various subsystems, and the finite energy and position resolution. The calorimeter should be designed optimally in such a way that the contribution to the missing energy from those is not larger than the amount of missing transverse energy carried off by the neutrinos from the heavy quark decays.[13]

The effect of the beam pipe is illustrated in FIGURE 9, where the effective missing energy cross section is shown for three different beam pipe apertures. The resulting curves are compared with the size of the contribution due to neutrinos.

The effect of the cracks was investigated by replacing "live" material with some "inert" material in the calorimeter. The resulting "missing" energy was calculated for two different fractional dead areas and again compared with the neutrino cross section. The results are displayed in FIGURE 10.

Finally, the effect of resolution was also studied. The conclusion was that the assumed calorimetric resolution, i.e., $\Delta E_{\text{had}} = 0.35\sqrt{E}$ and $\delta E_{\text{em}} = 0.15\sqrt{E}$, does not contribute at any significant level to an increase in events with missing p_T.

ELECTRONICS, COMPUTING, TRIGGER

Traditionally, these functions have been rather decoupled, but it is quite clear that for the SSC detectors they have to be integrated and designed in a coherent manner.[14] The data rates at the SSC will be staggering: in excess of 10^8 events/sec, with each event possibly encompassing about 10^6 bytes. Thus, even one event recorded every second, if recorded without any preprocessing, would result in writing on a 6250 BPI tape at its maximum possible speed (8×10^5 bytes/sec). Independent of the recording

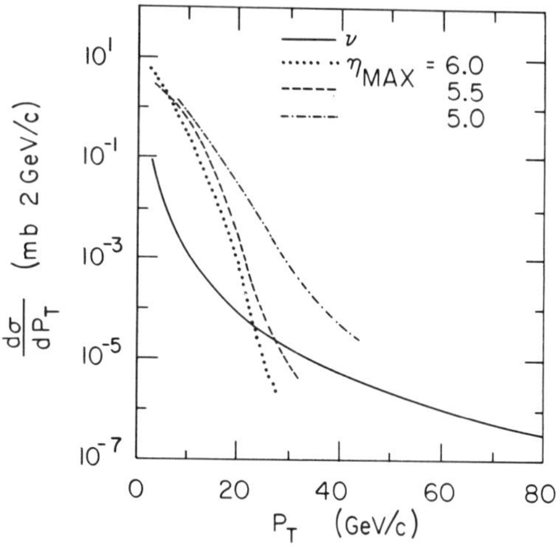

FIGURE 9. Contribution to the differential missing p_T cross section due to neutrino production from heavy quark decays (solid line) and due to incomplete coverage within beam holes of several sizes. The top quark mass is taken as 45 GeV/c^2 (from reference 13).

medium, however, the data rate is so large that a significant amount of on-line event preselection and processing will be required.

The fundamental difficulty in the trigger is the requirement of reducing the raw event by about 10^8. This is difficult, but probably not impossible technically. The real difficulty may, however, lie in doing this without losing any significant new physics. It is clear that the trigger and data acquisition architecture will require a considerable amount of flexibility built into it so that a variety of different topologies can be explored.

The extrapolations from the existing experiments, both of fixed target and colliding beam varieties, lead one to believe that 50 sec of IBM 3081K time will be required to process one event.[14] Thus, each 4π detector would require a computing capability that is equivalent to about 2½ times the projected Fermilab required computing power.

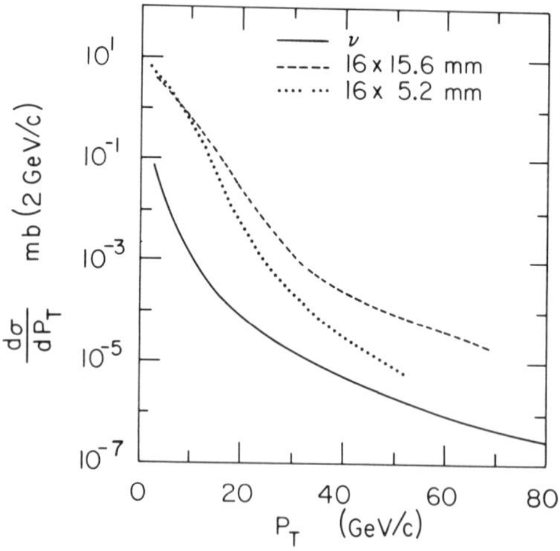

FIGURE 10. Differential missing p_T cross section due to azimuthally directed cracks in the calorimetry between $\theta = 30°$ and $\theta = 150°$. The cracks represent 1.3% (dotted) and 4.0% (dashed) of the area for the two cases shown. The cross section due to neutrinos from heavy quark decays is shown as a solid line. A beam hole of $\theta < 0.3°$ is used (from reference 13).

Very roughly, the present ideas about triggering rely on having several different trigger levels, each one of increasing complexity. Level 1 trigger should be deadtime-less and require no more than about 200 nsec for the decision. Some possibilities are a muon or electron of 5 GeV or higher, or some amount of missing energy. It is estimated that these requirements would generate a rate of about 10^5 Hz.

Additional, more sophisticated, hardware processing would be required at this stage to reduce the event rate to no more than 1 kHz. Such rates would make it feasible to have a bank of parallel processors select roughly one event/second that has the highest probability of being of interest. Clearly, this would require about 1000 processors in order to be capable of making a decision in 1 sec. The task is certainly formidable.

The schematic computer architecture being discussed, as well as the cost estimates, rely heavily on the assumption that significant technology progress will be made in the field in the next decade. Thus, for example, the estimate of the Snowmass 1984 Computing Group of about \$50 M/detector in computing costs assumed a gain of a factor of eight in compute power per dollar in the next five years.

There is no doubt that these ideas will evolve greatly during the next few years and that technology will progress. Today, however, the triggering and computing questions appear to present some of the most formidable problems for the SSC detectors.

SPECIAL PURPOSE DETECTORS

We shall limit this discussion to a few relatively obvious comments. First of all, there is no doubt that some experiments need special purpose, relatively small-scale detectors. It is highly likely that these experiments will also require special interaction region configurations. Some examples of such experiments are elastic pp scattering, searches for monopoles, and searches for new heavy long-lived particles.

In addition, there may be another class of experiments that may also need special purpose detectors, but larger in scope.[15] At least a part of their rationale is that they would be cheaper than the multipurpose 4π detectors. Some examples of such detectors would be diffractive dissociation, study of details of jets, and CP violation experiments. There have been some investigations in this area at Snowmass, but I do not believe that as yet a convincing case has been made for their necessity.

Finally, we should emphasize that because of its high rates, the SSC is inherently different from e^+e^- machines as far as its possible potential for special purpose experiments. Since the data-taking capability of the 4π detectors is way below the total SSC event rate, the general purpose detectors simply will not be able to do all the physics. Thus, in a certain way, SSC is halfway between e^+e^- storage rings and a hadronic fixed target situation. Whether this special situation will justify non-4π detectors is still an open question.

PLANS FOR THE FUTURE

Judging from the LEP experience, it is not too early to begin thinking about the detector problems. On the other hand, the machine testing and construction must have top priority.

Some of the obvious detector R&D that should commence soon must address such generic questions as radiation hardening of electronics, rapid and cheap computing, new calorimetric media and improvement of the existing ones, and new tracking techniques (e.g., scintillating fibers).

In parallel, work has to be done on realistic Monte Carlo calculations so that the effects of pileup on triggers and tracking become well understood. Furthermore, one should investigate to what extent the nth level accidentals could simulate (or provide serious backgrounds) for events of interest.

Finally, we should start thinking about new and innovative ways of solving the problem of detector and beam time approval mechanisms.

ACKNOWLEDGMENTS

The author would like to acknowledge the efforts of his many colleagues at the Snowmass workshop whose work provides the basis of this report. He would also like to thank the Organizing Committee of the Aspen Meeting for their tireless efforts that led to this very successful conference.

REFERENCES

1. PROCEEDINGS OF THE 1984 DPF SUMMER STUDY AT SNOWMASS. 1984.
2. ARNISON, G. *et al.* 1984. Phys. Lett. **139B:** 115; BAGNAIA, P. *et al.* 1984. Phys. Lett. **139B:** 105.
3. EICHTEN, E., I. HINCHLIFFE, K. LANE & C. QUIGG. 1984. Rev. Mod. Phys. **56:** 579.
4. ARNISON, G. *et al.* 1983. Phys. Lett. **122B:** 103; BANNER, M. *et al.* 1983. Phys. Lett. **122B:** 476.
5. BLOCK, M. & R. N. CAHN. 1985. Rev. Mod. Phys. **57:** 563.
6. THE REFERENCE DESIGNS REPORT FOR THE SUPERCONDUCTING SUPER COLLIDER. May 8, 1984.
7. ALPGARD, K. *et al.* 1981. Phys. Lett. **107B:** 315; 1983. Phys. Lett. **121B:** 209; ARNISON, G. *et al.* 1983. Phys. Lett. **123B:** 108.
8. FELDMAN, G. J., M. G. D. GILCHRIESE & J. KIRKBY. 1984. 4π detectors. Proceedings of the 1984 Summer Study at Snowmass.
9. BARNETT, B. A. *et al.* 1984. Muon identification and detection at the SSC. Proceedings of the 1984 Summer Study at Snowmass.
10. TING, S. C. C. 1984. Invited talk at the 1984 meeting of the DPF in Santa Fe, New Mexico.
11. YODH, G. B. *et al.* 1984. Electron identification. Proceedings of the 1984 Summer Study at Snowmass.
12. HANSON, G., D. MEYER *et al.* 1984. Charged particle tracking and vertex detection group summary report. Proceedings of the 1984 Summer Study at Snowmass.
13. GORDON, H. A. & P. D. GRANNIS. 1984. Calorimetry for the SSC. Proceedings of the 1984 Summer Study at Snowmass.
14. FIRESTONE, A. *et al.* 1984. Computing requirements at the SSC; SHAFER, R., D. HARTILL, D. MEYER & T. KONDO. 1984. Detector electronics group summary report. Proceedings of the 1984 Summer Study at Snowmass.
15. WITHERELL, M. S. 1984. Report of the specialized detector group. Proceedings of the 1984 Summer Study at Snowmass.

The Standard Model and Beyond[a]

WILLIAM J. MARCIANO

Department of Physics
Brookhaven National Laboratory
Upton, New York 11973

$SU(3)_C \times SU(2)_L \times U(1)$ STATUS REPORT

The "standard" $SU(3)_C \times SU(2)_L \times U(1)$ model of strong and electroweak interactions elegantly incorporates all the proven symmetries and successes of the quark model, quantum electrodynamics, and the Four-Fermi theory.[1] It correctly predicted weak neutral currents as well as the existence and properties of $W^{\pm}$, Z, and gluons. Only the predicted Higgs scalar boson remains undiscovered. At this time, there are no solid experimental results that cannot be accommodated by the standard model (at the 1 or 2 sigma level). Nevertheless, we do anticipate the emergence of new physics, beyond standard model expectations, that will hopefully provide guidance for theoretical advancement. Indeed, hints of some new phenomena may already be starting to appear in the CERN $\bar{p}p$ collider data (see section entitled MONOJETS).

As a means of surveying the status of the standard model, let me begin by giving what I consider to be the present "best" values of some $SU(3)_C \times SU(2)_L \times U(1)$ parameters:

$$\Lambda_{\overline{MS}}^{(4)} = 100 \begin{matrix} +100 \\ -50 \end{matrix} \text{ MeV} \tag{1.1}$$

$$\alpha^{-1}(m_W) = 127.70 \pm 0.30 + \frac{8}{9\pi} \ln\left(\frac{m_t}{36 \text{ GeV}}\right) \tag{1.2}$$

$$\sin^2\theta_W \equiv 1 - m_W^2/m_Z^2 = 0.220 \pm 0.006 \tag{1.3}$$

$$m_W = 82.2 \pm 1.8 \text{ GeV} \tag{1.4}$$

$$m_Z = 93.2 \pm 1.7 \text{ GeV} \tag{1.5}$$

$$\rho = 1.01 \pm 0.02 \tag{1.6}$$

$$m_e = 0.511 \times 10^{-3} \text{ GeV} \qquad m_d \simeq 9 \times 10^{-3} \text{ GeV} \qquad m_u \simeq 5 \times 10^{-3} \text{ GeV} \tag{1.7a}$$

$$m_\mu = 0.106 \text{ GeV} \qquad m_s \simeq 0.175 \text{ GeV} \qquad m_c \simeq 1.25 \text{ GeV} \tag{1.7b}$$

$$m_\tau = 1.78 \text{ GeV} \qquad m_b \simeq 4.5 \text{ GeV} \qquad m_t \simeq 45 \pm 15 \text{ GeV} \tag{1.7c}$$

[a]The submitted manuscript has been authored under contract no. DE-AC02-76CH00016 with the U.S. Department of Energy. Accordingly, the U.S. Governmemt retains a nonexclusive, royalty-free license to publish or reproduce the published form on this contribution, or allow others to do so, for U.S. Government purposes.

$$7\text{ GeV} \lesssim m_H \lesssim 1\text{ TeV}. \quad \mathbf{(1.8)}$$

In addition to the above quantities, there are four KM quark mixing angles (θ_1, θ_2, θ_3, and δ). The first three are constrained by β-decay, b lifetime, and $\Gamma(b \to u)/\Gamma(b \to c)$ measurements to lie in the ranges,[2]

$$|\sin \theta_1| = 0.221 \pm 0.002, \quad \mathbf{(1.9a)}$$

$$0.01 < |\sin \theta_2| \lesssim 0.15, \quad \mathbf{(1.9b)}$$

$$|\sin \theta_3| \lesssim 0.04, \quad \mathbf{(1.9c)}$$

while δ is more or less unconstrained. Future precision measurements of the b lifetime, $B^0 - \bar{B}^0$ oscillation, t-decays, ϵ'/ϵ, etc., and better theoretical estimates of strong interaction effects will hopefully pin down these mixing parameters. A deviation from unitarity in the KM matrix would be a signal for new physics (see section entitled BEYOND THE STANDARD MODEL).

Neutrino masses are additional parameters that can be accommodated easily in the standard model;[3] however, nonzero values naturally suggest new physics or at least guide us toward a better understanding of particle masses and hierarchies. Indeed, mass generation beyond the Higgs mechanism parametrization is perhaps the outstanding problem in elementary particle theory. Experimental bounds on neutrino masses are presently[3,4]

$$m_{\nu_e} < 6 \times 10^{-8}\text{ GeV}, \quad \mathbf{(1.10a)}$$

$$m_{\nu_\mu} < 0.25 \times 10^{-3}\text{ GeV}, \quad \mathbf{(1.10b)}$$

and

$$m_{\nu_\tau} < 0.07\text{ GeV}. \quad \mathbf{(1.10c)}$$

If neutrinos are massive, they will have mixing angles analogous to the KM matrix, which can (in principle) be determined by oscillation and decay experiments.[3,5]

In the remainder of this section, I will comment further on a few of these parameters and the role they play in testing the standard model.

$\underline{\Lambda_{\overline{MS}}^{(4)}}$: The effective four flavor QCD mass scale, $\Lambda_{\overline{MS}}^{(4)}$, is conventionally defined in terms of the $\overline{MS}$ (modified minimal subtraction) coupling, $\alpha_3(\mu) \equiv g_3^2(\mu)/4\pi$, via the two loop relationship of[6]

$$\Lambda_{\overline{MS}}^{(4)} \equiv \mu \exp\left[\frac{-6}{25}\left(\frac{\pi}{\alpha_3(\mu)} - \frac{77}{50}\ln\left(\frac{12\pi}{25\alpha_3(\mu)}\right)\right)\right], \; m_c \leq \mu < m_b. \quad \mathbf{(1.11)}$$

Radiative upsilon decays when used in conjunction with the Lepage-Mackenzie calculation,[7]

$$\frac{\Gamma(\Upsilon \to \gamma gg)}{\Gamma(\Upsilon \to ggg)} = \frac{4\alpha(1 + 2.2\,\alpha_3\,(0.157 M_\Upsilon)/\pi)}{5\alpha_3(0.157\,M_\Upsilon)}, \quad \mathbf{(1.12a)}$$

presently give[8]

$$\Lambda_{\overline{MS}}^{(4)} = 100 \pm 50 \text{ MeV} \qquad \text{(CUSB } \Upsilon \text{ \& } \Upsilon' \text{ data)}. \tag{1.12b}$$

Determinations of $\Lambda_{\overline{MS}}^{(4)}$ from $e^+e^- \rightarrow$ hadrons and deep-inelastic scattering tend to be larger,[9] but they entail bigger uncertainties. However, a recent analysis of energy-energy correlations by the MARK-J collaboration yields[9]

$$\Lambda_{\overline{MS}}^{(4)} = 150 \begin{matrix} +150 \\ -110 \end{matrix} \qquad \text{(MARK-J)}, \tag{1.13}$$

which certainly agrees with equation (**1.12b**). I therefore believe that the range for $\Lambda_{\overline{MS}}^{(4)}$ quoted in equation (**1.1**) represents a fair estimate.[10] Note that convention requires $\Lambda_{\overline{MS}}^{(N_F)}$ to depend on the number of effective flavors (a cause of some confusion). One expects[6]

$$\Lambda_{\overline{MS}}^{(6)} : \Lambda_{\overline{MS}}^{(5)} : \Lambda_{\overline{MS}}^{(4)} : \Lambda_{\overline{MS}}^{(3)} :: 27 : 63 : 100 : 130, \tag{1.14}$$

i.e., a significant variation.

$\alpha(\mu)$: The running electromagnetic coupling, $\alpha(\mu) = e^2(\mu)/4\pi$ (defined by $\overline{MS}$), given in equation (**1.2**) for $\mu = m_W \simeq 83$ GeV, is obtained from the usual fine structure constant, $\alpha = 1/137.035963$, via the perturbative relationship,[11]

$$\alpha^{-1}(\mu) = \alpha^{-1} - \frac{2}{3\pi}\sum_f Q_f^2 \ln(\mu/m_f) + \frac{1}{6\pi} + O(\alpha), \tag{1.15}$$

where the summation is over all charged fermions with mass, $m_f \leq \mu$. To incorporate strong interaction effects, the light quark contribution to equation (**1.15**) is obtained from a dispersive analysis of $e^+e^- \rightarrow$ hadrons. The 7.3% increase in $\alpha(\mu)$ in going from $\mu = 0$ to m_W is very important since it is the dominant radiative correction to $W^\pm$ and Z mass formulas. Thus, it is reassuring that QED tests at PEP and PETRA verify the running of $\alpha(\mu)$. Indeed, from a cursory examination of PETRA data,[12] I find

$$\alpha^{-1}(34.5 \text{ GeV}) \simeq 130 \pm 2, \tag{1.16}$$

which is in agreement with equation (**1.15**).

$\sin^2\theta_W$: There are two popular definitions of the renormalized weak mixing angle. The quantity, $\sin^2\theta_W(m_W)$, defined by $\overline{MS}$ at $\mu = m_W$, is useful for model-independent discussions and analysis of grand unified theories.[11] A second definition,[13]

$$\sin^2\theta_W \equiv 1 - m_W^2/m_Z^2, \tag{1.17}$$

is appropriate for the standard model with $\rho = 1$. By its very definition, $\sin^2\theta_W$ is well suited for discussions involving m_W and m_Z. These two definitions differ by $O(\alpha)$ corrections, which for $m_H \simeq m_Z$ are numerically small[14]:

$$\sin^2\theta_W \simeq 1.006 \sin^2\theta_W(m_W). \tag{1.18}$$

The success of the standard model is illustrated in TABLE 1, where I have given values of $\sin^2\theta_W$ (including radiative corrections) as determined by a wide variety of experiments with $|q^2|$ ranging from 0 to $\simeq 10^4$ GeV^2.

Some comments are in order. The ρ parameter was taken to be one in TABLE 1 (as implied by the Higgs doublet mechanism). The validity of this assumption is checked experimentally by comparing R_ν and $R_{\bar{\nu}}$ data. That procedure yields $\rho = 1.02 \pm 0.02$. Independently, measurements of m_W and m_Z when used in conjunction with[22]

$$\rho = \frac{m_W^2}{m_Z^2\,(1 - (38.65\ \text{GeV}/m_W)^2)} \qquad \textbf{(1.19)}$$

yield $\rho = 1.00 \pm 0.05$. The agreement of these two determinations of ρ provides strong support for the standard model, particularly since they entail very different momentum transfer domains, $|q^2| \simeq 20\ \text{GeV}^2$ and $|q^2| = 10^4\ \text{GeV}^2$, respectively.

Radiative corrections are very important in the above comparison of experiments. They reduce the value of $\sin^2\theta_W$ determined from R_ν data by about 5–6% and increase it by about 7% in the case of m_W and m_Z determinations. A test of the standard model at

TABLE 1. Values of $\sin^2\theta_W \equiv 1 - m_W^2/m_Z^2$ Obtained by a Variety of Experiments[a]

Exp. Measurement	$\sin^2\theta_W$	Comments
$R_\nu \equiv \sigma(\nu_\mu N \to \nu_\mu X)/\sigma(\nu_\mu N \to \mu X)$	0.217 ± 0.014	1981 World Ave.[15]
$\sigma(\nu_\mu e \to \nu_\mu e)/\sigma(\bar{\nu}_\mu e \to \bar{\nu}_\mu e)$	0.212 ± 0.023	1984 BNL & Charm[16]
$\overset{(-)}{\nu}_\mu$ N Scattering	0.224 ± 0.013	1984 Preliminary[17]
eD Asymmetry	0.218 ± 0.020	Reference 18
Atomic P.V.	0.22 ± 0.05	World Ave.[19]
μ–N Scattering	0.23 ± 0.08	Reference 20
$m_W = 82.2 \pm 1.8$ GeV	0.221 ± 0.010	UA1 & UA2 Ave.[21]
$m_Z = 93.2 \pm 1.7$ GeV	0.221 ± 0.010	UA1 & UA2 Ave.[21]
$1 - m_W^2/m_Z^2$	0.22 ± 0.04	Uncorrelated Errors
Weighted World Ave. $\sin^2\theta_W = 0.220 \pm 0.006$		

[a]The experiments assume $\rho = 1$ and the effects due to radiative corrections have been included.

the quantum loop level can be made by measuring the radiative correction, Δr, that appears in the mass formula,[11,13,22]

$$m_W = \frac{37.2804 \pm 0.0003\ \text{GeV}}{\sin\theta_W(1 - \Delta r)^{1/2}} = m_Z \cos\theta_W. \qquad \textbf{(1.20)}$$

The standard model predicts

$$\Delta r = 0.0696 \pm 0.0020 \qquad \textbf{(1.21)}$$

for $m_H \simeq m_Z$ and $m_t = 36$ GeV, where the quoted uncertainty comes from hadronic vacuum polarization effects. Experimental determination of Δr requires precise measurements of two of the three parameters, $\sin^2\theta_W$, m_W, and m_Z, used in conjunction with the derived formulas (from equation (**1.20**)]:

$$\Delta r = 1 - \left(\frac{37.28\ \text{GeV}}{m_W \sin\theta_W}\right)^2, \qquad \textbf{(1.22a)}$$

$$\Delta r = 1 - \left(\frac{74.56 \text{ GeV}}{m_Z \sin 2\theta_W}\right)^2, \tag{1.22b}$$

$$\Delta r = 1 - \frac{(37.28 \text{ GeV})^2}{m_W^2(1 - m_W^2/m_Z^2)}. \tag{1.22c}$$

Employing $m_W = 82.2 \pm 1.8$ GeV and $m_Z = 93.2 \pm 1.7$ GeV in equation (**1.22c**) gives

$$\Delta r = 0.074 \pm 0.158 \qquad \text{[from equation (1.22c)]}, \tag{1.23}$$

which is consistent with, but not yet a test of the standard model prediction in equation (**1.21**). One can do better by using $\sin^2\theta_W = 0.218 \pm 0.010$, obtained from neutral current data in TABLE 1 (excluding m_W and m_Z), along with the measured values of $m_W = 82.2 \pm 1.8$ GeV and $m_Z = 93.2 \pm 1.7$ GeV in equations (**1.22a** and **1.22b**). That procedure gives

$$\Delta r = 0.060 \pm 0.037 \qquad \text{[from equations (1.22a, b)]}. \tag{1.24}$$

If one accepts the above errors on $\sin^2\theta_W$, m_W, and m_Z, then when comparing the result in equation (**1.24**) and the prediction in equation (**1.21**), we see that the standard model has already been tested at the quantum loop level. In the future, both m_W and m_Z will be measured to ± 100 MeV. This will determine Δr to within $\pm$ 0.009 (for uncorrelated errors); or if the theoretical value of Δr is assumed, they can be used to determine $\sin^2\theta_W$ to within ± 0.0005. In addition, planned[23] $\nu_\mu - e$ and deep-inelastic $\nu_\mu - N$ scattering experiments, as well as Z decay asymmetry measurements at SLC, aim to independently measure $\sin^2\theta_W$ to within ± 0.005. Used in conjunction with the ± 100 MeV measurement of m_Z, this will determine Δr to within ± 0.015. One would like to use such measurements to constrain the Higgs mass, m_H. Unfortunately, the shift, δ, in Δr due to $m_H \gg m_Z$,

$$\delta \simeq \frac{11\alpha}{48\pi \sin^2\theta_W} \ln\left(\frac{m_H^2}{m_W^2}\right) - \frac{0.53\,\alpha^2}{128\pi^2 \sin^4\theta_W}\frac{m_H^2}{m_W^2}, \quad (m_H \gg m_W), \tag{1.25}$$

is small and quite insensitive to m_H.[22,24] A very large top quark mass can cause a more significant shift,[22,25]

$$\delta \simeq \frac{-3\alpha \cos^2\theta_W}{16\pi \sin^4\theta_W}\frac{m_t^2}{m_W^2}, \quad (m_t \gg m_W), \tag{1.26}$$

but the UA1 evidence for $m_t \simeq 45 \pm 15$ GeV indicates that such an effect is unlikely. Therefore, it seems that a clear deviation from the prediction in equation (**1.21**) would indicate contributions from new physics such as supersymmetry, compositeness, a fourth generation (see section entitled BEYOND THE STANDARD MODEL), etc.

I should also note that the Fermilab CCCFRR experiments[26] (averaged into the third entry of TABLE 1) give a relatively high value of $\sin^2\theta_W = 0.242 \pm 0.011 \pm 0.005$, which conflicts with some of the other determinations. In my opinion, this difference is probably due to additional theoretical and systematic uncertainties that result from combining $\nu_\mu N$ and $\bar{\nu}_\mu N$ data.

Are there experiments already that may be suggesting physics beyond the standard model? The anomalous UA1 monojets (see section entitled MONOJETS) could certainly be an important signal. In addition, the standard model may not be faring too well in its description of CP violation. The ϵ parameter $\simeq 2.3 \times 10^{-3}$ of the $K_L - K_S$ system is proportion to $\sin\theta_1$, $\sin\theta_2$, $\sin\theta_3$, and $\sin\delta$ in the standard model. As bounds on the KM angles [see equation (**1.9**)] have become more restrictive, accommodating the ϵ parameters has become more difficult.[27] Unfortunately, strong interaction uncertainties cloud this issue. In the case of neutrino mass and mixing, tritium β-decay and ν_e oscillation experiments are giving positive signals; however, they require independent confirmation before they can be taken seriously. Fortunately, several new experiments now under way should clarify the situation in the near future.[5] Finally, let me elaborate on what appears to be a disagreement between theory and experiment in the $e^+e^- \to \mu^+\mu^-$ forward-backward asymmetry, $A_{\mu\mu}$, measured at PETRA.[28] The standard model predicts[29]

$$A_{\mu\mu} \simeq \frac{-3}{8\sin^2 2\theta_W}\left(\frac{s}{m_Z{}^2 - s}\right) \quad (\mathbf{1.27})$$

or using equation (**1.20**),

$$A_{\mu\mu} \simeq \frac{-3\, m_Z^2\,(1 - \Delta r)}{8(74.56\ \mathrm{GeV})^2}\left(\frac{s}{m_Z{}^2 - s}\right). \quad (\mathbf{1.28})$$

Employing $\Delta r = 0.0696 \pm 0.0020$ and $m_Z = 93.2 \pm 1.7$ GeV, one obtains the theoretical predictions in TABLE 2.

TABLE 2. Comparison of the Average Experimental Asymmetry, $A_{\mu\mu}$, Measured at PETRA with Standard Model Predictions[a]

$\sqrt{s}$ (GeV)	$A_{\mu\mu}$ (Exp.)	$A_{\mu\mu}$ (Theory)
34.5	$-10.8 \pm 1.1\%$	$-8.7 \pm 0.05\%$
41.6	$-16.8 \pm 3.0\%$	$-13.6 \pm 0.1\%$

[a]Effects due to radiative corrections have been included.

The experimental asymmetry appears to be somewhat high (particularly the $\sqrt{s} = 34.5$ GeV measurement). Might this be a signal for new physics? One possibility could be a deviation from the standard model Δr value. Indeed, taking $m_Z = 93.2 \pm 1.7$ GeV and fitting equation (**1.28**) to the data yields

$$\Delta r = -0.16 \pm 0.10\ (e^+e^- \to \mu^+\mu^-), \quad (\mathbf{1.29})$$

which disagrees with the standard model prediction and also conflicts with the experimental determination of Δr in equation (**1.24**); thus, I do not consider this a viable solution. Toponium resonances will cause an enhancement in $|A_{\mu\mu}|$, but the effect is expected to be insignificant up until right before the resonance.[30] If, however, toponium (particularly a 1^{++} state) has an anomalously large coupling to $\ell^+\ell^-$ (perhaps due to its near degeneracy with the Z),[31] it could give rise to the observed asymmetry enhancement. That being the case, one would expect to see some effect in the total cross section for $e^+e^- \to \mu^+\mu^-$ (at about the 5–10% level). It will be interesting to keep an eye on both the asymmetry and total cross section, so as to see whether disagreement persists as more data is collected at PETRA.

HIGGS SCALAR AND η_t PRODUCTION

Discovery of the Higgs scalar is crucial for confirmation of the standard model. It is therefore important to ask how can it best be produced and detected? The answer depends on its mass since Higgs scalar production rates and branching ratios are extremely m_H dependent.[32] If $m_H < 60$ GeV, it should be observable at the Z resonance through the decays, $Z \rightarrow H\mu^+\mu^-$ or $H\gamma$. Higher masses up to $\simeq$100 GeV may be detectable at LEP II via $e^+e^- \rightarrow ZH$ if high luminosity is achieved. For $m_H > 100$ GeV, the Higgs scalar can be best produced by gluon-gluon fusion at a hadron-hadron collider.[33] It is my guess that the gluon-gluon fusion mechanism also provides a possible way of observing some toponium levels and a potentially powerful means of probing physics beyond the standard model; therefore, I will give a general analysis of this production mechanism.

The cross section for producing a boson, B, with spin, J, via gluon-gluon scattering is given by[34]

$$\sigma(\overset{(-)}{\mathrm{p}}\mathrm{p} \rightarrow B) = (2J + 1)\frac{\pi^2}{8m_B^3}\Gamma(B \rightarrow gg)\int_{\ln\sqrt{\tau}}^{-\ln\sqrt{\tau}} dy\, \tau F_G(\sqrt{\tau}e^y)F_G(\sqrt{\tau}e^{-y}), \quad \mathbf{(2.1)}$$

where $\Gamma(B \rightarrow gg)$ is the two-gluon decay rate of the B, $\tau = m_B^2/s$, and F_G is the proton's gluon distribution function. Using typical gluon distributions,[35] I find (approximately)

$$\sigma(\overset{(-)}{\mathrm{p}}\mathrm{p} \rightarrow B) = (2J + 1)\frac{\Gamma(B \rightarrow gg)}{1\text{ GeV}}(3 \times 10^{-30}\text{cm}^2)\exp\left[-m_B/11\text{ GeV}\right] \quad \mathbf{(2.2)}$$

for $\sqrt{s} = 540$ GeV and 60 GeV $\lesssim m_B \lesssim 150$ GeV. (For $\sqrt{s} = 630$ GeV, the cross section is about 50% larger.) The uncertainty in this cross section estimate is about a factor of five.

Given that the 1983 CERN $\bar{\mathrm{p}}\mathrm{p}$ collider run had an integrated luminosity of 1.3×10^{35} cm^{-2}, one expects that the number of B's produced, N_B, is given by

$$N_B \simeq 110(2J + 1)\frac{\Gamma(B \rightarrow gg)}{1\text{ GeV}}\exp\left[\frac{90\text{ GeV} - m_B}{11\text{ GeV}}\right]\text{events.} \quad \mathbf{(2.3)}$$

N_B runs from about 1700 events for $m_B = 60$ GeV and $J = 0$ to one event for $m_B = 140$ GeV and $J = 0$ per GeV of partial width into two gluons. Therefore, it seems that if any boson couples to two gluons with a width of $\approx$1 GeV and has a mass <140 GeV, it should have been produced at CERN.

What is $\Gamma(H \rightarrow gg)$? Unfortunately, it is predicted to be quite small in the standard model. For three generations of fermions, one finds[32]

$$\Gamma(H \rightarrow gg) \simeq \frac{5 \times 10^{-7}\, m_H^3}{m_W^2}. \quad \mathbf{(2.4)}$$

Thus, if $m_H = 90$ GeV, they should have produced less than 0.01 events in the 1983

CERN run. Clearly, a search for the Higgs scalar will require much higher energies and luminosities.

Although $\Gamma(H \to gg)$ is small, other 0^{++} or 0^{-+} states may couple more strongly to gluons. For example, in the $^1S_0$ state of toponium, η_t is expected to have a partial width of $\Gamma(\eta_t \to gg) \simeq 0.01$ GeV. Therefore, for $m_{\eta t}$ between 60 and 120 GeV, UA1 and UA2 should have between 17 and 0.1 η_t in their 1983 data. (Of course, $\Gamma(\eta_t \to gg)$ may be larger.) The anticipated branching ratio for $\eta_t \to \gamma\gamma$ is

$$B(\eta_t \to \gamma\gamma) = 8\alpha^2/9\alpha_3^2\,(m_{\eta_t}/4) \simeq 0.35\%, \tag{2.5}$$

which may be large enough to make $\eta_t \to \gamma\gamma$ observable in the future. Indeed, given the

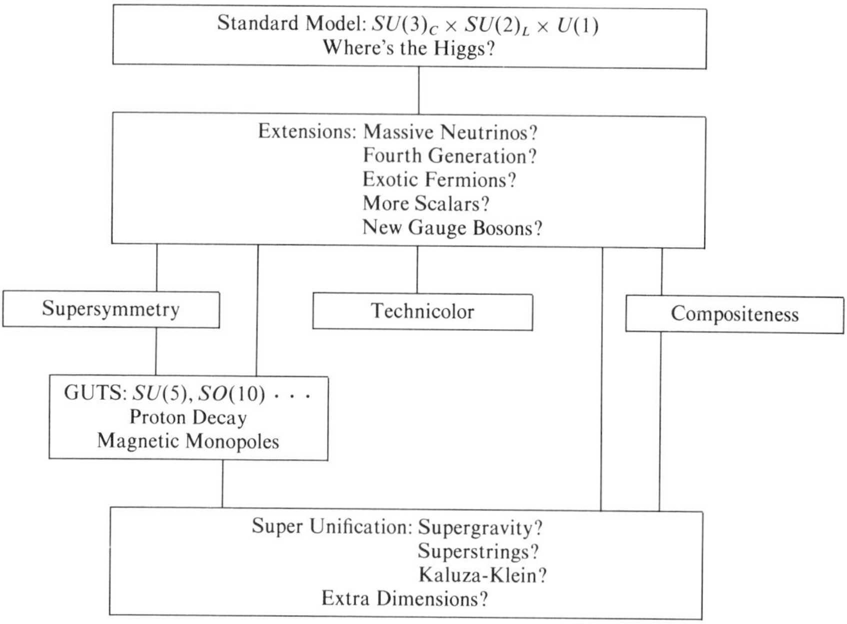

FIGURE 1. Directions in theoretical physics beyond the standard model.

increased energy and luminosity at the CERN collider, as well as uncertainties in the above analysis, I would not be too surprised if $\eta_t \to \gamma\gamma$ shows up in the 1985 run. A $\gamma\gamma$ event could also signal the decay of some more exotic particle such as a technipion; thus, it is an important type of event to search for.

BEYOND THE STANDARD MODEL

What new physics lies beyond standard model expectations? Listed in FIGURE 1 are some obvious generalizations and appendages to the standard model, as well as the types of theories actively being pursued. This illustration summarizes the main directions being explored. Although nature may have very different physics in store for us, the presently popular models are important in that they provide a framework for suggesting new experiments and evaluating their discovery potential. In that context, I will discuss in this section a few implications of these extensions.

Fourth Generation

The existence of a fourth generation of fermions is certainly a viable possibility. Indeed, some popular grand unified models and superstring theories seem to even require a fourth generation.[36] Of course, direct observation of the fourth generation leptons and quarks (e.g, via W or Z decay) would be the best that one could hope for. If, however, the quark and charged lepton masses are large, that may not be possible. Indirect indications of a fourth generation could then still come from precision measurements of W, Z, and top quark properties. The width of the Z can provide information on the number of neutrino species,[37]

$$N_\nu = 3 + \frac{\Gamma(Z \rightarrow \text{all}) - 2.8 \text{ GeV}}{0.178 \text{ GeV}}. \tag{3.1}$$

Already, by comparing W and Z production and decay mechanisms, UA2 has given the bound,[21]

$$N_\nu \leq 6 \text{ (95\% C.L.)}. \tag{3.2}$$

A fourth generation may also be indicated if the KM matrix is found to deviate from a unitary matrix after t decays are measured. This would imply fourth generation mixing. Mixing might also be observed through deviations in $\Gamma(W \rightarrow t + X)/\Gamma(W \rightarrow e\nu)$, from the prediction of

$$\frac{\Gamma(W \rightarrow tX)}{\Gamma(W \rightarrow e\nu)} = 3\left(1 - \frac{3}{2}\frac{m_t^2}{m_W^2} + \frac{1}{2}\frac{m_t^6}{m_W^6}\right). \tag{3.3}$$

Effects due to very heavy fermions could also show up as loop corrections in Δr or the ρ parameter. For example, a charged lepton with a mass of $m_L \gg M_W$ and a massless neutrino partner would change Δr to $0.0696 + \delta$ where[22,25]

$$\delta \simeq \frac{-\alpha}{16\pi}\frac{\cos^2\theta_W}{\sin^4\theta_W}\frac{m_L^2}{m_W^2}. \tag{3.4}$$

Employing the bound of $-0.047 \leq \delta < 0.027$ [from equation (**1.24**)] implies

$$m_L \lesssim 370 \text{ GeV}, \tag{3.5}$$

which is not yet a severe constraint. Similar bounds can be obtained using the ρ parameter. This effect can be even more important for quarks, since they come in three colors. In addition, supersymmetric multiplet partners (if they exist) will also contribute.[38] This illustrates the need for precise measurements of m_W, m_Z, and $\sin^2\theta_W$.

Additional Higgs Scalars

One may, in general, introduce many physical Higgs scalars into the standard model. Supersymmetric and grand unified theories are examples of extensions that require such fields. If some are relatively light (<100 GeV), they may be directly detected at high energy colliders. The expected signature is a relatively narrow resonance that decays into the most massive particles kinematically allowed. Indirect effects due to heavier Higgs scalars (or pseudoscalars) may be observable in rare decay searches (such as $K_L \to \bar{\mu}e$, $\pi\bar{\mu}e$, and π + missing energy) now getting under way at BNL or $\mu \to e\gamma$, $e\bar{e}e$, etc., at LAMPF. Other signals of Higgs scalar effects would be any sign of oscillations in the $D^0 - \bar{D}^0$ system or any unexpected contributions to electric and magnetic dipole moments. Finally, very heavy scalars implied by GUTS can lead to distinctive proton and neutron decay modes, such as p $\to K^+\bar{\nu}$ or $K^0\mu^+$ and n $\to K^0\bar{\nu}$. Many of these same effects can also come about in supersymmetric, technicolor, or composite models. Hence, experimental searches for such phenomena are well motivated and worth the effort.

Supersymmetry

Low energy supersymmetric theories are attractive in that they provide a natural mechanism for the cancellation of quadratic divergences in the Higgs scalar sector. For that reason, they more naturally accommodate mass hierarchies. Such theories predict the existence of supersymmetric partners for all observed elementary particles. Observing any of these partners would provide the first experimental evidence for supersymmetry in elementary particle physics.

Searches for charged scalar partners of the known fermions at PETRA have yielded lower bounds of O(20 GeV) for their masses.[39] Recently, however, supersymmetry advocates found something to cheer about. The UA1 monojet + missing energy events (see section on MONOJETS) were just what one might expect in some supersymmetry scenarios. Such events would correspond to the production of new particles that decay into photinos (low mass partners of the photon). Since photinos interact very weakly, they do not deposit energy in the detector; hence, the large missing energy.

Technicolor

Technicolor models provide a dynamical alternative to the Higgs mechanism.[40] Although the scale of such theories is O(1 TeV), they generally predict the existence of many relatively light pseudoscalars with masses <250 GeV. One of these, the techni-eta, η_T, is supposed to have a two-gluon fusion production cross section at least 50 times that of the Higgs scalar with the same mass.[41] It is expected to decay primarily back into two gluons or $t\bar{t}$. If, however, there is an enhancement of the $\eta_T \to Z$ + gluon mode, then a nice signal will be jet + missing energy corresponding to $Z \to \nu\bar{\nu}$ (just like the UA1 monojets).

Recently, Rubakov has made the interesting observation[42] that some massive techni-baryons (roughly >2 TeV) may violate ordinary baryon number in their decays via the 't Hooft mechanism.[43] Such a phenomenon would be interesting for supercollider experiments where $\sqrt{s} = 40$ TeV is expected. The signature of such decays would be high multiplicity events with at least 18 jets, 6 leptons, and several W, Z bosons.

Compositeness

The idea that quarks, leptons, $W^{\pm}$, Z, etc., may be bound states of more fundamental entities has been gaining in popularity as the number of "elementary particles" has increased.[44] In my opinion, no compelling composite model yet exists. However, there are quite general predictions associated with such scenarios. One should see deviations from standard model predictions, such as $\rho \neq 1$, axial-isoscalar neutral current effects, excited fermions, additional bosonic states, anomalous magnetic moments, etc. I will elaborate only on a few of these interesting, but as yet unobserved possibilities.

If the $W^{\pm}$ and Z are ${}^3S_1$ bound states of constituents that carry color, then one expects color octet excitations, $W^{*\pm}$ and Z^*. Their cross sections are enhanced by the factor of eight color degrees. The decays, $Z^* \rightarrow Z$ + gluon jet and $W^{*\pm} \rightarrow W^{\pm}$ + gluon jet, could explain the anomalous UA1 and UA2 events that seem to cluster at an invariant mass of around 150–160 GeV (see MONOJETS).[45] Such bosons could also help explain the $\Delta I = \frac{1}{2}$ rule in weak hadronic decays.

Spin-0 partners of the $W^{\pm}$ and Z, as well as their excitations, will also exhibit distinct phenomenology. The isoscalar spin-0 pseudoscalar partner of the Z, which I call the η_Z, is particularly interesting because it can be made by gluon-gluon scattering,[46] or if $m_{\eta_Z} < m_Z$, by the decay, $Z \rightarrow \eta_Z + \gamma$.[48] (I will not discuss the isovector pseudoscalars.) Possibly observable decays of the η_Z are $\eta_Z \rightarrow \gamma\gamma$, $\ell^+\ell^-\gamma$, $\nu\bar{\nu}\gamma$, or Jet-Jet-γ. The three $\ell^+\ell^-\gamma$ candidate events in the 1983 UA1 and UA2 data were somewhat suggestive of 90 GeV η_Z decays.[46] However, such events have not shown up in the 1984 data; most likely, then, the earlier events were $Z \rightarrow \ell^+\ell^-\gamma$ bremsstrahlung. In any case, $\gamma\ell^+\ell^-$ and $\gamma\nu\bar{\nu}$ (missing energy) events are anticipated signals for spin-0 bosons because such decays are not helicity suppressed.

My final comment on compositeness concerns anomalous magnetic moments, $a \equiv (g - 2)/2$. One expects composite structure to cause deviations from the standard model prediction at some level. For example, in the case of the proton and neutron, one finds

$$a_p = 1.79 \tag{3.6a}$$

and

$$a_n = -1.91. \tag{3.6b}$$

These large departures from $\alpha/2\pi$ (the Schwinger QED calculation) are indications of quark structure. Similarly, the $J = 1$ deuteron is found to have

$$a_D = -0.14. \tag{3.6c}$$

In the case of the known charged leptons, no such effects are seen. Indeed, strong experimental bounds,[48]

$$|\delta a_e| < 4 \times 10^{-10}, \tag{3.7a}$$

$$|\delta a_\mu| < 1 \times 10^{-8}, \quad \textbf{(3.7b)}$$

$$|\delta a_\tau| < 2 \times 10^{-2}, \quad \textbf{(3.7c)}$$

argue against substructure. To circumvent these bounds, one generally requires the mass scale of compositeness, Λ, to be large and masses of constituents to be zero. Then if the binding interaction respects chiral symmetry, the expected deviation, $\delta a_\ell \approx m_\ell^2/\Lambda^2$, is naturally small. Nevertheless, using the bound in equation **(3.7b)**, one finds

$$\Lambda > 1 \text{ TeV}. \quad \textbf{(3.8)}$$

A proposed BNL measurement[49] would lower δa_μ by a factor of 20 (or perhaps see an effect). Such an experiment would thus probe substructure at scales of 1–5 TeV.

If the $W^\pm$ boson is a composite state, then one should expect some deviation from $a_W = 0$ due to structure effects. A measurement of the W bosons' magnetic moment is thus of the utmost importance. Such a measurement may be possible at LEP II via W^+W^- pair production or in $p\bar{p} \to W\gamma$ at hadron-hadron colliders.[50]

MONOJETS

The UA1 collaboration has so far reported nine monojet events (Jet + missing energy > 40 GeV) in their 1983 and 1984 data. (Only about one-third of the 1984 data has been analyzed.) In addition, they have several two jet + missing energy events in the 1984 sample.[51] Such events may be harbingers of new physics beyond the standard model if they cannot be attributed to known backgrounds. Analysis by UA1 of the most likely backgrounds, $W \to \tau\bar{\nu}_\tau \to$ jet + missing energy and jet + Z production via QCD followed by $Z \to \nu\bar{\nu}$, were found to be small; however, those calculations should be independently scrutinized by phenomenologists. If this is new physics, there are already a number of speculations as to what it might be. A few scenarios are: (1) Heavy squark ($\simeq$50 GeV) production (supersymmetric scalar partners of quarks) decaying into quark + photino, with the photino giving rise to missing energy;[52] (2) Color excitations of the Z at $\simeq$150 GeV (as in some composite models), which decay into Z + gluon jet.[45] The subsequent decay, $Z \to \nu\bar{\nu}$, gives rise to missing energy; (3) Techni-eta or some other spin-0 color octet (at $\simeq$150 GeV) production via two gluons, followed by the decay into Z + gluon jet; (4) Z decay into two distinct neutral Higgs scalars with one decaying into a jet and the other going undetected as missing energy.[53]

The only comment I want to make concerns speculations (2) and (3), as well as the QCD background Z + jet production. If the missing energy comes from $Z \to \nu\bar{\nu}$, then we should be starting to see e^+e^- + missing energy events. The branching ratio, $\Gamma(Z \to e^+e^-)/\Gamma(Z \to \nu\bar{\nu}) \simeq \frac{1}{2}N_\nu$, is one-sixth or less. For $N_\nu = 3$, one expects 1.5 e^+e^- jet events for nine monojets. So far, no e^+e^- jet events (with high energy jets) have been observed. However, given the limited statistics, this is still a viable possibility. If e^+e^- events are observed, then in addition to possibly implying new physics, one can use the ratio of monojet and e^+e^- jet events to directly determine the number of neutrino species, N_ν.

GRAND UNIFICATION

Grand unified theories (GUTS) provide strong motivation for proton decay experiments and magnetic monopole searches. Using $\alpha \simeq 1/137$ and a range of $\Lambda_{\overline{MS}}^{(4)}$ values as input, one obtains the minimal $SU(5)$ model predictions in TABLE 3.[54] The predicted weak mixing angle,

$$\sin^2\theta_W = 0.218 + 0.006\, \ell n\,(100\text{ MeV}/\Lambda_{\overline{MS}}^{(4)}), \tag{5.1}$$

is in good agreement with the world average $\sin^2\theta_W = 0.220 \pm 0.006$ given in equation **(1.3)**. This represents an impressive success. Unfortunately, the predicted unification mass, m_X, where

$$m_X \simeq 1.3\, \Lambda_{\overline{MS}}^{(4)} \times 10^{15}, \tag{5.2}$$

implies a partial proton decay lifetime into $e^+\pi^0$ of

$$\tau(\text{p} \to e^+\pi^0) < 2.4 \times 10^{31}\text{ yr} \tag{5.3}$$

TABLE 3. Minimal $SU(5)$ Model Predictions as a Function of $\Lambda_{\overline{MS}}^{(4)}$

$\Lambda_{\overline{MS}}^{(4)}$ (MeV)	$\sin^2\theta_W$	m_W (GeV)	m_Z (GeV)	m_X (GeV)	$1/\Gamma(\text{p} \to e^+\pi^0)$ (yr)
25	0.226	81.3	92.3	3×10^{13}	$7.5 \times 10^{26\pm1.3}$
50	0.222	82.0	92.8	6.2×10^{13}	$7.5 \times 10^{27\pm1.3}$
100	0.218	82.8	93.6	1.3×10^{14}	$12 \times 10^{28\pm1.3}$
200	0.214	83.5	94.3	2.7×10^{14}	$12 \times 10^{29\pm1.3}$
400	0.210	84.3	94.9	5.5×10^{14}	$2.5 \times 10^{31\pm1.3}$

for

$$\Lambda_{\overline{MS}}^{(4)} \simeq 100^{+100}_{-50}\text{ MeV}.$$

This conflicts with the IMB experimental bound of[55]

$$\tau(\text{p} \to e^+\pi^0) > 1.2 \times 10^{32}\text{ yr} \tag{5.4}$$

and therefore seems to rule out the minimal $SU(5)$ model. The proton lifetime prediction can be increased if $\Lambda_{\overline{MS}}^{(4)}$ is larger (it scales approximately as the fourth power of $\Lambda_{\overline{MS}}^{(4)}$). More likely, new physics between m_W and m_X modifies the predictions. As that new physics is uncovered, it will either make grand unification less appealing or merely modify the above predictions. It is, therefore, important to have precise measurements of $\Lambda_{\overline{MS}}^{(4)}$ and $\sin^2\theta_W$ and to have ongoing searches for proton decay.

The large underground proton decay experiments are extremely important because they presently provide a window to physics on the 10^{14}–10^{15} GeV scale. Those detectors also allow searches for GUT magnetic monopoles via the reaction, $M + \text{p} \to M + e^+ + \pi$'s, neutron-antineutron oscillations, double proton decay ($\text{pp} \to e^+e^+$), cosmic ray neutrinos, etc. Given their obvious discovery potential and the fact that proton decay has not yet been observed, it is my belief that a new generation of detectors ten times larger than the IMB water detector should be a very high priority in the near future.

IN SEARCH OF EXTRA DIMENSIONS

Generalized Kaluza-Klein (KK) models offer the attractive possibility of unifying gravity with the other fundamental forces. The idea is to enlarge space-time to $4 + N$ dimensions such that the N extra dimensions form a small compact manifold with a mean radius of $R_{KK} \simeq 10^{-32}$ cm.[56] The $4 + N$ dimensional metric tensor, $g_{\mu\nu}(x)$, can then describe general relativity as well as gauge field interactions in our effective four-dimensional world, i.e., at energies $< 1/R_{KK} \simeq 10^{18}$ GeV.

Although Kaluza-Klein models presently exhibit serious theoretical flaws, the possibility of unifying gravity with other fundamental forces via additional dimensions is very appealing and is actively being pursued. It is therefore important to ask: Are extra dimensions a physical reality or merely a model building trick? If they are real, can we find experimental evidence for their existence? As yet, only general potentially observable effects have been suggested. I will comment on a few.

TABLE 4. Bounds on the Time Variation of Some Fundamental Quantities

Quantity Q	Bound on $\lvert\dot{Q}/Q\rvert$ (yr^{-1})	Method
G	1×10^{-11}	Solar System Measurements
α	1×10^{-17}	Geochemical
$\alpha^2 g_p m_e/m_p$	2×10^{-14}	Astrophysics

The size of the extra dimensions' R_{KK} may be contracting, expanding, or oscillating as a function of time,[57,58] i.e., $dR_{KK}/dt \neq 0$. (We know from red shifts that our ordinary three spatial dimensions are expanding.) If that is the case, it would give rise to time-varying fundamental "constants," as in Dirac's large numbers hypothesis.[59] One should therefore look for time variations of the gravitational constant, G, α, m_e/m_p, anomalous magnetic moment of the proton, etc. Although stringent bounds already exist (see TABLE 4), they should be pushed further.[58]

Perhaps the most likely scenario is the damped oscillation of R_{KK} about an equilibrium position. The amplitude of such oscillations would be damped by the radiation of energy into our three ordinary spatial dimensions.[58] In the early stages, this could have been the source of cosmic background radiation. In addition, this could have given rise to an effective inflationary universe scenario. Indeed, the rich structure of Kaluza-Klein cosmologies is only starting to be appreciated and explored. If the oscillation frequency is very large, then it has been suggested[60] that this mechanism can lead to isotropic ultrahigh energy ($\approx 10^{10}$–10^{16} GeV) cosmic ray production with a differential spectrum of $dN/dE \propto E^{-5/2}$. It would be nice to test this speculation.

Other consequences of Kaluza-Klein models could be baryon number violations and the existence of superheavy ($\simeq 10^{19}$ GeV) magnetic monopoles. Without a definite model, it is not possible to say how these phenomena should differ from GUT predictions. One does, however, anticipate that KK proton decay will be observable in the laboratory (i.e., $\tau_p < 10^{34}$ yr) only if it is induced by dimension five operators coming from scalar particle interactions. Hence, decays such as $p \rightarrow \bar{\nu}K^+$ or μ^+K^0 and $n \rightarrow \bar{\nu}K^0$ should dominate, just as in some supersymmetric GUT scenarios.[61] In the case of magnetic monopoles, their mass, hadronic properties, ability to catalyze proton decay, etc., may allow us to distinguish KK from GUT monopoles, but for now that is an open issue.[62]

The idea that there are more than four dimensions is more than 60 years old.[56] It has at one time or another fascinated almost all theoreticians. What we need now is a clever idea as to how we can prove or disprove experimentally the existence of extra dimensions.

CONCLUSIONS

The standard model is in very good shape. Discovery of the $W^{\pm}$ and Z, along with measurements of their masses and decay properties, provides the best confirmation of that theory to date. Indeed, we are at the point where the electroweak sector is already starting to be tested at the level of its quantum loop corrections. There still remain, of course, areas where experimental verification of the standard model is imperative. In the QCD sector, one would like to find and investigate glueballs as well as other predicted exotic states.[63] Direct evidence for the running of $\alpha_3(\mu)$, precise determination of $\Lambda_{\overline{MS}}^{(4)}$, and a better understanding of confinement and chiral symmetry breaking are also on the agenda. In the case of the latter, the relativistic heavy ion collider (RHIC) proposed by BNL[64] should prove very important.

At present, no clear departures from the standard model's electroweak sector have been found. The forward-backward asymmetry in $e^+e^- \rightarrow \mu^+\mu^-$ measured at PETRA is somewhat larger than expected, but more high energy data is needed before we can decide if it is due to some new phenomenon. Other interesting low energy experiments that not only test the standard model, but also provide possible windows to new physics, are neutrino mass and oscillation experiments, double β-decay, g–2 measurements, rare K and μ decays, searches for CP violation, etc. Fortunately, a variety of sophisticated experiments in those areas are planned or are already under way.

Perhaps the most important recent development in low energy experimental physics is the advent of big underground proton decay detectors. After more than a year of data taking, the IMB collaboration has set very constraining limits on a large number of proton and neutron decay modes. These bounds constrain GUTS and indicate the need for new physics between m_W and unification. Large underground detectors such as IMB are much more than single experiments; they are facilities that can also search for n-$\bar{n}$ oscillations, magnetic monopoles, cosmic ray neutrino emitters, etc.[54,64] I expect their role in elementary particle physics, astrophysics, and cosmology to steadily grow during the coming years.

The most exciting high energy experiments now under way are unquestionably UA1 and UA2 at CERN. Monojet events detected by the UA1 collaboration seem to be inconsistent with standard model backgrounds and have naturally created considerable excitement. Supersymmetry, technicolor, and composite model advocates all have competing explanations. Once again, we see how fresh experimental results can stir up theoretical creativity.

The next few years promise to be very interesting. Upgrading of the CERN $\bar{p}p$ collider combined with the start-up of TRISTAN, SLC, TeV I, and LEP will give the high energy community a wealth of facilities. I am sure that we can look forward to exciting new discoveries and surprises that will guide us in our aim to understand the laws of nature and the universe around us.

NOTES AND REFERENCES

1. For a review, see CHENG, T. P. & L. F. LI. 1984. Gauge Theory of Elementary Particle Physics. Clarendon Press. Oxford.
2. For reviews, see: CHAU, L. L. 1983. Phys. Rep. **95:** 1; LEUTWYLER, H. & M. ROOS. 1984. Z. Phys. **C25:** 91; BURAS, A. *In* Proceedings of the Workshop on the Future of Intermediate Energy Physics in Europe. H. Koch & F. Scheck, Eds.
3. For a review, see MARCIANO, W. 1981. Comments Nucl. Part. Phys. **9:** 169.
4. ABELA, R. *et al.* 1984. Phys. Lett. **146B:** 431; ALBRECHT, H. *et al.* 1985. Bari Conference.
5. See the talks by ROBERTSON, R., M. WITHERELL, B. VIGNON & P. VOGEL in this volume.
6. BARDEEN, W. A., A. BURAS, D. DUKE & T. MUTA. 1978. Phys. Rev. **D18:** 3998; BERNREUTHER, W. & W. WETZEL. 1983. Nucl. Phys. **B197:** 228; MARCIANO, W. 1984. Phys. Rev. **D29:** 580.
7. MACKENZIE, P. & G. P. LEPAGE. 1981. Phys. Rev. Lett. **47:** 1244.
8. SCHAMBERGER, R. D. *et al.* 1984. Phys. Lett. **138B:** 225.
9. DUKE, D. & R. ROBERTS. 1984. Rutherford preprint RAL-84-079; ADWA, B. *et al.* MARK-J COLLABORATION. 1984. MIT preprint 141.
10. LEPAGE, G. P. 1983. Proceedings of the 1983 Lepton-Photon Conference, Ithaca, New York.
11. MARCIANO, W. 1979. Phys. Rev. **D20:** 274; MARCIANO, W. & A. SIRLIN. 1981. *In* Proceedings of the Second Workshop on Grand Unification, Ann Arbor, MI. J. Leiveille, L. Sulak & D. Unger, Eds. Birkhauser. Boston, MA; Phys. Rev. Lett. **46:** 163.
12. MARCIANO, W. 1984. *In* Proceedings of the 4th Workshop on $\bar{p}p$ Collider Physics, Bern.
13. SIRLIN, A. 1980. Phys. Rev. **D22:** 971; MARCIANO, W. & A. SIRLIN. 1980. Phys. Rev. **D22:** 2695.
14. SARANTAKOS, S., A. SIRLIN & W. MARCIANO. 1983. Nucl. Phys. **B217:** 84.
15. SIRLIN, A. & W. MARCIANO. 1981. Nucl. Phys. **B189:** 442; LLEWELLYN SMITH, C. & J. WHEATER. 1981. Phys. Lett. **105B:** 486.
16. AHRENS, L. *et al.* 1985. Phys. Rev. Lett. **54:** 18; BERGAMA, F. *et al.* To be published; Talk by PANMAN, J. in this volume.
17. GEWENIGER, C. 1984. *In* Proceedings of Neutrino '84.
18. PRESCOTT, C. Y. *et al.* 1978. Phys. Lett. **77B:** 347.
19. FORTSON, E. N. & L. L. LEWIS. 1984. Phys. Rep. **113:** 289.
20. ARGENTO, A. *et al.* 1983. Phys. Lett. **120B:** 245; 1984. Phys. Lett. **140B:** 142.
21. See talks by CLINE, D. & C. CONTA in this volume.
22. MARCIANO, W. & A. SIRLIN. 1984. Phys. Rev. **D29:** 945. (Erratum. 1985. Phys. Rev. **D31:** 213.)
23. See talks by PANMAN, J. & C. PRESCOTT in this volume.
24. VAN DER BIJ, J. & M. VELTMAN. 1984. Nucl. Phys. **B231:** 205; VAN DER BIJ, J. 1984. NIKHEF preprint.
25. VELTMAN, M. 1977. Nucl. Phys. **B123:** 89.
26. See the talk by SCIULLI, F. in this volume.
27. See talks by LANGACKER, P. & J. CRONIN in this volume.
28. DAVIER, M. 1984. SLAC Summer Institute on Particle Physics.
29. MARK-J COLLABORATION. 1982. Phys. Rep. **87:** 77.
30. SEHGAL, L. M. 1983. Erice Lecture.
31. RENARD, F. M. 1979. Z. Phys. **C1:** 225; 1982. Nucl. Phys. **B197:** 237.
32. GORDON, H. *et al.* 1982. *In* Proceedings of the 1982 DPF Summer Study. R. Donaldson, R. Gustafson & F. Paige, Eds. Fermilab. Batavia, IL (and references cited therein).

33. GEORGI, H., S. GLASHOW, M. MARCHACEK & D. NANOPOULOS. 1978. Phys. Rev. Lett. **40:** 692.
34. EINHORN, M. B. & S. D. ELLIS. 1975. Phys. Rev. **D12:** 2007.
35. DUKE, D. & J. OWENS. 1984. Phys. Rev. **D30:** 49.
36. BAGGER, J., S. DIMOPOULOS & E. MASSO. 1984. Phys. Lett. **145B:** 211; SCHWARZ, J. H. 1982. Phys. Rep. **89:** 223.
37. ALBERT, D., W. MARCIANO, D. WYLER & Z. PARSA. 1980. Nucl. Phys. **B166:** 460.
38. BARBIERI, R. & L. MAIANI. 1983. Nucl. Phys. **B224:** 32; ELIASSON, E. 1984. Phys. Lett. **147B:** 65.
39. DAVIER, M. 1984. To be published in Proceedings of the 1984 SLAC Summer Institute.
40. For a review, see FARHI, E. & L. SUSSKIND. 1981. Phys. Rep. **74:** 277.
41. DIMOPOULOS, S., S. RABY & G. KANE. 1981. Nucl. Phys. **B182:** 77.
42. RUBAKOV, V. A. 1984. Nordita preprint; D'HOKER, E. & E. FARHI. 1984. Nucl. Phys. **B241:** 109.
43. 'T HOOFT, G. 1976. Phys. Rev. Lett. **37:** 8.
44. See, for example, ABBOTT, L. & E. FARHI. 1981. Phys. Lett. **99B:** 69.
45. BAUR, U. & K. H. STRENG. 1984. MPI-PAE 50/84 preprint.
46. MARCIANO, W. 1984. Phys. Rev. Lett. **53:** 975.
47. PECCEI, R. D. 1984. Phys. Lett. **136B:** 121; BAUR, U., H. FRITZSCH & H. FAISSNER. 1984. Phys. Lett. **135B:** 313; FRITZSCH, H. & G. MANDELBAUM. 1985. Phys. Lett. **105B:** 395.
48. PESKIN, M. E. 1981. *In* Proceedings of the 1981 International Symposium on Lepton and Photon Interactions. W. Pfeil, Ed.: 880. Bonn.
49. HUGHES, V. W. 1985. Yale University. Private communication.
50. MIKAELIAN, K. O., M. A. SAMUEL & D. SAHDEV. 1979. Phys. Rev. Lett. **43:** 746.
51. See the talk by ROHLF, J. in this volume.
52. See, for example, ELLIS, J. & M. SHER. 1984. CERN preprint; BARGER, V. *et al.* 1985. University of Wisconsin preprint.
53. GLASHOW, S. L. & A. MANOHAR. 1985. Phys. Rev. Lett. **54:** 528.
54. For a review, see GOLDHABER, M. & W. MARCIANO. 1984. BNL preprint-35787.
55. PARK, H. S. *et al.* IMB COLLABORATION. 1985. Phys. Rev. Lett. **54:** 22.
56. KALUZA, TH. 1921. Sitzungsber. Preuss. Akad. Wiss. Phys. Math. Kl. 966; KLEIN, O. 1926. Z. Phys. **37:** 895.
57. CHODOS, A. & S. DETWEILER. 1980. Phys. Rev. **D21:** 2167; FREUND, P. 1982. Nucl. Phys. **B209:** 146.
58. MARCIANO, W. 1984. Phys. Rev. Lett. **52:** 489 and references cited therein.
59. DIRAC, P. A. M. 1937. Nature **139:** 323.
60. KOIKAWA, T. & M. YOSHIMURA. 1984. Phys. Lett. **150B:** 107.
61. WEINBERG, S. 1982. Phys. Rev. **D26:** 287; SAKAI, N. & T. YANAGIDA. 1982. Nucl. Phys. **B197:** 5333.
62. See the talk by TURNER, M. in this volume.
63. See the talk by PARTRIDGE, R. in this volume.
64. RHIC AND QUARK MATTER. BNL proposal 51801.
65. See the talk by VAN DER VELDE, J. in this volume.

A Selective Experimental Review of the Standard Model[a]

E. D. BLOOM
Stanford Linear Accelerator Center
Stanford University
Stanford, California 94305

INTRODUCTION

Before discussing experimental comparisons with the "Standard Model" (S-M), it is probably wise to define more completely what is commonly meant by this popular term. This model is a gauge theory of $SU(3)_f \times SU(2)_L \times U(1)$ with 18 parameters. The parameters are α_s, α_{qed}, θ_W, M_W ($M_Z = M_W/\cos\theta_W$, and thus is not an independent parameter), M_{Higgs}; the lepton masses, M_e, M_μ, M_τ; the quark masses, M_d, M_s, M_b, and M_u, M_c, M_t; and finally, the quark mixing angles, θ_1, θ_2, θ_3, and the CP violating phase, δ. The latter four parameters appear in the quark mixing matrix as shown in FIGURE 1 for the Kobayashi-Maskawa[1] and Maiani[2] forms. Clearly, the present S-M covers an enormous range of physics topics and I can only lightly cover a few such topics in this report.

The measurement of R_{hadron} is fundamental as a test of the running coupling constant, α_s, in QCD. I will discuss a selection of recent "precision" measurements of R_{hadron}, as well as some other techniques for measuring α_s. QCD also requires the self-interaction of gluons. The search for the three gluon vertex may practically be realized in the clear identification of gluonic mesons. I will present a limited review of recent progress in the attempt to untangle such mesons from the plethora of $q\bar{q}$ states of the same quantum numbers that exist in the same mass range. The electroweak interactions provide some of the strongest evidence supporting the S-M that exists. Given the recent progress in this subfield and particularly with the discovery of the W and Z bosons at CERN,[3] many recent reviews[4] obviate the need for further discussion in this report. Finally, in attempting to validate a theory, one frequently searches for new phenomena that clearly would invalidate it. In that spirit, I will examine recent searches for new particle states.

THE MEASUREMENT OF α_s

In this section, I will discuss recent measurements of R_{hadron} that have systematic overall scale errors on the absolute value of R_{hadron} of $\pm 7\%$ or less. The most accurate measurements from the JADE[5] and MAC[6] collaborations are reported at the 2–3% level. The range in $\sqrt{s}$ of all the measurements is 5 to over 40 GeV, and thus the full combination of QCD and electroweak contributions to R_{hadron} must be considered.

[a]This work was supported by the Department of Energy, contract no. DE-AC03-76SF00515.

In the quark-parton model, the ratio, $R_{hadron} = \sigma_{hadron}/\sigma_{\mu\mu}$, of the total hadronic cross section to the lowest order muon pair cross section is given by summing over the available quark flavors, n_f,

$$R_0(s) = 3\left(\sum_{q=1}^{n_f} Q_q^2 - v_e Re(g(s)) \sum_{q=1}^{n_f} Q_q v_q + \frac{1}{4}(v_e^2 + a_e^2)|g(s)|^2 \sum_{q=1}^{n_f}(v_q^2 + a_q^2)\right), \quad (1)$$

where Q_q are the quark charges, v_i and a_i are the neutral current vector and axial vector coupling constants of the electroweak theory,[7] and

$$g(s) = \frac{1}{8\sin^2\theta_W\cos^2\theta_W}\left(\frac{s}{s - M_Z^2 + iM_Z\Gamma_Z}\right) = \frac{G}{\sqrt{2}}\frac{1}{4\pi\alpha}\frac{sM_Z^2}{s - M_Z^2 + iM_Z\Gamma_Z}. \quad (2)$$

It has been shown[8] that a rapidly converging perturbative expansion for R_{hadron} may be

Kobayashi – Maskawa

$$\begin{array}{c} \\ u \\ c \\ t \end{array}\begin{array}{ccc} d & s & b \\ \left(\begin{array}{c} c_1 \\ -s_1 c_2 \\ s_1 s_2 \end{array}\right. & \begin{array}{c} s_1 c_3 \\ c_1 c_2 c_3 - e^{i\delta} s_2 s_3 \\ -c_1 s_2 c_3 - e^{i\delta} c_2 s_3 \end{array} & \left.\begin{array}{c} s_1 s_3 \\ c_1 c_2 s_3 + e^{i\delta} s_2 c_3 \\ -c_1 s_2 s_3 + e^{i\delta} c_2 c_3 \end{array}\right) \end{array}$$

Maiani

$$\begin{array}{c} u \\ c \\ t \end{array}\left(\begin{array}{ccc} c_\beta c_\theta & c_\beta s_\theta & s_\beta \\ -s_\gamma c_\theta s_\beta e^{i\delta'} - s_\theta c_\gamma & c_\gamma c_\theta - s_\gamma s_\beta s_\theta e^{i\delta'} & s_\gamma c_\beta e^{i\delta'} \\ -s_\beta c_\gamma c_\theta + s_\gamma s_\theta e^{-i\delta'} & -c_\gamma s_\beta s_\theta - s_\gamma c_\theta e^{-i\delta'} & c_\gamma c_\beta \end{array}\right)$$

FIGURE 1. The quark mixing matrix in the Kobayashi-Maskawa and Maiani forms. c_i and s_i denote $\cos\theta_i$ and $\sin\theta_i$, respectively.

obtained from QCD:

$$R_{hadron}(s) = R_0(s)\left(1 + \frac{\alpha_s}{\pi} + C\left(\frac{\alpha_s}{\pi}\right)^2\right). \quad (3)$$

Calculated in the modified minimal subtraction ($\overline{MS}$) scheme, one has $C = 1.986 - 0.115n_f$. Additional corrections are needed for finite quark mass effects near $q\bar{q}$ thresholds.[8,9] The coupling constant, α_s, is predicted in QCD to run as a function of $\sqrt{s}$; the rate of the logarithmic decrease of α_s with increasing $\sqrt{s}$ is governed by $\Lambda_{\overline{MS}}$, the QCD scale parameter. Thus by determining α_s at one energy, the $\sqrt{s}$ dependence of R_{hadron}

predicted in QCD can be checked. Then at higher $\sqrt{s}$, the weak contribution to R_{hadron} can be measured in principle. The possibility of confronting theory with experiment depends on being able to measure R_{hadron} to within a few percent, given the present range of $\sqrt{s}$ available ($5 < \sqrt{s} < 45$ GeV). The difficulty of such measurements is

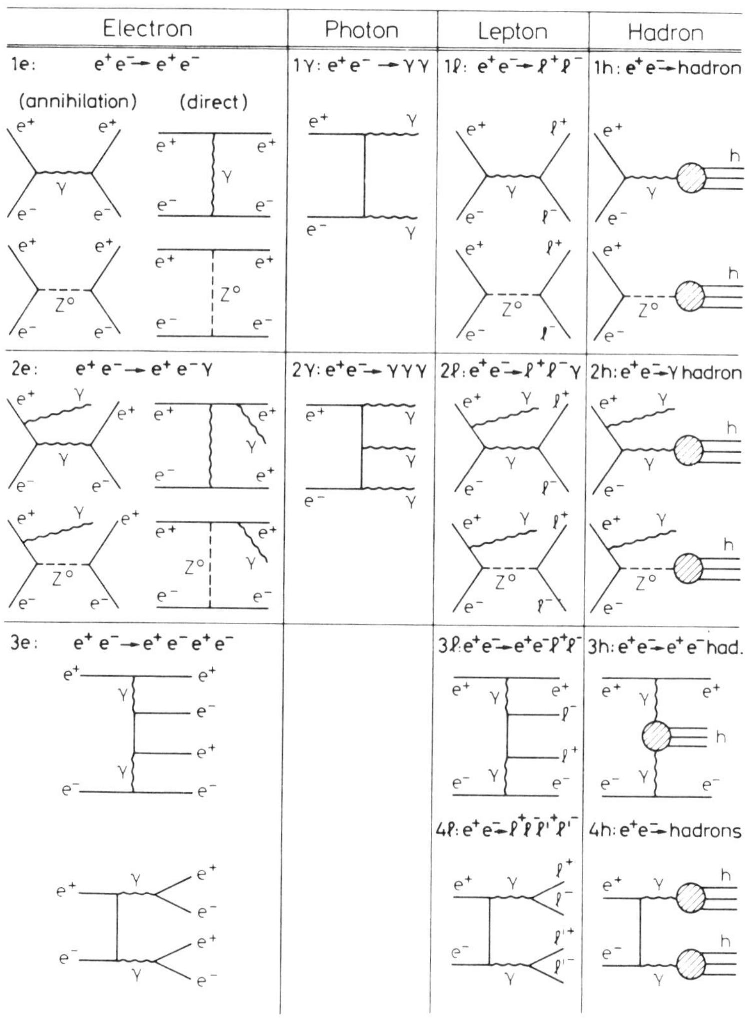

FIGURE 2. The various types of interactions seen from an e^+e^- initial state. Many of these reactions are serious backgrounds to R_{hadron} (the upper right cubby). See WU, S.L. 1984. *Phys. Reports* **107.**

partially illustrated in FIGURE 2, where a number of e^+e^- reactions are shown. Many of these reactions pose serious backgrounds to the process to be measured (in the cubby in the upper right of the figure). In addition, cosmic rays, beam gas interactions, and

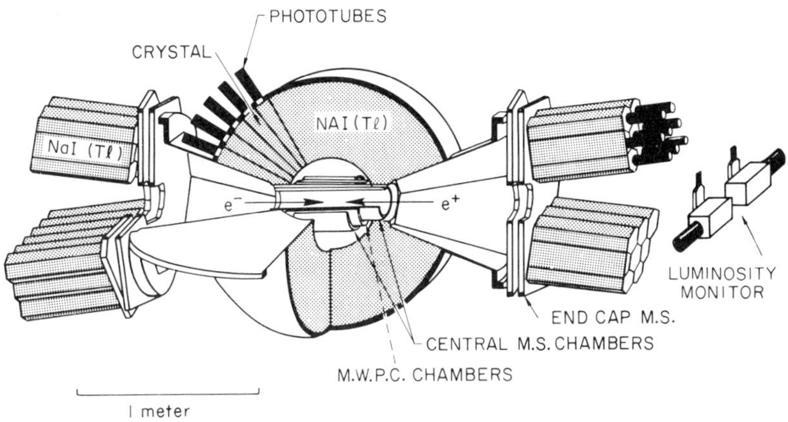

FIGURE 3. The generic R_{hadron} detector as illustrated by the Crystal Ball detector.

other machine-generated backgrounds can destroy the accuracy of an R_{hadron} measurement if not treated properly. Finally, the measurement of the integrated luminosity, typically using Bhabha scattering, is an important element in the accuracy of the measurement.

In order to best measure R_{hadron}, certain detector features must be present, and an example of the generic R_{hadron} detector is shown in FIGURE 3. Illustrated in the figure are the requirements of large solid angle (>90% of 4π), good hadron and electromagnetic calorimetry (over the entire solid angle), charged particle tracking (over most of the solid angle), and at least two independent measurements of the luminosity (large and small angle measurements, the latter with an independent apparatus), all needed for a proper R_{hadron} detector.

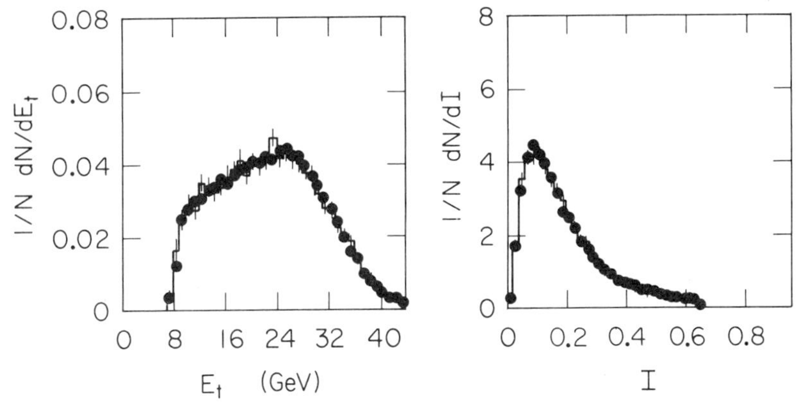

FIGURE 4. Comparison, in the MAC detector, of data (solid circles) and Monte Carlo (histogram) for distributions of transverse energy, E_t, and energy asymmetry, I.

FIGURE 4 illustrates the importance of a large solid angle. In this example from the MAC collaboration,[6] distributions in two (of a number of) variables used to extract the hadronic events are shown as obtained from data and Monte Carlo. The agreement is excellent, giving confidence that the hadronic event extraction efficiency is well understood. A large acceptance of the detector allows for a large hadronic efficiency and the ability to observe most of the event. Under such conditions, it is unlikely that

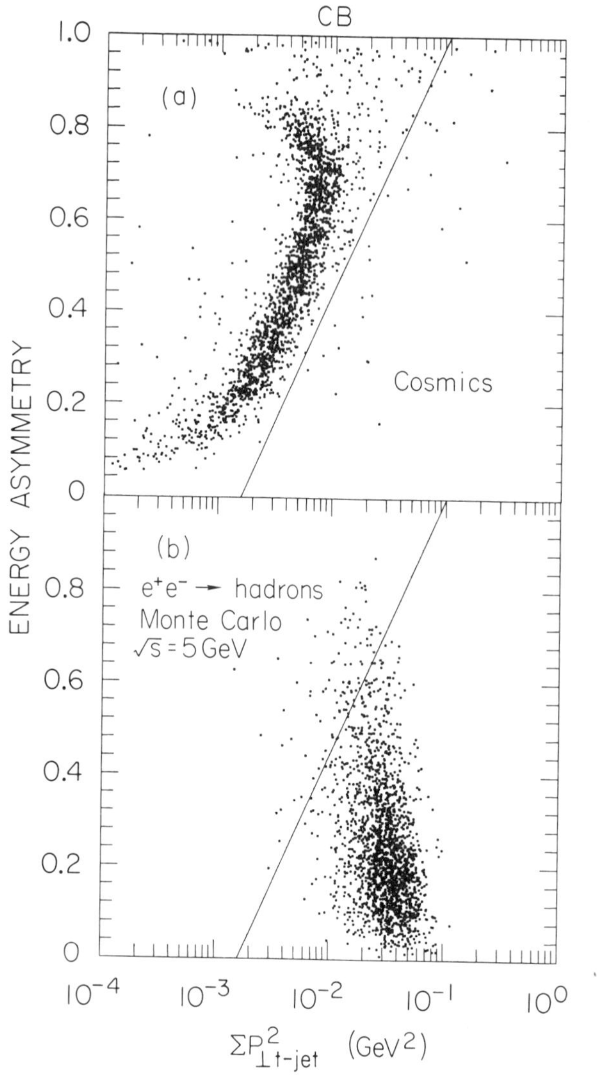

FIGURE 5. Comparison, in the Crystal Ball detector, of energy asymmetry for (a) cosmic rays and (b) Monte Carlo hadrons at $\sqrt{s} = 5$ GeV.

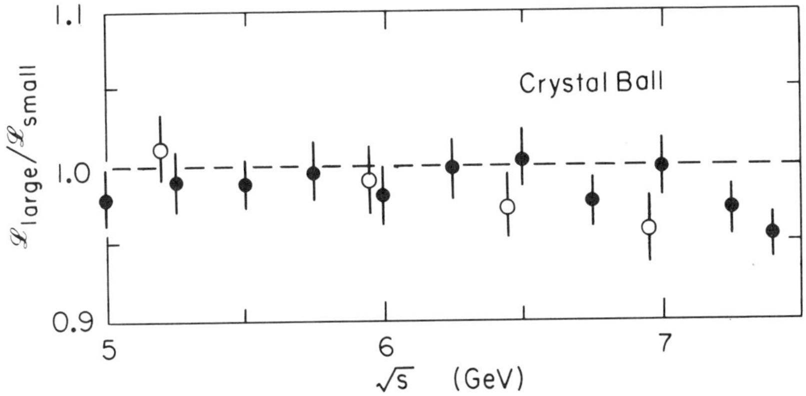

FIGURE 6. The ratio of large angle luminosity measurements to small angle in the Crystal Ball detector. The open circles are 1980 data and the closed circles are 1981 data.

the experimenters will be mislead by errant Monte Carlo calculations and even liberal error estimates on Monte Carlo parameters yield only a small contribution to the overall systematic error. FIGURE 5 illustrates the advantage of good calorimetry. This example from the Crystal Ball collaboration[10] demonstrates the ease of separating certain types of background. Using simple calorimetric variables obtained from only the energies and angles from each of the approximately 700 NaI (T1) elements of the detector, an almost totally complete separation of cosmic ray events (real data) and hadronic events (Monte Carlo) is achieved. The need for independent measurements of the luminosity is shown in FIGURE 6, also using results from the Crystal Ball. In observing the differences in the two measurements, one can with confidence estimate the systematic errors (overall and point-to-point) coming from the luminosity; in the case of the Crystal Ball, the systematic errors are $\pm 2.7\%$ overall, with an additional contribution from point-to-point.[10]

FIGURE 7 shows selected R_{hadron} values from experiments that have systematic errors less than 7%. The statistical errors are added in quadrature, with the quoted point-to-point systematic errors yielding the error bars shown in the figure. The rest of the (overall) systematic error of each experiment is given in the figure. Note that MAC reports a $\pm 2.3\%$ systematic error on their R_{hadron} value measured at $\sqrt{s}$ = GeV; this is the smallest error yet to be reported. FIGURE 8 repeats FIGURE 7 with the prediction of the simple parton model overlaid (electroweak effects are not included). The impression one obtained from FIGURE 8 is one of an increase in R_{hadron} at $b\bar{b}$ threshold and a general excess over the simple parton model over the entire range of $\sqrt{s}$ shown. These qualitative impressions are in agreement with the expectations of QCD; however, as we shall see, quantitative comparisons are not particularly supportive of QCD.

FIGURE 9 shows results to higher values of $\sqrt{s}$ for the Mark-J[11] and TASSO[12] collaborations. Only beyond $\sqrt{s}$ of 40 GeV may one hope to see an effect from the purely weak contribution to R_{hadron}; however, the errors are still too large for a useful determination of the Weinberg angle. Also, the overall systematic uncertainty on R_{hadron} of about $\pm 5\%$ for these experiments precludes the precise determination of α_s needed to obtain the QCD baseline from which the electroweak contribution can be

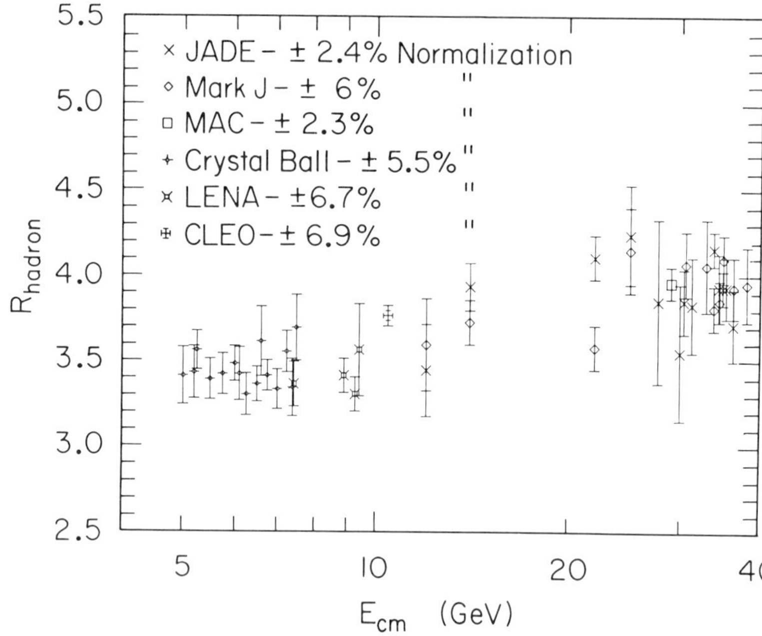

FIGURE 7. R_{hadron} for selected measurements with the overall systematic error of ±7%. The references to the various measurements are given in the text, except for LENA.[49]

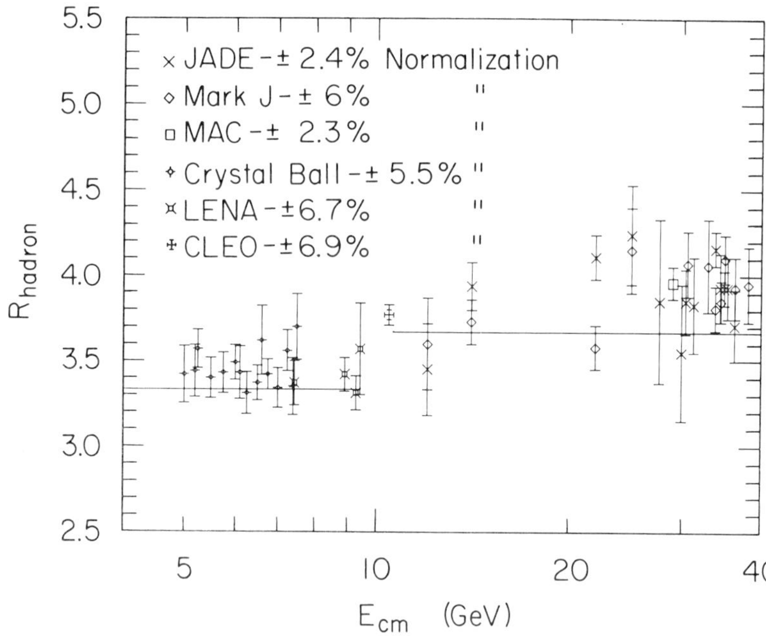

FIGURE 8. Same as for FIGURE 7, except the lines at $R_{hadron} = 3.333$ and 3.667 (from the simple parton model) shown for reference.

deduced at the high energies. FIGURE 10 demonstrates another difficulty; namely, at values of the Weinberg angle close to the presently accepted value of 0.23, the electroweak contribution to R_{hadron} (dominantly an interference term) has a zero. This makes the electroweak contribution even more difficult to detect in this energy range.

Of primary interest is how the value of R_{hadron} versus $\sqrt{s}$ quantitatively compares with the predictions of QCD. In order to attempt such a comparison for $\sqrt{s}$ below 40 GeV, a reliable determination of α_s at low energy is needed. Recently, the use of $\Upsilon(1S)$

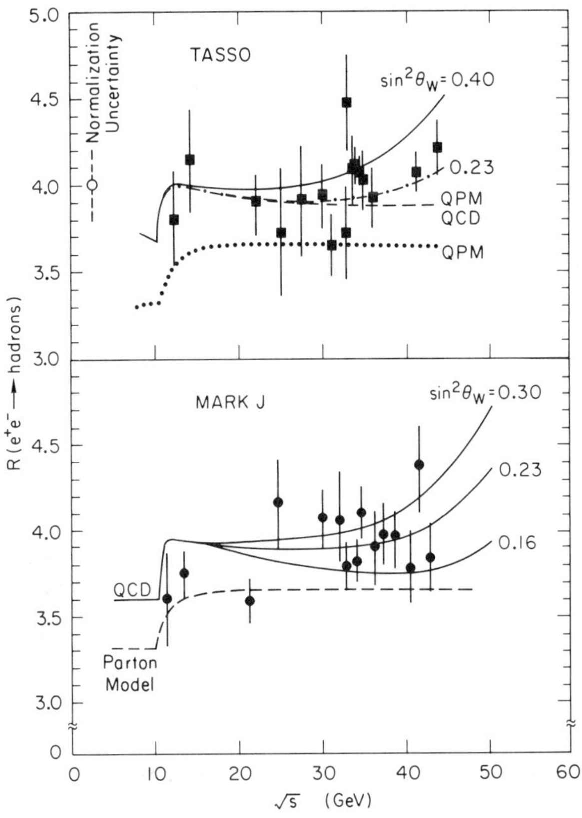

FIGURE 9. R_{hadron} from TASSO and Mark-J compared to full electroweak theory and ranges of $\sin \theta_W$.

decays has been considered to be reliable for such purposes. In particular, the calculation of Mackenzie and Lepage[13] that determines α_s from a simple ratio of well-measured $\Upsilon(1S)$ decays is particularly appealing:

$$\alpha_s^3(0.48 M_\Upsilon) = \frac{\Gamma_{had}(\Upsilon)}{\Gamma_{\mu\mu}(\Upsilon)} \alpha_{QED}^2 \frac{81\pi Q_b^2}{10(\pi^2 - 9)}, \tag{4}$$

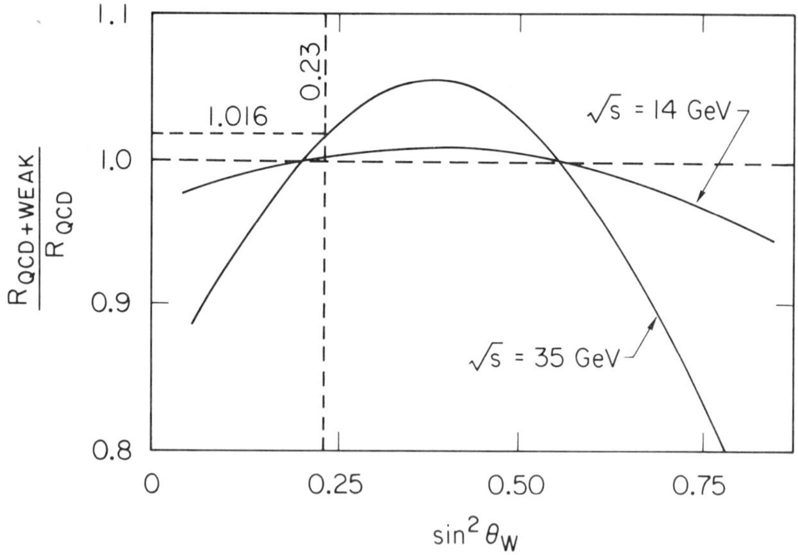

FIGURE 10. $R_{(QCD+WEAK)}/R_{QCD}$ versus $\sin^2\theta_W$ for $\sqrt{s} = 14$ and 35 GeV.

with α_s being defined to second order in the $\overline{MS}$ renormalization scheme by

$$\alpha_s(\sqrt{s}) \equiv \frac{12\pi}{(33 - 2n_f)\ln(s/\Lambda^2_{\overline{MS}})}\left[1 - \frac{462\ln\ln(s/\Lambda^2_{\overline{MS}})}{625\ln(s/\Lambda^2_{\overline{MS}})}\right]. \tag{5}$$

Application of equation (**4**) to the $\Upsilon(1S)$ data[14] yields the first value shown in TABLE 1 and $\Lambda_{\overline{MS}} = 118^{+16}_{-15}$ MeV. The other entries in TABLE 1 are obtained from R_{hadron} measurements as indicated in the table. That QCD is not well confirmed in its prediction of a decreasing α_s with increasing $\sqrt{s}$ is evident by examination of the table. Indeed, the MAC experiment alone shows almost a two standard deviation disagreement with the extrapolation of α_s from the $\Upsilon(1S)$ determination. However, it is clear that more data is needed, and with even smaller errors, to reliably check the QCD predictions.

TABLE 1. α_s from Various Sources

$(\sqrt{s})_{effective}$ (GeV)	Theory α_s	Experiment α_s	Comments
4.54	—	0.165 ± 0.005	Υ decay[14]
6	0.155	0.12 ± 0.11	R, (CB[10])
29	0.125	0.23 ± 0.06	R, no electroweak included (MAC[6])
32	0.123	~0.20	R, $\sin^2\theta_W = 0.23$ (JADE[5])
32	0.123	~0.18	R, $\sin^2\theta_W = 0.23$ (TASSO[12])

QUARKONIUM AND QCD

Another check of QCD of a qualitative nature is given by the comparison of the hadronic widths of the ^3P_J states of charmonium with those of bottomonium. FIGURE 11 shows the lowest order diagram in QCD perturbation theory describing the hadronic decay of the $^3P_0$ and $^3P_2$ states; two gluon emission is thought to mediate these decays. The diagram for the hadronic decay of the $^3P_1$ state is somewhat more involved.[15] Equations (6)–(8) give the lowest order QCD predictions for the decays,[15]

$$\Gamma(1^3P_2 \longrightarrow hadrons) = \frac{128}{5}\alpha_s^2 \frac{|\Psi_1'(0)|^2}{M_X^4} \tag{6}$$

$$\Gamma(1^3P_1 \longrightarrow hadrons) = \frac{128}{3\pi}\alpha_s^3 \frac{|\Psi_1'(0)|^2}{M_X^4} \ln\left(\frac{4m_q^2}{4m_q^2 - M_X^2}\right)\frac{m_f}{3} \tag{7}$$

$$\Gamma(1^3P_0 \longrightarrow hadrons) = 96\alpha_s^2 \frac{|\Psi_1'(0)|^2}{M_X^4}. \tag{8}$$

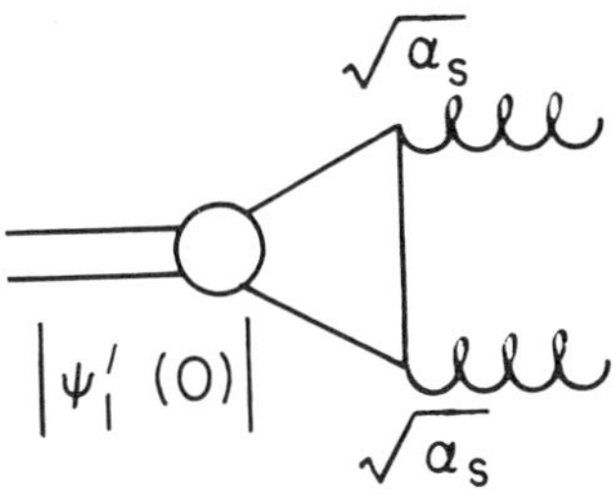

FIGURE 11. Lowest order diagram from QCD yielding $\Gamma_{had}(^3P_0)$ and $\Gamma_{had}(^3P_2)$.

In the above equations, the decay rates are all proportional to the derivative of the P-state wave function at the origin squared. Using a potential, $V(r)$, which yields a 1S–2S mass spacing independent of quark flavor, implies

$$V(r) \sim \ln(r), \tag{9}$$

and this in turn implies

$$|\Psi_1'(0)|^2 \sim M^{5/2}. \tag{10}$$

Taking all factors, we find

$$\frac{\Gamma_b(1^3P_2 \longrightarrow hadrons)}{\Gamma_c(1^3P_2 \longrightarrow hadrons)} \sim \left(\frac{\alpha_s(M_b)}{\alpha_s(M_c)}\right)^2 \frac{M_c^{3/2}}{M_b^{3/2}}. \tag{11}$$

TABLE 2 shows the numbers obtained from the experiments. The last four measurements in the table come from γ cascade decays of the $\Upsilon(2S)$ to the $\Upsilon(1S)$ and involve two assumptions. First, the lowest E_γ state observed is assumed to be the $^3P_2$ state, and the next lowest, the $^3P_1$ state (the spins of these states are yet to be measured). Second, the partial width for the decay via a γ from the ^3P_J state to the $\Upsilon(1S)$ is taken from theory, e.g., the CB uses Moxhay and Rosner.[19] (These are electric

TABLE 2. Experimental Determination of Hadronic Width of 1^3P_J States

State	System	Γ_{had} (MeV)	Comments
$^3P_0$	$c\bar{c}$	17 ± 3.5	CB[16]
$^3P_1$	$c\bar{c}$	<3.8(90%C.L.)	CB[16]
$^3P_2$	$c\bar{c}$	2.9 ± 2	CB[16]
$^3P_1$	$b\bar{b}$	$0.041^{+0.04}_{-0.02}$	CUSB[17]
$^3P_2$	$b\bar{b}$	$0.152^{+0.063}_{-0.050}$	CUSB[17]
$^3P_1$	$b\bar{b}$	$0.079^{+0.029}_{-0.019}$	CB[18]
$^3P_2$	$b\bar{b}$	$0.101^{+0.045}_{-0.028}$	CB[18]

dipole transitions in the theory.) The theory combined with experiment then gives

$$\Gamma_{had}(^3P_J) = \frac{Br^{exp}(\Upsilon' \rightarrow \gamma^3P_J) \times \Gamma_{E1}(^3P_J \rightarrow \gamma\Upsilon)}{[\,Br(\Upsilon' \rightarrow \gamma^3P_J) \times Br(^3P_J \rightarrow \gamma\Upsilon)\,]_{exp}}. \quad (12)$$

One can then compare theory and experiment in ratio for the $c\bar{c}$ and $b\bar{b}$ systems. Theory predicts [equation (**11**)]

$$\frac{\Gamma^{b\bar{b}}_{had}(^3P_2)}{\Gamma^{c\bar{c}}_{had}(^3P_2)} \sim 0.11, \quad (13)$$

while experiment gives 0.035 ± 0.028 for this ratio. Agreement is acceptable within the large errors and the assumptions made.

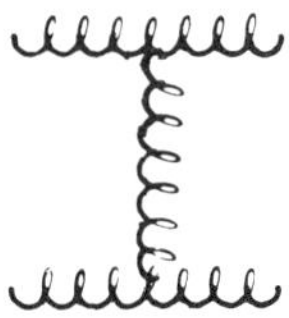

(a)

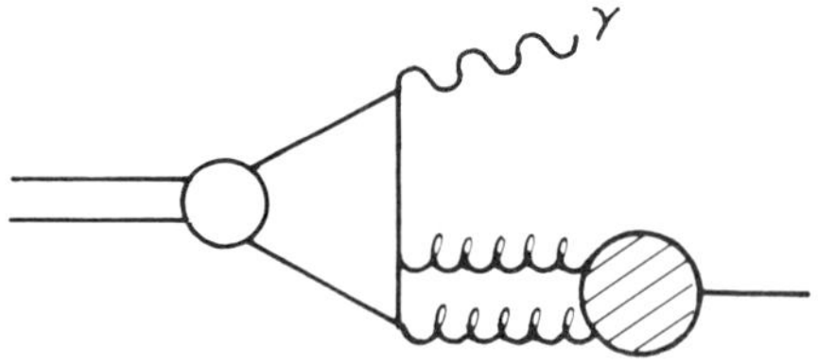

(b)

FIGURE 12. (a) Gluon-gluon interaction via gluon exchange predicted by QCD; (b) Possible gluonic meson production mechanism in quarkonium decays.

GLUONIC MESONS—THE SEARCH FOR THE THREE GLUON VERTEX OF QCD

One of the most distinctive features of QCD, its non-Abelian character, has yet to be verified experimentally. FIGURE 12a shows this feature realized in lowest order. The implication of the existence of such a process in lowest order is physically realized in the prediction by QCD of the existence of an extensive spectrum of colorless, flavorless bound states of two or more gluons. It is expected that the lower mass states of these gluonic mesons are bound states of mostly two gluons and their masses are predicted to lie in the range of 1 to 3 GeV/c^2. In this report, I will review briefly the status of two states that have for some time continued to hold the promise of containing a large gluonic meson component in their wave functions. These mesons are the $\iota(1440)$ and $\theta(1700)$, which have primarily been seen in the decay, $J/\psi \rightarrow \gamma + X$, depicted in FIGURE 12b. (However, the ι may have first been seen in $p\bar{p}$ annihilation at rest.[20]) FIGURE 13 briefly reviews the status of the ι. Part (a) of the figure indicates the level of information available as of 1982. The $K^+K^-\pi^0$ mass spectrum obtained from the decay of about two million J/ψ in the Crystal Ball[21] is shown with and without a $M_{K^+K^-}$ cut at the $\delta(980)$ mass. This data sample was used to first determine the J^P of the ι to be 0^-. The Mark II detector at SPEAR previously had seen this state in J/ψ decays, but was unable to measure its J^P; thus, the state was erroneously identified as the $J^P = 1^+$ meson $E(1420)$.[16,23] Recent results from the Mark III detector based on about 2.7 million J/ψ decays are shown in parts (b) and (c) of the figure; the $K_sK^{\pm}\pi^{\mp}$ and $K^+K^-\pi^0$ mass distributions are shown without a $K\bar{K}$ mass cut. This huge ι signal (and very little background) allows for a definitive J^P determination[24] of 0^-, thus confirming the result obtained by the Crystal Ball. All of the results are summarized in the table at the bottom of the figure. It should be noted that preliminary results from a recent BNL experiment, AGS #771,[25] have confirmed the 1^{++} $E(1420)$ in $p\pi$ and have identified this state in $p\bar{p}$ annihilation in flight; however, the experimenters report that the situation is more complicated for $p\bar{p}$ at rest, which is the same reaction where Baillon *et al.*[20] first reported a 0^- state at the same mass.

Is the ι a gluonic meson? From the point of view of many theorists, it has about the right mass and quantum numbers, and its big radiative decay rate from the J/ψ is also supportive of this interpretation. However, many other theorists disagree and classify the ι as a radially excited $q\bar{q}$ state. Seemingly, a way to settle the argument is to measure the radiative decays of the ι, since it is intuitive that a gluonic meson made from gluons having no charge should have a vanishingly small radiative decay, that is,

$$\frac{Br(\iota \rightarrow \gamma X)}{Br(\iota \rightarrow K\bar{K}\pi)} \ll 1\%. \tag{14}$$

FIGURE 14 shows that this is not likely the case. The Crystal Ball[26] and Mark III[27] have measured the decay, $J/\psi \rightarrow \gamma\gamma\rho^0$, and find a ratio of branching fractions (of what may be an ι-related $\gamma\rho^0$ final state) to $\iota \rightarrow K\bar{K}\pi$ of the order of 1%. The experimental situation is not clear, however, since the enhancement seen in $\gamma\rho^0$ (see figure) is broader than the ι. Evidence that part of this structure is the ι is based on a preliminary spin analysis (see reference 24 for a review of the present situation). The theoretical explanation for the large radiative decay of the ι, involving a gluonic meson, is mixing of the pseudoscalar mesons with a 0^- gluonic meson. The models in which the ι is mainly a $q\bar{q}$ state (radial excitation) easily accommodate a "large" radiative width; they would have difficulty explaining a "small" radiative width of the ι. The final status of the ι, though, is yet to be settled.

FIGURES 15 and 16 briefly review the status of the $\theta(1700)$. FIGURE 15 shows the signals seen in $J/\psi \rightarrow \gamma X$, where in FIGURE 15a, from the Crystal Ball,[28] X is $\eta\eta$, and in 16b, from the Mark II,[22] X is K^+K^-. Clear signals are seen in both detectors; however, in the Crystal Ball result, the statistics are limited, while for the Mark II, there is a large background. Also shown in FIGURE 15c and d is the $\pi\pi$ mass spectrum in both detectors; no clear signal is seen. Both detectors, with varying confidence, determined

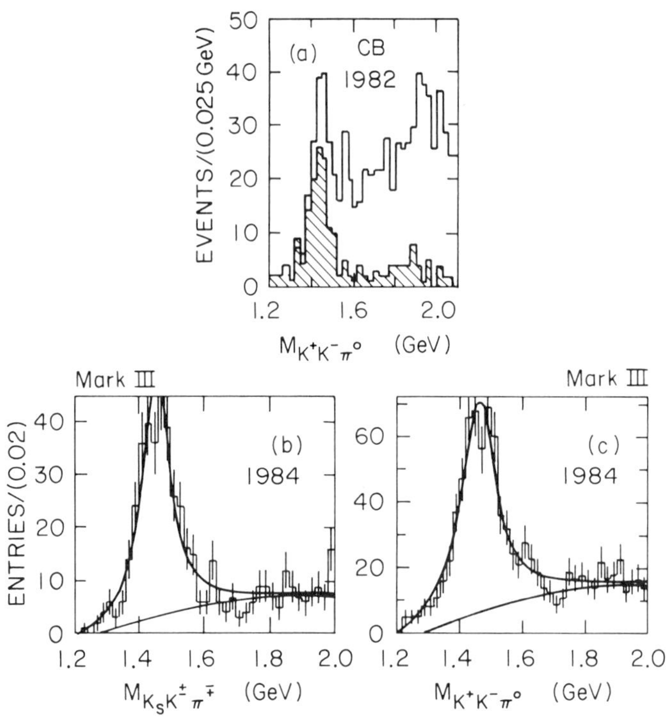

$\iota(1440)$ present experimental status.

Collaboration	Decay Mode	Mass (MeV)	Width (MeV)	$B(J/\psi \rightarrow \gamma\iota) \times B(\iota \rightarrow K\bar{K}\pi)$	J^{P}
Mark II	$K_S K^\pm \pi^\mp$	1440^{+10}_{-15}	50^{+30}_{-20}	$(4.3 \pm 1.7) \times 10^{-3}$	
Crystal Ball	$K^+K^-\pi^o$	1440^{+20}_{-15}	55^{+20}_{-30}	$(4.0 \pm 0.7 \pm 1.0) \times 10^{-3}$	0^-
Mark III	$K_S K^\pm \pi^\mp$	1456 ± 10	95 ± 7	$(5.0 \pm 0.5 \pm 0.7) \times 10^{-3}$	0^-
	$K^+K^-\pi^o$	1460 ± 10	103 ± 10		
DM2	$K_S K^\pm \pi^\mp$	1474 ± 15	76 ± 16		

FIGURE 13. Experimental status of ι. (a) Crystal Ball results; (b) and (c) Mark III results, with the table summarizing the status.

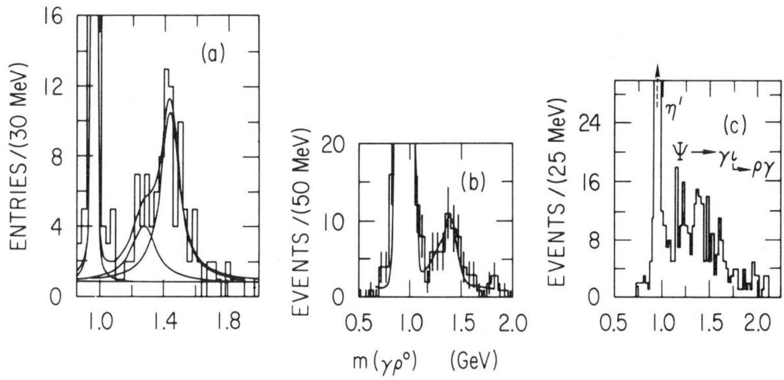

$J/\psi \rightarrow \gamma X(1400)$, $X(1400) \rightarrow \gamma\rho^o$ branching ratio. The mass and width of the 1400 MeV structure were respectively fixed at 1440 MeV and 55 MeV in the Crystal Ball fit.

Collaboration	$B(J/\psi \rightarrow \gamma X(1400))$ $\times B(X(1400) \rightarrow \gamma\rho^o)$
Crystal Ball	$(0.88 \pm 0.28 \pm 0.15) \times 10^{-4}$
Mark III	$(1.1 \pm 0.24 \pm 0.25) \times 10^{-4}$

FIGURE 14. Results on $J/\psi \rightarrow \gamma\gamma\rho$. (a) Mark III, (b) Crystal Ball, and (c) DM2, with the table summarizing the extraction of $\gamma\rho$ contribution at ι mass. The results in the table might correspond to an ι contribution (based on a spin analysis).

the J^P of the θ to be 2^+. These results are reviewed in the table that makes up the top of FIGURE 16, along with recent results from the Mark III detector[29] and the DM2 detector operating at DCI.[30] Parts (a) and (b) of the figure illustrate the data on which the Mark III tabular results are based. The data have clearly improved since 1982; good statistical information with little background is seen in the K^+K^- mass plot and a signal is evident in the $\pi\pi$ mass spectrum at the θ mass. Using this data, the Mark III was able to assign a J^P of 2^+ to the θ with high confidence, thus confirming the somewhat tentative assignments of the previous experiments.

Is the θ a gluonic meson? In my opinion, it has always been the best candidate seen in J/ψ radiative decays due to its quantum numbers and mass.[31] The case would be particularly strong if the $\gamma\rho\rho$ final state first seen by the Mark II from the J/ψ with an enhancement at $M_{\rho\rho}$ about 1700 MeV[22] had J^P of 2^+. The Mark III collaboration has recently presented evidence that this enhancement, which also has a large radiative branching ratio from the J/ψ, was dominantly $J^P = 0^-$. However, the observation of a $\pi\pi$ final state of the θ has helped the gluonic meson interpretation. Thus, as in the case of the ι, the situation remains unresolved.

Of course, the Υ system has hardly been explored when it comes to the question of gluonic mesons. The predicted production rates are quite small and the existing samples of data are about a factor of ten smaller than at the J/ψ; also, the detectors operating today are barely a match for this difficult region. However, as FIGURE 17 and the previous four figures illustrate, the progress at the J/ψ has been rapid and the surprises many; thus, the Υ may have even more to offer.

NEW PARTICLE SEARCHES

The Search for Right-Handed Currents

Beautiful precision experiments have been made on the muon decay, $\mu^+ \to \nu\bar{\nu}e^+$, particularly to search for deviations from the S-M and right-handed charged currents.

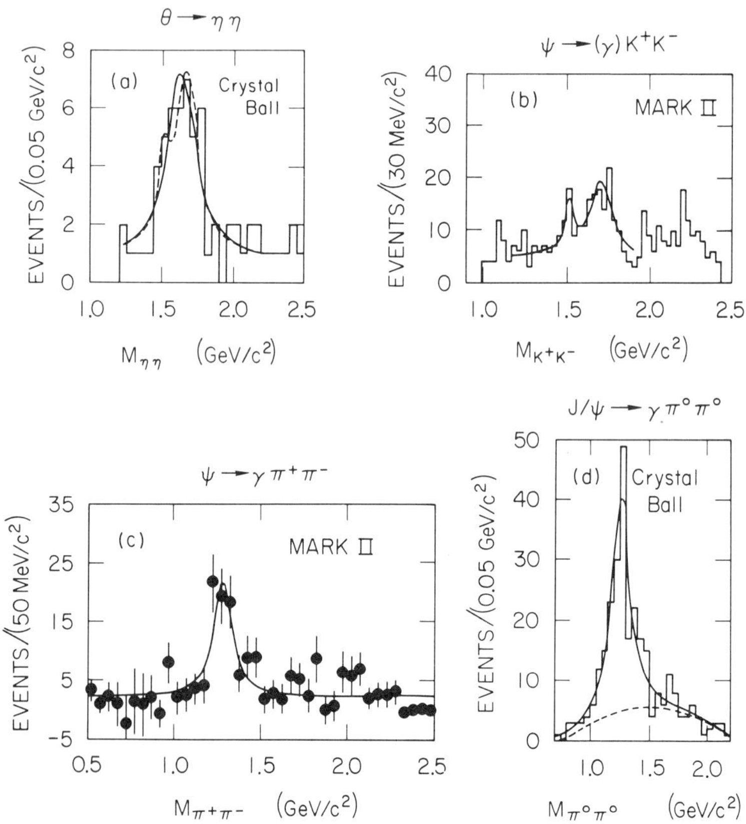

FIGURE 15. Results from 1982 on the $\theta(1700)$. (a) $\eta\eta$ final state, (b) K^+K^- final state, (c) no signal seen in $\pi^+\pi^-$ final state, and (d) no signal seen in $\pi^0\pi^0$ final state.

$\theta(1690)$ present experimental status. Upper limits are given at the 90% confidence level.

Collaboration	Decay Mode	Mass (MeV)	Width (MeV)	$B(\psi \to \gamma\theta) \times B(\theta \to X)$	J^P
Crystal Ball	$\eta\eta$	1670 ± 50	160 ± 80	$(3.8 \pm 1.6) \times 10^{-4}$	2^+ (95% CL)
	$\eta\eta'$			$< 2.1 \times 10^{-4}$	
	$\pi\pi$			$< 6 \times 10^{-4}$	
Mark III	K^+K^-	1720 ± 10	130 ± 20	$(4.8 \pm 0.6 \pm 0.9) \times 10^{-4}$	2^+ (99.9% CL)
	$\pi^+\pi^-$	1713 ± 15	—	$(1.6 \pm 0.4 \pm 0.3) \times 10^{-4}$	
	$\rho^o\rho^o$			$< 2.0 \times 10^{-4}$	
	$K\bar{K}\pi$			$< 2.5 \times 10^{-4}$	
	$K^+K^-\pi^+\pi^-$			$< 1.0 \times 10^{-4}$	
Mark II	K^+K^-	1708 ± 30	156 ± 60	$(6.0 \pm 0.9 \pm 2.5) \times 10^{-4}$	2^+ (80% CL)
	$\pi\pi$			$< 3.2 \times 10^{-4}$	
DM2	K^+K^-	1718 ± 29	233 ± 75		

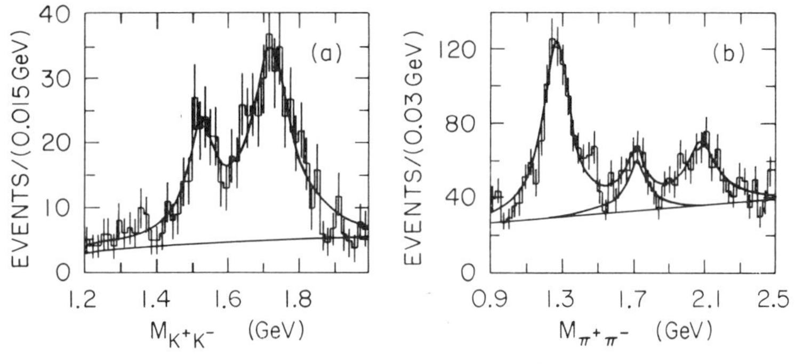

FIGURE 16. Present experimental status of $\theta(1700)$. (a) Mark III K^+K^- final state; (b) Mark III $\pi^+\pi^-$ final state with signal evident.

The experiments were performed at the TRIUMF meson factory by two somewhat overlapping groups—a Berkeley-SIN-TRIUMF group (B-S-T)[32] and a Berkeley-Northwestern-TRIUMF group (B-N-S).[33] What I am presenting in this report has been presented in more detail in reference 34.

The muon differential decay rate, averaged over the e^+ polarization and to the lowest nonvanishing order in M_e/M_μ, is given by

$$\frac{d^2\Gamma}{x^2 dx d \cos\Theta} \propto (3 - 2x) + (4\rho/3 - 1)(4x - 3) + 12\frac{M_{e^+}}{M_{\mu^+}x}(1 - x)\eta$$
$$-[(2x - 1) + (4\delta/3 - 1)(4x - 3)]\,\xi P_\mu \cos\theta, \quad \textbf{(15)}$$

where

$$x = \frac{E_{e^+}}{W},\ W = \frac{M_\mu^2 + M_e^2}{2M_\mu},$$

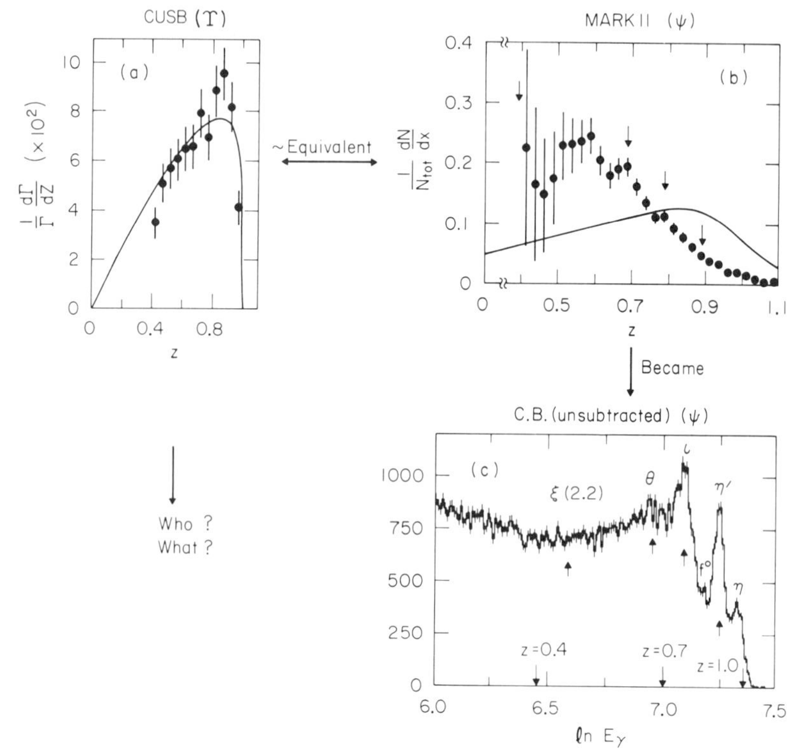

FIGURE 17. (a) The CUSB inclusive prompt photon spectrum from the Υ(1S), (b) the Mark II inclusive prompt photon energy spectrum from the J/ψ, and (c) the Crystal Ball inclusive (unsubtracted) photon energy spectrum from the J/ψ.

$\pi - \theta$ is the e^+ angle relative to the spin direction of the decaying μ^+, P_μ is the polarization of the μ^+, and ρ, η, ξ, and δ are the usual muon decay parameters[35] as calculated in the Standard Model.

The basic idea of the experiment I will discuss here is shown in FIGURE 18. The highly polarized muons are stopped in a magnetic field and allowed to decay. In one set of runs, the field is 1.1 T parallel to the muon polarization; in another set of runs, the field is 70 or 120 gauss perpendicular to the polarization direction. In the case shown in

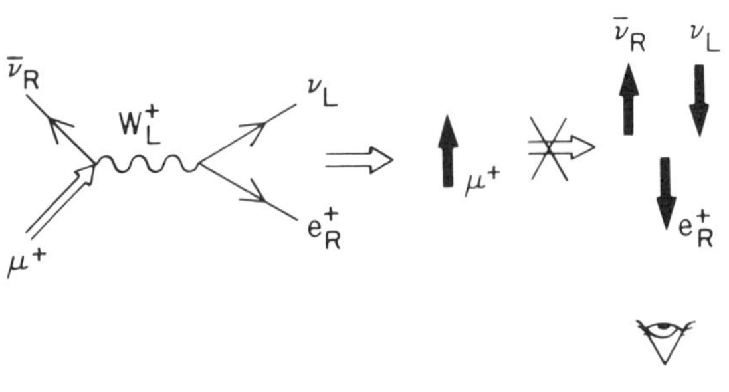

FIGURE 18. Muon decay via (V − A). Angular momentum conservation forces the decay rate to zero for positrons decaying backward relative to the muon spin direction. The presence of a (V + A) component allows such a backward decay.

FIGURE 18, the field is parallel to the polarization direction and the spectrometer is observing backward e^+ near $x = 1$. This decay cannot occur due to angular momentum conservation unless right-handed charged currents are contributing to the decay amplitude.

The experiments measured a number of parameters (sometimes in two ways) in order to obtain unprecedented limits on the mass of right-handed W's, under the assumption that a right-handed neutrino would be massless. FIGURE 19 shows some of the results of the experiments. The data presented in part (a) of the figure were used by the B-S-T collaboration to measure η. In the S-M, with $\eta = 0$, the preliminary results of the measurement[32] yield

$$\eta_{exp} = -0.087 \pm 0.097. \tag{16}$$

In the figure, the S-M, with η fitted, is compared to the measurement. The small x end of the spectrum is particularly useful in this measurement and the B-S-T group

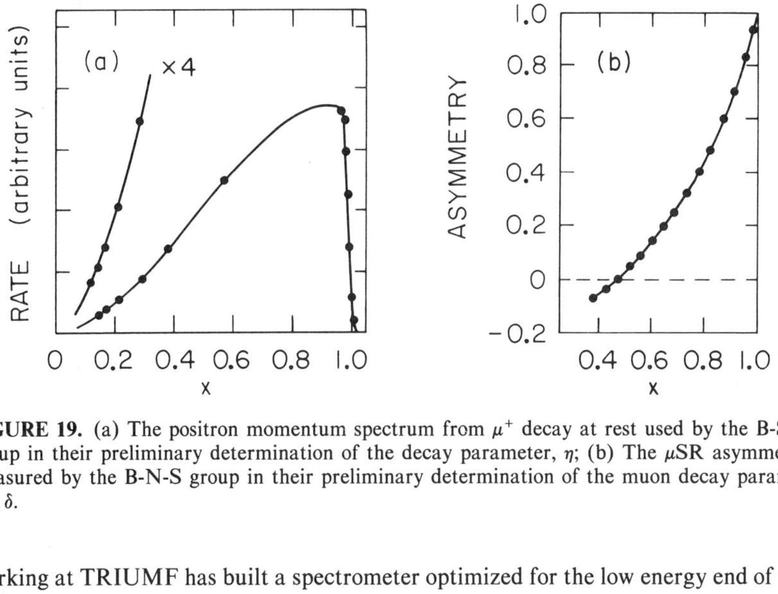

FIGURE 19. (a) The positron momentum spectrum from μ^+ decay at rest used by the B-S-T group in their preliminary determination of the decay parameter, η; (b) The μSR asymmetry measured by the B-N-S group in their preliminary determination of the muon decay parameter, δ.

working at TRIUMF has built a spectrometer optimized for the low energy end of the positron spectrum.

FIGURE 19b shows the data (and fit) used to extract a preliminary value for δ by the B-S-T collaboration. The S-M expects $\delta = 3/4$, and the fit, leaving ξP_μ free, but fixing ρ to the S-M value of $3/4$, yields the "very preliminary" value of

$$\delta = 0.748 \pm 0.004 \pm 0.003. \tag{17}$$

This is an improvement of almost a factor of two on the precision of the present world average.

FIGURE 20 shows the measurements of the positron endpoint spectrum (x near 1.0) by the B-N-S collaboration. Parts (a) and (b) of the figure compare the spectra in x near the positron endpoint for $B_\perp$ equal 70 or 100 gauss and $B_\parallel$ at 1.1 T, respectively. The effectively unpolarized spectrum in part (a) shows a characteristic sharp edge,

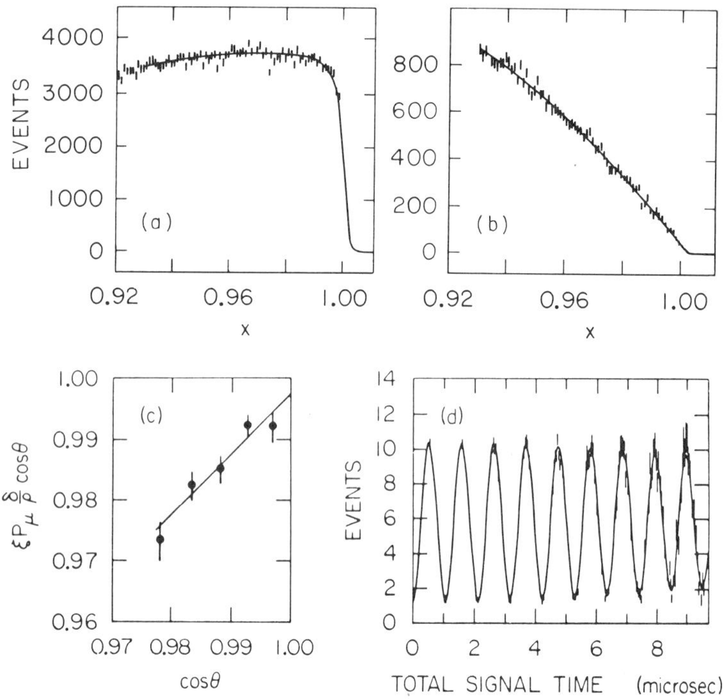

FIGURE 20. Distributions in reduced positron momentum, x, with the spin precessed due to (a) the presence of a 70 or 120 gauss transverse field and (b) a 1.1 T parallel field; (c) Fitted $(\xi P_\mu \delta/\rho)$ $\cos\theta$ for data like those in (b), divided into bins of $\cos\theta$; (d) The decay time spectrum for data with a transverse field [as in (a)], with the muon lifetime factored out.

which is used to calibrate the spectrometer. That edge almost vanishes completely when, as shown in part (b), the strong longitudinal field holds the muon spin nearly antiparallel to the positron direction. (There is a very small remnant edge left due to the finite acceptance of the spectrometer.[34]) In order to minimize the effect of the finite acceptance of the spectrometer, data like those of FIGURE 20b (but divided into bins of $\cos\theta$) are plotted in FIGURE 20c as $\cos\theta$ versus $(\xi P_\mu \delta/\rho)\cos\theta$. The intercept at $\cos\theta =$ 1 of the best fit line in the figure yields the limit[34] (which as we shall see is related to the limit on right-handed currents),

$$\xi P_\mu \delta/\rho > 0.9959 \text{ (90\% C.L.)}. \quad \textbf{(18)}$$

When $B_\perp$ is applied as for the data of FIGURE 19a, the muon spin rotates and $\langle P_\mu \rangle = 0$. (V − A) induces a maximal muon spin rotation (μSR) asymmetry in the e^+ decay rate. When the muon lifetime is factored out of the time spectrum, the sinusoidal modulation in FIGURE 20d is obtained. The amplitude of the pure (V − A) contribution would be reduced by a (V + A) admixture. This technique offers a measurement of $(\xi P_\mu \delta/\rho)$ that has rather different systematic biases than the previously described measurements.[34] Fitting the data of FIGURE 20d, the B-N-S collaboration obtains[34]

$$\xi P_\mu \delta/\rho > 0.9948 \text{ (90\% C.L.)}. \tag{19}$$

For $U(1) \times SU(2)_L \times SU(2)_R$, the W mass eigenstates, W_1 and W_2, need not be the same as the gauge bosons, W_L and W_R, and thus, there can be a mixing matrix of

$$\begin{bmatrix} W_1 \\ W_2 \end{bmatrix} = \begin{bmatrix} \cos\zeta & \sin\zeta \\ -\sin\zeta & \cos\zeta \end{bmatrix} \begin{bmatrix} W_L \\ W_R \end{bmatrix}. \tag{20}$$

If one defines a parameter, α, such that

$$\alpha \equiv M^2_{W_1}/M^2_{W_2}, \tag{21}$$

then ζ and α are related to the experimentally measured quantities by

$$\frac{\xi P_\mu \delta}{\rho} \simeq 1 - 2(2\alpha^2 + 2\alpha\zeta + \zeta^2) \tag{22}$$

with $M_{\nu_R} = 0$. One can then plot limits for ζ versus α for the B-N-S experiment under the assumption that $M_{\nu_R} = 0$, and this is done in FIGURE 21. Also shown in the figure is the only significant experimental constraint that does not depend on assumptions about the right-handed neutrino mass from CDHS.[36] The limit on ζ comes from the y distributions in νN and $\bar{\nu} N$ scattering where no right-handed neutrino need be produced. For a review of limits with other assumptions on the right-handed neutrino mass and from a number of experiments, see reference 34. Under somewhat restrictive assumptions, we thus find W_R mass limits of over 400 GeV from this B-N-S very low energy experiment. I find this a beautiful result.

The Search for Narrow States in the Radiative Decays of the J/ψ *and* Υ

First, I will discuss results from the Mark II detector at SPEAR on a narrow state at a mass of 2.2 GeV/c^2 in J/ψ radiative decays. This is really a brief status report on this state, called the ξ, since it was first reported in conferences in the summer of 1983.[37] For a complete status report, see the talk of R. Partridge in these proceedings.[38]

FIGURE 22 shows the signal as seen in 1983 data and the sum of 1983 and 1984 data. In the 1983 sample of about 1.8 million J/ψ decays, the signal is fitted as a 5 standard deviation (SD) effect; however, the sum of the 1982 and 1983 data, 2.6 millions J/ψ decays, shows a lower significance of 4.6 SD. The best estimate of mass and width are[37]

$$m_\xi = (2.218 \pm 0.003 \pm 0.010) \text{ GeV}/c^2$$

and

$$\Gamma \leq 40 \text{ MeV (95\% C.L.)}. \tag{23}$$

The state is best seen in $J/\psi \rightarrow \gamma K^+K^-$ and the product branching ratio is

$$B[J/\psi \rightarrow \gamma\xi(2.2)] \times B(\xi \rightarrow K^+K^-) = (5.7 \pm 1.9 \pm 1.4) \times 10^{-5}. \tag{24}$$

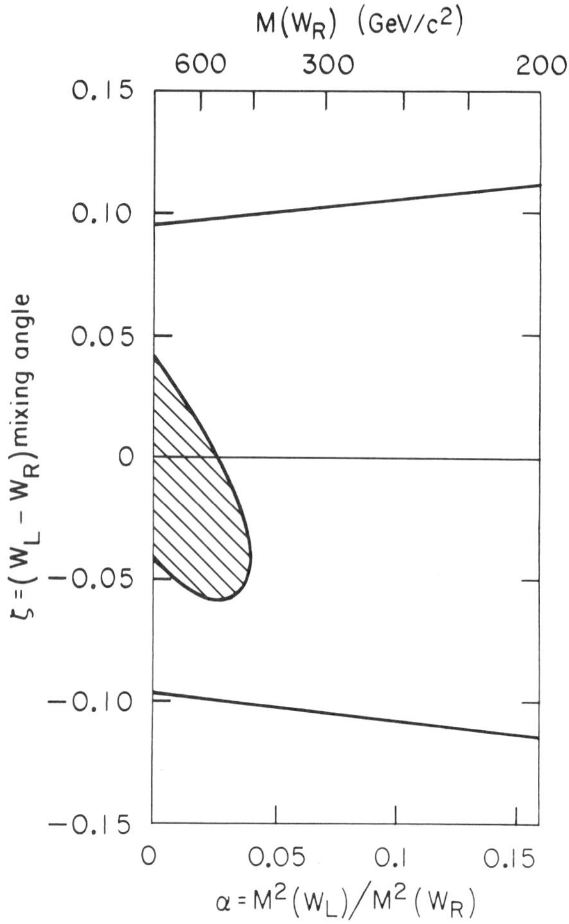

FIGURE 21. Experimental 90% C.L. limits in the $\alpha - \zeta$ plane for the B-N-S experiment, shaded area, and CDHS, region interior to bold lines.

The J^P of the state is not measured with the present data, but the most likely hypotheses are $J = 0$, 2, or 4.[37] FIGURE 23 shows the data taken in 1982 and 1983 separately. In parts (a) and (b) of the figure, the masses are allowed to be different in the fits to the two data samples and the fits show structures at different masses. If the mass is fixed to that seen in the 1983 data sample (see FIGURE 23d), the fit of FIGURE 23c results. In order to confirm the ξ, the Mark III collaboration hopes to obtain much more J/ψ data in the near future. It should be noted that DM2 at DCI has only reported upper limits on the ξ that are in mild conflict with the Mark III results.[39]

The possible existence of a narrow resonance in this mass range in J/ψ radiative decays has stimulated speculation that the ξ might be a non-S-M Higgs boson,[40] but at present more prosaic explanations[41] seem more likely.

Considerably more excitement was caused last summer by the announcement from the Crystal Ball collaboration of evidence for a narrow state in Υ radiative decays.[42]

The large mass of the proposed state, called ζ, at 8.3 GeV/c^2, suggested to a number of theorists the possibility that the state was a non-S-M Higgs candidate.[43] The state was observed by the Crystal Ball in two independent final state configurations—one of higher multiplicity and hadronic character and one of lower multiplicity, possibly $\tau\bar{\tau}$.

Given the numerous reports on the ζ over the past six months, I will review only briefly the results presented at the summer conferences[42] as an orientation to presenting recent results from a run of about 200k Υ events taken by the Crystal Ball at DESY last fall (1984) and results from the CUSB detector also from a fall run at CESR. These latest results are preliminary.

The first results came from a sample of about 100k Υ(1S) decays (10.4 pb^{-1}). FIGURE 24 shows the signal in the high multiplicity final state. The fit to the data, shown in the figure, yields a 4.2 SD effect at the ζ mass. FIGURE 25 shows the results from the low multiplicity final state. The fit to the data, shown in the figure, finds a 3.3 SD effect at a mass that is 10 MeV away from the fit in the high multiplicity case,

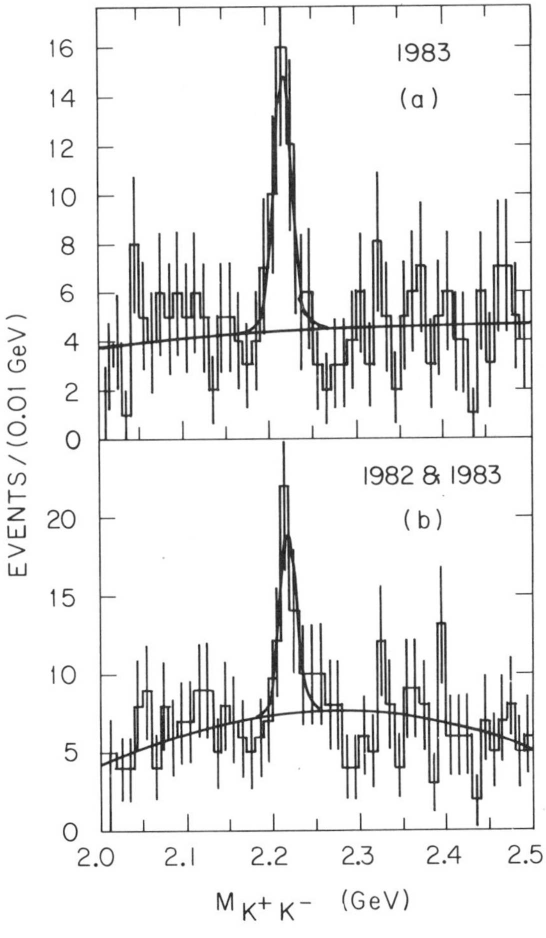

FIGURE 22. Mark III results for $J/\psi \to \gamma\xi$, $\xi \to K^+K^-$. (a) 1982 data alone; (b) 1983 and 1982 data combined.

which is well within the statistical error on the mass measurement. Taken as independent, the two final state configurations yield a signal of over 5 SD. The best estimate of the mass and width of the state is[42]

$$M_\zeta = (8.322 \pm 0.008 \pm 0.024)\ \mathrm{GeV}/c^2$$

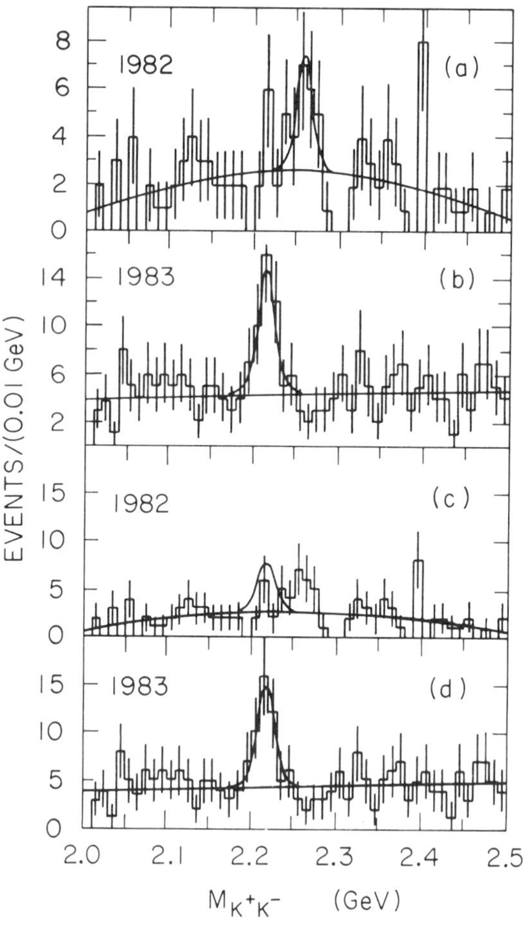

FIGURE 23. Mark III $\xi(2.2)$ results continued. (a) 1982 and (b) 1983, with the results fitted so as to allow the mass of the resonance to vary in the fit; (c) 1982 and (d) 1983, with the results fitted so as to keep the resonance mass fixed at the 1983 value.

and

$$\Gamma < 80\ \mathrm{MeV}\ (90\%\ \mathrm{C.L.}). \qquad \mathbf{(25)}$$

The branching ratio, $\Upsilon \rightarrow \gamma\zeta$, is somewhat model dependent since the manner in which the ζ decays can affect the efficiency for finding photons in the final state. This is reflected in the product branching ratio into hadrons for the ζ,

$$B[\Upsilon(1S) \rightarrow \gamma\zeta] \times B[\zeta \rightarrow \mathit{hadrons}] = (0.47 \pm 0.11 \pm 0.26)\%, \qquad \mathbf{(26)}$$

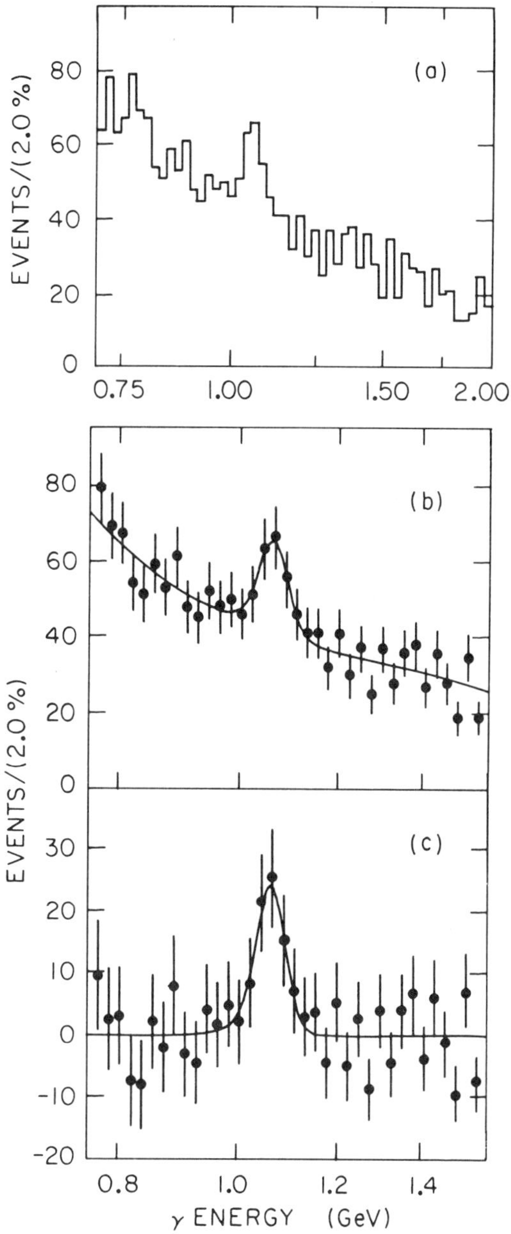

FIGURE 24. Crystal Ball results for $\Upsilon(1S) \rightarrow \gamma\zeta(8.3)$, $\zeta \rightarrow$ *hadrons*. (a) The inclusive photon spectrum after all cuts, including physics-oriented cuts; (b) The region of the ζ peak of (a), with fit shown as a solid line; (c) The same as (b) with the fitted background subtracted.

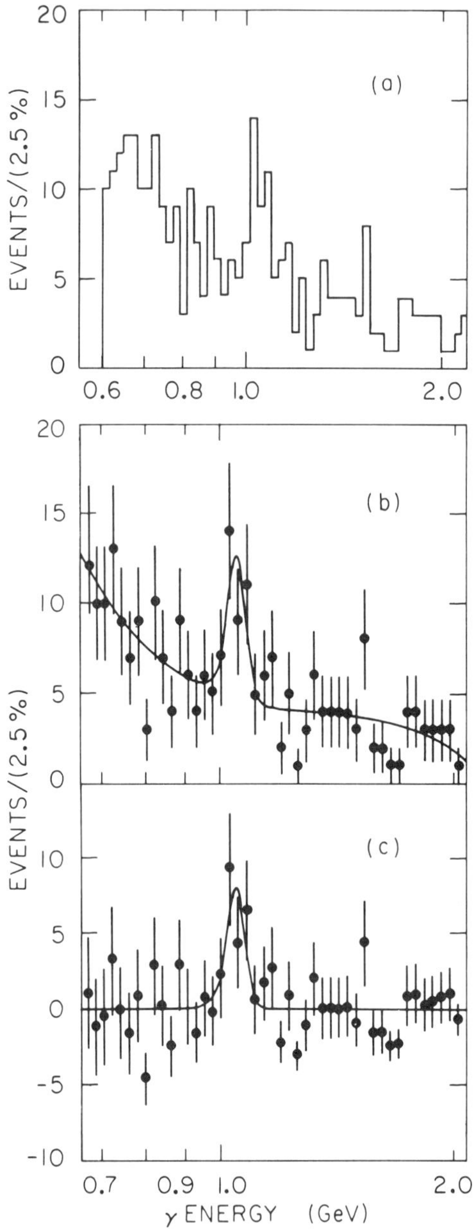

FIGURE 25. Crystal Ball results for $\Upsilon(1S) \rightarrow \gamma\zeta(8.3)$, $\zeta \rightarrow$ low multiplicity, $\tau^+\tau^-$ biased sample, including all cuts. (a) The inclusive photon spectrum; (b) The ζ peak region of (a), with fit shown as a solid line; (c) The same as (b) with fitted background subtracted.

where the first error is statistical and the second error is systematic with the bulk of this systematic error coming from uncertainties in the photon detection efficiency. Including the low multiplicity final state complicates matters even more, so the Crystal Ball collaboration prefers to give the result as

$$B[\Upsilon(1S) \to \gamma\zeta] \sim 0.5\%. \quad (27)$$

Given the excitement the announcement of the ζ caused, it was natural that a large effort was mounted to check the initial report from the Crystal Ball. This was done both at DESY and Cornell last fall (1984). The Crystal Ball obtained about 200 k more Υ decays in the detector (22 pb^{-1}) in a six-week run. The CUSB detector has obtained about 340 k more Υ decays in their detector (corresponding to about 22 pb^{-1} at CESR due to the narrower beam energy spread), making about 450 k events in total when adding in older data. In addition, 6 pb^{-1} were obtained at an energy just below the Υ in this run. CESR's run at the Υ continued for about 11 weeks. The results that I report here for both detectors, essentially, have been reported previously.[44,45]

In the interim between 1983 (100k Υ) and 1984 (200k Υ), the Crystal Ball detector underwent a major upgrade. A new tracking chamber system was installed which increased the number of proportional tube chamber layers from six to eight. Also, a new gas was used (Ar-CO_2-Methane), which stopped the chamber degradation with beam exposure that was previously plaguing the detector. This change necessitated a major restructuring of the on-line data acquisition hardware and software, along with the off-line analysis software. These changes had been in progress for some time, but the fall 1984 run was the first real physics data taken with the new system; of course, there were problems in the initial part of the run. The problems encountered allowed only a rapid analysis of the latter 60% of the data from this run and only for the high multiplicity final state. Most of the first 40% of the data has been recovered and will be reported on soon for the high multiplicity channel. The low multiplicity channel is more problematic since the backgrounds in the analysis are sensitive to the tracking chamber quality (not the case for the high multiplicity analysis). Thus, the dramatic improvement of the new chamber system over the old one has necessitated a total rethinking of the charged particle cuts in the low multiplicity analysis. The Crystal Ball collaboration hopes that this analysis will be completed by this summer (1985).

A comparison of the high multiplicity channel analysis from the 1983 run (100k Υ) and the last 60% of the 1984 run (about 125k Υ) is presented in FIGURES 26a and b. The new data obviously do not confirm the ζ. There is over a 4 SD difference at the ζ mass between the 1983 data and this preliminary analysis of the last 60% of the 1984 data. The Crystal Ball collaboration does not understand presently the origin of this difference; however, the potential physics impact of the ζ is so great that the experimenters have the burden of proof to show that the ζ signal reproduces in every valid data set.

The CUSB collaboration has also reported new results. Their result at the Leipzig Conference[46] was an upper limit of 0.2% (90% C.L.) for the product branching ratio, $B[\Upsilon \to \gamma\zeta] \times B\,[\zeta \to \mathit{hadrons}]$. This result was based on a sample of 112k Υ decays. However, this upper limit is calculated using a photon efficiency based on a model of the QCD process, $\Upsilon \to \gamma gg$, for Υ radiative decay to the ζ. If this model is used to obtain the branching ratio in the Crystal Ball (the 0.5% value in the Crystal Ball is obtained using a $c\bar{c}$ model for the hadronic decay of the ζ[42]), the value obtained is about 0.25%.

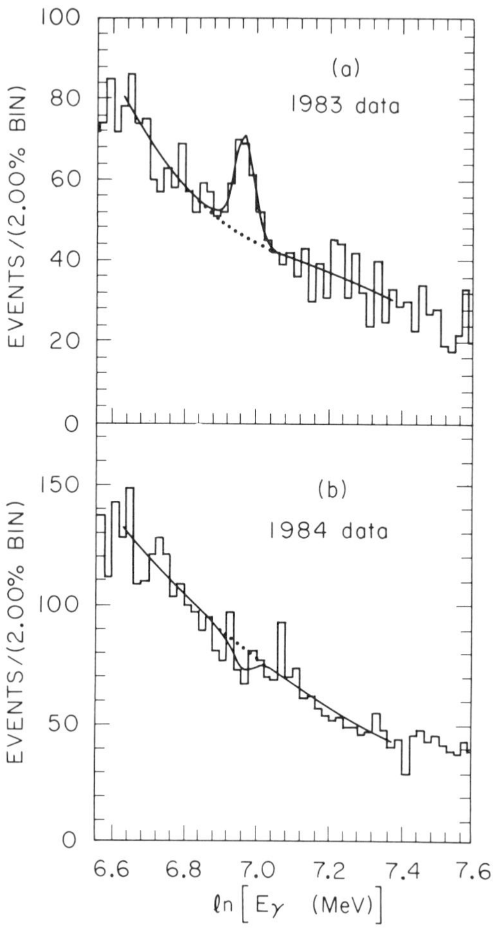

FIGURE 26. Crystal Ball results for $\Upsilon(1S) \rightarrow \gamma\zeta(8.3)$, $\zeta \rightarrow$ multihadrons. (a) The results from 1983 data, which were shown already in FIGURE 24a (100k Υ decays), and (b) the preliminary results from 1984 data, with no ζ peak seen (125k Υ decays).

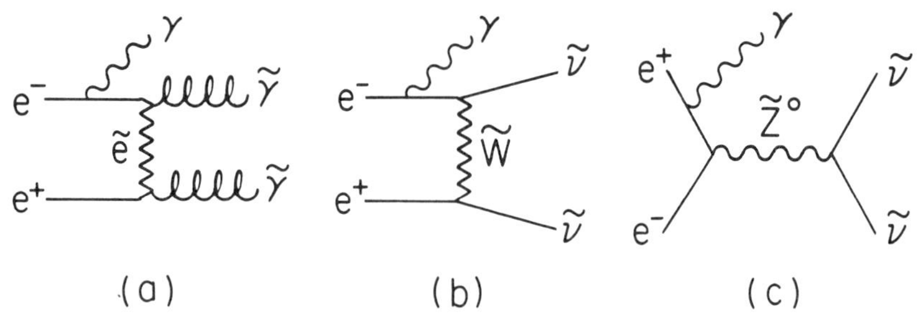

FIGURE 27. Diagrams used to calculate (a) $e^+e^- \rightarrow \gamma\tilde{\gamma}\tilde{\gamma}$; (b) and (c) $e^+e^- \rightarrow \gamma\tilde{\nu}\tilde{\nu}$.

The new preliminary analysis from the CUSB collaboration, using the new data (340k decays), gives an upper limit for the product branching ratio of 0.14%–0.2% (90% C.L.),[45] which depends on whether or not a new sector of the detector, made from BGO scintillator, was included in the analysis or not (lower limit without BGO sector). The same model for Υ decays, $\Upsilon \rightarrow \gamma gg$, was used to calculate the photon efficiency, so the CUSB upper limits are only in mild disagreement with the summer results[42] of the

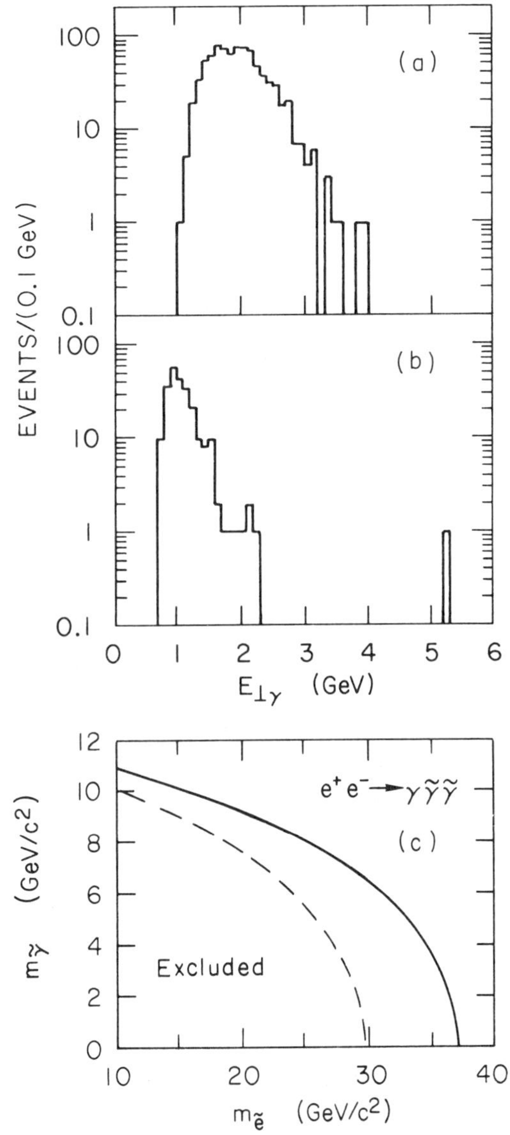

FIGURE 28. Results from the MAC collaboration on $\tilde{\gamma}$ production. (a) The observed $E_\perp$ spectrum for the first data sample of 36 pb^{-1} with θ_{veto} = 10° and a search region of $E_\perp > 4.5$ GeV; (b) The spectrum for the second data sample of 80 pb^{-1} with θ_{veto} = 5° and a search region of $E_\perp > 3.0$ GeV; (c) The lower limit for $M_{\tilde{e}}$ as a function of $M_{\tilde{\gamma}}$; the solid curve is for $M_{\tilde{e}_L} = M_{\tilde{e}_R}$; the dashed curve for $M_{\tilde{e}_L} \gg M_{\tilde{e}_R}$; the limits are at 90% C.L.

Crystal Ball. The most trouble for the ζ at this time is probably coming from the Crystal Ball itself.

The Search for Supersymmetric Particles

I report here recent limits obtained by the MAC collaboration[47] on the production of photinos, $\tilde{\gamma}$, and sneutrinos, $\tilde{\nu}$, in e^+e^- production at $E_{c.m.} = 29$ GeV. FIGURE 27 shows the diagrams tested for. The photino diagram has been calculated[48] and

$$\sigma(e^+e^- \longrightarrow \gamma\tilde{\gamma}\tilde{\gamma}) \sim \alpha^3 s/(M_{\tilde{e}})^4. \quad \textbf{(28)}$$

Candidate events are selected by the size of their "perpendicular energy," $E_\perp$. All that is seen in the detector is a single photon candidate with

$$E_\perp \geq (\sqrt{s} - E_\gamma) \sin\theta_{veto}, \quad \textbf{(29)}$$

where E_γ is the measured energy of the photon and θ_{veto} is the minimum angle covered by the detector. FIGURE 28 shows the result of the search. Part (a) shows the observed $E_\perp$ spectrum for the first data sample taken (36 pb^{-1}), with $\theta_{veto} = 10°$ and a search region of $E_\perp > 4.5$ GeV. Part (b) shows the observed $E_\perp$ spectrum for the second data sample taken (80 pb^{-1}), with $\theta_{veto} = 5°$ and a search region of $E_\perp > 3.0$ GeV. One event is observed in the latter data sample. This result leads to the upper limit for photino production versus $\tilde{e}$ mass, which is shown in part (c) of the figure. The experimental observations can also be turned into a limit on the sneutrino mass.[47] For $20 < M_{\tilde{W}} < 29$ GeV, $M_{\tilde{\nu}} > 10$ GeV/c^2.

CONCLUSIONS

(1) The running of the coupling constant, α_s, predicted by QCD is not seen presently by using measurements of R_{hadron}. However, the systematic errors on the measurements still allow a running α_s. It should be noted that when calculating α_s at the J/ψ and Υ using the technique of Lepage and Mackenzie,[13] a running α_s is also not observed, but is allowed within error.

(2) Evidence for the three gluon coupling predicted by QCD is still lacking. The search for gluonic mesons has continued to be inconclusive, even though candidate meson properties have been considerably refined.

(3) Searches for right-handed currents have so far proved negative, though impressive mass limits have been obtained. Searches for supersymmetric particles have also proved negative, with mass limits strongly coupled to accelerator energy. Much enthusiasm and inventiveness has been generated by the announcement of two possible new states, $\xi(2.2)$ and $\zeta(8.3)$; however, the existence of these states is presently uncertain. In the case of the ζ, theoretical interpretation outside the S-M played a very prominent role.

ACKNOWLEDGMENTS

I would like to thank Michael Peskin for interesting and useful discussions. I also thank Alfred Fridman and Andreas Schwarz for useful suggestions and a careful reading of the manuscript. Finally, I would like to thank the organizers of the first of what I hope will be a long series of marvelous and stimulating conferences.

REFERENCES

1. KOBAYASHI, M. & K. MASKAWA. 1973. Prog. Theor. Phys. **49:** 652.
2. MAIANI, L. 1977. Proceedings of the International Symposium on Lepton and Photon Interactions at High Energies, Hamburg 1977 (DESY, Hamburg 1977): 877.
3. ARNISON, G. *et al.* UA1 COLLABORATION. 1983. Phys. Rev. Lett. **122B:** 103; BANNER, M. *et al.* UA2 COLLABORATION. 1983. Phys. Lett. **122B:** 476.
4. STROVINK, M. 1984. LBL-18231; BÙRAS, A. J. 1984. MPI-PAE/PTh 46184.
5. BARTEL, W. *et al.* JADE COLLABORATION. 1979. Phys. Lett. **88B:** 171; 136; 1980. Phys. Lett. **91B:** 152; 1981. Phys. Lett. **100B:** 364; 1983. Phys. Lett. **129B:** 145.
6. FERNANDEZ, E. *et al.* MAC COLLABORATION. 1984. SLAC-PUB-3479. Submitted to Phys. Rev. D.
7. For a nice review of the development of this theory using the notation used in this report, see KIESLING, C. 1984. Recent Experimental Tests of the Standard Theory of Electroweak Interactions. Habilitationsschrift. Submitted to the Ludwig-Maximilians-Universität München.
8. DINE, M. & J. SAPIRSTEIN. 1979. Phys. Rev. Lett. **43:** 668; CHETYRKIN, K. G., A. L. KATAEV & F. V. TKACHEV. 1979. Phys. Lett. **85B:** 277; CELMASTER, W. & R. J. GONSALVES. 1978. Phys. Rev. Lett. **78:** 132.
9. SCHWINGER, J. 1973. Particles, Sources, and Fields, vol. II, chaps. 4 and 5. New York; APPELQUIST, T. & H. D. POLITZER. 1975. Phys. Rev. Lett. **34:** 43.
10. EDWARDS, C. *et al.* 1984. SLAC-PUB-3030.
11. ADEVA, B. *et al.* MARK-J COLLABORATION. 1983. MIT Report Number 131.
12. BRANDELIK, R. *et al.* TASSO COLLABORATION. 1979. Phys. Lett. **83B:** 261; 1980. Z. Phys. C. Part. Fields. **4:** 87; 1979. Phys. Lett. **88B:** 199; 1982. Phys. Lett. **113B:** 499.
13. MACKENZIE, P. B. & G. P. LEPAGE. 1981. Phys. Rev. Lett. **47:** 1244.
14. PARTICLE PROPERTIES DATA BOOKLET. 1984. Rev. Mod. Phys., vol. 56, no. 2, part II.
15. APPLEQUIST, T. *et al.* 1978. Ann. Rev. Nucl. Part. Sci. **28:** 387–499.
16. BLOOM, E. D. & C. W. PECK. 1983. Ann. Rev. Nucl. Part. Sci. **33:** 143–197.
17. TUTS, P. M. 1983. Proceedings of the Lepton Symposium, Cornell University. D. G. Cassel & D. L. Kreinick, Eds.
18. CRYSTAL BALL COLLABORATION. 1985. SLAC-PUB-3575.
19. MOXHAY, P. & J. ROSNER. 1983. Phys. Rev. **D28:** 1132.
20. BAILLON, P. *et al.* 1967. Nuovo Cimento **A50:** 393.
21. EDWARDS, C. *et al.* 1982. Phys. Rev. Lett. **49:** 259.
22. FRANKLIN, M. 1982. Ph.D. Thesis. SLAC-254; FRANKLIN, M. *et al.* 1983. Phys. Rev. Lett. **51:** 963.
23. DIONISI, C. *et al.* 1980. Nucl. Phys. **B169:** 1.
24. PERRIER, J. *et al.* MARK-III COLLABORATION. 1984. SLAC-PUB-3436.
25. PROTOPOPESCU, S. D. 1984. Proceedings of the Hadronic Session of the Nineteenth Rencontre de Moriond, La Plagne-Savoie-France, March 4–10, 1984, vol. 2. New Particle Production: 689.
26. EDWARDS, C. 1984. Ph.D. Thesis. CALT-68-1165.
27. PERRIER, J. 1984. Invited talk presented at the International Conference on Physics in Collision IV, Santa Cruz, California, August 22–24, 1984. SLAC-PUB-3436.
28. EDWARDS, C. *et al.* 1982. Phys. Rev. Lett. **48:** 458.
29. EINSWEILER, K. F. 1984. Ph.D. Thesis. SLAC-272.
30. AUGUSTIN, J. E. *et al.* 1984. Contributed paper to the XXII International Conference on High Energy Physics, Leipzig, July 19–25, 1984.
31. BLOOM, E. D. 1982. Invited talk presented at the 21st International Conference on High Energy Physics, Paris, France, July 26–31, 1982; J. Phys. **C3** (suppl. 12): 407.
32. BISTIRLICH, J., R. BOSSINGHAM, A. CHACON, C. CLAWSON, K. CROWE, T. HUMANIC, C. MARTOFF & C. MEYER. LBL/University of California, Berkeley; JANSEN, J. SIN;

BREWER, J., R. KEITEL, C. ORAM & M. SALOMON. University of British Columbia/TRIUMF.
33. BALKE, B., B. CARR, G. GIDAL, A. JODIDIO, K. A. SHINSKY, H. M. STEINER, D. P. STOKER, R. D. STROVINK & M. TRIPP. LBL/University of California, Berkeley; GOBBI, B. Northwestern; ORAM, C. J. TRIUMF.
34. STROVINK, M. 1984. Rapporteur talk presented at the XIth International Conference on Neutrino Physics and Astrophysics, Nordkirchen near Dortmund, FRG. LBL Report #18231.
35. SCHECK, F. 1978. Phys. Report **44C:** 187.
36. ABRAMOWICZ, H. *et al.* 1982. Z. Phys. **C12:** 225.
37. EINSWEILER, K. MARK-III COLLABORATION. 1983. SLAC-PUB-3202. Presented at the International Europhysics Conference on HEP, Brighton, England, July 20–27, 1983.
38. PARTRIDGE, R. 1984. Talk at this conference.
39. AUGUSTIN, J. E. *et al.* 1984. Contributed paper to the XXII International Conference on High Energy Physics, Leipzig, July 19–25, 1984.
40. HABER, H. E. & G. L. KANE. 1984. Phys. Lett. **135B:** 196.
41. GODFREY, S. *et al.* 1984. Phys. Lett. **141B:** 439.
42. PECK, C. CRYSTAL BALL COLLABORATION. 1984. SLAC-PUB-3380 and DESY 84-064; TROST, H. 1984. Contributed paper to the XXII International Conference on High Energy Physics, Leipzig, July 19–25, 1984; NICZYPORUK, B. 1984. SLAC Summer Institute on Particle Physics, July 23–August 3, 1984; COYNE, D. G. 1984. Invited talk presented at the Conference on Physics in Collision IV, University of Santa Cruz, Santa Cruz, California.
43. GLASHOW, S. L. & M. MACHACEK. 1984. Phys. Lett. **145B:** 302; PHAM, X. 1984. PAR/LPTHE-84/38; TU, T. 1984. BIHEP-TH-84-22; GLÜCK, M. 1984. DO-TH-84/23; SHIU, M. *et al.* 1984. HUTP-84/A068; GEORGI, H. *et al.* 1984. Phys. Lett. **149B:** 234; RODENBERG, R. 1984. Print-84-0843 (AACHEN); LANE, K. *et al.* 1984. DOE-ER-1545-342 and NSF-ITP-84-116; NANDI, S. 1984. DOE-ER-03992-563 (Texas); HABER, H. E. & G. L. KANE. 1984. UM TH 84-26; CLAVELLI, L. *et al.* 1984. IUHET-96a; PANTALEONE, Z. *et al.* 1984. SLAC-PUB-3439; WU, DAN-DI. 1984. KFK-TH 97.
44. BROCK, I. 1984. CRYSTAL BALL COLLABORATION. To be published in the Proceedings of the Meeting of the American Physical Society, Department of Particles and Fields, Santa Fe, New Mexico, October 31–November 3, 1984.
45. TUTS, M. 1984. CUSB COLLABORATION. To be published in the Proceedings of the Meeting of the American Physical Society, Department of Particles and Fields, Santa Fe, New Mexico, October 31–November 3, 1984.
46. FRANZINI, P. 1984. CUSB COLLABORATION. Question and answer period of Leipzig Meeting and private communication.
47. MAC COLLABORATION. 1985. Phys. Rev. Lett. **54:** 95.
48. WARE, J. & M. E. MACHACEK. 1984. Phys. Lett. **142B:** 300; KOBAYASHI, T. & M. KURODA. 1984. Phys. Lett. **139B:** 208; GRASSIE, K. & P. N. PANDITA. 1984. Phys. Rev. **D30:** 22.
49. NICZYPORUK, B. *et al.* LENA COLLABORATION. 1982. Z. Phys. **C15:** 299.

Brief Review of Neutrino Physics

FRANK SCIULLI

Department of Physics
Columbia University
New York, New York 10027

This is a review of some of the ongoing questions in neutrino physics. The audience should keep in mind that it may be biased by the perspective of this experimenter, who is actively investigating some of these very questions. In addition, some very important problems, including neutrino mass and oscillations, are not discussed here at all since they seem to be adequately covered in other sessions.

The most abundant reactions seen in high energy neutrino experiments are the charged current deep-inelastic processes:

$$\begin{pmatrix} \nu \\ \bar{\nu}_\mu \end{pmatrix} + N \rightarrow \begin{pmatrix} \mu^- \\ \mu^+ \end{pmatrix} + X. \qquad (1)$$

The corresponding processes induced by the neutral current and electromagnetic forces, respectively, are:

$$\begin{pmatrix} \nu_\mu \\ \bar{\nu}_\mu \end{pmatrix} + N \rightarrow \begin{pmatrix} \nu_\mu \\ \bar{\nu}_\mu \end{pmatrix} + X \qquad (2)$$

and

$$\begin{pmatrix} \mu^- \\ \mu^+ \end{pmatrix} + N \rightarrow \begin{pmatrix} \mu^- \\ \mu^+ \end{pmatrix} + X. \qquad (3)$$

Here N represents a nucleon and X represents a system of hadrons with appropriate quantum numbers. The standard notation[1] for the description of these processes is shown in TABLE 1.

STRUCTURE FUNCTIONS

As outlined in TABLE 1, measurement of the deep-inelastic cross sections provides the structure functions, $F_2(x, q^2)$, $xF_3(x, Q^2)$, $R(x, Q^2)$, which are related to the quark densities in the nucleon as shown. The best statistical precision among these experiments is with those using complex targets (about which we will have more to say later), and, hence, the most accurate measurements are for the nearly isoscalar case.

Quark-Parton Model (QPM) Tests

The historical tests of the quark-parton model are in very good shape. These are described in the following subsections.

Mean-Square Charge Test

This requires that the values of F_2 measured in electromagnetic and neutrino scattering be related by the mean-square quark charge:

$$F_2^{\text{em}} = \langle q^2 \rangle F_2^{\text{cc}}. \qquad \textbf{(4)}$$

Here, if we neglect the strange-charm asymmetry in the sea, $\langle q^2 \rangle = 5/18$. FIGURE 1 shows the comparison of the x-dependence of F_2 from two neutrino[2,3] and two muon[4,5] scattering experiments using iron nuclei as targets. The muon scattering values have been multiplied by 18/5 and three of the data sets have been arbitrarily renormalized by the factors shown to give the best agreement with the fourth.

TABLE 1. Formulae and Notation of Deep-Inelastic Scattering

$x \equiv Q^2/2M\nu$	E = beam (neutrino) energy
$y \equiv \nu/E$	E_μ = outgoing lepton (muon) energy
	$\nu = E - E_\mu$
	$Q^2 = 2EE_\mu(1 - \cos\theta_\mu)$
	$s = M^2 + 2ME$
Scaling two-quark model	$\begin{cases} F_2(x, Q^2) = q(x) + \bar{q}(x) + 2k(x) \\ 2xF_1(x, Q^2) = q(x) + \bar{q}(x);\ R(x, Q^2) = 2\dfrac{k(x)}{q(x) + \bar{q}(x)} \\ xF_3(x, Q^2) = q(x) - \bar{q}(x) \end{cases}$
Neutrino	$q(x) \equiv u(x) + d(x) + s(x) + c(x)$
	$\bar{q}(x) = \bar{u}(x) + \bar{d}(x) + \bar{s}(x) + \bar{c}(x)$
Muon-electron	$q^{\text{em}}(x) = 4/9\,u(x) + 1/9\,d(x) + 1/9\,s(x) + 4/9\,c(x)$
	$\bar{q}^{\text{em}}(x) = 4/9\,\bar{u}(x) + 1/9\,\bar{d}(x) + 1/9\,\bar{s}(x) + 4/9\,\bar{c}(x)$
Neutrino	$\dfrac{d^2(\sigma^\nu + \sigma_{\bar{\nu}})}{dx\,dy} = \dfrac{G^2 s}{\pi}\left\{\left(1 - y - \dfrac{Mxy}{2E}\right) F_2^{\text{wk}}(x, Q^2) + 1/2\, y^2\,[2xF_1^{\text{wk}}(x, Q^2)]\right\}$
	$\dfrac{d^2(\sigma^\nu - \sigma^{\bar{\nu}})}{dx\,dy} = \dfrac{G^2 s}{\pi}[y(1 - 1/2\,y)\,xF_3^{\text{wk}}(x, Q^2)]$
Muon-electron	$\dfrac{d^2\sigma^{\text{em}}}{dx\,dy} = \dfrac{4\pi\alpha^2}{Q^4} s\left\{\left(1 - y - \dfrac{Mxy}{2E}\right) F_2^{\text{em}}(x, Q^2) + 1/2\, y^2\,[2xF_1^{\text{em}}(x, Q^2)]\right\}$
	$\dfrac{2xF_1}{F_2} \equiv \left(1 + \dfrac{4M^2x^2}{Q^2}\right)\Big/(1 + R);\ 0 \geq R \geq \infty$

As can be seen, there are normalization differences of about ±8% internal to the neutrino experiments and about ±3% internal to the muon experiments. To within these precisions, the mean-square charge test works well.

These normalization differences are slightly larger than those that the experiments tabulate. Therefore, because the total charged current cross section can be directly measured with neutrino beams, a comparison of such measurements will show overall systematic discrepancies. We can see in FIGURE 2 that most of the neutrino structure function difference can be traced directly to the overall flux normalization, which is common to structure function and cross section measurements.[6]

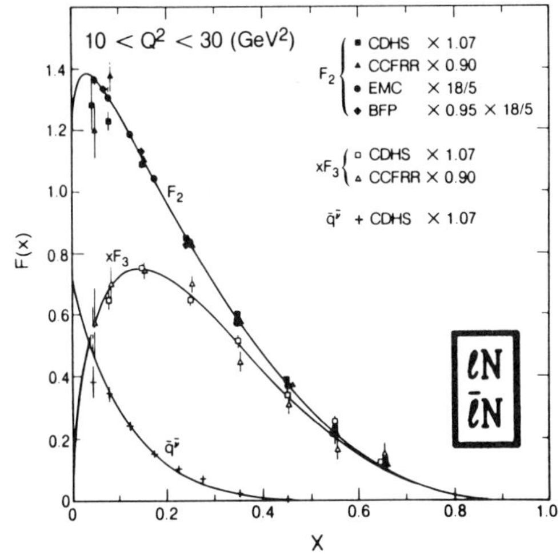

FIGURE 1. Structure function measurements from iron targets.

Valence-Quark Test (Gross–Llewellyn-Smith Sum Rule)

Another important test of the QPM is the test of the number of valence quarks,[7] as measured by xF_3,

$$\int_0^1 F_3 dx = \int_0^1 \frac{xF_3}{x}\, dx = 3\left(1 - \frac{\alpha_s}{\pi}\right),$$

up to first order in the QCD corrections,[8] where α_s is the quark-gluon coupling. At presently available Q^2, the sum rule should have the value of about 2.8. Three measured values have been published:

CCFRR[3]: 2.83 ± 0.20,

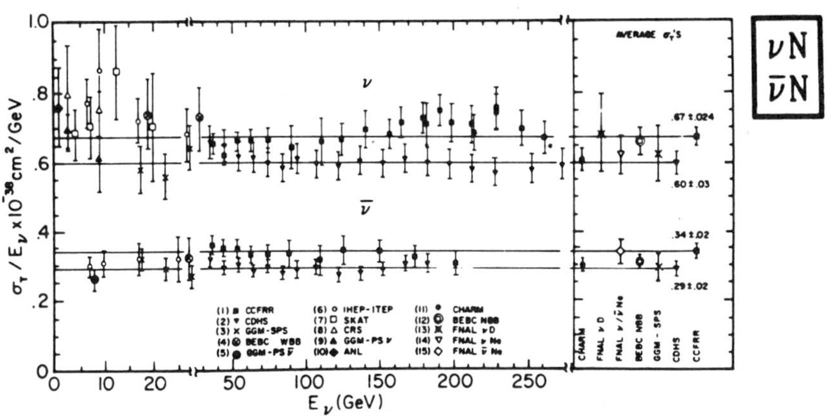

FIGURE 2. Total cross section measurements of (anti)neutrinos on nucleons.

CHARM[9]: 2.66 ± 0.41,

CDHS[10]: 3.2 ± 0.5.

It is clear that the experimental values are in very good agreement with the prediction. The CCFRR result[3] is shown, both differentially and integrally, in FIGURE 3. The differential spectrum is consistent with a dependence close to $\sqrt{x}$ at small x.

NC versus CC Structure Functions

Even raw differential cross sections are expected to be very similar for neutral current and charged current neutrino x-distributions. The differences, from the presence of V + A currents in the NC case, are rather small. Several groups have measured these distributions[11] and have found no demonstrable differences.

EMC Effect

An important discovery, named for the experimental group[12] that first saw it in muon scattering, has been demonstrated quite clearly. FIGURE 4 shows the original data and the SLAC/MIT data that corroborated the effect[13] for $x > 0.2$. The x-dependence for iron differs by $\pm 15\%$ from that expected from deuterium measurements. No Q^2 dependence on atomic weight at fixed x has been seen. A recent experiment by a Rochester/SLAC/Am U group has measured[14] the dependence on A from several nuclei at SLAC and they again corroborate the effect for $x > 0.2$. No

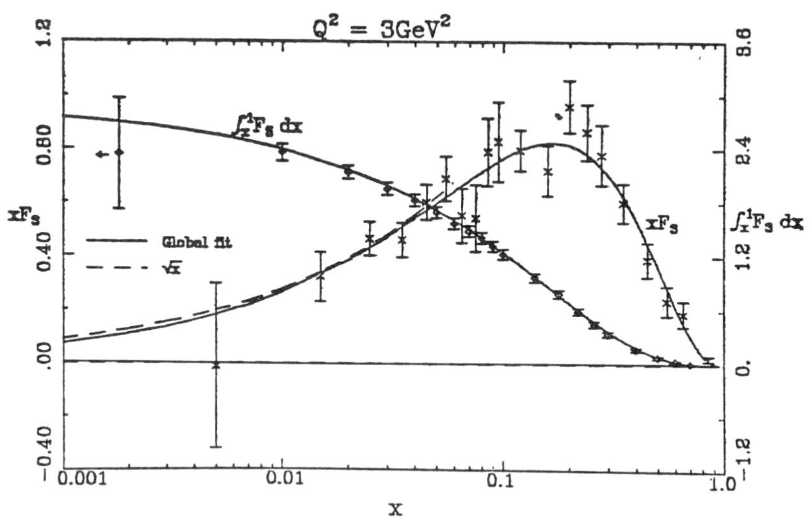

FIGURE 3. Measurement by CCFRR of $xF_3(x)$ and the integral.

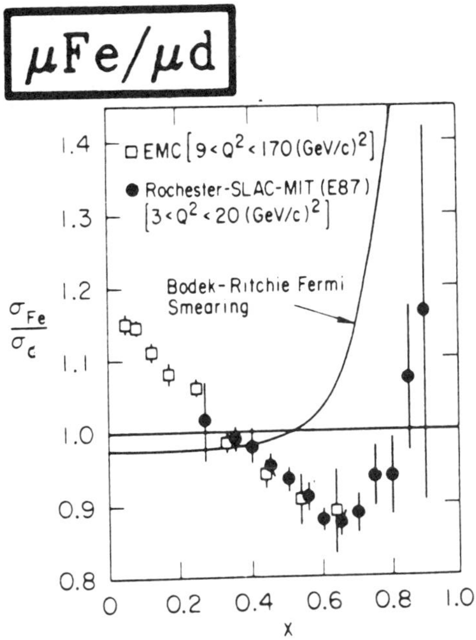

FIGURE 4. Comparison of charged lepton scattering from iron and deuterium.

effect was seen at smaller x, but this might be due to the fact that they measure a combination of F_2 and R at small x.

Many ideas for explaining this effect would expect a big effect at small x, particularly from the sea quarks. One might expect the integrated antiquark component to be bigger by 40% or more in iron than in deuterium. Measurement of the sea is in the province of neutrino interactions, but there is some concern about combining data from different experiments. TABLE 2 gives results to date on searching for this increase in the antiquark sea.

We must conclude that the dramatic increase in the sea expected in many models has not yet been seen.

Q^2 *Dependence of Structure Functions (and Comparison with QCD)*

Since the earliest days of deep-inelastic scattering (~1968), it has been known that structure functions had some small (scale-breaking) dependence on Q^2 as well as the dramatic dependence on x. The very beautiful theory of Quantum Chromodynamics (QCD) predicts a specific pattern of scale-breaking in high energy deep-inelastic scattering that is not unlike the observed pattern. There is good experimental evidence that the very low Q^2 scale-breaking is not simply described by perturbative QCD. Our hope is that the higher Q^2 range over which data is available is appropriate for testing and measuring QCD properties. One very fortunate aspect of this endeavor is that there is quite good agreement among many experiments on the Q^2 dependence of structure functions, as seen in FIGURE 5, for iron.[2,3] An unfortunate aspect is that there are insufficient statistics where it matters.

TABLE 2. Results on Searching for an Increase in the Antiquark Sea

Group	$\overline{Q}_A/Q_D$	Comments
BEBC[15] (WA 24)	0.98 ± 0.16	Track sensitive D_2 target in heavy liquid
BEBC[16] (WA 59/24)	0.89 ± 0.06 1.11 ± 0.06	All Q^2 } Compare two different runs High Q^2 } with D_2 and heavy liquid
CDHS[17]	1.05 ± 0.09 ± 0.06	From H_2 target data in front of Fe detector, select high y ($\langle Q^2 \rangle = 7$ GeV^2)

The leading order expression for the quark-gluon coupling constant of QCD is

$$\alpha_s(Q^2) = \frac{4\pi}{(11 - \frac{2}{3}n_f)\log Q^2/\Lambda^2},$$

where n_f is the number of extant quark flavors and Λ is the scale parameter of the theory. Naively, one might expect that $\Lambda^{-1} \approx$ scale of strong interactions $\approx$ typical radius of bound hadrons $\approx m_\pi^{-1}$. Theoretical attempts are in progress to calculate this parameter precisely. There are, of course, other experimental techniques for extracting Λ, which were summarized by LePage at the Cornell Conference.[18] That author

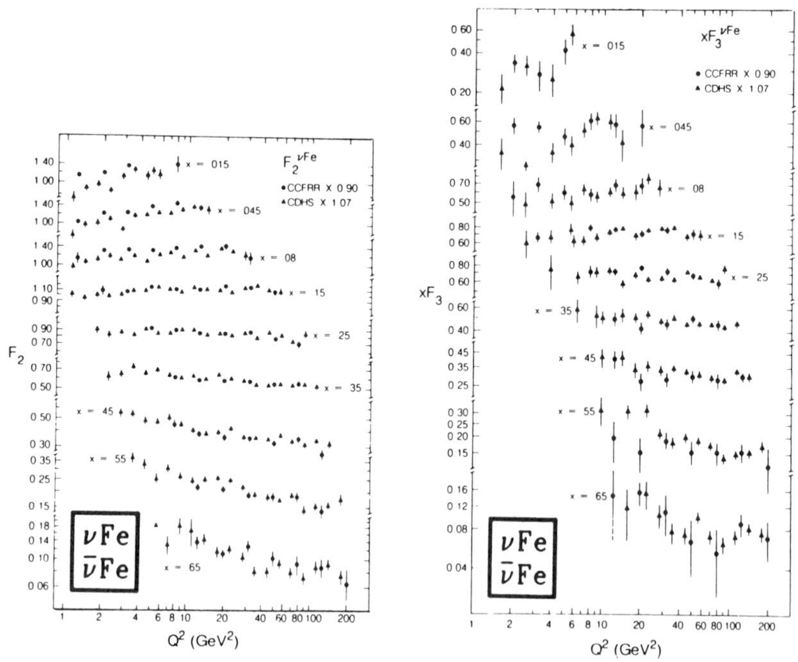

FIGURE 5. Structure functions from two neutrino experiments.

expresses the opinion that the ratio of gluonic to radiative decay rates of the Υ and Υ are trustworthy ways of measuring $\Lambda_{\overline{MS}}$, which is the second order Λ in the $\overline{MS}$ renormalization scheme. Such measurements, it is stated, suggest that $\Lambda_{\overline{MS}}$ is in the range, 50–200 MeV. It should be noted that the equivalent Q^2 for these processes is about 5 GeV^2. Also, for the ψ decay, with $Q^2 \simeq 2$ GeV^2, the spectrum of single photons does not follow the QCD prediction,[19] whereas the Υ decay spectrum does conform to the shape prediction. The best values quoted by Schamberger *et al.*[20] are:

$$\Upsilon \text{ decay} \qquad \Lambda_{\overline{MS}} = 116 \begin{matrix} +105 \\ -\ \ 57 \end{matrix} \text{ MeV},$$

$$\Upsilon' \text{ decay} \qquad \Lambda_{\overline{MS}} = \ \ 80 \begin{matrix} +195 \\ -\ \ 59 \end{matrix} \text{ MeV}.$$

Of course, an important part of showing that we have a viable theory with predictive power is that we can measure the same value for a parameter in many different processes. In deep-inelastic scattering, these values of Λ have come from the behavior of F_2, $\bar{q}$, etc. (singlet analysis), on the one hand, and from xF_3 (non-singlet analysis), on the other.

Singlet Analysis

The Q^2 dependence of $F_2\,(x, Q^2)$ is given, in first order, by:

$$\frac{dF_2(x, Q^2)}{d \ln Q^2} = \frac{\alpha_s(Q^2)}{\pi} \int_x^1 \left[F_2(z, Q^2) P_{qq}\left(\frac{x}{2}\right) + G(z, Q^2) P_{qg}\left(\frac{x}{z}\right) \right] \frac{dz}{z},$$

where the P's are known functions describing the probability of finding a quark given an initial quark or gluon, respectively.

There is a similar equation for the evolution of the gluon distribution. We must extract simultaneously $\alpha_s(Q^2)$ and the correct function for $G(x, Q^2)$ from the data on $F_2(x, Q^2)$. Typically, one assumes a form for the gluon distribution at a specific $Q^2 = Q_0^2$:

$$G(x, Q_0^2) = A_G(1 - x)^{\eta_G}(1 + \gamma_G x).$$

Several authors[21,22] have pointed out an important difficulty in these analyses: there is a very strong correlation between Λ (or α_s), on the one hand, and η_G or γ_G, on the other. These latter parameters largely determine the "softness" of the gluon distribution. In one case,[22] the second order fits give $\Lambda_{\overline{MS}} \approx 200 \pm 100$ MeV with a soft gluon distribution and a comparably good fit for $\Lambda_{\overline{MS}} \simeq 350 \pm 100$ MeV with a hard gluon distribution. In first order, one finds[23] essentially equal confidence for $\Lambda \simeq 200$ MeV with a soft gluon distribution and for $\Lambda \approx 400$ MeV with a hard gluon distribution. The two gluon distributions in question are shown in FIGURE 6 for $Q^2 = 4$ GeV^2. The question of comparing the evolved gluon distribution at very high energies (from $\bar{p}p$, for example) is a moot point since the two gluon distributions evolve to essentially the same distribution at $Q^2 = 10{,}000$ GeV^2 (also shown in FIGURE 6).

Non-Singlet Analysis

Neutrino scattering permits measurement of $xF_3(x, Q^2)$, which has a relatively simple evolution:

$$\frac{dxF_3(x, Q^2)}{d \ln Q^2} = \frac{\alpha_s(Q^2)}{\pi} \int_x^1 \frac{dz}{z} \left[xF_3(x, Q^2) P_{qq}\left(\frac{x}{z}\right) \right].$$

It does not involve the gluon distribution and has the further advantage that the extraction of xF_3 does not depend on R. The principal difficulty is that the values of xF_3 come from differences between neutrino and antineutrino differential cross sections. This, for example, makes the statistical error on xF_3 considerably larger than that of F_2.

TABLE 3 shows a selected tabulation of fits made on high statistics data in non-singlet analysis. (Note that some of the fits were not made by the experimenters.) The first entry, BEBC, originally made a very low Q^2 cut and determined Λ_{Lo} to be 740 MeV. A later analysis, with a substantially higher cut on Q^2, found Λ to be three times smaller. This may be taken as evidence that data with $Q^2 \lesssim 2$ GeV^2 are suspect for comparisons with perturbative QCD, which is a very similar situation to that prevailing in the γ decay discussed earlier.

The trick of using F_2 at large x, justifiable if there are not $\bar{q}$ or R contributions, improves the statistical precision, but for the cases shown it systematically gives uncomfortably larger values of Λ. There is no strong evidence that there is any

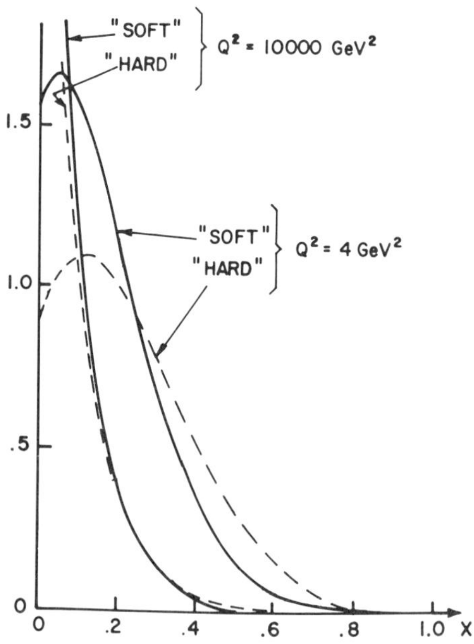

FIGURE 6. Gluon distributions consistent with present data at low and high Q^2.

TABLE 3. Fits Made on High Statistics Data in Non-Singlet Analyses

Expt	Beam/ Target	Cuts (GeV^2)	Λ_{Lo} (MeV)	α_{Lo} ($Q^2 = 30$)	$\Lambda_{\overline{MS}}$ (MeV)	Ref
BEBC	$\nu, \bar{\nu}/$ H Liq	$Q^2 > 0.15$	740 ± 50	0.166		25
		$Q^2 > 1.5$	245^{+130}_{-145}	0.107		26
CDHS	$\nu, \bar{\nu}/$ Fe	$Q^2 > 2$, $W^2 > 11$			200^{+200}_{-100}	2
		F_2 for $x > 0.3$	275 ± 80		300 ± 130	
		$Q^2 > 5$ $W^2 > 10$	131 ± 99	0.089	336 ± 131	22
		F_2 for $x > 0.4$	227 ± 77		284 ± 80	
CHARM	$\nu, \bar{\nu}/$ $CaCO_3$	$Q^2 > 3$	$187^{+130}_{-160} \pm 70$		$310 \pm 140 \pm 70$	27
		$Q^2 > 10$	185^{+130}_{-120}			
CCFRR	$\nu, \bar{\nu}/$ Fe	$Q^2 > 5$ $W^2 > 10$	88^{+163}_{-78}	0.080	193^{+272}_{-136}	3
		F_2 for $x > 0.4$	266^{+114}_{-104}	0.110		
EMC	$\mu/$Fe	$Q^2 > 5$ $x > 0.25!$	$125^{+22}_{-20}{}^{+114}_{-70}$		$173^{+29}_{-27}{}^{+158}_{-97}$	28
		$Q^2 > 5$ $x > 0.25!$	133 ± 23		163 ± 26	22

dependence on Q^2 cuts made above a few GeV^2 within the experimental precision. It has been noted[24] that there may be some systematic shifts due to W^2 cuts, but more statistics will be needed[22] to quantify this for cuts much more than 10 GeV^2. For completeness, we show in the table results from analysis of a muon experiment that attempted a non-singlet analysis for $x > 0.25$. Aside from the concern over using F_2 in this region, it has been pointed out[22] that small systematic normalization differences between runs at different energies can substantially change the result.

The only published values from xF_3 data for $\Lambda_{\overline{MS}}$, made by the experimenters, are from three groups:

$$\text{CDHS: } (Q^2 > 2 \text{ GeV}^2) \qquad \Lambda_{\overline{MS}} = 200^{+200}_{-100} \text{ MeV}$$

$$\text{CCFRR: } (Q^2 > 5 \text{ GeV}^2) \qquad \Lambda_{\overline{MS}} = 193^{+272}_{-156} \text{ MeV}$$

$$\text{CHARM: } (Q^2 > 3 \text{ GeV}^2) \qquad \Lambda_{\overline{MS}} = 310 \pm 140 \pm 70 \text{ MeV}$$

While these values are larger than those extracted from radiative γ decays, the errors on each are too large to claim inconsistency at present. These measurements are largely limited by the size of the event sample. We really require an increase in statistics by at least an order of magnitude to determine Λ to much higher precision. An extension of the Q^2 region will also be helpful. The run beginning now at Fermilab, supplemented by planned runs in 1987, should be very useful in addressing these questions.

Measurements of R (σ_L/σ_T)

The magnitude of R is very difficult to obtain with precision. On the other hand, there is a very specific prediction from QCD. Since $R = F_L/(F_2 - F_L)$, the prediction is, in leading order,

$$F_L(x, Q^2) = \frac{\alpha_s(Q^2)}{2\pi} x^2 \int_x^1 \frac{dz}{z^3} \left\{ \frac{8}{3} F_2(z, Q^2) + 4n_f \left(1 - \frac{x}{z}\right) zG(z, Q^2) \right\},$$

or that the average value of R is $\langle R \rangle \approx 0.35\, \alpha_s$.

TABLE 4. Neutral Current Measurements Distinguishing between Lepton-Lepton and Lepton-Quark Interactions

Process/ Measurement	Lepton-Lepton	Lepton-Quark
	$\nu_\mu e \rightarrow \nu_\mu e$	$\nu_\mu N \rightarrow \nu_\mu X$
NC Rate	CHARM: 0.215 ± 0.032 ± 0.013	CHARM: 0.208 ± 0.016
{NC or CC} Rate	BNL: 0.209 ± 0.029 ± 0.013	CCFRR: 0.242 ± 0.012
	$e^-e^+ \rightarrow \mu^-\mu^+$	$eD_2 \rightarrow eX$
NC Amplitude / EM Amplitude	{Petra & PEP}: 0.18 ± 0.02	SLAC: 0.215 ± 0.015

Measurements at higher Q^2 have been made by the CHARM,[29] CDHS,[30] CCFRR,[31] EMC,[16] and BFP[32] groups. One may conclude that R is not substantially larger than predicted, but precision comparisons are yet to be made.

NEUTRAL CURRENTS

The Standard Model gives us very strong predictive capabilities for comparison to measurements of neutral current phenomena. There is a single unknown parameter, $\sin^2\theta_w$, that should be the same (after small correction for electroweak effects) whether obtained from lower energy measurements or from measurements of the vector boson masses. Of the low energy tests, we may distinguish between lepton-lepton interactions, on the one hand, and lepton-quark interactions, on the other. Also, the measurements may consist of comparing two weak rates or of comparing a neutral current amplitude with a weak amplitude. These are illustrated in TABLE 4 with some results obtained in each case.

While there are no discrepancies larger than two standard deviations from the best

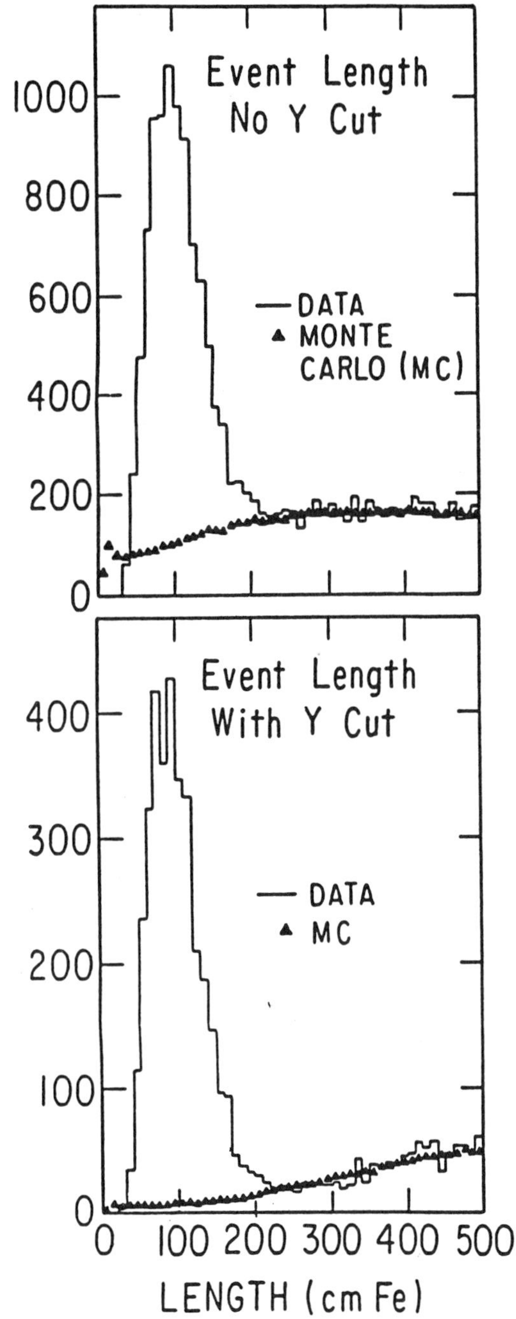

FIGURE 7. Length distribution of neutrino data showing the effect of Y-cut.

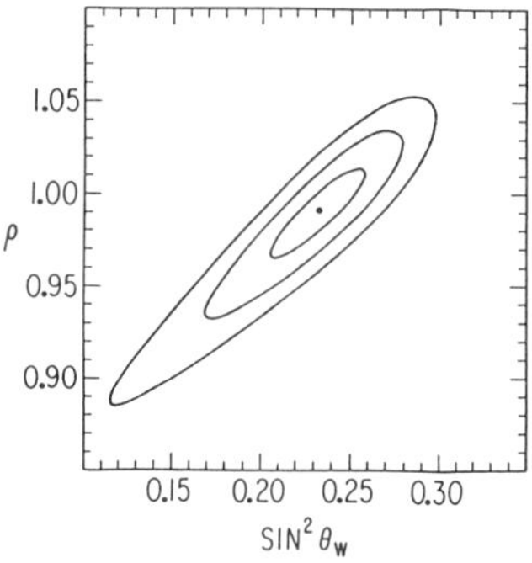

FIGURE 8. Correlation of ρ and $\sin^2\theta_w$ from the CCFRR experiment.

value obtained from the UA1 and UA2 boson masses, $\sin^2\theta_w = 0.220 \pm 0.01$, there are a few reasons to be alert. The asymmetry measurement from the $\mu^+ \mu^-$ events at PEP and PETRA uniformly give lower values. In addition, recent precise measurements of the neutrino neutral current rate are slightly higher.

The recent value measured by the CCFRR group[33] is the highest of those tabulated. An evaluation of the sources of systematic error indicates that this is about 0.005, which is less than half the statistical error. The largest single source of systematic error in such experiments is the uncertainty in the subtraction of charged current background with low energy stopping muons or wide angle (also of low energy) muons that exit the apparatus. All experiments with calorimeters as targets must have such a background, unless they can demonstrate clearly that muons can be tracked through the hadron shower with very high efficiency. In order to reduce this uncertainty, the CCFRR experiment has made a cut in the data that reduces by a factor of two the total data sample, but which reduces the background by about an order of magnitude. The cut is made on events with large values for the ratio of hadron energy to neutrino energy (available in a dichromatic beam); such charged current events have lower muon energy and larger muon angle. FIGURE 7 shows the effect on the distribution in length, which is the total distance in the calorimeter over which charged particles penetrate. The neutral currents are visible as a large peak at small length values, while charged currents, with a penetrating muon, have a more continuous distribution. The upper figure shows the data without making the additional cut. The reduction in the background, and hence the smaller uncertainty, is clearly evident from the lower figure.

The result of the two parameter fit to ρ and $\sin^2\theta_w$ is shown in FIGURE 8. The one standard deviation limits are:

$$\rho = 0.991 \pm 0.025 \pm 0.008,$$

$$\sin^2\theta_w = 0.234 \pm 0.026 \pm 0.010.$$

The one parameter fit, for $\rho = 1$, is

$$\sin^2\theta_w = 0.242 \pm 0.011 \pm 0.005.$$

MULTIMUONS

The opposite-sign dimuon data are used to extract the strange component of the nucleon wave function. There is a consensus at present that the $\bar{s}$ quarks carry about half the momentum carried by the $\bar{u}$ and $\bar{d}$ quarks. The more controversial question has to do with the same-sign dimuons, which have no simple accepted interpretation in the standard model.

The strongest evidence for the existence of this phenomenon comes from three results published in 1981 by the HPWFOR,[34] CFNRR,[35] and CHARM[36] groups. The second of these came from data taken by our group during an engineering shakedown run in Lab E at Fermilab, using a wideband beam. More recent data[37] by our group (CCFRR) taken in a narrow-band beam still shows evidence for same-sign events, as will be discussed, but the rates are not as high; this will also be discussed further.

The worst difficulty in understanding this phenomenon is due to background from charged current events with a second muon from the decay of a π or K either in the first

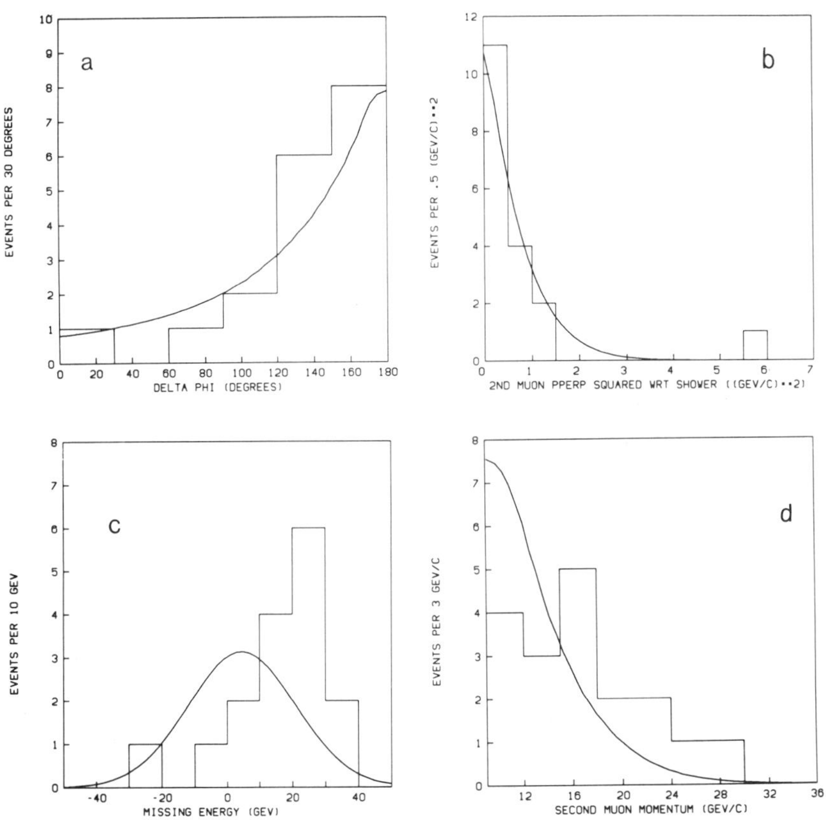

FIGURE 9. Distributions of kinematic variables from same-sign dimuons (see text).

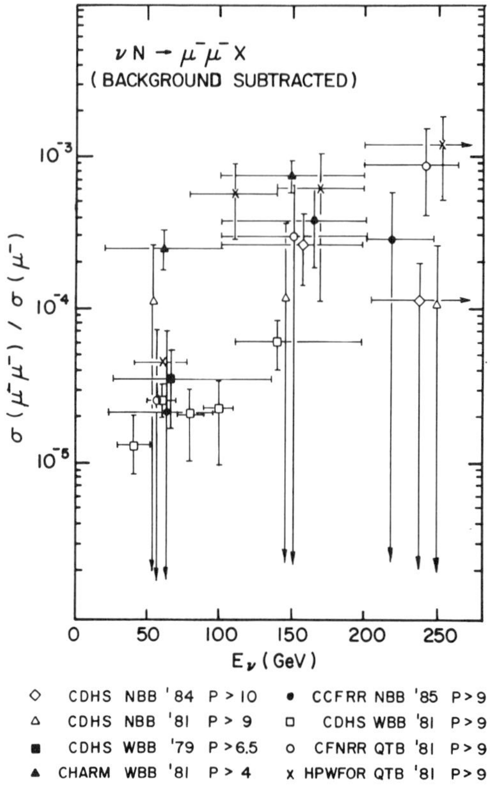

FIGURE 10. Recent data on rates for same-sign dimuons.

or succeeding generation in the hadron shower. Our estimate is that this could account for about half of the observed 18 events in the CCFRR experiment. The background is calculated to about 20% accuracy. Possible interpretations of the remaining events are D^0-$\overline{D}^0$ mixing and associated charm production. These share with π and K decay the qualitative feature that the second muon would be strongly correlated with the direction of the hadron shower. The data tend to have this property. FIGURE 9a shows the distribution in the azimuthal angle, ϕ, between the first muon and the second muon, with the neutrino direction taken as the z-axis. Clearly, the second muon is preferentially on the opposite side from the first muon.

This feature is even more clearly seen in FIGURE 9b, which is the distribution in the square of p_T^{min}. This is the transverse momentum between the hadron momentum vector and that of the muon in the configuration (of which there are two possible) that minimizes it. This parameter, then, defines the second muon. One event is at a high value of nearly $6(\text{GeV}/c)^2$, but the other 17 are low.

FIGURE 9c shows the distribution in missing energy in these events, presumably carried by a final state neutrino. The solid curve is the expected distribution if all the events were from π decay (which is a factor of two higher than the calculated level).

The width of this distribution is dominated by resolution in beam and event energy measurement. The means of the distributions differ. The data give

$$\langle E^{data}_{missing} \rangle = 15.1 \pm 4.2 \pm 1.0 \text{ GeV},$$

where the first error is statistical and the second is systematic. The missing energy expected from π or K decay is

$$\langle E^{\pi,K\ decay}_{missing} \rangle = 3.5 \pm 0.5 \text{ GeV}.$$

Hence, the missing energy is about 2.7 standard deviations higher than expected if all the events were from meson decays.

FIGURE 9d shows the distribution in momentum of the second muon. Again, the distribution is stiffer than that expected from π or K decay. The data give a mean of $\langle p^{data}_{\mu 2} \rangle = 16.8 \pm 1.3 \pm 0.9$ GeV, while the expected average from background is $\langle p^{decay}_{\mu 2} \rangle = 12.5 \pm 1.0$ GeV, which is about two standard deviations lower.

The rates measured for this process are shown in FIGURE 10. The rates from the recent CCFRR experiment do not show a dramatic increase at the highest energies, but the statistical precision is still poor. Concern in this regard caused a reexamination of the data from the CFNRR run.[35] The data points plotted reflect this: two of the twelve events were removed since they did not pass criteria presently imposed and the background increased somewhat. The rate still appears higher at high energy, but it is not inconsistent with the lower points due to the large error bars. Overall from this group, there are several effects at the two to three standard deviations level, in rates, in mean energies, etc., which give credence to the prompt same-sign phenomena. This is not satisfactory; the run beginning now at Fermilab could provide more than an order of magnitude more data to examine that would hopefully reach to energies higher by a factor of two. This clearly requires that the proton intensities planned for the Tevatron be realized.

It should be noted that the strongest indication by far for the existence of these events is from the CHARM experiment.[36] However, it also should be noted that this experiment sets a high level even at low energy ($\sim 1.5 \times 10^{-4}$ for $E_\nu < 100$ GeV), where there is a conflict with another experiment in the μe channel. This experiment[38] gives a 90% confidence limit of 3×10^{-5} with an average neutrino energy of $\langle E_\nu \rangle = 50$ GeV.

The question of the existence and, if so, the rate for the same-sign dimuons remains open. More information, therefore, is clearly needed.

REFERENCES

1. FISK, H. E. & F. SCIULLI. 1982. Ann. Rev. Nucl. Sci. **32:** 499.
2. CDHS: ABRAMOWICZ, H. *et al.* 1983. Z. Phys. **C17:** 282.
3. CCFRR: MACFARLANE, D. B. *et al.* 1984. Z. Phys. **C26:** 1.
4. EMC: AUBERT, J. J. *et al.* 1981. Phys. Lett. **105B:** 315, 322.
5. BFP: CLARK, A. R. *et al.* 1983. Phys. Rev. Lett. **51:** 1826.
6. WOHL, C. G. *et al.* 1984. Rev. Mod. Phys. **56:** 51.
7. GROSS, D. J. & C. H. LLEWELLYN-SMITH. 1969. Nucl. Phys. **B14:** 337.
8. BEG, M. A. B. 1975. Phys. Rev. **D11:** 1165.
9. CHARM: JONKER, M. *et al.* 1982. Phys. Lett. **109B:** 133.
10. CDHS: DE GROOT, J. G. H. *et al.* 1979. Phys. Lett. **82B:** 292.
11. BALTAY, C. *et al.* 1980. Phys. Rev. Lett. **44:** 916; JONKER, M. *et al.* 1983. Phys. Lett.

128B: 117; IGO-KEMENES, P. 1983. Proceedings of the International Europhysics Conference on High Energy Physics, Brighton, England (July 1983).
12. AUBERT, J. J. *et al.* 1983. Phys. Lett. **123B:** 275.
13. BODEK, A. *et al.* 1983. Phys. Rev. Lett. **50:** 1431; **51:** 534.
14. GOMEZ, J. *et al.* 1984. Proceedings of the DPF Meeting, Santa Fe, New Mexico (October 1984).
15. PARKER, M. A. *et al.* 1983. Proceedings of the 1983 International Symposium on Lepton and Photon Interactions at High Energy, Cornell University (Contribution #212).
16. COOPER, A. M. *et al.* 1983. Proceedings of the International Europhysics Conference on High Energy Physics, Brighton, England (July 1983). Summarized by K. Rith.
17. MERLO, J. P. *et al.* 1983. Proceedings of the International Europhysics Conference on High Energy Physics, Brighton, England (July 1983). Summarized by K. Rith.
18. LEPAGE, G. P. 1983. Proceedings of the 1983 International Symposium on Lepton and Photon Interactions at High Energy, Cornell University.
19. ABRAMS, G. S. *et al.* 1980. Phys. Rev. Lett. **44:** 114.
20. SCHAMBERGER, R. D. *et al.* 1984. Phys. Lett. **138B:** 225.
21. BARNETT, R. M. & D. SCHLATTER. 1982. Phys. Lett. **112B:** 475.
22. DEVOTO, A. *et al.* 1983. Phys. Rev. **D27:** 508.
23. DUKE, D. W. & J. F. OWENS. 1984. Phys. Rev. **D30:** 49.
24. BARNETT, R. M. 1982. Phys. Rev. Lett. **48:** 1657.
25. BOSETTI *et al.* 1978. Nucl. Phys. **B142:** 1.
26. FITZE *et al.* 1981. Proceedings Neutrino '81 (Hawaii), vol. I, p. 344.
27. BERGSMA, F. *et al.* 1983. Phys. Lett. **123B:** 269; 1984. CERN EP/84-143.
28. AUBERT, J. J. *et al.* Phys. Lett. **114B:** 291.
29. BERGSMA, F. *et al.* 1984. Phys. Lett. **141B:** 129.
30. ABRAMOWICZ, H. *et al.* 1983. Z. Phys. **C17:** 283; 1982. Z. Phys. **C15:** 19.
31. See reference 3 and MACFARLANE, D. 1983. Ph.D. Thesis Caltech (Nevis preprint R1287); PUROHIT, M. V. 1983. Ph.D. Thesis Caltech (Nevis preprint R1298).
32. See reference 5 and MEYERS, P. D. 1983. Ph.D. Thesis Berkeley (LBL preprint 17108).
33. CCFRR: REUTENS, P. *et al.* To be published in Phys. Lett.
34. HPWFOR: TRINKO, T. *et al.* 1981. Phys. Rev. **D23:** 1889.
35. CFNRR: NISHIKAWA, K. *et al.* 1981. Phys. Rev. Lett. **46:** 1555. (Erratum. 1985. Phys. Rev. Lett. **54:** 1336.)
36. CHARM: JONKER, M. *et al.* 1981. Phys. Lett. **107B:** 241.
37. SMITH, W. *et al.* 1984. SLAC Summer Institute.
38. SCHAFFER, A. C. 1984. Columbia Ph.D. Thesis (Nevis preprint 250).

Constraints on Particle Physics from Cosmology[a]

DAVID N. SCHRAMM AND JANE C. CHARLTON

Enrico Fermi Institute
University of Chicago
Chicago, Illinois 60637

INTRODUCTION

Cosmology and particle physics have become symbiotic in their relationship. In the past, developments in physics have been used to explain astrophysics problems. Recently, cosmology also has been able to place constraints on particle properties and these constraints can be tested by experiment. Thus, the flow of information at the interface of particle physics and cosmology is no longer just one-way. (Astronomy is no longer a parasite of physics.) Many examples of the interchange will be described in this review.

The timeline of cosmology is rapidly filling in as later events find their explanations in earlier events. FIGURE 1 depicts the timeline of the early universe, with an emphasis on the occurrence of phase transitions. We shall see that phase transitions in the early universe produce observables that can be used as evidence of the occurrence of a transition or as a way of working out the details of the transition. In this review, we will mention what is known about each epoch and show how it might constrain the particle models. Since a great deal of effort is devoted currently to the study of the dark matter problem, special emphasis will be placed on this issue. This study of dark matter and galaxy formation will allow us to draw upon much of what was discussed in earlier epochs. This review will draw heavily on a previous review[1] we prepared.

QUANTUM GRAVITY

We begin by discussing the least understood epoch of the universe, the era of quantized gravity. Before the Planck time, 10^{-43} sec, at energies exceeding 10^{19} GeV, it is expected that gravity is unified with the other forces. The physics of this time period is not well described because no reliable quantum field theory for gravity exists. It has been suggested by Stephen Hawking that, at the Planck time, all of space-time is a foam of mini black holes, each with a mass of around 10^{19} GeV. These black holes will evaporate and reform on the Planck time scale of 10^{-43} seconds. In analogy to other gauge coupling constants, it may be that the gravitational constant varies at temperatures greater than its characteristic scale of 10^{19} GeV. On an even more speculative note, it is hoped that the expansion rate, H_0, and the deceleration rate, q_0, are

[a]The work in this paper (Enrico Fermi Institute preprint no. 85-21) was supported in part by NSF grant no. AST-83-13128 and DOE grant no. DE-AC02-80ER10773 A004.

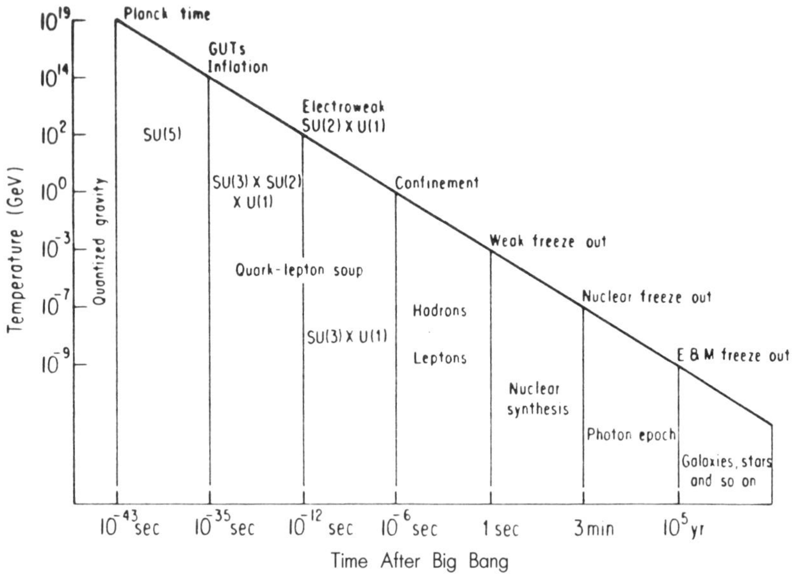

FIGURE 1. The timeline of the early universe, with an emphasis on the occurrence of phase transitions.

observables produced at the "freeze-out" of quantum gravity. The basis for this hope is the many examples of observables produced when an interaction decouples from the background. Perhaps further study of the quantum gravity epoch will add H_0 and q_0 to the growing list of those properties of the universe, once thought to be initial conditions, that now have explanations, just as study of the GUTs epoch has yielded the cosmologically observed baryon to photon ratio.

To unify gravity with the other three interactions in a consistent theory seems to require supersymmetry (SUSY), which puts fermions and bosons on an equal footing. The application of SUSY to gravity is known as supergravity. Most theories of supergravity behave better and appear far less ad hoc if they have more than a 3 + 1 dimension space-time. The extra dimensions are not observed today, so the theories operate on the assumption that the extra dimensions are compactified on the scale of the Planck length. Perhaps this will lead to an explanation of the origin of our 3 + 1 dimensions. The most promising supergravity theory to date, which attempts to unify everything (Theory of Everything ≡ T.O.E.), involves superstrings in ten dimensions.[2] It has been shown[2] that the two optimal T.O.E.'s, $O(32)$ and $E(8) \times E(8)$, (both of which contain $SU(3) \times SU(2) \times U(1)$), are the only known theories that can contain all known particles including chiral fermions. They are finite so no renormalization problems (including the small cosmological constant) exist and they are anomaly free. The low energy ($\lesssim 10^{19}$ GeV) phenomenology of these theories remains to be determined, but a T.O.E. is no longer beyond conception. Perhaps the most severe constraints on the T.O.E. will be cosmological, including dark matter and large-scale structure arguments.

GUT

In the epoch following the Planck time, the strong, weak, and electromagnetic forces remain unified until the energy drops below $\sim 10^{14}$ GeV at 10^{-35} seconds. Above this energy, the Grand Unified Theory (GUT) gauge particles, X and Y, which carry the grand unified force, exist. The X's and Y's would be able to interconvert quarks and leptons. Thus at the GUTs epoch, there would be nonconservation of baryon number and lepton number, which are quantities that are conserved at lower energies. The nonconservation of baryon number led to the GUTs' prediction of proton decay. Experiments to measure the proton lifetime are a way of exploring which quantum field theory describes the GUTs epoch correctly. The simplest GUT considered is $SU(5)$.[3] $SU(5)$ predicts proton decay with a lifetime of $\sim 10^{30}$ years by the dominant mode, $p \rightarrow e + \pi^0$. As we will see, even before the experiments[4] failed to find this mode, cosmologists already doubted $SU(5)$ from baryosynthesis arguments. Although $SU(5)$ was the simplest GUT, another class of GUT, supersymmetric or (SUSY) GUTs, have

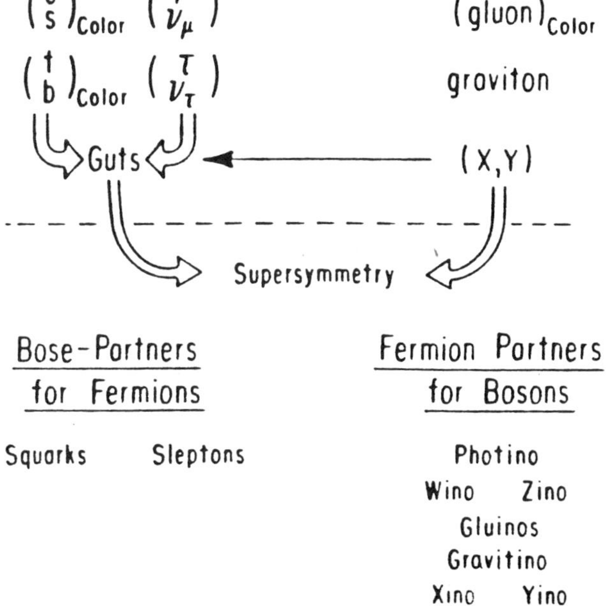

FIGURE 2. A summary of the particle types including their supersymmetric partners.

a different and perhaps, as mentioned above, a more significant appeal. In SUSY, each fermion (boson) is required to have a boson (fermion) partner. New families of particles were created since existing particles did not match up correctly. FIGURE 2 is a summary of the particle types including their supersymmetric partners. The bulk of these SUSY particles are not expected to be seen until the energies are above $M_Z \sim 100$ GeV. Thus, future TeV accelerators may potentially discover a whole zoo of new particles and confirm the concept. As mentioned at this meeting,[5] the monojets at CERN already may be the first experimental assault on SUSY. It is expected that the lowest mass SUSY particle will be stable and thus will survive once it freezes out in the early universe.

In addition to SUSY pointing the way to unification of GUTs with gravity, it also attempts to explain the mass differences between the various gauge bosons. Masses of gauge particles are produced by the Higgs field during phase transitions. We will describe such transitions further when discussing inflation. SUSY models predict proton decay via the mode, $p \rightarrow \mu^+ K^0$, with a somewhat larger range in lifetimes than the 10^{30} years required by minimal $SU(5)$. SUSY favors μK or $\mu\eta$ because, in SUSY, the decay goes via a Higgs particle that couples to mass, which favors the most massive quark that is less massive than the proton, i.e., the strange quark. The IMB experiment is not particularly sensitive to decay via this mode, but it does have possible candidates.[4] The experiment, NUSEX, conducted at Mont Blanc Tunnel, and the Kamioka experiment also have candidate events that favor SUSY. New proton decay detectors sensitive to μK and $\mu\eta$ modes should give a definite result and it is tentatively possible that the proton lifetime is on the order of a few times 10^{31} years, in accordance with the predictions of SUSY, but the few events seen to date are not sufficient to prove anything.

Baryogenesis

One of the most attractive characteristics of GUTs is that they are able to yield a net n_b/n_γ for the universe.[6] Direct observations of cosmic rays and of the γ-ray background, as well as n_b/n_γ being $\gg 10^{-18}$, demonstrate a net baryon asymmetry. Nucleosynthesis arguments, to be discussed later in this paper, constrain the baryon to photon ratio to the range of $(4\text{–}7) \times 10^{-10}$. Sakharov[7] outlined the three ingredients needed for baryogenesis: (1) baryon nonconservation, (2) C and CP violation, and (3) departure from thermal equilibrium. Baryon nonconservation is clearly required if an initially symmetric universe is to develop a net n_b/n_γ. Unless both C (charge conjungation) and CP (charge conservation combined with parity) are violated, B (the net baryon number) is zero since it changes sign under C and CP. Departure from thermal equilibrium is required since, in thermal equilibrium, the baryon density depends only on the temperature and the particle mass, which is the same for a particle and an antiparticle. We have already seen that baryon nonconservation follows from GUTs and that C and CP violation are known to occur (although their violation during baryon nonconserving interactions remains to be measured in proton decay). An upper limit for the magnitude of CP violation can be found from a measurement of the electric dipole moment of the neutron since the two quantities are related.[8] Thermal equilibrium is maintained only if reaction rates are much greater than the Hubble constant. As the universe expands and cools, reaction rates drop to values lower than expansion rates and processes drop out of equilibrium. Detailed calculations of specific GUTs models[9,10] have yielded a net baryon to photon ratio in the observed range. It was noted that minimal $SU(5)$ made too small a baryon to photon ratio[11] unless an additional generation, a richer Higgs sector, or an axion U_1 field was added.[6] Thus, cosmology has constrained GUTs physics and the constraint has been supported subsequently by experiment.

One important consequence of the GUTs determination of n_b/n_γ as a unique function of temperature was the requirement that primordial fluctuations must be adiabatic.[12,13] Isothermal fluctuations were ruled out since $n_\gamma = constant$ implies $n_b = constant$. We will return to this point later.

Inflation

Another consequence of the GUTs transition is the generation of magnetic monopoles. When the group $SU(5)$ is broken to yield the $U(1)$ subgroup of electromagnetism, monopoles are produced.[14,15] The number of monopoles produced should be quite high, comparable to the number of baryons produced.[16,17] Observations clearly rule out the existence of this many monopoles. Some adjustment is needed to solve this problem. This adjustment leads to the theory of inflation, which also was found to solve many long-standing initial condition problems in cosmology. In particular, inflation solves the horizon problem, the flatness problem, and the clumping problem. The horizon problem deals with the question of why the universe is so smooth on large scales. Measurements of the dipole anisotropy of the 3K background give $\delta T/T \sim 10^{-3}$ and once the local velocity is accounted for and higher moments are examined, it is found that $\delta T/T < 2 \times 10^{-5}$ at the decoupling of radiation from matter at $T \sim$ 3000K.[18] At that time, opposite parts of the sky were apparently causally disconnected even more so than at the present. Why is the universe so homogeneous and isotropic if separate regions were causally disconnected, that is, each outside the others' horizon? The second problem, the flatness problem, is finding an explanation for how the cosmological density parameter, $\Omega = \rho/\rho_{crit}$, became so finely tuned to one. Although we don't know whether Ω is within an order of magnitude or so of unity today, it is known to evolve on a time scale of $\Omega/|\dot{\Omega}| \sim H^{-1}$. Therefore, at the Planck time, Ω should evolve on a dynamical time scale of $\sim 10^{-43}$ seconds. By the present time, then, it would have evolved to ∞ or 0, depending on whether $\Omega > 1$ or $\Omega < 1$, unless it was 1.000 . . . to almost 60 decimal places. Finally, why does the universe contain density perturbations ("bumps") such as stars, galaxies, and clusters of galaxies on a small scale if the universe is so smooth on a large scale?

In 1981, Guth[19] proposed the theory of inflation as a solution to these problems. Although buds of inflation existed earlier,[20–22] Guth was the first to put all the pieces together and to realize that inflation could solve many cosmological problems. The idea is based on the existence of a nonzero vacuum energy produced by the Higgs field in GUTs. When the radiation density is no longer higher than the energy density of this false vacuum, the universe enters a phase in which the scale factor grows exponentially, a de Sitter phase. When symmetry is broken in the GUTs phase transition, the energy in the Higgs field produces the masses of the gauge particles, the vacuum energy goes to zero, inflation stops, and the universe returns to normal expansion. The horizon problem is solved since all of the present universe came from a small region that was causally connected before inflation. Inflation also solves the monopole problem. Since each horizon prior to inflation would contain less than one monopole, less than one monopole generated in the GUTs transition would exist in the universe today. Inflation would make the scale factor, $R \longrightarrow \infty$, so the curvature term, k/R^2, would be suppressed and the universe would be flat, $\Omega = 1$. Finally, bumps could be generated in the phase transition as different domains mixed to create our present space. The problem with this original inflation theory is that the phase transition is unable to go to completion and have the universe leave the de Sitter phase because the bubbles of the broken symmetry phase inflate away from each other faster than they expand.

This problem was solved in the theory of new inflation proposed by Linde[23] and Albrecht and Steinhardt.[24] In this theory, the transition is a smooth Coleman-Weinberg type in which, during the vacuum-dominated era, the universe is evolving towards the true vacuum. The result of this kind of a transition is that our universe is a single bubble. As in the original inflation, the horizon, flatness, and monopole problems are solved. However, because we must live in a single bubble, all perturbations must be generated within this bubble. Numerous investigators[25] found that fluctuations could be generated with the Harrison-Zeldovich spectrum,[26] which put equal power on all scales as observed. Standard GUTs, though, resulted in $\delta\rho/\rho \gtrsim 10$, which is a factor of 10^5 more than the observed limits required by 3K anisotrophy.[25] Such large fluctuations would collapse to black holes rather than to galaxies. Possible solutions to this problem are supersymmetry (SUSY), supergravity, or extra Higgs sectors.[27–30] These all require what presently seems to be ad hoc tuning. Despite the fact that the details of inflation have yet to be worked out, it is viewed with a great deal of certainty since it has been so successful in solving the cosmological problems discussed above. In fact, Gott[31] has argued that such a de Sitter phase expansion is required by geometry alone if we are to achieve the known initial conditions of the universe. Therefore, something like inflation must occur and Ω today must be unity to a high accuracy to avoid having us live at a special epoch.

Finally, the GUTs transition may produce strings rather than bubbles if the vacuum in the model has the correct topology. This is analogous to phase transitions in crystallization, which produce filaments. Strings will later be discussed as a possible solution to the dark matter problem. It is interesting that the current, in vogue, T.O.E.'s break via phase transitions that produce such cosmic strings.[32] These cosmic strings contain a core of "unified flux" where the strong and electroweak interactions are symmetric.

QUARK LEPTON SOUP

After the GUTs transition, we enter a somewhat more familiar epoch, that of the strong and electroweak interactions. In this epoch, above 100 GeV, the weak and electromagnetic interactions remain unified as predicted over a decade ago by Weinberg, Salam, and Glashow and verified in 1983 by the discoveries of the intermediate vector bosons (W^+, W^-, and Z^0) at CERN.[33] Quarks and leptons are no longer interconverted since baryon number is conserved and the universe exists as a "soup" of quarks and leptons. At 100 GeV (10^{-12} sec), the Weinberg-Salam transition occurs and the symmetry between the weak and the electromagnetic forces [$SU(2) \times$

$U(1)$] is broken. The next event is the quark-hadron symmetry transition at $\sim$1 GeV (10^{-6} sec). Although the quark-hadron transition (quark confinement) and chiral symmetry breaking (quarks obtaining mass) seem to be quite different physical processes, calculations have shown that they probably occur at the same time. It is thought that the transition is first order, although the final word on this subject has not been given. Details on this subject can be found in the *Quark Matter '83* proceedings.[34]

The possibility of black hole formation at both the electroweak and the quark-hadron-chiral symmetry transitions has been discussed; the former by Crawford[35] and the later by Crawford and Schramm[36] and Schramm and Olive.[37] These black holes would be of planetary mass and thus would not explode by the Hawking process in the age of the universe. They also form before big bang nucleosynthesis so they are not limited by the baryon density limits as later black holes are. Production of such black holes would require a first order phase transition. The electroweak transition does not seem to be capable of producing many black holes.[35] The quark-hadron-chiral transition, however, may generate planetary mass black holes. Such black holes will be of interest when discussing the dark matter problem. Another option at the quark-hadron transition is the production of quark nuggets.[38] Nuggets are stable clumps of more than three quarks. In fact, Witten argues that such clumps could be as massive as asteroids or small planets and would thus behave cosmologically very similar to the planetary mass black holes. However, recent arguments[39] do not favor the formation of such large quark lumps. In any case, cosmological density constrains the efficiency of the production of any relic from the early universe, including black holes and nuggets. It will be interesting to see whether heavy-ion collisions exploring the quark-hadron region find any trend towards large clumping.

WEAK FREEZE-OUT

In the early universe, neutrinos are produced by neutral-current weak interactions via reactions of the type:

$$e^+e^- \leftrightarrow \nu_i + \bar{\nu}_i \ (i = e, \mu, \tau).$$

This equilibrium is maintained until the temperature drops below a few MeV. At lower temperatures, the weak interaction rate is too slow to keep up with the expansion of the universe so neutrinos are decoupled from the hadron-lepton soup. These neutrinos would have a temperature of 2K today, compared to 3K for photons, because the decoupled neutrinos are not heated by e^+e^- annihilation at $T \leq m_e/3$ as the photons are.

Soon after the neutrinos decouple, the charged-current weak interactions,

$$p + e^- \leftrightarrow n + \nu_e, \quad n + e^+ \leftrightarrow p + \bar{\nu}_e, \quad n \leftrightarrow p + e^- + \bar{\nu}_e,$$

are no longer able to maintain chemical equilibrium between neutrons and protons. The equilibrium ratio of n/p is given by $\exp(-\Delta m/T)$. When the rates of the charged-current weak interactions become small compared to the universal expansion rate, the n/p ratio essentially freezes out. Since nearly all neutrons are incorporated into ^{4}He, the value of n/p will determine the abundance of ^{4}He produced in the nucleosynthesis period.

Nucleosynthesis begins with the production of deuterons via the reaction, n + p $\leftrightarrow$ d + γ. At energies greater than 0.1 MeV, the abundance of deuterons is very small because energetic photons rapidly dissociate any deuterons that are formed. This deuterium bottleneck to nucleosynthesis breaks at $T \approx 0.1$ MeV and the epoch of nucleosynthesis begins.

NUCLEOSYNTHESIS

When the deuterium abundance becomes significant, reactions with n, p, and d occur, which lead to the synthesis of D, ^{3}He, ^{4}He, and a small amount of ^{7}Li. The agreement of the predictions of primordial nucleosynthesis with the observed abundances is one of the profound successes of the big bang theory. We shall begin by discussing the dependence of the various abundances on the baryon to photon ratio, n_b/n_γ, the neutron half-life, $\tau_{1/2}$, and the number of neutrino flavors, N_ν. These dependences will then allow us to examine the constraints placed on n_b/n_γ, N_ν, and Ω_b, the baryon density parameter. For a detailed current review of nucleosynthesis, see Yang *et al.*[40] and the references therein. It is interesting to note how the initial success in the late 1960s of big bang nucleosynthesis, in enabling an understanding of how roughly a quarter of the universe is ^{4}He, has improved spectacularly as detailed studies of the noncosmological synthesis and destruction of the other light elements converged on what was required from the big bang. Therefore, this gave agreement between predictions and observations over an abundance range of nearly ten orders of magnitude.[40]

As mentioned in the previous section, the production of ^{4}He depends upon the value of n/p, which is determined by a balance of the weak interaction rate with the expansion of the universe. The weak interaction rate varies inversely with the neutron half-life. The expansion rate increases with an increasing number of light particle species and thus with increasing N_ν. Finally, if n_b/n_γ is large, the deuterium bottleneck is broken sooner, so n/p is higher and more ^{4}He can be produced.

The abundances of D and ^{3}He are much smaller than that of ^{4}He. Nuclear reactions tend to burn both of these elements to ^{4}He since it is more tightly bound. The surviving abundances are determined by a competition between the reaction rates and the expansion rate of the universe. The higher the value of n_b/n_γ, the more nucleons are available to react with D or ^{3}He and so the lower the abundances of the remaining D and ^{3}He.

Few elements heavier than ^{4}He are produced primordially because the expansion rate overtakes the rate of reactions producing heavier elements. The absence of stable nuclei at mass-5 and mass-8 inhibits the production of heavier nuclei. Only small amounts of ^{7}Li are synthesized via the reaction, ^{4}He + ^{3}H $\rightarrow$ ^{7}Li + γ, at lower nuclear abundance ($n_b/n_\gamma \leq 3 \times 10^{-10}$) and from the decay of ^{7}Be produced in the reaction, ^{3}He + ^{4}He $\rightarrow$ ^{7}Be + γ, at higher densities. The curve of ^{7}Li/H versus n_b/n_γ goes through a minimum at $n_b/n_\gamma \approx 3 \times 10^{-10}$.

All of the abundances discussed above can be used to place various constraints on n_b/n_γ. Since the deuteron is barely bound and is thus easily destroyed in stars, the observed abundance of deuterium provides a lower limit to the primordial abundance. Since the abundance decreases with increasing n_b/n_γ, the observed value can be used to place an upper limit on n_b/n_γ. A lower limit on n_b/n_γ can be obtained using the observed abundances of D and ^{3}He. Nearly all the deuterium in stars is burned to ^{3}He. Since stars are not producers of ^{3}He and since both ^{3}He and D are produced in greater amounts at low n_b/n_γ, it follows that D and ^{3}He taken together can be used to set a lower limit for n_b/n_γ, which avoids overproduction of ^{3}He. Similar observations can be made for ^{7}Li and ^{4}He. All are in agreement with a conservative estimate of $n_b/n_\gamma = (3\text{–}10) \times 10^{-10}$ and a best value of $n_b/n_\gamma = (4\text{–}7) \times 10^{-10}$.

As mentioned earlier, the abundance of ^{4}He depends upon the number of neutrino flavors as well as $\tau_{1/2}$ and n_b/n_γ. Since we have a constraint on n_b/n_γ and a value of $\tau_{1/2} = 10.6$ min $\pm$ 0.2, we can limit the number of neutrino species if given a range of possible values for the primordial ^{4}He abundance, Y_p. An upper limit for Y_p from observations of objects with low heavy element abundances is 0.25. It is more difficult to bound Y_p from below, but such a bound is unnecessary for the argument. (See Yang *et al.*[40] and the references therein.) The conclusion of such studies[40] is that $N_\nu < 4$ with a best fit value of $N_\nu = 3$. $N_\nu = 4$ is allowed only if $Y_p \geq 0.253$. Predictions on N_ν can be verified[41] by high energy particle physics experiments currently being conducted to determine the width of Z^0 at CERN and those that will be done in the near future at SLAC, Fermilab, and LEP. This[42] is the first time that cosmology has made a prediction about fundamental physics that can be tested at a colliding beam machine. In fact, the S$\bar{\text{P}}$PS, LEP, SLC, and TEV-I will be checking the big bang to earlier times than is possible with any telescope.

The density parameter is related to the value of n_b/n_γ by

$$\Omega_b = 3.53 \times 10^{-3} h_0^{-2} (T_0/2.7\text{K})^3 n_b/n_\gamma,$$

where h_0 is the Hubble constant in units of 100 km/sec/Mpc. The conservative range is $0.011 \leq \Omega_b \leq 0.19$ using the values of n_b/n_γ quoted above and allowing T_0 to range from 2.7K to 3.0K. A best value range is $0.014 \leq \Omega_b \leq 0.14$. If the upper limit on T_0 is reduced to 2.78K as argued by Partridge,[43] then the best limit is $\Omega_b < 0.10$. If $\Omega = 1$, as predicted by inflation, then we see that the universe must be dominated by nonbaryonic matter. If constraints on the age of the universe are used,[44] the range tightens to $0.03 \leq \Omega_b \leq 0.10$.[40] The implications of this, including the possible identities of the dark matter, will be discussed later.

ELECTRICITY AND MAGNETISM FREEZE-OUT

When the temperature of the universe dropped below ~1eV at ~10^5 years, two important things happened. First, since 1eV is approximately the binding energy of electrons in atoms, at this time electrons combined with nuclei to form atoms. Free electrons that had been capable of scattering were no longer abundant, so the universe became transparent. The second event was that the universe became matter dominated, that is, the energy density in radiation dropped below that in matter. The significance of this is that density perturbations could begin growing without being smeared out by the radiation. The photons at 10^4K that were able to propagate freely after the E&M freeze-out are observed today as the 3K blackbody radiation.

DARK MATTER

We have seen that inflation seems to require that $\Omega = 1$, while nucleosynthesis requires the density in baryons to be in the range of $0.03 \leq \Omega_b \leq 0.10$. This leads us to the conclusion that the universe is dominated by nonbaryonic matter. This is one of the three distinct cosmological dark matter problems. First, though, we will describe the other two problems—dark halos of galaxies and galaxy formation. Correlation functions and galaxy clustering also will be reviewed briefly. Then possible solutions to the problems shall be discussed.

Halos

The problem of dark matter in halos of galaxies has been well established and described in detail.[45,46] Characteristic masses of galaxies can be found using the simple dynamical relationship, $M \sim v^2r/G$, where v is the orbital velocity and r is the separation distance. For spiral galaxies, a typical value for M is $\sim 10^{11}\ M_\odot$ with a mass to luminosity ratio, $M/L \sim 10\ h_0$. In our galaxy, this value is approximately twice that accounted for by matter visible in the optical, such as stars, gas, dust, etc., in the disk.[47] This dark matter problem in our galaxy, however, probably does not have a cosmological origin. When we look at larger scales such as binaries and small groups, the characteristic mass increases by $\sim$10, while the light/galaxy does not change. The resulting M/L is approximately 100 h_0. As the scale increases, M/L seems to increase, although uncertainties in the data are large. Values of M/L for large clusters and superclusters range from $\sim$100 h_0 to $\sim$500 h_0. This problem of dark halos has been referred to as the "missing light problem" because the problem is not mass that is missing, but the existence of nonluminous mass.[48]

A value of M/L can be related to the density parameter if it is assumed to be an average M/L for the universe. To do this, we must introduce the average luminosity density, $\mathbf{L} = 2 \times 10^8\ h_0L_\odot/\mathrm{Mpc}^3$, as determined by Kirshner, Oemler, and Schecter.[49] The implied matter density is given by $\rho = M/L \times \mathbf{L}$. The result for the largest scale gives an Ω in the range of 0.07–0.4. In theory, it is possible that this dark halo problem could be solved by baryonic matter since there is an overlap in allowed Ω. It is known that nonoptical baryons provide at least a partial solution to the problem since some are seen in the X-ray region.[50] It is difficult to find a type of baryonic object that is not excluded by other considerations.[51] Jupiters and low mass stars can in principle work, but observed stellar initial mass functions indicate that they do not come close to providing a complete solution unless extreme production occurs at a very low mass with no high mass tail. The number of stellar mass black holes required to solve the problem would require the existence of heavy elements produced in supernovae in excess of the abundances observed, unless black holes can form without supernova ejecta. This latter possibility would be consistent with an initial production of very massive stars, $M > 100\ M_\odot$. It is important to note that some of the dark matter must be baryonic since $\Omega_b \gtrsim 0.03$, whereas $\Omega_{visible}$ is <0.01. It is intriguing that $\Omega \sim 0.1$ is still consistent with all dynamical and baryonic arguments.[52]

Note also that even on the largest scale, that of superclusters, $\Omega \lesssim 0.4$ is the largest value observed. The only way to reach $\Omega = 1$ (the value required by inflation) is to distribute the bulk of the dark matter on even larger scales than those of giant superclusters.

Galaxy Formation

For galaxy formation to occur, a density fluctuation, δn_b, in the baryon density, n_b, is required. In principle, the fluctuation can arise via adiabatic or isothermal modes or from seeds formed at earlier times. If planetary mass black holes were formed in the quark-hadron transition, they may serve as seeds, first clustering baryons and then exploding.[54] Introducing such seeds requires the introduction of isothermal-like fluctuations on small scales, even those less than the horizon size at the quark-hadron transition. We saw, when discussing GUTs and baryogenesis, that primordial fluctuations must be adiabatic. Isothermal fluctuations cannot persist through the GUTs epoch and thus must be confined to scales smaller than the horizon during the epoch when their production occurred. Thus, isothermals seem to be relegated to possibly serving as Ostriker-Cowie[53] seeds and then only if a phase transition[36] or an other physical process can produce them.

In adiabatic fluctuations, there is a connection between the variations in baryon density, $\delta n_b/n_b$, and the fluctuations in the 3K background, $\delta T/T$, since baryons are coupled to the radiation field. Naively, $\delta\rho/\rho \sim 3\delta T/T$ and detailed calculations showed that $\delta\rho/\rho$ does track $\delta T/T$.[55,56] Recent observations set the limit on the 3K background anisotropy as $\delta T/T \leq 2.1 \times 10^{-5}$ at the E&M decoupling, $T \sim 3000$K. The naive adiabatic requirement for density perturbations at decoupling is thus $\delta n_b/n_b \lesssim 6 \times 10^{-5}$. If we assume linear growth, this would imply $\delta n_b/n_b \leq 6 \times 10^{-2}$ today at $T \sim 3$K. This is much smaller than the observed value of $\delta\rho/\rho \sim 1$ on scales up to at least clusters of galaxies. (Nonlinear growth can occur when $\delta\rho/\rho \geq 1$, so some objects with $\delta\rho/\rho \gg 1$ can easily be explained.) For $\delta\rho/\rho$ to be this large, it requires growth to begin earlier than the time when baryonic matter decouples from the radiation. Thus, nonbaryonic matter is needed. Detailed calculations show that $\Omega \sim 1$ (at least $\Omega > 0.4$) is required to be consistent with $\delta\rho/\rho \sim 1$.

Galaxy Clustering

The two-point correlation function applied to galaxies by Peebles[57] and his co-workers is an invaluable method for describing the distribution of matter. The two-point correlation function, $\xi(r)$, is defined as the excess probability over random for an object to be a distance, r, from another object. It is found that ξ_{gg} is proportional to $r^{-1.8}$. On large scales, the correlation function no longer has this r dependence, but decreases and may be negative for $r \geq 20/h_0$ Mpc.[58] The correlation function for clusters, ξ_{cc}, has the same r dependence, $r^{-1.8}$, but it is ~20 times larger in magnitude than that for galaxies.[59,60] In fact, the greater the richness of the cluster, the greater the amplitude of the correlation function. The cluster-cluster correlation function is nonzero up to scales of at least 100 Mpc. At first glance this seems difficult to understand, especially since gravitational clustering should be less, not more, effective at larger scales. However, Kaiser[61] and Fry[62] have pointed out that if clusters are merely the 3σ peaks in the galaxy distribution, then the factor of twenty can be understood and ξ_{cc} is proportional to ξ_{gg}. Unfortunately with this approach, if $\xi_{gg} < 0$, then ξ_{cc} should be <0 at the same r, but ξ_{cc} seems to be positive where ξ_{gg} is negative. Even if the Davis *et al.*[58] data reflects a local statistical fluctuation, there is no adiabatic model that does not have $\xi_{gg} < 0$ for r below 100 Mpc. An alternative way of looking at the cluster-cluster correlation is using a renormalized or dimensionless[63] approach that uses the average separation distance between the objects as units for r rather than using the same r for clusters and galaxies. The result is that the correlation amplitude for clusters is about three times weaker than that for galaxies and all clusters, regardless of richness, have approximately the same amplitude. The constancy of the cluster amplitude with scale tells us that clustering may be a scale-free process and not due merely to gravitational clustering. The increased amplitude for galaxies over this scale-free amplitude may simply be due to the enhanced gravitational clustering on small scales. Thus, the real question may be, "what is the scale-free process?", not "what is the dark matter?". Note that a scale-free process produces a constant velocity dispersion about the Hubble flow, which is opposed to the increasing velocity dispersion with scale normally associated with $\Omega \sim 1$.

A scale-free process with the correlation function proportional to $r^{-1.8}$ is known as a fractal,[64] with a dimension of $D = 1.2$. Such a fractal is close to the linear $D = 1$ case. In fact, Geller[65] has argued that the power of the correlation function is sufficiently uncertain, so -2 cannot be ruled out. Therefore, $D = 1$ is a possibility. Hierarchical clustering (building from small-scale to large) can give us such behavior if clusters are biased selections of peaks.[61,62] However, as mentioned above, the positivity of ξ_{cc} at large scales argues against such a model. An Ostriker-Cowie[53] explosive galaxy formation scenario can yield a scale-free process if the density of the explosion producing seeds is sufficiently high so that the resulting explosions percolate.[66] The fractal dimension of such a percolated system depends on the geometry of the explosions (e.g., jets or spheres) and the geometry of the material in which the percolation operates (pancakes, filaments, or uniform gas). In principle, such a mechanism can produce a $D \sim 1.2$ fractal. Another scenario that can produce a near linear fractal is the cosmic string idea.[29] If galaxy and cluster formation occur along fragmented cosmic strings, then D near unity is possible. Such string-induced formation can also lead to a nonrandom distribution of galaxies and holes.[67] Let us now look at how the need for such a scale-free process affects the dark matter problem.

Possible Solutions

We have seen that inflation requires $\Omega \sim 1$, which leads to the conclusion that the universe is dominated by nonbaryonic matter. It is difficult, but not impossible to explain dark matter in halos without invoking nonbaryonic solutions. Finally, the discrepancy between perturbations in matter density and temperature either implies existence of nonbaryonic matter so that perturbations in matter can begin growing before decoupling or that the perturbations do not couple with the matter and radiation until after recombination.

Single particle nonbaryonic candidates are described as "hot," "warm," or "cold" matter following Bond and Szalay.[68] Cold matter is nonrelativistic at the time it decouples from matter, so it can condense on small scales. The smallest scales that can collapse when particle i first dominates the mass density of the universe is the effective Jeans mass:

$$(M_J)_i = 3 \times 10^{18} \frac{M_\odot}{m_i^2(eV)}.$$

Hot matter is relativistic at decoupling and it remains relativistic until shortly before it becomes the dominant matter of the universe. The Jeans mass for hot matter is large, so large cluster scales form first and eventually fragment to form smaller scales. It has been argued that warm matter behaves essentially the same as hot matter except that it decouples before the neutrinos decouple; thus, it has a lower temperature than the neutrinos at present.

Massive neutrinos with $10\text{eV} \leq m_\nu \leq 25\text{eV}$ for the most massive neutrino eigenstate are the least exotic of the hot matter solutions. Hot matter solutions give the large-scale structure, but they have difficulty making galaxies. Phase-space arguments prevent hot matter from solving the dark halo problem, particularly for dwarf spheroidals. Thus, hot matter scenarios require galactic halos to be dark baryons. They have galaxies forming after $z \sim 1$, which is in conflict with observations of quasars with $z \gtrsim 3.5$. These difficulties are surmounted with isothermal-like fluctuations, either strings or exploding seeds. Various other light "inos" and some types of shadow matter also fall under the classification of hot matter. (Shadow matter[69] is the matter associated with the other $E(8)$ in the $E(8) \times E(8)$ T.O.E., where the first $E(8)$ yields our $SU(3) \times SU(2) \times U(1)$ universe.)

The many examples of cold matter include neutral heavy leptons, various heavy "inos," other types of shadow matter, axions, quark nuggets, and the planetary mass black holes that have been discussed in connection with the quark-hadron-chiral symmetry transition. Black holes above planetary mass would still be in the form of baryons at the time of nucleosynthesis, while less massive black holes would evaporate via the Hawking process. Cold matter has major advantages[70,71] in that it is able to explain galaxy formation and is able to solve the problem of dark matter in halos. The serious flaw in cold matter models is that they place all the mass on small scales. If $\Omega \sim 1$ as predicted by inflation, then $\Omega_{cluster} \sim 1$ is in conflict with observations of $\Omega_{cluster} \leq 0.4$.[72] One way around the dilemma is to have cold matter and to additionally postulate that the bulk of the clumped baryonic plus cold matter does not contain any shining baryons. Thus, this would mean that galaxies for some reason only occur in certain clumps.[70] Such biasing is akin to the biasing for clusters in the ξ_{cc} versus ξ_{gg}, except here the bulk of the universe is not shining and the light is not tracing mass in an unbiased way. Note that the key difference between the hot and cold models is where the dark baryons end up.

All warm particles that have been suggested have the same difficulties as either the hot or cold matter. Hybrid cold particle and hot particle models also fail because the hot particles, while relativistic, damp out growth of the cold density fluctuations.[44,73,74] Models with a nonzero cosmological constant, Λ, or a gravity differing from $1/r^2$ are even more bizarre and will not be discussed here.

Various other speculative solutions have also been proposed. If a cold or warm particle was to decay to a hot one after galaxy formation, the dark matter problems could be solved.[73,75–78] Unfortunately, it requires a finely tuned particle for which there is no other evidence.

If we invoke a scale-free galaxy-cluster formation process, then the decay scenario is no longer needed and biasing runs into the trouble of negative ξ_{gg} with positive ξ_{cc} observed. The only surviving processes are percolated explosions or cosmic strings. Neither of these has a problem with galaxy formation, but neither yields $\Omega = 1$ unless hot matter is also added.

Scale-free structure enables the hot scenario to again thrive, but with it comes the need for dark baryons in halos. It is interesting that the two optimal solutions, strings and percolated explosions, each involve phase transitions in the early universe; strings at the GUTs epoch or seeds from the quark-hadron transition. Also, both require dark baryonic halos and both require hot matter to reach $\Omega = 1$. The closest viable alternative is cold matter plus biasing, whose success requires that ξ_{cc} really be negative at large r. As bizarre as it may seem, the large-scale structure of the universe may be telling us about primordial phase transitions.[79]

CONCLUSIONS

When two branches of science are able to pool their resources and stimulate each other, an exciting and productive era of scientific study results. At the interface of particle physics and cosmology, such an era is beginning. Many examples of invaluable interchange between these two fields have been described in this review. We will reiterate some of the more significant examples:

(1) Cosmologists have used abundances of elements synthesized in the big bang to constrain the range of values for n_b/n_γ to $(3\text{–}10) \times 10^{-10}$. The baryon to photon ratio given by the minimal $SU(5)$ field is smaller than the lower bound of this constraint. This gave cosmologists reason to doubt minimal $SU(5)$ before proton decay experiments were conducted and confirmed this doubt.
(2) The idea of inflation is based on the existence of a vacuum energy that was proposed in connection with GUTs. Inflation is able to solve long-standing problems in cosmology such as the horizon problem, the flatness problem, and the clumping problem.
(3) Big bang nucleosynthesis arguments have enabled cosmologists to find a best fit value of three for the number of neutrino flavors, with four as an upper limit. If the best fit value is correct, we expect no more quarks after top to be found in experimental particle physics. Measurement of the width of the Z^0 will soon test this constraint.
(4) The large-scale structure of the universe appears to be scale-free. Such scale-free behavior may be indicative of a phase transition in the early universe.
(5) The bulk of the matter in the universe is probably nonbaryonic.

Many other entries for this list come rapidly to mind. Rather than enumerating them, let us just remember that this young interdisciplinary field is growing at a phenomenal rate with new interfaces being found with great frequency.

ACKNOWLEDGMENTS

We would like to thank F. Accetta, M. Davis, J. Frieman, E. Kolb, K. Olive, G. Steigman, A. Szalay, M. Turner, and N. Vittorio for useful discussions.

REFERENCES

1. CHARLTON, J. C. & D. N. SCHRAMM. 1984. *In* Proceedings of the Third International School of Gravitation and Cosmology. J. M. Salim, Ed. Rio de Janiero, Brazil.
2. SCHWARZ, J. 1984. Physica **124A:** 543 (and references therein).
3. GEORGI, H. & S. GLASHOW. 1974. Phys. Rev. Lett. **32:** 438.
4. VAN DER VELDE, J. 1985. *In* Aspen Winter Physics Conference. M. Block, Ed. N.Y. Academy of Sciences. New York.
5. KANE, G. & D. HAGILAN. 1985. *In* Aspen Winter Physics Conference. M. Block, Ed. N.Y. Academy of Sciences. New York.
6. KOLB, E. & M. TURNER. 1983. Ann. Rev. Nucl. Part. Sci. **33:** 645 (and references therein).
7. SAKHAROV, A. D. 1967. Pis'ma Zh. Eksp. Teor. Fiz. **5:** 32. (1967. JETP Lett. **5:** 24.)
8. ELLIS, J., M. K. GAILLARD, D. V. NANOPOULOS & S. RUDAZ. 1981. Phys. Lett. **99B:** 101.
9. FRY, J. N., K. OLIVE & M. TURNER. 1980. Phys. Rev. **D22:** 2952.
10. KOLB, E. W. & S. WOLFRAM. 1980. Nucl. Phys. **B172:** 224.
11. ELLIS, J., M. GAILLARD & D. V. NANOPOULOS. 1979. Phys. Lett. **88B:** 320.
12. TURNER, M. & D. N. SCHRAMM. 1978. Nature **279:** 303.
13. PRESS, W. 1980. Phys. Scr. **21:** 702.
14. T'HOOFT, G. 1974. Nucl. Phys. **B79:** 276.
15. POLYAKOV, A. M. 1974. JETP **20:** 194.
16. ZELDOVICH, Y. & A. KHLOPOV. 1979. Phys. Lett. **79B:** 242.
17. PRESKILL, J. 1979. Phys. Rev. Lett. **43:** 1365.
18. USON, J. & D. WILKINSON. 1984. *In* Proceedings, Inner Space/Outer Space Conference. M. Turner & E. Kolb, Eds. University of Chicago Press. Chicago, Illinois.
19. GUTH, A. 1981. Phys. Rev. **D23:** 347.
20. GLINER, E. B. & I. G. DYNIKOVA. 1975. Sov. Astron. Lett. **1:** 93.
21. SATO, K. 1981. Mon. Not. R. Astr. Soc. **195:** 467.
22. KAZANAS, D. 1980. Astrophys. J. **241:** L59.
23. LINDE, A. D. 1982. Phys. Lett. **108B:** 389.
24. ALBRECHT, A. & P. STEINHARDT. 1982. Phys. Rev. Lett. **48:** 1220.
25. GUTH, A. 1984. *In* Proceedings, Inner Space/Outer Space Conference. M. Turner & E. Kolb, Eds. University of Chicago Press. Chicago, Illinois.
26. ZELDOVICH, Y. & I. NOVIKOV. 1983. Relativistic Astrophysics: Volume II-Cosmology. University of Chicago Press. Chicago, Illinois.
27. ELLIS, J., D. NANOPOULOS, K. OLIVE & K. TAMVAKIS. 1983. Phys. Lett. **118B:** 335.
28. PI, S. Y. 1984. Nucl. Phys. **B252:** 127.
29. VILENKIN, A. 1985. Phys. Rep. **121:** 1.
30. STEINHARDT, P. 1984. *In* Proceedings, Inner Space/Outer Space Conference. M. Turner & E. Kolb, Eds. University of Chicago Press. Chicago, Illinois.
31. GOTT, R. 1982. Nature **295:** 304.
32. WITTEN, B. 1985. Phys. Lett. **149B:** 351.
33. ARNISON, G. 1983. Phys. Lett. **129B:** 273.
34. LUDLAM, T. W. & H. E. WEGNER, Eds. 1984. Proceedings of the Brookhaven Conference on Quark Matter, 1984. Nucl. Phys. **A418.**
35. CRAWFORD, M. 1985. Phys. Rev. **D31:** 1956.
36. CRAWFORD, M. & D. N. SCHRAMM. 1982. Nature **298:** 538.
37. SCHRAMM, D. N. & K. OLIVE. 1984. Nucl. Phys. **A418:** 289.
38. WITTEN, E. 1984. Phys. Rev. **D30:** 272.
39. APPLEGATE, J. & C. HOGAN. 1985. Preprint Caltech. Pasadena, California.

40. YANG, J., M. TURNER, G. STEIGMAN, D. N. SCHRAMM & K. OLIVE. 1984. Astrophys. J. **281:** 493.
41. SCHRAMM, D. N. & G. STEIGMAN. 1984. Phys. Lett. **141B:** 337.
42. STEIGMAN, G., D. N. SCHRAMM & J. GUNN. 1977. Phys. Lett. **66:** 202.
43. PARTRIDGE, B. 1985. *In* Proceedings, Third Rome Conference. Nuovo Cimento. In press.
44. FREESE, K. & D. N. SCHRAMM. 1983. Nucl. Phys. **B233:** 167.
45. FABER, S. & J. GALLAGHER. 1979. Ann. Rev. Astron. Astrophys. **17:** 135.
46. PEEBLES, P. J. E. 1980. *In* Physical Cosmology. J. Audouze, R. Balian & D. N. Schramm, Eds. Reidel. Amsterdam.
47. BAHCALL, J. *In* Proceedings, Inner Space/Outer Space Conference. M. Turner & E. Kolb, Eds. University of Chicago Press. Chicago, Illinois.
48. SCHRAMM, D. N. & G. STEIGMAN. 1981. Astrophys. J. **243:** 1.
49. KIRSHNER, R., G. OEMLER & R. SCHECTER. 1979. Astrophys. J. **84:** 95.
50. LEA, S., J. SILK, E. KELLOGG & S. MURRAY. 1973. Astrophys. J. **184:** L105.
51. HEGYI, D. & K. OLIVE. 1983. Phys. Lett. **126B:** 28.
52. GOTT, J., J. GUNN, D. N. SCHRAMM & B. TINSLEY. 1974. Astrophys. J. **194:** 543.
53. OSTRIKER, J. & L. COWIE. 1980. Astrophys. J. **243:** L127.
54. FREESE, K., R. PRICE & D. N. SCHRAMM. 1984. Astrophys. J. **275:** 405.
55. VITTORIO, N. & J. SILK. 1984. *In* Proceedings, Inner Space/Outer Space Conference. M. Turner & E. Kolb, Eds. University of Chicago Press. Chicago, Illinois.
56. BOND, J. R. & G. EFSTATHIOU. 1984. *In* Proceedings, Inner Space/Outer Space Conference. M. Turner & E. Kolb, Eds. University of Chicago Press. Chicago, Illinois.
57. PEEBLES, P. J. E. 1982. The Large-Scale Structure of the Universe. Princeton University Press. Princeton, New Jersey.
58. DAVIS, M. 1984. *In* Proceedings, Inner Space/Outer Space Conference. M. Turner & E. Kolb, Eds. University of Chicago Press. Chicago, Illinois.
59. BAHCALL, N. & R. SONIERA. 1983. Astrophys. J. **270:** 20.
60. KLYPIN, A. & A. KHLOPOV. 1983. Sov. Astron. Lett. **9:** 41.
61. KAISER, N. 1985. Astrophys. J. **282:** 374.
62. FRY, J. N. 1983. Private communication.
63. SZALAY, A. & D. N. SCHRAMM. 1985. Nature **314:** 718.
64. MANDELBROT, B. 1977. The Fractal Geometry of Nature. Freeman Press. San Francisco, California.
65. GELLER, M. 1985. *In* Proceedings, Third Rome Conference. Nuovo Cimento. In press.
66. CHARLTON, J. & D. N. SCHRAMM. 1985. In preparation.
67. TUROK, N. & D. N. SCHRAMM. 1984. Nature **312:** 598.
68. BOND, J. R. & A. SZALAY. 1983. Astrophys. J. **274:** 443.
69. KOLB, E., D. SECKEL & M. TURNER. 1985. Preprint Fermilab. Chicago, Illinois.
70. DAVIS, M., G. EFSTATHIOU, C. FRENK & S. WHITE. 1985. Astrophys. J. Suppl. **57:** 241.
71. BLUMENTHAL, G. R., S. FABER, J. PRIMACK & M. REES. 1984. Nature **311:** 517.
72. GUNN, J. E., B. W. LEE, I. LERCHE, D. N. SCHRAMM & G. STEIGMAN. 1977. Astrophys. J. **223:** 1015.
73. DAVIS, M., M. LECAR, C. PRYOR & E. WITTEN. 1981. Astrophys. J. **250:** 423.
74. DEKEL, A. & M. WEST. 1984. Astrophys. J. **279:** 1.
75. OLIVE, K., D. SECKEL & E. VISHNIAC. 1984. Preprint Fermilab. Chicago, Illinois.
76. TURNER, M., G. STEIGMAN & L. KRAUSS. 1984. Phys. Rev. Lett. **52:** 2090.
77. GELMINI, G., D. N. SCHRAMM & J.W.F. VALLEE. 1985. Phys. Lett. **146B:** 311.
78. OLIVE, K., D. N. SCHRAMM & M. SREDNICKI. 1984. Preprint Fermilab. Chicago, Illinois.
79. SCHRAMM, D. N. 1985. *In* Proceedings, Third Rome Conference. Nuovo Cimento. In press.

The LEP Program

JAMES V. ALLABY

CERN CH-1211

Geneva 23, Switzerland

INTRODUCTION

LEP is an electron-positron collider 26 kms in circumference. Initially, four interaction regions are to be equipped with experiments, although there is the potential for eight crossing points with four-bunch operation. Before reviewing the experiments, though, it is useful to recall some basic facts about LEP.

The collider is located underground between the outskirts of Geneva and the Jura mountains (see FIGURE 1.1). The major part of the tunnel in which LEP will be installed will be bored by machine in the "molasse" (a kind of sandstone) that lies at depths of several tens of meters below the surface of the ground in this region, which is formed of glacial moraine. However, about one-eighth of the circumference lies under the foothills of the Jura and here the tunnel must pass through the limestone rock where a boring machine cannot be used. Since the surface of the ground is itself not flat over this huge area, the plane of the machine is not horizontal, but slightly inclined to minimize the distance from the surface to the interaction regions; hence, the cost of the access shafts.

Even with this trick, the interaction regions are typically about 100 meters below the surface. An artist's impression of a possible interaction region is shown in FIGURE 1.2. Three access shafts will be provided at these regions: one for access to the machine itself; one to allow the large experiments to be installed; and one, of smaller diameter, to allow the physicists to intervene at the detector from their control room on the surface.

The planning for LEP sees first beam collisions at the end of 1988 and all the four approved experiments (ALEPH, DELPHI, L3, and OPAL) plan to be installed by that date, which is when the machine should provide electrons and positrons of up to 50 GeV energy. Operation at this energy will use conventional radio-frequency accelerating cavities. However, research and development of suitable superconducting cavities is progressing very well and it is believed that further energy increases will occur adiabatically by the gradual addition of such cavities in the available straight sections and the eventual replacement of the original copper cavities by superconducting ones. The final operating energy thus attainable should be above the W-pair threshold.

The four LEP experiments are under construction by large collaborations of physicists and engineers, representing many institutes and laboratories from around the world. In this report, individual names will not be given, but they can be found in the appropriate Technical Proposal.

After an all too brief review of the physics at LEP, the four approved LEP experiments will be described in separate sections of this paper. Since it is impossible to describe everything in all detail, an attempt is made to pick out novel techniques in the

FIGURE 1.1 The location of LEP.

experiments. The choice is the author's and he would like to apologize to any constructors of components of a LEP detector that are either not described, or worse, described incorrectly. In particular, no description of data acquisition systems or computing is given, nor of the small-angle tagging systems for luminosity monitoring.

PHYSICS AT LEP

The initial physics operation will certainly be at the Z^0 peak. There the event rate will be high, which is important for the start-up of new detectors. At the energy of the Z^0, the maximum luminosity of LEP will be 6×10^{31} cm^{-2} s^{-1}. For the purposes of rates quoted here, a luminosity of one-tenth of the maximum will be used, i.e., 6×10^{30} cm^{-2} s^{-1}. The total rate at the peak is expected to be $\approx$2600 times the QED μ-pair rate ($R \approx 2600$ after radiative corrections), so that in an operating year of 2500 hours, 1.5 million Z^0 events will be detected in the full solid angle for an integrated luminosity of 54 pb^{-1}.

In the first year of operation, with the above assumptions, more than 50,000 $\mu^+\mu^-$ pairs will be produced, together with a similar number of e^+e^- pairs inside the fiducial volume of the detectors. Aside from their intrinsic interest, these events will be ideal to calibrate and understand fully the LEP detectors.

Furthermore, there is a wide range of physics topics that can be addressed at the Z^0 peak, some of which will be discussed below, so there is every indication that the physicists at LEP will wish to operate at or near the peak for at least the first two years.

Initial Physics

Some of the obvious topics of interest are:

(a) Z^0 mass and width: Mass determination to 0.1–0.2 GeV seems feasible, allowing a precise determination of the Weinberg angle (±0.001 in $\sin^2\theta$). The width measurement can be used to put limits on the number of neutrino families. The beam energy spread in the Z^0 region is expected to be $\approx$100 MeV and the energy uncertainty to be $\approx$20 MeV.

(b) Electroweak asymmetry near the pole: The μ-pair reaction, $e^+e^- \Longrightarrow \mu^+ + \mu^-$, has a forward-backward asymmetry passing through zero at the pole. Thus, the magnitude of the asymmetry can be used to locate the exact position of the Z^0 pole and hence determine the Weinberg angle. With 50,000 μ-pairs, an error of 0.005 can be achieved. Radiative corrections on the pole change the asymmetry by $\leq$0.02 and this effect can be calculated or removed by cuts.

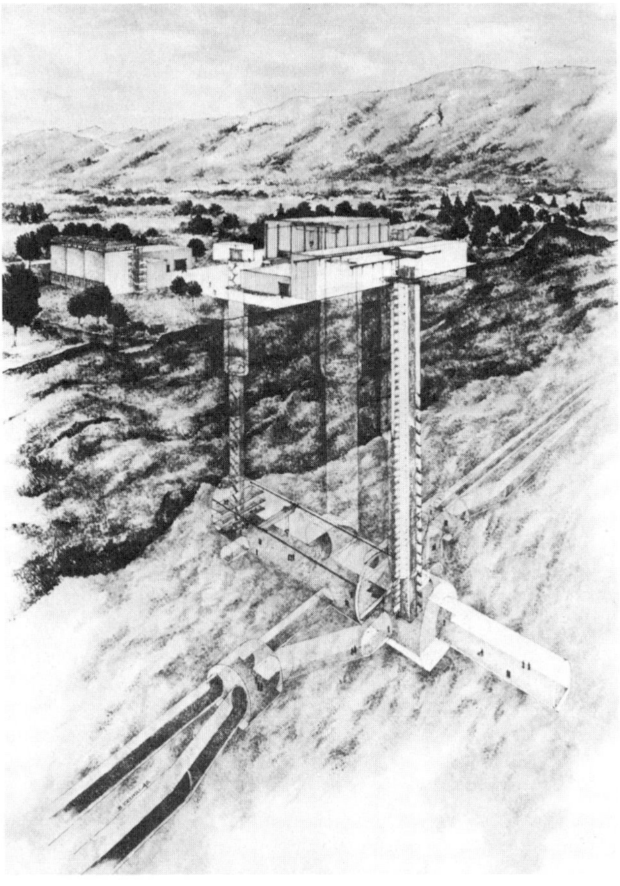

FIGURE 1.2. An artist's impression of an interaction region.

(c) $e^+ + e^- \Longrightarrow e^+ + e^-$: This reaction has contributions (see FIGURE 2.1) from photon exchange in the t-channel (a), Z^0 exchange in the s-channel (b), and interference between photon exchange in the s- and t-channels (c). Above the pole, interference between photon exchange in the t-channel and Z^0 exchange in the s-channel (d) becomes important. Below 20°, photon exchange dominates and the angular region of 10°–20° can be used to determine the beam luminosity, whereas the backward region is pure Z^0 exchange.

(d) The unexpected: One always hopes that high-energy e^+e^- collisions might succeed in "breaking the bag" and thus liberate free quarks. Supersymmetry might reveal itself by producing spin-0 partners of ordinary leptons and quarks. Scalar lepton pairs from Z^0 decay would decay rapidly to an ordinary lepton pair plus noninteracting photinos or goldstinos carrying large missing transverse momentum. New levels of matter (preons, rishons, etc.) might be produced. Excited leptons could be seen decaying by radiative decay to normal leptons.

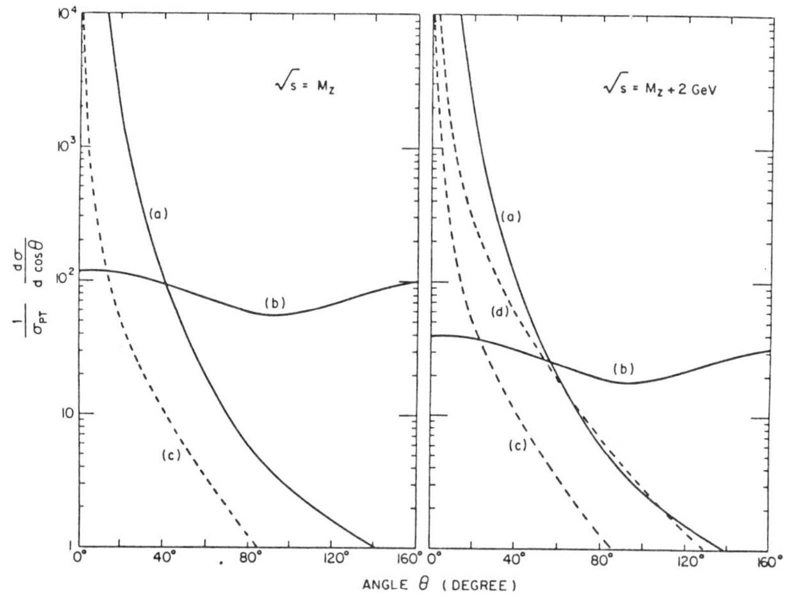

FIGURE 2.1. The contributions to $e^+e^- \rightarrow e^+e^-$ at and near the pole (see text).

Jets

The study of hadronic events (Jet Physics) will be important on the Z^0 peak. With the luminosity assumptions above, ≈200,000 events per year are expected for each quark flavor. In order to study the behavior of LEP detectors in this environment,

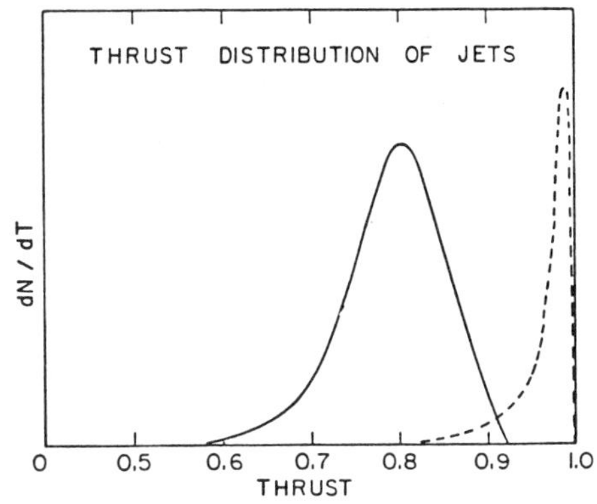

FIGURE 2.2. The thrust distributions for light (dashed line) and top (solid line) quarks.

Monte Carlo simulations are extensively used, such as the Lund model. A key element is flavor identification. Selection of top-quark pair production can be made by cuts on thrust. The expected thrust distribution for light-quark jets (dashed line) and top-quark jets (solid line) is shown in FIGURE 2.2 ($m_t = 25$ GeV). It can be seen that a thrust cut of $T < 0.9$ will select top-quark jets, whereas a selection of $T > 0.9$ will retain the light-quark jets (including *b*-quarks).

To select *b*-quark jets from the light-quark sample, one then will demand at least one identified prompt lepton, either $e^{\pm}$ or $\mu^{\pm}$. Finally, those leptons coming from *b*-decay will be selected by cuts in the transverse momentum of the leptons with respect to the jet axis.

FIGURE 2.3 shows the expected distribution of this transverse momentum for leptons from charm decay (dashed line) and beauty decay (solid line). With a selection of $p_T > 1$ GeV$/c$, the *b*-decays will be selected with 55% efficiency, whereas only 7% of the charm decays will survive the cut.

If we put in the efficiencies for electron and muon identification and for the cuts, we find that from an initial sample of 140,000 *b*-quark pairs, 16,000 events identified

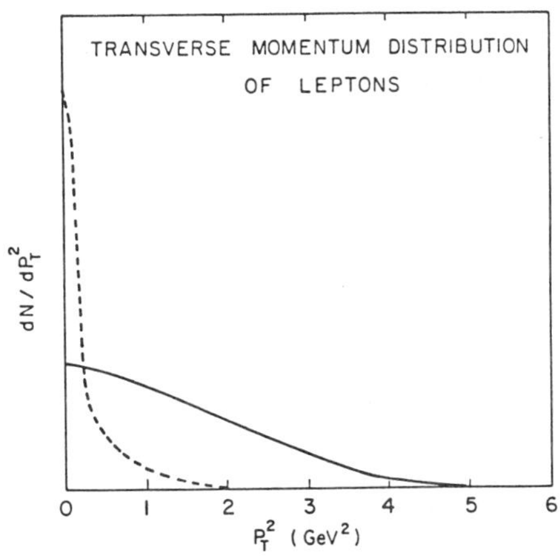

FIGURE 2.3. The transverse momentum distribution of leptons from charm decay (dashed line) and beauty decay (solid line).

by electron decay ($p_T > 1$ GeV/c) are retained containing 12,600 b-pairs, 2500 charm decays, and ≈900 symmetric background. For the muon sample ($p_T > 1$ GeV/c, $p > 5$ GeV/c), 8500 events are retained containing 6850 b-pairs, 1150 charm decays, and ≈500 symmetric background. Thus, we identify ≈19,500 of the original 140,000 b-quark pair events with ≈20% background. Further selection criteria will depend on details of the specific detector. For example, the identification of $K^{\pm}$ mesons in the charm decay, possible in the DELPHI detector, could reduce the charm contamination.

The background in the b-quark pair sample can be serious in the case of asymmetry measurements. If we take the electron sample above and evaluate the measured asymmetry, assuming the true asymmetry to be 0.22, we find $A_{meas} = 0.147 \pm 0.008$. The systematic error of the correction to the asymmetry due to the charm contamination would dominate over the statistical error. In this case, one would prefer to apply the same semileptonic decay selection to the opposite jet. In the example above, by selecting the semileptonic decay in both jets and by combining e^+e^-, $\mu^+\mu^-$, μ^+e^-, and $e^+\mu^-$, one would obtain a pure sample of 900 b-pairs and the error would be dominated

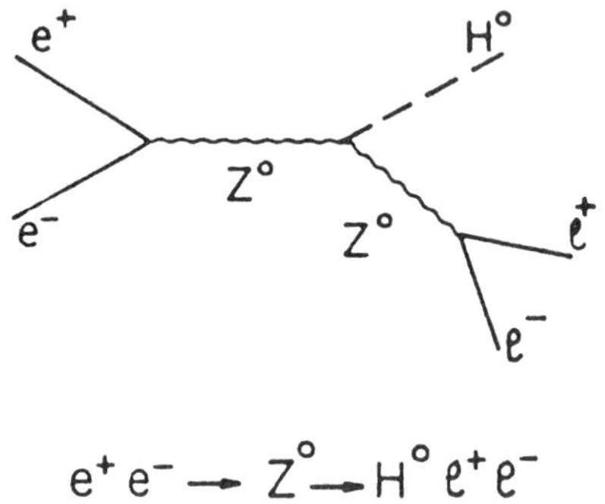

FIGURE 2.4. The dominant graph for Higgs production.

by statistics, thus yielding a background-free measurement of the asymmetry, $A_{meas} = 0.220 \pm 0.032$.

Higgs Search

This is clearly one of the "hot" topics of LEP physics that surely will be attacked by all the experiments. The most promising reaction to detect a neutral Higgs is

$$e^+ + e^- \Longrightarrow Z^0 \Longrightarrow H^0 + \ell^+ + \ell^-,$$

or at energies above $M_H + M_Z$,

$$e^+ + e^- \Longrightarrow H^0 + Z^0, \quad Z^0 \Longrightarrow \ell^+ + \ell^- \ (\ell = e \text{ or } \mu).$$

The first reaction occurs on the Z^0 pole producing the H^0 with a virtual Z^0. The second is an example where the energy is higher and a real Z^0 is produced. The diagram for this reaction is shown in FIGURE 2.4.

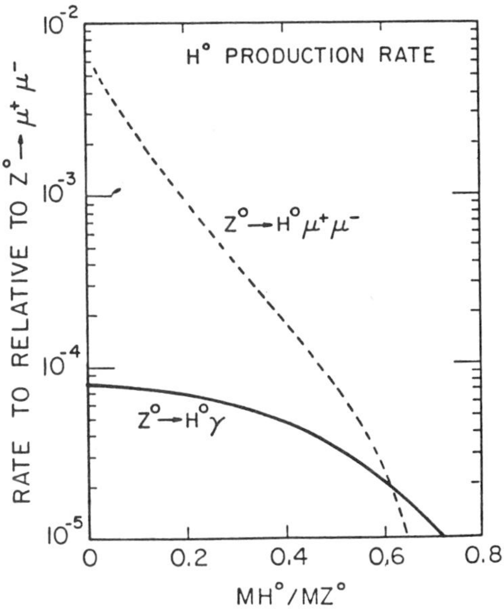

FIGURE 2.5. The H^0 rate of production on the Z^0 pole as a function of the mass ratio, M_H/M_Z.

The rate for production on the pole is shown in FIGURE 2.5 as a function of the mass ratio, M_H/M_Z. The lepton pair tends to take the highest mass kinematically possible. The method is to search for a mass peak in the missing mass to two well-measured leptons. The background comes from semileptonic decay of heavy quarks. FIGURE 2.6 shows simulation results for three different Higgs masses. For $M_H < 40$ GeV, a clear separation from background is obtained. If the Higgs mass is higher, it is more favorable to go to an energy of 164 GeV. There, the cross section for the reaction producing a real Z^0 is 3 pb. The simulation results in this case, for a Higgs mass of 50 GeV, are shown in FIGURE 2.7.

Final identification of the Higgs events will require further selection. The H^0 will decay dominantly into two jets with an isotropic distribution. Hence, a cut in the

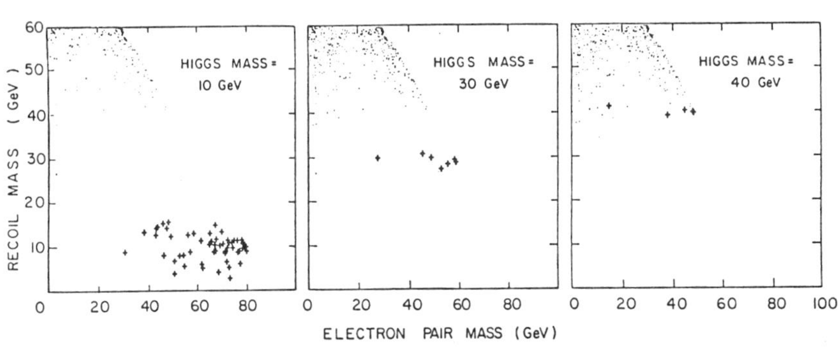

FIGURE 2.6. The events and backgrounds for Higgs search (see text).

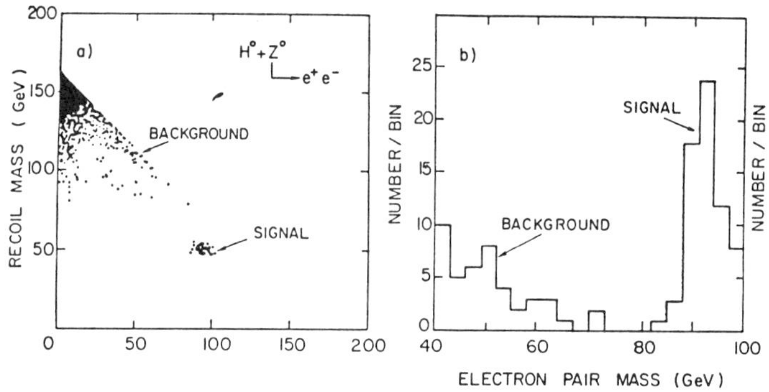

FIGURE 2.7. The signal and background for higher energy search.

momentum of the lepton (>10 GeV/c) and on its p_T with respect to the jet axis (>8 GeV/c) may allow one to single out the Higgs events. If the top quark is heavy (e.g., > 35 GeV), the dominant background will then be produced by top-quark pairs, which both decay semileptonically to produce the high momentum lepton used in the Higgs selection and which can also have high transverse momentum. This background could make it impossible to see the Higgs on the Z^0 peak.

This has been a small sample of the physics one might see at LEP. We will look now at the details of the four approved detectors to see how they prepare for this exciting era.

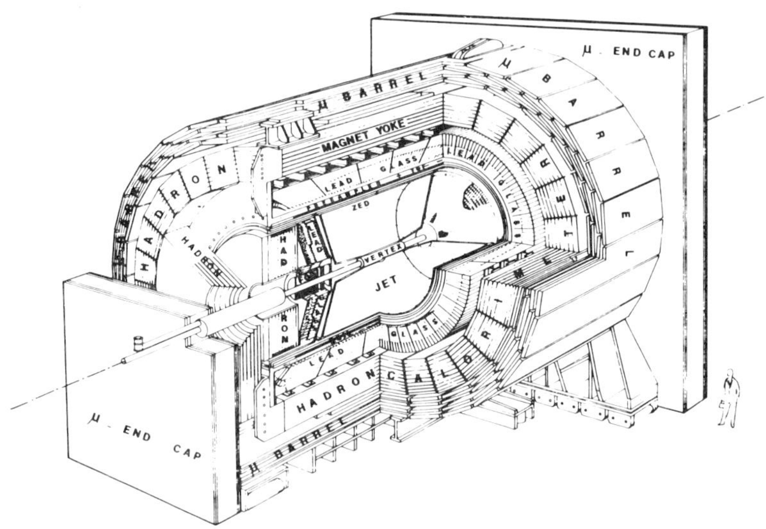

FIGURE 3.1. The overall view of the OPAL detector.

OPAL

This detector, whose name represents Omni-Purpose Apparatus for Lep, is probably the most conventional of the four LEP detectors. Many of its design features are based on proven techniques used in the JADE detector at PETRA. An overall view of the OPAL detector is shown in FIGURE 3.1.

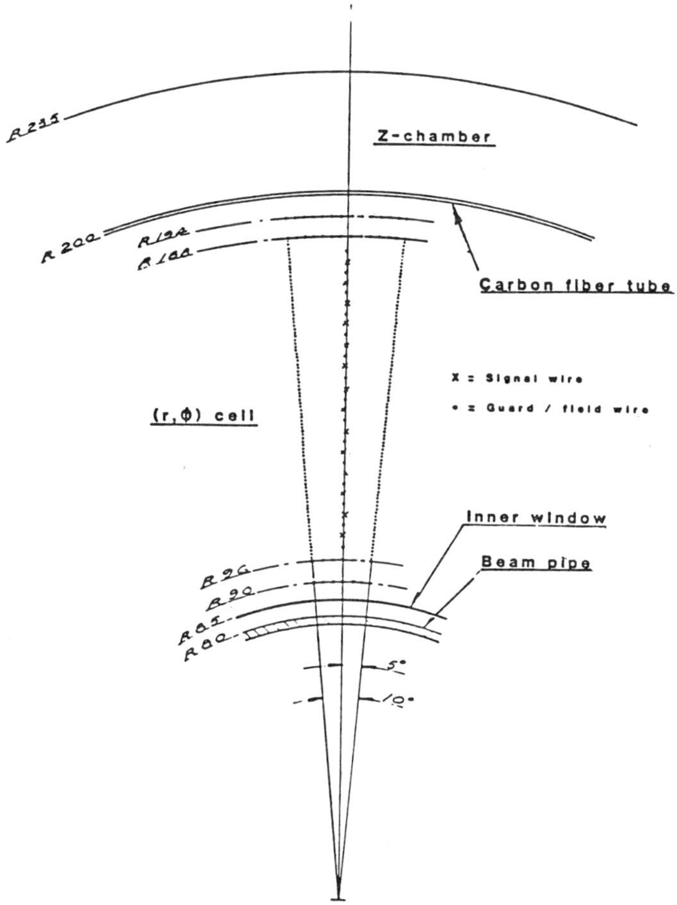

FIGURE 3.2. The OPAL vertex detector.

The OPAL detector measures charged tracks with a vertex detector and a central tracking chamber, both designed to operate at 2–4 bar pressure. The electromagnetic calorimeter is outside the coil and is followed by a hadron calorimeter and muon chambers.

The vertex detector surrounds the beam pipe, is of an active length of one meter, and covers the radial space from 85 mm to 235 mm. This small drift chamber, shown in FIGURE 3.2, should provide high precision (r, ϕ) measurements ≈ 50 μm and a z measurement better than 1 mm. A z-coordinate readout from the (r, ϕ) cells based on the pulse propagation time difference will provide z information for triggering with a precision $\approx 2\%$ of the wire length. The outer 35 mm will contain a device to provide precise z information, which is still under study.

The central track detector of OPAL is a jet chamber with wires four meters long. In order to assure that the design is sound, a full-scale prototype containing two complete sectors has been built. FIGURE 3.3 shows this prototype jet chamber. Tracking accuracies between 110 μm and 150 μm have been obtained for drift distances up to 16 cm (as shown in FIGURE 3.4a) and the relative z-resolution is $\approx 1\%$ (as shown in FIGURE 3.4b). By sampling the ionization of a particle up to 160 times, a relative particle identification by dE/dx of 3–4% is obtained, thus giving π/e separation (as shown in FIGURE 3.4c).

The track detection of OPAL is completed by separate Z-chambers located outside the jet chamber, which are in the form of 24 azimuthal chambers that are each 4.5 meters long in the z-direction and contain eight bidirectional drift cells that are each 56 cm long. A prototype chamber at atmospheric pressure has been tested in magnetic fields of four and ten kilogauss.

A solenoidal magnetic field of four kilogauss is provided by a conventional (warm) coil of 2.25 m radius. The Collaboration plan to replace this with a superconducting coil that will give up to a ten kilogauss field at a later stage.

FIGURE 3.3. The prototype jet chamber (OPAL).

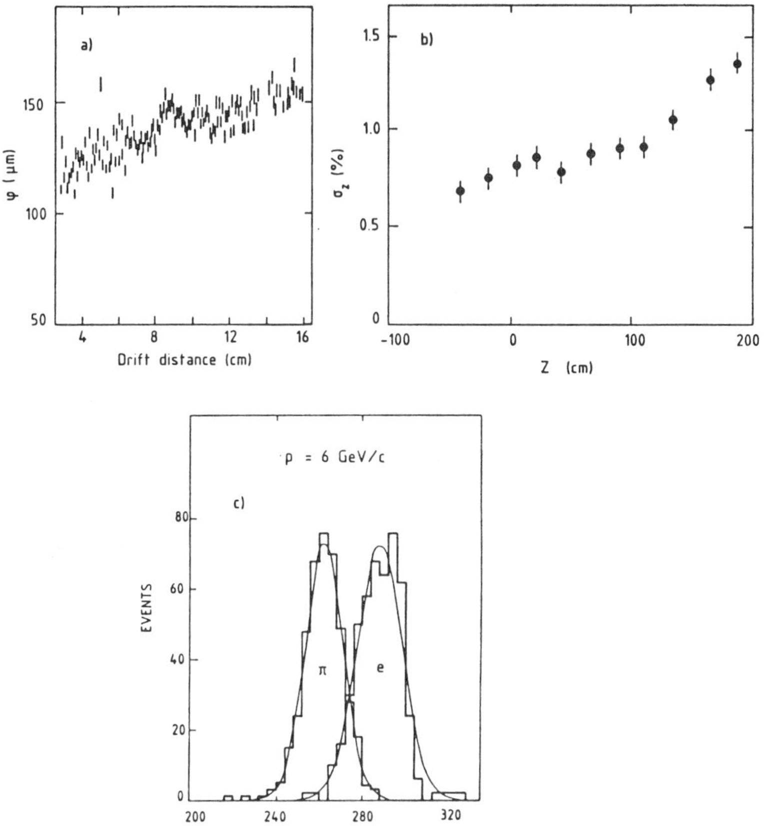

FIGURE 3.4. The results on jet chamber prototype (see text).

Immediately outside the coil, time-of-flight counters will be installed. These are followed immediately by the electromagnetic calorimeter, which is composed of lead glass blocks in a pointing geometry as shown in FIGURE 3.5. Each block, of thickness 23.9 X_0, will be viewed by a Hamamatsu R2238 photomultiplier tube. In order to maintain spatial resolution for electromagnetic showers behind the coil, the calorimeter is preceded by a "presampler" that comprises two layers of Iarocci tubes (each offset by half a cell) equipped with a cathode strip readout.

Hadron calorimetry is provided by interleaving the 10 cm thick iron plates of the solenoid return yoke with carbon-coated plastic tubes operating in the limited streamer mode.

Muon identification will be obtained by four layers of drift chambers outside the iron return yoke. Most of the chambers will be 10.4 m long and 1.2 m wide and will contain two drift cells, each with a maximum drift length of 30 cm. Resolution in the transverse drift direction of ≈ 1 mm will be obtained by conventional timing techniques. Studies have been made of specially shaped cathode pads to improve the z-resolution

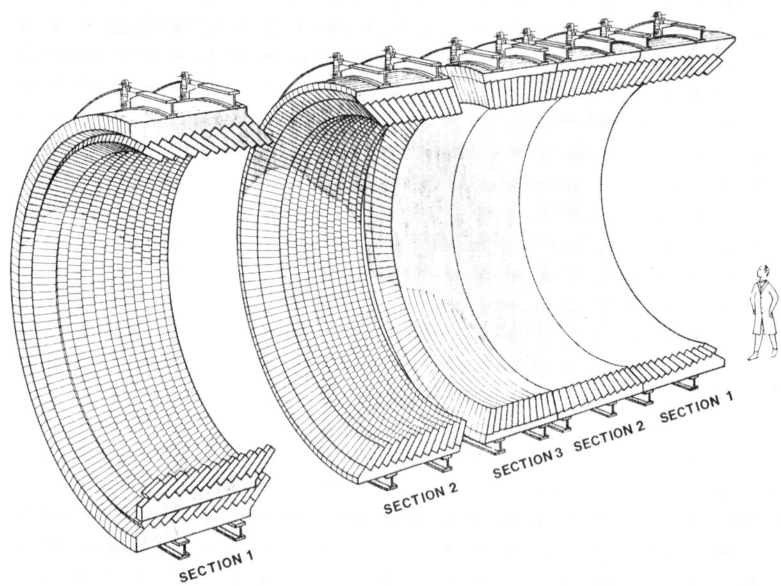

FIGURE 3.5. The OPAL electromagnetic calorimeter (see text).

obtained by the usual charge division or the time difference measurement from the wire signals.

The principle is shown in FIGURE 3.6, which indicates the form of the cathode pads above and below the wire. The charge, Q_A, induced on the diamonds and Q_B on the triangles is measured with a conventional ADC so that the ratio, $\Delta = (Q_A - Q_B)/(Q_A + Q_B)$, can be formed off-line. Two such ratios, Δ_t and Δ_b, can be formed for the top and bottom sets of pads, which are displaced relative to each other along the wire direction by a quarter of the repeat distance, so that Δ_t and Δ_b uniquely determine the longitudinal coordinate, modulo the repeat distance.

The endcaps are equipped similarly with electromagnetic and hadron calorimeters

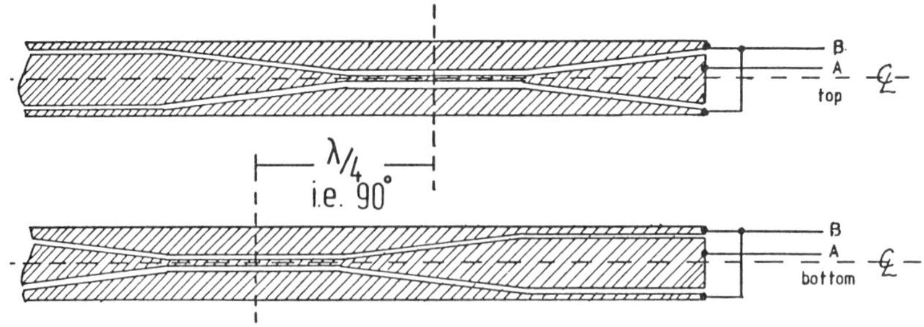

FIGURE 3.6. The cathode pad structure of the OPAL muon chambers (see text).

as well as forward muon chambers. The readout of the lead glass blocks of the forward electromagnetic calorimeter will be by vacuum phototriodes that have been shown to tolerate the magnetic field when its axis is $< \pm 30°$ with respect to the normal to the photocathode.

ALEPH

The ALEPH detector is technically more ambitious than OPAL and is based on a large superconducting solenoid (5.3 m in diameter and 6.4 m long) to provide a

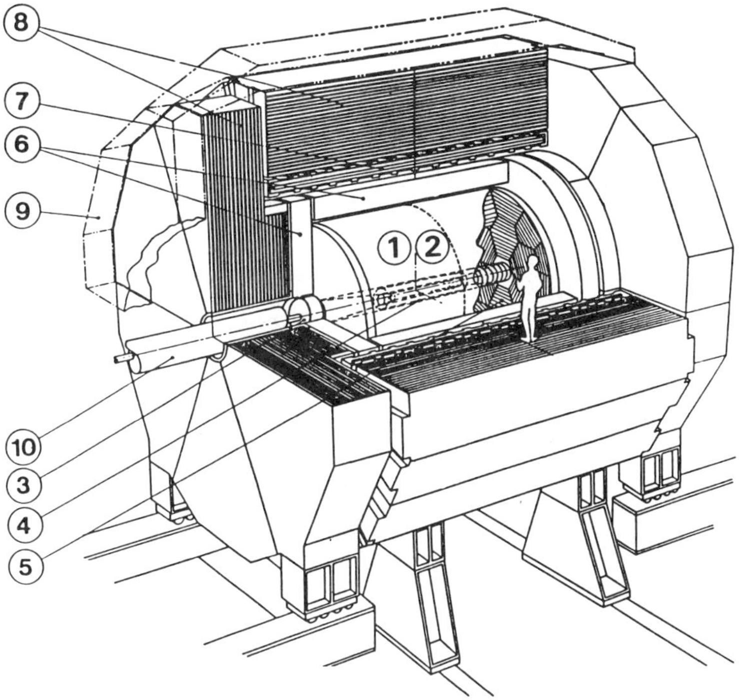

FIGURE 4.1. The overall view of the ALEPH detector: 1 = beam pipe; 2 = minivertex detector; 3 = luminosity monitor; 4 = inner chamber; 5 = Time Projector Chamber (TPC); 6 = electromagnetic calorimeter; 7 = superconducting coil; 8 = hadron calorimeter; 9 = muon detector; 10 = superconducting quadrupole.

magnetic field of 15 kgauss. The central detector is a large Time Projection Chamber (TPC), 3.6 m in diameter and 4.8 m long, and this should lead to optimum momentum measurements on charged tracks.

The overall layout is shown in FIGURE 4.1. As with OPAL, the beam pipe is surrounded by an inner track chamber whose main function is to provide trigger information on charged tracks. Precision of 100 μm is the goal in the (r, ϕ) projection. This detector will not be described in detail.

The TPC of ALEPH is a key component of the detector and although it is based on the PEP4 TPC, it contains some novel features. The endplates are divided into 18 sectors of three different types and they are arranged to give a radial "zigzag" boundary between sectors, thus ensuring that straight tracks never fall completely on a radial boundary. Furthermore, the pad structure is in the form of 21 concentric circles, each pad having a radial length of 30 mm and a width of 6.5 mm. Each endplate contains 20,500 pads and 3240 wires. The use of circular pad rows eliminates, for stiff tracks, a term in the resolution function caused by fluctuations in the size of the avalanche on each wire and proportional to the angle between the track and the pad axis. FIGURE 4.2 shows a sketch showing the dimensions of the ALEPH TPC, and details of the endplates are shown in FIGURE 4.3.

Extensive tests have been carried out on a prototype (called TPC90) using laser beams and cosmic rays in a magnetic field up to 12 kgauss. Tests of various schemes for

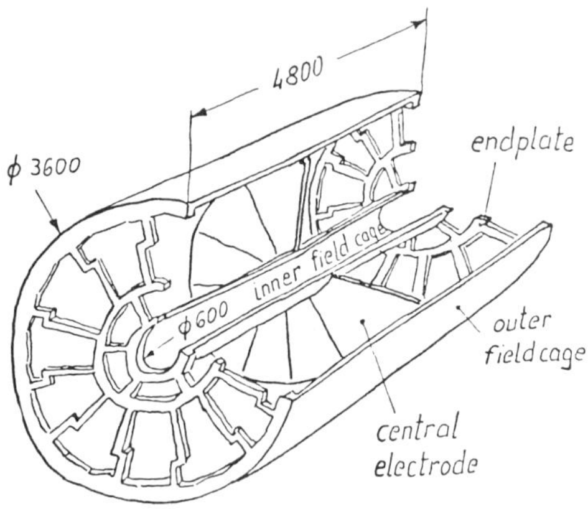

FIGURE 4.2. The dimensions of the ALEPH TPC.

"gating" the TPC have been investigated, as well as tests of production techniques for the field cage, etc.

Immediately outside the TPC, the electromagnetic calorimeter is located. ALEPH have chosen to use a lead/wire-chamber sandwich technique for their electromagnetic calorimeter. The chambers are made from aluminum extrusions with readout by cathode pads. Pads from consecutive planes are connected to form towers pointing to the intersection region and are summed in three depth layers corresponding to the first 4 X_0 (10 layers), the central 9 X_0 (23 layers), and the last 9 X_0 (12 layers). Lead sampling of 2 mm is used for the first and central layers, and of 4 mm for the last layer. The same technique is used for the endcaps.

The lateral size of each tower as seen along its axis is typically 3 $\times$ 3 cm^2. The segmentation in azimuth angle is $\approx 1°$ and is similar in polar angle near 90°.

The chambers have been tested with the usual argon + ethane mixture. However, it was found that the resolution in a magnetic field was degraded quite severely with this mixture. The use of a Xe(80%) + CO_2(20%) mixture restored the resolution found with argon + ethane in the absence of a magnetic field.

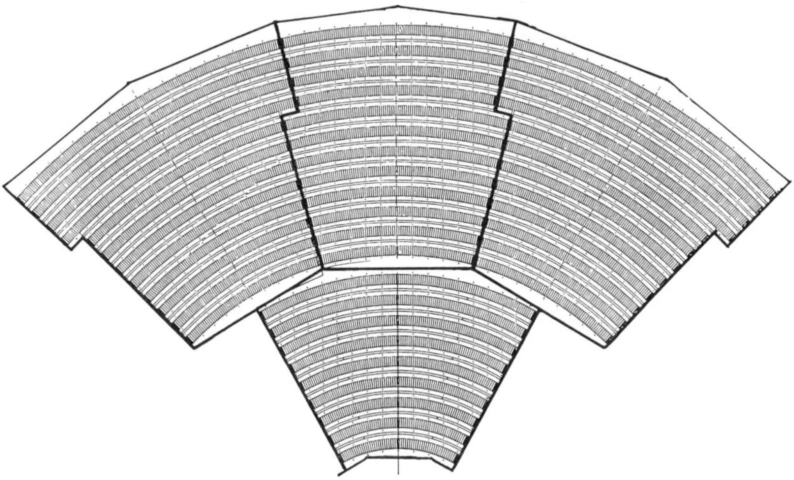

FIGURE 4.3. A diagram of 4 of the 18 sectors for each endplate of the ALEPH TPC, which shows the arrangement of some of the 22,000 readout pads.

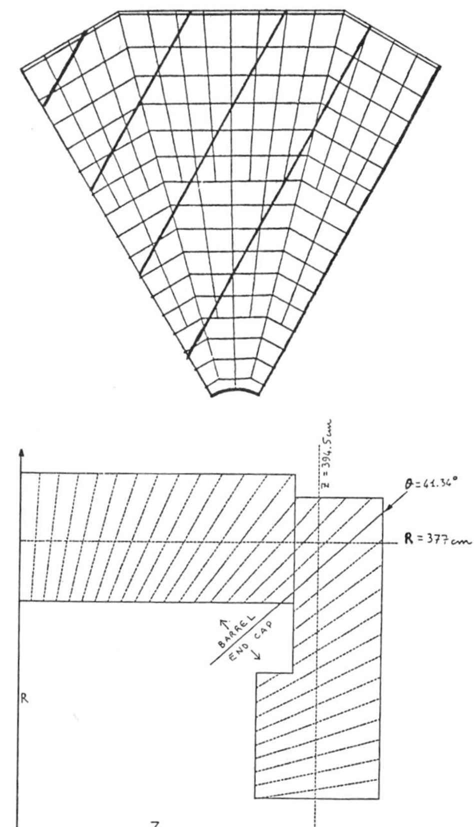

FIGURE 4.4. The tower structure of the ALEPH hadron calorimeter (see text).

Energy resolution of $18\%/\sqrt{E}$ was measured at 5 and 10 GeV for electrons. Typical spatial resolution of 2–3 mm for the centroid of a shower has been obtained.

Outside the superconducting coil, the return yoke is instrumented as a hadron calorimeter and muon detector. The iron is in the form of 5 cm plates and planes of streamer tubes are interleaved between subsequent plates. A total of 23 plates (of which the last is 10 cm thick) make up the total iron thickness of 120 cm. Each tube layer is equipped with pad readouts on one side for integrated energy flux measurement and with strips parallel to each tube on the other side, which yield digital information as the basic tool for muon identification.

The pad readout is organized into projective towers, each tower being longitudinally divided into two parts. The azimuthal and polar angle subdivision is shown in FIGURE 4.4. Resolution of $\approx 85\%/\sqrt{E}$ is expected with good linearity up to 30 GeV.

For safety reasons, the usual argon + isobutane mixture of streamer tubes cannot be used. Instead, the safer mixture of 14% argon, 56% CO_2, and 30% *n*-pentane will be used.

External to the magnet—both in the barrel and the endcap regions—two double layers of streamer tubes are installed to detect muons traversing the full iron and to measure their direction.

L3

The L3 detector is a special purpose device whose aims are to measure photons, electrons, and muons with a resolution of $\Delta p/p \approx 1\%$, to measure hadron jet energies with good resolution, and to measure vertices with high precision.

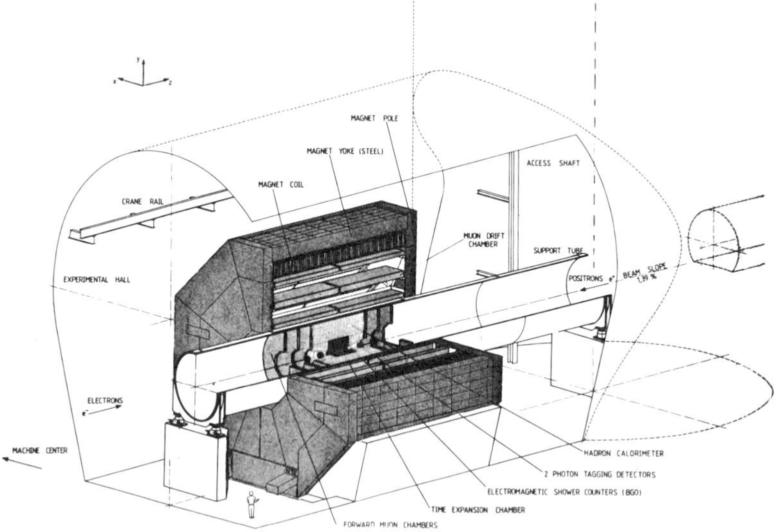

FIGURE 5.1. The overall view of the L3 detector (see text).

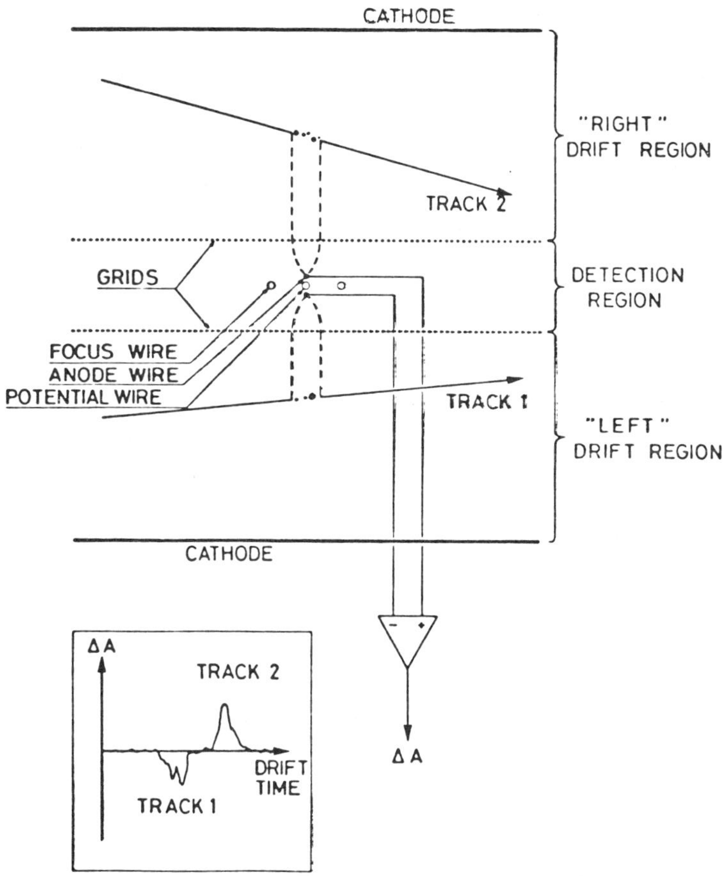

FIGURE 5.2. The principle of the TEC (Pattern Recognition Configuration) (see text).

The overall view of the L3 detector is shown in FIGURE 5.1. Charged particles from the interaction are tracked in a Time Expansion Chamber (TEC) out to $r = 50$ cm with high spatial and double-track resolution. Electrons and photons will shower in a cylindrical array of Bismuth Germanate Oxide (BGO) of thickness 22 X_0. Hadrons that start to shower in the BGO will continue into the hadron calorimeter extending out to $r = 217$ cm. Only muons will penetrate into the muon spectrometer where they will be momentum analyzed in a magnetic field of five kilogauss by three layers of chambers extending out to $r = 568$ cm. The magnetic field is provided by a warm coil of octagonal shape constructed from welded aluminum plates. The inner radius of the aluminum conductor is 5.93 m and the outer radius is 6.82 m.

The TEC vertex chamber uses a low drift velocity in order to detect the shape of the anode signal and uses its centroid to define the drift-time value rather than the leading edge of the pulse. In addition, double anode wires are used to resolve the left-right ambiguity as shown in FIGURE 5.2. The aim is to obtain a position resolution of $\leq 40\ \mu$m and a double-track resolution of ≤ 500 μm to permit lifetime measurements on short-lived particles and momentum measurements on charged tracks. FIGURE 5.3 shows two sectors of the TEC chamber. Prototype work has demonstrated the viability of the TEC concept.

The electromagnetic calorimeter of L3 will be constructed from BGO crystals. The number of segments in the barrel region (42°–90°) has been fixed to be 160 in azimuth and 48 in polar angle, while the geometry of the endcaps is not yet final. The light from each BGO crystal will be detected by a photodiode. Intensive prototype work is under way, including tests of a 5 × 5 matrix of tapered crystals in a beam, and industrial methods for cutting and polishing the crystals have been determined. A sketch of this calorimeter is shown in FIGURE 5.4.

Hadron energy that is not deposited in the BGO (one absorption length) will be detected by a fine-grained calorimeter containing four absorption lengths of depleted

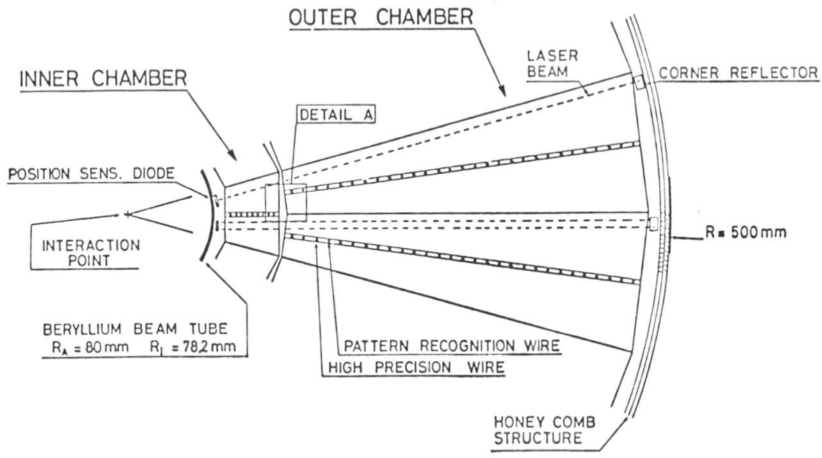

FIGURE 5.3. A schematic drawing of two TEC sectors.

uranium and instrumented with wire chambers. Any hadron energy that leaks beyond the uranium will be detected in the muon filter, which comprises two absorption lengths of copper and is instrumented with streamer chambers.

The barrel hadron calorimeter will be constructed from nine rings, each 50 cm wide, which can roll inside the muon filter on rails (see FIGURE 5.5). Each ring contains 16 separate modules arranged azimuthally on a frame. Tests of the chambers are under way.

The muon detector of L3 comprises two halves, each containing eight octants fixed on a torque tube. One octant has already been constructed to study the effects of loading, etc., since precise alignment is required. Each octant contains three types of momentum chambers and four "Z" chambers that form the covers of the inner and outer momentum chambers. A sketch of an octant showing the assembly technique is shown in FIGURE 5.6. Prototypes of all these chambers have been constructed and are

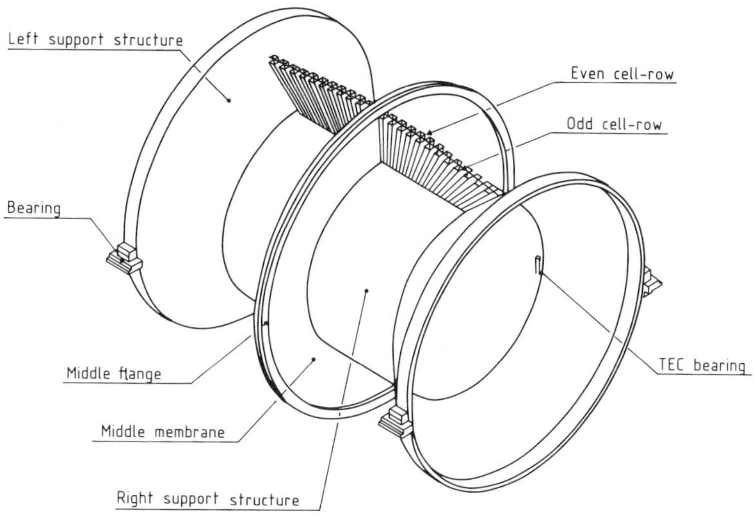

FIGURE 5.4. A sketch of the L3 BGO calorimeter.

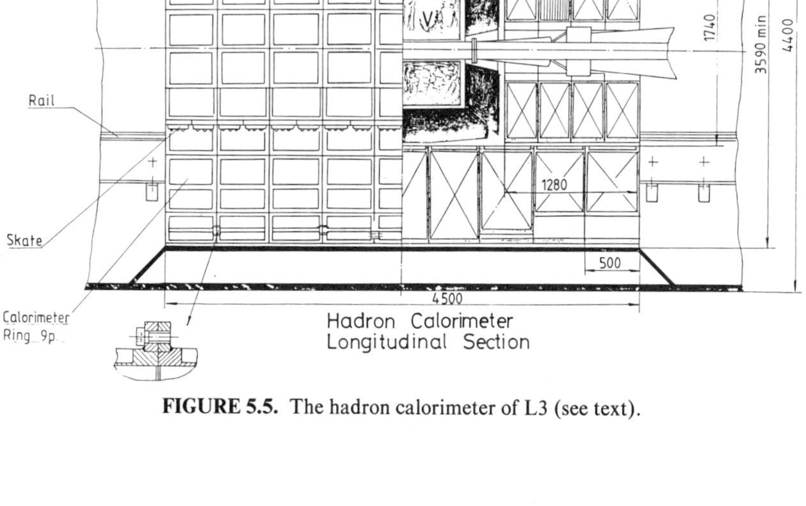

FIGURE 5.5. The hadron calorimeter of L3 (see text).

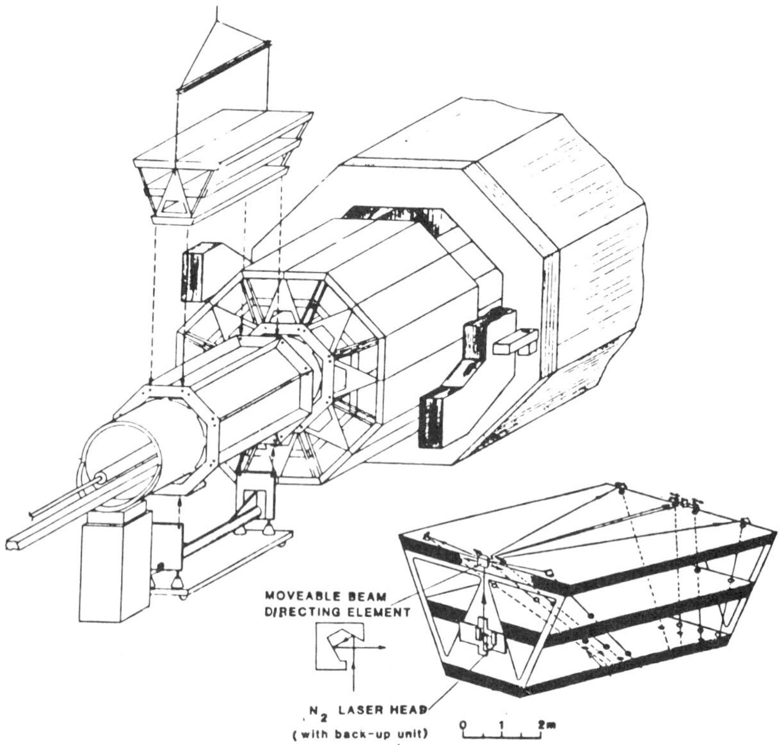

FIGURE 5.6. A sketch of the L3 muon detector.

under test. First tests have yielded a single wire resolution of 195 μm, which is better than the design goal.

DELPHI

This detector, whose name represents DEtector with Lepton, Photon, and Hadron Identification, is an ambitious detector in which, in addition to the general features found in large solenoid detectors such as OPAL or ALEPH, hadron identification is obtained over a large part of the full solid angle and over a good momentum range by means of special ring-imaging Cherenkov counters.

A sketch of the DELPHI detector is shown in FIGURE 6.1. Immediately surrounding the beam pipe is a vertex detector. This device, based on silicon μ-strips, is under development in the Collaboration, but has not been fully designed yet. It is planned for implementation after the start-up of LEP and will not be described further here.

Outside the vertex detector is the inner detector (which comprises a small jet chamber between 12 and 22 cm in radius, segmented azimuthally into 24 segments), which will provide a track trigger in the (r, ϕ) plane. The novel feature of this chamber lies in the arrangement of the drift field such that the drifting electrons have a constant angular velocity towards the detection plane. Tests have shown that even for tracks producing the longest drift times within the 15° sector, the spread in arrival time of the signals is within ±50 ns, which leads to a very powerful track trigger. Outside of this is a trigger layer comprising five layers of axial wires read out by cathode strips to yield z-coordinates. In addition to providing fast trigger information, the inner detector forms the first of a series of track chambers used in DELPHI that provide 24 point measurements, each with ≈100 μm precision.

FIGURE 6.2 shows the longitudinal view of DELPHI. In the barrel region, charged tracks are first measured by the inner detector, then by a TPC covering the annular region from 30 cm radius to 120 cm radius and with an active half-length of 130 cm, and finally by an outer detector at about 200 cm radius comprising five staggered layers of 1.8 cm square drift tubes, each 4.8 m long in the z-direction, operating in the limited streamer mode. In the forward region, good momentum measurement capability is maintained by the use of two packs of forward drift chambers, chambers A and B, respectively.

The DELPHI TPC is similar in dimensions and construction to the PEP4 TPC. Extensive prototype work over the past three years on endplate construction, pad design, field cage construction, gating techniques, and hybridized preamplifier and shaper electronics, along with careful comparison between the use of CCDs versus flash ADCs, cannot possibly be covered in this report. The two end-plates will be divided into six sectors as in PEP4, but the 16 pad rows will be circular. The contribution to the resolution due to $E \times B$ and track angle effects has been studied for different pad geometries. FIGURE 6.3 shows Monte Carlo simulation results for straight pad rows (a) without corrections for wire pulse-height and (b) with these

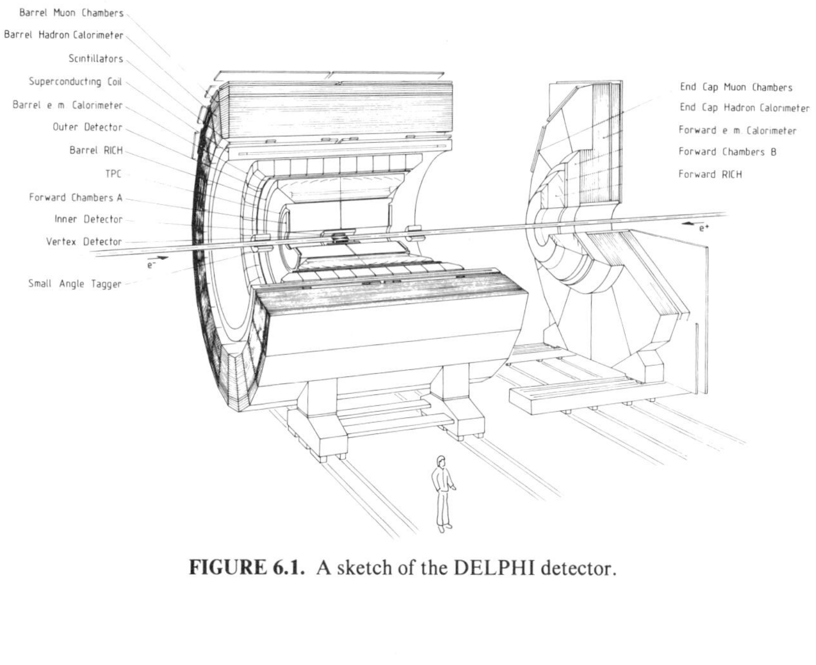

FIGURE 6.1. A sketch of the DELPHI detector.

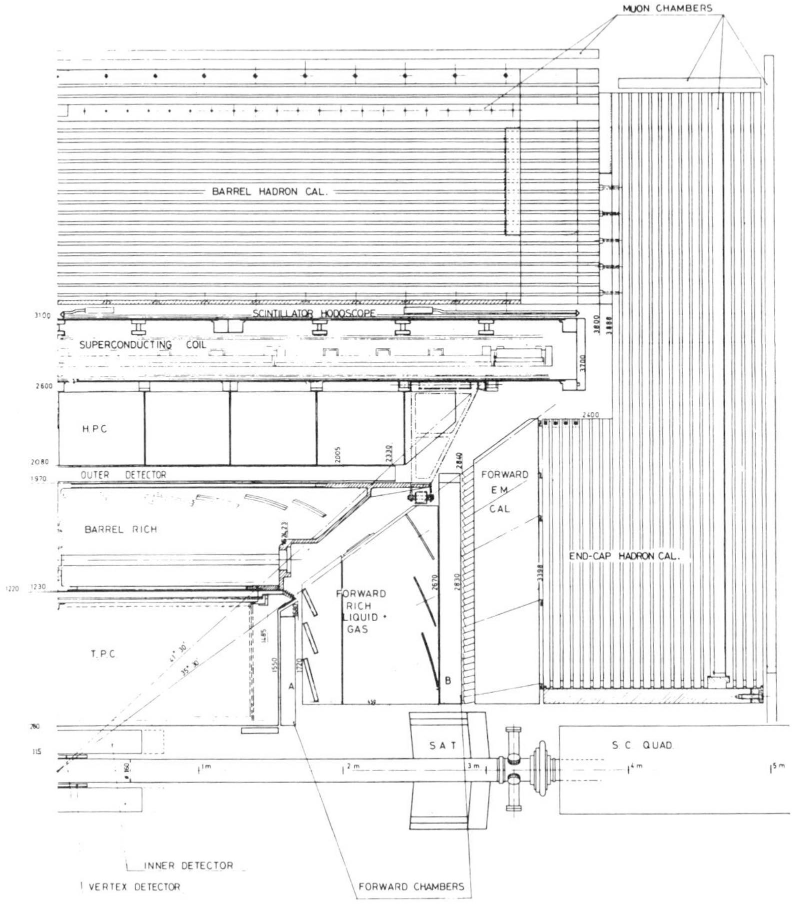

FIGURE 6.2. The longitudinal view of the DELPHI detector.

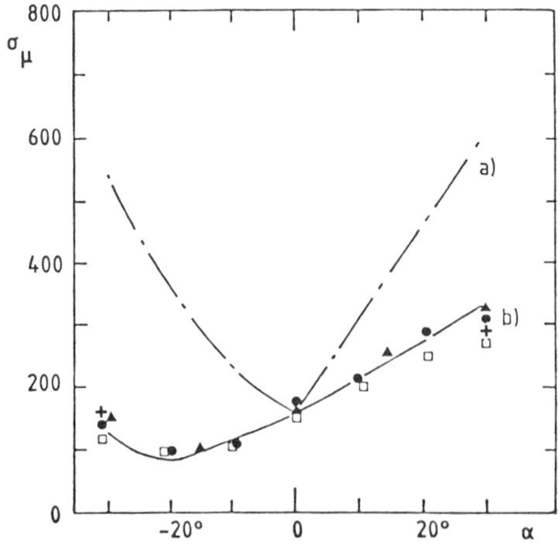

FIGURE 6.3. The contribution of track angle to TPC resolution (see text).

corrections, and for circular rows of pads with the dimensions of 8×8 mm^2 (●) and 32×8 mm^2 (□) without corrections. Data points are shown from test beam measurements on prototypes with straight pad rows (▲) after pulse-height correction and with circular pad rows (+) without correction.

FIGURE 6.4. The DELPHI TPC endplate prototype (see text).

The DELPHI TPC will use short pads (8 mm along the radius) rather than long pads, such as chosen by ALEPH, since the simulation shows that the (r, ϕ) resolution is only slightly worse for high momentum particles, but is substantially better at low momenta.

A full-scale mechanical prototype of a sector plate has been constructed (see FIGURE 6.4) that showed a positioning accuracy of ≤ 10 μm. The readout will use a newly developed 8-bit 15 MHz flash ADC, rather than CCDs.

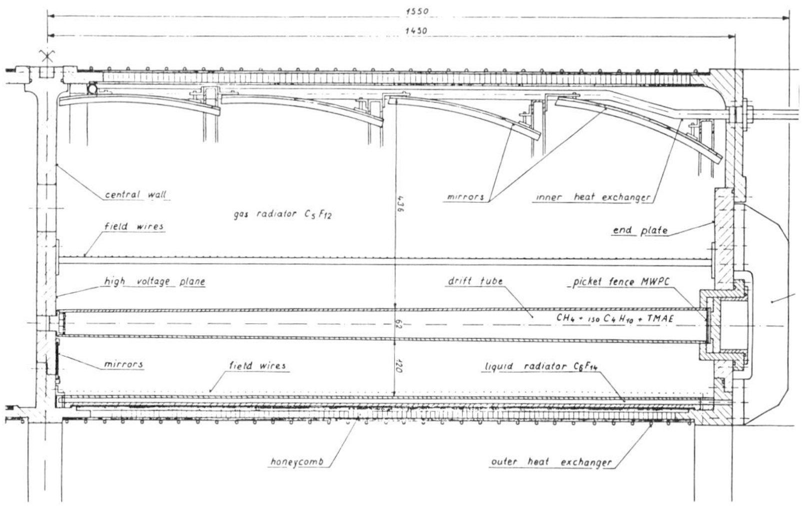

FIGURE 6.5. A drawing of the DELPHI Barrel RICH (see text).

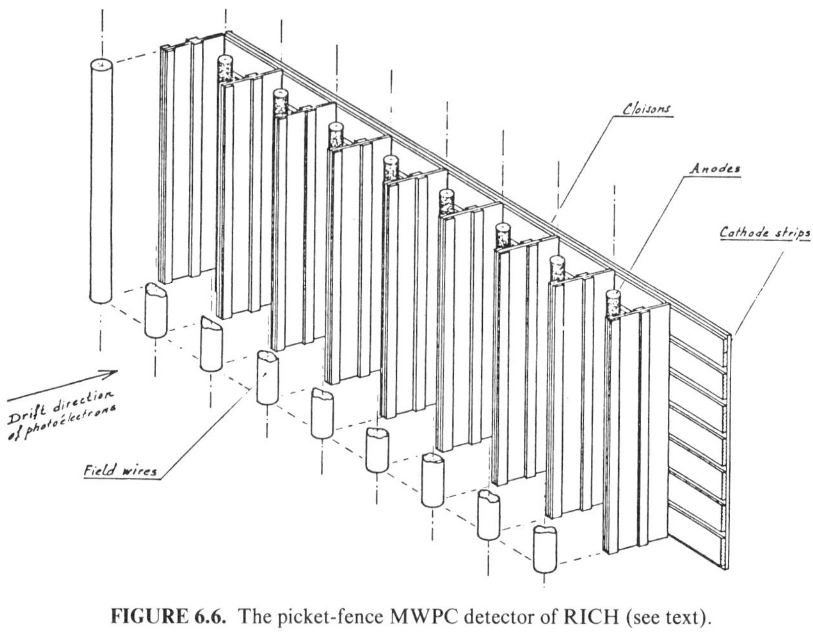

FIGURE 6.6. The picket-fence MWPC detector of RICH (see text).

Charged particles emerging from the TPC are identified by velocity measurement in ring-imaging Cherenkov (RICH) counters. The Barrel RICH occupies the annular region from a radius of 123 cm to a radius of 197 cm and covers the angular region from 41.5° to 90°. A drawing of the Barrel RICH is shown in FIGURE 6.5.

Particles first traverse a thin (1 cm) liquid radiator cell containing freon (C_6F_{14}), and if their velocity is above the threshold, Cherenkov light is emitted in a narrow cone that emerges through the quartz of the radiator cell and enters one or more of the drift tube cells, also through quartz windows. Inside each drift tube, the gas contains sufficient TMAE to absorb the Cherenkov photons by photoionization within the first 2–3 cm. Since there is a uniform drift field inside the tubes, the ionization electrons

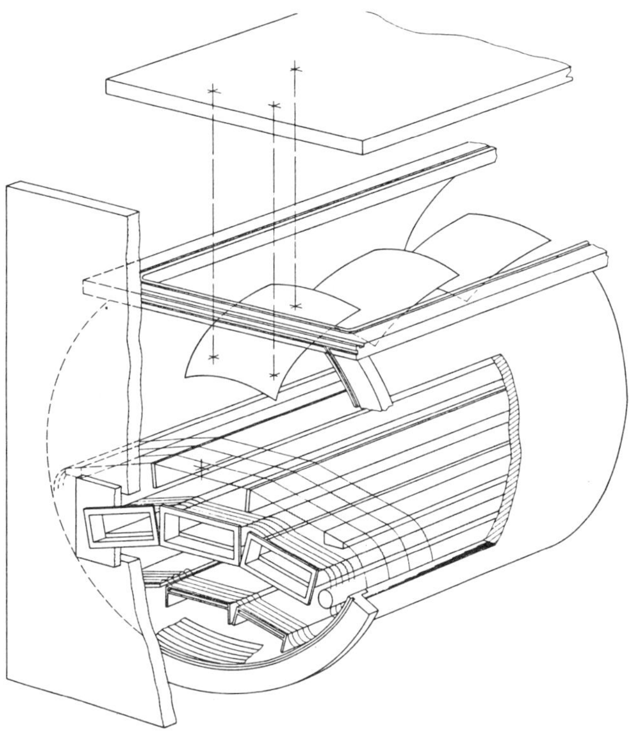

FIGURE 6.7. The Barrel RICH prototype (see text).

drift along the tube to its end where they are detected by a picket-fence MWPC, in which each wire is optically separated from its neighbor, as shown in FIGURE 6.6. The hit anode wire gives the (r, ϕ) coordinate and the time of arrival of the signal gives the z-coordinate of the detected Cherenkov photon. The cathode strip barycenter gives the radial (depth) coordinate at which the ionization occurred.

Higher momentum particles that emerge from the liquid radiator can pass through the drift tube region and into the gas radiator region containing another freon (C_5F_{12}), where, with sufficient velocity, they will also radiate Cherenkov photons. Since about 45 cm of radiator gas is traversed, the light must be focused by parabolic mirrors onto the drift tubes, thus passing through quartz windows on the outer face and producing photoionization electrons close to that face. They will drift down to the MWPCs in the same way as before. The depth coordinate measured by the cathode strips allows

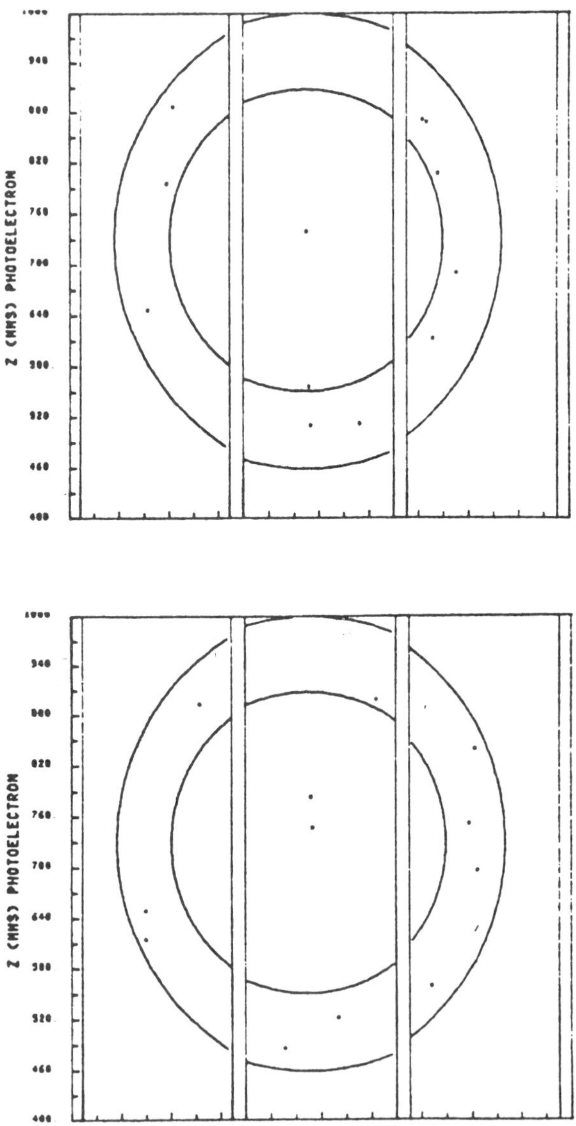

FIGURE 6.8. The events from the liquid radiator using 10 GeV/c pions.

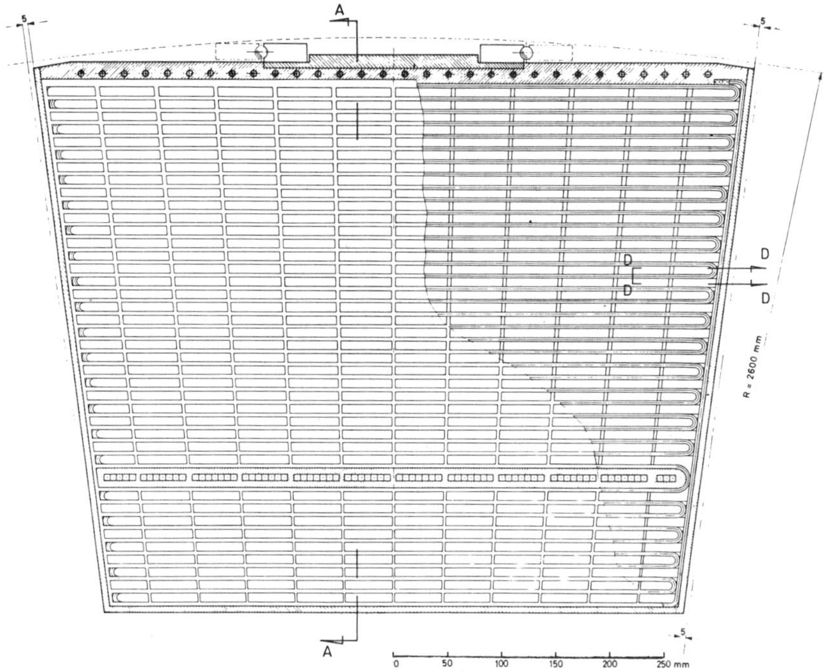

FIGURE 6.9. The cross section of an HPC module (see text).

separation between electrons from the two kinds of Cherenkov rings. With the radiator gas at 1.3 atmospheres pressure, ≈11 photoelectrons should be obtained from a relativistic particle, with all of them in one drift tube due to the focusing of the parabolic mirrors.

During the past 18 months, tests have been carried out using the prototype sketched in FIGURE 6.7, which is to scale, but represents one-eighth of the full azimuth of the final Barrel RICH. As can be seen from the figure, the prototype contains two liquid radiator cells, three drift tubes, and the parabolic mirrors all in a highly uniform electric field.

A Cherenkov ring image from the liquid radiator will fall typically on three adjacent drift tubes and will produce ≈20 photoelectrons. The actual number detected should reach ≈16, provided optimum operating conditions are maintained. Examples of events containing about ten photoelectrons measured with the Barrel RICH prototype using 10 GeV pions are shown in FIGURE 6.8.

Charged particles emitted at <35.5° pass through the forward RICH. It is similar in principle to the Barrel RICH, but the drift tube technique is not applicable because of the magnetic field. Hence, the photon detector is divided into twelve independent TPC units, each covering 30° in azimuth. Time does not permit a more detailed description of this detector.

Electromagnetic calorimetry in the barrel region of the DELPHI detector is provided by a novel device known as the High-density Projection Chamber (HPC). This is a lead/gas sampling calorimeter in which the ionization electrons are drifted along the sampling slots to proportional chamber detectors at the end of each module. The construction is modular with each module covering 15° in azimuth and being divided into eight annular regions in the z-direction. The cross section of a module of the HPC is shown in FIGURE 6.9.

FIGURE 6.10. The assembled lead structure of the HPC prototype.

The drift field is established in the HPC by using a lead-wire construction in which each wire carries a different voltage. A photograph of a prototype of an HPC module is shown in FIGURE 6.10.

The attraction of the HPC lies in its spatial and angular resolution at an affordable cost. For example, prototype measurements have demonstrated the angular resolution for reconstruction of the axis of an electromagnetic shower to be ≈ 40 mrad/ $\sqrt{E}$ in the drift direction, which is in agreement with shower simulations. The energy resolution depends on the final optimization of the lead structure, but it will be $\approx 18\%/\sqrt{E}$.

The forward electromagnetic calorimeter of DELPHI will be made from lead glass blocks read out by vacuum phototriodes. The detail of construction will not be given here. The magnetic field of DELPHI will be 12 kilogauss, which is provided by a superconducting solenoid with a radius of 2.75 m.

The iron return yoke, which is in the form of 5-cm laminations, is used to provide hadronic calorimetry by the insertion of streamer tube detectors, which are read out in a projective tower geometry by cathode pads grouped together to give four samplings in depth.

Two planes of muon chambers complete the DELPHI detector, with the first plane being inside the return yoke and the second being outside. The chambers are flat drift chambers, ≈ 4 m long, that have a maximum 10 cm drift distance to a central anode wire. The precision is ≈ 1 mm in the drift direction and the z-coordinate will be measured to ≈ 10 mm by delay-line readout or by charge division.

ACKNOWLEDGMENTS

This talk would not have been possible without the cooperation of the spokesmen of the four LEP experiments who have provided most of the material for the discussions in this paper. The discussion on the physics at LEP (the second section) is based principally on the work of the DELPHI Collaboration. The author would like to express his sincere thanks to his colleagues for their generous help and takes full responsibility for any errors of fact in the report.

The SLC Program[a]

CHARLES Y. PRESCOTT

Stanford Linear Accelerator Center
Stanford Univesity
Stanford, California 94305

INTRODUCTION

The SLC proposal was conceived as a prototype of a new class of particle accelerators, specifically the linear colliders.[1] The desires for higher energy e^+ e^- collisions were based on the highly successful e^+ e^- physics programs at existing machines and the promise of exciting physics experiments at higher energies. Scaling of storage ring parameters to reach higher energies led to designs of very large machines and estimates of very costly projects. The SLAC Linear Collider proposal was an economical step to achieve the next increment in energy that seemed appropriate, namely that needed to reach the Z^0.

To achieve economically the desired luminosity, comparable to that possible in a scaled-up storage ring, the small beam size and short duty cycle capabilities of a linac were exploited. Modifications of the existing SLAC linac were required, including damping rings to reduce beam emittance, higher power klystrons to increase the accelerating gradient, and beam transport lines to split the e^+ and e^- beams and bring them into collision. The SLC proposal was approved in stages, with final approval of construction in early 1984. This report summarizes the status as of January 1985. The first beams are scheduled for January 1987.

THE STATUS OF THE MACHINE

FIGURE 1 shows schematically the plan for the SLC. The various components include the existing linac structure upgraded for higher energy and better control of the beams, a new gun and booster, two damping rings to provide small beam emittances, a positron source, beam transport arms, a final focus section, and the experimental hall and detectors.

Tests of the injector and one of the two damping rings have been under way for over a year. The purpose of the damping rings are to reduce the beam emittance normally delivered by a linac gun and an injector to values low enough to be focused to small spots. At design performance, the south damping ring is expected to have two bunches of 5×10^{10} electrons circulating diametrically opposed. The design of the lattice is optimized for fast damping of electron orbits through quantum fluctuations. The damping time is three milliseconds. After circulating in the damping rings for the linac interpulse period (nominally 5.3 milliseconds at the linac pulse rate of 180 Hz), the two

[a]This work was supported by the Department of Energy, contract no. DE-AC03-76SF00515.

bunches are extracted and sent into the linac, along with an e^+ bunch from the other damping ring. The first e^- bunch and the e^+ bunch are sent to the experiments, while the second e^- bunch generates e^+'s for a future e^+ bunch.

In February 1984, single bunches of 1×10^{10} electrons were accelerated into the south damping ring, extracted, and accelerated through ten sectors (one-third of the linac). Invariant emittance, momentum spread, and bunch length were all within

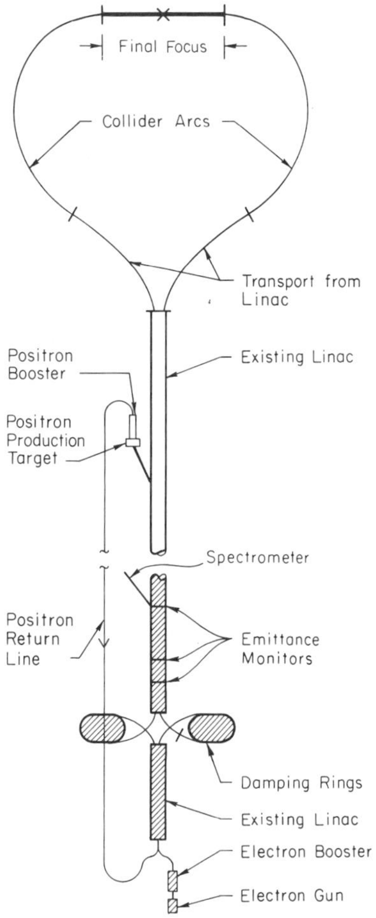

FIGURE 1. A schematic layout of the SLC main components.

design specifications. This demonstration of the fundamental principles represented a major step in the project and satisfied a milestone to be passed before conventional construction started.

Since February 1984, construction of the second damping ring on the north side of the linac has begun and improvements to the injection and extraction lines have been done. Beam tests in the fall of 1984 demonstrated that two electron bunches of 5×10^{10}

electrons in each bunch were accelerated up to the south damping ring. Bunch charges of 4×10^{10} electrons were circulated in the damping ring, but not extracted. No emittance measurements were made on these bunches. The immediate goals for the spring 1985 running are to demonstrate that two bunches can be injected and extracted from the damping ring, with 2×10^{10} electrons in each bunch, and accelerated through ten sectors. Emittance and momentum spread will be measured at that point. The north damping ring construction is well under way and operation of that ring on e^+ beams is scheduled for March 1986.

Conventional construction for the SLC is well along. The connections between the linac switchyard and the collider tunnels were completed last summer. The arc tunnels are close to completion, with access available in March 1985 and July 1985 for the north arc (e^- beam) and the south arc (e^+ beam), respectively. The next steps for the arc tunnels are to provide a surveyed coordinate system and to install cabling and pipes for the magnets, followed by the installation of the magnet supports and then the magnets themselves. The magnet installation is scheduled to begin in September 1985.

The construction of the interaction hall for the experiments has begun. Preliminary excavation started last summer (1984). The construction of the hall itself is under way, with completion expected in February 1986.

Installation of the first detector (Mark II) is presently scheduled to begin in spring 1986. Other technical components for the collider include the positron source, new klystrons, arc magnets and final focus magnets, and a polarized electron gun. The positron source includes a pulsed kicker and beam line to extract the second e^- bunch, a target, a 200 MeV booster, and a return line to bring the e^+'s back to the front of the accelerator. The kicker and beam line are expected to be tested toward the end of the spring running in May 1985. The return line consists of a simple vacuum pipe, which is mounted on the ceiling of the linac tunnel. The vacuum pipe and its quadrupoles and steering magnets will be installed between June and October 1985. The 200 MeV booster will be tested in March 1986, and shortly thereafter, e^+'s are scheduled to be sent back to the front of the linac, accelerated through sector 1, and injected into the north damping ring. The final step, acceleration of e^+'s down the linac, is scheduled for June 1986.

Upgrading the klystrons from 38 megawatt units to 50 megawatt units is required to achieve sufficient energy for Z^0 production. The linac presently has a complement of 240 klystrons. The energy upgrade requires replacing approximately 180 of the klystrons with 50 megawatt tubes. Production of the 50 megawatt tubes has not been easy. SLAC must design and build them since no supplier of these tubes exists. Early tests of these higher power units revealed a problem with vacuum windows failing due to the higher electric fields present. The problem has been solved by splitting the microwave plumbing at the output into two waveguides and two windows, which then recombine into one waveguide. Fifteen such tubes have been completed at SLAC and are being installed in sectors 9 and 10 for tests under realistic accelerator conditions. Lifetimes of the cathodes, a critical parameter, will be measured during these tests.

Magnets for the collider arcs and final focus are under construction at SLAC. The magnets combine bending and quadrupole fields into one element. The basic design allows the beam to follow the local terrain so that tunnels and the experimental hall remain near to the local ground surface. Installation of the magnets begins in September 1985.

The polarized electron gun exists in prototype form. An earlier version of this device was operated successively on the linac in 1978 and 1979. It is a laser-driven photoemission source with a cathode of gallium arsenide. Polarizations of 40% have been achieved and currents in excess of 5×10^{10} electrons in a single bunch are achieved readily with modest laser power. Plans to replace the gallium arsenide cathode with other similar materials are expected to provide considerable higher polarization. The polarized source is not scheduled to operate during the first year of the SLC operation.

Final focus and steering are essential in order to reach the SLC luminosity goals. Without focusing, the luminosity would be about 5×10^{26} cm^{-2} sec^{-1} at full current. At this luminosity, the SLC would yield 1.6 Z^0's per day, which isn't too bad by today's standards, but far below what is possible. Focusing the linac beams from elliptical spots of 250 μm $\times$ 30 μm to round spots of 1 to 2 μm diameters can raise the luminosity to well above 1×10^{30} cm^{-2} sec^{-1}. The function of the final focus is to provide a small spot size in the presence of a $\pm\frac{1}{2}$% momentum spread. Control of the aberrations leads to a complex set of magnets, requiring 500 feet of space, 6 bending magnets, 25 quadrupoles, and 8 sextupoles for each of the two incoming beams. In addition, the extraction and dumping of used beam and the steering of beams into collision are part of the final focus section. Installation of these components is scheduled for late 1986.

Mechanical stability of the arc magnets and final focus magnets is essential. Tolerances of $\pm$0.1 mm are required. Seismic noise has been shown not to be a problem, but thermal effects and floor settlement must be accommodated. Automatic alignment mechanisms are being designed for the final focus section.

Beam steering and beam finding techniques are important to bring micron-sized beams into collision. The techniques envisioned incorporate several steps to go from the millimeter scale to the micron scale. They are:

(i) For gross steering, beam position monitors, which sense the proximity of charge to an electrode, are suitable down to $\pm$100 μm. Such devices are currently in use in the linac;

(ii) Below the 100-μm level, it is proposed that momentary raster scans over approximately 100 μm $\times$ 100 μm areas in steps of 10-μm sizes be used. A single scan requires less than one second;

(iii) When the two incoming beams pass within 10 μm, they begin to interact through the magnetic fields, causing deflections. Detectors downstream sense the presence of synchroton radiation emanating from the beam-beam interaction. Gamma rays in the 1-MeV range signal the nearby passage of the other beam. The flux of soft γ's, the energy spectrum, and the angles of the γ's tell the control system the direction and the amount to adjust offsets to bring the beams into full collision. The final 10-μm adjustments will depend on sensing the magnitude and direction of the γ's from the beam-beam interaction ("beamstrahlung"). In head-on collisions, the beamstrahlung photon energies reach 100 MeV and scatter outward at angles up to 2.5 mrad.

Luminosity of the SLC is expected to reach 6×10^{30} cm^{-2} sec^{-1} at the design intensity of 5×10^{30} e^+'s and e^-'s. The linac pulse rate is 180 Hz. The spot size is expected to be 1.8 μm diameter and 1.5 mm long. Divergence of the incoming beam is 0.3 mr and grows to 2.5 mr after collision with the other beam. All beamstrahlung photons lie inside the 2.5 mr after collision with the other beam, so detector components are mostly free of this source of backgrounds. At full luminosity of 6×10^{30} cm^{-2} sec^{-1}, each beam is focused strongly in passing through the other. The expected pinch effect is important to the SLC physics and will be studied carefully because of the implications for future collider projects. At a luminosity of 6×10^{30} cm^{-2} sec^{-1}, a production of 3×10^6 Z^0's per year is expected, which is based on an operation efficiency of one-quarter of the calendar year.

THE STATUS OF THE DETECTORS

Through a series of deliberations beginning in 1982 and extending into 1984, the SLAC Experimental Program Advisory Committee (EPAC) selected two detectors for the SLC experimental program. The first of these, the Mark II, was selected from several proposals to upgrade existing PEP detectors. The ground rules established by the EPAC were that the first detector be an upgrade, that the modifications be tested at PEP, and that the detector be ready for SLC turn-on in early 1987. The EPAC further decided that a second detector should be built from "grounds-up," thus optimizing its capabilities for Z^0 physics. Desirability to incorporate 4π hadron calorimetry and that the second detector should be ready for data in calendar 1988 were recommended. The single proposal for a new detector was the SLD, which was approved in the spring of 1984.

The Mark II Detector

The Mark II collaboration consists of eight institutions[2] of approximately 67 physicists, which is expected to grow to 100 by the time of beam turn-on, plus students. The proposal is a significant upgrade of the existing Mark II. It will incorporate the following new components: a new coil, an end-door retraction mechanism, an endcap electromagnetic calorimeter, a new central tracker, a time-of-flight scintillator and photomultiplier system, new FASTBUS electronics, a six-layer trigger chamber, a luminosity monitor, and a vertex detector. The only components not changed are the iron return yoke, the liquid argon calorimeter, and the muon detector. These changes are nearing completion, with the testing at PEP to be under way in October 1985. A move to the SLC interaction region is planned in the spring or early summer of 1986. FIGURE 2 shows a quadrant cross section of the upgraded Mark II. The new endcap calorimeter, combined with the liquid argon modules and the luminosity monitor, provide nearly 4π coverage for electromagnetic interactions. The endcap calorimeters are under construction at LBL and are expected to be tested in February of 1985 at SLAC. Installation of the endcaps is planned for April 1985. The endcaps are a lead-drift tube sandwich device, which is somewhat like the Mark III endcap design. They consist of 36 layers of aluminum tubes and $\frac{1}{2}$ X_0 lead sheets ganged together in quasi-tower geometry and segmented in depth three times. The first 8 layers are ganged together into $3 \times (X\text{–}Y\text{–}U\text{–}V)$ groups, followed by 12 layers ganged again in $X\text{–}Y\text{–}U\text{–}V$ geometry, followed by 16 layers of $X\text{–}Y$ geometry. The total system consists of 1300 channels; resolution with argon-ethane gas operating in a saturated avalanche mode is expected to be $\approx 17\%$ $\sqrt{E}$. The angular coverage is 18° to 45°.

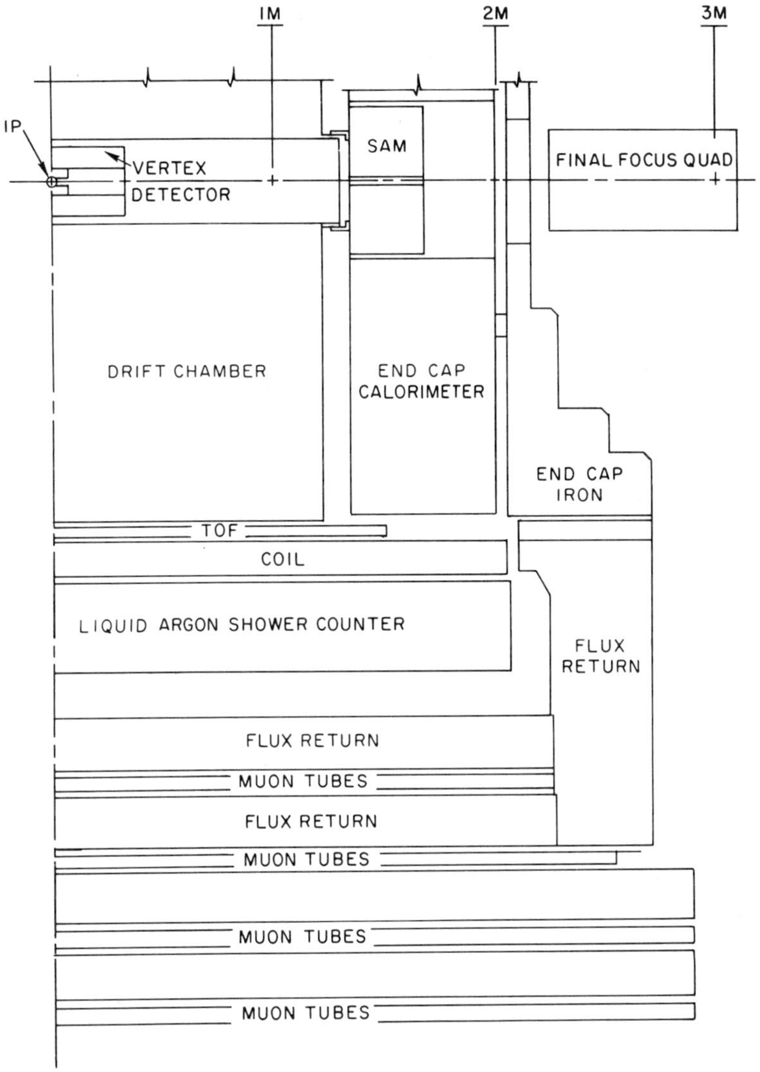

FIGURE 2. A quadrant section of the Mark II upgraded detector.

The Mark II central tracker consists of 12 layers of vector cells, each containing six sense wires. The inner radius of the tracker is 19 cm, the outer radius is 148 cm, and the length is 2.5 m. The magnetic field during running will be 5 kG. The system contains 5832 sense wires, processed in FASTBUS by multihit electronics. Resolution with argon–CO^2–CH_4 gas is expected to be below 200 μm, with a 2-mm two-track separation.

The drift chamber will measure dE/dx to augment time-of-flight and electromagnetic calorimetry for e/π separation. Construction of this drift chamber is complete. Tests are under way, with cosmic ray tests in the detector scheduled for April 1985.

Several vertex detector schemes are being considered for Mark II: a precision drift chamber capable of 40 μm resolution at four atmospheres, a silicon strip detector, and CCDs. None of these will be tested at PEP and a choice of which to use will be made in the summer of 1987.

The SLD Detector

The SLD detector will be constructed by a collaboration of 22 institutions[3] consisting presently of approximately 100 physicists, but which is expected to grow considerably before data begins. This detector goes beyond the Mark II capabilities in the following areas: hadron calorimetry and full hadron particle identification over nearly 4π solid angle, and tracking and muon identification over 4π solid angle. The detector was approved by the SLAC EPAC in May 1984 and achieved DOE funding approval in November 1984. The rate of funding has not yet been established. Construction will begin in the fall of 1985.

FIGURE 3 shows a cross section of a quadrant of the SLD. Access to the interior systems is achieved by retraction of the large doors, which withdraw the endcap calorimeters, Cerenkov ring imaging devices, and endcap tracking devices.

Tracking in the SLD consists of a vertex detector and a central and endcap drift chamber system. Resolutions for single tracks in the vertex and drift chambers are expected to be 5 μm and 100 μm, respectively, with two-track separations of 40 μm and 2 mm. Tests of prototypes have confirmed these performance values. Processing of raw data from these devices will be done in fast front-end processors in FASTBUS. Momentum resolution in the central region is expected to have the form, $\Delta p/p = \sqrt{(.0092)^2 + (.0013p)^2}$, in a 6 kG field.

Particle identification in the SLD is based on Cerenkov rings in a Cerenkov ring imaging device (CRID). Full $e/\pi/k/p$ separation is expected over the full momentum range of particles. Muons are identified at the back of the iron calorimeter. The CRID system will cover nearly the full 4π solid angle. The system will achieve complete particle identification at all momenta by combining a liquid radiator (FC-72) and a gas radiator (isobutane) into a single detector. Cerenkov photons are collected in rings in a photon detector consisting of isobutane, methane, and small amounts of the photosensitive compound, tetrakis-dimethyl-amino-ethylene (TMAE). With suitable electric fields, photoelectrons drift toward a detector of sense wires at the ends of the CRID volume. The arrival times can be translated into drift distances from the detectors. Circles of photoelectrons can be identified; the radii serve to identify the particle types. Prototypes of the CRID system have demonstrated rings from both liquid and gas radiators for π's at 11 GeV/c. Performance from the liquid radiator has achieved theoretical expectations and results from a gas radiator are expected in the spring of 1985.

Calorimetry in the SLD is based on a depleted uranium–lead tile–liquid argon device, followed by an iron-gas calorimeter incorporating the Iarocci streamer tube technology. The liquid argon calorimeter contains 20 X_0 in the electromagnetic section and 3.6 λ in the combined electromagnetic and hadronic sections. The liquid argon

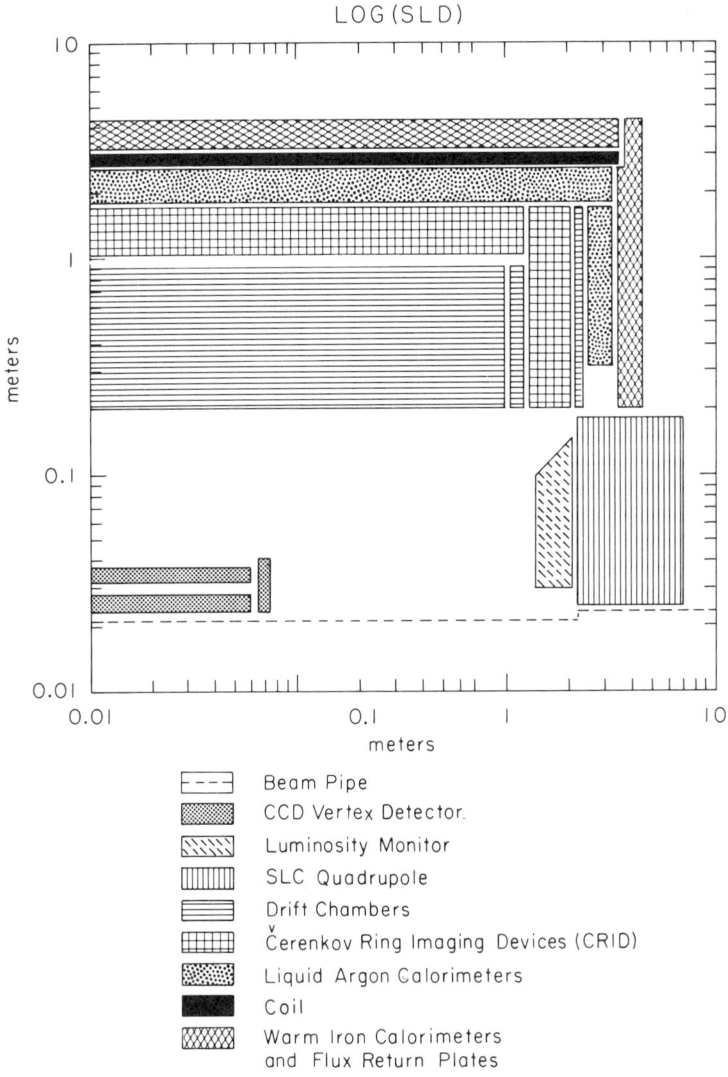

FIGURE 3. A quadrant view of the SLD detector on a log scale. Traditional detector schematics give a short shift to the precision components near the collision point. This logarithmic sketch strikes a different balance.

calorimeter will be inside a conventional aluminum coil of $\approx 0.5\ \lambda$, followed by the iron-gas calorimeter of $\approx 5\ \lambda$ thickness. Muon identification is incorporated in the outer two layers of the iron-gas system. The choice of the uranium and lead for the liquid argon calorimetry combines the desirable features of high density for compactness, fission compensation for better energy resolution, liquid argon for calibration stability and ease of readout segmentation, and tower geometry for simplicity in the analysis. Tower sizes are 33 × 33 mr in the electromagnetic section and 66 × 66 mr in the hadronic portion. Design goals for resolution are $\sigma(E) \approx 8\% \sqrt{E}$ (electromagnetic) and $\approx 50\% \sqrt{E}$ (hadronic). Tests of prototypes are under way, with results to be reported in the next month or two.

THE PHYSICS GOALS

The electroweak parts of the standard model today appear to work wherever they have been tested. The field of particle physics is facing an important branch in the progress of understanding. On the one side are predictions of grand unification that see no new physics beyond the 100 GeV scale. The three forces explain all phenomena out to the GUTS scale of 10^{15} GeV. The Higgs boson(s) remains to be found. On the other side are many models of interesting processes with mass scales below 1 TeV. Hints of new physics such as supersymmetry, technicolor, compositeness, or other processes may be seen already in the collider experiments. The physics program of the SLC concentrates on production of Z^0's for study. In the decays of this object lie detailed information on weak interactions and potential new physics up to the 1-TeV energy

TABLE 1. The Z^0 Physics Program Demands[a]

Physics Topic	Z^0's 10^4	10^5	10^6	Vertex	Tracking	EM Calor.	Particle ID	Had. Calor.	Muon ID	4π Ω
Mass	●				●	●				
Width	●				●	●				
ν Counting		●			●	●				●
Lepton Couplings		●			●	●			●	
Heavy Quark Couplings		●		●	●		●			
Jets	●			●	●	●	●	●		
Higgs Search			●	●	●	●		●		
Hadronic Studies			●	●	●	●	●	●	●	
Top	●			●	●	●	●	●		
Toponium			●	●	●	●				
CP Violation			●	●	●	●	●	●	●	
Polarization Studies		●		●	●	●	●	●	●	●
Rare Z^0 Decays			●	●	●	●	●	●	●	●
SUSY		●			●	●		●	●	●
$SU(2)_L \times SU(2)_R \times U(1)$			●		●	●			●	●
Compositeness		●			●	●			●	●
Flavor-changing N. C.'s			●	●	●	●	●	●	●	

[a]This table characterizes the demands placed on the machine and detectors. It is a list of physics topics that may be of interest. For each topic, an estimate of the required luminosity, in terms of the number of produced Z^0's, is given. Estimates of the detector capabilities that are desirable or necessary are also given. The table is intended to be an overview. The list of topics is surely incomplete and the demands of the machine and detector may change as new ideas and techniques develop.

scale. The large counting rates available at the peak of the Z^0 provide the source of data we need to probe particle interactions out to the TeV scale.

TABLE 1 is a long list of physics topics that have been discussed in various contexts relating to Z^0 physics. It is not intended necessarily to be a complete list. The purpose of this list here is to characterize the Z^0 program in terms of the demands on the machine and the demands on the detectors. Not all of the listed topics wil be of equal importance or interest, but that may be difficult to judge today. TABLE 1 characterizes each topic according to the luminosity required, which is translated into the equivalent number of Z^0's produced. The design goal of 6×10^{30} cm^{-2} sec^{-1} yields an estimated 3×10^6 Z^0's in a calendar year, with an operation at about one-quarter efficiency. Many physics topics do not require that many Z^0's. A machine that operates at 10^{29}, for example, can provide enough data to study many of these questions. TABLE 1 is an attempt to estimate the demands placed on the machine.

The second part of TABLE 1 looks at the same topics for the detectors. Some of these topics place little demand on detector capability, while others require considerable performance from a detector. For example, processes with final state neutrinos are best studied with calorimetry that has 4π coverage. Heavy quark processes can be separated from other processes with a vertex detector and hadron particle identification. Particle identification is of significant use in studies of new mesons and baryons. The discovery potential of large detectors will be best in the detectors providing the most complete and clean particle information. TABLE 1 provides two simple one-dimensional projections of a complex set of conditions and circumstances associated with Z^0 physics. Surely some of these topics will lead to new physics, new experiments, and many talks and publications in the coming years.

CONCLUSIONS

The SLC program is an accelerator experiment and a physics experiment. The progress in the accelerator experiment has been rapid, with injector and damping ring components working, conventional construction on schedule, and technical components in production. Accelerator studies will investigate beam-linac and beam-beam interactions with application to the design of future linear colliders. The physics experiments start in 1987 to study Z^0 properties and to look for new physics effects. Detectors to fully exploit the potential physics are under construction.

ACKNOWLEDGMENTS

This report summarizes the work of a great number of people. I would like to thank the following individuals for their assistance in assembling the material: R. Steining, R. Erickson, J. Clendenin, S. Ecklund, D. Getz, and J. Faust helped with the accelerator status and construction schedules; G. Feldman, J. Dorfan, J. Kadyk, M. Breidenbach, D. Leith, and D. Hitlin provided information on the detectors; and numerous colleagues, too many to name, have contributed to studies of the physics goals.

NOTES AND REFERENCES

1. RICHTER, B. 1984. Large Linear Colliders. SLAC-PUB-3371.
2. California Institute of Technology, Colorado, Lawrence Berkeley Labs, Hawaii, John Hopkins, Michigan, UC Santa Cruz, and SLAC.
3. California Institute of Technology, Cincinnati, Colorado, Columbia, Frascati, Illinois, Massachusetts Institute of Technology, Northeastern, Rutherford-Appleton Labs, Perugia, Pisa, Triumf, Northridge, University of British Columbia, Ferrara, UC Santa Barbara, Tennessee, Vanderbilt, Victoria, SLAC, Wisconsin, and Washington.

High Energy Electroweak Parameters and Experimental Data from $\bar{p}p$ Collider Runs[a]

DAVID B. CLINE

Physics Department
University of Wisconsin
Madison, Wisconsin 53706

INTRODUCTION AND DEFINITIONS

The discovery of the W and Z particles will make it possible to measure the high energy electroweak parameters, ρ and $\sin^2 \theta_w$, with great accuracy.[1,2] This also will provide a test of the radiative corrections. In addition, the decays of W and Z provide a laboratory for the observation of new particles. The first example is the observation of the t-quark through[3]

$$W \rightarrow t + \bar{b}.$$

In addition, the decay,

$$Z \rightarrow \nu\bar{\nu},$$

provides a sensitive test for the number of neutrinos in nature.[4] Also, the decay,

$$W \rightarrow L + \nu_L,$$

with W decaying into a new heavy lepton, could "discover" the next generation.[5] Furthermore, the decay of W into old leptons (e, μ, τ) can be used to test for the universality of leptons, which might lead someday to a better understanding of this universality.

Finally, $\bar{p}p$ colliders produce copious numbers of b-quarks and we will show that the present UA1 data of same-sign dimuons can limit B^0-$\bar{B}^0$ mixing parameters.

In TABLE 1, we give the definition of the high energy parameters that determine the success of the $SU(2) \times U(1)$ theory. One goal of the extensive improvement program for the UA1 detector is to measure these parameters using W and Z particles from the $\bar{p}p$ collider.[6]

W AND Z PRODUCTION AND DETECTION IN THE UA1 AND UA2 EXPERIMENTS—1983 AND PRELIMINARY 1984 DATA

The 1983 $\bar{p}p$ collider run at CERN yielded a significant number of W and Z events analyzed by both the UA1 and UA2 groups. During the 1984 run, the number of W

[a]These $\bar{p}p$ collider runs took place in 1983 and 1984.

TABLE 1. High Energy Electroweak Parameters

1. Spin W, Z; $(V - A)/(V + A)$ components, $\sin^2 \theta_{w,\rho}$
2. $M_W = \left[\dfrac{\pi\alpha}{\sqrt{2}\, G_\mu \sin^2 \theta_w (1 - \Delta r)}\right]^{1/2}$
3. $M_Z = M_w/\cos\theta_w = \dfrac{77.3 \pm 0.08 \text{ GeV}}{\sin^2 \theta_w}$
4. Γ_W, Γ_Z
5. N_ν: number of neutrino families
6. Mass heavy sequential lepton, M_t quark

and Z events had increased by about a factor of 2.5, thus bringing the total of W and Z events to 0(300)(W) and 0(30)(Z).[7]

The first goal of the study of the W and Z is to identify the largest possible number of expected decay modes. In TABLES 2 and 3, we give a summary of the decay channels that have been identified so far.

PARAMETERS OF THE W AND Z PARTICLES AND $SU(2) \times U(1)$ VALUES AND RADIATIVE CORRECTIONS TESTS

The 1983 run parameters of the W and Z particles from the UA1 experiment are given in TABLE 4. In TABLE 5 are given preliminary results from the UA2 experiment (1984) run. Note the excellent agreement with the low energy $SU(2) \times U(1)$ parameters. The next step will be to make precise measurement of these parameters that test the theory at the loop calculation level and, hence, this will test the radiative corrections. A comparison of the parameters measured by the UA1 and UA2 groups is given in TABLE 6 and is shown in FIGURE 1.

The radiative corrections are expressed as Δr where the value of the G_μ weak coupling constant can be expressed as

$$\frac{G_\mu}{\sqrt{2}} = \frac{g^2}{8M_W{}^2\,[1 - \Delta r]},$$

TABLE 2. W Decays

Decay Mode	Identification Status
$W \to e + \nu_e$	Yes
$W \to \mu + \nu_\mu$	Yes
$W \to \tau + \nu_\tau$	Yes
$W \to L \to \nu_L$	Limit only
$W \to u + \bar{d}$	No (indirectly by $u + \bar{d}$, W production)
$W \to c + \bar{s}$	Most likely
$W \to t + \bar{b}$	Most likely
$W \to H^0 + e + \nu$	No
$H^0 + \mu + \nu$	
$H^0 + q + \bar{q}$	

TABLE 3. Decays of Z

Decay Mode	Identification Status
$Z \rightarrow e^+e^-$	Yes
$Z \rightarrow \mu^+\mu^-$	Yes
$Z \rightarrow \nu_i\bar{\nu}_i$	Possibly (high $P_\perp$ gluon recoil)
$Z \rightarrow \tau\bar{\tau}$	Yes ($\tau \rightarrow$ hadrons, $\bar{\tau} \rightarrow \bar{\ell} + \nu + \nu$)
$Z \rightarrow u + \bar{u}$	No ($u\bar{u} \rightarrow Z^0$ indirectly)
$Z \rightarrow d + \bar{d}$	No ($d\bar{d} \rightarrow Z^0$ indirectly)
$Z \rightarrow s + \bar{s}$	No
$Z \rightarrow c + \bar{c}$	Yes ($c \rightarrow \mu$)
$Z \rightarrow b + \bar{b}$	Yes ($b \rightarrow \mu$)
$Z \rightarrow t + \bar{t}$	No (M_t, Phase-space)
Exotic Decays $Z \rightarrow \mu\mu H^0 \ldots$	No

with g being the $SU(2) \times U(1)$ coupling constant. The radiative correction, Δr, is given by a sum of terms,

$$\Delta r = \Delta r^{(1)} + \Delta r^{(2)} + \Delta r^{(3)},$$

where $\Delta r^{(1)}$ comes from corrections due to gauge bosons, $\Delta r^{(2)}$ due to leptonic effects, and $\Delta r^{(3)}$ due to hadronic effects. The best estimate for Δr is

$$\Delta r = 0.0696 \pm 0.002.$$

Δr is related to the M_W and M_Z parameters as

$$\Delta r = 1 - \frac{[37.28 \text{ GeV}]^2}{\left(\dfrac{M_W^2}{M_Z^2}\right)(M_Z^2 - M_W^2)}.$$

Therefore, a measurement of $M_Z - M_W$ and $[M_W/M_Z]$ will allow determination of Δr. The theoretical estimate for Δr (or $M_Z - M_W$) depends on one other parameter (such as $\sin^2\theta_W$, for example). However, the precise absolute measurement of M_Z and a precise measurement of $M_Z - M_W$ are sufficient to tie down the theory completely. As

TABLE 4. Measured Properties of IVBs (UA1)

	$W^\pm$	Z^0	
Mass	80.9 ± 2.8 GeV/c^2	93.9 ± 2.9 GeV/c^2	
Full Width (90% C.L.)	< 7 GeV	< 8.5 GeV	
Observed Decays	$e^\pm\nu$ (68) $\mu^\pm\nu$ (14)	e^+e^- (4) $\mu^+\mu^-$ (5)	e: $\sin^2\theta_w = 0.226 \pm 0.008 \pm 0.014$; $\rho = 0.96 \pm 0.05$
$\sigma \cdot B$ $\sqrt{s} = 540$ GeV $p\bar{p}$	0.53 ± 0.08 nb ($\pm$ 0.09)	58 ± 21 pb ($\pm$ 9)	μ: $\sin\theta_w = 0.228 \pm 0.04$ / 0.03; $\rho = 1.15 + 0.22 - 0.28$
Spin	1	---	

TABLE 5. Data from the UA2-1984 Run[a]

$\sigma_z^e = 0.11 \pm 0.04 \pm 0.02$ nb
$\sigma_w^e = 0.53 \pm 0.1 \pm 0.1$ nb
QCD $\Gamma_W/\Gamma_Z = (9.3 \pm 9)\ \sigma_e/\sigma_w^e$
$\Gamma_W = 2.77$ GeV/c^2
$\Gamma_Z < 2.6 \pm$ GeV/c^2 (90%)
$\Gamma_Z < 3.1 \pm 0.3$ GeV/c^2 (95%)
$N_{Z\to e^+e^-} = 16$ ('83 + '84 data)
$N_{W\to e\nu} \approx$ ('83 + '84 data)
$M_W = 83.1 \pm 1.9 \pm 1.3$ GeV/c^2
$M_Z = 92.7 \pm 1.7 \pm 1.4$ GeV/c^2

[a]See reference 8.

an example, if $M_Z = 93.8 \pm 0.3$ GeV, the theoretical values for $M_Z - M_W$ have an uncertainty of ~70 MeV (i.e., once M_Z has been determined to 300 MeV).[8] Many new effects can cause a larger deviation of $M_Z - M_W$ (Δr). For example, in TABLE 7 (Marciano[9]), there is shown the dependence of $M_Z - M_W$ on the t-quark mass. Note that M_t from 36–60 GeV causes a shift of 30 MeV. A 1-TeV Higgs boson gives comparable changes as well.

THE RATIO OF *W* AND *Z* WIDTHS AND LIMITS ON THE NUMBER OF NEUTRINO FAMILIES

There are theoretical uncertainties in the predicted properties due to the unknown number of lepton and neutrino families; indeed, the Z^0 decays provide an important source of decay into all neutrino families,

$$Z^0 \longrightarrow {}_i\Sigma(\nu_i + \bar{\nu}_i),$$

and these decays in turn increase the Z^0 width by ~180 MeV for each new family. At present, three families are known (ν_e, ν_μ, ν_τ). The decays of the $\overline{W}$ are less sensitive to the number of neutrinos, but they will depend on the existence of heavy leptons and heavy quarks (see FIGURE 2). In time, we can expect to understand the bulk of the W decays and, therefore, the W width will be "exactly" calculable. In contrast, the Z^0 width is not calculable unless the neutrino decay can be observed directly, which is a formidable, if not impossible task. A simple technique has been proposed so as to

TABLE 6. Comparison of the Parameters Measured by the UA1 and UA2 Groups

	UA1	UA2	Combined
M_W	80.9 ± 2.8	$83 \pm 1.9 \pm 1.3$	81.8 ± 2.6
M_Z	93.9 ± 2.9	$92.7 \pm 1.7 \pm 1.4$	93.3 ± 2.7
$\sin^2\theta_w$[a]	0.226 ± 0.016	$0.216 \pm 0.01 \pm 0.007$	0.221 ± 0.013
ρ[b]	0.96 ± 0.05	1.02 ± 0.06	0.99 ± 0.05

[a]$\sin^2\theta_w = 0.221 \pm 0.01$.
[b]$\rho = 0.99 \pm 0.05$.

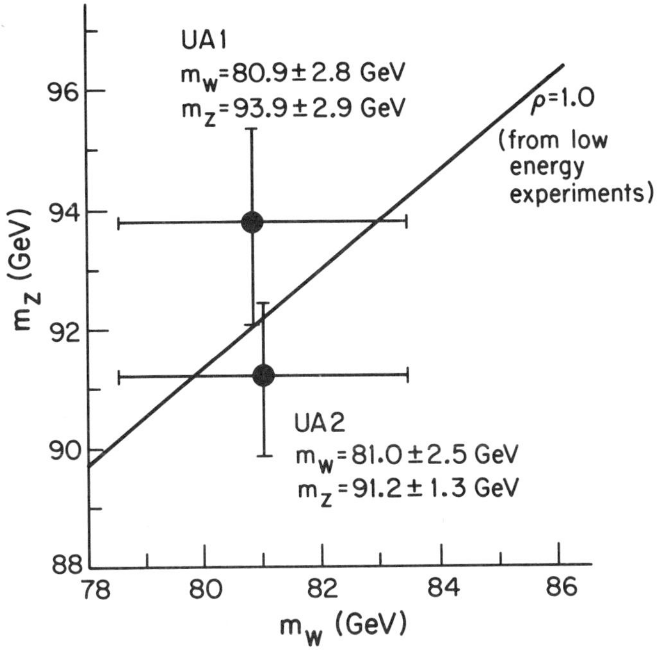

FIGURE 1. A comparison of the M_Z and M_W as measured by the UA1 and UA2 groups.

estimate the Z^0 width from the production rates of W and Z events and to explore the limits of this technique.[10]

We assume that the absolute rates of $W \rightarrow \ell + \nu_\ell$ and $Z \rightarrow \ell + \bar{\ell}$ and the ratio of production cross sections, σ_Z/σ_W, are described within the standard model. If so, then the ratio of the W and Z total widths can be "measured" by forming the ratio,

$$\frac{N_{Z \rightarrow \ell\ell}}{N_{W^- \rightarrow \ell\nu_{\ell j}}} = \left(\frac{\Gamma_{Z \rightarrow \ell + \bar{\ell}}}{\Gamma_{W \rightarrow \ell + \nu_\ell}}\right) \cdot \left(\frac{\Gamma_W}{\Gamma_Z}\right) \cdot \frac{\sigma_Z}{\sigma_W}\frac{Q^2}{Q^2},$$

TABLE 7. Dependence of $M_Z - M_W$ on the t-Quark Mass[a]

M_t(GeV)	$\Delta\Gamma$	A(GeV)	$M_Z - M_W$
20	0.0699	38.66	10.79
36	0.0696	38.65	10.79
60	0.0714	38.69	10.82
83	0.0623	38.50	10.86
100	0.0558	38.37	10.56
150	0.0379	38.01	10.27
200	0.0167	37.60	9.96
240	−0.0036	37.21	9.68
Tree level	0	37.28	9.73

[a] Above evaluated for $\sin^2\theta_W = 0.217$.

where $N_{Z\to\ell\bar{\ell}}$ and $N_{W\to\ell\nu_\ell}$ refer to the corrected number of Z and W events into the partial channels, $\ell\bar{\ell}$ and $\ell\nu_\ell$, respectively, σ_Z and σ_W refer to the production cross sections, and Q^2 is the momentum squared of the recoiling W and Z particles. Γ_W and Γ_Z are the total widths of the W and Z, respectively. The theoretical predictions for σ_Z and σ_W follow from the coupling of the W and Z to the $u\bar{u}$, $d\bar{d}$, and $u\bar{d}$ quark-antiquarks in the beam particles.

The calculation of $\Gamma_{W\propto\ell\nu}$ and $\Gamma_{Z\to\ell\bar{\ell}}$ are straightforward in the standard model,

$$\Gamma_{W\to\ell\nu_\ell} = \frac{G\,M_W^3}{6\sqrt{2}\pi},$$

$$\Gamma_{Z\to\ell\bar{\ell}} = \frac{G\,M_W^3}{12\sqrt{2}\pi}(1 - 4s^2 + 8s^4),$$

where the calculations for the Γ_W and Γ_Z total widths depend on the number of leptonic neutrino flavors, the t-quark mass, the Weinberg angle of $s^2 = \sin^2\theta_w$, and the mass of the W and Z. We use the M_W, M_Z, and s^2 determined from the UA1 experimental data. The ratio of Γ_W/Γ_Z can be calculated reliably on the basis of the assumed t-quark mass and the number of neutrino flavors. A very precise measurement of Γ_W/Γ_Z and the

	Factor in Branching Ratio
$Z^0 \to \nu_e\bar{\nu}_e$	$2\times e^+e^-$ decay
$\to \nu_\mu\bar{\nu}_\mu$	$2\times\mu^+\mu^-$ decay
$\to \nu_\tau\bar{\nu}_\tau$	$2\times\tau\bar{\tau}$ decay
$\to \nu_{new}\bar{\nu}_{new}$	$N_{\text{new neutrino}} \times 2\times e^+e^-$ decay

Total branching fraction into neutrinos

$= (2+2+2)\times e^+e^-$ fraction $+ N_{\text{new neutrino}} \times 2\times e^+e^-$ fraction

$$\text{Ratio of}\left[\frac{Z\to\nu\bar{\nu}}{Z\to e^+e^-}\right] = \left[6 + 2N_{\text{new neutrino}}\right]\text{decays}$$

FIGURE 2. The decays of the Z^0 into neutrino families.

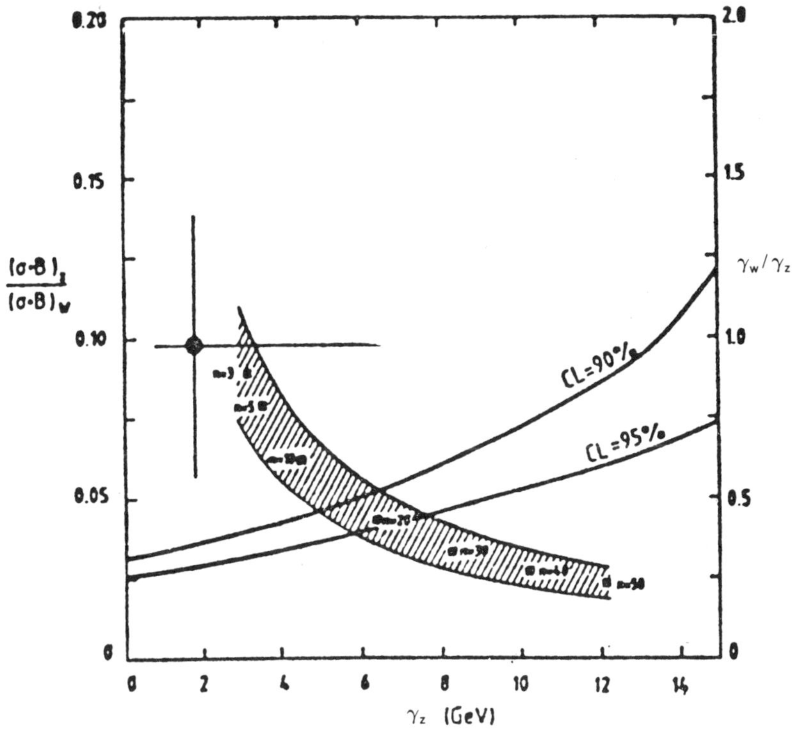

FIGURE 3. The ratio of rates of $Z \to e^+e^-$ to $W \to e\nu$ and the N_ν, number of neutrino families in nature, from the UA1 experiment (crosshatched curve). The ratio is plotted against γ_z (the width of the Z^0) and on the other side, it is plotted against the ratio of the W and Z full widths.

knowledge of the t-quark mass allow for a determination of N_ν, which is the number of neutrino families in nature (see FIGURE 3).

In the limit that $u_p = d_p = \bar{u}_{\bar{p}} \gg s, \bar{s}$, the rate of cross sections becomes

$$\left[\frac{\sigma_Z}{s_{W^+} + \sigma_{W^-}}\right]\bar{p}p = \frac{\left[\frac{1}{4} - \frac{2s^2}{3} + \frac{8s^4}{9}\right]\int_{\tau_Z}^{1} dx \frac{[u(x)\,\bar{u}(x)]}{x} + \left[\frac{1}{4} - \frac{s^2}{3} + \frac{2}{9}s^4\right]\int_{\tau_Z}^{1} dx \frac{d(x)\bar{d}(x)}{x}}{\int_{\tau_W}^{1} dx \left\{\frac{[u(x)\bar{d}(x)]}{x}\right\} \cos^2\theta_c + \left[\frac{u(x)\bar{s}(x)}{x}\right]\sin^2\theta_c},$$

which gives

$$\left[\frac{1}{2} - s^2 + \frac{10}{9}s^4\right]\frac{1}{\cos^2 d_c}\left(\frac{\tau_Z}{\tau_W}\right) = 0.27$$

if we neglect the strange quark contribution and the different x dependence of u and d quarks. Thus, it is plausible that the ratio of $[\sigma_Z/\sigma_{W^+} + \sigma_{W^-}]$ can be calculated reliably for $\bar{p}p$ collisions since the dominant contribution comes from the $SU(2) \times \nu(1)$ parameters and the kinematic factors. We expect corrections to higher order QCD processes to scale approximately like the first order calculations.

We conclude that for $\bar{p}p$ collisions, the ratio of

$$\left(\frac{\sigma_Z}{\sigma_{W^+} + \sigma_W}\right)[Q^2]$$

is most likely to be calculated reliably. As $Q^2 \rightarrow 0$, the Drell-Yan $q\bar{q}$ annihilation graphs should give the dominant production cross section. Higher order QCD processes should give rise to a Q^2 distribution for the W and Z, which has been observed already in the UA1 data.

We believe that the ratio, $\sigma_Z/\sigma_{Z^+} + \sigma_{W^-}$, can be estimated more reliably than the individual cross sections since many of the uncertainties drop out of the ratio calculations; the ratio depends mainly on the semiweak coupling of the $u\bar{u}$, $d\bar{d}$, and $u\bar{d}$ to the Z and W status.

In the standard model with a t-quark mass of 40 GeV and three generations of neutrinos, the ratio of the W to Z width is predicted exactly to better than a few percent. If there are additional decays of the W or Z into new types of particles, such as

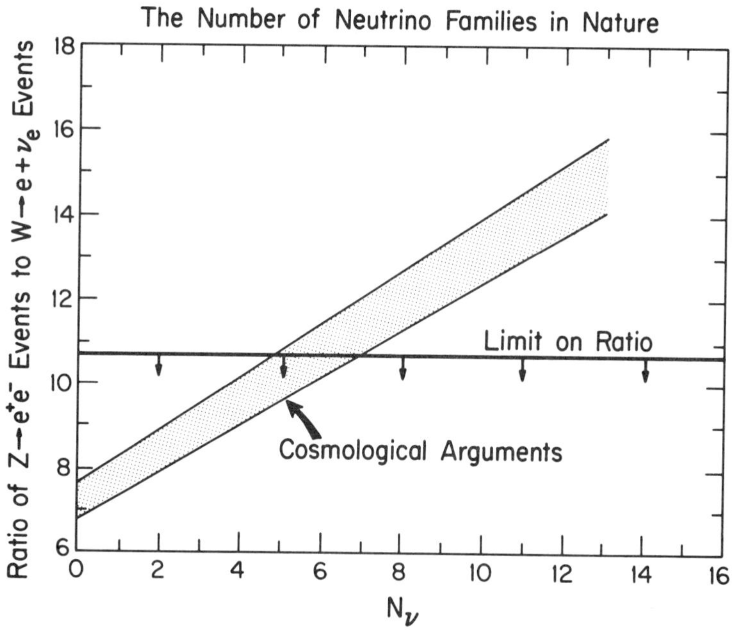

FIGURE 4. A summary of the present limits on the ratio, Γ_W/Γ_Z, and the theoretical uncertainties due to the measurement errors on the $SU(2) \times U(1)$ parameters.

TABLE 8. Initial Test of e-μ-τ Universality Using W Decays

$\frac{\Gamma_{W\to\mu\nu}}{\Gamma_{W\to e\nu}} = 1.24 \pm 0.6^a$
$\frac{\Gamma_{W\to\tau\nu\tau}}{\Gamma_{W\to e\nu}} = 0.96 \pm^b$

[a] 1984. Phys. Lett. **134B**.
[b] UA1 paper in preparation.

supersymmetric particles, the ratio of widths probably will depart from these predictions. In FIGURE 4, we report a result of Barger *et al.* that summarizes the present limits on the ratio of Γ_W/Γ_Z and the theoretical uncertainties due to the measurement errors on the $SU(2) \times U(1)$ parameters. If we take the viewpoint that the theoretical limits on N_ν from big bang cosmology are correct, then the ratio of Γ_W/Γ_Z is predicted quite precisely.

The primordial abundance of ^{4}He is very sensitive to the number of neutrino flavors since, in the early universe, additional species of light neutrinos resulted in a faster universal expansion rate, thus leaving more neutrinos to be bound in ^{4}He. The number of species inferred from the primordial ^{4}He abundance also depends weakly on the exact neutron half-life and the total density of baryons in the universe. Yang *et al.* have shown that the number of families is less than or equal to four (see reference 4). Their calculations strongly disfavor even five families, which is a remarkably precise deduction for a cosmologically based calculation. Present collider data are also in agreement with this result so far.

e-μ-τ UNIVERSALITY TESTS WITH W AND Z DECAYS

The existence of a very high mass particle that decays into $\ell + \nu_e$ gives another possibility to test for e-μ-τ universality. Hopefully, precise measurements of the relative branching ratios might someday give a clue to the differences (if any) between the structure of these leptons. We report in TABLE 8 the present values of the ratio of these decay modes. Precise theoretical predictions of the effects of the radiative corrections would be useful in order to determine the possible significance of future measurements in observing real differences between e, μ, and τ particles.

QCD CORRECTIONS TO W AND Z PRODUCTION AND COMPOSITE MODELS

High transverse momentum Z^0 production can occur through the processes,

$$q + \bar{q} \rightarrow Z^0 + g,$$

$$q + g \rightarrow Z^0 + q,$$

as well as processes with soft multigluon radiation. The subsequent decays of

$$Z^0 \longrightarrow \nu_e\bar{\nu}_e$$
$$\nu_\mu\bar{\nu}_\mu$$
$$\nu_\tau\bar{\nu}_\tau$$
$$\nu_L\bar{\nu}_L$$
$$\bullet\ \bullet$$

give events with large missing energy. These processes give a background for the search for "new physics," i.e.,

$$W^+ \longrightarrow L^+ + \nu_L \quad \text{(reference 5)}$$
$$\llcorner\!\!\longrightarrow q + \bar{q} + \bar{\nu}_L.$$

[Supersymmetric (gluinos, squarks)
monojets from unknown processes.]

For this reason, it is important to understand the plausible upper limits on the rates for the above reactions in the UA1 experiment.

We have carried out calculations of these processes with the ISAJET program. Undoubtedly, the calculations of these processes are uncertain by a factor of at least two and, in addition, the effects of the detector (trigger bias and smearing) are not known precisely. Nevertheless, the ISAJET Monte Carlo program includes these processes and an elaborate detector simulation program has been written to match the

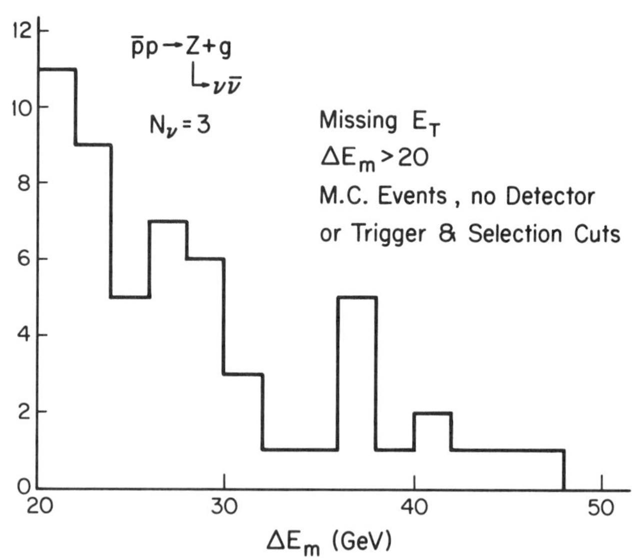

FIGURE 5. The raw ΔE_m distribution above 20 GeV/c.

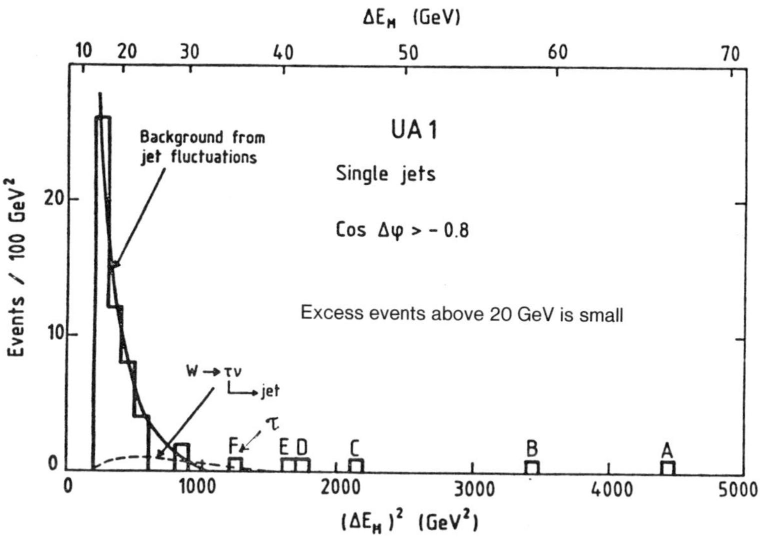

FIGURE 6. The missing $P_{\perp}$ distribution for the monojet events.

Monte Carlo events. The same program can be used to calculate

$$
\begin{array}{l}
q + (\bar{q}) \rightarrow Z_0 + \text{jet} \ldots \\
q + g \qquad \hookrightarrow e^+ e^- \\
\qquad\qquad \hookrightarrow \mu^+ \mu^- .
\end{array}
$$

There is already some hint in the data that this process occurs due to the "flat" $P_{\perp z}$ distribution of the Z^0 events.

The ISAJET Monte Carlo program generates 1000 events of the type,

$$
\bar{p}p \rightarrow Z^0 + \text{jet} + \ldots \quad \hookrightarrow \nu\bar{\nu}.
$$

We assume only three generations. If more generations are assumed (such as, six), then the number of missing $E_{\perp}$ events will increase by another factor of two. Thus, it is possible to put limits on the number of lepton generations by this technique of using the monojet events.

The raw ΔE_m distribution above 20 GeV/c is shown in FIGURE 5 and displays a long tail. These events then were run through detector simulation, trigger simulation, and reconstruction programs. Note that the effect of the trigger threshold and calorimeter smearing is to flatten out the distribution in FIGURE 5.[12]

In FIGURE 6, we show the missing $P_{\perp}$ distribution for the monojet events.[13] The ISAJET calculations cannot reproduce the long flat tail above 40 GeV of missing energy. Nevertheless, the events between 20 and 40 GeV that are not τ-like would arise from Z + gluon. A careful study of these events could provide an independent measurement of the $Z^0 \rightarrow \nu\bar{\nu}$ branching ratio. In addition, very careful measurements of the W + gluon or Z + gluon production at very high Q^2 could be used to test for compositeness of the W and Z. Thus, this is a challenge for the next generation of running at CERN and FNAL.

SEARCH FOR B^0 MIXING WITH SAME-SIGN DIMUON EVENTS

By now, it is well known that B_d^0-$\bar{B}_d^0$ mixing is expected to be very small. However, since B_s^0-$\bar{B}_s^0$ mixing proceeds through a Wts vertex, the effect can be much larger (see FIGURE 7). The mixing is related also to the mass of the t-quark, which is due to the loop diagrams in the figure. There are several uncertainties in attempting to calculate the production of same-sign dileptons from B^0-$\bar{B}^0$ mixing:

(1) $\tau_{B_d^0}$, $\tau_{B_u^\pm}$, and $\tau_{B_s^0}$ are individually unknown.
(2) The specific leptonic branching fractions of B_d^0, $B_u^\pm$, and B_s^0 also are not known.
(3) For the $D^\pm$ and D^0 systems, the lifetimes appear to differ, thus affecting the semileptonic branching ratio.

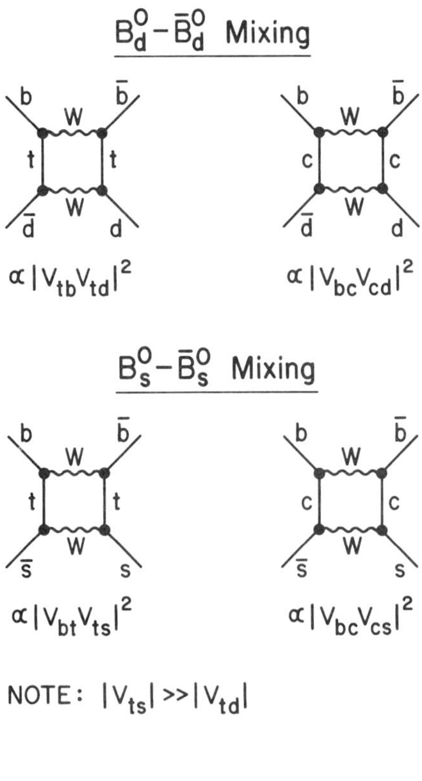

FIGURE 7. B_d^0-$\bar{B}_d^0$ and B_s^0-$\bar{B}_s^0$ mixing.

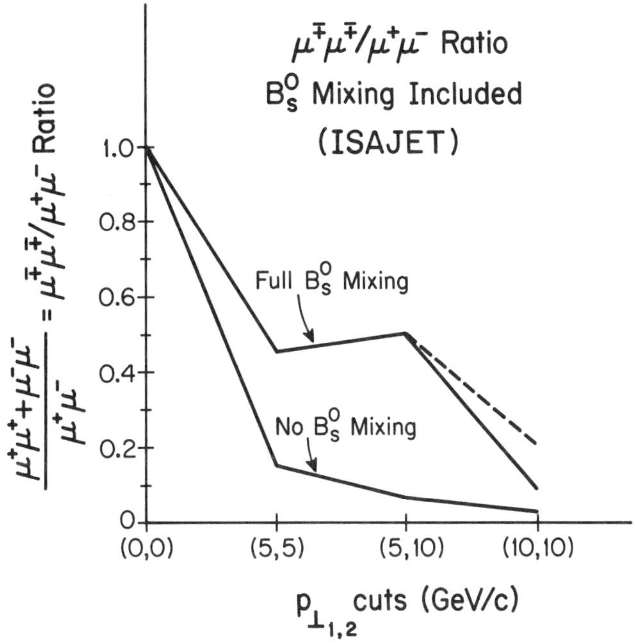

FIGURE 8. The results of the full mixing and no mixing cases for B_s^0.

(4) The production of B_s^0 and $\bar{B}_s^0$ requires the pickup of a strange particle in a gluon-gluon produced $b\bar{b}$ reaction.

We have modified ISAJET to include B^0 mixing. Briefly, the mixing formalism gives for $e^+e^- \rightarrow b\bar{b}$,

$$\frac{\text{Same-Sign}}{\text{Opposite-Sign}} = \frac{2\gamma}{(1+\gamma)^2},$$

$$\text{where } r = \frac{(\Delta m)^2 + (\Delta\Gamma/2)^2}{2\Gamma^2 + (\Delta M)^2 - (\Delta\Gamma/2)^2}.$$

For the b system, we expect

$$\frac{\Delta\Gamma}{\Delta M} \ll 1.$$

Thus,

$$r \approx (\Delta M/\Gamma)^2/2 + \left(\frac{\Delta M}{\Gamma}\right)^2$$

and full mixing occurs for $\Delta M \approx \Gamma$.

However, for the production of B_s^0 in $\bar{p}p$ collisions, the situation is more complex since we may produce combinations of $B^{\pm}B_s^0$, $B^{\pm}B_d^0$, $B_d^0B_s^0$, $B_s^0\bar{B}_s^0$, etc. Thus, a Monte Carlo approach is more useful. Furthermore, the pickup of strange particles may be $P_\perp$ dependent.

Unfortunately, the CP violation is expected to be extremely small in the B_s^0-$\bar{B}_s^0$ system, even though the mixing may be relatively large. We have used two procedures (ISAJET was modified to include mixing):

(1) A large sample of unbiased events was run. This produced a few events with high $P_\perp$ muons.
(2) The generation program was biased to give high $P_\perp$ muons to obtain better statistics.

In both procedures, the several $P_\perp$ cuts and the cases of full mixing, 5% mixing, and no B_s^0 mixing were investigated. The results are shown in FIGURE 8 (biased). For the unmixed case, the ratio of $(\mu^+\mu^+) + (\mu^-\mu^-)/\mu^+\mu^-$ falls to 0.04 for $P_{\perp 1,2} > 10$ GeV/c. The addition of full mixing increases the ratio, even at a $P_{\perp 1,2}$ of (5, 5) GeV/c. Thus, the measurement of this ratio in the $P_\perp$ region of (5–10) GeV/c can be used to detect B_s^0-$\bar{B}_s^0$ mixing. This effect is still much larger than the effect expected from $W \rightarrow t + \bar{b}$ production.

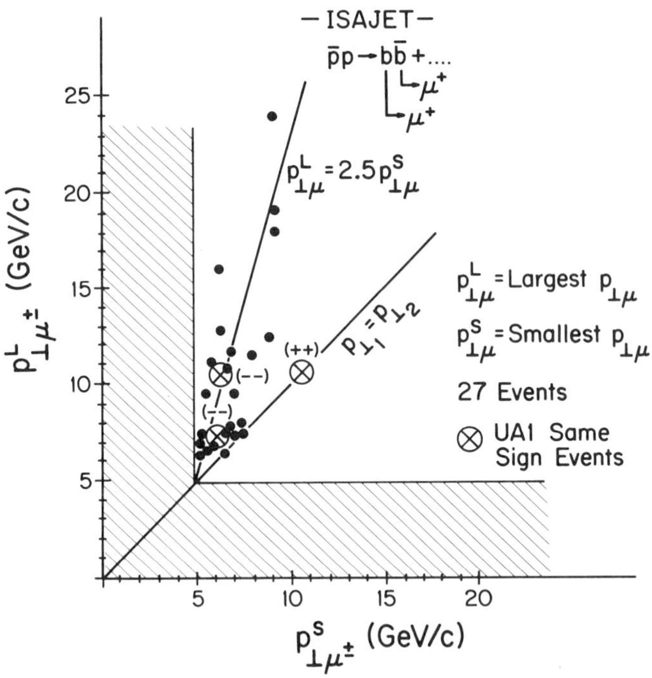

FIGURE 9. An illustration of how high $P_\perp$ same-sign dimuons can be used to distinguish the mixing case from the flavor cascade case ($P_{\perp 1} \sim P_{\perp 2}$) using the ISAJET Monte Carlo calculations.

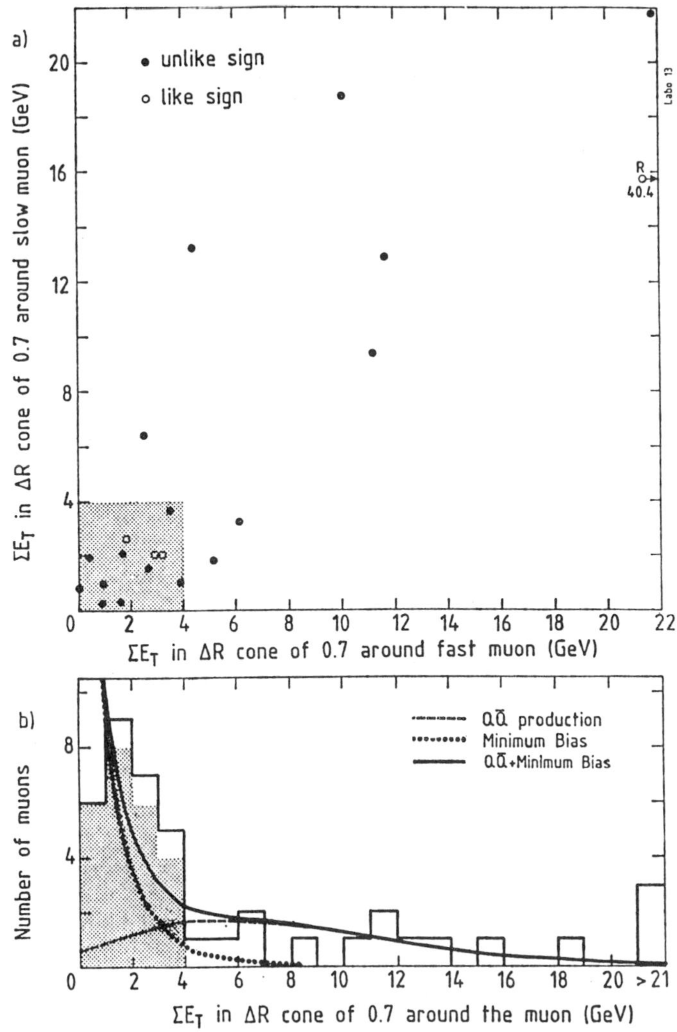

FIGURE 10. The hadronic energy associated with the two leptons.

In FIGURE 9, we illustrate how the study of high $P_\perp$ same-sign dimuons can be used to distinguish the mixing case from the flavor cascade case ($P_{\perp_1} \sim P_{\perp_2}$) using the ISAJET Monte Carlo calculations.

The 1983 UA1 dimuon data includes a class of events with large hadronic energy near one or both leptons.[14] These events are most likely to come from $b\bar{b}$ production. FIGURE 10 shows the hadronic energy associated with the two leptons. For the events that are clearly $b\bar{b}$-like productions, only one same-sign event out of eight opposite-sign events occurs. This event is also a candidate for $Z^0 \rightarrow b\bar{b}$. The ratio of ⅛ is in good agreement with the no mixing case.

ACKNOWLEDGMENTS

I wish to thank the many members of the UA1 experiment who have explained many aspects of the data to me.

NOTES AND REFERENCES

1. ARNISON, G. *et al.* 1983. Phys. Lett. **122B:** 103; ARNISON, G. *et al.* 1983. Phys. Lett. **126B:** 398.
2. BANNER, M. *et al.* 1983. Phys. Lett. **122B:** 476; BAGNANA, P. *et al.* 1983. Phys. Lett. **129B:** 130.
3. ARNISON, G. *et al.* 1984. Phys. Lett. **147B:** 222.
4. See, for example, TURNER, M. 1980. *In* Proceeding Telemark Neutrino Mass Miniconference. V. Barger & D. Cline, Eds.: 73. University of Wisconsin report.
5. CLINE, D. & C. RUBBIA. 1983. Phys. Lett. and CERN EP/83-61.
6. UA1 improvement program proposal no. SPSC/83-48/84-72.
7. The 1984 run provided a net integrated luminosity of 270 nb^{-1}.
8. See the UA2 report from this meeting.
9. SIRLIN, A. & W. MARCIANO. 1985. Private communication.
10. This method was proposed by HALZEN, F. & K. MURSULA. 1983. Phys. Rev. Lett. **51:** 857; CLINE, D. & J. ROHLF. CERN report (unpublished); HIKASA, K. 1984. Phys. Rev. **D29:** 1939.
11. BARGER, V. & F. HALZEN. 1985. University of Wisconsin preprint.
12. CLINE, D. & M. MOHAMMADI. 1984. UA1 TN 1984.
13. ARNISON, G. *et al.* 1984. Phys. Lett. **134B:** 469.
14. ARNISON, G. *et al.* 1985. Submitted to Phys. Lett.

Precision Measurements of the Electroweak Mixing-Angle

J. PANMAN

CERN CH-1211
Geneva 23, Switzerland

INTRODUCTION

After the discovery of the weak neutral current,[1] a lot of effort has been spent to test the prediction of the Glashow-Salam-Weinberg model, which states that the coupling in all neutral current phenomena depends on one single parameter, $\sin^2 \theta$. Indeed, a unique value of this parameter can explain the couplings measured in many different processes, namely, the leptonic and semileptonic neutrino scattering, the asymmetries in electron-nucleon, electron-positron, and muon-nucleon interactions, the parity violating effects in atomic transitions, and the masses of the W and Z bosons.

This prediction is based on the Born approximation of the theory.[2] In the corrections to the Born terms, different processes pick up different correction terms. Hence, measurements of $\sin^2 \theta$ made with sufficient precision and extracted from the data using the simple zero order approximations should show differences for different reactions. It is this feature that provides a test of the gauge nature of the theory.

The present status of measurements of $\sin^2 \theta$ are summarized briefly in the first section. Then it is shown that the effects of the first order corrections to the Born approximation are within reach of a new round of experiments. Some of the experiments aiming at such high precision measurements are described in some detail in the latter two sections.

LOW ENERGY TESTS

Many different experiments have been performed to test the low energy predictions of the theory. These tests include the weak isospin structure, the value of ρ, and the observation that $\sin^2 \theta$ is process independent.

Isomultiplet Structure and Generation Universality

The left-handed and right-handed isospin assignments, T^3_L and T^3_R, as postulated in the standard model, and the resulting vector and axial vector couplings, g_V and g_A, are given in TABLE 1. In determining the deviation from the standard model, the left-handed isospin is taken from theory. Thus, all deviations are attributed to a nonzero right-handed isospin value, which is predicted to vanish in the standard model. The results are summarized in TABLE 2.[3] For the light quarks, one has to add a systematic uncertainty of the size of the statistical one since the result is obtained from

TABLE 1. Weak Isospin Assignments and Couplings

	T_L^3	T_R^3	g_A	g_V
ν	1/2	0	1/2	1/2
e, μ, τ	−1/2	0	−1/2	$-1/2 + 2\sin^2\theta$
u, c, t	1/2	0	1/2	$1/2 - 4/3\sin^2\theta$
d, s, b	−1/2	0	−1/2	$-1/2 + 2/3\sin^2\theta$

a combined fit of many different processes. No serious deviations from the model are found. This evidence confirms the isomultiplet structure of the model. By comparing the couplings of the various generations, one obtains good agreement, which confirms the assumption of generation universality.

The Value of ρ and sin² θ

In many different processes, a value of $\sin^2\theta$ was derived. According to the lowest order predictions of the standard model, all these experiments should be described by a single value of the mixing angle. The experimental status is given in TABLE 3. From the table, it is clear that at this level of precision no discrepancies have been found.

Another possible test of the standard model is the measurement of ρ. The ρ parameter describes the overall strength of the neutral current (NC) couplings with respect to the charged current (CC) couplings. In the standard model, ρ is equal to unity (to lowest order) and is a consequence of the symmetry breaking mechanism with doublets in the Higgs sector. Indeed, experimental evidence favors values of ρ close to unity as shown in TABLE 3.

NEXT TO LOWEST ORDER TESTS

The standard model, being based on gauge symmetries, is in principle calculable to all orders, and its predictions, therefore, can be tested by experiment. One expects that effective values of $\sin^2\theta$ obtained in different processes are modified in a different way by higher order corrections. Therefore, these higher order corrections can be tested if sufficiently precise measurements of this parameter are available.

TABLE 2. Axial Vector Coupling and T_R^3 Determination

	g_A	T_R^3
u	0.56 ± 0.02	−0.05 ± 0.02 ± syst.
d	−0.41 ± 0.04	−0.09 ± 0.04 ± syst.
s		$-0.6 < T_R^3 < 0.3$ (90% C.L.)
c	0.63 ± 0.20	−0.13 ± 0.20
b	−0.55 ± 0.13	0.05 ± 0.13
e	−0.51 ± 0.04	0.01 ± 0.04
μ	−0.52 ± 0.05	0.02 ± 0.05
τ	−0.44 ± 0.07	−0.06 ± 0.07

Direct Mass Measurements

The discovery of the intermediate vector bosons, W and Z, and a direct measurement of their mass at the Sp$\bar{\rm p}$S collider[4] opened a new way of checking the standard model.

The relevant equations have been given in reference 5. The masses are given by

$$M(W) = A/\sin\theta \qquad M(Z^0) = 2A/\sin 2\theta, \tag{1}$$

with $A = (37.2810 \pm 0.0003)$ GeV$/\sqrt{(1 - \Delta r)}$ and the correction term, $\Delta r = 0.0696 \pm 0.0020$. By rewriting the equations, one obtains a formula for $\sin^2\theta$ that is independent of the absolute calibrations of the mass measurement:

$$\sin^2\theta = 1 - M(W)^2/M(Z^0)^2. \tag{2}$$

The radiative corrections can be measured by using

$$\Delta r = 1 - \frac{(37.2810\ \text{GeV})^2}{M(W)^2[1 - M(W)^2/M(Z^0)^2]}.$$

TABLE 3. Measurements of $\sin^2\theta$ and ρ

Process	$\sin^2\theta$	Condition	ρ
νN	$0.223 \pm 0.007 \pm 0.01$	$\rho = 1$	1.01 ± 0.02
eD	$0.215 \pm 0.015 \pm 0.005$	$\rho = 1$	
e^+e^-	0.18 ± 0.02	$M(Z^0) = 93 \pm 2$ GeV	1.17 ± 0.09
$\nu_\mu e$	$0.22 \pm 0.03 \pm 0.01$	ρ indep.	1.09 ± 0.14
μC	$0.23 \pm 0.07 \pm 0.04$	$\rho = 1$	
Atomic Parity violation	$0.205 \pm 0.034 \pm 0.024$	ρ indep.	
νp	0.182 ± 0.023	$\rho = 1$	
W and Z masses	0.22 ± 0.01	ρ indep.	1.00 ± 0.036

In the preceding equations, ρ was taken from the standard model; if one releases this condition, one obtains

$$\rho = \frac{M(W)^2}{M(Z^0)^2(1 - A^2/M(W)^2)},$$

which can be compared with the measurement.

In TABLE 4, the results of the direct mass measurements are compared to the predictions of the model for $\sin^2\theta = 0.217 \pm 0.014$. Good agreement is observed at the level of accuracy of the experiments.

Future Improvements

Although many tests have been performed, none of them has yet reached sufficient precision to determine the validity of the first order corrections. In order to recall the situation, one sees that

$$\Delta r = 0.0696 \Longrightarrow \Delta(M(Z^0)) = 3.3 \text{ GeV}$$

$$\Delta \sin^2\theta = 0.014 \Longrightarrow \Delta(M(Z^0)) = 2.4 \text{ GeV}$$

$$\text{from UA1 and UA2} \Longrightarrow \Delta(M(Z^0)) = 2.5 \text{ GeV},$$

thus demonstrating the need for higher precision measurements.

The future improvements to the luminosity of the collider (ACOL project) are expected to boost the integrated luminosity from 100 nb^{-1} to 10,000 nb^{-1}. The improvement to be expected in the measurements of $\sin^2\theta$ can be estimated from the two equations by which $\sin^2\theta$ can be determined from the masses:

(1) From $\sin^2\theta = 1 - M(W)^2/M(Z^0)^2$: In this expression, the absolute mass scale drops out and probably other systematic effects are small. One could expect an improvement of the order of the improvement in the statistical error. Hence, a precision of $\Delta \sin^2\theta = \pm 0.004$ can be expected.

TABLE 4. Comparison with Direct Mass Measurements

Quantity	UA1	UA2	Standard Model with $\sin^2\theta =$ 0.217 ± 0.014
$M(W)$ (GeV)	80.9 ± 1.5 ± 2.4	83.1 ± 1.9 ± 1.3	83.0 + 2.9/− 2.7
$M(Z^0)$ (GeV)	95.6 ± 1.5 ± 2.9	92.7 ± 1.7 ± 1.4	93.8 + 2.4/− 2.2
$M(Z^0) - M(W)$ (GeV)	14.7 ± 2.1 ± 0.4	9.6 ± 2.5 ± 0.2	10.8 ± 0.5
$\sin^2\theta$ [eq. (**2**)]	0.284 ± 0.035	0.196 ± 0.035	0.217 ± 0.014
Δr	0.252 ± 0.072 ± 0.045	0.051 ± 0.173 ± 0.030	0.0696 ± 0.0020
$\sin^2\theta$ [eq. (**1**)]	0.228 ± 0.008 ± 0.014	0.216 ± 0.010 ± 0.007	0.217 ± 0.014
ρ	0.928 ± 0.038 ± 0.016	1.006 ± 0.043 ± 0.01	1.00

(2) From $\sin^2\theta = A^2/M(W)^2$: For higher statistics, the result from this expression will be dominated by the scale error on the mass measurement. If one assumes that the absolute calibration of these big calorimeters can be performed to 1%, one derives a limiting accuracy of $\Delta \sin^2\theta = \pm 0.004$.

Having these future improvements to the direct mass measurements in mind, these can serve as a goal to be reached by the $\sin^2\theta$ measurements in other processes. At present, there are several attempts to reach a precision of $\Delta \sin^2\theta = \pm 0.005$ in neutrino scattering; some of them will be discussed in the following sections.

HIGH PRECISION SEMILEPTONIC NEUTRINO MEASUREMENTS

It was shown by Llewellyn-Smith that for isoscalar targets, the major contributions to the NC and CC cross sections are related by assuming isospin invariance alone.[6] The

TABLE 5. Comparison of High Precision R_ν Measurements

	CDHS[7]	CHARM[8]
Beam	160 GeV NBB	160 GeV NBB
Statistics CC	250,000	120,000
NC	80,000	40,000
E_H-cut	10 GeV	2 GeV
p_μ-cut	$\approx$ 3 GeV	1 GeV
Method	event-length	muon-recognition
Total corr.	$\approx$ 25%	$\approx$ 15%
$\Delta \sin^2 \theta^a$	0.005	0.005

[a] These uncertainties do not include common theoretical errors.

following equation for the differential cross sections applies:

$$d^2\sigma_{NC}^{\nu(\bar{\nu})}/dxdy = (1/2 - \sin^2\theta + 5/9\sin^4\theta)\, d^2\sigma_{CC}^{\nu(\bar{\nu})}/dxdy + (5/9\sin^4\theta)\, d^2\sigma_{CC}^{\bar{\nu}(\nu)}/dxdy \quad (3)$$

Integrating equation (3) over x and y, one obtains a relation containing only ratios of total cross sections. The measurement sensitive to $\sin^2\theta$ is the ratio of neutral-current over charged-current total cross sections for neutrinos, R_ν.

At present, two groups at CERN are analyzing data taken in 1984 to measure R_ν with high precision. An upgraded version of the narrow-band beam is used at 160 GeV central momentum, with nearly a factor of two more flux than previously. The method

FIGURE 1. Photograph of the CHARM detector.

used by the two groups is completely different and a comparison of the results will provide a good test of the systematic uncertainties involved. The experimental parameters are summarized in TABLE 5. In the following sections, the CHARM experiment will be described in some detail.

The Experimental Setup

The CHARM neutrino detector is a fine-grain calorimeter that is followed by an iron spectrometer with a toroidal magnetic field. It is surrounded by a magnetized iron frame. The calorimeter has a sampling corresponding to one radiation length, with scintillators, proportional drift tubes, and streamer tubes as detecting elements. It is described in detail elsewhere.[9] A view of the CHARM-I detector is shown in FIGURE 1. The fiducial mass used in this exposure was 87 tons.

Data were taken in the 160 GeV narrow-band beam (NBB), which was developed by Grant and Maugain[10] as a high flux version of the CERN NBB optics. It satisfies the conditions for a precision measurement of R_ν, namely,[11]

(i) A sufficiently high average neutrino energy, which is needed for efficient pattern recognition in the CHARM detector.
(ii) Low background contributions that are measurable with high accuracy.
(iii) The possibility to measure the relative $\bar{\nu}/\nu$ flux normalization accurately.
(iv) A calculable neutrino energy spectrum.
(v) A high event rate.

In addition to these points, it possesses a well-defined energy-radius correlation of the neutrino flux. This makes it possible to obtain y-distributions of neutral current interactions.

Event Classification and Corrections

Interactions in the CHARM detector are classified by an automatic pattern recognition program on an event-by-event basis. Selection criteria are applied that attempt to identify optimally the physical processes and to minimize the corrections needed to relate the visible cross sections to the physical cross sections.

Neutrino interactions are defined as events without entering charged tracks. A neutrino event is called a charged current interaction if it contains a muon originating from the event vertex. The muon has to be visible over a range corresponding to an energy-loss of at least 0.67 GeV. Its total range has to exceed 1 GeV energy-loss. All other neutrino event candidates are classified as neutral current interactions. Only those events that have their vertex inside the fiducial volume and that have a shower energy of at least 2 GeV are analyzed. Off-line representations of an NC and CC event in the CHARM-detector are shown in FIGURES 2 and 3.

The following corrections have to be applied to obtain the CC and NC event numbers:

The classification efficiency of the automatic pattern recognition is known to be better than 99.5%. In addition, a flag is generated by the program to tag situations

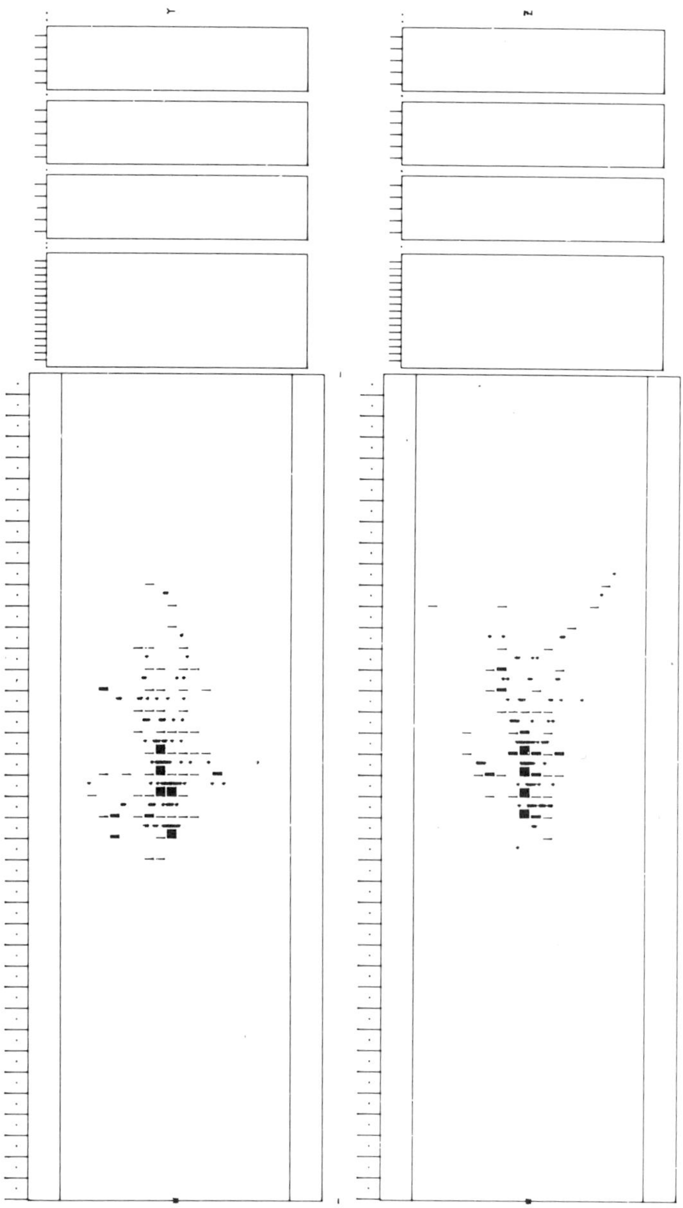

FIGURE 2. Off-line representation of a neutral current event recorded in the CHARM detector.

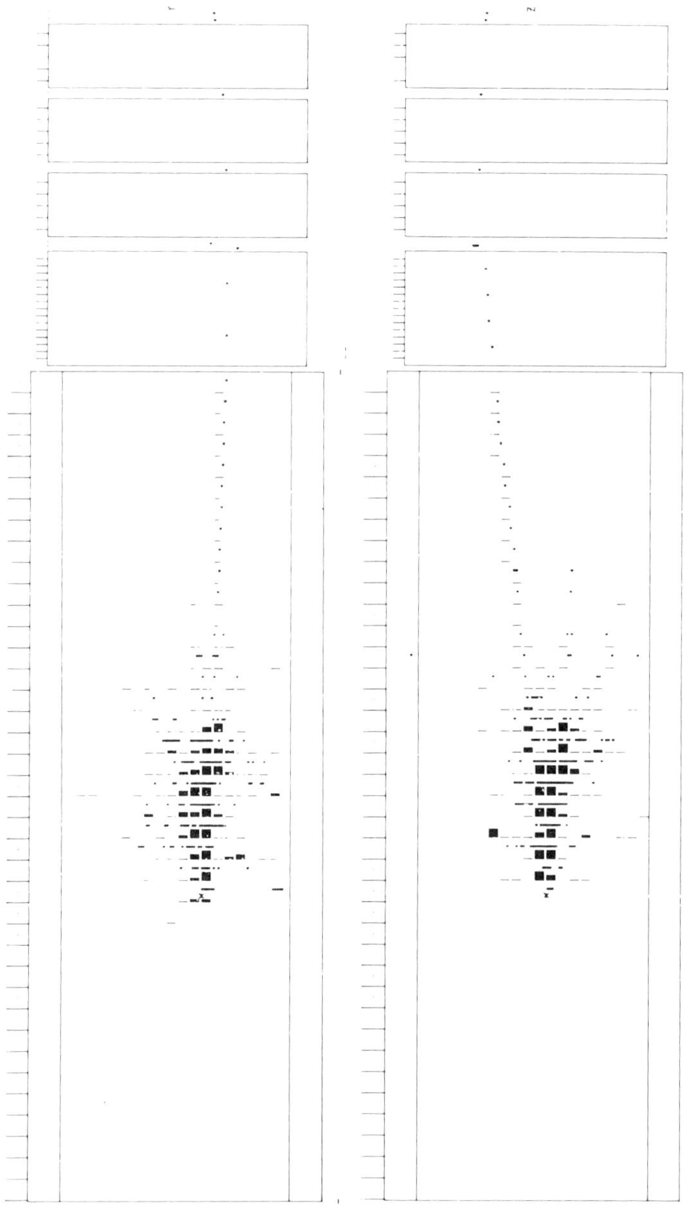

FIGURE 3. Off-line representation of a charged current event recorded in the CHARM detector.

where the automatic procedure could fail. A visual inspection of these flagged events (~4000), together with a scan of a randomly selected sample of events, reduces the error introduced by classification errors to $\Delta R_\nu/R_\nu \leq 0.1\%$.

CC events in which the primary muon cannot be identified are classified as NC. The majority of these muons have an energy less than 1 GeV. A smaller contribution to the loss of CC events is caused by muons of more than 1 GeV either leaving the sides of the detector before their energy loss is sufficiently large or by muons obscured by the hadronic shower. The total correction induced by all sources of unidentified muons is of the order of 2% of the CC event rate, which induces an uncertainty amounting to $\Delta R_\nu/R_\nu \simeq 0.3\%$.

A small fraction of NC events contains a track that fulfills the requirements of a primary muon of a CC event. Therefore, these events are classified as CC events. This background is caused by decays of pions or kaons in the shower or by punch-through tracks. The size of this correction is of the order of 2%, giving $\Delta R_\nu/R_\nu \simeq 0.2\%$.

Both the NC and CC interactions of electron neutrinos originating from K_{e3}-decays in the beam are classified as NC events. The contribution of these events amounts to

TABLE 6. Error Sources in the Measurement of R_ν

Error Source	% Error in R_ν
Statistics	0.7%
Classification inefficiencies	0.1%
WB background + cosmic	0.3%
μ recognition losses	0.3%
π/K decays in the shower	0.2%
K_{e3} decays	0.4%
Total systematic error	0.7%
Total error	1.0%

~5% of the NC candidates (~2% in the $\bar{\nu}$ exposure). With a measurement of the K/π ratio with an accuracy of ~2 ÷ 3%, this correction can be calculated to ±5% of its value, corresponding to $\Delta R_\nu/R_\nu \simeq 0.4\%$.

The flux from the decays before the decay tunnel is not well calculable and has to be subtracted using data obtained with the entrance of the decay tunnel blocked by an absorber. The uncertainty introduced by this wide-band (WB) background amounts to $\Delta R_\nu/R_\nu \simeq 0.3\%$.

The uncertainty introduced by the cosmic ray background during normal data taking is small because sufficient data can be taken in periods without the beam to measure the correction and because the correction is rather small.

The final experimental precision obtained in the measurement of R_ν is summarized in TABLE 6. The uncertainty in r is dominated by the event statistics (1.5%), the WB background subtraction (1.5%), and the relative normalization error of the $\bar{\nu}$ to the ν flux (2.0%). A 3% accuracy in r is therefore expected. The accuracy of the measurements of R_ν and r as given above is sufficient to ensure an experimental precision of $\Delta \sin^2 \theta = \pm 0.005$.

Theoretical Uncertainty in sin² θ

At this level of experimental precision, it is worthwhile to consider the uncertainties introduced by the interpretation of R_ν in terms of $\sin^2\theta$. The main contributions to the theoretical error can be classified as follows:

(i) Uncertainties in the Kobayashi-Maskawa (K.M.) mixing matrix.
(ii) The momentum weighted content of s and c quarks in the nucleon and the threshold effects in c-production.
(iii) Higher twist terms.

A careful analysis[6] of the uncertainties shows that, using equation (**3**), the uncertainty is $\Delta\sin^2\theta(\text{theory}) \simeq 0.01$. If in addition it is assumed that the matrix is almost diagonal, this error is reduced to ± 0.005. An additional uncertainty due to higher twist terms is estimated to be smaller than 0.005.[6]

However, if one uses other methods such as the Paschos-Wolfenstein relation,

$$R^- = \frac{\sigma^{\nu}_{\mathrm{NC}} - \sigma^{\bar\nu}_{\mathrm{NC}}}{\sigma^{\nu}_{\mathrm{CC}} - \sigma^{\bar\nu}_{\mathrm{CC}}} = (1/2) - \sin^2\theta,$$

one obtains a smaller theoretical uncertainty by canceling the sea contributions. This method requires more antineutrino data to obtain the same statistical accuracy. Since the event rates in antineutrino beams are usually a factor of six lower per incident proton, the small statistics in the antineutrino measurements determine the final accuracy.

The Measurement of sin² θ Using the y*-Distributions*

An alternative method was proposed that relied on comparisons of the y-distributions of neutral-current and charged-current neutrino and antineutrino interactions.[12] The uncertainties proportional to the strange quark component in the nucleon can be removed by exploiting the additional information provided by a simultaneous measurement of the NC and CC y-distributions of neutrino and antineutrino scattering. The normalization of the ill-known terms can be introduced as free parameters. Corrections for the s-quark contributions to the NC cross sections are treated by the substitution,

$$d\sigma_{\mathrm{NC}}/dy \rightarrow d\sigma_{\mathrm{NC}}/dy - (g_s^2 \cdot s) \cdot f(y), \qquad (\mathbf{4})$$

where $(g_s^2 \cdot s)$ is a free parameter and $f(y)$ is obtained using a specific model, e.g., in the scaling limit,

$$f(y) \rightarrow 2[1 + (1 - y)^2]. \qquad (\mathbf{5})$$

The free parameter, $(g_s^2 \cdot s)$, is interpreted as being the product of the coupling of the weak neutral current to the strange quark, g_s^2, and the strange quark content of the nucleon, s. This model can be improved by using parametrizations including quantum-chromodynamic (QCD) type corrections. In the charged current sector, a similar correction is made, which is based on dimuon measurements:

$$d\sigma_{\rm CC}/dy \rightarrow d\sigma_{\rm CC}/dy - B^{-1}\, d\sigma_{{\rm di}-\mu}/dy, \tag{6}$$

in which B (the average branching ratio to muons of charm states produced in this process) is a free parameter. This procedure takes care of threshold effects and removes any need for a priori knowledge on the product of the element, $|U_{cs}|^2$, and the absolute strange quark component in the nucleon.

The corrections that remain to be applied involve the knowledge of the mixing elements, $|U_{ud}|$, $|U_{us}|$, and $|U_{cd}|$, and the contribution of charm quarks to the NC and CC cross sections. According to Llewellyn-Smith,[6] these corrections are calculable to within an uncertainty of $\Delta \sin^2 \theta = \pm 0.004$. No assumptions on the shape of the parton distributions other than equation (**5**) or its equivalent have to be made.

This method improves on previous determinations of $\sin^2 \theta$ from the differential cross sections, $d\sigma/dy$, because it uses the quark-parton model in correction terms only and, therefore, scaling violations and threshold effects are taken into account intrinsically. The influence of higher twist effects can be shown to be small ($\Delta \sin^2 \theta < \pm 0.005$).[6]

TABLE 7. $\Delta \sin^2 \theta$ with Different Methods

Error Type:	From R_ν	From P.W. Relation	From $d\sigma/dy$
Statistical	0.0035	0.008	0.006
Systematic	0.0035	≈0.0035	≈0.0035
Theoretical	0.01	0.003	0.004

The Measurement of the y*-Distribution*

The fine granularity of the CHARM calorimeter not only permits a powerful separation of NC and CC induced events, but also it permits a precise measurement of the vertex and energy of hadronic showers.[13] These features, together with the energy-radius correlation of the NBB, make a measurement of the y-distribution of NC events possible.[14]

The statistical accuracy of $\sin^2 \theta$ as obtained from the y-distributions is determined mainly by the neutrino statistics, but also some antineutrino data are needed. A fit to the y-distributions, including the contributions of strange quarks to the CC and NC cross sections as free parameters, provides a statistical uncertainty of $\Delta \sin^2 \theta$ (statistical) $= \pm 0.006$. The experimental systematic uncertainty is estimated to be $\Delta \sin^2 \theta$ (systematic) $= \pm 0.003$.

TABLE 7 summarizes the different errors in the various methods for an exposure yielding 100,000 neutrino CC events and 10,000 antineutrino CC events.

NEUTRINO-ELECTRON SCATTERING

The advantage of using a purely leptonic process to determine $\sin^2\theta$ is shown clearly by the discussion on theoretical errors in the semileptonic case. This process is fully understood theoretically and the measurement is a purely experimental problem. The cross sections of muon-neutrino electron scattering can be written in terms of the vector and axial vector coupling of the electron, g_V^e and g_A^e, as follows:

$$\sigma(\nu_\mu e \rightarrow \nu_\mu e) = G^2 m_e E_\nu/2\pi \,\{(g_V^e + g_A^e)^2 + (1-y)^2 (g_V^e - g_A^e)^2\}$$

$$\sigma(\bar{\nu}_\mu e \rightarrow \bar{\nu}_\mu e) = G^2 m_e E_\nu/2\pi \,\{(g_V^e - g_A^e)^2 + (1-y)^2 (g_V^e + g_A^e)^2\},$$

where $y = E_e/E_\nu$, the ratio of the recoil electron energy and the incident neutrino energy, and m_e is the electron mass. The low cross section of the elastic scattering of muon-neutrinos on electrons compared to the inclusive semileptonic cross section, $\sigma(\nu_\mu e) \approx 10^{-4}\,\sigma(\nu_\mu N)$, presents a formidable experimental problem. The experimental tools involve a good electron-pion separation, some electron-photon separation, and a measurement of the recoil electron angle, θ_e, in large-mass detectors.

A sensitive measurement of $\sin^2\theta$ can be obtained using the ratio of neutrino and antineutrino cross sections:

$$R = \frac{\sigma(\nu_\mu e)}{\sigma(\bar{\nu}_\mu e)} = \frac{1+\eta+\eta^2}{1-\eta+\eta^2}, \tag{7}$$

where $\eta = 1 - 4\sin^2\theta$. From equation (7), it follows that $\Delta\sin^2\theta \simeq (1/8)\,\Delta R/R$ for $\sin^2\theta$ close to 0.25. The use of this ratio also has experimental advantages if it is measured with a single apparatus, since:

(i) the efficiencies for finding electrons cancel;
(ii) some of the systematic uncertainties in the background subtraction cancel;
(iii) no absolute normalization of the neutrino fluxes is needed since only the ratio of neutrino and antineutrino fluxes enters.

Measurements of $\sin^2\theta$ using the method of the ratio of the cross sections were first obtained by the CHARM collaboration.[15] The results are summarized in the following section. A measurement with similar precision was reported recently by the E734 collaboration at Brookhaven.[16] On the basis of the experience with the CHARM-I detector, a new dedicated detector is being built, with the aim to provide a precision of $\Delta(\sin^2\theta) \simeq 0.005$.

The CHARM-I ν_μe Scattering Experiment

The CHARM collaboration has obtained 83 $\nu_\mu e$ scattering events and 112 $\bar{\nu}_\mu e$ scattering events in the CERN wide-band beam (WBB).

Events of electromagnetic nature are selected on the basis of their shower properties, rejecting the background by a factor of $\simeq$300. Events were selected requiring:

(i) a small energy deposition in the first scintillator plane (E_F) not exceeding seven minimum ionizing particles;

(ii) the number of wires hit in the proportional and streamer tube planes following the vertex to be small;
(iii) a small width of the shower, as shown in FIGURE 4.

The measurement of the recoil electron angle reduces the backgrounds further by exploiting the kinematics of the process, $E_e \theta_e^2 \leq 2m_e$, which provides a signal peaking in the forward region, while the backgrounds have much broader angular distributions. The results are shown in FIGURE 5. The remaining backgrounds are composed of ν_e quasi-elastic interactions and ν_μ neutral current interactions, including coherent π^0 production. The relative amount of the two background components was evaluated from the study of the E_F distribution shown in FIGURE 6; they are different for each of the showers initiated by electrons, photons, and neutral pions.

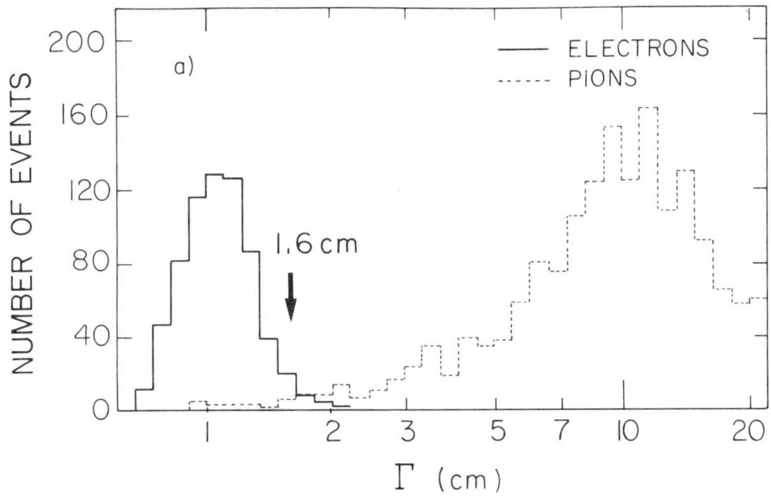

FIGURE 4. Width, Γ, of showers induced by electrons and pions measured in a test beam with the CHARM-I detector.

The neutrino and antineutrino fluxes were measured by analyzing events induced by quasi-elastic charged-current scattering on nucleons[17] and the events induced by inclusive neutral- and charged-current processes on nucleons.[18] The ratio of the normalized numbers of $\nu_\mu e$ and $\bar{\nu}_\mu e$ events is found to be

$$R_{\text{exp}} = N(\nu_\mu e)/N(\bar{\nu}_\mu e) \times F = \sigma(\nu_\mu e)/\sigma(\bar{\nu}_\mu e) = 1.26 + 0.41/-0.28.$$

The flux factor, F, was obtained by taking the average of both flux determinations. The relation between R_{exp} and $\sin^2 \theta$ given by equation (7) is shown in FIGURE 7. One obtains:

$$\sin^2 \theta = 0.215 \pm 0.032 \text{ (stat)} \pm 0.012 \text{ (syst)}.$$

In FIGURE 8, the allowed areas in the g_V-g_A plane are shown. Four regions are

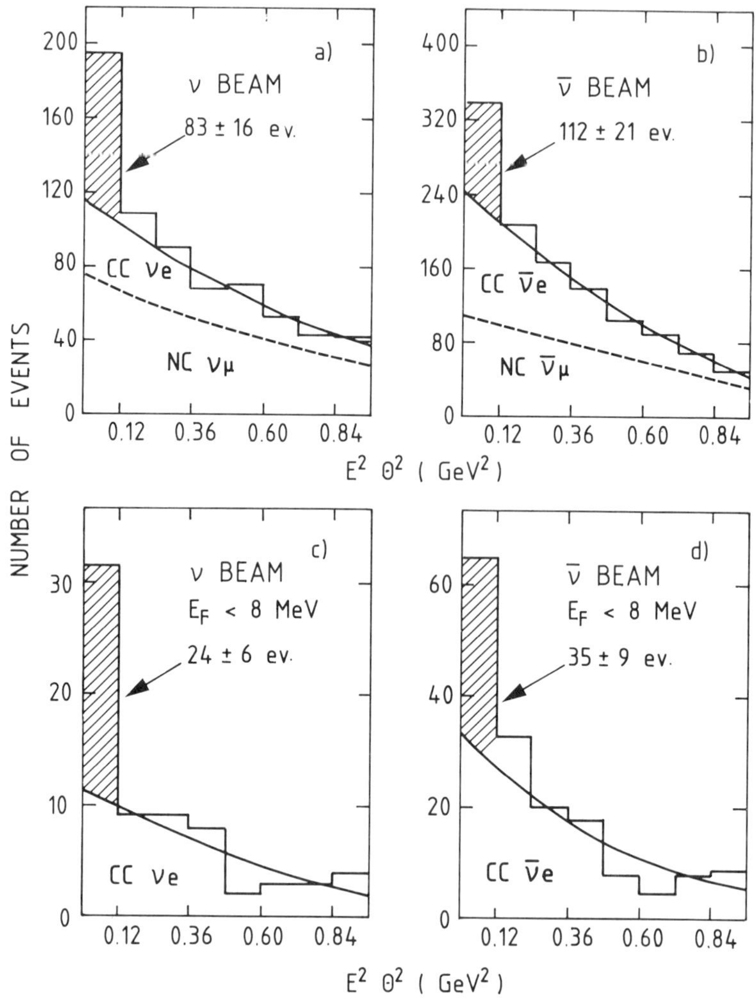

FIGURE 5. $E^2\theta^2$ distributions for (a) neutrino, (b) antineutrino events; in (c) and (d), the additional condition of $E_F < 8$ MeV is applied to (a) and (b), respectively.

allowed by $\nu_\mu e$ and $\bar{\nu}_\mu e$ scattering. Combining these regions with electron-positron annihilation data and electron-neutrino-electron scattering results, one solution survives:

$$g_A^e = -0.54 \pm 0.05 \text{ (stat)} \pm 0.06 \text{ (syst)}$$

$$g_V^e = -0.08 \pm 0.07 \text{ (stat)} \pm 0.03 \text{ (syst)}.$$

Using the absolute value of the cross section determinations, the relative strength of the neutral- and charged-current coupling constants is found to be

$$\rho = 1.09 \pm 0.09 \text{ (stat)} \pm 0.11 \text{ (syst)},$$

which is in agreement with the prediction of the standard model.

The CHARM-II Detector

The CHARM-II detector is being built at CERN and is approved for an exposure corresponding to 2000 events in each channel, thus aiming at an accuracy of $\Delta \sin^2 \theta = 0.005$.[20] This detector is expected to be fully operational in 1986. The techniques used here are extrapolated from the experience obtained with the CHARM-I detector. The main improvements with respect to the CHARM-I detector are:

(1) A finer granularity, both longitudinally and laterally, in order to achieve a higher efficiency for selecting electron showers, as well as a better angular resolution for electron showers by a factor of three, thus reducing the total background by a factor of five.
(2) A fivefold increase in fiducial mass.
(3) An optimized performance of the wide-band neutrino beam.

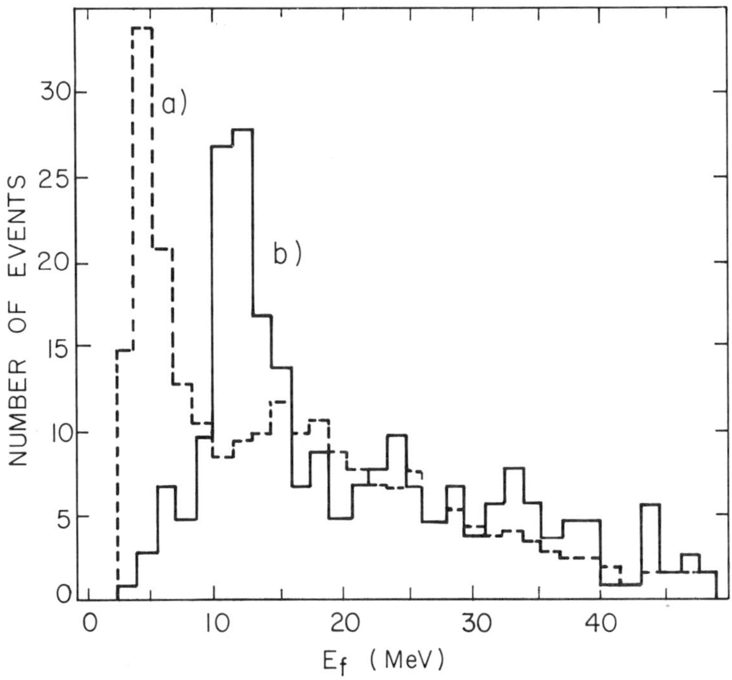

FIGURE 6. Distribution of E_F (energy deposition in the first scintillator plane following the vertex) for (a) electrons and (b) π^0.

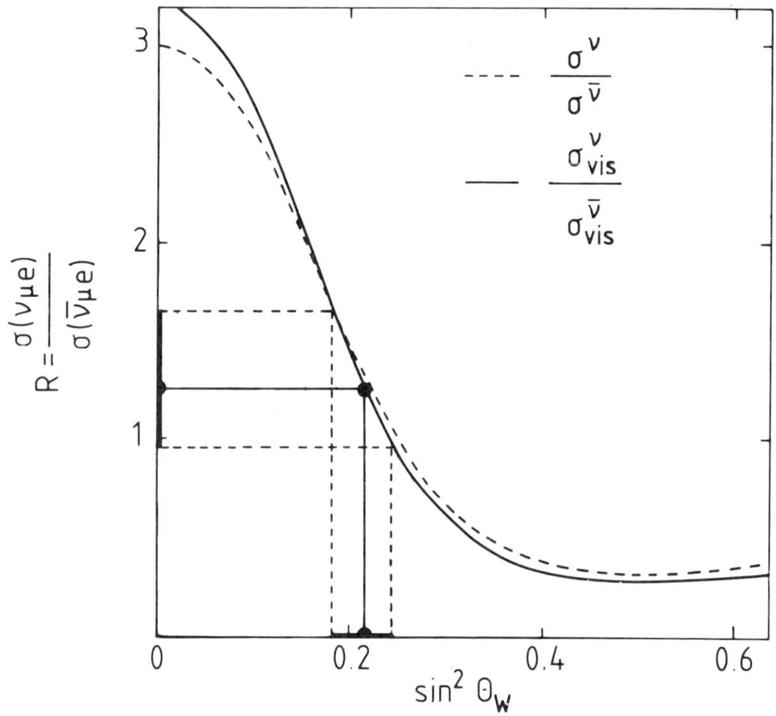

FIGURE 7. Relation between $R = \sigma(\nu_\mu e)/\sigma(\bar{\nu}_\mu e)$ and $\sin^2\theta$ with the measured value of R and the corresponding value of $\sin^2\theta$.

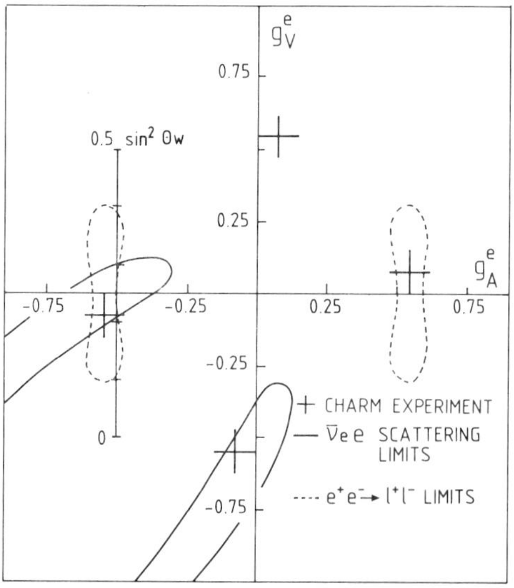

FIGURE 8. Values of g_A^e and g_V^e, the neutral current coupling constants, obtained from measurements of $\sigma(\nu_\mu e)$ and $\sigma(\bar{\nu}_\mu e)$ by the CHARM collaboration.[15] The limits from forward-backward asymmetry in $e^+e^- \rightarrow \ell^+\ell^-$ (see, for example, reference 3) and from $\sigma(\bar{\nu}_e e)$ measurements[19] select a unique solution.

The structure of the detector is sketched in FIGURE 9. It consists of 420 equal units of $3.7 \times 3.7\ m^2$ surface area, each composed of a 5 cm thick glass plate and of a plane of streamer tubes with 1 cm wire spacing. They are read out by the wires and by cathode strips of 2 cm spacing, which are perpendicular to the wires. The wires are read out digitally to obtain unambiguous information about the track multiplicity near the vertex. The signal from the strips is read out with analogue electronics and provides a ± 2 mm accuracy of the centroid position of a track, along with a precise measurement of the position of the shower core. A simulation of this detector by Monte Carlo methods gives an electron shower angular resolution of $\sigma(\theta) \sim 16\ \text{mrad}/\sqrt{E/\text{GeV}}$, which is similar to the natural angular spread of recoil electrons in this process. The signal to background ratio is expected to improve by a factor of five. An exposure with 1.5×10^{19} protons will yield ~2000 events in each channel.

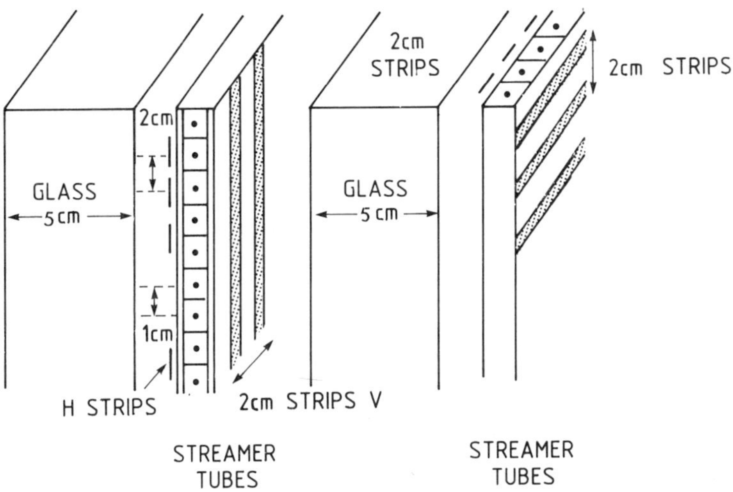

FIGURE 9. Sketch of the structure of the CHARM-II detector.

Monitoring of the neutrino flux and of the beam composition using quasi-elastic events has been studied in detail by the CHARM-I experiment. A total flux ratio error of $\pm 2\%$ is estimated, which leads to a total error of $\pm 4\%$ on R and a corresponding experimental error of $\Delta \sin^2 \theta = \pm 0.005$. There is no theoretical uncertainty in extracting this result from the data.

SUMMARY

The present experimental status supports the low energy predictions of the standard model. However, to obtain a meaningful test of the loop-level corrections, it is necessary to improve the experimental precision. This precision will be reached by two semileptonic neutrino-scattering experiments performed in 1984 at CERN. However, the present theoretical uncertainties in the interpretation of these experiments are still too large. A determination of $\sin^2\theta$ using a purely leptonic process does not suffer from these uncertainties. The CHARM-II detector is being built at CERN and is aiming at a precision in $\sin^2\theta$ that will give meaningful tests of the first order corrections of the theory. These neutrino experiments will match the precision reached in the near future with the direct mass measurements obtained at the Sp$\bar{p}$S collider at CERN, upgraded with the ACOL project.

REFERENCES

1. HASERT, F. J. *et al.* 1973. Phys. Lett. **B46:** 138.
2. GLASHOW, S. L. 1961. Nucl. Phys. **22:** 579; SALAM, A. & J. WARD. 1964. Phys. Lett. **13:** 168; WEINBERG, S. 1964. Phys. Rev. Lett. **D5:** 1264.
3. PANMAN, J. 1984. XIth International Conference on Neutrino Physics and Astrophysics, Dortmund (and references therein).
4. ARNISON, G. *et al.* 1983. Phys. Lett. **B122:** 103; Phys. Lett. **B126:** 398; Phys. Lett. **B129:** 273; BANNER, M. *et al.* 1983. Phys. Lett. **B122:** 476; BAGNAIA, P. *et al.* 1983. Phys. Lett. **B129:** 130.
5. MARCIANO, W. 1983. Proceedings of the International Lepton and Photon Symposium, Cornell; MARCIANO, W. & A. SIRLIN. 1983. Testing the standard model by precise determinations of *W* and *Z* masses. BNL preprint.
6. LLEWELLYN-SMITH, C. H. 1983. Contribution to Fixed Target Workshop, CERN (1982); Nucl. Phys. **B228:** 205–215.
7. DYDAK, F. *et al.* 1983. CERN/SPSC/83-49.
8. BERGSMA, F. *et al.* 1984. CERN/SPSC/84-1.
9. DIDDENS, A. N. *et al.* 1978. Nucl. Instr. Methods **157:** 35; JONKER, M. *et al.* 1982. Nucl. Instr. Methods **200:** 183.
10. GRANT, A. & J. M. MAUGAIN. 1983. CERN/EF/BEAM 83-2.
11. PANMAN, J. 1982. Contribution to the SPS Fixed Target Workshop, CERN.
12. PANMAN, J. 1984. Phys. Lett. **B154:** 221–223.
13. ABT, I. *et al.* 1983. Nucl. Instr. Methods **217:** 377–381.
14. JONKER, M. *et al.* 1981. Phys. Lett. **B102:** 67.
15. JONKER, M. *et al.* 1982. Phys. Lett. **B117:** 272; BERGSMA, F. *et al.* 1984. Phys. Lett. **B147:** 481.
16. AHRENS, L. A. *et al.* 1985. Phys. Rev. Lett. **54:** 18.
17. JONKER, M. *et al.* 1980. Phys. Lett. **B93:** 203; BERGSMA, F. *et al.* 1983. Phys. Lett. **B122:** 465.
18. JONKER, M. *et al.* 1981. Phys. Lett. **B99:** 265.
19. REINES, F., H. S. GURR & H. W. SOBEL. 1976. Phys. Rev. Lett. **37:** 315. (The limits shown in FIGURE 8 have been evaluated by KRENZ, W. 1982. RWTH Aachen, Pitha 82/26.)
20. BUSI, C. *et al.* 1983. CERN/SPSC/83-24 and 83-37.

Radiative Corrections in the $SU(2)_L \times U(1)$ Theory[a]

A. SIRLIN

Department of Physics
New York University
New York, New York 10003

INTRODUCTION

The fact that the modern gauge theories of electroweak interactions are renormalizable has stimulated considerable interest in the evaluation of radiative corrections to physical observables. The main objective of this work is the verification of the theory at the level of its quantum corrections. If successful, such an enterprise could become the counterpart of the extraordinary comparisons between theory and experiment carried out in QED since the late forties. There is also a second, more speculative, aim in these studies. It may occur that some of the radiative corrections are sensitive to very short distance phenomena. In that case, accurate comparisons with experiment may give subtle, but significant clues about new physics and exotic phenomena.

TRADITIONAL WEAK INTERACTIONS

Loop effects are important in many weak interaction processes. For example, there are processes, such as $K_L \rightarrow \mu^+ + \mu^-$, that are forbidden in lowest order and are roughly estimated theoretically by considering higher order corrections. There are also some allowed processes in which the experimental precision is so great that inclusion of the corrections becomes imperative. A classical example is the study of the spectrum and angular distribution in muon decay,[1,2] in which the radiative corrections have played an essential role in achieving agreement with the predictions of the local V-A theory (or more generally, the local theory with two-component neutrinos). However, as the photonic corrections responsible for these effects are known to be convergent to all orders in α (but lowest order in G_μ),[3] it perhaps can be argued that the accurate agreement between theory and experiment in this case represents a success of the local V-A theory allied with QED, rather than the modern electroweak theories. The situation is markedly different in another old problem, namely, the analysis of the universality of the weak interactions.[4] In modern parlance, we can formulate the problem as follows: is the experimentally determined Kobayashi-Maskawa matrix unitary? Using the accurate data from superallowed Fermi transitions, μ decays, and $\Delta Q = \Delta S = 1$ semileptonic processes, and ignoring loop effects, one finds $|U_{ud}|^2 + |U_{us}|^2 > 1$ by about 3.2%. Thus, neglect of the radioactive corrections leads to

[a]This work was supported in part by the National Science Foundation grant no. PHY-8116102.

a sharp disagreement between experiment and one of the most general properties of the $SU(2)_L \times U(1)$ theory, namely, the unitarity of the K-M matrix. The question is then whether inclusion of higher order effects can rescue the theory from such disagreement. In the Standard Model (SM), there is a theorem that suggests that this indeed may happen. It says that in semileptonic processes mediated by $W^{\pm}$, the large m_Z behavior of the O(α) corrections can be written as[5]

$$\Delta P = \overset{0}{P}\left[\frac{3\alpha}{2\pi}(1 + 2\overline{Q})\ln\left(\frac{m_Z}{\mu}\right) + \ldots\right], \tag{1}$$

where $\overset{0}{P}$ is the zeroth order probability expressed in terms of the μ-decay coupling constant, G_μ, μ is an energy scale characteristic of the process, and $\overline{Q}$ is the average charge of the quark isodoublet. For semileptonic processes, with the usual quark changes, we have $\overline{Q} = 1/6$. For μ decay, the analogue of the quark isodoublet is (ν_μ, μ) and $\overline{Q} = -1/2$; thus, the large $\ln(m_Z/\mu)$ terms are absent in this case. Using Current-Algebra and Short-Distance Expansion techniques, one obtains for the decay probability of the superallowed Fermi transitions[4]:

$$\Delta P = \overset{0}{P}\left[\frac{3\alpha}{2\pi}\ln\left(\frac{m_Z}{2E_m}\right) + \ldots\right], \tag{2}$$

where E_m is the endpoint energy of the positron. Note that we have included only the dominant logarithm. The argument of the logarithm in equation (**2**) is O(2×10^4); as a consequence, inclusion of the radiative corrections decreases $|U_{ud}|^2$ by $\simeq$3.7%. In this way, one finds $|U_{ud}| = 0.9735 \pm 0.0016$. Combining with $|U_{us}|^2 = 0.221$[6] and recalling $|U_{ub}|^2 < 10^{-4}$, we have $|U_{ud}|^2 + |U_{us}|^2 + |U_{ub}|^2 = 0.997 \pm 0.004$, so that agreement between theory and experiment is restored. In the local V-A theory, the effective cutoff was infinite rather than m_Z; as a consequence, while the corrections to μ decay were finite, those involving β decay were divergent.[1-3] Thus, both the renormalizability of the theory and the actual sign and magnitude of the corrections have been found to be crucial in establishing the phenomenological tenability of the SM.

THEORETICAL PREDICTION OF m_W, m_Z

Recently, several theoretical groups have studied a new generation of problems involving the theoretical predictions of m_W and m_Z (and their interdependence) and the accurate determination of $\sin^2\theta_W$.[b] I will follow the approach of a program carried out in collaboration with W.J. Marciano. The methodology of this approach is based to a large extent on a "Simple Renormalization Framework"[9] and the Current Algebra Formulation of Radiative Corrections.[4] It is well known that in the SM, there are a number of important tree-level relations:

$$\cos\theta_W = m_W/m_Z, \tag{3a}$$

[b] Detailed reports and references can be found in references 7 and 8.

$$e = g \sin \theta_W, \tag{3b}$$

$$\frac{G_\mu}{\sqrt{2}} = \frac{g^2}{8m_W^2}. \tag{3c}$$

In the Simple Renormalization Framework of reference 9, equations (**3a, b**) remain exact to all orders, with the understanding that m_W and m_Z are the physical masses of W and Z, and e is the conventionally defined charge of the positron. Thus, equations (**3a**) and (**3b**) can be regarded as the precise definitions of $\cos\theta_W$ and g, respectively, in terms of the three fundamental observables, e, m_W, and m_Z. On the other hand, equation (**3c**) is corrected to O(α). To determine the size of this effect, one must study the corrections to the muon decay lifetime. It is convenient to define G_μ by

$$\frac{1}{\tau_\mu} = \frac{G_\mu^2 m_\mu^5}{192\pi^3} f\left(\frac{m_e}{m_\mu}\right)\left[1 + \frac{3}{5}\frac{m_\mu^2}{m_W^2}\right]\left[1 + \frac{\alpha}{2\pi}\left(\frac{25}{4} - \pi^2\right)\left(1 + \frac{2\alpha}{3\pi}\ln\left(\frac{m_\mu}{m_e}\right)\right)\right], \tag{4a}$$

$$f(x) = 1 - 8x^2 + 24\, x^4 \ln x + 8x^6 - x^8, \tag{4b}$$

which, aside from the very small term, $(3/5)\,(m_\mu^2/m_W^2)$, are the relevant formulae of the local V-A theory.[10] Insertion of the experimental value of τ_μ leads to $G_\mu = (1.16635 \pm 0.00002) \times 10^{-5}\ \text{GeV}^{-2}$. Evaluating the corrections to τ_μ in the SM, one finds[9]

$$\frac{G_\mu}{\sqrt{2}} = \frac{g^2}{8m_W^2(1 - \Delta r)}. \tag{5a}$$

The radiative correction, Δr, is written as $\Delta r = \Sigma_{i=1}^3 \Delta r^{(i)}$. The term $\Delta r^{(1)}$, which represents contributions from the bosonic sector, is a complicated function of $\sin^2\theta_W$ and $\xi \equiv m_H^2/m_Z^2$, and it turns out to be relatively small: $\simeq -0.22\%$ for $\xi = 0$, 0.30% for $\xi = 1$, and 1.18% for $\xi = 100$. On the other hand, $\Delta r^{(2)}$ and $\Delta r^{(3)}$, which represent leptonic and hadronic contributions, are quite large:

$$\Delta r^{(2)} = \frac{2\alpha}{3\pi}\left[\sum_{1=e,\mu,\tau} \ln\left(\frac{m_Z}{m_1}\right) - \frac{5}{2} + \frac{3}{8}\frac{(2s^2 - 1)\ln c^2}{s^4}\right] \tag{5b}$$

$$\Delta r^{(3)} = -e^2\, Re\Pi_{\gamma\gamma}^{r(h)}(m_Z^2) + \frac{3\alpha}{4\pi}\frac{(2s^2 - 1)}{s^4}\ln c^2 \tag{5c}$$

where $s^2 \equiv \sin^2\theta_W$, $c^2 \equiv \cos^2\theta_W$, and $\Pi_{\gamma\gamma}^{r(h)}(m_Z^2)$ is the hadronic contribution to the renormalized vacuum polarization function evaluated at m_Z^2. In equations (**5b, c**), we have neglected terms of $O(m_f^2/m_W^2)$, but the complete expressions are known.[11,12] The large additive logarithms, anticipated by Marciano,[13] make $\Delta r^{(2)}$ and $\Delta r^{(3)}$ dominant. The appearance of logarithms of fermion masses can be traced to equation (**3b**) and the fact that e represents an amplitude evaluated at an "exceptional" invariant momentum $(q^2 = 0)$. For $\sin^2\theta_W = 0.217$, $m_H = m_Z$, and $m_t = 36$ GeV, one finds[14]

$$\Delta r = 0.0696 \pm 0.0020, \tag{5d}$$

where the error is associated with uncertainties in the hadronic contribution.

Recently, bounds for the contribution of the first five flavors to $\Delta r^{(3)}$ have been derived using a method of constrained functional minimization, in which the constraint parameters are nonperturbative QCD condensates determined in a self-consistent way from the value of the hadronic contribution to the muon anomaly.[15] The results turned out to be very close in magnitude and in estimation of the errors as those employed in references 9 and 14. Equation (**5d**) is not sensitive to small variations in $\sin^2\theta_W$, m_H, and m_t. For this reason, we may take that result as the "standard" value for Δr. Instead, if m_t and m_H are allowed to vary over the wide range of 20 GeV $\leq m_t \leq$ 60 GeV and 10 GeV $\leq m_H \leq$ 1 TeV, respectively, one obtains the (approximately symmetric) result of $\Delta r = (7.3 \pm 0.8) \times 10^{-2}$. Recently, the detailed results for Δr given in references 9 and 11 have been confirmed to $O(\alpha)$.[16] Renormalization group arguments and theorems on cancellation of mass singularities have been used to show that the terms of $O(\alpha^n \ln{}^n m)$ (where m is a small generic fermion mass) are contained in equation (**5a**), with the provision that one replaces $\Delta r \rightarrow \Delta r - e^4 Re\Pi_2^{(r)}(m_Z^2)$, where the last term represents the fourth order contribution to vacuum polarization.[17] Neglecting such a small term, we find from equation (**5a**),

$$m_W = \left(\frac{\pi\alpha}{\sqrt{2}G_\mu}\right)^{1/2} \frac{1}{\sin\theta_W}\left[\frac{1}{1-\Delta r}\right]^{1/2} = \frac{(38.65 \pm 0.04)\ \text{GeV}}{\sin\theta_W}, \tag{6}$$

with $\sin\theta_W$ defined according to equation (**3a**). Another popular definition, which appears naturally in GUT predictions, is $\sin^2\theta_W(m_W)$, introduced by modified minimal

TABLE 1. Comparison of Recent m_W and m_Z Data and 1981 Theoretical Predictions

	UA1 Electron Data	UA2 Electron Data	Weighted Average Including μ Data	1981 Prediction
m_W(GeV)	80.9 ± 1.5 ± 2.4	83.1 ± 1.9 ± 1.3	82.1 ± 1.7	$83.0^{+2.9}_{-2.7}$
m_Z(GeV)	95.6 ± 1.4 ± 2.9	92.7 ± 1.7 ± 1.4	93.0 ± 1.7	$93.8^{+2.4}_{-2.2}$

subtraction. The analytic relation with $\sin^2\theta_W$ is given to $O(\alpha)$ in reference 18. Numerically, $\sin^2\theta_W = 1.006 \sin^2\theta_W(m_W)$ for $m_H \simeq m_Z$.

To predict m_W and m_Z, one needs $\sin\theta_W$ in equation (**6**). The most precise values at present are obtained from the ν_μ deep inelastic scattering and the e-D asymmetry. In the case of $R_\nu = \sigma_T(\nu_\mu + N \rightarrow \nu_\mu + X)/\sigma_T(\nu_\mu + N \rightarrow \mu^- + X)$, equation (**1**) tells us that the denominator is renormalized by a factor of $1 + (\alpha/\pi)\ln(m_Z^2/\mu^2)$, which is not present in the numerator. This leads to an effective $\rho^2(\nu_{\mu;h}) = 0.983$, which, together with other effects, lowers $\sin^2\theta_W$ by $\simeq 5\%$.[19] Using the 1981 data, this led to[19,20]

$$\sin^2\theta_W = 0.217 \pm 0.014\ (R_\nu), \tag{7a}$$

$$\sin^2\theta_W = 0.218 \pm 0.020\ (e - D). \tag{7b}$$

Inserting this value in equations (**3a**) and (**6**) led to the 1981 prediction of m_W and m_Z, which is compared with recent UA1 and UA2 data[21] in TABLE 1. If the radiative corrections had been ignored both in equation (**6**) and in the extraction of $\sin\theta_W$ from experiment, the theoretical predictions would have been $m_W \simeq 78.2$ GeV and $m_Z \simeq 89.0$ GeV, which are about 4.8 GeV smaller.

However, now we can invert the process and compute $\sin\theta_W$ from m_W and m_Z. For example, inserting $m_W = 82.1 \pm 1.7$ GeV and $m_Z = 93.0 \pm 1.7$ GeV (see TABLE 1) into equation (**6**) and $\sin(2\theta_W) = 77.30\ \text{GeV}/m_Z$, respectively, leads to a weighted average of

$$\sin^2\theta_W = 0.222 \pm 0.007\ (\text{from } m_W, m_Z). \qquad \textbf{(8)}$$

Eliminating $\sin\theta_W$ between equations (**3a**) and (**6**) leads to relations describing the m_W and m_Z interdependence:[14,22]

$$m_Z = m_W/(1 - A^2/m_W^2)^{1/2} \qquad \textbf{(9a)}$$

$$m_W = \frac{m_Z}{\sqrt{2}}[1 + (1 - 4A^2/m_Z^2)^{1/2}]^{1/2} \qquad \textbf{(9b)}$$

where

$$A = \left(\frac{\pi\alpha}{\sqrt{2}G_\mu}\right)^{1/2}(1 - \Delta r)^{-1/2} = 37.281\ \text{GeV}\ (1 - \Delta r)^{-1/2}. \qquad \textbf{(9c)}$$

We also can determine in principle the radiative correction, Δr, from m_W and m_Z:

$$\Delta r = 1 - \frac{(37.281\ \text{GeV})^2}{m_W^2(1 - m_W^2/m_Z^2)}. \qquad \textbf{(9d)}$$

A quantity of particular interest is $m_Z - m_W$ as a function of m_Z.[14] For $m_Z = 93.8 \pm 2.5$ GeV and $A = 38.65$ GeV, one predicts that $m_Z - m_W = 10.8 \pm 0.6$ GeV. About 1 GeV, i.e., 10% of $m_Z - m_W$ for a given m_Z, arises from the quantum corrections.

A useful criterion to signal the possible need to alter the theory is to consider the quantity,

$$\frac{1}{(1 - \Delta r)^2} = \left[\frac{m_W}{37.281\text{GeV}}\right]^4\left[1 - \frac{m_W^2}{m_Z^2}\right]^2, \qquad \textbf{(10)}$$

which represents the corrections to the muon decay rate when expressed in terms of the fundamental parameters, g and m_W. As the large logarithmic effects have been taken into account by the renormalization group arguments to all orders, it would be very difficult to explain large departures from the "standard" value of 1.1552. As an illustration, if $m_W = 81.0$ GeV and $m_Z = 95.0$ GeV with great accuracy, we would have $(1 - \Delta r)^{-2} = 1.66$, which would indicate strongly the need to modify the theory at the tree level.

EXOTIC CONTRIBUTIONS

If the deviations are small, they could signal the existence of exotic contributions to Δr. Indeed, massive particles may affect the value of this quantity. One possible way to change Δr, involving large m_t, would be ruled out if the result of 30 GeV $\lesssim m_t \lesssim$ 60 GeV is confirmed. Other interesting mechanisms include: (1) the possible effect of new fermion generations, (2) large m_H, and (3) contributions from supersymmetric partners.

It is convenient to write $\Delta r = (\Delta r)_{st} + \delta$, where $(\Delta r)_{st}$ stands for the standard value of equation (**5d**) and δ stands for the new exotic contributions. If the up-and-down masses, m_i and m_j, of an additional quark generation satisfy $m_i^2/m_j^2 = 1 + \epsilon$ ($\epsilon \ll 1$), then asymptotically for large $\eta_i \equiv m_i/m_W$ and $\eta_j \equiv m_j/m_W$, we have

$$\delta = -\frac{\alpha}{4\pi}\frac{c^2}{s^4}\,|\,U_{ij}\,|^2\,\frac{(m_i - m_j)^2}{m_W^2}, \tag{11}$$

which is an expression proportional to the Veltman correction[23] to the ρ parameter. For general values of m_i and m_j, δ has a complicated mathematical structure and has been studied in detail in reference 12 by a combination of analytic and numerical methods. The general conclusion is that an additional generation can give contributions to Δr ranging from large negative to small positive values. Positive contributions are attained only in part of the $\zeta_i, \eta_j \leq 1$ region and, for $\zeta_{i,\eta_j} \geq 1$, only in a narrow strip parallel to the diagonal $\zeta_i = \eta_j$. The upper bounds from the quark contribution are attained closely to the line, $1 = \zeta_i + \eta_j$, which corresponds to the threshold for the process, $W \rightarrow q_i + \bar{q}_j$. Numerical values for the upper bounds to the contribution of an additional generation of quarks and leptons are given in TABLE 2. For $m_Z = 93.80$ GeV, these potential positive effects can at most increase $m_Z - m_W$ by $\lesssim 130$ to 180 MeV. It is noteworthy that massive degenerate fermions do not fully decouple to $O(\alpha)$, but give small positive contributions.[12,24] As we move away from the diagonal, δ becomes negative. As suggested by equation (**11**), for $m_i - m_j$ equal to about a few times m_W, these negative contributions are large in magnitude and can cancel $(\Delta r)_{st}$, thus decreasing the theoretical prediction for $m_Z - m_W$ for a given m_Z. Thus, Δr is very sensitive to isodoublet splittings of such an order of magnitude.

For large m_H, one has asymptotically

$$\delta = \frac{11}{48\pi}\frac{\alpha}{s^2}\ln\left(\frac{m_H^2}{m_Z^2}\right) + \ldots \tag{12}$$

For $m_H^2/m_Z^2 < 100$, the increase in Δr is < 0.0088. Recently, van der Bij and Veltman have uncovered corrections of order, $(\alpha/\pi s^2)^2\,(m_H^2/m_W^2)$, to the ρ parameter and the vector boson mass shifts.[25,26] From an examination of the coefficients of $\ln\,(m_H^2/m_Z^2)$ and m_H^2/m_W^2, the authors argue that their results suggest a breakdown of perturbation theory for $m_H \gtrsim 3$ TeV. On the other hand, even for $m_H = 3$ TeV, the Higgs corrections remain quite small relative to the zeroth order contributions and rather small in comparison with $(\Delta r)_{st}$. It is worthwhile to point out that there are theoretical scenarios that lead to small upper bounds for m_H.[27–29] In particular, the assumptions and arguments of reference 29 lead to $m_H \lesssim 125$ GeV.

TABLE 2. Upper Bounds to the Contribution to Δr from an Additional Fermion Generation according to Whether or Not the "Up" Quark is Restricted to Be Heavier and the Neutrino Massless

	Heavier "Up" Quark	No Restriction on Quarks
Massless Neutrinos	0.754×10^{-2}	0.883×10^{-2}
Massive Neutrinos	0.900×10^{-2}	1.029×10^{-2}

TABLE 3. Comparison of UA1 and UA2 Results with Theoretical Expectations

	UA1 Electron Data	UA2 Electron Data	SM with $\sin^2\theta_W = 0.217 \pm 0.014$
m_W(GeV)	$80.9 \pm 1.5 \pm 2.4$	$83.1 \pm 1.9 \pm 1.3$	$83.0 \begin{smallmatrix} +2.9 \\ -2.7 \end{smallmatrix}$
m_Z(GeV)	$95.6 \pm 1.4 \pm 2.9$	$92.7 \pm 1.7 \pm 1.4$	$93.8 \begin{smallmatrix} +2.4 \\ -2.2 \end{smallmatrix}$
$m_Z - m_W$ (GeV)	$14.7 \pm 2.1 \pm 0.4$	$9.6 \pm 2.5 \pm 0.2$	10.8 ± 0.5
Δr	$0.252 \pm 0.070 \pm 0.044$	$-0.025 \pm 0.211 \pm 0.032$	0.0696 ± 0.0020
ρ	$0.928 \pm 0.036 \pm 0.016$	$1.025 \pm 0.051 \pm 0.009$	1
$\sin^2\theta_W = \left(\frac{38.65 \text{ GeV}}{m_W}\right)^2$	$0.228 \pm 0.008 \pm 0.014$	$0.216 \pm 0.010 \pm 0.007$	0.217 ± 0.014
$\sin^2\theta_W = 1 - m_W^2/m_Z^2$	0.284 ± 0.034	0.196 ± 0.047	0.217 ± 0.014

Additional contributions arising in supersymmetric models have been studied recently.[30–32] As in the case of higher fermion generations, dramatic modifications of the radiative corrections occur only by allowing large mass splittings in supersymmetric isomultiplets (which do not occur in models with breaking induced by supergravity) or for large m_t. For $0 < m_t \lesssim 100$ GeV, one finds that in a class of supergravity unified models, the supersymmetric corrections vary in the range of $-0.02 < (\Delta r)_{\text{sup}} < 0.01$, with the lower and upper bounds corresponding to the thresholds, $m_Z = 2\, m_{\tilde{W}}$ and $m_W = m_{\tilde{W}} + m_{\tilde{Z}}$, respectively. ($\tilde{W}$ and $\tilde{Z}$ denote winos and zinos.) Thus, these effects are small relative to $(\Delta r)_{st}$ and can lead to changes of at most a few hundreds MeV in the $m_W - m_Z$ predictions.

For the future, likely areas of interest include the detailed study of the forward-backward asymmetry and the longitudinal polarization asymmetry in $e^+e^- \rightarrow \mu^+\mu^-$ near the Z^0 resonance, in addition to the search for the electroweak contributions to $(g_\mu - 2)/2$.

TABLE 3 summarizes the comparison of the theoretical expectations of the SM with recent information obtained in the UA1 and UA2 experiments. Although there is good agreement within error bars, clearly more accurate data are needed to verify the SM at the level of its quantum corrections or to uncover new physics.

REFERENCES

1. KINOSHITA, T. & A. SIRLIN. 1959. Phys. Rev. **113:** 1652.
2. BERMAN, S. M. 1958. Phys. Rev. **112:** 267.
3. BERMAN, S. M. & A. SIRLIN. 1962. Ann. Phys. (N.Y.) **20:** 20.
4. SIRLIN, A. 1978. Rev. Mod. Phys. **50:** 573.
5. SIRLIN, A. 1982. Nucl. Phys. **B196:** 83.
6. LEUTWYLER, H. & M. ROOS. 1984. Z. Phys. **C25:** 91.
7. LYNN, B. W. & J. F. WHEATER, Eds. 1984. Radiative Corrections in $SU(2)_L \times U(1)$. World Scientific Publishing Company, Singapore.
8. BEG, M. A. B. & A. SIRLIN. 1982. Phys. Rep. **88:** 1.
9. SIRLIN, A. 1980. Phys. Rev. **D22:** 971.

10. Roos, M. & A. Sirlin. 1971. Nucl. Phys. **B29:** 296.
11. Marciano, W. J. & A. Sirlin. 1980. Phys. Rev. **D22:** 2695.
12. Bertolini, S. & A. Sirlin. 1984. Nucl. Phys. **B248:** 589.
13. Marciano, W. J. 1979. Phys. Rev. **D20:** 274.
14. Marciano, W. J. & A. Sirlin. 1984. Phys. Rev. **D29:** 945.
15. Verzegnassi, C. 1984. Trieste University preprint.
16. Consoli, M., S. Lo Pestri & L. Maiani. 1983. Nucl. Phys. **B229:** 474.
17. Sirlin, A. 1984. Phys. Rev. **D29:** 89.
18. Sarantakos, S., W. Marciano & A. Sirlin. 1983. Nucl. Phys. **B217:** 84.
19. Marciano, W. J. & A. Sirlin. 1981. Nucl. Phys. **B189:** 442.
20. Llewellyn Smith, C. & J. F. Wheater. 1982. Nucl. Phys. **B208:** 27.
21. Radermacher, E. 1984. CERN-EP/84-41.
22. Sirlin, A. Contribution to reference 7, page 155.
23. Veltman, M. 1977. Nucl. Phys. **B123:** 89.
24. Veltman, M. 1980. Phys. Lett. **91B:** 95.
25. van der Bij, J. J. & M. Veltman. 1984. Nucl. Phys. **B231:** 205.
26. van der Bij, J. J. 1984. Nucl. Phys. **B248:** 141.
27. Cabibbo, N., L. Maiani, G. Parisi & R. Petronzio. 1979. Nucl. Phys. **B158:** 295.
28. Callaway, D. J. E. 1984. Nucl. Phys. **B233:** 189.
29. Beg, M. A. B., C. Panagiotakopoulos & A. Sirlin. 1984. Phys. Rev. Lett. **52:** 883.
30. Eliasson, E. 1984. Phys. Lett. **147B:** 65.
31. Schwarzer, K. H. G. 1984. University of Oxford preprint: 40/84.
32. Lynn, B. W. 1984. SLAC-PUB-3358.

Charm and Heavy Flavor Lifetimes[a]

W. T. FORD

Department of Physics
University of Colorado
Boulder, Colorado 80309

INTRODUCTION

The flavor-changing transitions among quarks or among leptons are described, according to the standard electroweak theory, by their coupling to the charged weak bosons, $W^{\pm}$. The Feynman diagrams corresponding to the decay of a heavy quark or lepton to a lighter one are copies of the diagram that describes muon decay, $\mu^- \rightarrow \nu_\mu e^- \bar{\nu}_e$, in which $\mu^- \rightarrow \nu_\mu$ with the emission of a virtual W^-, which in turn produces the $e^-, \bar{\nu}_e$ pair. The assumption of a universal weak coupling for all fundamental fermions immediately leads to the prediction that, in the absence of flavor mixing or final-state mass corrections to phase-space, the formula for the electronic partial decay rate of a free lepton or quark is the same as that for the muon. Thus, for a heavy lepton, L (e.g., τ), we expect

$$\frac{1}{\tau L} = \frac{1}{B_e} \frac{G_F^2 m_L^5}{192\pi^3}, \tag{1.1}$$

where G_F is the Fermi constant and B_e is the branching ratio to $\nu_L e^- \bar{\nu}_e$. The value predicted by equation (**1.1**) for τ_τ is $(2.82 \pm 0.18) \times 10^{-13}$ sec, with the uncertainty coming from $B_e = 0.176 \pm 0.011$.[1]

The quarks inside hadrons are, of course, not free; however, when one of the quarks, Q, is heavy, one may hope that the effects of binding are small and so we proceed to write down the analog of equation (**1.1**) for this case:

$$\frac{1}{\tau_Q} = \frac{1}{B_e} \frac{G_F^2 m_Q^5}{192\pi^3}. \tag{1.2}$$

Here, the light quark is assumed to behave as a spectator to the heavy quark decay. The branching ratio, B_e, can be estimated by parton counting: the W^- radiated by a decaying charm quark can create pairs of (ν_e, e), (ν_μ, μ), or (u, d), with weight, 3, for the color sum, which leads to $B_e \simeq 1/5$. The predicted charm lifetime then is $\tau_c \simeq 8 \times 10^{-13}$ sec.

The charm decay is Cabbibo-allowed; in the general case, we need to add to the right-hand side of equation (**1.2**) a factor, $\Sigma_i |V_{q_iQ}|^2$, for the kinematically allowed q_i, where V_{q_iQ} is the corresponding Cabbibo-Kobayashi-Maskawa matrix element. In the absence of K–M suppression, the lifetime for bottom would be $\tau_b \simeq 2 \times 10^{-15}$ sec.

In the following sections, we will discuss the status of lifetime measurements of the

[a]This work was supported by the Department of Energy, contract no. DE-AC02-81ER40025.

charmed D^0, D^+, and F^+ mesons and Λ_c^+ and A^+ baryons, the status of the bottom hadron mixture produced at PEP and PETRA (presumably mostly B^0, B^+), and the status of the τ lepton.

CHARM PARTICLE LIFETIMES

New results are available as of this conference for the D^0 and D^+ mesons from the Mark II and Hybrid Facility Collaborations at SLAC.

Mark II

The Mark II detector, operated at PEP, which provided e^+e^- collisions with a center-of-mass energy of 29 GeV, features a large volume drift chamber, with a high resolution vertex chamber near the beam, surrounded by liquid-argon electromagnetic calorimeters. Tracks are reconstructed from measurements starting ten cm from the interaction point, whereas the decay path length is typically $\lesssim$ 1 mm, so a priori assignment of tracks to secondary vertices is not possible. This and the lack of particle identification mean that simple effective mass distributions of candidate daughter particles would be overwhelmed with combinatoric background.

The D^0's are isolated through the decay chain of

$$D^{*+} \longrightarrow D^0\pi^+ \quad \hookrightarrow K^-\pi^+. \tag{2.1}$$

Here, the difference in mass between D^{*+} and D^0 exceeds the pion mass by only a few MeV. Thus, when the K,π combinations that are consistent with D^0 are constrained to the D^0 mass and combined with other pions, the mass difference of $M(D^0, \pi^+) - M(D^0)$ is found to peak very sharply around 145 MeV. This is so because the fractional uncertainty of the mass difference is roughly that of the momentum measurement. With the further requirement of $z > 0.6$, where z is the energy of the D^* divided by the beam energy, the background is about 15%.

The Mark II Collaboration earlier published a result[2] based upon 27 events of reaction (**2.1**). Recently, they have analyzed[3] a sample for this decay mode augmented to 39 events and added 35 decays of $D^0 \rightarrow K^-\pi^+\pi^0$. For the latter decay mode, it was necessary to sharpen the resolution on the "bachelor" pion angle by constraining it to the D^0 decay vertex (rather than the less well-known production vertex). Monte Carlo studies were used to demonstrate that this approximation does not result in any significant bias against events with long decay times. The proper lifetime distribution for the 74 decays of both modes is shown in FIGURE 1a. Although the measurement of decay distances is resolution-limited in this experiment, a clear displacement of the distribution from zero can be seen. A maximum-likelihood fit to the distribution gives the mean life value of

$$\tau(D^0) = (4.5^{+0.9}_{-0.8} \pm 0.5) \times 10^{-13}\ \text{sec}\ .$$

A more difficult measurement with the Mark II is the D^+ lifetime via the decay chain of

$$D^{*+} \rightarrow D^+\pi^0 \quad \llcorner\!\!\rightarrow K^-\pi^+\pi^+. \tag{2.2}$$

Since the π^0 is almost at rest in the D^{*+} frame, its laboratory energy has a nearly fixed ratio to that of the D^{*+}. This fact is used to improve the mass resolution and pull out a

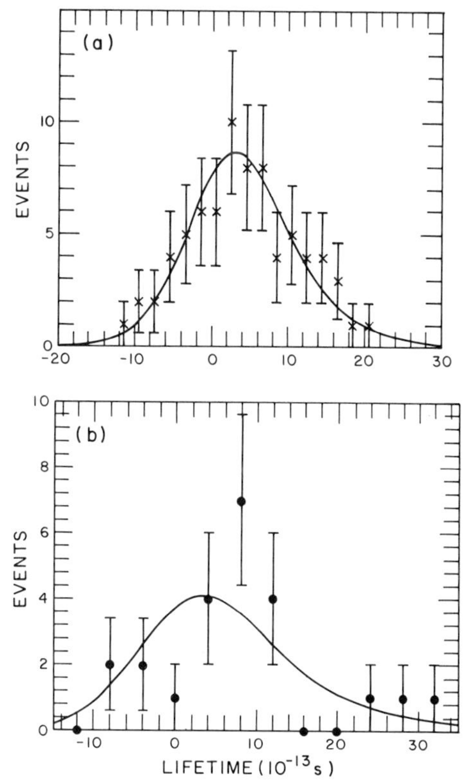

FIGURE 1. (a) Lifetime distribution from Mark II for the decays, $D^0 \rightarrow K\pi$ and $D^0 \rightarrow K\pi\pi^0$; (b) Lifetime distribution for D^+ decays. The curves are the results of maximum-likelihood fits.

sample of 23 events, including 7 ± 2.6 background events, with which to measure the D^+ lifetime. The proper decay time distribution is shown in FIGURE 1b. The fit gives

$$\tau(D^+) = (8.5^{+3.4}_{-2.5} \pm 0.5) \times 10^{-13} \text{ sec.}$$

The ratio of the two lifetimes measured in this experiment is

$$\tau(D^+)/\tau(D^0) = 1.9^{+1.0}_{-0.8} \pm 0.5.$$

SLAC Hybrid Facility

The SLAC Hybrid Facility group have completed processing of their series of exposures in the 20-GeV backscattered laser beam, referred to as BC-72, BC-73, and BC-75. The 40-inch bubble chamber was equipped with extra high resolution cameras and the chamber operating conditions adjusted to optimize the point measurement resolution. Particle identification was facilitated by tracking chambers, Čerenkov counters, and a wall of lead glass downstream of the bubble chamber. This system permits measurement of an individual decay flight path with error much less than the path length, except at the smallest vertex separation distances. Cuts are made on the decay path and on the largest track impact parameter to remove the bias against short path lengths that occurs because the vertices tend to merge in these cases and be missed in the scanning.

Because of missing π^0's, only about one-third of the events are fully constrained. For those that are not, the momentum of the D is found to be well approximated by the scaling,

$$p_D = p_{vis}(m_D/m_{vis}).$$

There are 47 D^+, 46 D^0, and 5 ambiguous decays. The lifetime distributions are shown in FIGURE 2, which lead to the results,[4]

$$\tau(D^+) = (9.2 \pm 1.5 \pm 0.5) \times 10^{-13}\ \text{sec},$$

$$\tau(D^0) = (6.1 \pm 1.1 \pm 0.4) \times 10^{-13}\ \text{sec}, \ {}^{+0.6}_{-0.3} \pm 0.1.$$

$$\tau(D^+)/\tau(D^0) = 1.5$$

We shall see below that the value for $\tau(D^0)$ from this experiment is the largest of those presently reported, although it is by no means inconsistent with the world average. The authors have called attention[5] to one event in their sample that is fully constrained and well measured, and with a very low probability of being a background event, for which the proper lifetime is 55×10^{-13} sec. The statistical probability for this observation is a few percent if the true mean life is around 6×10^{-13} sec, but only a few hundredths of a percent if the current world average value of $\sim 4 \times 10^{-13}$ sec is assumed.

Other Recent Measurements

Some further experiments that have reported recently on charm lifetimes or have work in progress are discussed in the following paragraphs and are included in the summary of results below.

CERN experiment NA11 of the ACCMOR Collaboration is among the first to apply silicon microstrip detectors for precision vertex reconstruction. Resolution of 10 μm transverse to the beam is achieved, leading to the typical longitudinal resolution for the vertex position of $\sim$150 μm. A lifetime value for the F^+ has been published,[6] based on three fully constrained events, and results for D^0 and D^+ have appeared in conference reports.[7]

The European Hybrid Spectrometer collaboration have used the small hydrogen bubble chamber, LEBC, to study π^- production of charm. Lifetime values have been presented for D^0 mesons,[8] for D^+ mesons, and for the Λ_c^+ baryon,[7,9] based on exposures designated NA16 and NA27.

The Omega spectrometer has been used in experiment WA58 at CERN to study charm photoproduction. Events were produced in a stack of 0.6-mm-thick emulsion pellicles set at a grazing angle (5°) to the beam. Results have been published[10] for D^0, D^+, and Λ_c^+, and more data are being analyzed.

Besides those of Mark II and BC-75, the other high-statistics measurement of $\tau(D^0)$ comes from Fermilab E531. This is a neutrino production experiment employing

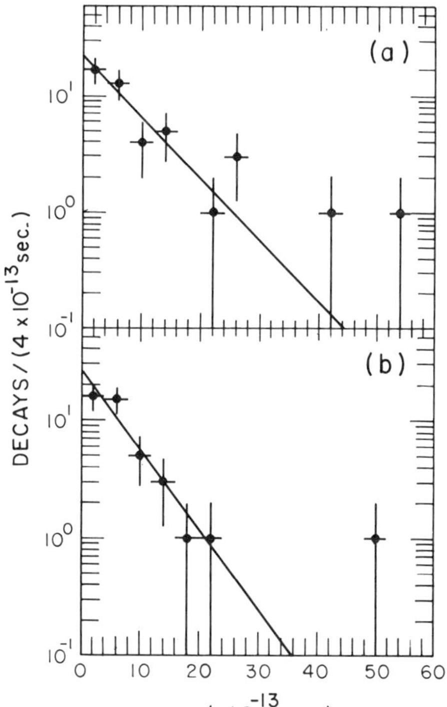

FIGURE 2. Lifetime distributions from BC-72/73/75 for (a) D^+ and (b) D^0 decays. The curves are the results of maximum-likelihood fits.

nuclear emulsion. The data analysis is still in progress; preliminary results were presented at the 1983 Cornell Conference.[11]

Finally, the CERN hyperon beam experiment, WA62, recently[12] found the A^+ (csu) baryon state in $\Lambda K^-\pi^+\pi^+$ at 2460 MeV. The effective mass plot is shown in FIGURE 3a for events satisfying the primary and secondary vertex quality cuts. There are ~53 events in the peak over a background of ~59 events. The lifetime is extracted from a fit to the vertex separation distribution obtained by subtracting the mass sideband region from the peak region, which is shown in FIGURE 3b. The result is

$$\tau(A^+) = 4.8^{+2.9}_{-1.8} \times 10^{-13} \text{ sec.}$$

Charm Lifetime Summary

FIGURES 4, 5, and 6 summarize the results of the several experiments for $\tau(D^0)$, $\tau(D^+)$, $\tau(F^+)$, and $\tau(\Lambda_c^+)$. The error bars represent the quadrature sum (σ) of statistical and systematic error for each experiment. The "combined" result in each figure refers to the weighted average, in which the weight for experiment i is given by $1/\sigma_i^2$. Where errors are asymmetric, the sign of the residual with respect to the average is used to choose σ_i. This weighting scheme assumes Gaussian errors; on the other hand, a lifetime experiment with negligible resolution error and bias measures an exponential distribution, for which the proper weight is simply the number of events, N_i (at least for large N_i). Equivalently, in this case one may use the reciprocal square of the fractional error, σ_i/τ_i. The actual experiments fall somewhere between the Gaussian and pure exponential limits (e.g., Mark II is resolution limited, WA58 has corrections for scanning biases against long decay lengths, etc.). We find that in the case of the current measurements of $\tau(D^0)$, it makes some difference which weighting scheme is used. Therefore, both results are shown for the world average to indicate the range of uncertainty from these considerations.

In general, the agreement among measurements for each particle is good. The $\tau(D^+)$ measurement of WA58 was omitted from the average because it differs from the average by more than two standard deviations and because the experiment appears to have limited sensitivity to the larger values found by the other experiments.

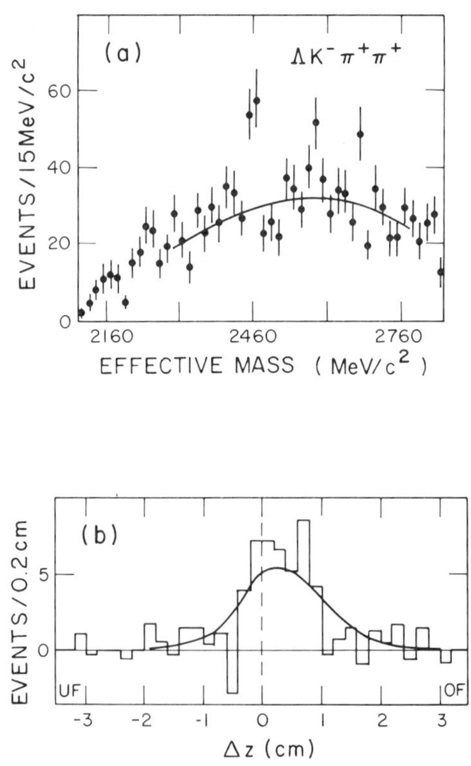

FIGURE 3. (a) Effective $\Lambda K^-\pi^+\pi^+$ mass plot for quality vertex events; (b) Background-subtracted decay length distribution for A^+ region.

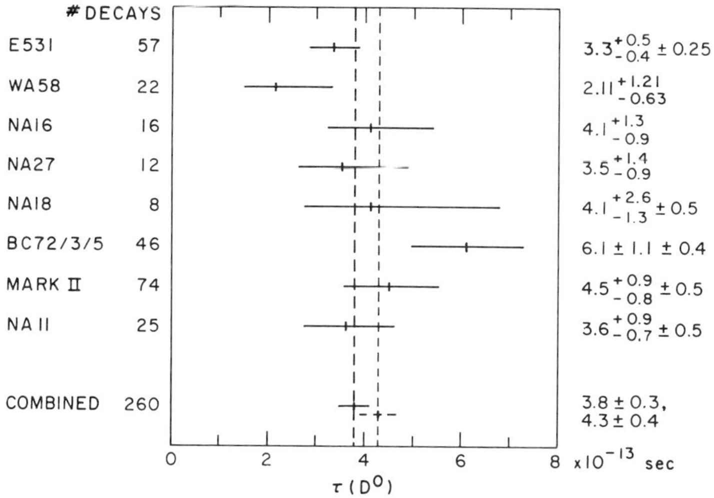

FIGURE 4. $\tau(D^0)$ for all experiments.

Conclusions for Charm Lifetimes

In FIGURE 7 are shown the lifetimes for the several charm states, along with the prediction of (**1.2**), the naive spectator picture. There is agreement within a factor of two or three with the predicted absolute value of the lifetime for charm (which is really quite uncertain because of the strong dependence in (**1.2**) on the mass of the charm quark), but the expectation that all the lifetimes should be the same clearly is not met. In particular, the D^+ has a substantially longer lifetime than the others (except perhaps Λ^+). Corrections to the phase-space for final-state particle masses turn out to be similar for the different states because the mass difference between charm and strange quarks accounts for most of the difference in particle masses.

For the D^+, D^0 ratio, we find

$$\tau(D^+)/\tau(D^0) = 2.4 \pm 0.3. \tag{2.3}$$

The lifetime ratio can also be inferred from the ratio of semielectronic branching ratios since $\Gamma_e(D^+) = \Gamma_e(D^0)$ at least for the Cabibbo-allowed component. The Mark III collaboration have reported[13]

$$B_e(D^+) = 17 \pm 3 \pm 3\%,$$

$$B_e(D^0) = 6 \pm 3 \pm 2\%,$$

$$\frac{B_e(D^+)}{B_e(D^0)} = \frac{\tau(D^+)}{\tau(D^0)} = 2.78^{+0.86+0.27}_{-0.60-0.42}, \tag{2.4}$$

which is in agreement with the ratio (**2.3**) from direct lifetime measurements.

Modifications to the naive spectator model of FIGURE 8a include[14] (in FIGURE 8b) hard gluon corrections (which, however, have little effect on the relative lifetimes among charm states) and consideration of Pauli suppression. FIGURES 8c and 8d show

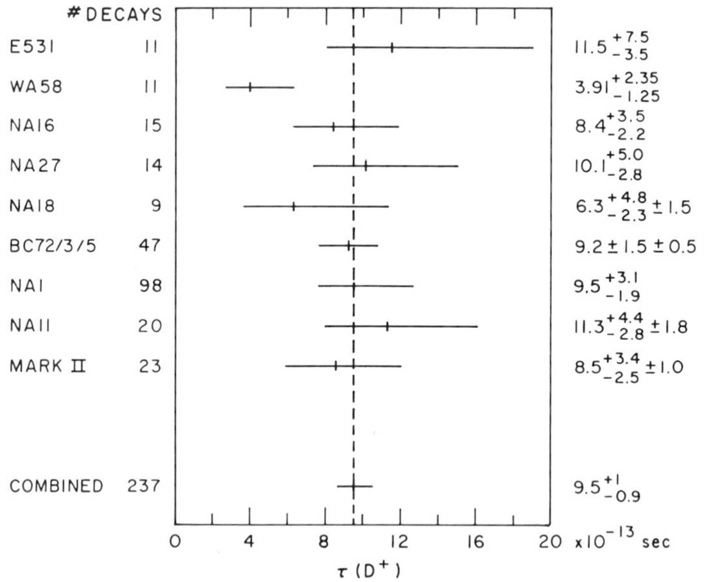

FIGURE 5. $\tau(D^+)$ for all experiments.

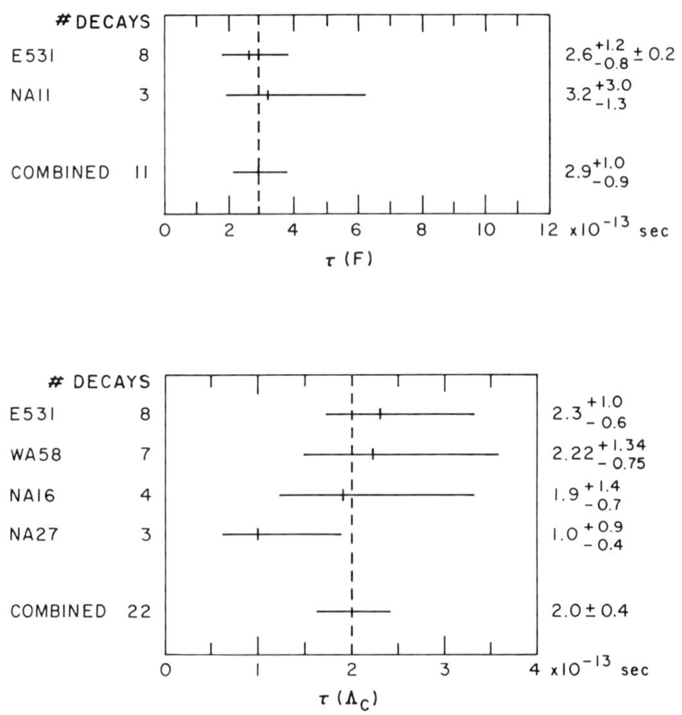

FIGURE 6. (a) $\tau(F^+)$ for all experiments; (b) $\tau(\Lambda_c^+)$ for all experiments.

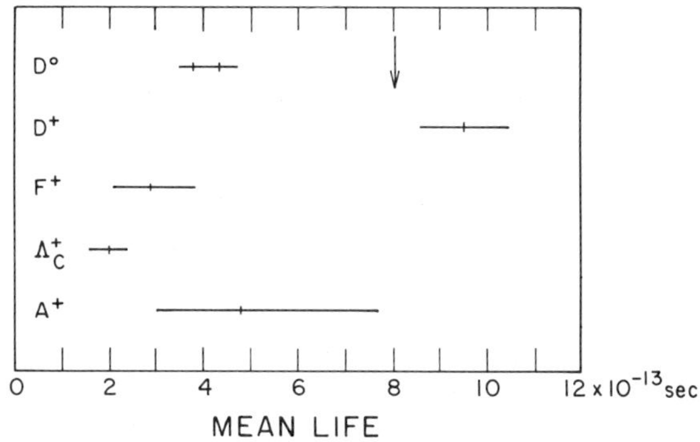

FIGURE 7. World averages for the charmed particles. The arrow indicates the prediction of equation (**1.2**).

spectator diagrams for D^+ and D^0, in which two identical light quarks appear for D^+, but not for D^0. This effect accounts for only ~20% increase in $\tau(D^+)$.[14]

FIGURE 8e shows an exchange diagram for D^0 that has no counterpart in D^+ since the $\bar{d}$ cannot absorb the W^+. Similarly, the annihilation diagram of FIGURE 8f for the F^+ is Cabbibo-allowed, while its counterpart for D^+ (FIGURE 8g) is Cabbibo-suppressed. These may account for the higher D^0 and F decay rates. However, if the decay is really controlled by the parton interactions, one expects the exchange and annihilation graphs to be suppressed by helicity conservation (as in $\pi \rightarrow e\nu$), by small wave function overlap, and in the case of the annihilation diagram, by color factors, compared with the spectator graph in which the light quark from decay gets together with its color-matched spectator. Resolution of these problems requires information beyond the lifetime data, including study of specific exclusive decay modes.[13]

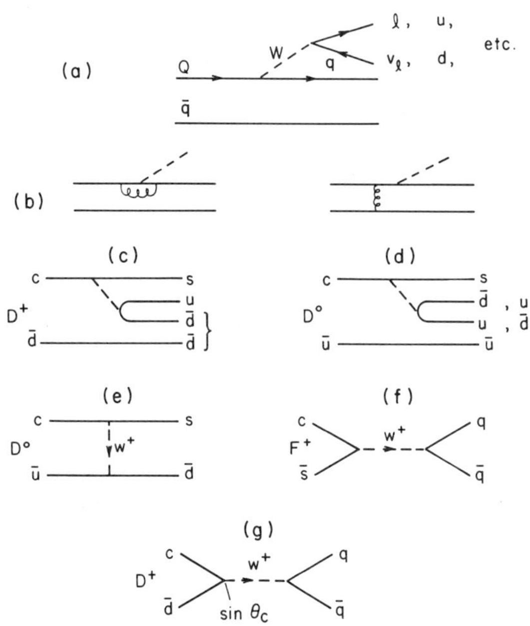

FIGURE 8. Parton Feynman graphs describing charm decays: (a) spectator model; (b) typical hard gluon corrections; (c), (d) spectator diagrams for D^+ and D^0; (e) exchange diagram for D^0; (f), (g) annihilation diagrams for F^+ and D^+.

BOTTOM PARTICLE LIFETIMES

Fixed-target experiments have not yet isolated signals corresponding to production of bottom particles. Specific states have been reconstructed in e^+e^- at CESR running on the $\Upsilon(4S)$, where they are produced nearly at rest. Contending with higher multiplicities, the experimenters at PEP and PETRA have obtained samples enriched in bottom production by tagging the relatively massive bottom particles either with leptons from the semileptonic decays or with event shape. The tracks associated with decay products from bottom particles reveal the separation between the production point and the decay vertex (via the track impact parameter in most analyses) for lifetimes greater than a few tenths of a picosecond. The surprise is that bottom particles indeed live long enough for these experiments to see an effect. The first measurements by MAC and Mark II in the summer of 1983 were followed in 1984 by contributions from three other groups, as well as updates to the original measurements. New at this conference is a further update of the MAC measurement.

The production of bottom hadrons in e^+e^- annihilation is via creation of a b quark-antiquark pair, followed by fragmentation of the quarks into jets of hadrons. It has been found[15] that the bottom hadron in the jet carries, on average, about 80% of the quark momentum. The lepton from the B decay usually goes forward relative to the B direction; by assuming it always goes forward, we can define the signed impact parameter, δ, to be positive (negative) if the lepton track extrapolates to a crossing point downstream (upstream) of the production point. The production point is taken in these experiments as the center of the beam crossing region and is therefore known within about $\pm 500\mu$m in the horizontal direction and 50–100μm in the vertical. To obtain the mean life, τ_b, from the data, the observed distribution of the events in δ is compared with the calculated probability distribution function,

$$P(\delta)d\delta = [f_b P_b(\tau_b, \delta) + f_c P_c(\delta) + f_{b\to c} P_{b\to c}(\tau_b, \delta) + f_{bg} P_{bg}(\tau_b, \delta)]d\delta, \qquad (3.1)$$

by maximum-likelihood or other means. In equation (**3.1**), the subscripts b, c, $b \to c$, and bg refer to direct b leptonic decays, c leptonic decays, cascade decays, and misidentification background, respectively. The last term depends upon τ_b since the hadron that faked a lepton may have come from b decay; in fact, this component of the hadronic events is used by TASSO to measure τ_b (see below).

MAC

The MAC detector is a nearly 4π electromagnetic-hadronic calorimeter enclosing a solenoid coil and drift chamber for charged-particle tracking. The hadron calorimeter contains magnetized steel plates so that, in conjunction with outer drift chambers, it serves also as a muon spectrometer. With this arrangement, it is possible to obtain samples of multihadron events containing muons or electrons with high efficiency and low bias.

The calorimetric energy flow is used to determine the jet (thrust) axis. The signature for bottom production and semileptonic decay is the appearance of a muon or electron with a large value of the component of momentum perpendicular to the thrust axis ($p_\perp$). For lepton (p, $p_\perp$) greater than (2, 1.5) GeV, the sample contains about 53%

direct b decays, 5% cascade decays, 13% charm background events, and 29% background events from lepton misidentification.

The present event sample is from an integrated luminosity of ~210 pb^{-1}, which is an increase of about 40% compared with the measurement reported previously[16] by this group. FIGURE 9a gives the lepton impact parameter distribution for the 505 events. Each event entered into this histogram has been weighted by the reciprocal square of its assigned error (see FIGURE 9b), which averages about 530 μm. The mean value of δ

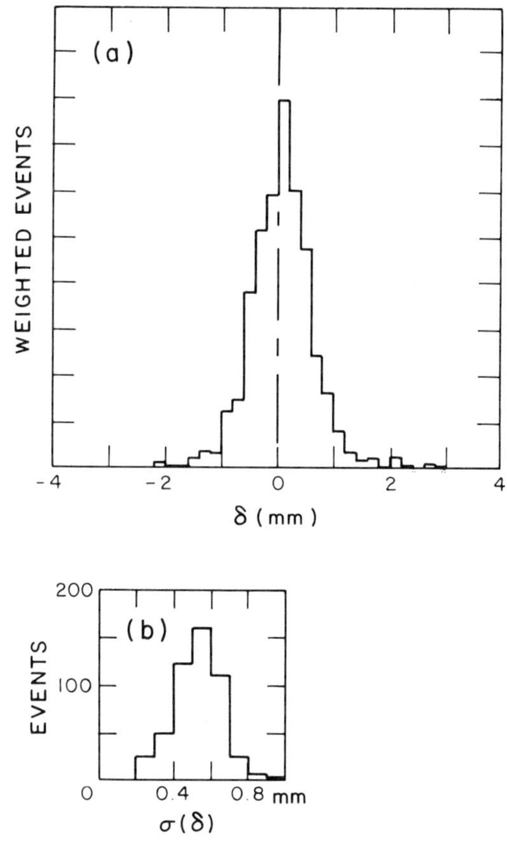

FIGURE 9. (a) Impact parameter distribution of b decay events from MAC; (b) Distribution of the errors in δ.

(actually the median, to minimize the effect of tails) is 70 ± 22 ± 10 μm. Taking the average of both sides of equation (**3.1**), we find

$$\langle \delta \rangle = f_b \alpha c \tau_b + \text{(background correction)}, \quad \textbf{(3.2)}$$

where the Monte Carlo gives $f_b = 0.53$, $\alpha = 0.45$, and the background corrections add up to $(9 + 4.3\tau_b)\mu$m. The lifetime is

$$\tau_b = 0.81 \pm 0.28 \pm 0.17 \text{ psec.}$$

The MAC group expects to improve the measurement of τ_b with data from a recently installed precision vertex chamber. This device is built around a vacuum pipe of only 3.5 cm radius and consists of six layers of thin-walled tube chambers operated at a pressure of four atmospheres.

Mark II

An analysis similar to that just described has been performed by the Mark II Collaboration. Electrons and muons are identified, respectively, by the liquid argon calorimeters and muon chambers placed outside the magnet return yoke. The momentum and $p_\perp$ cuts on the lepton are >2 and >1 GeV, respectively. The precision vertex detector gives a determination of the extrapolated track displacement near the origin of $\simeq 100$ μm, which is much smaller than the beam size except for events that are oriented so the beam is seen edge on. The 282 events used in the analysis are those having a δ error less than 310 μm, roughly half of the total. A likelihood fit to the δ distribution of these events gives the result,[17]

$$\tau_b = 0.85 \pm 0.15 \pm 0.20 \text{ psec.}$$

The Mark II group has also analyzed the data another way. For the events selected by the criteria on the lepton described above, the tracks are sorted into two hemispheres divided by the plane perpendicular to the thrust axis. A vertex is formed for each hemisphere since it is not possible in general to resolve the primary, *b*-decay, and cascade *c*-decay vertices. The separation of this hemispheric vertex from the beam center is larger (by an angle factor) than the impact parameter, so the size of the beam is less important. A Monte Carlo calculation is relied on to relate the resulting distribution to τ_b. The result from this analysis is

$$\tau_b = 1.25^{+0.26+0.35}_{-0.19-0.39} \text{ psec.}$$

DELCO

The DELCO detector at PEP has an open arrangement of Čerekov counters around cylindrical drift chambers, surrounded by drift chambers and shower counters. With superior electron identification, it is possible to relax the electron momentum cut to >1 GeV. The *b*-enriched sample is made by requiring $p_\perp > 1$ GeV, and it turns out that the average impact parameter is larger for a given τ_b than for the MAC or Mark II cuts. The average impact parameter error is about 400 μm for 60 events. The average δ of 215 ± 81 μm leads to the result,[18]

$$\tau_b = 1.16^{+0.37}_{-0.34} \pm 0.23 \text{ psec.}$$

JADE

The JADE experiment operates at PETRA, which has provided e^+e^- collisions over a range of center-of-mass energies with the largest accumulation of data around 35 GeV. The detector consists of a central tracking chamber of the "jet" type (i.e., having cells with multiple sense wires, long drift distances, and multihit readout), surrounded by lead glass electromagnetic calorimeters and a three-layer filter with drift chambers for muon identification. Muons and electrons having $p_\perp > 0.9$ GeV are required for the b-enriched sample. The impact parameters are measured with an average uncertainty of 480 μm. Fits to the δ distributions give $\tau_b = 1.8$ psec with $\langle \delta \rangle$ of 282 ± 66 μm for muons and 457 ± 107 μm for electrons. A second method entails selecting events having an identified muon, assigning a weight representing the probability that the event comes from $b\bar{b}$ (according to the aplanarity and the muon $p_\perp$), and making the distribution of δ for all tracks in the event. From both analyses combined, the result is[19]

$$\tau_b = 1.8^{+0.5}_{-0.3} \pm 0.4 \text{ psec.}$$

TASSO

The TASSO detector at PETRA has a large drift chamber and solenoid for charged-particle tracking, electromagnetic calorimeters covering about two-thirds of the azimuth, and Čerenkov counters plus shower counters for the remainder. Absorbers and tracking chambers for muon identification are placed beyond the calorimeters in each region. A precision inner drift chamber was added in 1983. With it, the track extrapolation component to σ_δ is about 100μm.

The selection of heavy flavor is accomplished by applying several event shape cuts to the multihadron event sample. Events are chosen that look like two fat quark jets with minimal gluon radiation. The procedure raises the b fraction in the events to 32%. Since all tracks having $p > 1$ GeV contribute, the measurement has relatively high statistics (at some cost in model-dependence). A fit to the distribution of impact parameters gives[20]

$$\tau_b = 1.83^{+0.38+0.37}_{-0.35-0.34} \text{ psec.}$$

Summary of Bottom Lifetime Measurements

The results for τ_b from all of the experiments are summarized in FIGURE 10. Included as dashed or dotted bars in the figure are some earlier results from experiments that have since reported on augmented data samples. The six recent results are quite consistent with each other ($\chi^2 = 5.8$ for five degrees of freedom) and confirm the large value of $\tau_b \simeq 1$ psec. It has been observed[17] that different values of τ_b would be found by experiments that include nonleptonic decays (such as TASSO) from that measured with semileptonic decays alone, if the B^+ and B^0 have different lifetimes (as do D^+ and D^0). This is because the charge state with longer lifetime has a larger semileptonic branching ratio; thus, that state is overrepresented in the leptonic decay sample. This may be an interesting issue for future, more precise experiments.

FIGURE 10 shows the result of averaging the current measurements (counting the two Mark II results as independent). The error on this world average is probably an underestimate since each of the experiments has a large error with a substantial systematic component, Monte Carlo calculations are relied on to measure the sample purities, sensitivities, biases, etc., and some of the assumptions in the particle

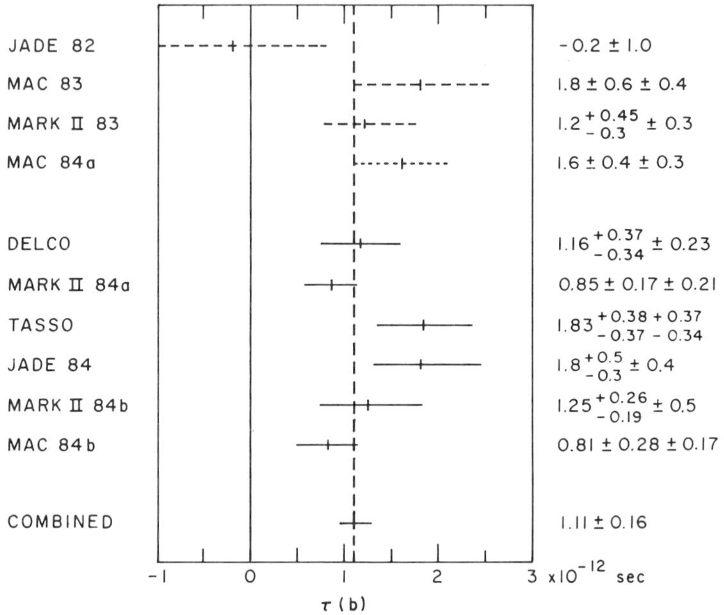

FIGURE 10. Summary of τ_b measurements. Dotted and dashed bars denote earlier results now superceded.

production modeling are not entirely independent among the experiments. There is a clear need for more precision work on this problem.

Mixing in b *Decays*

Putting the details into equation (**1.2**), we can relate the b lifetime to the Kobayashi-Maskawa mixing matrix elements, V_{ub} and V_{cb}:[21]

$$\frac{1}{\tau_b} = \frac{1}{B_e} \frac{G_F^2}{192\pi^3} m_b^5 \left[I_c f_c |V_{cb}|^2 + I_u f_u |V_{ub}|^2 \right] \tag{3.3}$$

Here, V_{cb} or V_{ub} is the mixing amplitude for the weak charged current transition between b-quarks and c- or u-quarks. I is the final-state mass correction to phase-space and f is a QCD radiative correction factor. I and f are functions of the mass ratios, $m_{c,u}/m_b$.

The small value, measured at CESR,[21] of the ratio,

$$R_{uc} \equiv \frac{\Gamma(b \to u)}{\Gamma(b \to c)} = \frac{I_u f_u |V_{ub}|^2}{I_c f_c |V_{cb}|^2} = 2.13 \frac{|V_{ub}|^2}{|V_{cb}|^2}, \tag{3.4}$$

suggests use of the form,

$$\frac{1}{\tau_b} = \frac{1}{B_e}\frac{G_F^2}{192\pi^3} m_b^5 I_c f_c |V_{cb}|^2(1 + R_{uc}). \tag{3.5}$$

The measurements give $R_{uc} < 0.04$. A fit to the lepton endpoint spectrum from semileptonic b decay gives the product, $m_b^5 I_c = 1430 \pm 60$ GeV^5. The QCD radiative correction has been calculated[22] as a function of m_c/m_b, leading to the value,[21] $f_c = 0.857$. The semileptonic branching ratio[23] is $11.6 \pm 0.5\%$. Putting these numbers into equation (**3.5**), together with the world average of $\tau_b = 1.1 \pm 0.2$, we find finally

$$|V_{cb}| = 0.050 \pm 0.005,$$
$$|V_{ub}| < 0.010.$$

These results represent a much smaller degree of mixing of the third generation of quarks with the lighter ones compared with the mixing between the first two generations. Implications for the Kobayashi-Maskawa picture of weak mixing with CP violation are discussed in the talk at this conference by Langacker.

τ LEPTON LIFETIME

Measurements of the τ lifetime are summarized in FIGURE 11. The recent precise results from TASSO,[24] Mark II,[17] and MAC[25] are combined to give the average value,

$$\tau_\tau = 2.95 \pm 0.25 \times 10^{-13}\text{ sec},$$

which is in good agreement with the universality prediction from equation (**1.1**).

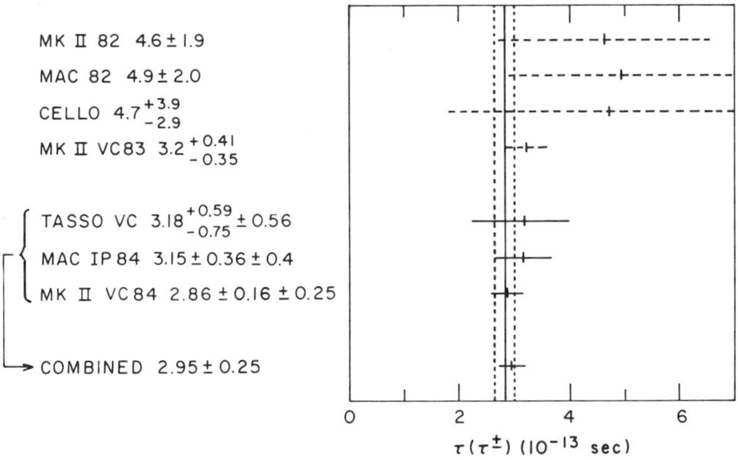

FIGURE 11. Measurements of the τ lifetime. The universality prediction from equation (**1.1**) with its uncertainty is indicated by the vertical lines. The notations "VC" and "IP" mean "with vertex chamber" and "impact parameter method," respectively.

CONCLUSIONS

We have seen that τ-μ universality is obeyed within ~8%. In the charm system, lifetimes are in rough agreement with universality at the parton level (within uncertainties in quark masses, etc.). On the other hand, the observed inequalities of

$$\tau(D^+) > \tau(D^0, F^+),$$

$$\tau(A^+) \gtrsim \tau(\Lambda_c^+),$$

indicate either the necessity for exchange and/or annihilation diagrams in the explanation of these decay rates or the inadequacy of the parton plus short-distance QCD description, or both. Information from exclusive decay modes is required to pursue these questions further.

The bottom lifetime has now been measured by five experiments and is confirmed to be approximately 10^{-12} sec. This implies only about 5% mixing of the third with the second generation (even less of the third with the first), to be compared with $\sin\theta_c = 0.23$. Outstanding problems include establishment of the value, presently bounded only from above, of the u-b mixing, and separate B^0 and B^+ lifetime measurements as a probe of the composite structure of the bottom system, which one hopes because of the larger quark mass will be more easily understood than the charm system in terms of the parton-level physics in short-distance approximation.

ACKNOWLEDGMENTS

It is a pleasure to acknowledge helpful discussions with M. Bosman, T. DeGrand, J. Jaros, K. Moffeit, N. W. Reay, and J. G. Smith.

NOTES AND REFERENCES

1. BLOCKER, C. A. *et al.* 1982. Phys. Lett. **109B:** 119.
2. YELTON, J. M. *et al.* 1984. Phys. Rev. Lett. **52:** 2019.
3. GLADNEY, L. 1985. Private communication.
4. MOFFEIT, K. 1985. Private communication.
5. ABE, K. *et al.* 1984. SLAC-PUB-3493.
6. BAILEY, R. *et al.* 1984. Phys. Lett. **139B:** 320.
7. KLANNER, R. 1984. Talk given at the International Conference on High Energy Physics, Leipzig.
8. AGUILAR-BENITEZ, M. *et al.* 1984. Phys. Lett. **146B:** 266.
9. REUCROFT, S. 1985. Private communication.
10. ADAMOVICH, M. I. *et al.* 1984. Phys. Lett. **140B:** 119, 123.
11. REAY, N. W. 1983. *In* Proceedings of the 1983 International Symposium on Lepton and Photon Interactions at High Energies. D. G. Cassel & D. L. Kreinick, Eds.: 244–274. Cornell University. Ithaca, New York; see also USHIDA, N. *et al.* 1983. Phys. Rev. Lett. **51:** 2362.
12. BIAGI, S. F. *et al.* 1985. Phys. Lett. **150B:** 230.
13. HITLIN, D. 1984. Talk given at the SLAC Summer Institute and Topical Conference.
14. See, for example, the review by RÜCKL, R. 1984. *In* Flavor Mixing in Weak Interactions. Ling-Lie Chau, Ed.: 681–716. Plenum. New York.

15. Scc, for example, FERNANDEZ, E. *et al.* 1983. Phys. Rev. Lett. **50:** 2054; NELSON, M. E. *et al.* 1983. Phys. Rev. Lett. **50:** 1542.
16. FORD, W. T. 1984. *In* Flavor Mixing in Weak Interactions. Ling-Lie Chau, Ed.: 77–89. Plenum. New York.
17. JAROS, J. 1984. Talk given at the "Physics in Collision" conference, Santa Cruz, California.
18. KLEM, D. E. *et al.* 1984. Phys. Rev. Lett. **53:** 1873.
19. STEFFEN, P. 1984. Talk at the International Conference on High Energy Physics, Leipzig.
20. ALTHOFF, M. *et al.* 1984. Phys. Lett. **149B:** 524.
21. LEE-FRANZINI, J. 1984. *In* Flavor Mixing in Weak Interactions. Ling-Lie Chau, Ed.: 217–238. Plenum. New York.
22. CABBIBO, N. & L. MAIANI. 1978. Phys. Lett. **79B:** 109.
23. STONE, S. 1983. *In* Proceedings of the 1983 International Symposium on Lepton and Photon Interactions at High Energies. D. G. Cassel & D. L. Kreinick, Eds.: 215. Cornell University. Ithaca, New York.
24. ALTHOFF, M. *et al.* 1984. Phys. Lett. **141B:** 264.
25. FERNANDEZ, E. *et al.* 1984. Phys. Rev. Lett. **54:** 1624.

Rare ψ Decays[a]

RICHARD PARTRIDGE

Institute for Particle Physics
University of California
Santa Cruz, California 95064

INTRODUCTION

Slightly more than ten years have passed since the ψ was discovered, yet the study of ψ decays continues to be an active and fruitful area of research. One reason for such longevity is that each successive experiment has increased their sensitivity over previous experiments either by improving detection efficiency or by increasing statistics. This has allowed the observation and, in some cases, detailed studies of rare ψ decays. Branching ratios of $\approx 10^{-4}$ are now routinely studied, while certain decay channels are beginning to show interesting effects at the 10^{-5} level. Future experiments at the Beijing Electron Positron Collider (BEPC) have the potential for increasing sensitivities by one or two orders of magnitude, thus enabling many interesting studies impossible with current data samples.

Since the theme of this conference is electroweak physics, I will first examine the extent to which ψ decays can be used to study electroweak phenomena. These phenomena tend to be extremely rare at the ψ, where strong and electromagnetic decays dominate; however, the prospect of greatly increased sensitivity using BEPC makes this an interesting topic to consider. The remainder of this work is devoted to the more traditional task of using the ψ to study quarks, gluons, and the properties of the strong interaction. Of particular interest is the study of radiative ψ decays, where a number of new particles have been discovered. Recent results regarding two of these particles, the $\theta(1700)$ and $\iota(1450)$, will be discussed, as well as a study of the quark content of the η and η' using decays of the ψ to vector-pseudoscalar final states.

ELECTROWEAK PHYSICS AT THE ψ

There are three types of electroweak effects that one might try to look for at the ψ: weak decays, such as $\psi \rightarrow \nu\bar{\nu}$; electroweak interference effects; and the existence of new fundamental couplings, such as the Higgs coupling. This chapter examines the feasibility of searching for these rare effects in ψ decays.

Weak Decays

The ψ can couple to the weak neutral current through the annihilation diagram shown in FIGURE 1d. A simpleminded estimate of the branching ratio for decays of this

[a]This work was supported in part by the Department of Energy, contract no. DE-AM03-76SF00034.

type is given by

$$BR(\psi \rightarrow Z^0 \rightarrow X) \sim \left(\frac{M_\psi}{M_Z}\right)^4 BR(\psi \rightarrow e^+e^-) \sim 10^{-7}. \tag{1}$$

In a year of running, the Beijing Electron Positron Collider should be able to produce $\approx 10^8\ \psi$ (100 Mψ), a small handful of which will have weak decays.

The most interesting of the weak decay modes is $\psi \rightarrow \nu\bar{\nu}$, which can in principle be used to count the number of light neutrino generations. The branching ratio for this

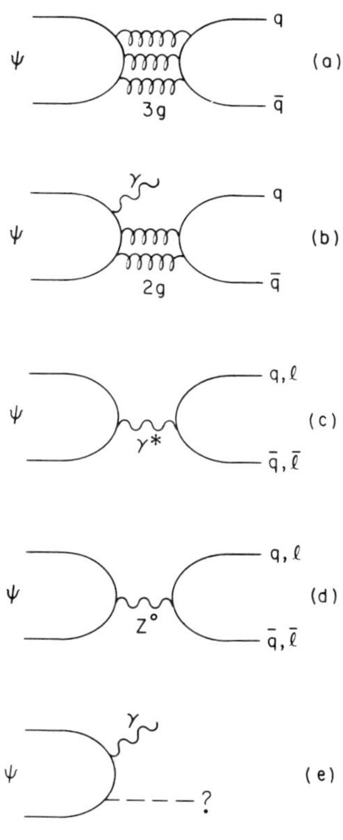

FIGURE 1. Feynman diagrams for ψ decay: (a) three gluon decay; (b) radiative decay; (c) electromagnetic decay; (d) weak decay; (e) a new fundamental coupling.

decay can be calculated in the standard model[1] and is given by

$$BR(\psi \rightarrow \nu\bar{\nu}) = N_\nu \frac{(3/8 - \sin^2\theta_W)^2}{8\sin^4\theta_W \cos^4\theta_W}\left(\frac{M_\psi}{M_Z}\right)^4 BR(\psi \rightarrow e^+e^-)\,, \tag{2}$$

where N_ν is the number of light neutrino generations. Taking $\sin^2\theta_W = 0.22$, equation (**2**) gives $BR(\psi \rightarrow \nu\bar{\nu}) = 10^{-8}\,N_\nu$. One way to "see" the $\nu\bar{\nu}$ decay mode is to look for the decay sequence of $\psi' \rightarrow \pi^+\pi^-\psi$, $\psi \rightarrow$ *nothing*. Unfortunately, various backgrounds make it unlikely that this decay mode can be seen unless $N_\nu > 10^5$. Such a large number of light neutrino generations has already been ruled out by the observation of a narrow Z^0. Note that this decay mode is much more interesting for heavy quarkonium systems because the rate increases as the fourth power of the mass.

Part of the problem with looking for the decay of $\psi \longrightarrow \nu\bar{\nu}$ is that it doesn't have an unmistakable signature in a detector. Probably the most distinctive signature for a weak ψ decay is the C-parity violating $\psi \longrightarrow \phi\phi$ decay. The branching ratio for this mode has been estimated[2] to be $BR(\psi \longrightarrow \phi\phi) \leq 10^{-8}$, which is too small even for BEPC. Thus, it seems unlikely that weak decays of the ψ will ever be observed.

Electroweak Interference

An alternative way to see the effects of the electroweak interaction is to search for electroweak interference. Electromagnetic decays of the type shown in FIGURE 1c also should have a small contribution from the weak decay shown in FIGURE 1d. For example, the decay of $\psi \longrightarrow \mu^+\mu^-$ can be mediated by either a photon or a Z^0, which leads to an interference term in the decay distribution. Unfortunately, the interference term does not give a forward-backward asymmetry unless the helicity of the μ is measured,[3] and even then the asymmetry is only $\approx 0.1\%$.

New Fundamental Couplings

Understanding the symmetry breaking mechanism that gives masses to the gauge bosons and fermions is one of the most exciting challenges facing high energy physics today. In this section, I will briefly examine how the ψ can be used to search for the new fundamental couplings that must result from symmetry breaking.

In the standard model, symmetry breaking is accomplished by the Higgs mechanism, which results in one or more scalar Higgs bosons. The Higgs couples to the charm quark, so if $M_H < M_\psi$, then the Higgs would be produced in radiative ψ decays by the diagram shown in FIGURE 1e. In the minimal standard model (one Higgs), the Linde-Weinberg bound[4] constrains the Higgs mass to be considerably heavier than the ψ mass, $M_H > 7$ GeV, which seems to rule out seeing the Higgs in ψ decays. However, this bound does not apply to more complicated models that have two or more Higgs doublets, so it is clearly worthwhile to look for the Higgs in radiative ψ decays.

The signature for a low-mass Higgs ($M_H \approx 1$–3 GeV) is an extremely narrow state that has spin 0 and decays primarily to final states involving K mesons. The MARK III experiment has evidence[5] for a new narrow state, the $\xi(2220)$, that decays to K^+K^-. The possible existence of such a state has generated considerable interest and many suggestions for what it could be, including the possibility that it is a Higgs boson.[5,6] I want to emphasize that there are a large number of plausible explanations for the ξ, including the possibility that it is a statistical fluctuation or an ordinary quark state, which are at least as likely as the Higgs interpretation. However, since the Higgs interpretation has not yet been ruled out by the data, I will take this opportunity to update the status of the ξ.

The MARK III K^+K^- mass spectrum for the channel $\psi \longrightarrow \gamma K^+K^-$ is shown in FIGURE 2. A clear peak with a statistical significance of 4.6 standard is seen near 2.2

GeV. Fitting a Breit-Wigner line shape to the peak yields the following parameters:

$$M_\xi = (2218 \pm 10)\ \text{MeV},$$

$$\Gamma_\xi < 40\ \text{MeV},$$

$$BR(\psi \rightarrow \gamma\xi) \cdot BR(\xi \rightarrow K^+K^-) = (3.8 \pm 1.6) \times 10^{-5}. \tag{3}$$

The experimental situation regarding the ξ is somewhat murky at the moment, but it should become much clearer in the near future. The DM2 collaboration is analyzing a large sample of 8.6 Mψ; preliminary results[7] fail to confirm the ξ. The MARK III is currently running on the ψ and will obtain an additional 2–3 Mψ to check the earlier result. If the ξ is confirmed, substantially more data will be needed to measure its spin

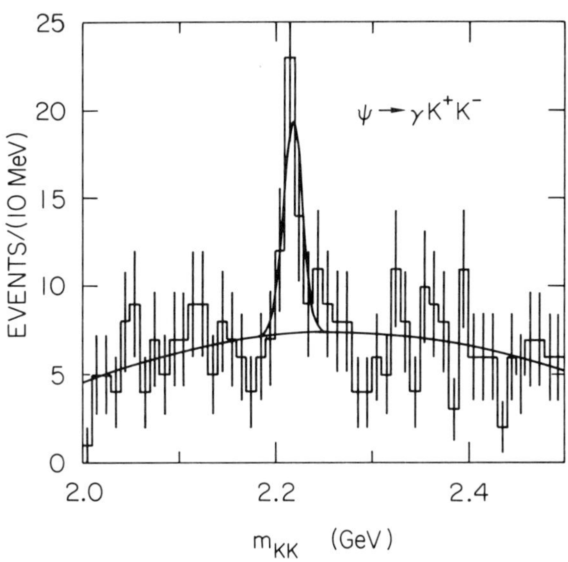

FIGURE 2. The MARK III K^+K^- mass distribution in the vicinity of the $\xi(2220)$ for the decay of $\psi \rightarrow \gamma K^+K^-$.

and branching ratios to other decay channels, and to improve the limit on its width; such information will play an important role in establishing the nature of the ξ.

The Higgs mechanism is probably the weakest link in the standard model, which has prompted considerable interest in alternative mechanisms for symmetry breaking. Whatever the mechanism, there must be new fundamental couplings to the quarks that might be observable in radiative ψ decays if the appropriate mass scale is low enough. One of the more popular alternatives to the standard model is supersymmetry, which requires that each fundamental fermion and gauge boson have a supersymmetric partner. If the supersymmetric partner to the gluon, the gluino, has a low mass, it is possible that "gluinonium," a bound state of two gluinos, would then be produced in radiative ψ decays.[8] The authors of reference 8 argue that the apparent absence of such a state in ψ decays is evidence that the gluino, if it exists, must have a mass greater than

$\approx$1 GeV. Note that unlike limits from beam dump experiments, this limit is independent of the squark mass or the gluino lifetime.

RADIATIVE DECAYS OF THE ψ

During the past few years, a large number of previously unobserved states have been seen in radiative ψ decays. These include $\iota(1450)$, $\theta(1700)$, one or more states decaying to $\eta\pi\pi$, and one or more states decaying to $\rho\rho$ and $\omega\omega$. The abundance of new states in radiative ψ decays and their relatively large decay rates are probably indications that a new spectroscopy is emerging. Understanding the nature of these new states is one of the principal goals of current research into ψ physics.

Conventional wisdom dictates that hadronic ψ decays proceed through the three gluon diagram shown in FIGURE 1a, while radiative decays proceed through the two gluon diagram shown in FIGURE 1b. One of the reasons radiative ψ decays attract so much interest is that they allow one to study the two gluon system, with the energy of the photon "tuning" the mass of the two gluon system. In principle, the two gluon system can resonate to form a new type of matter that is made of gluons and held together by gluons: a "glueball." Such an unusual object can occur only in theories where the gauge fields can couple to themselves, which is a fundamental property of non-Abelian gauge theories like QCD. Thus, the identification of one or more glueball states would unambiguously confirm a fundamental property of QCD, a theory that has very few unambiguous tests.

One of the problems with identifying glueballs is that there are no direct tests to determine whether or not a particle is a glueball. The situation is further complicated by the possibility that quark states and glueballs can mix with each other, further blurring the distinction between the two. Since it is hard to prove a new particle is a glueball and not an ordinary meson, four quark state, hybrid, etc., we need to learn as much as possible about any glueball candidates we find. Relevant information includes the spin of the candidate, its branching ratios for various decay modes (including electromagnetic decay modes), and identifying the isoscalar members of nonets with the same quantum numbers as the candidate. Making as many of these measurements as possible is one of the goals of the present ψ analysis effort.

In the next two chapters, I will focus on recent measurements by the MARK III of the properties of the $\iota(1450)$ and the $\theta(1700)$; these states are difficult to fit into quarkonium nonets and are strong glueball candidates.

STATUS OF THE θ(1700)

At the present time, the $\theta(1700)$ has only been observed in radiative ψ decays. The first observation was by the Crystal Ball[9–11] in studies of $\psi \rightarrow \gamma\eta\eta$; a second decay mode, $\theta \rightarrow K^+K^-$, was later seen by the Mark II.[12] Both of these experiments studied the decay angle distributions and found that a spin-parity assignment of 2^+ was preferred over 0^+, but neither experiment had sufficient statistics to make a definitive determination.[13]

Spin Analysis of the θ(1700)

The MARK III has recently studied[5] the decay sequence of $\psi \rightarrow \gamma\theta$, $\theta \rightarrow K^+K^-$ and performed a spin analysis of the θ. FIGURE 3 shows the presence of well-separated $f'(1525)$ and $\theta(1700)$ signals in the K^+K^- mass distribution. For a spin-parity determination, there are three relevant angles: θ_γ, the polar angle of the photon with respect to the beam direction; θ_K, the polar angle of the K^+ in the KK rest frame; and ϕ_K, the azimuthal angle of the K^+ in the KK rest frame. The distributions for these angles are shown in FIGURE 4, along with the acceptance corrected predictions for spin 0 and spin 2. The figure shows that spin 2 is clearly favored over spin 0; an analysis of the decay distributions shows that the confidence level for this assignment is 99.9%.

Observation of θ(1700) → ππ

The MARK III has also looked for the decay of $\theta \rightarrow \pi^+\pi^-$. FIGURE 5 shows the $\pi^+\pi^-$ mass distribution in $\psi \rightarrow \gamma\pi^+\pi^-$. A large peak corresponding to the $f(1270)$ is seen, as well as smaller peaks near 1700 MeV and 2100 MeV.[14] The solid curve in FIGURE 5 shows the result of a fit incorporating three noninterfering Breit-Wigner line shapes. The mass and width of the peak near 1700 MeV are consistent with the θ, so we are probably seeing the decay of $\theta \rightarrow \pi^+\pi^-$. Note that interference effects could be significant in this channel, especially if the state near 2100 MeV has spin 2. TABLE 1 summarizes the currently known properties of the θ.

Interpretation of the θ(1700)

The tensor nonet is already well established with the $f(1270)$ and $f'(1525)$ filling the isoscalar slots, so what is the θ? The list of suspects includes a radial or orbital excitation, a four quark state, a $q\bar{q}g$ hybrid, and a glueball. I will briefly review each of these possibilities.

There are a number of reasons why the θ is probably not a radial excitation of the f' or an $L = 3$ $s\bar{s}$ state. First, the mass of the θ is probably too low for such a state.[15] Furthermore, nonets in this mass region have nearly ideal mixing, which would lead one to expect a much lower rate for $\theta \rightarrow \pi\pi$. We would also need to explain the apparent absence of the nonstrange partner of the θ, which would presumably decay to $\pi\pi$. The available evidence seems to weigh heavily against the θ being a $q\bar{q}$ state.

The properties of four quark states are not presently known, but it is generally expected they will be very wide unless they are below threshold for their "fall-apart" decay mode. One possibility is that the θ is a $(u\bar{u} + d\bar{d})s\bar{s}$ state, which is nominally below threshold for favored decays such as $K^*\bar{K}^*$ or $\phi\omega$. However, such a state should have an extremely small $\pi\pi$ decay rate and should exhibit a strongly distorted line shape due to the opening of the $K^*\bar{K}^*$ channel. Thus, it seems unlikely that the θ is a four quark state.

Hybrids may provide a rich new spectroscopy of $q\bar{q}g$ nonets, but there is currently no experimental evidence for such states. Bag model calculations[16] indicate the 2^{++} $s\bar{s}g$ hybrid should lie in the mass range of 2.1–2.5 GeV, which is inconsistent with the 1.7

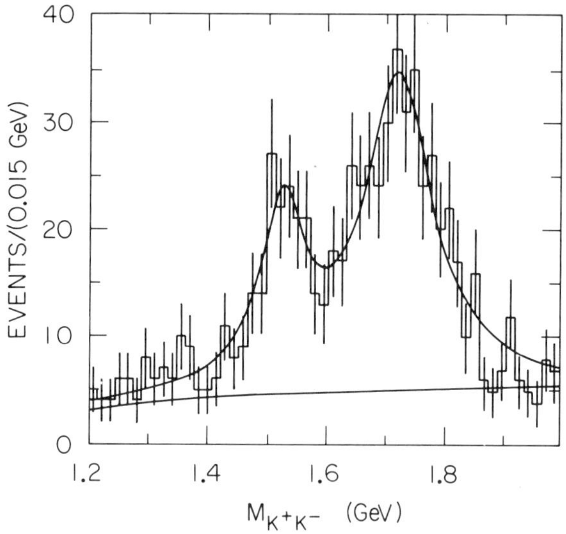

FIGURE 3. The MARK III K^+K^- mass distribution in the vicinity of the $\theta(1700)$ for the decay of $\psi \to \gamma K^+K^-$.

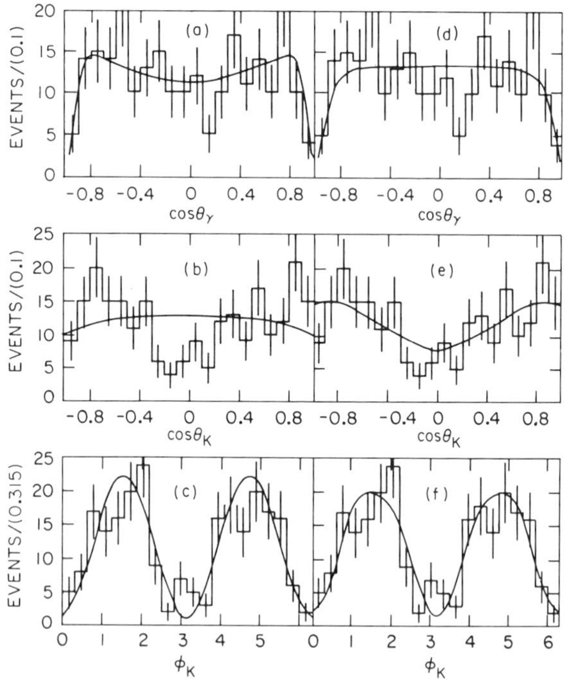

FIGURE 4. Decay angle distributions for the $\theta(1700)$ from the MARK III experiment. The solid curves in (a)–(c) show the acceptance corrected distributions expected for spin 0; (d)–(f) show the distributions expected for spin 2.

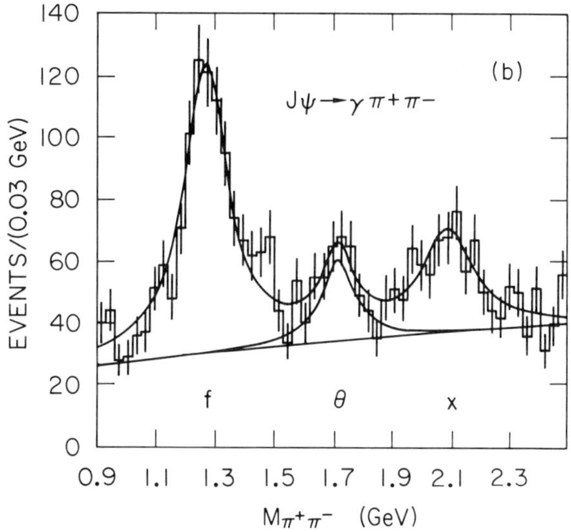

FIGURE 5. The MARK III $\pi^+\pi^-$ mass distribution for the decay of $\psi \rightarrow \gamma\pi^+\pi^-$.

GeV mass of the θ. Furthermore, it would be very surprising to have such a strong signal for an excited (TM gluon) hybrid without there being some evidence for the four ground state (TE gluon) hybrid nonets.

The properties of the θ are generally consistent with being a glueball. It is produced in radiative ψ decays with a high rate, as expected for a glueball. Since glueballs are $SU(3)$ flavor singlets, the θ should decay to KK, $\pi\pi$, and $\eta\eta$ in the ratio of 4:3:1 if the flavor symmetry is preserved. The observed ratio is 4:1.0 ± 0.4:1.6 ± 0.8, which is in disagreement with the $SU(3)_f$ prediction for the $\pi\pi$ rate. It is well known that flavor symmetry is badly broken in this mass region, so the above result should not be too surprising. The remarkable feature is that the $\pi\pi$ decay mode is observed at such a large rate. This strongly suggests that the θ is not an $s\bar{s}$ or $s\bar{s}X$ state because the decay to $\pi\pi$ would then be OZI suppressed.

TABLE 1. Properties of the $\theta(1700)$

Experiment	Decay Mode	Mass (MeV)	Width (MeV)	$BR(\psi \rightarrow \gamma\theta) \cdot BR(\theta \rightarrow X)$ $(\times 10^{-4})$
Crystal Ball[9–11]	$\eta\eta$	1670 ± 50	160 ± 80	3.8 ± 1.6
	$\eta\eta'$			<2.1
	$\pi\pi$			<6
Mark II[12]	KK	1708 ± 30	156 ± 60	12 ± 6
	$\pi\pi$			<3.2
MARK III[5]	KK	1720 ± 10	130 ± 20	9.6 ± 2.2
	$\pi\pi$	1713 ± 15	—	2.4 ± 0.7
	$\rho\rho$			<6
	$\bar{K}K\pi$			<2.8
	$K^*\bar{K}^*$			<4.5
DM2[7]	KK	1718 ± 29	233 ± 75	

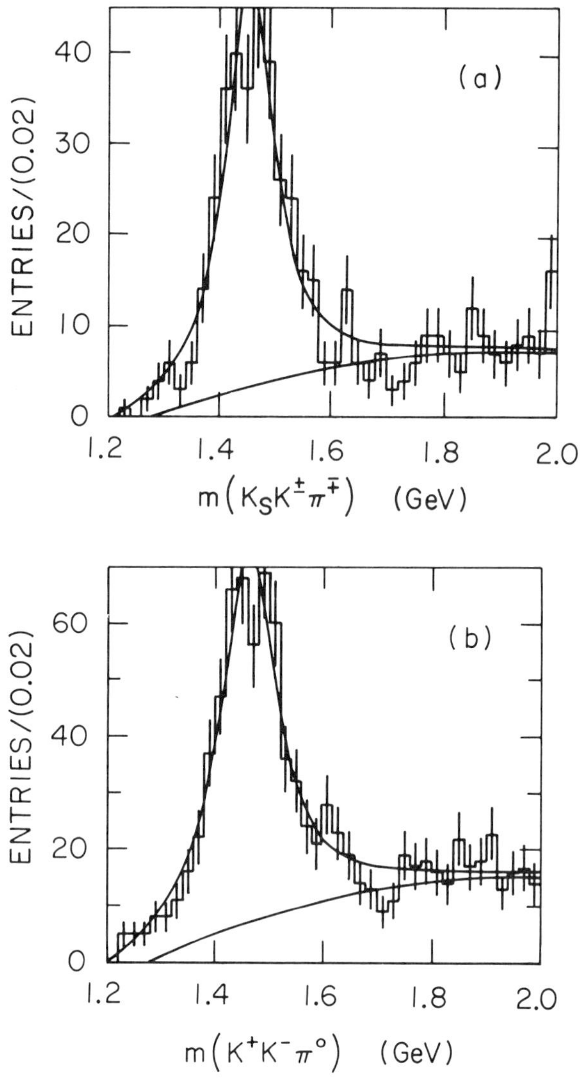

FIGURE 6. The MARK III $KK\pi$ mass distribution for the decay of $\psi \rightarrow \gamma KK\pi$: (a) The $K_S K^{\pm}\pi^{\mp}$ distribution; (b) the $K^+K^-\pi^0$ distribution.

From the above evidence, I conclude that the θ is most likely a glueball. However, this interpretation is based more on the improbability of the alternatives than on the measured properties of the θ itself. Further checks of the glueball hypothesis can be made by measuring quantities sensitive to the quark content of the θ, such as $\Gamma(\theta \rightarrow \gamma\gamma)$, and quantities that demonstrate flavor-symmetric couplings, such as $BR(\psi \rightarrow \omega\theta)$ and $BR(\psi \rightarrow \phi\theta)$.

STATUS OF THE ι

The $\iota(1450)$ may have first been seen by early $\bar{p}p$ experiments,[17] but the excitement began when the decay of $\psi \rightarrow \gamma\iota$, $\iota \rightarrow K\bar{K}\pi$ was observed by the Mark II with a surprisingly large branching ratio.[18] The Mark II tentatively identified this structure as the $E(1420)$, an axial vector state with properties similar to those observed by the Mark II. However, a partial wave analysis by the Crystal Ball found that the ι decays predominately through the 0^- $\delta\pi$ isobar.[19]

Since the δ decays to $\eta\pi$ as well as KK, it was expected that the ι would also be seen in the $\eta\pi\pi$ channel. The subsequent failure[20] to find the ι in this channel probably indicates that the ι decay mechanism is more complicated than the mechanisms considered in the Crystal Ball analysis. Another difficulty with this analysis is the presence of ι events in the phase-space term, which should not occur if the decay is correctly modeled by a $\delta\pi$ isobar. These difficulties led the MARK III to analyze the spin of the ι using a method that is only sensitive to the spin of the ι, and not its decay mechanism.[21]

Spin Analysis of the $\iota(1450)$

The MARK III has analyzed the ι in both the $K^+K^-\pi^0$ and $K_SK^\pm\pi^\mp$ decay modes. FIGURE 6 shows clear ι signals in each channel with very little background. The spin of the ι can be determined by studying the angular distribution of the plane formed by the $KK\pi$ system in the ι rest frame.[22] Three angles are needed in the spin analysis (see FIGURE 7): θ_γ, the angle of the photon with respect to the beam direction; β, the polar angle of the normal to the decay plane in the ι rest frame; and α, the azimuthal angle of the decay plane normal in the ι rest frame. For spin 0, the cos β distribution will be flat, while cos θ_γ will have a $(1 + \cos^2\theta_\gamma)$ distribution. For spin 1, the distributions depend on an unknown ratio of helicity amplitudes; however, there is no value of this ratio that will give distributions similar to spin 0.

FIGURE 8 shows the distributions for cos β and cos θ_γ in the $K_SK^\pm\pi^\mp$ decay mode. The cos β distribution is flat (FIGURE 8a), as is the detector acceptance for this angle (FIGURE 8b). The acceptance corrected cos θ_γ distribution is in good agreement with

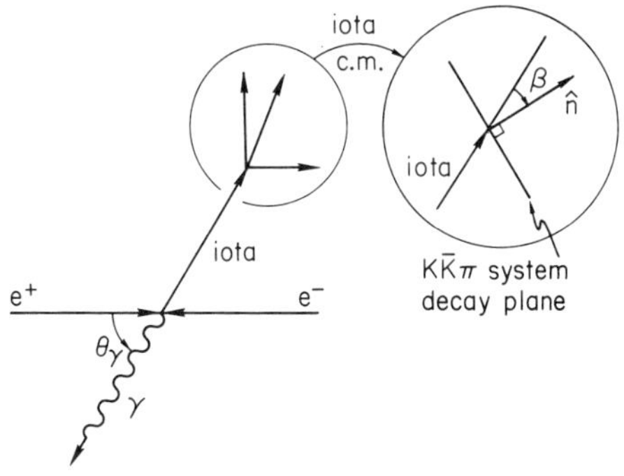

FIGURE 7. Definition of the angles used in the ι spin analysis.

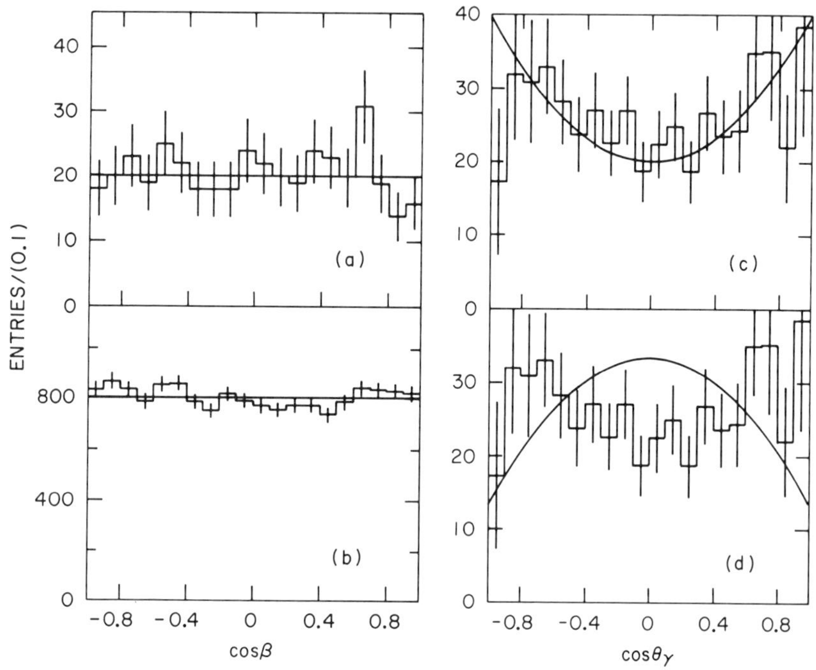

FIGURE 8. Angular distributions for the MARK III spin analysis of $\iota \rightarrow K_S K^{\pm}\pi^{\mp}$: (a) The observed cos β distribution; (b) the detector acceptance for cos β; (c) the acceptance corrected cos θ_γ distribution with a solid curve showing the prediction for spin 0; (d) same as (c), except the curve shows the distribution expected for spin-parity 1^+.

the prediction for spin 0 (FIGURE 8c) and is clearly inconsistent with the spin 1 prediction (FIGURE 8d).[23] A full-blown likelihood analysis of the decay distributions reaches the same conclusion for both the $K_S K^{\pm}\pi^{\mp}$ and $K^+K^-\pi^0$ channels: spin 0 is favored over spin 1 by an enormous factor.

Evidence for $\iota(1450) \rightarrow \gamma\rho$

The ι may also have been observed[7,11,21] in the decay sequence of $\psi \rightarrow \gamma X$, $X \rightarrow \gamma\rho^0$. FIGURE 9 shows that Crystal Ball, DM2, and MARK III all have a peak in the $\gamma\rho$ mass distribution at $\approx$1400 MeV that is $\approx$100–200 MeV wide. Given the shape of the peak and its surprisingly large width, it is quite possible that two (or more) particles contribute to the observed signal. There is also some evidence[11,21] that background events may populate the low-mass side of the peak. With these difficulties in mind, the Crystal Ball and MARK III have fit the peak under various assumptions and find that $BR(\psi \rightarrow \gamma X) \cdot BR(X \rightarrow \gamma\rho) \simeq (1\text{–}2) \times 10^{-4}$.

With the present statistics, it is not possible to positively identify the peak in FIGURE 9, but the number of possibilities is rather limited. The partial width to $\gamma\rho$ for a particle in this mass region is expected to be at most a few MeV.[15,24,25] The observed peak has a total width of at least 100 MeV, so whatever produces this peak should have $BR(X \rightarrow \gamma\rho) \leq 5\%$ and should be copiously produced in radiative ψ decays: $BR(\psi \rightarrow \gamma X) \geq 2 \times 10^{-3}$. Between the Crystal Ball and MARK III, all of the radiative decay channels expected in this mass region have been studied and only three states have large branching ratios: the $f(1270)$, the $\iota(1450)$, and the structure[26] near 1380 MeV observed in $\psi \rightarrow \gamma\eta\pi\pi$. The $f(1270)$ may substantially influence the low-mass region, but by itself cannot account for the observed peak. The $\gamma\rho$ mass distribution has been checked for consistency with the $\iota(1450)$ and is fully compatible with this interpretation if there is a low-mass contribution from another channel or background. The $\gamma\rho$ signal is also compatible with the poorly understood structure near 1380 MeV observed in the $\eta\pi\pi$ channel; further work is needed to identify the nature of this structure and to check for consistency with the $\gamma\rho$ signal.

If the $\gamma\rho$ signal is due to decays of the $\iota(1450)$, the decay angle distributions should agree with the predictions for spin-parity 0^-. Two angles are particularly good indicators: $\theta_{\gamma 2}$, the polar angle of the secondary photon in the $\gamma\rho^0$ rest frame; and $\theta\pi$, the polar angle of the π^+ in the ρ^0 rest frame. For spin 0, cos $\theta_{\gamma 2}$ should have a uniform distribution, while cos θ_π should have a $\sin^2 \theta_\pi$ distribution. FIGURE 10 shows that the data are consistent with the predictions for 0^-. Note that spin 1 can have the same distributions for these two angles for a particular choice of helicity amplitudes; to definitively rule out spin 1 requires an analysis of the full decay angle distribution and a substantial increase in statistics.

Interpretation of the ι(1450)

A comparison of the $E(1420)$ and $\iota(1450)$ properties is given in TABLE 2. It is now clear that these must be different particles; not only do they have different spin, but the mass and width measurements are also incompatible. The $E(1420)$ is probably the axial vector $s\bar{s}$ state, but the interpretation of the $\iota(1450)$ is currently unsettled. There

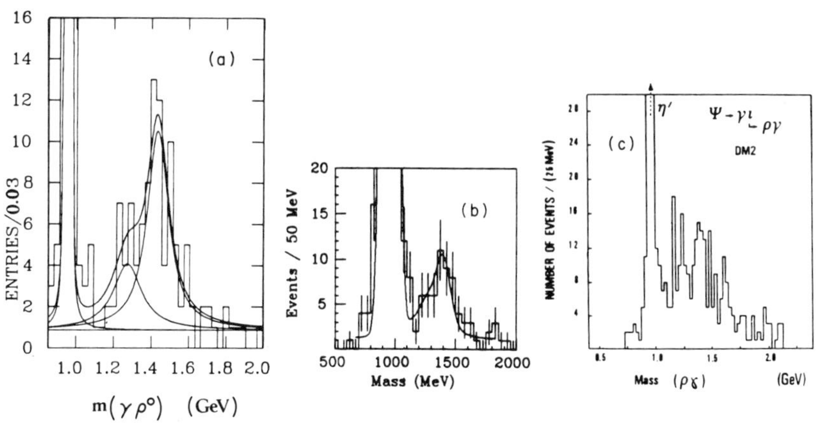

FIGURE 9. The $\gamma\rho^0$ mass distribution for the decay of $\psi \rightarrow \gamma\gamma\rho^0$: (a) MARK III; (b) Crystal Ball; (c) DM2.

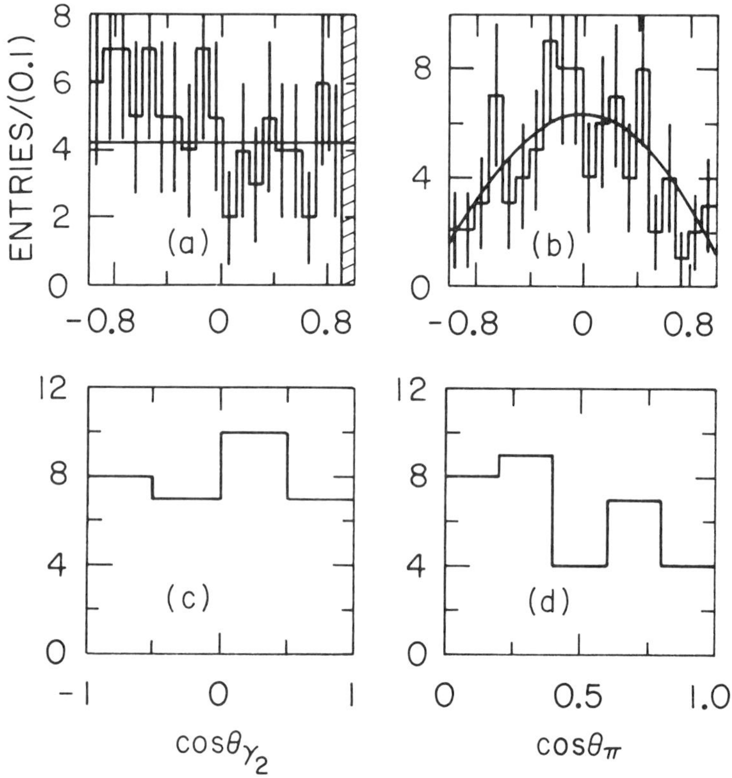

FIGURE 10. Decay angle distributions for the decay of $\psi \rightarrow \gamma\gamma\rho^0$: (a) and (b) MARK III distributions; (c) and (d) Crystal Ball distributions.

is the usual list of suspects: radial excitation, glueball, four quark state, and hybrid. To further complicate matters, the ι may be operating in disguise by mixing with a quark state (see the next chapter for a discussion of possible ι–η' mixing).

The $\iota(1450)$ is unlikely to be a four quark state since the quarks would be in an orbitally excited state and should have an extremely large width for the fall-apart decay mode.[27] It would also be surprising if either a hybrid or four quark state is so copiously produced in radiative ψ decay: $BR(\psi \rightarrow \gamma\iota) \geq (5.0 \pm 0.8) \times 10^{-3}$, which is one of the largest radiative branching ratios that has been observed.

The most likely explanations for the $\iota(1450)$ are a radial excitation or a glueball. Arguments have been given for why a radial excitation should not be so copious in

TABLE 2. Comparison of $E(1420)$ and $\iota(1450)$

Particle	Mass (MeV)	Width (MeV)	$BR(\psi \rightarrow \gamma X) \cdot BR(X \rightarrow KK\pi)$	J^{PC}
$E(1420)$	1418 ± 10	52 ± 10	?	1^{++}
$\iota(1450)$[21]	1458 ± 10	98 ± 15	$(5.0 \pm 0.8) \times 10^{-3}$	0^{-+}

radiative ψ decays,[27] as well as arguments for why it should.[25] The radially excited pseudoscalar nonet is rather poorly understood and it is quite possible the ι could fit in as one of the isoscalars. While the possible observation of $\iota \to \gamma\rho$ may seem to contradict the glueball hypothesis, glueballs are expected to mix with quark states, thus giving the "glueball" radiative decay modes. The large branching ratio for producing the ι supports the glueball interpretation and the decay to both $KK\pi$ and $\gamma\rho$ suggests flavor-symmetric couplings.

Since both the radial excitation and glueball hypothesis appear to be viable, it is important to search for additional evidence to identify the nature of the ι. Probably the most important task is to acquire a better understanding of the radially excited pseudoscalar nonet. In addition, the quark content of the $\iota(1450)$ can be studied by measuring $\Gamma(\iota \to \gamma\gamma)$ and $BR(\iota \to \gamma\phi)$, while measuring $BR(\psi \to \omega\iota)$ and $BR(\psi \to \phi\iota)$ would check for flavor-symmetric couplings.

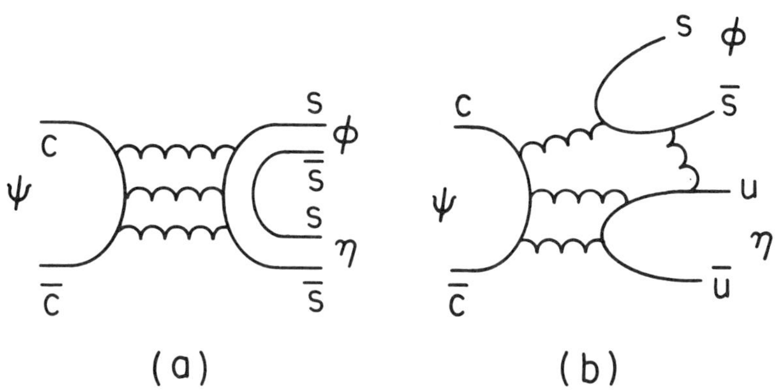

FIGURE 11. Diagrams for the decay of $\psi \to \phi\eta$: (a) The three gluon diagram; (b) the doubly disconnected diagram.

WHAT IS THE η'?

The MARK III has used decays of the ψ to a vector and a pseudoscalar meson to study the quark content of the pseudoscalar nonet.[28] Included in this study are measurements of the branching ratios for the following ψ decay modes: $\rho\pi$, $K^*\overline{K}$, $\omega\eta$, $\omega\eta'$, $\phi\eta$, $\phi\eta'$, $\rho\eta$, $\rho\eta'$, and $\omega\pi^0$. FIGURE 11a shows the three gluon diagram for a typical vector-pseudoscalar decay, $\psi \to \phi\eta$. Note that this diagram exhibits flavor correlations: only the $s\bar{s}$ component of the η contributes to the decay (the vector meson nonet is assumed to be ideally mixed in this analysis). A similar diagram exists for $\psi \to \omega\eta$; in this case, only the $(u\bar{u} + d\bar{d})$ component of the η will contribute. Diagrams that do not exhibit flavor correlation, such as FIGURE 11b, are doubly disconnected and are expected to be suppressed relative to the singly disconnected three gluon diagram.[29]

To relate the vector-pseudoscalar branching ratios to the quark content, we use a technique similar to one developed by Rosner[30] for radiative decays of the pseudoscalar mesons. Following Rosner, we define the η and η' wave functions as follows:

$$|\eta\rangle = x_\eta \left|\frac{u\bar{u} + d\bar{d}}{\sqrt{2}}\right\rangle + y_\eta|s\bar{s}\rangle,$$
$$|\eta'\rangle = x_{\eta'} \left|\frac{u\bar{u} + d\bar{d}}{\sqrt{2}}\right\rangle + y_{\eta'}|s\bar{s}\rangle, \tag{4}$$

where for the moment we assume no other states mix with the η or η'. Using $SU(3)$ symmetry, we can relate the x and y parameters to various ratios of vector-pseudoscalar branching ratios. For example,

$$\frac{BR(\psi \to \omega\eta)}{BR(\psi \to \rho^0\pi^0)} = x_\eta^2 \times \text{Phase-Space Factor}. \tag{5}$$

The MARK III has developed a model for vector-pseudoscalar decays that takes the $SU(3)$ predictions and incorporates electromagnetic decays and strange-quark mass effects.[28] This model has been fit to the observed branching ratios, yielding the following quark content for the η and η':

$$|x_\eta| = 0.64 \pm 0.06,$$
$$|y_\eta| = 0.84 \pm 0.12,$$
$$|x_{\eta'}| = 0.34 \pm 0.05,$$
$$|y_{\eta'}| = 0.72 \pm 0.12. \tag{6}$$

For a simple mixing of the η and η', we should have $x^2 + y^2 = 1$ and $|y_\eta| = |x_{\eta'}|$. Using the quark content determined above gives $x_\eta^2 + y_\eta^2 = 1.1 \pm 0.2$, which indicates that the η wave function is almost entirely made up of quarks. However, $x_{\eta'}^2 + y_{\eta'}^2 = 0.63 \pm 0.18$ suggests that there may be another state that mixes with the η', while $|x_{\eta'}| \ll |y_\eta|$ indicates that a simple mixing of the η and η' is incorrect.

The above results are displayed graphically in FIGURE 12. This figure shows that the η is consistent with a nearly pure octet state, but the η' appears to be missing a substantial fraction of its nonstrange quarks and is inconsistent with the predictions for a 10° mixing angle. If the above model for vector-pseudoscalar decays is correct, the η' must be mixing with another pseudoscalar state.

Do $\iota(1450)$ and η' Mix?

Since the η' is produced in radiative ψ decays with a large branching ratio, whatever state mixes with it must also be produced in radiative ψ decays with a reasonably large branching ratio. The obvious candidate for a pseudoscalar with a large radiative branching ratio in ψ decays is the $\iota(1450)$. In this scenario, the ι must contain the missing nonstrange quark content of the pseudoscalar nonet: $x_\iota^2 = 1 - x_\eta^2 - x_{\eta'}^2 = 0.47 \pm 0.08$. We can calculate the mixing contribution to the partial width $\Gamma(\iota \to \gamma\rho)$ using $SU(3)$ symmetry:

$$\Gamma(\iota \to \gamma\rho) = 3x_\iota^2 \left(\frac{M_\omega(M_\iota^2 - M_\rho^2)}{M_\iota(M_\omega^2 - M_\pi^2)}\right)^3 \Gamma(\omega \to \gamma\pi^0). \tag{7}$$

Using the value for x_ι calculated above yields a mixing contribution to the $\gamma\rho$ partial width of 3.1 ± 0.5 MeV. This demonstrates that even if the $\iota(1450)$ is a "glueball," it can still have a substantial width for radiative decays. Under the assumption that the $\gamma\rho$ signal near 1400 MeV is due to the $\iota(1450)$, the MARK III has measured $\Gamma(\iota \rightarrow \gamma\rho) = 1.9 \pm 0.7$ MeV, which is in agreement with the predicted contribution from mixing. Although such agreement does not prove that the η' and ι mix, it is an important consistency check for what I believe is an attractive explanation for the observed properties of the η'.

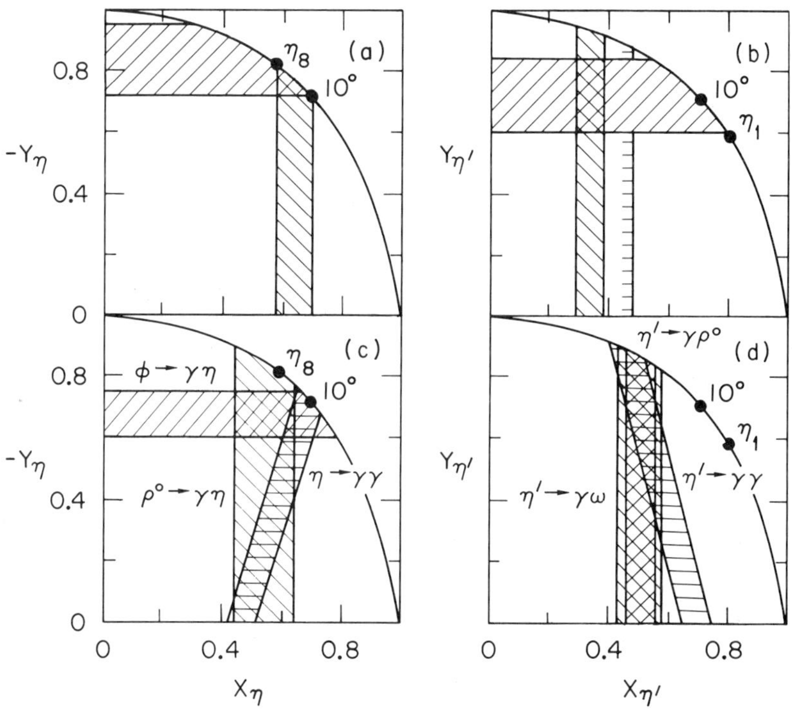

FIGURE 12. Quark content of the η and η': (a) and (b) from vector-pseudoscalar decays; (c) and (d) from radiative decays.

CONCLUSIONS

Althought the ψ is proving to be an extremely useful tool for studying the strong interaction, the prospects for observing electroweak effects in ψ decays are rather poor. It is unlikely that weak decays or electroweak interference will be observed at the ψ. However, there is a chance that new fundamental couplings (such as the Higgs) will be observed if the mass scale for electroweak symmetry breaking is low enough.

The large number of new states observed in radiative ψ decays suggests that we are seeing one or more new spectroscopies. Progress has been made in understanding the nature of two of these states, the $\theta(1700)$ and $\iota(1450)$. New results strongly suggest that the $\theta(1700)$ is a glueball state, but more work is necessary to determine whether the $\iota(1450)$ is a radial excitation, glueball, hybrid, or something even more exotic.

The ψ is also proving to be an excellent place to study light-quark spectroscopy; future study in this area has enormous potential for giving a better understanding of this complicated subject.

Having recently concluded the first decade of ψ physics, it is time to look ahead to what the next ten years will bring. The DM2 collaboration has acquired the world's largest sample of ψ decays, 8.6 Mψ, while the MARK III expects to have $\approx$5.5 Mψ by the summer. The analyses of these large data samples should provide a new round of interesting discoveries. In the longer term, BEPC has the opportunity to acquire a very large data sample (perhaps 100 Mψ), which would allow rare processes to be studied in detail. Although the first decade is a tough act to follow, there is still much to learn about how quarks and gluons can be put together to form particles. There is no better place to do this than at the ψ.

ACKNOWLEDGMENTS

I would like to thank Howard Haber and my colleagues in the MARK III collaboration for their discussions and contributions to the results presented here. I would also like to thank Norman Amos, Martin Block, Ruth Goldhor, and Shekhar Shukla for the tremendous job they did in organizing the premier of the Aspen Winter Physics Conference Series.

NOTES AND REFERENCES

1. ELLIS, J. 1978. *In* Proceedings of Summer Institute on Particle Physics. M. Zipf, Ed.: 69. Stanford Linear Accelerator Center. Stanford, California.
2. GOLDMAN, T. & H. E. HABER. 1983. Los Alamos preprint LA-UR-83-2323.
3. ISGUR, N. 1975. Phys. Rev. **D11:** 3236.
4. LINDE, A. D. 1976. JETP Lett. **23:** 64; WEINBERG, S. 1976. Phys. Rev. Lett. **36:** 294.
5. EINSWEILER, K. 1984. Ph.D. thesis. Stanford University. SLAC report no. 272.
6. HABER, H. E. & G. L. KANE. 1984. Phys. Lett. **135B:** 196; WILLEY, R. S. 1984. Phys. Rev. Lett. **52:** 585; GODFREY, S., R. KOKOSKI & N. ISGUR. 1984. Phys. Lett. **141B:** 439.
7. SZKLARZ, G. 1984. *In* Proceedings of the XXIIth International Conference on High Energy Physics, vol. 1. A. Meyer & E. Wieczorek, Eds.: 180. Akademie der Wissenschaften der DDR. Leipzig, Germany.
8. GOLDMAN, T. & H. HABER. 1984. Los Alamos preprint LA-UR-84-634.
9. EDWARDS, C. *et al.* 1982. Phys. Rev. Lett. **51:** 859.
10. SCHARE, D. L. 1982. *In* Field Theory in Elementary Particles. B. Kursunoglu & A. Perlmutter, Eds.: 239. Plenum Press. New York.
11. EDWARDS, C. 1984. Ph.D. thesis. California Institute of Technology. CALT-68-1165.
12. FRANKLIN, M. E. B. 1982. Ph.D. thesis. Stanford University. SLAC report no. 254.
13. The θ is allowed to have the following J^{PC} values: 0^{++}, 2^{++}, 4^{++}, etc. Spins greater than two are not expected in the 1700 MeV mass region and have not been included in spin analyses of the θ.
14. The mass and width of the peak near 2100 are consistent with the $h(2030)$, but such an identification must be considered tentative pending a spin determination.
15. GODFREY, S. & N. ISGUR. 1984. Mesons with Chromodynamics. University of Toronto preprint.
16. CHANOWITZ, M. S. & S. R. SHARPE. 1983. Phys. Lett. **132B:** 413.

17. BAILLON, P. *et al.* 1967. Nuovo Cimento **A50:** 393.
18. SCHARRE, D. L. *et al.* 1980. Phys. Lett. **97B:** 329.
19. EDWARDS, C. *et al.* 1982. Phys. Rev. Lett. **49:** 259.
20. EDWARDS, C. *et al.* 1983. Phys. Rev. Lett. **51:** 859; TOKI, W. 1983. *In* Proceedings of Summer Institute on Particle Physics. P. M. McDonough, Ed.: 471. Stanford Linear Accelerator Center. Stanford, California.
21. RICHMAN, J. D. 1985. Ph.D. thesis. California Institute of Technology. CALT-68-1231.
22. BERMAN, S. & J. JACOB. 1965. Phys. Rev. **139B:** 1023.
23. The ratio of helicity amplitudes was chosen to give a flat cos β distribution, as observed in the data.
24. DONOGHUE, J. F. 1984. Phys. Rev. **D30:** 114; PALMER, W. F. & S. S. PINSKY. 1983. Phys. Rev. **D27:** 2219.
25. BARNES, T. 1984. Rutherford Lab. preprint RAL-84-046.
26. BECKER, J. J. 1984. Ph.D. thesis. University of Illinois at Champaign-Urbana.
27. CHANOWITZ, M. S. 1981. *In* Particles and Fields–1981: Testing the Standard Model. C. A. Heusch & W. T. Kirk, Eds.: 85. American Institute of Physics. New York.
28. PERRIER, J. 1984. To be published in Proceedings of Physics in Collision IV. Santa Cruz, California. SLAC-PUB-3436.
29. A different conclusion is reached by PINSKY, S. S. 1984. Los Alamos preprint LA-UR-84-3208.
30. ROSNER, J. L. 1983. Phys. Rev. **D77:** 1101.

Measurements of Neutrino Mass[a]

R.G.H. ROBERTSON

Physics Division
Los Alamos National Laboratory
Los Alamos, New Mexico 87545

In this review, we shall consider "direct" experimental information on neutrino mass as derived from the study of nuclear and elementary-particle weak decays. Neutrino oscillations, double β decay, and searches for heavy neutrinos admixed with lighter ones will not be discussed since they are given extensive treatment by others.[1] Similarly, only cursory consideration will be given to cosmological limits on neutrino mass.

TRITIUM BETA DECAY

The ITEP Experiments

As is well known, a group at the Institute for Theoretical and Experimental Physics (ITEP) in Moscow has reported[2-4] evidence for a massive electron antineutrino (ν_β) from a long experimental campaign. The basic method is a careful study of the beta spectrum from tritium decay:

$$\frac{dN}{dE} = G_F^2 \frac{m^5c^4}{2\pi^3\hbar^7} \cos^2\theta_c \,|M_{if}|^2\, F(Z_f, E)\, \mathrm{p}\, E\, (W_0 - E)\, [(W_0 - E)^2 - m_\beta^2 c^4]^{1/2}, \quad \mathbf{(1)}$$

where M_{if} is the nuclear matrix element (for ^{3}H decay, $|M_{if}|^2 \sim 5.55$) and $F(Z_f, E)$ is the Fermi Coulomb distortion factor.[5]

$$F(Z_f, E) \sim x(1 - e^{-x})^{-1}\, (1.002037 - 0.001427 v/c),$$

where

$$x = 2\pi Z_f \alpha/(v/c).$$

The function, $F(Z_f, E)$, differs from unity by about 20% for ^{3}H near the endpoint. To simplify notation, we refer to neutrino masses as m_β, m_μ, and m_τ, and make no further distinction between neutrino and antineutrino mass. The total energy is W_0. Since actual experiments are usually done in the center-of-momentum frame, the observed maximum energy of the betas is less by about 3 eV, the energy of nuclear recoil.

This expression applies to the decay of a bare nucleus, and when atomic electrons are also present, it is necessary to express the total decay probability as a sum of individual branches to all possible final states. Ching and Ho[6] have given the general expression, which is indistinguishable (at the level of about 10^{-3}) from simply

[a]This work was performed under the auspices of the U.S. Department of Energy.

replacing W_0 by

$$W_n = W_0 + V_n$$

in equation (**1**), where $V_n = E_i - E_{fn}$, the difference between atomic electron binding energies in the initial and final states.

For the free tritium atom, one can calculate both the branching and the excitation energies to great accuracy (see Williams and Koonin[7]) (FIGURE 1). Such a calculation can probably also be done for T_2, although even in this elementary molecule the situation is vastly more complex because excited states in the ^{3}He daughter are now in

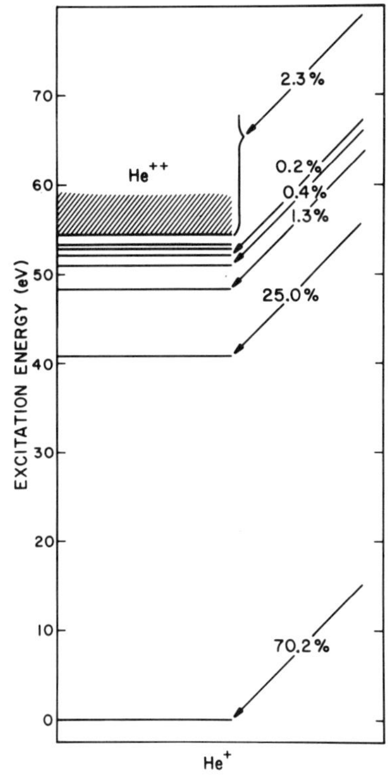

FIGURE 1. Branching in the decay of free atomic tritium to various final states in ^{3}He.

the continuum and have a finite width. For example, Martin and Cohen[8] find that 58% of the decays lead to the ^{3}HeT$^+$ ground state, 26% lead to excited ^{3}HeT$^+$ states (which can dissociate or decay radiatively), and the remaining 16% goes to the continuum. For a molecule as complicated as valine, which is used by the ITEP group, a detailed calculation is very difficult, especially for the continuum. Kaplan, Smutnyi, and Smelov[9] have reported large-basis calculations for valine.

Much has been made of this "final-state" problem and it does indeed appear to be a serious limitation on how accurately a neutrino mass can be determined. Nevertheless, a model-independent statement about a lower limit to the mass can be made by attempting to fit the data with expression (**1**), the "nuclear" case. Both the endpoint energy, W_0, and the neutrino mass, m_β, are treated as free parameters. Actual spectra are a superposition of many spectra with different endpoints, W_n, ranging up to a maximum, W_0. Thus, lumping all constants and $F(Z_f, E)$ into a single parameter, A, gives

$$dN/dE = \Sigma_n A\,\lambda_n \mathrm{p}E\,(W_n - E)^2[1 - m_\beta^2/(W_n - E)^2]^{1/2},$$

where λ_n is the weight of the nth state. Consider now the case for $m_\beta = 0$, expand W_n, and carry out the summations:

$$dN/dE = A\mathrm{p}E\,(W_0 - E + \overline{V}_n)^2[1 + (\overline{V_n^2} - \overline{V}_n^2)/(W_0 - E + \overline{V}_n)^2].$$

This spectrum looks like a simple Fermi spectrum with an extrapolated endpoint, $W_0 + \overline{V}_n$, and a "neutrino mass" squared of $-2(\overline{V_n^2} - \overline{V}_n^2)$, provided one does not inspect the endpoint region with good resolution. However, $\overline{V_n^2} - \overline{V}_n^2$ is always ≥ 0, which causes an excess of counts in the endpoint region, rather than a deficiency as would result from an actual neutrino mass. The most conservative assumption that can be made is $\overline{V_n^2} - \overline{V}_n^2 = 0$ (i.e., only one final state). If, under that assumption (the "nuclear" case), one still sees a deficiency of counts in the endpoint region, then it can only be due to neutrino mass.

This is a very powerful result that permits the use of complex sources to determine whether the neutrino has mass. Although there are innumerable other kinds of experimental errors that could subvert the conclusion, final-state effects cannot affect this model-independent lower limit.

Similar arguments can be made about the effect of instrumental resolution. Let the resolution function be $R(E, E')$, the normalized probability of observing an energy, E, when the actual energy is E'. We assume that $R(E,E')$ is sharply peaked around $E \sim E'$ and therefore we neglect the effect of the term, pE, in the spectral distribution. Then,

$$dN/dE \sim A\mathrm{p}E \int \{(W_0 - E' + V_n)^2 - m_\beta^2/2 + \sigma_V^2\} R(E,E')dE',$$

where $\sigma_V^2 = \overline{V_n^2} - \overline{V}_n^2$. Thus

$$dN/dE = A\mathrm{p}E\,\{(W_0 - \overline{E}_R + \overline{V}_n)^2 - m_\beta^2/2 + \sigma_V^2 + \sigma_R^2\},$$

where

$$\overline{E}_R = \int_0^{W_0} E'R(E,E')dE'$$

and

$$\sigma_R^2 = \int_0^{W_0} (E')^2 R(E,E')dE' - \overline{E}_R^2.$$

Again, this expression will be valid close to, but not at, the endpoint. One can see that resolution plays the same role as final states, and that in the absence of knowledge of $R(E,E')$, one could make the conservative assumption that $\sigma_R^2 = \sigma_V^2 = 0$. If there were still a deficiency of counts at the endpoint, that could only be due to neutrino mass.

Unfortunately, no experiment has sufficiently good resolution (yet) to obtain useful results under this assumption.

We now briefly review the history of the ITEP measurements: the first report[10] of data from the tritium experiment in the Tret'yakov toroidal spectrometer[11] gave, curiously, an upper limit of 35 eV (90% C.L.). By 1981, a great deal more data had been accumulated and the result,

$$14 < m_\beta < 46 \text{ eV} \qquad (99\% \text{ C.L.}),$$

was published.[2] (The earlier data were included in this.) The lower limit was established in the model-independent assumption of one final state. However, Simpson[12] pointed out that, in measuring the resolution function, the natural linewidth of the ^{169}Yb conversion lines had been overlooked and the resolution was therefore better than assumed. As described above, this has the effect of reducing the derived neutrino mass and the best-fit value fell from 34.3 to 28.0 eV, based on measurements by the ITEP group and by Bennett *et al.*[13] of the ^{169}Yb linewidths. After this correction, the model-independent (one-final-state) lower limit went to zero. The original analysis for the best-fit value made use of the atomic approximation and the ITEP group claims[4] that reanalysis with valine distributions brings m_β back up to 36.4 eV because $\sigma_{\text{valine}} > \sigma_{\text{atom}}$. There seems to be no basis for this claim in published calculations,[9] which would indicate that $\sigma_{\text{valine}} < \sigma_{\text{atom}}$ and, therefore, this gives an even lower value for m_β, 21eV.

After 1981, the resolution was improved from about 45 to about 20 eV by setting up a potential gradient along the source to compensate for finite source size. The background was reduced 20-fold by accelerating the betas by 4 kV into the clear region above the tritium endpoint (based on, we believe, a suggestion[14] of the Los Alamos group.) New results were reported in 1983 at Cornell[3] (and at Brighton). The best-fit value (valine distribution) was 33.0 eV and a "model-independent" lower limit of 20 eV was claimed. However, the spectrum could not be well fit by any theoretical description (FIGURE 2). Two massive neutrinos were proposed, but did not really alleviate the problem.

There were two major errors in the 1983 analysis. The "model-independent" lower limit was derived by examining the last 40–50 eV of the spectrum and claiming that, because this region was free of final states (other than the ground state), the total energy release and therefore the neutrino mass could be derived unambiguously. That was incorrect, though, because not only are there many states in this region,[9] but the derived endpoint lay outside the region chosen. This erroneous claim should not be confused with the type of model-independent limit discussed earlier. The second error was in relating the measured line shapes of ^{169}Yb conversion lines to the instrumental resolution. It was incorrectly assumed that the conversion lines have symmetric, Lorentzian shapes and that the observed low energy tails were entirely instrumental. In fact, conversion lines are always accompanied by lower energy shake-off satellites and by degraded electrons that have scattered in the source backing (see, for example, Freedman[15]). This assumption led to the bizarre spectrum distortions observed. Thus, the results reported[3] at the Cornell and Brighton meetings can be disregarded.

Finally, at the 1984 meeting in Leipzig, the ITEP group reported[4] new data with what appears to be a substantially correct analysis and a genuine model-independent lower limit. Three runs are shown in FIGURE 3. Chi-squared per degree of freedom is essentially equal to one. The best-fit value (valine distribution) is 34.8 ± 1.9 eV and the

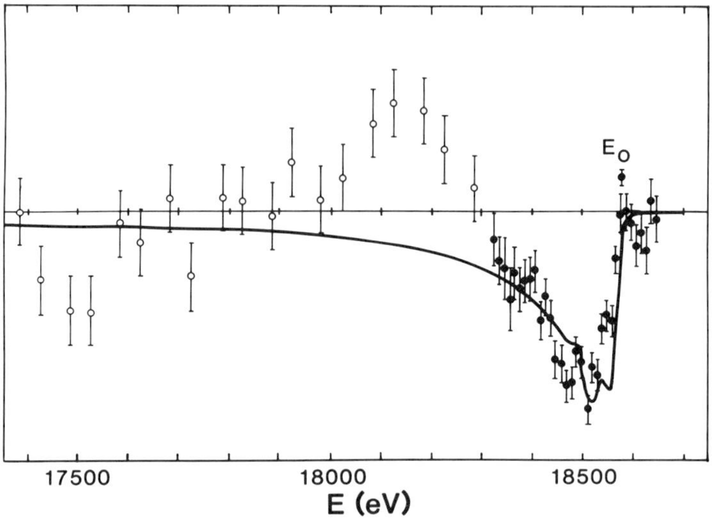

FIGURE 2. ITEP data of 1983, expressed as deviations from a curve with $m_\beta = 0$. The solid line is the best fit for a single neutrino with $m_\beta = 33$ eV. (Units on the ordinate are not given.)

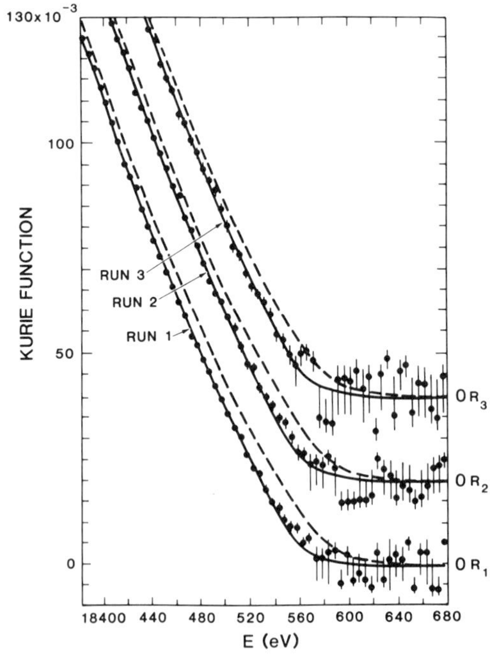

FIGURE 3. ITEP data of 1984. The solid curves are for the best-fit value of 35 eV and the dashed curves are for $m_\beta = 0$. Runs with three separate sources are shown.

model-independent lower limit ("nucleus" distribution) is 9 eV (90% C.L.). Until a detailed account of the new work is published, we can only speculate on whether problems may still remain.

In our opinion, the most serious concern is that Kurie plots are never straight (for an exception, see Simpson[5]). The reasons are as follows:

(1) The efficiency of the spectrometer may vary with energy. By accelerating the betas to a fixed energy, the variation of detector efficiency can be eliminated, but it is replaced by a variation in spectrometer acceptance.
(2) The energy calibration may be nonlinear. This may be due to instrumental problems or binning errors (Bennett *et al.*[1]).
(3) Scattering in the source leads to an excess of low-energy counts.
(4) Nuclear recoil loss changes from a maximum of 3 eV at the endpoint to smaller average losses further down in the spectrum.[16]

One is extremely sensitive to errors in the extrapolated endpoint energy; depending on details of the spectrum, an endpoint error of 10 eV (due to spectral distortions rather than absolute error) can account entirely for a 35 eV neutrino mass. As has been pointed out by Bennett *et al.*,[13] a proper treatment of energy calibration and spectrum curvature is thus important. We have some reassurance on this point in that the ITEP group gets essentially the same results from the last 330 eV of the spectrum as from the last 1690 eV in the 1984 data set.[4]

The ^{3}He-T Mass Difference

Although not required for deriving the neutrino mass, the ^{3}He-T mass difference clearly provides an important constraint. The final state spectrum used fixes $\overline{V}_n$ and therefore must give the correct endpoint energy. Also, experimental problems might be revealed through an endpoint discrepancy. Finally, using a fixed value for the endpoint energy can in principle improve the accuracy with which m_β can be derived. FIGURE 4 shows that the situation is not very satisfactory. The data of Bergkvist[17] and of Tret'yakov *et al.*[2] have been corrected for changes in calibration energies and the datum of Simpson[5] is not included because of probable systematic errors (Dixon *et al.*[20]). There is good agreement between the ITEP data[2,4] and the Ion Cyclotron Resonance (ICR) result of Lippmaa *et al.*,[21] which lends some support to the ITEP claims. However, other ICR data (not shown) are not in agreement and most likely suffer from systematic errors. Further independent mass measurements are obviously desirable.

Other Tritium Experiments

At this writing, we are aware of 17 groups involved in (or intending to be involved in) tritium beta-decay measurements of m_β. For reasons of brevity, we will simply tabulate our information (see TABLE 1), with the disclaimer that some of it may now be inaccurate and/or out-of-date.

One experiment[22] that is under development at Los Alamos National Laboratory is to study the spectra of free atomic and molecular tritium. With these sources, one can

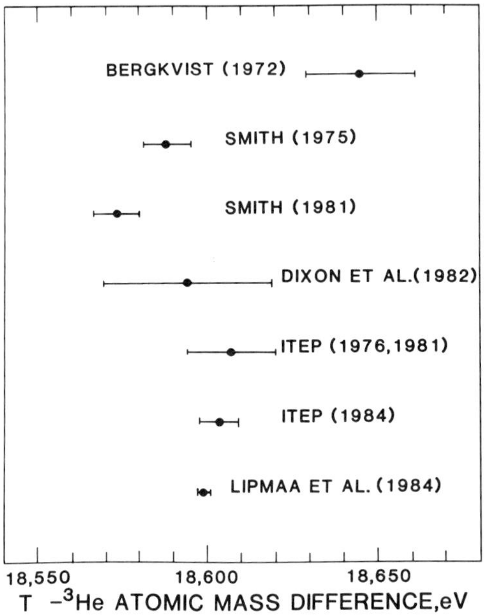

FIGURE 4. Difference between atomic masses of ^{3}H and ^{3}He. Data are from Bergkvist,[17] Smith,[18] Smith,[19] Dixon *et al.*,[20] the ITEP group,[2,4] and Lippmaa *et al.*[21]

eliminate uncertainties associated with the final-state distribution. For reasons of greater familiarity, we will describe this experiment in more detail.

The apparatus, shown in FIGURE 5, consists of three principal sections: a source region, a pumping restriction and extraction region, and a toroidal magnetic spectrometer. The basic concept of the experiment is to measure the beta-decay spectrum of an

TABLE 1. Tritium Experiments

Investigator	Location	Spectrometer	Source	Resolution (eV)
Lyubimov	ITEP	Toroidal Magnetic	Valine-T	20
Bergkvist	Stockholm	Iron-core Magnetic	Al-T	25
Kundig	Zurich	Toroidal Magnetic	C-T	12
Jelley	Oxford	Cylind. Mirror	Na-palmitate-T	15
Fackler	Livermore	Retarding E-S	T_2 Solid	1.3
Stoeffl	Livermore	Toroidal Magnetic	T,T_2 Gas	6
Boyd	Ohio State	Retarding E-S	T_2 Solid	12
Graham	Chalk River	$\pi\sqrt{2}$ Magnetic	Ti-T	18
Bowles, HR	Los Alamos	Toroidal Magnetic	T,T_2 Gas	30
Ohshima	Tokyo	$\pi\sqrt{2}$ Magnetic	?	30
Clark	IBM	Retarding E-S	?	?
Hughes	Yale	?	T_2 Gas	?
Kalbfleisch	Oklahoma	?	?	?
Moore	McGill	Retarding E-S	?	?
Lobashov	Moscow	Retarding E-S	T,T_2 Gas	8
Simpson	Guelph	Solid-State Si	Si-T	250[a]
Hansen	Berkeley	Solid-State Ge	Ge-T	250

[a]Experiment completed (reference 5); $m_\beta \leq 65$ eV (95% C.L.).

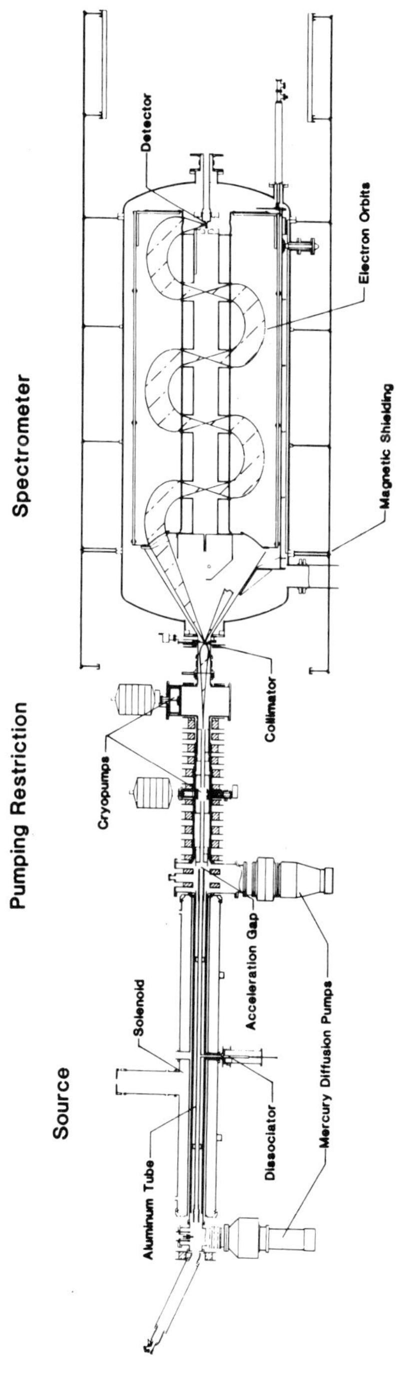

FIGURE 5. Cross section of the Los Alamos atomic tritium experiment. The overall length of the apparatus is 16 meters.

intense gaseous atomic tritium source. An extended source is required to reach a sufficient source intensity. Placing this source in a strong axial magnetic field with a pinch at one end allows one to extract most of the betas from the tritium decays into the extraction region. The pumping restriction and extraction region serves two purposes: to stop tritium molecules from reaching the spectrometer and to focus the betas emerging from the source region into the spectrometer. The toroidal spectrometer analyzes the momentum of the betas, allowing a precise measurement of the tritium energy spectrum near the endpoint.

The source region consists of a 4-m long, 3.8-cm diameter highly polished aluminum tube that is pumped by Hg diffusion pumps located at both ends. An RF dissociator produces the atomic tritium gas that enters the tube at the midpoint between the two pumps. The tube is cooled to a temperature of about 120 kelvin to minimize recombination of the atomic gas. The expected integrated source intensity will be 10^{15}–10^{16} atoms/cm^2 of 85–90% pure atomic tritium.

Sapphire standoffs allow the aluminum tube to be thermally regulated, but electrically isolated. Because of this, the potential of the tube can be varied up to 20 kV. Hence, the betas from the source region can be differentiated from background tritium decays that occur elsewhere in the system. Another advantage of varying the source potential is that it is possible to step the source voltage instead of the spectrometer magnetic field.

The source tube is enclosed in a superconducting solenoid that produces a field on the order of 1.5 kG with a small axial gradient. The electrons are constrained to spiral about the field lines, thus allowing 93% of the betas from tritium decays in the source region to enter the extraction region.

The pumping restriction and extraction region is a differential pumping system that guides electrons to the focus, while limiting tritium contamination of the spectrometer to 2.5 mCi/day. The accelerated beta particles from the source are focused to a collimator located at the object of the spectrometer. The collimator limits the acceptance of the spectrometer so that no betas that emerge from or strike the source walls enter the spectrometer. The extraction efficiency has been measured to be 25% with a ^{169}Yb source and is in good agreement with ray-tracing calculations.

The design of the toroidal spectrometer, an improved version of an earlier design by Tret'yakov,[11] has been described previously.[23] The resolution of the spectrometer is determined mainly by source diameter and is calculated to be about 30 eV FWHM for 25 keV electrons and a 1 cm diameter source. A two-axis cosine coil active magnetic shielding system surrounds the spectrometer and cancels the ambient magnetic field to the necessary 20-milligauss level. The detector is a single-wire proportional counter.

Understanding the system's characteristics and behavior is critical to obtaining unambiguous results. There is a variety of diagnostics to ensure that we understand our system. Off-line diagnostics include an electron gun to study orbits, wall effects, and resolution. Several solid sources and a gaseous ^{83m}Kr line source will also be used to study system resolution and extraction. On-line diagnostics include a detector to monitor total source activity, a UV laser to determine the molecular fraction of the source gas, and a residual gas analyzer to monitor gas composition.

The construction of the experimental apparatus is nearly complete. Data taking should commence during the next year. We expect a count rate of 1 Hz in the last 100 eV below the endpoint and an ultimate sensitivity to neutrino mass better than 10 eV.

^{163}Ho AND ^{158}Tb

In 1981, Bennett[24] and de Rujula[25] urged that a study of internal bremsstrahlung (IB) emission in electron capture (EC) be undertaken to look for spectral distortions near the endpoint caused by neutrino mass. Usually, IB is a relatively weak process, but under the right conditions,[6] a resonant enhancement process makes it practical for use in neutrino-mass searches. Just as in tritium beta decay, one searches for the loss of phase-space near the endpoint caused by neutrino mass.

The technique has been applied successfully to the EC decay of ^{193}Pt, from which a limit on m_β of 500 eV was obtained.[26] This is certainly the best limit obtained specifically for the electron neutrino, although CPT invariance requires that the limits obtained for $\bar{\nu}_\beta$ must also apply to ν_β. The decay Q-value for ^{193}Pt is 56.8 keV, and it appears that the ^{163}Ho might be a more favorable case because of a much lower Q-value, 2.3 ± 1.0 keV. After a great deal of experimental effort in determining the Q-value and half-life of this isotope, it appears that it will not be possible to achieve a result competitive with the ITEP measurements. In part, this is due to experimental difficulties, but there are more fundamental problems of interpretation. Although it was claimed by de Rujula[25] that atomic and molecular problems were insignificant, that argument considered only the low probability for shake-up and shake-off. A more serious difficulty discovered by Bennett[27] is that the tails of distant resonances, which lie close to the valence level and even in the continuum, have substantial effects on the IB spectrum shape. The effects probably cannot be calculated to sufficient accuracy. Another problem is the difficulty in measuring the instrumental response to sufficient accuracy; as Bennett observed,[27] the energies of X rays following electron capture and photoionization are slightly different because of differences in screening potentials in the two cases.

A second promising candidate was found by Raghavan,[28] who reported that the decay of ^{158}Tb proceeded to the 1187-keV level of ^{158}Gd by orbital K-capture and emission of a neutrino with an energy of only 156 ± 17 eV. A massive neutrino would alter the line shape of subsequent X rays and would even alter the total K-capture rate to this level by a few percent. Unfortunately, Raghavan's value for the energy release is in substantial conflict with mass measurements made by Burke[29] (by nuclear reactions) and by Barber[30] (by direct mass spectroscopy). Recently, Thomas *et al.*[31] have repeated the coincidence experiments of Raghavan with refined apparatus and find no evidence for the reported K-capture decay. It thus appears that ^{158}Tb does not have the reported low energy K-capture branch.

LIMITS ON MASSIVE NEUTRINOS FROM COSMOLOGY

There is abundant evidence from the dynamics of galaxies that the universe is filled with nonluminous matter.[32] While the precision with which the total mass can be determined is limited, the successful inflationary models and general philosophical considerations suggest that the universe is flat (closed, but not bound). The abundances of ^{2}H, ^{3}He, ^{4}He, and ^{7}Li all indicate that the baryon density is no more than 10% of that needed for closure and that the remaining 90% must therefore be nonbaryonic. The initial ITEP reports of a mass for ν_β between 14 and 46 eV were greeted with enthusiasm because a neutrino mass in this range could close the universe exactly. However, this appealing picture quickly ran into difficulties with the damping-out of the primordial density fluctuations needed to trigger the formation of galaxies. Relativistic neutrinos free-stream away from density knots and it is only at a late epoch that a 30 eV neutrino would have become nonrelativistic. By that time the scale of fluctuations is very large, and not only would one find galaxies to be much younger than they actually are, but the clustering of galaxies would be quite different from that observed. Briefly, a neutrino massive enough to produce the appropriate clustering would overclose the universe.

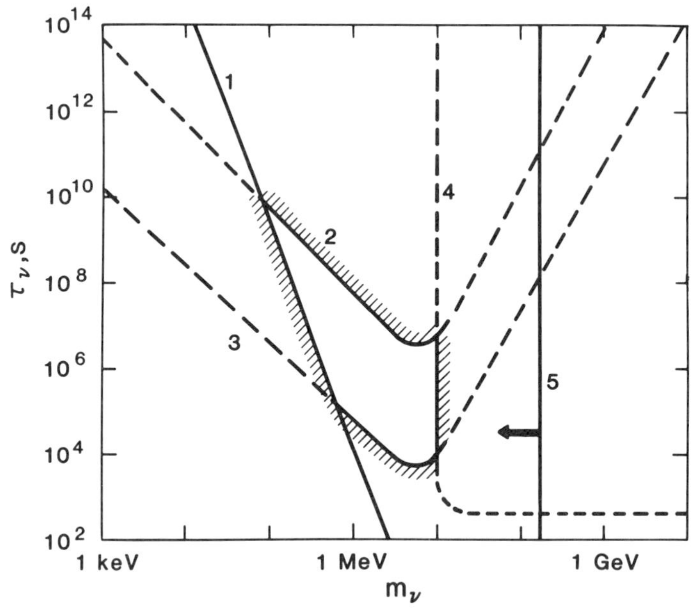

FIGURE 6. Constraints on the mass and lifetime of a "heavy" weakly decaying neutrino. See text for an explanation of the lines.

Hut and White[33] have reviewed the question of whether one can, using only particles known to exist, reach a cosmology consistent with observation. They conclude that if one neutrino is massive enough to close the universe and is stable, then another must be substantially more massive (in order to catalyze clustering early enough) as well as unstable (to avoid overclosing the universe). Furthermore, the only possible instability is of the form,

$$\nu_H \rightarrow \nu_L l l,$$

where l is a neutrino or charged lepton. FIGURE 6 shows more quantitatively how restrictive the limits on the properties of the heavy neutrino are. Boundary 1 is set by the plausible assumption that the coupling for neutrino decay will not be stronger than

G_F; Boundary 2 is the condition that the heavy neutrinos should have decayed at least by the time quasars formed; Boundary 3 is the lower limit of lifetime that permits clustering on the right scale; and Boundary 4 roughly (and conservatively) indicates limits established by Krauss[34] and Lindley[35] that rescue deuterium from destruction by inverse-Compton photons from the charged lepton decay products. The line 5 is the recently established[36] upper limit on m_τ (164 MeV). The experimental limit[37,38] on m_μ (0.25 MeV) almost eliminates it as a candidate and there is relatively little room even for m_τ. One has

$$0.1 < m_\tau < 10 \text{ MeV and } 10^3 < \tau_\tau < 10^9 \text{ s}.$$

Central to this conclusion is the belief that the mechanism of galaxy formation and clustering is adequately understood.

CONCLUSIONS

The present knowledge of the masses of the three known types of neutrino may be summarized as follows:

(a) ν_β: From recent measurements of the tritium beta spectrum,[4]

$$20 < m_\beta < 45 \text{ eV} \qquad \text{(model dependent)}$$

$$9 < m_\beta \qquad \text{(model independent);}$$

(b) ν_μ: From measurements[37,38] of the energy release in $\pi^+ \longrightarrow \mu^+\nu_\mu$,

$$m_\mu < 250 \text{ keV (90\% C.L.);}$$

(c) ν_τ: From observation[36] of the decay, $\tau \longrightarrow 3\pi^\pm\pi^0\,\nu_\tau$,

$$m_\tau < 164 \text{ MeV (95\% C.L.).}$$

REFERENCES

1. Witherell, M.; Declais, Y.; Vogel, P.; Schramm, D. 1985. This volume.
2. Lubimov, V. A., E. G. Novikov, V. Z. Nozik, E. F. Tret'yakov & V. S. Kosik. 1980. Phys. Lett. **94B:** 266; Kozik, V. S., V. A. Lyubimov, E. G. Novikov, V. Z. Nozik, E. F. Tret'yakov, V. S. Kozik & N. F. Myasoedov. 1981. Zh. Eksp. Teor. Fiz. **81:** 1158.
3. Boris, S., A. Golutvin, L. Laptin, V. Lubimov, V. Nagovizin, E. Novikov, V. Nozik, V. Soloshenko, I. Tichomirov & E. Tret'yakov. 1983. Proc. Intern. Europhysics Conf. on High Energy Physics, Brighton, 1983 (Rutherford Appleton Laboratory, 1983), p. 386, and preprint. Also presented at 1983 Intern. Symp. on Lepton and Photon Interactions, Cornell.
4. Lyubimov, V. A. 1984. Proc. XXII Intern. Conf. on High Energy Phys. (Leipzig).
5. Simpson, J. J. 1981. Phys. Rev. **D23:** 649.
6. Ching, C. R. & T. H. Ho. 1984. Phys. Rep. **112:** 1.
7. Williams, R. D. & S. E. Koonin. 1983. Phys. Rev. **C27:** 1815.
8. Martin, R. L. & J. S. Cohen. 1984. Private communication.
9. Kaplan, I. G., V. N. Smutnyi & G. V. Smelov. 1983. Zh. Eksp. Teor. Fiz. **84:** 833.
10. Tret'yakov, E. F., N. F. Myasoedov, A. M. Apalikov, V. F. Konyaev, V. A. Lyubimov & E. G. Novikov. 1976. Izv. Akad. Nauk SSSR, Ser. Fiz. **40:** 2026.
11. Tret'yakov, E. F. 1975. Izv. Akad. Nauk SSSR, Ser. Fiz. **39:** 583.
12. Simpson, J. J. 1983. Proc. Intern. Conf. ICOMAN '83 (Frascati, Italy 1983).

13. BENNETT, C. L., A. B. MCDONALD, P. T. SPRINGER, T. E. CHUPP & M. L. TATE. 1985. Phys. Rev. **C31:** 197.
14. ROBERTSON, R. G. H., T. J. BOWLES, J. C. BROWNE, A. G. LEDEBUHR, R. MILLS, M. MALEY, D. KNAPP, D. TAYLOR & J. WHEATLEY. 1981. Annual Report, Michigan State University Cyclotron Laboratory, p. 78.
15. FREEDMAN, M. S. 1974. Annu. Rev. Nucl. Sci. **24:** 209.
16. REPKO, W. W. & C. E. WU. 1983. Phys. Rev. **C28:** 2433.
17. BERGKVIST, K. E. 1972. Nucl. Phys. **B39:** 371.
18. SMITH, L. G. & A. H. WAPSTRA. 1975. Phys. Rev. **C11:** 1392.
19. SMITH, L. G., E. KOETS & A. H. WAPSTRA. 1981. Phys. Lett. **102B:** 114.
20. DIXON, W. R., W. F. DAVIDSON, R. S. STOREY & D. M. REHFIELD. 1982. Phys. Lett. **108B:** 265.
21. LIPPMAA, E., R. PIKVER, E. SUURMAA, J. PAST, J. PUSKAR, I. KOPPEL & A. TAMMIK. 1985. Phys. Rev. Lett. **54:** 285.
22. KNAPP, D. A., T. J. BOWLES, J. C. BROWNE, T. H. BURRITT, J. S. COHEN, J. A. HELFFRICH, M. P. MALEY, R. L. MARTIN, R. G. H. ROBERTSON & J. F. WILKERSON. 1984. Los Alamos National Laboratory Preprint LA-UR-84-3980. Proceedings of Neutrino Mass Miniconference. Telemark, Wisconsin. V. Barger & D. Cline, Eds. (1985). World Scientific. Philadelphia.
23. BOWLES, T. J., R. G. H. ROBERTSON, M. MALEY, J. C. BROWNE, T. BURRITT, M. STELTS, J. F. WILKERSON, J. HELFFRICH, D. KNAPP & A. G. LEDEBUHR. 1983. Am. Inst. Phys. Conf. Proc. **114:** 212.
24. BENNETT, C. L. 1981. Phys. Lett. **107B:** 19.
25. DE RUJULA, A. 1981. Nucl. Phys. **B188:** 414.
26. JONSON, B. 1983. Nucl. Phys. **A396:** 479.
27. BENNETT, C. L. 1984. Bull. Am. Phys. Soc. **29:** 666.
28. RAGHAVAN, R. S. 1983. Phys. Rev. Lett. **51:** 975, 1395.
29. BURKE, D. G. 1984. Phys. Rev. **C29:** 2339.
30. BARBER, R. C. 1984. Private communication.
31. THOMAS, J., J. LOSECCO, F. DEJONGH, B. YIP, J. GIMLETT, T. ALTZITZOGLU & R. A. NAUMANN. 1985. To be submitted for publication.
32. FABER, S. M. & J. S. GALLAGHER. 1979. Annu. Rev. Astron. Astrophys. **17:** 135.
33. HUT, P. & S. D. M. WHITE. 1984. Nature **310:** 637.
34. KRAUSS, L. M. 1984. Phys. Rev. Lett. **53:** 1976.
35. LINDLEY, D. 1979. Mon. Not. R. Astron. Soc. **188:** 15P.
36. MATTEUZZI, C. *et al.* 1984. Phys. Rev. Lett. **52:** 1869.
37. ABELA, R., M. DAUM, G. H. EATON, R. FROSCH, B. JOST, P-R. KETTLE & E. STEINER. 1984. Phys. Lett. **146B:** 431.
38. ANDERHUB, H. B. *et al.* 1982. Phys. Lett. **114B:** 76.

Status of Double Beta Decay

MICHAEL S. WITHERELL
Department of Physics
University of California, Santa Barbara
Santa Barbara, California 93106

INTRODUCTION

In recent years, there has been a resurgence of interest, both experimental and theoretical, in the physics of double beta decay. In order to understand this, it is useful to review the physics involved. Double beta decay can occur when an even-even nucleus, energetically unable to decay to the neighboring odd-odd nucleus, decays instead by the process,

$$(A, Z) \rightarrow (A, Z + 2) + 2e^- + 2\bar{\nu}_e.$$

This process, called two-neutrino double beta decay, $(\beta\beta)_{2\nu}$, is an ordinary second order weak interaction with a correspondingly long lifetime. In addition, the more interesting neutrinoless mode, $(\beta\beta)_{0\nu}$, may also occur:

$$(A, Z) \rightarrow (A, Z + 2) + 2e^-.$$

Neutrinoless double beta decay is allowed only if the lepton number is not conserved and the neutrino is a Majorana particle (self-conjugate). In addition, $(\beta\beta)_{0\nu}$ decay requires that the neutrino helicity not be exactly -1. Otherwise, the emitted neutrino would have the wrong helicity to be reabsorbed at the second weak vertex.

At first glance, the existence of $(\beta\beta)_{0\nu}$ seems doubly unlikely since two of the standard properties of the weak interaction must be violated. However, the two violations are not independent. There is no observable distinction between Dirac and Majorana neutrinos if the neutrinos are massless and the weak interaction is pure V-A. In other words, all tests of lepton conservation for the neutrino (such as nonobservation of $\bar{\nu}_e + \text{n} \rightarrow e^- + \text{p}$) are explained by helicity alone.

The reason that the Majorana nature of the neutrino has received a great deal of attention lately is that in grand unified theories, lepton number is violated (as well as baryon number). It is usual in these theories to have a Majorana mass term and it is natural for $m_\nu \ll m_e$. Thus, if the neutrino has mass, it may well be a Majorana particle. The only experiment that is sensitive to Majorana neutrino masses as small as a few eV is neutrinoless double beta decay.

The two-neutrino mode of double beta decay has a lifetime that should be calculable and it is expected to be in the range of 10^{19}–10^{23} years. If the neutrinoless mode were not suppressed, it would be favored over two neutrinos by a phase-space factor of about 10^6. The amplitude for $(\beta\beta)_{0\nu}$ is proportional to the admixture of the wrong helicity state of the neutrino. If the decay is allowed because of a Majorana neutrino mass, then the lifetime is proportional to m_ν^{-2}.

DOUBLE BETA DECAY EXPERIMENTS

There are two types of experiments to study double beta decay—geochemical experiments and those designed to observe the decay directly. In the geochemical experiments, one looks for a daughter nucleus, which is a noble gas, occluded in rocks of known age. The live time in such an experiment is about one billion years. The drawback is that those experiments can only measure the total rate of double beta decay and will not be sensitive to $(\beta\beta)_{0\nu}$ if it has the smaller decay rate. The most notable geochemical experiment is one that measures the ratio of $t_{1/2}(^{128}\mathrm{Te})/t_{1/2}(^{130}\mathrm{Te})$. This ratio is about $10^{3.8}$ if both are dominated by the two-neutrino mode, but only $10^{1.9}$ if the neutrinoless decay dominates. Possible systematic effects in the absolute rates cancel in this ratio. The experimental result is consistent with pure 2ν decay.[1] However, the absolute ratio is much lower than that calculated for the two-neutrino mode.

In direct experiments, most of the work in the last two years has been on the decay, $^{76}\mathrm{Ge} \rightarrow {}^{76}\mathrm{Se} + 2e^-$. In early $\beta\beta$ experiments, the thin sources needed for low background limited the amount of source material. Germanium detectors, with the active nucleus distributed as an 8% component throughout, solve this problem. The detectors have an energy resolution of about 0.1%, which gives a powerful background rejection when looking for the line from the two electrons in neutrinoless decay. In addition, the germanium is highly purified and therefore relatively free of natural radioactivity. Finally, the energy release in the decay is 2040.7 ± 0.5 keV,[2] which is high enough for a reasonable rate and for low backgrounds. Thus, germanium makes an excellent source/detector for $(\beta\beta)_{0\nu}$ decay. One simply looks for a peak in the energy spectrum at 2.04 MeV with as little background as possible.

To reach the sensitivity desired in ^{76}Ge experiments, a large amount of active germanium is required. For example, a half-life of 10^{24} years corresponds to one count in three years for a 130 cm^3 detector. Thus, all experiments now in progress involve multiple counter arrays and great efforts have been taken to reduce background. If you mount a standard germanium detector aboveground and count, the background rate at 2 MeV is about 10^{-3} counts/keV-hour-cm^3 of Ge. A 1200 cm^3 array counting for 100 hours would give 4000 counts within the resolution line, so one is looking to suppress background by a factor of about 10^4. The most difficult cosmic ray background is from neutrons produced in muon interactions, which in turn produce capture gammas. The other source of background is natural radioactivity, which can come from the surrounding matter, from radon gas, or from impurities in the apparatus. Gamma rays from either source can Compton scatter in the germanium, thus depositing energy in the energy region around the possible 2.04 MeV peak.

As an example of the steps taken to reach the background level needed for this experiment, I will use the UCSB-LBL experiment.[3] The detectors are completely surrounded by a 6″ thick box of NaI counters, thus providing an efficient Compton shield. This in turn is shielded from the surroundings by a neutron absorber and lead of low radioactivity. A minimum of inert material is used inside the shield, all of low atomic number, to minimize absorption of scattered gammas. Finally, it is necessary to go underground to achieve the lowest possible background level. The background levels achieved in this and similar underground ^{76}Ge experiments are the lowest reached in any counting environment at energies around 1 MeV. The experience and results are especially interesting for other experiments looking for rare signals at 1 MeV and below, such as electronic detection of solar neutrinos.

RESULTS FROM ^{76}Ge EXPERIMENTS

The original experiment to study double beta decay in germanium detectors was the Milan group, which sets a limit, $t_{1/2} > 5 \times 10^{21}$ years.[4] The results of recent experiments are shown in TABLE 1. Detector fiducial volumes are given and all backgrounds are given in counts per keV per 1000 hours per 100 cm^3 of Ge. The limits are at 68% confidence, all using a consistent technique. All, but the Guelph/Aptec/Kingston experiment and the Milan experiment, achieved these results aboveground. The Milan group, by virtue of their very long running time, has the best present limit, 1.2×10^{23} years.

All of these groups are continuing to improve the sensitivity of their experiments. The Battelle-USC group has achieved a background level of 0.15 operating in the Homestake mine. They are building a new array of total volume of 1400 cm^3 for the next version. The Caltech group has moved their detector to the Gotthard tunnel in Switzerland, thus lowering the background to about 0.2, and are building an 1100 cm^3 array. The Osaka group has moved their detector to the Kamioka mine. The Guelph group, running in a salt mine, is adding more detectors to make a 600 cm^3 array. UCSB-LBL has installed 600 cm^3 in an underground site at the Oroville, California dam. The background level is now about 0.15 and an additional 600 cm^3 of detectors will complete the detector. Finally, the Milan group continues to collect data with their 250 cm^3 detector in the Mont Blanc tunnel.

TABLE 1. Results from ^{76}Ge $\beta\beta$ Experiments

Group	Reference	Volume (cm^3)	Background	Live Time (10^3 hrs.)	$t_{1/2}$ Limit (10^{22} yrs.)
Battelle-USC	5	125	6	4.0	2
Caltech	6	90	1.8	3.8	2
Osaka	7	160	1.6	2.0	2
Guelph/Aptec/Kingston	8	194	1.5	1.6	3
UCSB/LBL	3	300	0.5	1.6	5
Milan	9	116	1.4	14.5	12
		138	0.3	4.7	

This present round of experiments should push the limit on $t_{1/2}$ for $(\beta\beta)_{0\nu}$ to (0.5–1) $\times 10^{24}$ years over the next year. For example, 900 cm^3 of germanium with background of 0.15 reaches the present limit of 10^{23} years in about three weeks of running.

LIFETIME LIMITS AND NEUTRINO MASS

There are two problems that confuse the conversion of $t_{1/2}$ to an effective Majorana neutrino mass. Multiple Majorana neutrinos coupling to the electron can lead to cancellations that reduce the decay rate. There is also some uncertainty in the calculation of the nuclear matrix element. The calculations agree reasonably well and they give upper limits on the neutrino mass of about 5 eV for lifetime limits of around 10^{23} years.[10–12] There is a serious discrepancy, however, between lifetimes measured by geochemical means and the relevant calculations. The geochemical limit[13] for ^{82}Se is about six times longer than that calculated. Moe and Lowenthal reported a much shorter lifetime in a laboratory experiment, but an ongoing experiment by Moe using a TPC will soon be reporting a much improved result. Until the experimental situation is clarified, it is not clear how reliable the calculations are.

If the neutrino is massless, then by using the Los Alamos calculations,[10] the present limits on an admixture of RHC are about $\eta < 10^{-5}$. It should also be noted that the double beta decay results also set a limit on mixing if the neutrinos are Majorana, given by $|m_1 \cos^2\theta - m_2\cos^2\theta| < 5$eV, where the sign depends on the relative CP eigenvalue of the two Majorana neutrinos. Such a limit is more stringent than neutrino oscillation experiments for $m_2 > 100$ eV, or more stringent than the searches for peaks in $\pi \rightarrow e\nu$ decay or kinks in nuclear beta decay at higher mass.

CONCLUSIONS

The most sensitive test of lepton number violations in the neutrino system is neutrinoless double beta decay. In particular, the limit for the half-life of this process in ^{76}Ge has been pushed from 5×10^{21} years to about 10^{23} years. Within another year, another factor of five to ten in improvement can be expected. The present limit corresponds to a Majorana neutrino mass of around 5 eV, if present calculations are right. To reduce the uncertainty in the calculations, the lifetime for two-neutrino double beta decay must be well measured in a nucleus in which the calculation of the matrix element should be reliable.

To go beyond the limit of about 10^{24} years requires a breakthrough in experimental technique. Many groups are investigating the use of large volume wire chambers filled with ^{136}Xe, which has a 2.5 MeV energy for double beta decay. A novel approach has been suggested by Fiorini and Niinikoski,[14] using low temperature calorimetry. One of these, or another new approach, will be necessary to carry the search for lepton number violation in neutrinos further.

REFERENCES

1. KIRSTEN, T. 1982. Proceedings, Science Underground 396 (Los Alamos).
2. ELLIS, R. J., B. J. HALL, G. R. DYCK, G. A. LANDER, K. S. SHARMA, R. C. BARBER & H. E. DUCKWORTH. 1984. Phys. Lett. **136B:** 146.
3. CALDWELL, D. O., R. M. EISBERG, D. M. GRUMM, D. L. HALE, M. S. WITHERELL, F. S. GOULDING, D. A. LANDIS, N. W. MADDEN, D. F. MALONE, R. H. PEHL & A. R. SMITH. 1985. Phys. Rev. Lett. **54:** 281.
4. FIORINI, E., A. PULLIA, G. BERTOLINI, F. CAPPELANI & G. RESTELLI. 1973. Nuovo Cimento **13A:** 747.
5. AVIGNONE, F. T., R. L. BRODZINSKI, D. P. BROWN, J. C. EVANS, W. K. HENSLEY, J. H. REEVES & N. A. WOGMAN. 1983. Phys. Rev. Lett. **50:** 721.
6. FORSTER, A., H. KWON, J. K. MARKEY, F. BOEHM & H. E. HENRICKSON. 1984. Phys. Lett. **138B:** 301.

7. Ejiri, H., N. Takahashi, T. Shibata, Y. Nagai, K. Okada, S. Nakayama & N. Kamikubota. 1984. Osaka University preprint.
8. Simpson, J. J., P. Jagam, J. L. Campbell, H. L. Malm & B. C. Robertson. 1984. Phys. Rev. Lett. **53:** 141.
9. Bellotti, E., O. Cremonesi, E. Fiorini, C. Liguori, A. Pullia, P. Sverzellati & L. Zanotti. To be published in Phys. Lett.
10. Haxton, W. C., G. J. Stephenson & D. Strottman. 1981. Phys. Rev. Lett. **47:** 153; 1982. Phys. Rev. **D25:** 2360; Haxton, W. C. & G. J. Stephenson. Prog. Part. Nucl. Phys. In press.
11. Doi, M., T. Kotani, H. Nishiura & E. Takasugi. 1983. Prog. Theor. Phys. **69:** 602.
12. Zamick, L. & N. Auerbach. 1985. Phys. Rev. **C26:** 2185.
13. Moe, M. U. & P. D. Lowenthal. 1980. Phys. Rev. **C22:** 2186.
14. Fiorini, E. & T. O. Niinikoski. 1984. Nucl. Instrum. Methods **224:** 83.

Neutrino Oscillation Studies with Reactor Antineutrinos[a]

P. VOGEL

Physics Department
California Institute of Technology
Pasadena, California 91125

Neutrino oscillations will occur if at least some neutrinos are massive and if the weak eigenstates (the neutral partners of the charged leptons) are not identical with the mass eigenstates (the particles whose propagation in vacuum is characterized by a single exponential function). In analyzing experimental results, one often makes the simplifying assumption that it is sufficient to consider mixing of only two neutrinos, so that consequently the mixing can be characterized by a single mass splitting, $\Delta m^2 = | m_1^2 - m_2^2|$, and by a single mixing angle, θ. In such a case, the probability that a neutrino created in a given weak interaction eigenstate will remain in the same eigenstate after traveling the distance, L, is given by

$$P(E_\nu, L, \Delta m^2, \theta) = 1 - \sin^2(2\theta) \cdot \sin^2\left(\frac{1.27 \cdot \Delta m^2 \cdot L}{E_\nu}\right), \tag{1}$$

where E_ν is in MeV, L in meters, and Δm^2 in eV^2.

Nuclear reactors are intense sources of electron antineutrinos with energies up to about 8 MeV and the $\bar{\nu}_e$ flux can be determined for distances up to $\sim$100 m. By studying the variation of the $\bar{\nu}_e$ flux with distance and energy, one obtains sensitivity to rather small values of Δm^2 ($\Delta m^2 > 0.01$ eV^2). At the present time, no other terrestrial neutrino source reaches similar sensitivity. The determination of the mixing angle, θ, depends on the statistical and systematic uncertainties of the $\bar{\nu}_e$ flux; in practice, sensitivity to $\sin^2(2\theta) > 0.1$ has been achieved.

The inverse neutron beta decay, $\bar{\nu}_e + p \rightarrow n + e^+$, is the usual detection reaction. The cross section of this reaction is determined by the neutron lifetime, and up to the small recoil terms, $\sim E_\nu/M_{\text{nucleon}}$, the positron and $\bar{\nu}_e$ energies are related by the relation,

$$E_{\bar{\nu}} = E_{e^+} + M_n - M_p. \tag{2}$$

Thus, the measurement of the yield of positrons with a given energy at several distances from the reactor core allows one to determine the values (or limits) of the oscillation parameters, Δm^2 and $\sin^2(2\theta)$.

Two series of high statistics experiments of this kind have been performed so far. The first of them, by the Caltech-SIN-TU Munich collaboration, uses the power reactor at Goesgen (Switzerland) and reported results for $L = 37.9$ m[1] and 45.9 m.[2] An earlier measurement at the ILL reactor in Grenoble with $L = 8.76$ m,[3] performed by

[a]This work was supported by the U.S. Department of Energy under contract no. DE-AT03-81ER40002.

the Caltech-TU Munich-ISN Grenoble collaboration with an essentially identical detector, could be combined with the Goesgen result. No evidence for neutrino oscillations have been seen. The second experiment,[4] performed at the Bugey (France) reactor at $L = 13.6$ m and 18.3 m, sees an indication for neutrino oscillations.

In this paper, I shall discuss the compatibility of these two results. (See references 1–4 for the description of the detectors, data taking, and analysis.)

THE GOESGEN RESULTS

The observed positron spectra at 38 m and 46 m in units of counts per MeV and hour are shown in FIGURE 1. The background spectra, measured with reactor off, have been subtracted and several corrections (positron annihilation, bremsstrahlung, neutrino interaction in the target walls, etc.) have been applied. The full line in FIGURE 1 represents the "no oscillation" spectrum based primarily on the electron spectrum measurements of Schreckenbach *et al.*[5] The construction of this "no oscillation" spectrum is discussed below.

The Goesgen experiment is being continued at $L = 65$ m and the results will be available in the near future. The sensitivity of the 65-m positron spectrum to the oscillation parameters suggested by the Bugey group is illustrated in FIGURE 2.

The existing results can be analyzed in several ways, with various amounts of reliance on the measured or calculated reactor antineutrino spectra.

The interpretation of the two Goesgen measurements (38 m and 46 m) does not require any assumptions about the $\bar{\nu}_e$ spectrum. The fuel composition of the reactor is accurately known at all times and it differed by only small amounts (1.3% difference in the spectrum) during the two measuring cycles. The ratio of the integral positron

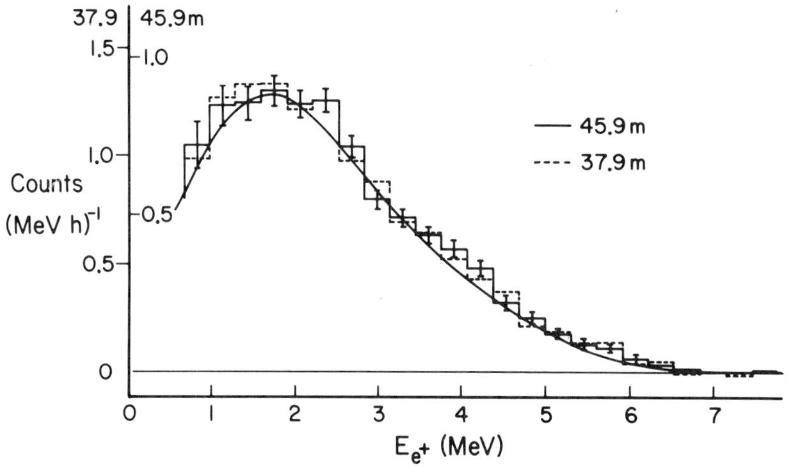

FIGURE 1. Positron spectra obtained by subtracting reactor-off from reactor-on spectra for the 38-m and 46-m experiments at Goesgen. The solid curve represents the predicted positron spectrum assuming no neutrino oscillations.

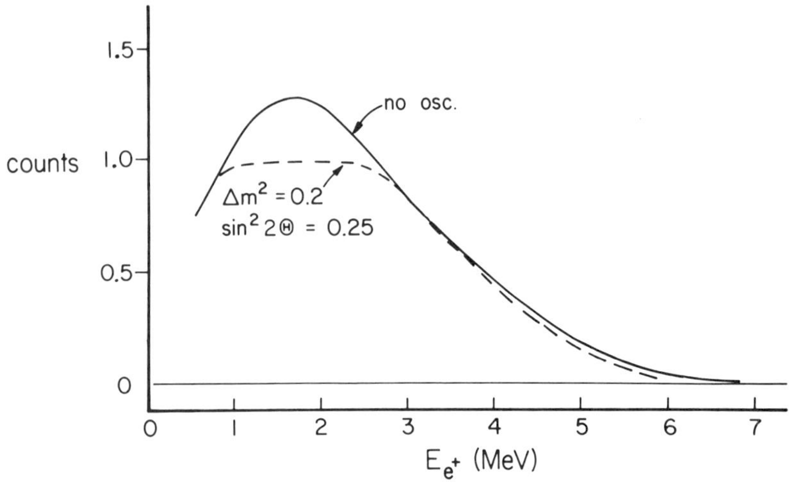

FIGURE 2. Illustration of the expected results at 65 m distance for the oscillation parameters suggested by the Bugey group.

yields, corrected for the trivial solid angle difference, is $Y(46)/Y(38) = 1.01 \pm 0.03 \pm 0.02$, which is compatible with the "no oscillation" hypothesis. The exclusion plot is shown in FIGURE 3 (dashed line). There is an overlapping region with the set of oscillation parameter values suggested by the Bugey experiment. This "allowed" region is centered around $\Delta m^2 = 0.2$ eV and $\sin^2(2\theta) = 0.25$.

INCLUSION OF THE ILL DATA

The ILL reactor uses enriched U-235 and only its fission contributes to the $\bar{\nu}_e$ spectrum. Therefore, we must know how different the antineutrino spectra of the two reactors are, or what their ratio actually is. This ratio is shown in FIGURE 4a, which is based on the measured spectra[5] of the main fuels, U-235 and Pu-239, and on the calculated spectra[6,7] of the secondary fuels, U-238 and Pu-241. It turns out that the ratio of the $\bar{\nu}_e$ spectra of different fissioning elements is less sensitive to the calculational details. This agrees better with the measured values than the spectra themselves. That conclusion is illustrated in FIGURE 4b, which is based only on calculations. One can see that the ILL and Goesgen spectra differ only by 4–15% and that various ways of determining this ratio give very similar results. When ILL and Goesgen data are combined, one obtains the dotted curve in FIGURE 3. The analysis[2] includes 3% uncertainty in the spectrum shape and 7% uncertainty in the overall normalization. There is a definite disagreement with the Bugey results, with no overlap of their respective 90% and 95% C.L. contours.

Note that inclusion of the ILL data does not depend on the absolute value of the detection reaction cross section (that is, on the neutron lifetime); it depends only weakly on the neutrino spectrum, as described above, and it also depends only on the relatively small change in the detector efficiency. (The detector has been moved and modified between the ILL and Goesgen runs, so an absolute efficiency calibration has been performed at both sites.)

INCLUSION OF THE REACTOR NEUTRINO SPECTRUM

If the reactor neutrino spectrum and the detector efficiency and response are known, the oscillation analysis can be extended and an equivalent of the $L = 0$ result can be obtained. For a power reactor, one has to add proper time averages of the spectra of various fuels (whose contribution to the reactor power is illustrated in FIGURE 5). The absolute spectra of the main fuels, U-235 and Pu-239, are needed, while the secondary fuels, Pu-241 and U-238, can be treated as described above. The various purely theoretical treatments of the neutrino spectrum[6,7] depend on assumptions regarding the beta decay of the rare short-lived fission fragments and also on the large amounts of experimental beta decay data, some of which may be unreliable.[8] Thus, it is preferable to use directly experimental spectra. The procedure is as follows: One measures the electron spectra associated with fission of the corresponding actinide element. An example[5] is shown in FIGURE 6 for the case of U-235. Because the electrons and electron antineutrinos are emitted simultaneously in each beta decay and because they together carry the full available decay energy, there is a one-to-one

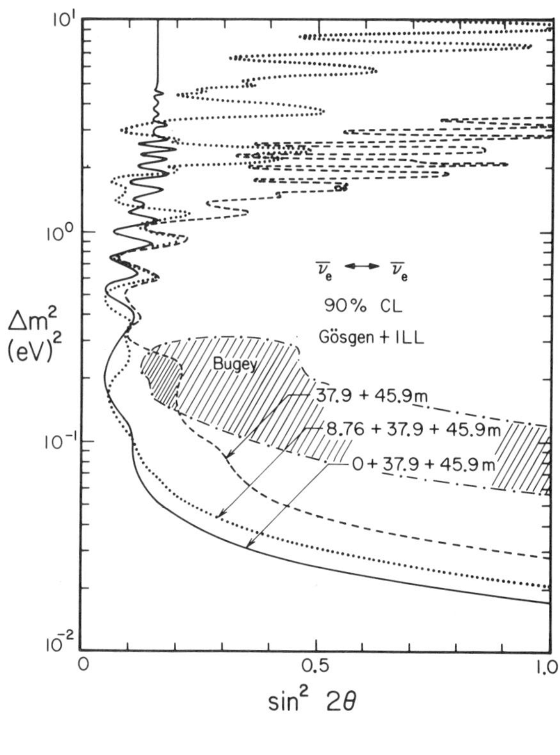

FIGURE 3. Exclusion plot obtained from the Goesgen and ILL experiments at 90% confidence level. The curve labeled 37.9 + 45.9 m refers to the limit obtained by using the 37.9 m and 45.9 m data only, for example. The 0-m label represents the on-line beta spectroscopic measurement (see text). The excluded region is to the right of these three curves. The allowed region of the Bugey experiment is inside the dot-and-dashed contour.

correspondence between the electron and $\bar{\nu}_e$ spectra. Thus, the experimentally determined electron spectrum can be converted into an antineutrino spectrum[5,6] essentially without loss of accuracy.

Note, in passing, the steep decrease of the spectrum in FIGURE 6. It is actually remarkable that nuclear structure calculations are able to describe at all a spectrum changing by four orders of magnitude. It is also clear why uncertainties increase with increasing energy.

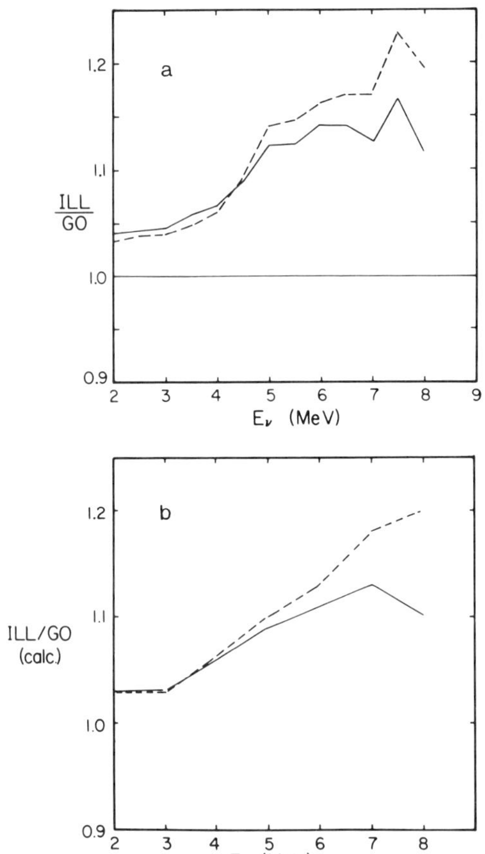

FIGURE 4. The ratio of the ILL and Goesgen neutrino spectra. The upper part (a) is based on the measured spectrum[5] of U-235 and Pu-239 and on the calculated spectrum of Pu-241 and U-238. The lower part (b) is based on calculations only. Full lines use the calculations of reference 6 and dashed lines use reference 7.

Let us enumerate the assumptions made when one wants to include the "no oscillation" spectrum in the analysis. First, one uses the electron spectra (and their errors) of Schreckenbach *et al.*[5] These spectra differ from some of the other ones obtained elsewhere under less favorable experimental conditions. However, there is no reason to assume that such a difference implies large uncertainties in the measured spectra. Secondly, one uses the experimental neutron lifetime and the experimental

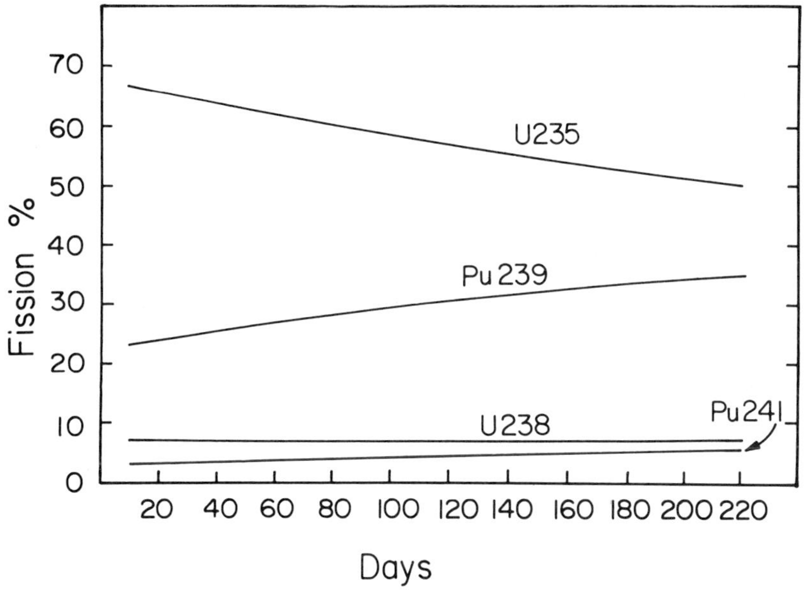

FIGURE 5. Relative contributions of the fissioning isotopes to the total number of fissions as a function of time since the beginning of the yearly refueling cycle (Goesgen reactor).

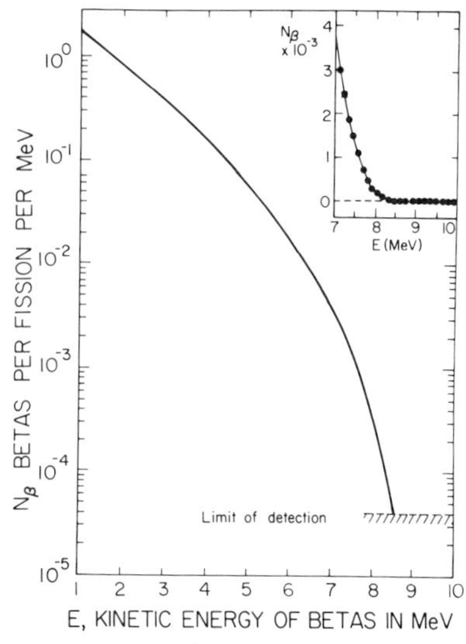

FIGURE 6. Measured electron spectra of fissioning U-235 according to reference 5. Error bars are indicated in the insert.

detection efficiency and response. The corresponding error bars are included in the analysis. Thirdly, one uses the spectra of the secondary fuels as described above. When all that is done, one obtains the solid curve in FIGURE 3, which again is in contradiction with the Bugey result.

FURTHER APPLICATIONS: NEUTRINO DECAY

The oscillation experiments described above lose their sensitivity if Δm^2 becomes large, which is due to the smearing of the oscillating part in equation (**1**). Indeed, only inclusion of the $L = 0$ "no oscillation" spectrum makes it possible to say anything at all about this regime. However, there are other ways to study the effect of massive neutrinos as first suggested by Shrock.[9] (The limits on the mixing angle obtained by various methods are reviewed in reference 10.)

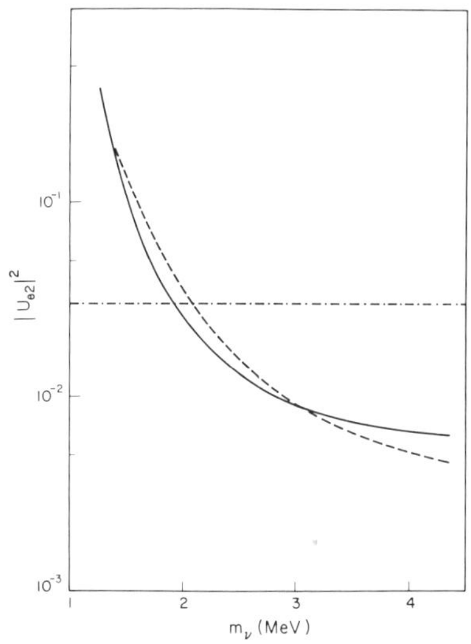

FIGURE 7. Limits of the mixing parameter, $|U_{ei}|^2 = \sin^2\theta$. The solid curve is for the threshold of 0.6 MeV (2 counts/sec limit) and the dashed curve is for the threshold of 2.0 MeV (0.5 counts/sec limit). The horizontal line is the 68% C.L. limit for the neutrino oscillation search.

The study of reactor neutrinos could also be used in this context. Assume that a heavy neutrino with $m_\nu > 2m_e$ couples with a small mixing angle (subdominant coupling) to the electron. Such a neutrino will be produced in all beta decays having a sufficient decay energy and, thus, it will also be produced in a nuclear reactor. Therefore, it will be a part of the reactor neutrino beam. In the detector used in the oscillation studies, the heavy neutrinos can decay by the process,

$$\nu \longrightarrow e^+ + e^- - \nu'$$

(ν' is the light neutrino dominantly coupled to electrons), and the kinetic energy of the e^+, e^- pair will be detected. Thus, the signature of such a heavy neutrino is an excess of the singles counts with reactor on. Experimentally, no such excess is observed at the level of 2 counts/sec for the 0.6 MeV energy threshold and the level of 0.5 counts/sec for the 2 MeV threshold at the Goesgen reactor. The interpretation of this result is only sketched here. (Details are presented in reference 11.)

The mixing angle, $|U_{ei}|^2 = \sin^2\theta$, appears in the problem twice. This includes once in the production where the probability of producing the heavy neutrino is proportional to $|U_{ei}|^2$ and depends on m_ν as well. Also, the decay rate in the rest frame of the heavy neutrino is again proportional to $|U_{ei}|^2$ and to m_ν^5 in analogy to the muon decay. Combining the production and decay rates and taking into account the kinematics of the situation,[11] one obtains the limits shown in FIGURE 7, which is better than the extrapolation of the oscillation results. Other restrictions on the mixing angle at this heavy neutrino mass range are based on the ^{76}Ge double beta decay and on the lack of the interplanetary positron flux associated with the decay of solar neutrinos;[12] additional assumptions must be made, however, in these cases.

ACKNOWLEDGMENT

It is a pleasant duty to thank Professor F. Boehm for his help in preparing this report.

REFERENCES

1. VUILLEUMIER, J-L. *et al.* CALTECH-SIN-TUM COLLABORATION. 1982. Phys. Lett. **114B:** 298.
2. GABATHULER, K. *et al.* CALTECH-SIN-TUM COLLABORATION. 1984. Phys. Lett. **138B:** 449.
3. KWON, H. *et al.* CALTECH-ISN-TUM COLLABORATION. 1981. Phys. Rev. **D24:** 1097.
4. CAVAIGNAC, J-F. *et al.* 1984. Phys. Lett. **148B:** 387.
5. SCHRECKENBACH, K. *et al.* 1981. Phys. Lett. **99B:** 251; VON FEILITZSCH, F. *et al.* 1982. Phys. Lett. **118B:** 162.
6. VOGEL, P. *et al.* 1981. Phys. Rev. **C24:** 1543.
7. KLAPDOR, H. V. *et al.* 1982. Phys. Rev. Lett. **48:** 127; Phys. Lett. **112B:** 22; METZINGER, J. 1984. Thesis. Heidelberg.
8. BEHR, J. A. & P. VOGEL. 1983. Nucl. Phys. **A411:** 199.
9. SHROCK, R. E. 1980. Phys. Lett. **96B:** 159; 1981. Phys. Rev. **D24:** 1232, 1275.
10. BOEHM, F. & P. VOGEL. 1984. Ann. Rev. Nucl. Part. Sci. **34:** 125.
11. VOGEL, P. 1984. Phys. Rev. **D30:** 1505.
12. TOUSSAINT, D. & F. WILCZEK. 1981. Nature **289:** 777.

Free Neutron-Antineutron Oscillations[a]

SERGIO P. RATTI

Dipartimento di Fisica Nucleare e Teorica
dell' Universitá
Istituto Nazionale di Fisica Nucleare
Sezione di Pavia
Via Bassi, 6 I27100 Pavia, Italy

INTRODUCTION

The present report is a new analysis of the scenario existing at Icoban-84.[1] The first experiment has been completed at Grenoble[2f] and the lower limits for $n\bar{n}$ transitions in nuclear matter have been given.[3]

Experimentation for the detection of possible neutron-antineutron oscillations using free neutrons produced by nuclear reactors has been in progress in Europe since the early eighties. A collaboration between the Institute von Laue-Langevin (ILL), the Rutherford-Appleton Laboratory, CERN, the University of Sussex, and the University of Padua[2] have completed an experiment at the 57 Megawatt Reactor in Grenoble. A collaboration between the Universities of Pavia, Rome (I and II), and CESNEF-Milan[4] is installing an experiment at the Triga Mark II, ¼ Megawatt Reactor at the University of Pavia. Contemporarily, a new experiment is being proposed at Grenoble, while another experiment using neutrons produced by dumping an intense proton beam of the new Moskow Meson Factory has been proposed by the Moskow group.[5]

At the best of my knowledge, no development has utilized the idea proposed by H. Yoshiki[6] of using 0.1 meV neutrons stored in totally reflecting neutron bottles for as long as their lifetime.

Finally, it seems that the experiments proposed in the United States have been postponed by the Adair Committee.[7]

PHYSICS SCENARIO

The neutron-antineutron oscillation phenomenon, although a pure experimental question, is more similar to the neutrino oscillation[8a] than to the $k\bar{k}$ case.[8] On the basis of this and several other different motivations,[9] this phenomenon is now strictly connected to the Grand Unification Theories.[10]

In 1973, Pati and Salam[10a] proposed to unify the strong, electromagnetic, and weak interactions by gauging a symmetry group associated with a multiplet containing quarks and leptons. Their model does not allow proton decay unless the quarks have a unitary charge. In 1974, Georgi and Glashow[10b] proposed an elegant Grand Unification Theory with a sensational sentence: "We present a series of hypotheses and

[a]This work was supported by the Ministero della Pubblica Istruzione.

speculations leading inescapably to the conclusion that $SU(5)$ is the group of the world—that all elementary particle forces (strong, weak, and electromagnetic) are different manifestations of the same fundamental interaction involving a single coupling strength: the fine structure constant. Our hypotheses may be wrong and our speculations idle, but the uniqueness and simplicity of our scheme are reasons enough that it be taken seriously." The proposed $SU(5)$ is a group that does not allow $n\bar{n}$ oscillations and predicts proton lifetimes of the order,

$$t(\text{proton}) = 10^{(29\pm2)} \text{ years.}$$

S.L. Glashow proposed $SU(5)$ on several occasions.[10c] Marshak and Mohapatra[10e,f] made proposals alternative to $SU(5)$; in particular,[10e–m] it seems that $SO(10)$ is a simple group that contains $SU(2)$ and $U(1)$ as subgroups. Its higher rank compared to $SU(5)$ implies that the $SO(10)$ group can accommodate intermediate mass scales between M_W and M_X, which are the two mass scales arising in minimal $SU(5)$ Grand Unification. One of these, M_C, if confined between 100 and 10,000 TeV, could lead to observable $n\bar{n}$ transition times of the order of 10–100 megasecs. The theoretical models leave a rather large uncertainty in the values expected for $t(n\bar{n})$.

Pati, Salam, and Strathdee[10o] put forward alternative routes for a spontaneous breaking of $SU(16)$ that allow for the coexistence of proton decay and $n\bar{n}$ oscillations. Therefore, a direct measurement of the free $n\bar{n}$ transition time is a measurement able to discriminate among several possible Grand Unification Theories.

The first generation of proton decay experiments measured a limit,[11]

$$t(\text{proton}) > 10^{32} \text{ years,}$$

at 90% C.L. for the decay, $p \rightarrow e^+ + \pi^0$, which is in disagreement with the $SU(5)$ prediction by more than one order of magnitude.

Neutron-antineutron oscillations are an interesting example of $(\Delta B = 2, \Delta L = 0)$ transitions, in addition to the obvious process of

$$NN \rightarrow \text{pions.}$$

The $n\bar{n}$ transition is related to a typical time,

$$1/t\ (n\bar{n}) = \langle \bar{n}\,|M|\,n \rangle,$$

where the matrix element, M, is defined by a proper effective Lagrangian.[12] The transition time is defined as

$$t(n\bar{n}) = \hbar/dm.$$

The $n\bar{n}$ mass splitting, dm, may depend to first order on the corresponding Hamiltonian, H′,[10d] through

$$1/t\ (n\bar{n}) = dm = \langle \bar{n}/H'/n \rangle \sim \sqrt{\Gamma(\Delta B = 2) \cdot M},$$

where $\Gamma\ (\Delta B = 2)$ is the width of the $\Delta B = 2$ process and M is the nucleon mass.

In the quark-lepton picture,[10] the simplest term in the effective Lagrangian able to induce a $\Delta B = 2$ process is a six quark term:

$$L = M_x^{-5}\,(qqq \cdot \overline{qqq}) + h.c.$$

Observation of $n\bar{n}$ oscillations up to 10^8 s will open up new physics in the mass range of 10–10,000 TeV.[10d–m]

Information on $n\bar{n}$ oscillations has been extensively obtained[13] using approaches based on the nuclear instability associated with the $n\bar{n}$ process, which is followed by antineutron annihilation with a surrounding nucleon inside the nuclear matter.

Lower limits can be derived from experimental lower bounds for the nuclear lifetime obtained from data collected at large underground proton decay setups.[3,13h] Such an approach is under constant theoretical refinement,[13e,g;14] which improves the treatment of the nuclear aspect of the problem. Furthermore, $n\bar{n}$ oscillation is not the only $\Delta B = 2$ process possible inside nuclear matter.[13i]

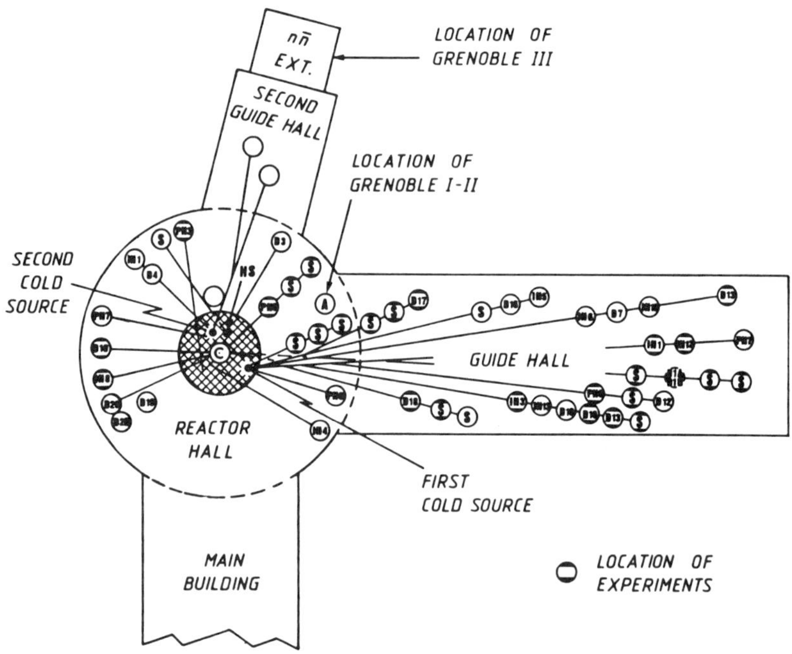

FIGURE 1. Sketch of the von Laue-Langevin (ILL) Laboratory. The location of Grenoble I–II is indicated as (A). The new annex to host Grenoble III is also indicated.

Therefore, there seems to be several problems to solve.[14–16] For instance, using the same data on energy shifts and widths of p-atomic levels, Dover *et al.*[13e] fit a real part of the antineutron potential, $V = 86$ MeV, and an imaginary part, $W = 161$ MeV, and Wong *et al.*[15] fit $V = 367$ MeV and $W = 34$ Mev. Such an indirect way of determining $t(n\bar{n})$ from nuclear instability[13g,14] rests on two unproven assumptions:

(a) that the $n\bar{n}$ transition mass inside the nucleus is the same as for free neutrons;[14a]

(b) that the nuclear instability is essentially a single nucleon process leading to antineutron conversion, with the remaining nuclear matter acting as a medium to generate an average field.

Sparrow[14b] pointed out that all published calculations use the Schrödinger equation for nucleons and antinucleons. The use of a relativistic field theory implies the existence of strong scalar and vector fields that require a Dirac equation treatment even at zero energy. The relativistic approach differs from the nonrelativistic ones since the strong fields enhance the lower components in the nuclear matter. Therefore, the nuclear decay rates due to $n\bar{n}$ oscillations could be very sensitive to such lower component contributions.

Thus, although in the context of a Schrödinger nucleon treatment the model dependence of the $n\bar{n}$ oscillation time seems to be small, a treatment based on Dirac nucleons accounting for strong scalar and vector fields appears to be needed.[14b]

Finally, Basecq and Wolfenstein[16] pointed out that some ($\Delta B = 2$) operators favor other [$NN \rightarrow$ mesons] processes in nuclei, instead of $n\bar{n}$ oscillations even in vacuum. Other details are also questioned by several authors.[14c–i] It is, therefore, at least natural to assume that experiments on free $n\bar{n}$ oscillations and estimates inferred from nuclear instability are independent searches of which both are of crucial character.

In this paper, I shall compare four experiments:

(1) Grenoble I–II[2]: the completed experiment giving $t(n\bar{n}) > 1$ megasecs.
(2) Pavia–Rome[4]: aiming at a sensitivity of 10 megasecs.
(3) Grenoble III[2c,d,17,18]: aiming at 100 megasecs.
(4) Moskow (Meson Factory)[5]: a real large experiment able in principle to reach a limit as high as 1–10 gigasecs.

THE EXPERIMENT COMPLETED AT ILL (GRENOBLE I AND II)[b]

The first measurement of the transition time, $t(nn)$, for free neutrons has been attempted by the CERN, ILL, Padua, RAL, and Sussex collaboration[2] at the 57 Megawatt Research Reactor at ILL in Grenoble. It was measured in two phases—the first one[2b] giving a limit,

$$t(n\bar{n}) > 0.1 \text{ megasecs},$$

and the second one[2f] giving

$$t(n\bar{n}) > 1 \text{ megasecs}.$$

In both phases (Phase I and II), cold neutrons are transported to the experimental area (FIGURE 1) by a totally reflecting guide, 3×20 cm^2 in cross section, that has a 25-m radius of curvature to kill direct gamma rays from the reactor core. The beam intensity in this experiment is $I = 1.5$ billion n/s with an average speed of $v = 160$ m/s. FIGURE 1 shows a schematic arrangement of the experimental halls. (This figure is

[b]For sake of brevity, I shall neither elaborate the phenomenology nor mention the possibility of performing experiments that use alternating driving magnetic fields, which was recently proposed (reference 19) for the investigation of the $n\bar{n}$ oscillation phenomenon (see reference 4d).

useful also for the discussion of the section entitled GRENOBLE III.) The neutrons propagate free from the Earth's magnetic field interaction along a straight section, 4.7 m long, followed by a diverging vessel, 2.7 m long. The vessel terminates in a ^{6}LiF beam dump, 54 cm in diameter and 0.3 cm thick, that is glued to an aluminum foil, which is 0.3 cm thick. The time of flight from their "last" collision inside the guide to the target is evaluated to be

$$t = 0.027 \text{ sec.}$$

During Phase I,[2b] the ^{6}LiF beam dump was used as the target for the antineutrons that were possibly created by the $n\bar{n}$ transitions. The detector used was a Pb-scintillator sandwich calorimeter (FIGURE 2a).

Phase II differs from Phase I essentially because the calorimeter has been replaced by a set of limited streamer tubes that surround the target on all sides, except for the side through which the beam enters the apparatus (FIGURE 2b). All planes of limited streamer tubes are sandwiched between 5-mm Al plates, except for the last ten planes of the forward chambers, which are sandwiched between 5-mm Fe plates. Eighteen planes are placed in the forward direction and four planes are placed all around the target so as to cover 70% of 4π.[2c,e,f]

The detector is completed by two logically separated systems of scintillation counters. The forward system, used for the trigger, is made of a front hodoscope (H_x, H_y) plus a set of counters; a lateral system is composed of a barrel hodoscope. Both hodoscopes are used for track reconstruction.

The whole apparatus is covered by a system of 66 anticoincidence counters that provide a veto signal against cosmic rays. For the rest, the apparatus is the same as that of Phase I.

In 3.8 megasecs, only one event, shown in FIGURE 3, has been found with its vertex inside the target area allowed for an $\bar{n}$ annihilation by the beam dimensions. No events have been found in that area; neither in 2.2 megasecs of running time with the condition of "reactor on - field on," nor in 3.1 megasecs with the reactor off.

The number of background events has also been evaluated from the total number of events with vertex inside the ^{6}LiF target. Since the antineutron annihilation vertices are expected only in about one-fourth of the total target area and since nine events have been found, the number of background events is estimated to be (2.3 ± 0.8). The observed antineutron annihilation candidate is considered to be consistent with the expected cosmic ray background.

After about 100 days of effective running time, the apparatus has been further improved: a thin carbon foil, 0.125 mm thick, has been placed inside the vacuum tank, ~15 cm before the ^{6}LiF absorber, and a set of scintillator counters has been added between the lateral chambers and the Fe shield.

Data have been taken with the new experimental conditions for 5.6 megasecs with the reactor on and the magnetic field off, for 2.3 megasecs with the reactor on and the magnetic field on, and for 2.1 megasecs with the reactor off. No more $\bar{n}$-annihilation candidate events have been found.

In conclusion, no evidence has been found for $\bar{n}$ annihilations in 9.4 megasecs effective running time in Grenoble (I and II). Having estimated from test run data the antineutron detection efficiency to be 0.3, the authors[2f] set a lower limit on the $n\bar{n}$

transition time:

$$t(n\bar{n}) > 1 \text{ megasec}$$

at a 90% confidence level.

THE PAVIA EXPERIMENT[c]

The experiment that the collaboration between the Universities of Roma (I and II), the Enrico Fermi Center (CESNEF) of the Milano Politechnique, and the University of Pavia is performing at the 250 Kilowatt Research Reactor of the Laboratory for Applied Nuclear Energy (LENA) at the University of Pavia (which is a search for free $n\bar{n}$ oscillations) is presently the only experiment in progress after the completion of Grenoble (I and II). Presented at ICOMAN-82,[4a] a number of parameters have been updated on several occasions.[4c–e]

The reactor used is by far the smallest source of neutrons proposed for $n\bar{n}$ oscillations experiments. However, the dimensions of the thermal column allows one to gain in surface what is lost in neutron flux.[d]

A new annex has been added to the laboratory (FIGURE 4). The opening of windows and access doors in the building required the certification of the National Agency (ENEA), which is the legal supervisor of all nuclear plants.

In spite of the fact that the building is not yet formally delivered and the nuclear plant is not yet formally certified, the experiment is running with about one-fourth of the detector. The neutron flight path (FIGURE 4a), in a vacuum tube, will have a residual magnetic field on the order of some milligauss for a length of 16.5 m [the length of the neutron channel (FIGURE 4b), ~1.2 m in diameter], which corresponds to a flight time of $t = 7.5$ milliseconds. The μ-metal shielding is being extended to the target and beyond to increase possibly the time of flight to $t = 8.4$ ms.

Repeated measurements with the multiple foil attenuation method give a current on the target (~1.3 m diameter) of about 10^{11} thermal neutrons, which comes directly from a plane source that is ideally reproduced at the entrance window of the vacuum pipe.

An unexpected photon background is being cured by adding a Pb frame around the original Bi-graphite gamma attenuator. However, it is placed at the entrance of the vacuum pipe (FIGURE 4b), which consists of two vertical slabs, 40 cm wide, 3.5 cm

[c]Coauthors of this section of the paper are: G. Bressi (1), E. Calligarich (1), R. Cardarelli (3), M. Cambiaghi (1), M. Conversi (2), G. de Zorzi (2), R. Dolfini (1), A. Gigli Berzolari (1), A. Lanza (1), G. Liguori (1), F. Massa (2), F. Mauri (1), A. Piazzoli (1), S.P. Ratti (1), R. Santonico (3), D. Scannicchio (1), F. Sebastiani (2), M. Terrani (4), P. Torre (1), and D. Zanello (2). 1 = Dipartimento di Fisica Nucleare e Teorica dell' Universitá and INFN di Pavia; 2 = Dipartimento di Fisica, "G. Marconi" Universitá, "La Sapienza"—INFN–Sezione di Roma; 3 = Dipartimento di Fisica, Universitá di Roma II Tor Vergata, and INFN–Sezione di Roma; 4 = Centro Studi Nucleari ed Esperienze "E. Fermi"–Politecnico di Milano.

[d]Under contract no. 2188-83-09TL-ISP-I, the authors acknowledge that the first five meters of the tunnel and the beam shutters have been built by Ccr Ispra.

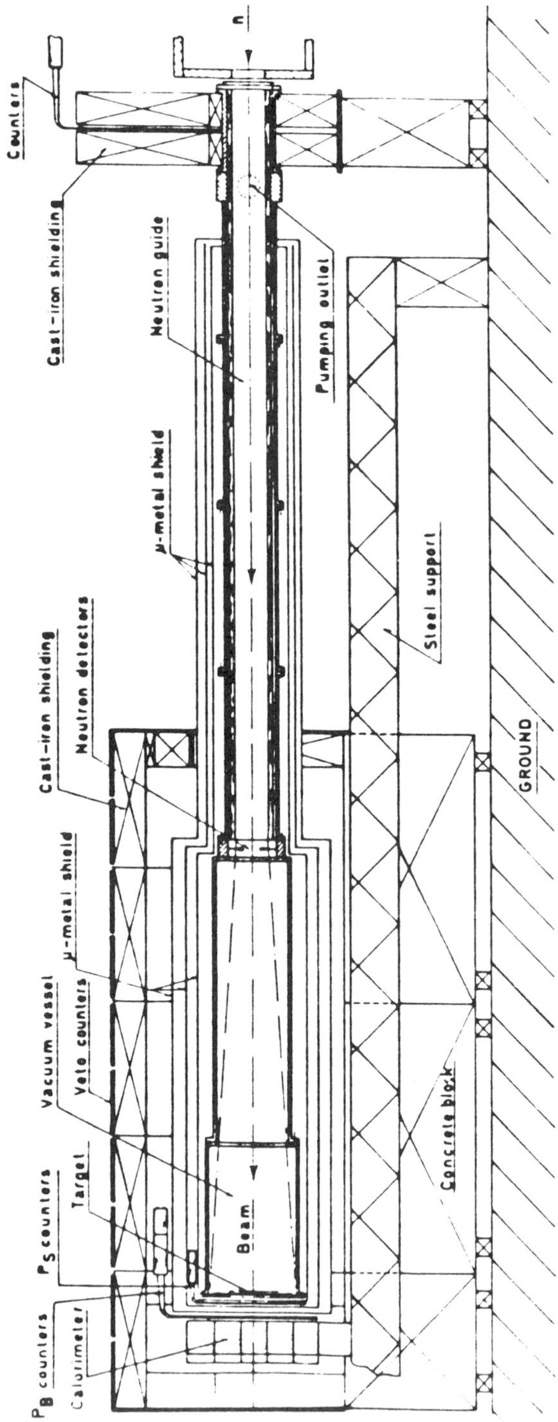

FIGURE 2a. The calorimeter-detector in Grenoble I.

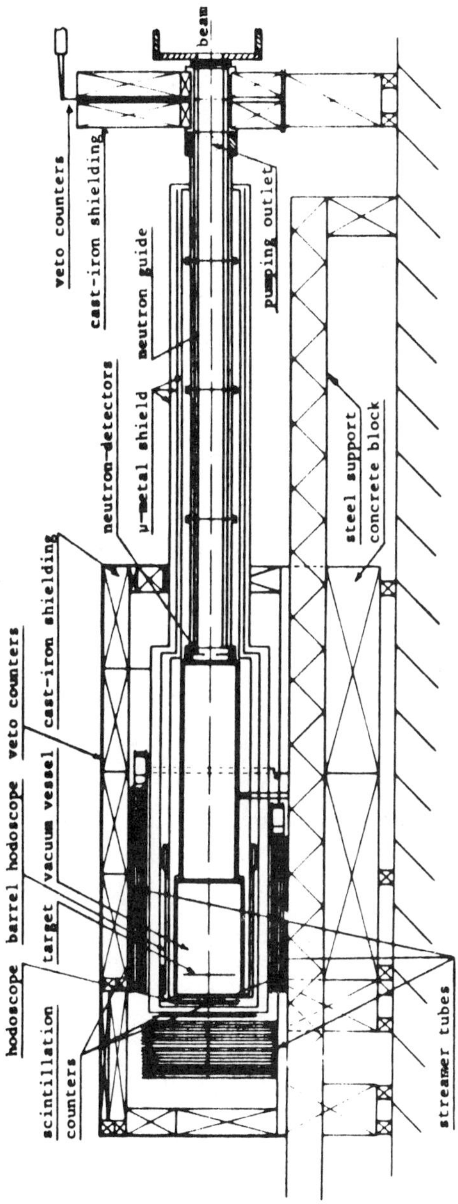

FIGURE 2b. The final setup in Grenoble II.

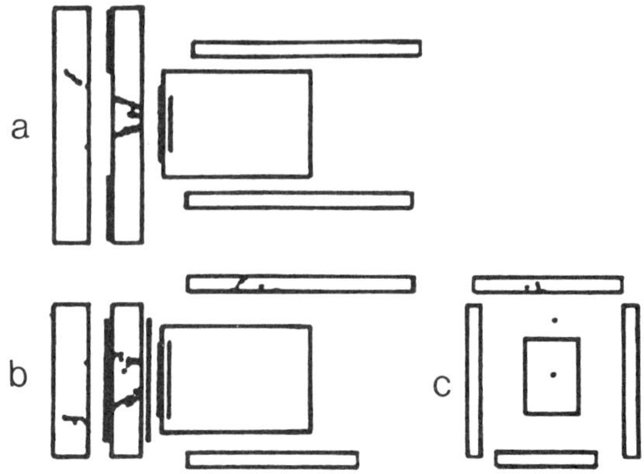

FIGURE 3b. Three views of the Grenoble I candidate: (a) horizontal: (b) vertical, parallel to the beam axis; and (c) vertical, orthogonal to the beam axis.

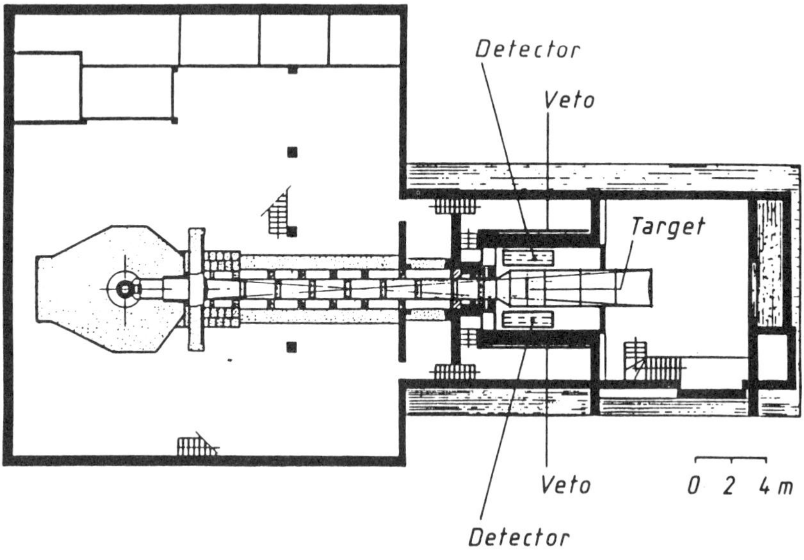

FIGURE 4a. The Pavia experiment: map of the Reactor Building and the new annex.

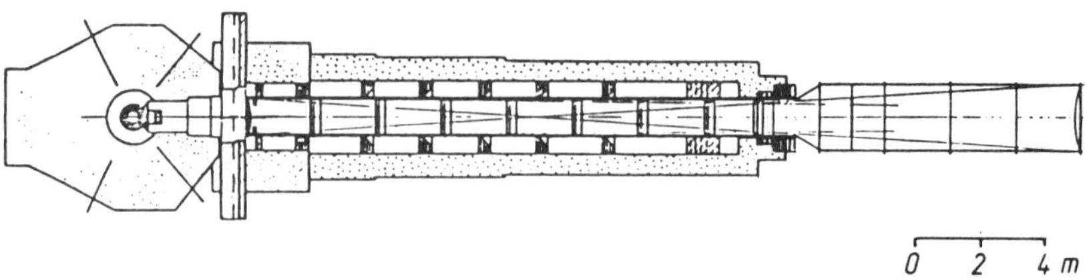

FIGURE 4b. The Pavia experiment: horizontal cut of the beam transport system. The Bi-graphite attenuator at the reflector and the additional Pb slabs at the vacuum tube entrance are indicated.

thick, that do not depress, so far, the neutron flux. The measured beam profile at the target region, shown in FIGURE 5, reproduces the Monte Carlo calculated shape of the beam.

The target is intended to be a set of two grafoil sheets, 125 microns thick, about 40 cm apart, that are located 18.5 m from the entrance window of the vacuum pipe and give permanent control of the signal to background ratio.

Cross sections of the whole detector are sketched in FIGURE 6. As already described,[4a] it is a vertex detector associated with a fine grain calorimeter that is made

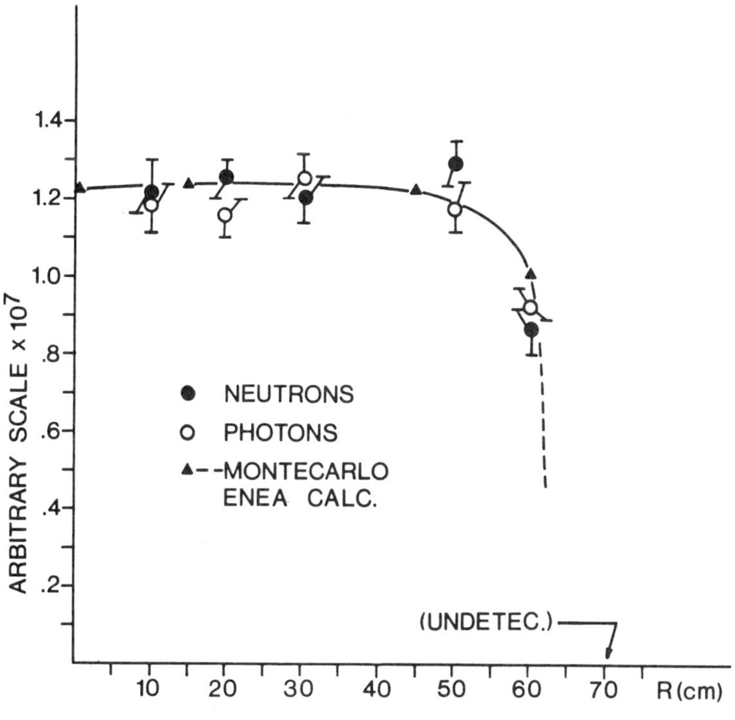

FIGURE 5. Measured neutron and photon fluxes at the target position as compared to Monte Carlo calculations (arbitrary scale).

of plastic flash chamber modules sandwiched with 0.5-cm Pb or 2.0-cm Fe absorber-converter plates for a total of ten radiation lengths.

The calorimeter, near completion, is made of four quadrants covering about 90% of 2π around the vacuum tube and about 65% of the total solid angle. On three sides, a concrete wall, 80 cm thick, separates the detector from the double-layer $120 \times 2\ m^2$ RPC veto system (FIGURE 6a).

Plastic material Flash Chambers, PFCs,[20a] alternated with metal plates for shower sampling and assembled in a flash chamber calorimeter, constitute the cheapest fine grain calorimeter that can be built using present technologies. They are already in use for large dimension experiments.[21a,b] In addition, systematic experimentation was carried out[21c] to improve the performances of PFCs for our experiment.

The spurious hits due to the overall background have been experimentally investigated, even in the present situation, which is not yet optimal (FIGURE 7). The firing probability per single cell, in percent, for 3.0-m cells, is (see the straight lines in FIGURE 7):

$$p(f) = (1.45 \pm 0.25)\ [t/\text{microsec}]\ [P/100\ \text{kwatts}]$$

for FCs close to the beam pipe (open points in FIGURE 7) and

$$p(f) = (0.33 \pm 0.10)\ [t/\text{microsec}]\ [P/100\ \text{kwatts}]$$

for FCs after the trigger counters (black points in FIGURE 7); t is the memory time in microseconds and P is the Reactor's power in kW. We are still confident that a further improvement of at least a factor of three can be achieved.

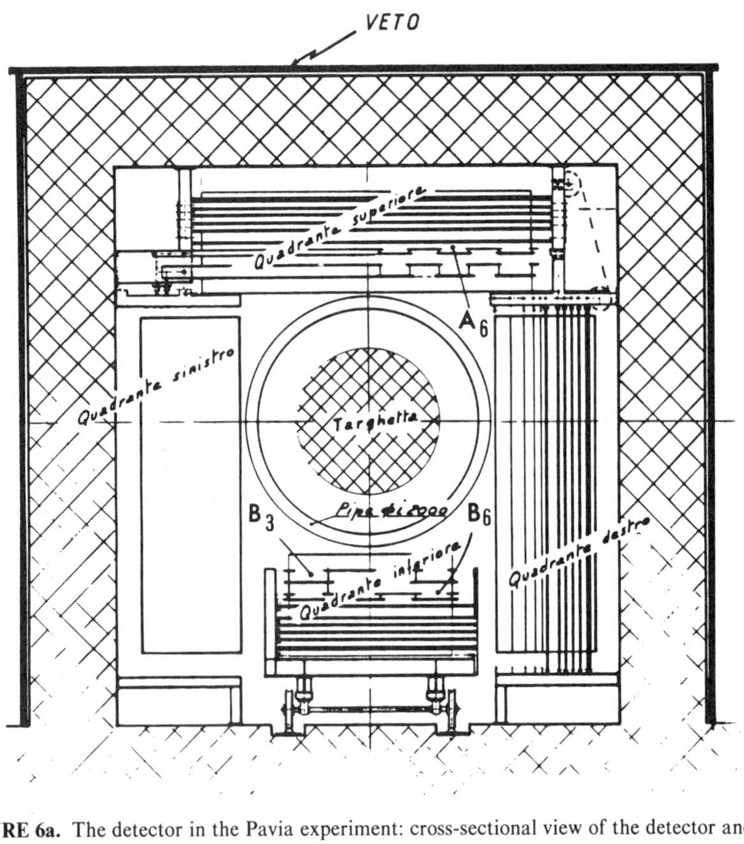

FIGURE 6a. The detector in the Pavia experiment: cross-sectional view of the detector and the veto.

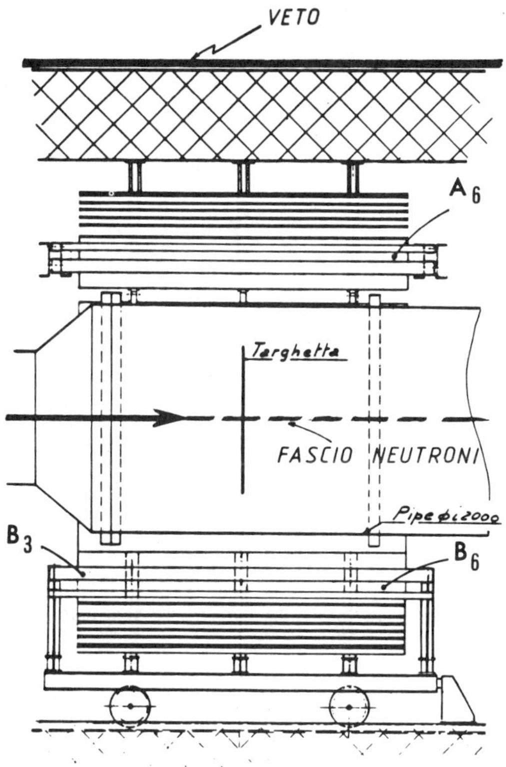

FIGURE 6b. The detector in the Pavia experiment: longitudinal cut in the target region.

The veto system is made of Resistive Plate Counters (RPC) recently developed by Santonico and Cardarelli.[22] They are fast, wireless, DC-operated counters made of parallel electrodes of high resistivity ($\sim 10^{11}$ ohm $\times$ cm), between which an argon-butane gas mixture [(70/30)%] flows at atmospheric pressure. At any particle crossing, they yield pulses of ~0.5 volts, 2–3 ns rise-time, on inductive pickup aluminum-strip lines, ~50 ohm impedance, which are placed on the external counter walls. They have been described on several occasions.[1,22c]

The trigger system is provided by three layers of 11 plastic scintillators per quadrant, 27 cm $\times$ 270 cm and 4 cm thick (six counters per plane for the bottom quadrant).

TABLE 1 summarizes the trigger efficiency situations with or without an additional RPC layer, at a background single counting rate of $S = 500$ kHz (first line) or $S = 100$ kHz (second line).

The trigger efficiency, $\sqrt{E_{TR}} = 0.7$, seems to be a reasonable assumption for our experiment. We assume, though, in our experiment:

(1) a corrected neutron current on target: $\dot{N}_c\,(n) = 10^{11}$ n/s;

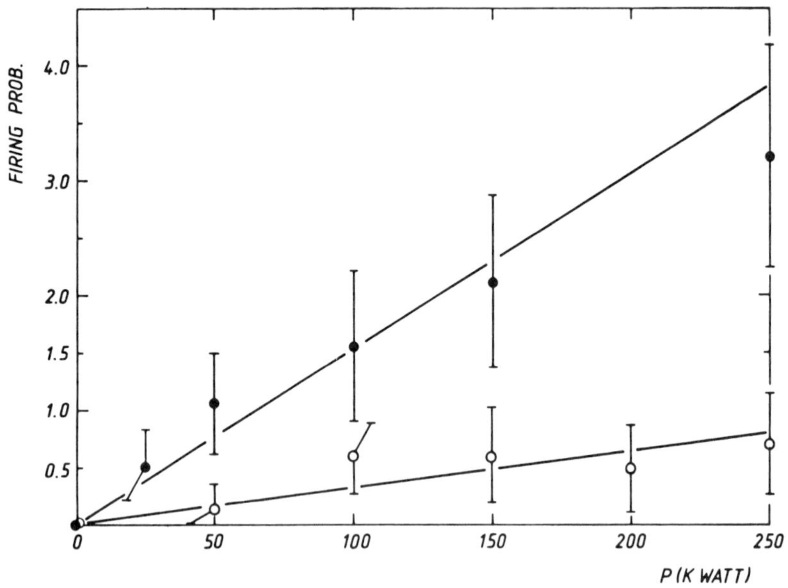

FIGURE 7. Spurious hit multiplicity in the flash chambers: Measured firing probability versus reactor power.

(2) a total running time equivalent to the running time of two solar years ($\Delta T = 2.5 \times 10^{11}$ sec);
(3) a neutron time of flight, $t = 7.5$ ms;
(4) imposing $N(\bar{n}) = 1$ and defining $t(n\bar{n}) = t\ \sqrt{N_c \Delta T}$.

In doing so, we find a total of $N_c = 2.5 \times 10^{18}$ neutrons and may quote as a reasonable transition time limit,

$$t(nn) \sim 10 \text{ megaseconds.}$$

The cosmic ray background has been estimated in detail by M. Tavani and D. Zanello[23] on the basis of the amount of material encountered by the known cosmic ray fluxes at sea level. The cosmic ray neutrons' fluxes and energy spectra at sea level are reported in TABLE 2. About 90% have energies of 1.0 GeV $< E <$ 3.0 GeV and an angular spectrum of $dn/d\phi = (1 - \cos^8\phi)$.

TABLE 1. Trigger Efficiencies, ϵ_{TR}(%)

Single Background Rate (s^{-1})	$S_1 S_2 S_3$ *T3* Structure: Two Tracks	$S_1 S_2 S_3$ *T3* Structure: Three Tracks	$S_1 S_2 S_3 R4$ *T4* Structure: Two Tracks	$S_1 S_2 S_3 R4$ *T4* Structure: Three Tracks
5×10^5	.33	.27	.34	.32
10^5	.61	.44	.65	.49

TABLE 2. Cosmic Ray Fluxes and Energy Spectra at Sea Level

K.E. (GeV)	ϕ (cm^{-2} S^{-1} $ster^{-1}$)	%
1–2	6.0×10^{-5}	78.2
2–3	1.1×10^{-5}	14.3
3–4	0.4×10^{-5}	5.0
4	0.2×10^{-5}	2.5
TOT	7.7×10^{-5}	—

These neutrons also suffer the following attenuation lengths:

$$L_1 = 47.1 \text{ cm concrete;}$$

$$L_2 = 21.9 \text{ cm Iron;}$$

$$L_3 = 23.7 \text{ cm Lead;}$$

$$L_4 = 47.6 \text{ cm Aluminum.}$$

The cosmic ray background is the number of neutron interactions in the 2-cm thick Al-vacuum pipe or in the target during a conventional running time assumed here as $\Delta T = 10^{11}$ s to simulate a real $\bar{n}$ annihilation event. Assuming as an ambiguity region a cylinder ± 25 cm wide around the target, there is a total of:

$N(P)$ = 400,000 neutrons on the vacuum pipe (P) during 10 Ms;
$N(T)$ = 50,000 neutrons on the target (T) during 10 Ms;
$N_i\,(P)$ = 13,000 interactions in (P) during 10 Ms;
$N_i\,(T)$ = 14 interactions in the target (T) during 10 Ms.

The details on the number of interactions suffered by the cosmic ray neutrons in the different parts of the apparatus are summarized in TABLE 3. It is worth mentioning that the 1.2-m heavy concrete as "ceiling" of the apparatus is essential. The total number of interactions is kept very low.

We also assume, pessimistically, that in our calorimeter:

(a) we have measurements errors: ± 0.5 cm for charged particles (the size of the flash cell) and ± 2.0 cm for photons;
(b) we reconstruct the interaction vertex and estimate the number of "misleading

TABLE 3. Number of Neutrons in $\Delta T = 10^7$ sec

Number of Events	Crossing: 2m Conc. plus Calorimeter	Crossing: 2m Conc., but not Calorimeter	Crossing: Less than 2m Conc.; (no) Calorim.
Inter. in Concrete	4.6×10^7	2.5×10^6	1.9×10^6
Inter. in Fe } Calorim.	1.9×10^5	—	—
Inter. in Pb } Calorim.	3.2×10^4	—	—
Inter. in Tank Al	1.0×10^4	5.1×10^2	2.6×10^3
Inter. in Top Part of Calorimeter	2.4×10^5	1.5×10^4	6.5×10^4

events" that simulate an interaction coming from a point having a distance, d, from the geometrical axis of the vacuum tube ($d < 70$ cm) under our trigger conditions.

The data are collected in TABLE 4. Assuming the total number of misleading events to be

$$n_{sim} = n_{tot} \times f_{tr} \times f_{ric} \times (E_\gamma)^{n_\gamma},$$

where

n_{sim} = interactions in the tube misreconstructed in the target,
n_{tot} = total number of events (for each reaction channel),
f_{tr} = fraction of events giving a trigger,
f_{ric} = fraction of events giving a vertex on the tube,
E_γ = 0.7 trigger efficiency per single photons, and
n_γ = number of photons needed for geometrical reconstruction,

we find in $\Delta T = 10$ megaseconds that $n_{sim} = (2.1 \pm 1.4)$ events, of which 1.6 are interactions in the vacuum pipe within ± 25 cm from the target position along the beam.

It is found that 98% of the event's charged secondaries or photons cross only one quadrant of the detector and that 2% cross no more than two quadrants. Therefore, events having particles crossing three or more quadrants would be typical of only $\bar{n}$ annihilations.

Systematic studies are in progress. In particular, is the study on the reconstruction of the cosmic ray interactions in the vacuum pipe to investigate the separation between interactions in the target and interactions in the vacuum pipe.

GRENOBLE III

As a possibility for the future, a new experiment is plausible at Grenoble. It is intended to be an "ultimate experiment" for the detection of free $n\bar{n}$ oscillations for neutrons produced by a nuclear reactor. It is aiming at a sensitivity of the order of 0.1–1.0 gigasecs. The Grenoble Laboratory has called for letters of intent and it is prepared to make available a new experimental area in an annex to be built in the near future (see FIGURE 1).

Thermal neutrons[10] can be transported to the experimental area by a new curved guide, which is 60 m long and has a radius of $R = 5000$ m. Its neutron current will be about 300 billion n/s at an average temperature of $T \sim 15°$K.

The "interaction free" flight path would be the final part of the beam transport line plus a 30-m straight section with a 3-mrad divergence, which is enclosed in a conical vessel with a final aperture diameter of $\phi \sim 1.3$ m. It is maintained at a pressure, P, of a few nanobars and has a residual magnetic field of $B < 0.10$ milligauss to provide a flight time of $t \sim 0.1$ s. A carbon target, 125 microns thick, could have a cross-sectional area of 70 cm × 100 cm, which is far enough from the vacuum pipe.

Two tentative detector setups[2d,9,10] have been circulated so far. For the first one, G. Fidecaro[2b,d] proposed an "exercise" aiming at a maximal selectivity of an apparatus

TABLE 4. Track Data

Topology		Error along the Beam Direction: Δx (cm) Target	Δx (cm) Tank	Fraction of Mislead. Events ($R < 700$ cm)
Three Tracks on Three Different Quadrants	Three Charged	3.7	4.4	
	2CH + 1 Photon	8.9	10.0	6.3×10^{-3}
	1CH + 2 Photons	12.0	14.5	1.1×10^{-2}
	Three Photons	14.0	16.8	2.6×10^{-2}
Three Tracks on Two Quad.	Three Charged	3.5	4.2	2.6×10^{-4}
	2CH + 1 Photons	8.0	10.6	1.2×10^{-2}
	1CH + 2 Photons	9.8	12.6	3.7×10^{-2}
	Three Photons	10.8	16.0	5.8×10^{-2}
Only Two Tracks	—	5.5	—	—

"able to reach solid conclusions on the basis of very few events in the lucky circumstance in which candidate events would show up." The suggested detector (FIGURE 8) is designed to obtain a drastic improvement in the measurement of the energy release in the annihilation process.

It consists of an Inner Time Projection Chamber System (TPC) operating in a magnetic field, which is provided by a superconducting solenoid surrounded by an electromagnetic calorimeter. The TPCs (or an equivalent detector of the same family, i.e., drift or multiwire chambers) would provide momentum and ionization of the charged particles, as well as their geometrical coordinates.

Energy and geometrical coordinates of the photons from π^0 decay would be provided by the electromagnetic calorimeter. Due to the presence of a magnetic field giving the momentum of the charged particles, there would be no need for a hadron

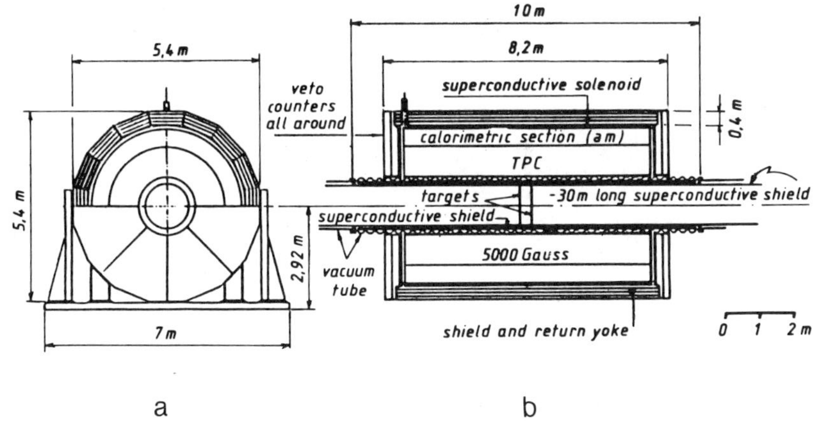

FIGURE 8. A possible design of a detector for Grenoble III as proposed in references 2b and 2d: (a) cross-sectional view; (b) longitudinal cut in the target region.

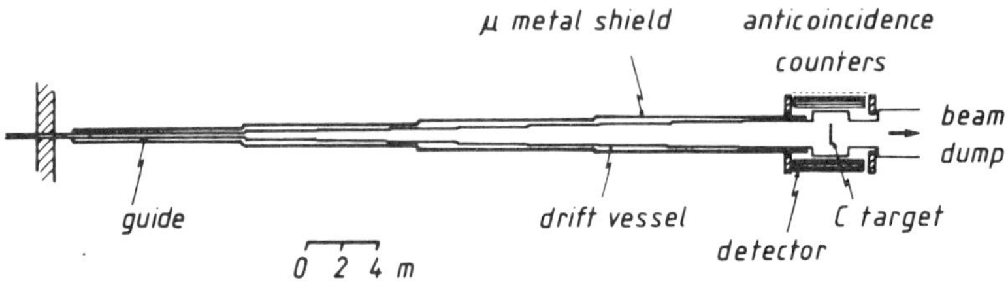

FIGURE 9a. Possible design of a detector for Grenoble III proposed in references 2c, 2e, and 18: beam transport system.

calorimeter. The magnetic field is confined inside the detector by a superconducting shield placed inside the vacuum pipe (FIGURE 8b). This would be preferable to the multilayer μ-metal shield that was used in the previous Grenoble experiments. In addition, the use of honeycomb-structured materials for the vacuum pipe would reduce the multiple scattering suffered by the produced particles before they enter the detector.

K. Green[17] has circulated a "feasibility study" aiming at limits as high as 10^9 without describing particular detectors. However, M. Baldo Ceolin[2e] and G. Puglierin[2c] designed a detector surrounding a target with maximal solid angle coverage (FIGURE 9), which is split into four quadrants placed azimuthally around the target (as done in the Pavia experiment). In a recent proposal,[10] the detector is composed of:

(a) a charged particle vertex detector (ten planes per quadrant) with essentially no absorptive material;
(b) a neutral particle vertex detector (ten planes per quadrant) placed after a 0.5-cm Pb gamma converter and interleaved with 0.5-cm Al plates;
(c) a final calorimeter designed to absorb most, if not all, of the energy emitted in

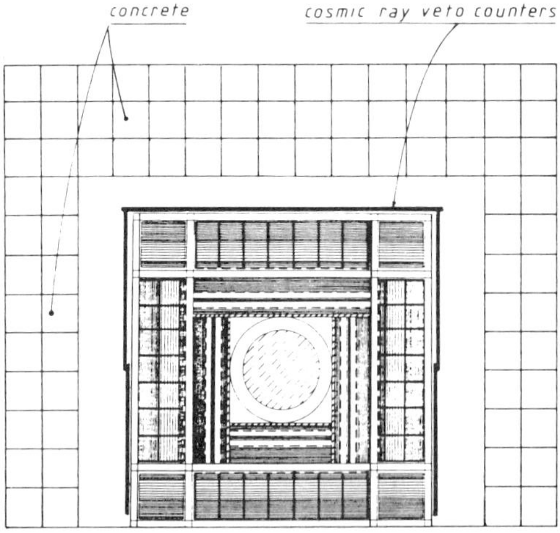

FIGURE 9b. Possible design of a detector for Grenoble III proposed in references 2c, 2e, and 18: view of the detector and veto.

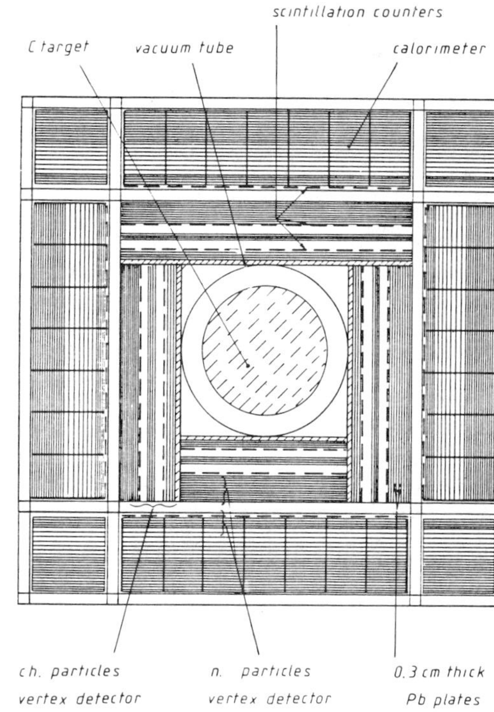

FIGURE 9c. Possible design of a detector for Grenoble III proposed in references 2c, 2e, and 18: structural detail of the detector.

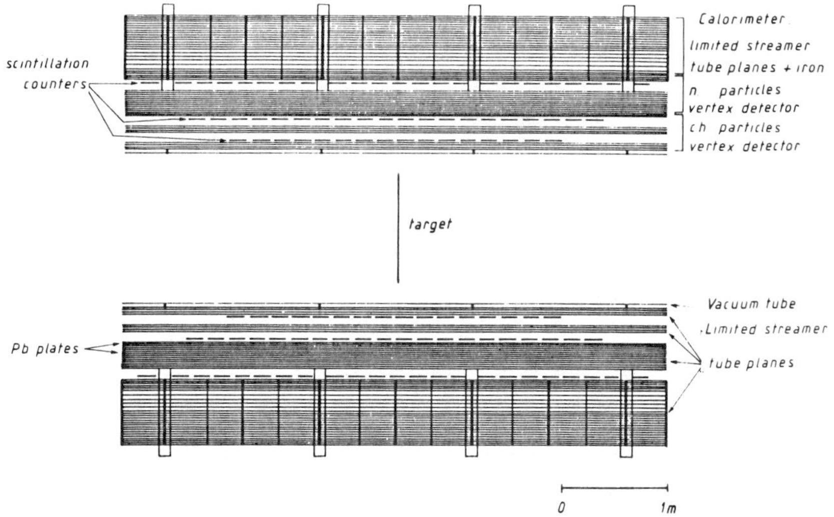

FIGURE 9d. Possible design of a detector for Grenoble III proposed in references 2c, 2e, and 18: longitudinal cut of the detector in the target region.

the $\bar{n}N$ annihilations, which is made of 16 planes of limited streamer tubes interleaved with Fe absorbers of increasing thickness.

The free propagation region (FIGURE 10) is a diverging guide ($\delta \sim 3$ mrad), 30 m long, followed by a conical vessel ($\delta \sim 15$ mrad) evacuated to a pressure of $P = 10$ microtorr.

The "interaction free" path is enclosed by the neutron guide and the first 30 m of the vessel, and both are shielded against external magnetic fields to values of $B < 0.1$ milligauss.

The conical vessel has a definite shape since the gravitational effect on the neutron mass cannot be neglected. In fact, the fall-down, $z = 1/2gv^2$, turns out to be some centimeters and a proper Li absorber is installed for the purpose.

Grenoble III is still at the stage of design and it seems proper to say that it might

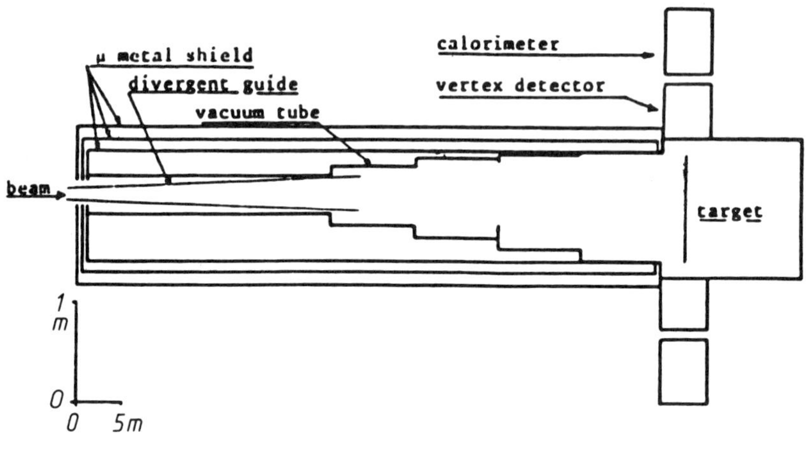

FIGURE 10. Cross section of the conical vessel (Grenoble III) in the target region. The asymmetric structure to cure the gravitational fall-down of the neutrons and the "ad hoc" ^{6}LiF absorber is indicated.

become a reality around 1986–1987. A collaboration is now being established between the Universities of Padua, Pavia, Heidelberg, and the ILL Laboratory.

THE EXPERIMENT AT THE MOSKOW MESON FACTORY (PROPOSAL)

V. Kuzmin and collaborators[5] have presented on several occasions a real gigantic experimental possibility by which one theoretically could reach sensitivities in $t(n\bar{n})$ as high as 5.0–10.0 gigasecs at the Moskow Meson Factory recently completed at the Institute for Nuclear Research.

The general view of the experimental layout is sketched in FIGURE 11. The installation begins at a location where a full proton beam of the meson factory is dumped; it extends for over 60 meters and is about 10 meters underground. It includes:

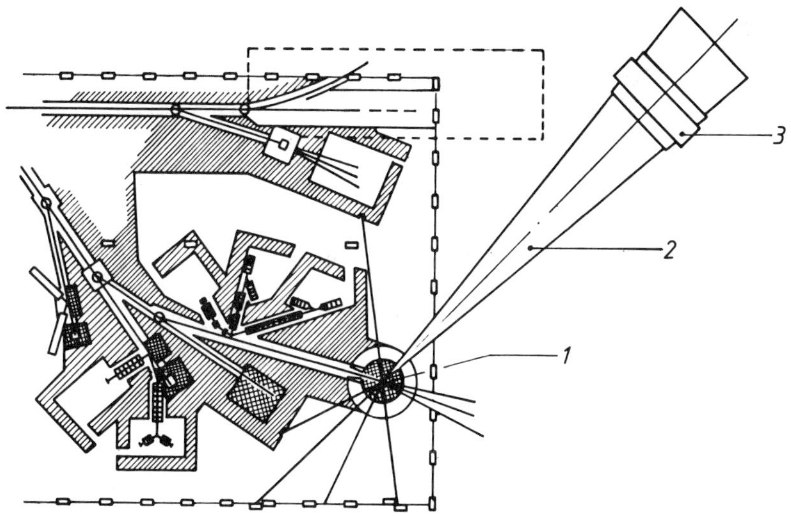

FIGURE 11. Overall sketch of a possible experimental layout at the Moskow Meson Factory (from reference 5): (1) neutron source; (2) neutron guides; (3) detectors system.

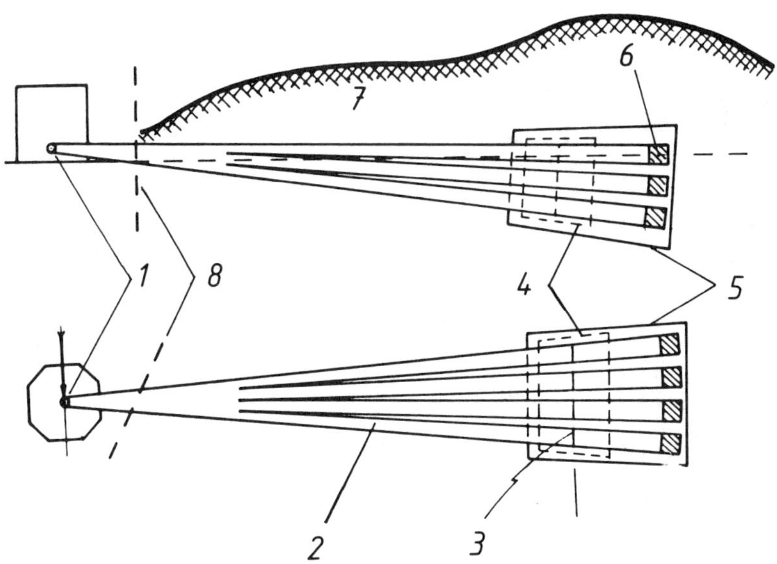

FIGURE 12. Detailed sketch of the pulsed neutron source: (1) the kinac's target; (2) the hydrogen moderator; (3) the deuterium moderator; (4) the proton beam; (5) the neutron port; (6) the target container; (7) the cold moderator container; (8) the iron shield.

the neutron source (FIGURE 12); 12 neutron pipes (FIGURE 13); and a detector-veto system (FIGURE 14) combining detectors of $\bar{n}N$ annihilation products and cosmic ray event vetoes.

The whole installation, except for the pulsed neutron source, is located outside the main hall of the Meson Factory. The detector system is placed into a special pavilion. Flight pipes and the detector hall are shielded from cosmic rays by as much as 20 m of water equivalent material.

The neutron source of the Meson Factory is intended to be the one devoted to a wide range of investigations both in nuclear and in solid-state physics. It is sketched in FIGURE 12; it consists of a target stopping a 0.5-mA proton beam, which is cooled with heavy water and surrounded by a liquid deuterium moderator facing a large aperture neutron port. The use of an ^{238}U beam stopper would provide 7×10^{15} n/s at the moderator surface, while PuO^2 would give 3×10^{16} n/s; a Pb stopper would guarantee 3×10^{15} n/s. The current of cold neutrons in the target region is quoted to be 6×10^{12} n/s.

A large vacuum pipe (FIGURE 13) at about 25 m from the deuterium moderator splits into 12 independent cylindrical channels that extend through an additional 32 meters to reach 12 target foils, 1.6 m in diameter, and which continue for an additional 10 to 15 meters until they hit 12 beam stoppers in order to minimize backscattering background. The gaps between the 12 pipes in the target region are supposed to allow the installation of 1-m thick track detectors. The Earth's magnetic field is suppressed to about 10 milligauss by a passive shielding.

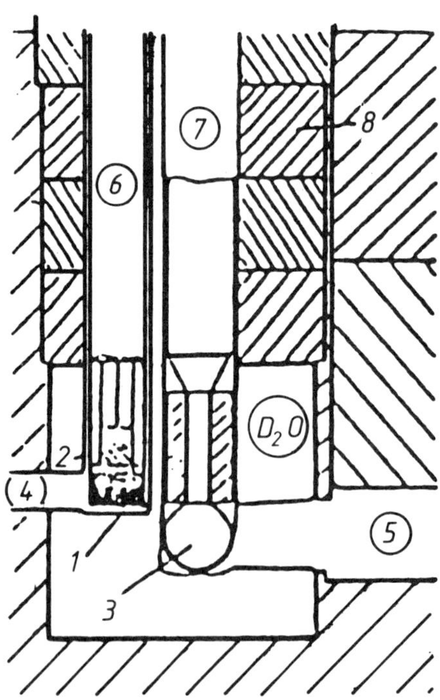

FIGURE 13. The neutron flight system: (1) the neutron source; (2) the neutron guides; (3) the system of 12 targets; (4) the location of the 12 detectors (dashed lines); (5) the veto system; (6) the final beam stoppers; (7) the magnetic shield; (8) the closing walls of the experimental halls.

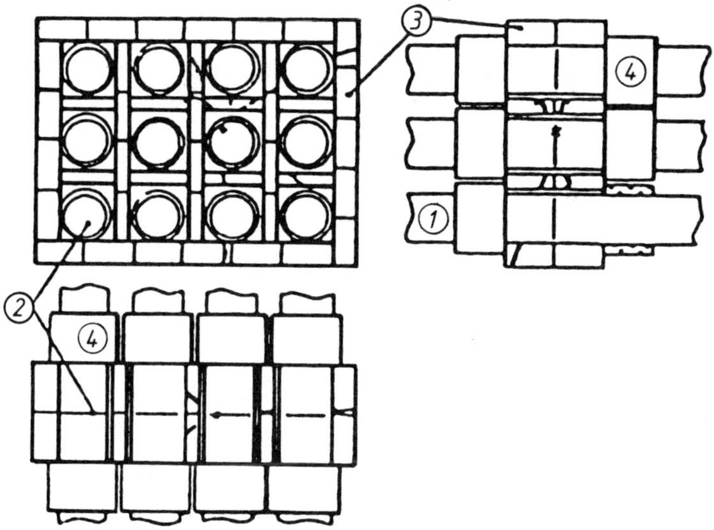

FIGURE 14. Details of the detector system: (1) the neutron flight tubes; (2) the 12 targets; (3) the track's detector system; (4) the scintillator calorimeter-collars. The series of dotted tracks indicate a simulated annihilation event.

At a residual gas density of d = 10^{14} cm$^{-3}/t$ (where t is the neutron time of flight), neutron dephasing becomes important. The total volume of about 1600 cubic meters with a wall surface of 5000 square meters is evacuated to about 0.1 millipascal by a unique vacuum system. The setup is rather complex; for a detailed description, the reader should refer to reference 5a. Briefly, the systems (FIGURE 14) allocated around the 12 target foils consist of two parts:

(i) a track detector (equivalent, in a sense, to the calorimeter of the Pavia experiment) located azimuthally around the target and made of four quadrants;

(ii) a scintillator calorimeter made of two sandwiches located up-beam and down-beam to cover about 75% of 4π.

The "azimuthal part" of the detector, though, has to give a space resolution in the location of the interaction vertex of 10 cm. It is made of 24 "coordinate planes" given by multiwire Heiger counters (see reference 5) that are interlayed by low density metal converters. The "up-beam – down-beam" calorimeters are made of liquid scintillators interlayed by heavy (300 g/m^2) absorbers. The designed weight of the detector is about 100 tons; the liquid scintillator volume is 40 m^3 and the number of the readout channels is about 300. To reach the designed result with only 1 m of distance between the vacuum pipes is undoubtedly a severe technical task.

Also, rejection against cosmic rays provided by the ~20 m.w.e. of ground is complemented by a layer of scintillators (~400 m^3) extending over a surface of about 1500 m^2, located behind a 1-m concrete wall surrounding the target-detector hall.

In conclusion, V. Kuzmin and collaborators hope to achieve a sensitivity as high as gigasecs. Forcing the parameters of both the linac and the neutron source, the limit can be in principle raised to approximately $t(n\bar{n}) \sim 10$ gigaseconds.

CONCLUSIONS

An experiment has been attempted to detect $n\bar{n}$ oscillations rather fast. It gave a limit of $t(n\bar{n}) > 1$ megasec.

The Pavia experiment is aiming at $t(n\bar{n}) = 10$ megasecs with an all inclusive cost of about one million dollars.

Grenoble III might have an overall cost of about ten million dollars and should be able to attempt a limit of $t(n\bar{n}) = 0.1$ gigaseconds.

The Moskow experiment, still a proposal, multiplies the detector by about ten and the beam transport problems by about five, which is due to the addition of the cost of a cold neutron-pulsed source. I would guess that such a gigantic experiment might be done at an all-inclusive cost not far from one hundred million dollars.

ACKNOWLEDGMENTS

The author is indebted to the members of the Pavia-Roma Milano (CESNEF) collaboration for providing help and data. Useful discussions with M. and G. Fidecaro, as well as some conversations with V. Kuzmin are gratefully acknowledged. The author thanks M. Baldo Ceolin for a lengthy discussion of the manuscript. He is also indebted to G. Bonaschi for the careful drawing or redrawing of the many figures.

NOTES AND REFERENCES

1. Ratti, S. P. 1984. ICOBAN-84. D. Cline, Ed.: 30.
2. (a) Fidecaro, G., M. Fidecaro, L. Lanceri, A. Marchioro, W. Mampe, M. Baldo Ceolin, F. Mattioli, G. Puglierin, C. J. Batty, K. Green, H. Prosper, P. Sharman, J. M. Pendlebury & K. Smith. 1980. ILL-5 proposal; (b) Fidecaro, G. 1981. Proceedings of Neutrino-81; Department of Physics and Astrophysics, Honolulu, p. 246; (c) Puglierin, G. 1983. Proceedings of H.E.P.-83; Brighton (U.K.). J. Guy & C. Costain, Eds.: 381; (d) Fidecaro, G. 1983. *N$\bar{n}$* oscillation experiments. Preprint CERN-EP/83-102: Fourth Workshop on Grandunification. University of Pennsylvania (April 1983); (e) Baldo Ceolin, M. 1983. *N$\bar{n}$* oscillations. *In* Workshop on Reactor Based Fundamental Physics. ILL, Grenoble (November '83); 1983. International Conference on Unified Theories and Their Experimental Tests. Edited by Ac.Naz.Lincei, p. 127; 1984. Vth International Symposium on Capture Gamma Ray Spectroscopy and Related Topics. Knoxville, Tennessee. To appear; (f) Fidecaro, G., M. Fidecaro, L. Lanceri, A. Marchioro, W. Mampe, F. Mattioli, G. Puglierin, C. J. Batty, K. Green, H. B. Prosper, P. Sharman, J. M. Pendlebury & K. F. Smith. 1985. Phys. Lett. **156B:** 122.
3. (a) Cherry, N. J., K. Lande, C. K. Lee, R. I. Steinberg & B. Cleveland. 1983. Phys.Lett. **50:** 1354; (b) Battistoni, G., E. Bellotti, G. Bologna, P. Campana, C. Castagnoli, V. Chiarello, A. Ciocio, D. C. Cundy, B. d'Ettorre Piazzoli, E. Fiorini, P. Galeotti, E. Iarocci, C. Liguori, G. Mannocchi, G. P. Murtas, P. Negri, G. Nicoletti, P. Picchi, M. Price, A. Pullia, S. Ragazzi, M. Rollier, O. Saavedra, L. Satta, L. Trasatti & L. Zanotti. 1983. Phys. Lett. **133B:** 454; (c)

JONES, T. W., R. M. BIONTA, G. BLEWITT, C. B. BRATTON, B. G. CORTAZ, S. ERREDE, G. W. FOSTER, W. GAJEWSKI, K. S. GANEZER, M. GOLDHABER, T. J. HAINES, D. KIELCZEWSKA, W. R. KROPP, J. G. LEARNED, F. LEHMANN, J. M. LOSECCO, H. S. PARK, F. REINES, J. SCHULTZ, E. SHUMARD, D. SINCLAIR, H. W. SOBEL, J. L. STONE, L. R. SULAK, R. SVOBODA, J. C. VAN DER VELDE & C. WUEST. 1984. Phys. Rev. Lett. **52:** 720.

4. (a) BRESSI, G., E. CALLIGARICH, R. CARDARELLI, G. CECCHET, F. CINGOLI, A.CIRILLO, M. CONVERSI, G. DE ZORZI, L. DICIACCIO, A. GIGLI, G. LIGUORI, F. MASSA, E. ORVINI, G. PERLINI, A. PIAZZOLI, S. RATTI, R. SANTONICO, F. SEBASTIANI, M. TERRANI, P. TORRE & D. ZANELLO. 1982. ICOBAN-82. V. S. Narashiman, P. Roy, K. V. L. Sharma & B. V. Sreekantan, Eds.: 229. Pramaña Suppl.; (b) RATTI, S. P. 1981. *In* Weak Interaction as Probes of Unification. A.I.P. Conf. Proc., n. 72. G. B. Collins, L. N. Chang & J. R. Ficenec, Eds.: 183. A.I.P.; (c) BRESSI, G., E. CALLIGARICH, R. CARDARELLI, A. CESANA, F. CINGOLI, A. CIRILLO, M. CONVERSI, G. DE ZORZI, L. DICIACCIO, R. DOLFINI, A. GIGLI, A. LANZA, G. LIGUORI, F. MASSA, E. ORVINI, A. PIAZZOLI, S. P. RATTI, R. SANTONICO, F. SEBASTIANI, M. TERRANI, P. TORRE, D. ZANELLO & L. ZANELLO. 1982. Proc. of The VIIth European Conference of Triga Reactor Users (Istanbul), p. 5–1 and p. 5–15; (d) RATTI, S. P. 1983. Hadronic Journal **6:** 1123; 1983. ICOMAN-83. E. Bellotti & S. Stipcich, Eds.: 212. Lab. Naz. Frascati; 1984. ICOMAN-84. D. Cline, Ed.: 30; (e) BRESSI, G., E. CALLIGARICH, R. CARDARELLI, M. CAMBIAGHI, M. CONVERSI, G. DE ZORZI, R. DOLFINI, A. GIGLI, A. LANZA, G. LIGUORI, F. MASSA, F. MAURI, A. PIAZZOLI, S. P. RATTI, R. SANTONICO, F. SEBASTIANI, M. TERRANI, P. TORRE & D. ZANELLO. 1985. Progress report of the Pavia experiment on free $n\bar{n}$ oscillations. Report DFNT 85/2.
5. ILJINOV, A. S., M. V. KAZARNOSKY, V. A. KUZMIN, E. A. MONICH, M. E. SHAPOSHNIKOV, YU. YA. STAVISSKY & B. E. STERN. 1983. ICOMAN-83. E. Bellotti & S. Stipcich, Eds.: 179. Lab. Naz. Frascati (and references therein); 1982. ICOBAN-82. V. S. Narashiman, P. Roy, K. V. L. Sharma & B. V. Sreekantan, Eds.: 179. Pramaña Suppl.
6. YOSHIKI, H. 1980. Check of neutron oscillations using ultra cold neutrons. KEK preprint 80-10 (September 1980) and references therein.
7. MARSHAK, R. E. 1982. Private communication. The last presentation was done by ANDERSON, H. L. *In* 3rd Workshop on Grandunification. A. Jaffe & D. Ruelle, Eds.: 392. Birkhausen (and in reference 4d).
8. (a) CHANG, L. N. & N. P. CHANG. 1980. Phys. Rev. Lett. **45:** 1540; (b) GELL-MANN, M. & A. PAIS. 1955. Phys. Rev. **97:** 1387.
9. (a) SACHAROV, A. 1967. Pis'ma Zh. Eksp. Teor. Fiz. **5:** 32; JEPT Lett. **6:** 236; (b) KUZMIN, V. 1970. Pis'ma Zh. Eksp. Teor. Fiz. **12:** 335; JEPT Lett. **12:** 228.
10. (a) PATI, J. & A. SALAM. 1973. Phys. Rev. **D8:** 1240 (The first suggestion of baryon-lepton symmetry is as old as 1959, which was put forward by GAMBA, A., R. E. MARSHAK & S. OKUBO. 1959. Proc. Natl. Acad. Sci. **45:** 88); 1973. Phys. Rev. Lett. **31:** 661; 1974. Phys. Rev. **D10:** 275; (b) GEORGI, H. & S. L. GLASHOW. 1974. Phys. Rev. Lett. **32:** 438; (c) GLASHOW, S. 1980. *In* Astrophysics and Elementary Particles, Common Problems, p. 11. Accademia Naz. dei Lincei, Roma 1980; 1979. *In* Proceedings of Neutrino '79, p. 518; 1980. *In* Proceedings XIth International Symposium On Multiparticle Dynamics. E. DeWolf & F. Verbeure, Eds.: 701. Bruges; (d) WILCZEK, F. & A. ZEE. 1979. Phys. Rev. Lett. **43:** 1571; (e) MARSHAK, R. E. 1982. Proc. Nat. Acad. Sci. **79:** 3371 (see also: Report VPI-HE 81/3); (f) MARSHAK, R. E. & R. N. MOHAPATRA. 1980. Phys. Lett. **91B:** 222; **94B:** 183; Phys. Rev. Lett. **44:** 1316; 1984. Rep. VPI-84/10 (October 5, 1984); (g) TOSA, Y., R. E. MARSHAK & S. OKUBO. 1983. Phys. Rev. **D27:** 444; (h) MARSHAK, R. E. 1983. Proceedings European Conference on Electroweak Effects at High Energy, Erice. To be published; (i) GIPSON, J. M. & R. E. MARSHAK. 1984. Report VPI HEP 84/5 (September 9, 1984); 1984. VPI HEP 84/13 (October 30, 1984); (j) MOHAPATRA, R. N. & G. SENJANOVIC. 1982. Report BNL 31720 IC/82/89; (k) PARIDA, M. K. & M. CVETIC. 1984. Report VPI HEP 84/6; (l) CHANG, D., R. N. MOHAPATRA & M. K. PARIDA. 1984. Phys. Rev. Lett. **52:** 1072; Phys. Rev. **D30:** 1052; (m) CHANG, D., R. N. MOHAPATRA, J. M. GIPSON, R. E. MARSHAK & M. K. PARIDA. 1984. Report VPI HEP 84/9 (October 2, 1984); (n) A useful reference to all papers published prior to 1981 is the collection of papers published by MOHAPATRA, R. N. & D. CHANG. 1981. Gauge

Theories of Fundamental Interactions. World Scientific; (o) PATI, J., A. SALAM & J. STRATHDEE. 1981. Nucl. Phys. **B185:** 445; (p) LUST, D., A. MASIERO & M. RONCADELLI. 1982. Phys. Rev. **D25:** 3096; 1984. Z. Phys. Part. Fields. **C21:** 327; (q) KUO, T. K. & S. T. LOVE. 1980. Phys. Rev. Lett. **45:** 93; (r) FUJIMOTO, Y. & Z. ZHIYONG. 1982. ICTP (Trieste) Report IC/82/60.

11. See: ALLISON, W. W. M. 1984. *In* ICOBAN-84. D. Cline, Ed.: 323; LOVELESS, J. ibid. p. 1; KOSHIBA, M. ibid. p. 69; see also VAN DER VELDE, J. 1986. These proceedings.
12. RAO, S. & R. SHROCK. 1982. Phys. Lett. **116B:** 238.
13. (a) SANDERS, P. G. H. 1980. J. Phys. G: Nucl. Phys. **6:** 1616; (b) CHETYRKIN, K. G., M. V. KAZARNOVSKY, V. A. KUZMIN & M. E. SHAPOSHNIKOV. 1981. Phys. Lett. **99B:** 358; (c) COWSIK, R. & S. NUSSINOV. 1981. Phys. Lett. **101B:** 237; (d) ARAFUNE, J. & O. MIYAMURA. 1981. Prog. Theor. Phys. **66:** 661; (e) DOVER, C. B., A. GAL, J. M. RICHARD. 1983. Phys. Rev. **D27:** 1090; (f) RIAZZUDIN, R. 1982. Phys. Rev. **D25:** 885; (g) ALBERICO, W. M., A. BOTTINO & A. MOLINARI. 1982. Phys. Lett. **114B:** 266; 1983. Preprint TH 3537-CERN (February 1983); ALBERICO, W. M., J. BERNABEU, A. BOTTINO & A. MOLINARI. 1984. Nucl. Phys. **A429:** 445; (h) ELLIS, R. 1983. CERN-Preprint EP/83-26. Unpublished; (i) VERGADOS, J. D. 1982. Phys. Lett. **118B:** 107.
14. (a) KABIR, P. K. 1983. Phys. Rev. Lett. **51:** 231; (b) SPARROW, D. A. 1984. Effects of the nuclear medium on the observation of baryon number violation, University of Pennsylvania Preprint; (c) HORWITZ, L. P. & E. KATZNELSON. 1983. Phys. Rev. Lett. **50:** 1184; (d) FLEMING, G. N. 1983. Phys. Lett. **125B:** 287; (e) LAPAGE, G. P., T. M. YAN & D. R. YENNIE. 1983. Phys. Rev. Lett. **51:** 1599; (f) CAHILL, K. 1983. Phys. Rev. Lett. **51:** 1600; (g) WHEATHER, J. & R. PEIERLS. 1983. Phys. Rev. Let. **51:** 1601; (h) LANE, A. M. 1983. Phys. Lett. **99A:** 359; (i) FONDA, L. & G. C. GHIRARDI. 1983. Phys. Lett. **131B:** 309.
15. WONG, C. Y., A. K. KERMAN, G. R. SATCHLER & A. D. MACKELLAR. 1984. Phys. Rev. **D29:** 574.
16. BASECQ, J. & L. WOLFENSTEIN. 1983. Nucl. Phys. **B224:** 21.
17. GREEN, K. 1984. Feasibility study of an experiment to search for *nn* oscillations to a limit of 10 gigaseconds. Preprint.
18. HEIDELBERG-PADOVA COLLABORATION. 1984. A sensitive search for $n\bar{n}$ transitions at the ILL Reactor. Proposal DP84/16 (21.1.84).
19. (a) ARNDT, R. A., V. B. PRASAD & R. RIAZUDDIN. 1982. Phys. Rev. **D24:** 1431; (b) ZOVKO, N. & W. P. TROWER. 1982. Phys. Rev. **D25:** 3088; (c) PUSCH, G. D. 1983. Nuovo Cimento **74A:** 149.
20. (a) CONVERSI, M. & L. FEDERICI. 1978. Nucl. Instrum. Methods **161:** 219; (b) FEDERICI, L., M. NARDI, F. CERADINI & M. CONVERSI. 1978. Nucl. Instrum. Methods **151:** 103; (c) CONVERSI, M. & F. LA CAVA. 1979. Nucl. Instrum. Methods **161:** 219.
21. (a) TAYLOR, F. E., T. OHSUGI, J. K. WALKER, D. MILLER & D. BUCHHOLZ. 1978. IEEE Trans. Nucl. Sci. **NS-25:** 312; 1980. **NS-27:** 30; (b) CHEN, H. H. 1977. U.C.I.-Neutrino n. 20, U.C.I. 10P19-119; (c) BRESSI, G., P. TRESPI, M. CAMBIAGHI, E. D'USCIO & F. MASSA. 1983. Hadronic Journal **6:** 1330; SANTONICO, R. 1983. ibid. p. 1321.
22. (a) SANTONICO, R. & R. CARDARELLI. 1981. Nucl. Instrum. Methods **187:** 377; 1982. Test of a double resistive plate counter under the Gran Sasso Tunnel. Report IFGM-INFN-Roma n. 788 (February 1982); (b) SANTONICO, R. 1981. Proceedings of GUD Workshop (University of Rome), p. 105.
23. TAVANI, M. & D. ZANELLO. 1983. Cosmic ray background estimate for the *Nn* experiment. IFGM-University of Rome-INFN Rome, Internal Report n. 805.

Experimental Status of Proton Decay

J. C. VAN DER VELDE

Physics Department
University of Michigan
Ann Arbor, Michigan 48109

Question: How can we even be here?
Answer: Baryon number is not conserved.

This is a seemingly paradoxical answer to the long-standing mystery of why there appears to be a preponderance of matter over antimatter in the universe. The basic interaction that produced baryons in the early instants after the "big bang" did not conserve baryon number. Therefore, when the last baryon-antibaryon annihilations occurred, "we" were left with an excess of baryons.

As I write this, I look out over the mountains of Aspen. There is a lot of baryon number sitting there. Do the anti-mountains of anti-Aspen exist somewhere? Apparently not.

There are other theoretical reasons for believing that the quarks inside baryons should be able to change directly into leptons via a new interaction and leave behind no baryon number. Hence, it would be nice if the hypothesis of baryon number nonconservation could be confirmed and quantified by direct measurement. This is what has motivated the experimenters to delve into that which I will describe here.

We begin with a list of six current experiments having data on nucleon decay, which is shown in TABLE 1. The first column gives the fiducial mass from those candidate events that are accepted. The first three experiments use tracking calorimeters in which the principal mass is in the form of iron plates. The last three are of the water Cherenkov type. The analyzed exposure and number of events found within the fiducial volume are given for each experiment in the last two columns. A more detailed description of these experiments can be found in the recent review by Perkins.[1]

Contained event rates per kilotonne-yr of exposure are shown in FIGURE 1. One sees that the rates are generally in good agreement at a value of ~ 100 events/kt-yr. This is approximately the rate expected[2,4] due to atmospheric neutrino interactions above a threshold neutrino energy of about 300 MeV. The Kamioka experiment has a somewhat lower threshold than the others and sees a correspondingly higher rate. The rate at KGF is expected to be a little lower because of their lower geomagnetic latitude. The HPW group has only searched for events with ≥ 2 muon decays and therefore sees a much smaller rate.

The rates shown in FIGURE 1 have not been corrected for detection efficiency. Typical efficiencies for nucleon decays and neutrino interactions are in the range of 50–90%.

The arrows at 6 (60) events/kt-yr represent the rate of events expected if the nucleon lifetime were 10^{32} (10^{31}) yr, assuming 100% detection efficiencies and equal rates for protons and neutrons. From this, one sees immediately that one is faced with a formidable neutrino background problem in searching for nucleon decays at a lifetime level of 10^{32} years or longer.

TABLE 1. Current Proton Decay Experiments[a]

Experiment	Fiducial Mass	Magnetic Latitude	Analyzed Exposure	Contained Events
Kolar G.F.	70 tonne Fe	2° N	210 tonne-yrs	16
NUSEX	100 tonne Fe	50° N	220 tonne-yrs	22
FRÉJUS	450 tonne Fe	50° N	90 tonne-yrs	9
IMB	3300 tonne H_2O	52° N	3800 tonne-yrs	401
KAMIOKANDE	880 tonne H_2O	27° N	660 tonne-yrs	89
HPW	560 tonne H_2O	50° N	380 tonne-yrs	5[b]

[a]TABLE 1 is an updated version of a similar one recently presented by LoSecco, J. 1984. Invited paper at the APS-DPF annual meeting, Santa Fe, New Mexico.
[b]$2\mu + 3\mu$ only.

The small circles represent the rate at which the various experiments find events that are consistent with one or more nucleon decay modes. These events are usually referred to as "candidates." Given an event with more than just a single track, a detector with a less-than-ideal resolution, and some amount of imagination, one can usually find some nucleon decay mode that fits it. Hence, one should be somewhat wary about the term "candidate" at this point. So far, the candidates do not show the clustering about a dominant decay mode that one would expect for real nucleon decays.

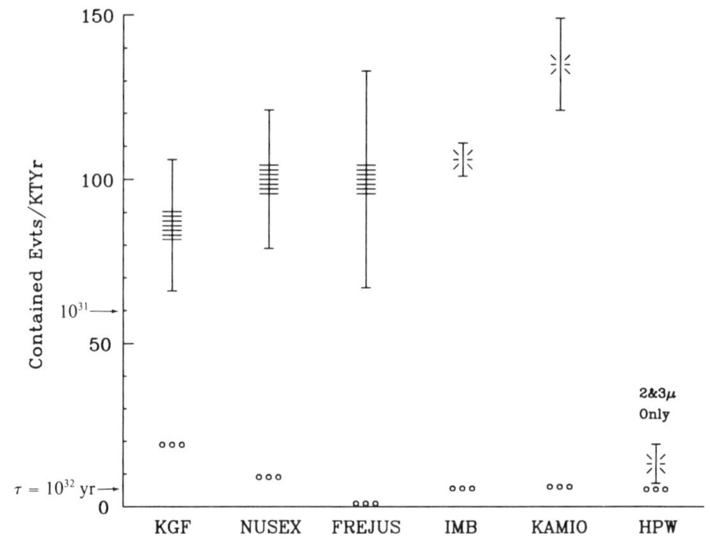

FIGURE 1. Contained event rates for six current nucleon decay detectors. The first three points are from iron-plate tracking devices and the second three are from water Cherenkov detectors. The overall rates (top points with error bars) are in reasonable agreement with each other and with the expectations based on atmospheric neutrino fluxes. Arrows at the left represent the total expected event rate from the decay of protons and neutrons (assuming they had the same lifetime). The small circles represent reported rates of "candidate" events for nucleon decay. (See text.)

In setting lower limits on lifetime/branching ratios (τ/B) for various decay modes, most experimenters have taken the conservative attitude that all of the candidates for that particular mode are real. Background estimates for various modes are still fairly crude, so that, generally, background has not been subtracted from the candidate samples in calculating limits. Quoted lower limits on τ/B are, therefore, generally quite conservative.

As can be seen from FIGURE 1, candidates are being found at a level of 5% to 10% of the contained events. This is just about the rate that atmospheric neutrino interactions start to mimic nucleon decays. We illustrate the problem in FIGURE 2. A proton with Fermi momentum can decay into $\mu^+\pi^0$ with an included angle of 150° or more. This configuration can be mimicked by the reaction,

$$\nu_e + \mathrm{n} \rightarrow \mathrm{n} + e^- + \pi^+,$$

where the e^- makes a shower and the π^+ decays into a μ^+. Making a cut of $\theta > 150°$ on

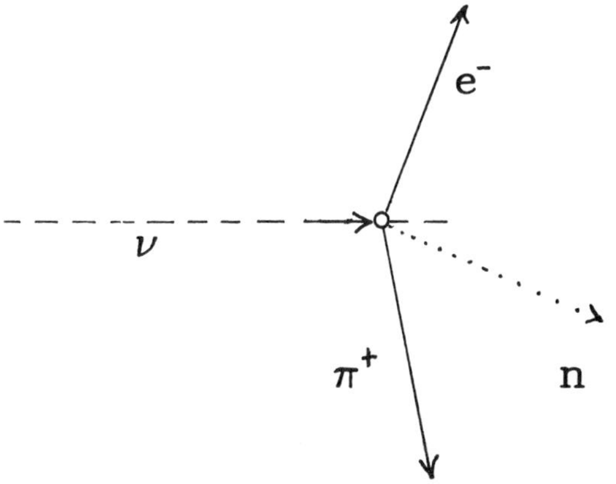

FIGURE 2. A typical low energy single pion production reaction by an atmospheric electron neutrino. Such wide-angle two-track events are a serious source of background in (earthbound) nucleon decay detectors if the nucleon lifetime is $>10^{32}$ yrs.

such two-track events saves ~90% of the decays from a nucleus like oxygen and rejects a large fraction, but not all, of the neutrino background around 1 GeV. (We will give a more precise estimate later in the paper.)

Of course, a wide-angle two-track event can be mimicked by a single track that scatters through a small angle if one does not know the track directions. Hence, given the present limits on the nucleon lifetime and the present knowledge of neutrino background, it is essential to be able to measure track directionality. This is straightforward for the water Cherenkov detectors, but more difficult for tracking calorimeters since they must rely on ionization measurements, shower profiles, multiple scattering, etc.

Understanding the properties of neutrino background has become the watchword for nucleon decay experimenters. It will also be the watchword for the remainder of this paper.

We now give a brief review of the present results from the six experiments listed. At the end we will say something about experiments planned for the near future.

KGF RESULTS

Collaborating Institutions:

Tata Institute of Fundamental Research
Osaka City University

Detector Characteristics:

Fiducial mass = 70 tonnes
10 × 10 cm proportional tubes with 0.23 cm iron walls
1.2 cm iron plates
Depth = 7600 mwe
Ionization measurement gives some track directionality

Results:

210 tonne-yr exposure
16 contained events
4 nucleon decay candidates

The KGF detector is comprised of two crossed horizontal stacks of large iron-walled proportional tubes, 6 m × 4 m in one view and 4 m × 4 m in the other. The

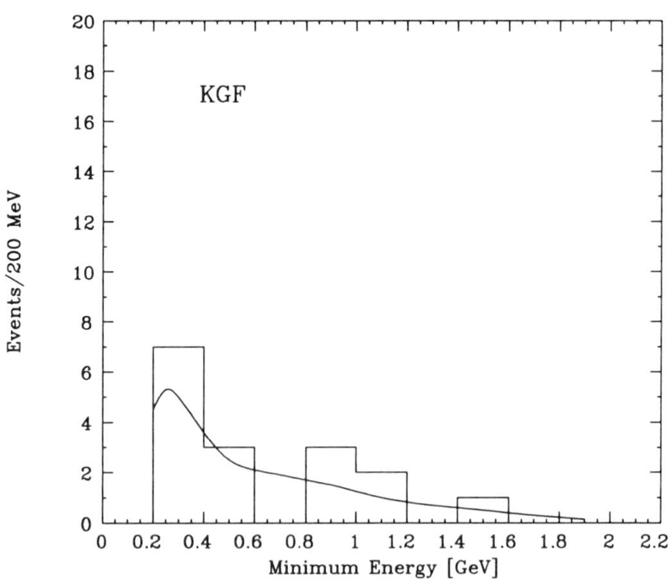

FIGURE 3. Energy distribution of KGF contained events. The curve is the KGF estimate of neutrino background.

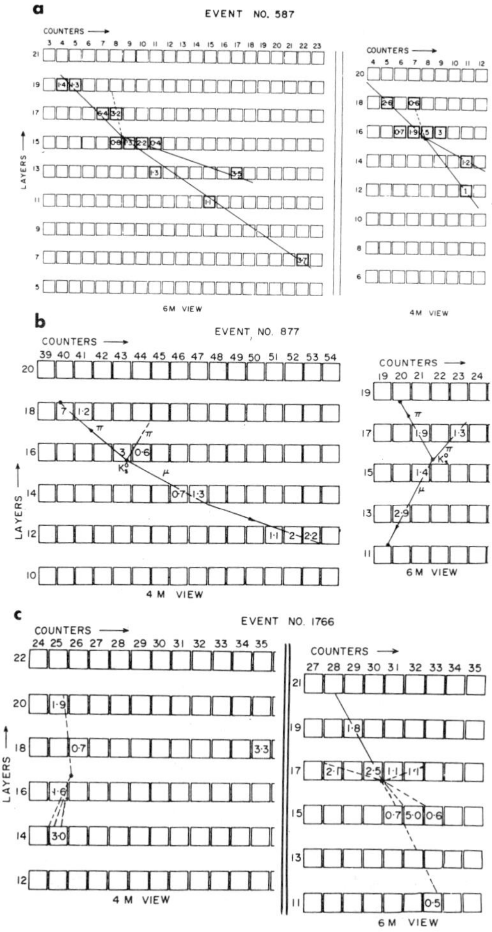

FIGURE 4. (a) KGF candidate for $P \rightarrow e^+\pi^\circ$. Widely spaced hits indicate an electromagnetic shower. (b) KGF candidate for $P \rightarrow \mu^+(K^\circ \rightarrow \pi^+\pi^-)$. Numbers in boxes indicate the ionization level in the associated proportional tube. (c) KGF candidate for $N \rightarrow \nu\eta^\circ$, $N \rightarrow e^+\rho^-$, or $P \rightarrow e^+\pi^\circ$.

experiment was the first dedicated nucleon decay apparatus to turn on, in recent times, and has now logged more than three years of livetime. The visible energy spectrum of 16 contained events is shown in FIGURE 3. The estimated background due to neutrino interactions is also shown, along with the position of four nucleon decay candidates. Orthogonal views of three of the candidates are shown in FIGURES 4a, b, and c. The ionization level recorded by each tube is also indicated. The experimenters make use of this information to get track directionality. Corrections for the amount of gas traversed by a hypothesized track can be a significant factor in the ionization information.

As can be seen from FIGURE 4, the spatial resolution and number of tubes firing in this pioneering experiment appear to be minimal, which is due to the tube size and amount of iron between tubes. One would hope to have more information content per event in order to make a convincing case for the discovery of nucleon decay in the midst of the neutrino background, especially given what is now known about the rate.

NUSEX RESULTS

Collaborating Institutions:

Frascati, Milano, Torino, CERN

Detector Characteristics:

Fiducial mass = 100 tonnes
9 mm × 9 mm limited streamer tubes, bidirectional readout
1.0 cm iron plates
No ionization information
Depth = 5000 mwe

Results:

220 tonne-yr exposure
22 contained events
2 nucleon decay candidates

This detector has better spatial resolution than that of KGF due to the smaller tube size and thinner iron plates. The two nucleon decay candidates are shown in FIGURES 5a and b.

The NUSEX experimenters determine the neutrino background from the results of running a test module of the detector in a neutrino beam at CERN. In principle, such tests can be made quite realistic for tracking calorimeter detectors because of their modular nature. In practice, difficulties arise in trying to match the atmospheric neutrino spectrum and ν_e, $\bar{\nu}_e$, ν_μ, $\bar{\nu}_\mu$ content with an accelerator beam. Also since tracking detectors do not have isotropic response, it is difficult to obtain sufficient statistics at a sufficiently large number of incident angles.

The NUSEX group calculates the background for event #1 to be 0.07 ± 0.05 events. They have also put a test module into a gamma beam at an incident angle of 45° to help them calculate the background for event #2 to be 0.18 ± 0.06 events.

FRÉJUS RESULTS

Collaborating Institutions:

Aachen, Orsay, Palaiseau, Saclay, Wuppertal

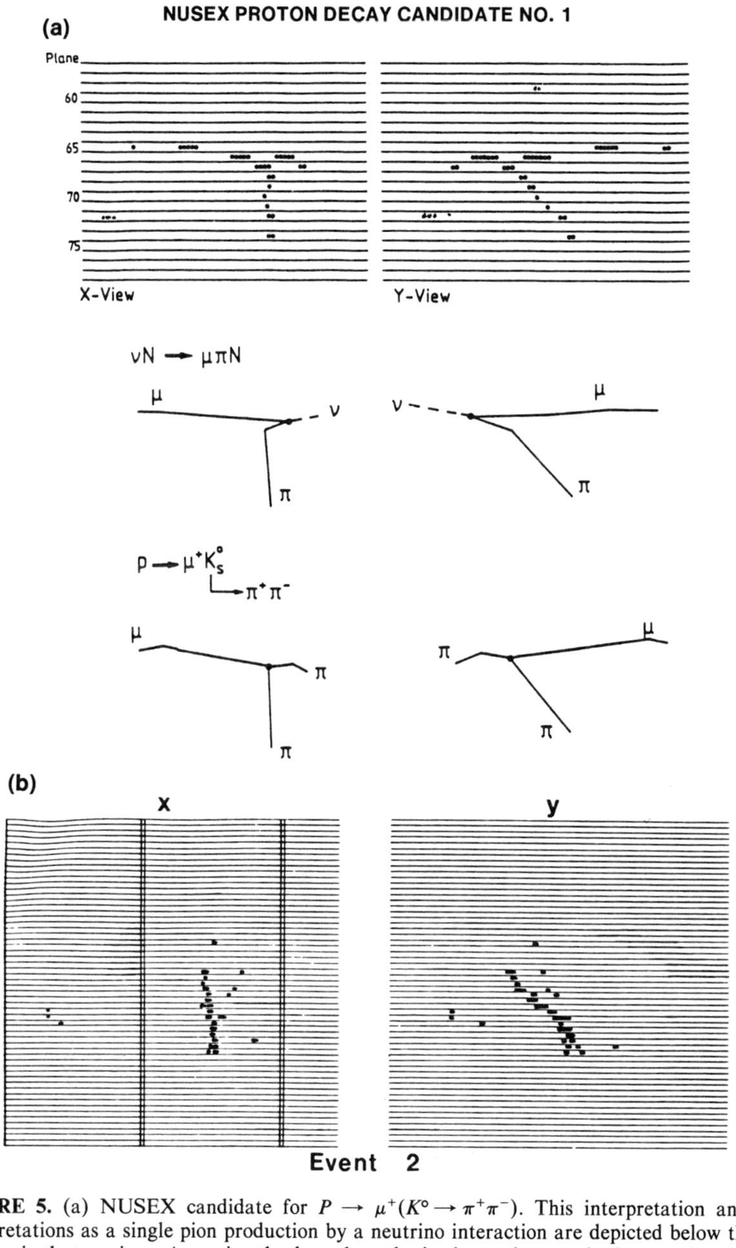

FIGURE 5. (a) NUSEX candidate for $P \rightarrow \mu^+(K^\circ \rightarrow \pi^+\pi^-)$. This interpretation and the interpretations as a single pion production by a neutrino interaction are depicted below the hit patterns in the two views. Assuming the decay hypothesis, the total energy in the event is 1.0 ± 0.2 GeV, the invariant mass of K° is 550 ± 80 MeV, and the momentum imbalance is $\Delta P = 0.4 \pm 0.2$ GeV/c. (b) NUSEX candidate for $P \rightarrow e^+\pi^\circ$. Assuming two showering tracks, the total energy is 1.1 ± 0.2 GeV and $\Delta P = 0.25 \pm 0.15$ GeV/c.

Detector Characteristics:

Fiducial mass = 450 tonnes
6 mm × 6 mm flash tubes triggered by 15 mm × 15 mm Geiger tubes
3 mm thick iron plates
No ionization information
Depth = 4200 mwe

Results:

90 tonne-yr exposure
9 contained events
0 nucleon decay candidates

This detector has excellent spatial resolution because of the thin plates and small flash tubes. It can also detect muon decays with an efficiency of ~70%. Early results give them nine contained events at the rate expected for neutrino interactions. None of the events appears to be of an unusual variety. FIGURE 6 shows an 800 MeV muon that starts and stops inside the detector. The good spatial resolution allows one to see the multiple scattering of tracks like this.

HPW RESULTS

Collaborating Institutions:

Harvard, Purdue, Wisconsin

Detector Characteristics:

Fiducial mass = 560 tonnes
Water Cherenkov type
700 5-inch phototubes distributed in cylindrical volume
Walls lined with mirrors for increased light collection
PWC anti surrounds tank
Depth = 1500 mwe

Results:

380 tonne-yr exposure
5 contained events (≥ 2 muon decays only)
2 nucleon decay candidates

Until now, the experimenters have only used their data to look for events with two or more muon decays. This, along with their system of PWC tubes, more easily allows them to filter out cosmic ray muon background. They make a further cut at 300 photoelectrons, which corresponds roughly to accepting events with Cherenkov equivalent energy (E_c) $\leq$ 450 MeV. The 2-μ sample so selected is shown in FIGURE 7.[3] The sample is further reduced to four contained events by scanning away entering showers and electronic noise, as shown in FIGURE 7. A similar procedure is applied to the 3-μ sample, which leaves them with one contained event having 92 MTD hits ($\approx$ photoelectrons) in the primary vertex trigger.

Separate vertices for the primary trigger and muon decay positions are determined with an accuracy of $\approx \pm$ 1 m for each event. These data are shown in TABLE 2. Events 2 and 3 are chosen as possible nucleon decay candidates because their "Isotropy" variable makes them look more isotropic (proton-decay-like) than the general neutrino background events. The vertex locations for the two candidates are depicted in FIGURES 8a and b.

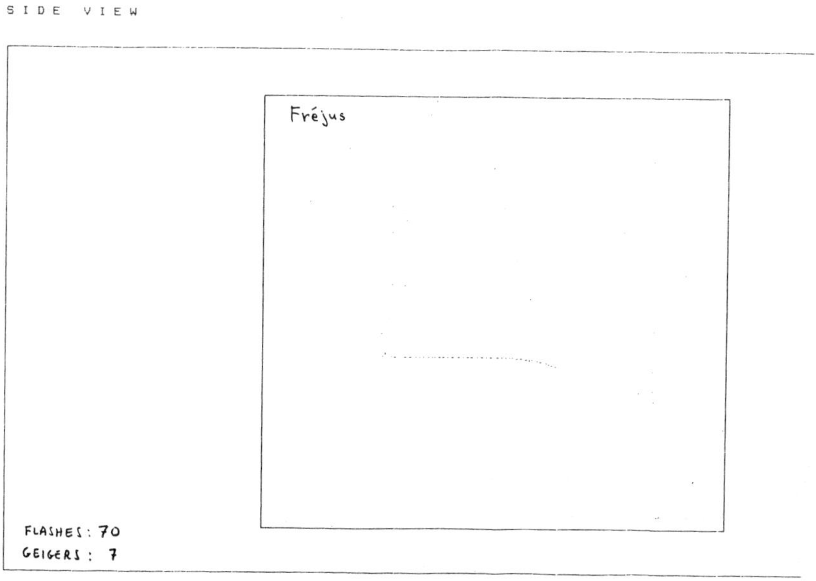

FIGURE 6. Event in the new Fréjus detector. An 0.8 GeV horizontal muon, probably made by a neutrino interaction, starts and stops deep inside the detector volume.

The experimenters calculate the total neutrino background for the five events to be ~0.3 events. This calculation is in reasonable agreement with IMB results of five events with $\geq$ 2 muons having E_c < 450 MeV since the IMB data are based on ten times the exposure.

It seems clear that the five HPW contained events cannot be due to neutrino interactions. The experimenters, though, suggest that they may be interesting: perhaps multimuon/pion modes of nucleon decay.

There seems to be a problem with all five of the events in that the number of MTD hits associated with the primary vertex (trigger 1) does not indicate enough Cherenkov energy [$E_c \approx$ 1.5 (MTD hits)]. That is, one expects factors of three to ten times more

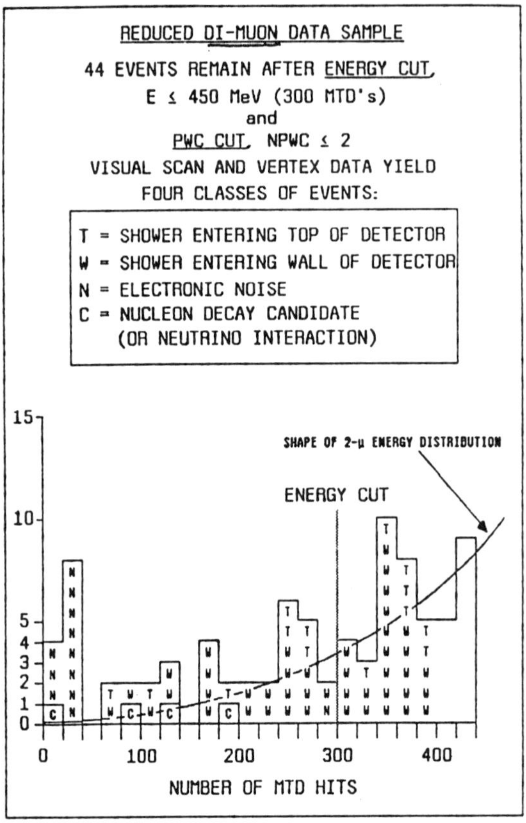

FIGURE 7. HPW sample of events with two muon decays. The energy cut (300 MTD hits) is indicated, along with the scanning results on each event.

TABLE 2. HPW Contained Events

	Trig.[a]	MTDs[b]	Iso.[c]	Vertex (x, y, z)	μ-Decay Time (μs)
Event 1 [1429–11,137]	1	162	.34	−1.6, 2.9, 1.3	
	2	22		−1.5, −1.5, 1.5	0.6
	3	39		−1.5, 2.6, 1.6	1.5
Event 2 [2122–5236]	1	17	.36	3.2, −1.7, 0.2	
	2	42		3.3, −2.1, −2.0	1.2
	3	38		1.1, −3.5, −1.5	2.6
Event 3 [2133–16,773]	1	81	.21	3.1, −2.3, −0.1	
	2	32		3.5, −1.0, 2.5	0.9
	3	25		1.5, −2.5, 2.5	2.4
Event 4 [2489–2499]	1	132	.48	2.3, −2.0, 1.6	
	2	14		3.0, −2.0, 0.0	0.9
	3	33		2.4, 0.3, 2.3	1.9
Event 5 [2920–32,257]	1	92	.40	2.8, −2.5, 2.3	
	2	24		3.5, −3.5, 2.5	2.4
	3	40		2.0, −2.0, 2.5	3.2
	4	15		2.5, 0.5, 3.5	4.3
1 PWC HIT ⟶				−2.3, −6.3, ---	

[a]Trigger number.
[b]Number of MTD hits.
[c]Isotropy.

light to be recorded if muons or other charged particles are to propagate from the primary vertex to the decay vertices. The experimenters attribute this inconsistency to possible errors in the individual vertex location estimates, i.e., the vertices are really much closer together than indicated by the positions given in TABLE 2.[3]

One can ask if there is any evidence for HPW type events in the other detectors. (Only the water Cherenkov experiments so far have low enough thresholds and high enough muon detection efficiencies to see such events.) If one takes the number of MTD hits in TABLE 2 at face value, then three of the five events are below the present

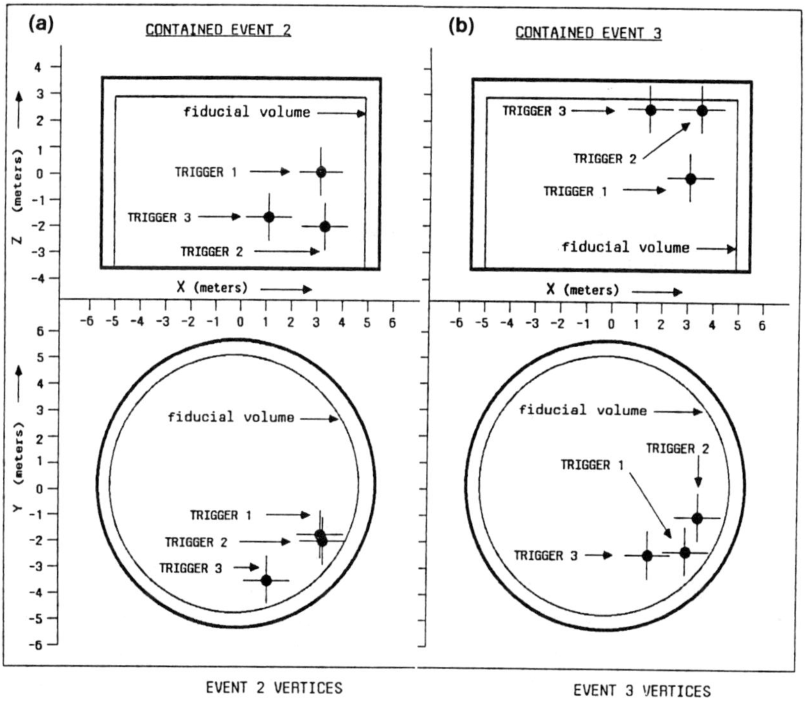

FIGURE 8. (a, b) HPW event candidates for nucleon decay. Projections of the event vertices are shown in top and side views. Trigger 1 is the primary vertex and triggers 2 and 3 are the positions of the two muon decays. More details for these and three other contained events can be found in TABLE 2.

IMB threshold of $E_c = 180$ MeV (≈ 120 MTD hits in HPW). With ten times the HPW exposure, IMB sees ten contained events with 2-μ decays and none with 3-μ decays. Only one of these events has an "Isotropy" of ≤ 0.5. It has $E_c = 350$ MeV. We will take a closer look at it later (in FIGURE 15a). Hence, whereas IMB might have expected to see $\sim 20 \pm 14$ HPW type events, only one is seen.

In principle, the KAMIOKANDE experiment can detect events with several muon decays at E_c values of ≤ 50 MeV. Hence, it should be able to check the HPW results on a possible multimuon signal concentrated below $E_c = 100$ MeV. At present, I don't have the appropriate information from KAMIOKANDE in order to make the comparison. KAMIOKANDE does, however, have almost twice the exposure of HPW.

KAMIOKANDE RESULTS

Collaborating Institutions:

University of Tokyo (Department of Physics, International Center for Elementary Particle Physics, and Institute for Cosmic Ray Research), National Lab for High Energy Physics (KEK), and University of Niigata

Detector Characteristics:

Fiducial mass = 880 tonnes
Water Cherenkov type
1000 20-inch phototubes on surface of cylinder
20% area coverage
Depth = 2700 mwe

Results:

660 tonne-yr exposure
89 contained events
4 nucleon decay candidates

Preliminary results on the energy spectrum of the KAMIOKANDE events are shown in FIGURE 9. Only 65 of the 89 events, those that appear to be just a single track (pseudo-elastic), are shown. The energy spectrum agrees, at least approximately, with the expectation from atmospheric neutrinos, although the atmospheric predictions for the total rate and ν_μ/ν_e ratio are not very well determined below 500 MeV.[4] It is significant that KAMIOKANDE sees events with less than 100 MeV of Cherenkov energy. This is an area totally unexplored by any other current (or previous) detector.

In addition to the 65 pseudo-elastic events in FIGURE 9, the KAMIOKANDE group has 24 contained events with ≥ two tracks. If the tracks (i.e., Cherenkov cones) are well isolated from one another, the distinction between showering (S) or nonshowering (M) tracks can be made. The group lists four candidates consistent with one or more nucleon decay modes.[5]

The first of these is the rather unusual event shown in FIGURE 10a. It has three distinct tracks, as shown on the unfolded cylinder. The size of the small circles represents the pulse height, or photoelectron count, of the lit phototubes. The event also has a possible (but not definite) delayed pulse from a $\mu \rightarrow e$ decay (shown in inset). The event can be interpreted as a muon (weakest track) plus two showers arising from $K^\circ \rightarrow 2\pi^\circ$, $\eta^\circ \rightarrow 2\gamma$, or simply two nonresonant gammas or electrons. The event is nearly planar, having $\Sigma\theta_{ij} = 340°$. The background from atmospheric neutrinos is estimated as ≤ 0.3 events at 90% C.L. One problem in interpreting the event as a nucleon decay is the relatively high momentum imbalance. The two showering tracks have a net momentum of about 485 MeV/c, which when added to the momentum of the "muon" track, leaves 380 MeV/c net momentum. This is rather large for Fermi momentum from the oxygen nucleus. One would need to see some similar events with lower net momentum in order to make a case for this one.

The second event (FIGURE 10b) is interpreted as possibly having five tracks. The minimum energy for the event is 600 MeV. There is no evidence of $\mu \rightarrow e$ decay. Of course, a neutrino interaction having five primary tracks at this energy would be very unusual. On the other hand, charged pions are very likely to scatter in water at these

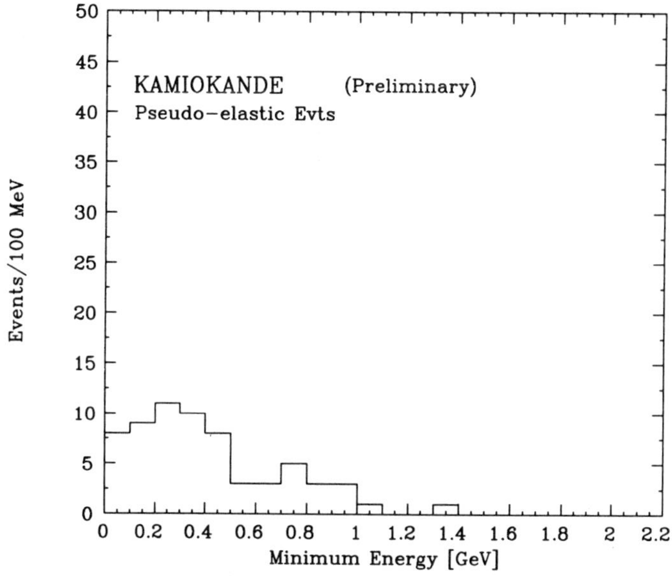

FIGURE 9. Energy distribution for 65 KAMIOKANDE contained events that appear to be pseudo-elastic neutrino interactions. Note that this detector can record events below 100 MeV.

energies so that one track can appear as two. The estimated neutrino background for this topology is ≤ 0.8 events at 90% C.L. If the Kamioka group interprets the event as $p \rightarrow e^+\omega^o$ ($\omega^o \rightarrow \pi^+\pi^-\pi^o$), it gets a proton mass of 880 MeV (consistent with 940 within errors) and a net momentum of 250 MeV/c.

The third KAMIOKANDE candidate (FIGURE 10c) is a wide-angle ($\theta = 166°$) two-track event with an energetic (535 MeV/c) showering track and a low momentum muon or pion of about 200 MeV/c. The invariant mass of the pair is about 720 MeV with a momentum imbalance of more than 300 MeV/c. In order to fit a nucleon decay mode, one needs to assume a third particle below Cherenkov threshold, thus making decays such as $N \rightarrow \mu^+\rho^-$ or $p \rightarrow \nu\rho^+$ possible. The event has a definite $\mu \rightarrow e$ decay.

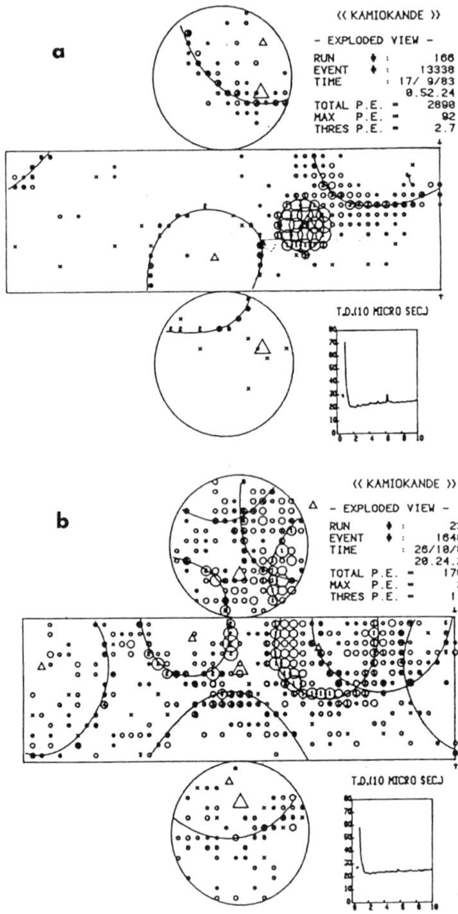

FIGURE 10a,b. KAMIOKANDE nucleon decay candidates. The small circles represent PMT hits projected onto an unfolded cylinder with the size proportional to light level. The large circles are drawn by the experimenters to guide the eye. Inset at lower right shows summed light output for the whole detector and indicates the presence or absence of stopping muon decays.

The fourth candidate (FIGURE 10d) also has a $\mu \longrightarrow e$ decay and is a wide-angle (146°) two-track event with both tracks being nonshowering. The tracks have 404 and 203 photoelectrons, respectively. Assuming a $\pi\mu$ pair, one gets an invariant mass of 650 MeV and a net momentum of 230 MeV/c. Hence, it is also necessary in this case to invoke an unseen particle in order to fit nucleon decay. Then modes such as $p \longrightarrow \mu^+\rho^\circ$, μ^+k°, or $N \longrightarrow \nu\rho^\circ$ are possible. The background is calculated as ≤ 0.3 events at 90% C.L.

IMB RESULTS

Collaborating Institutions:

University of California Irvine, University of Michigan, Brookhaven National Laboratory, California Institute of Technology, Cleveland State University, Fermilab, University of Hawaii, Lawrence Livermore Laboratory, University College, Warsaw University, University of Illinois

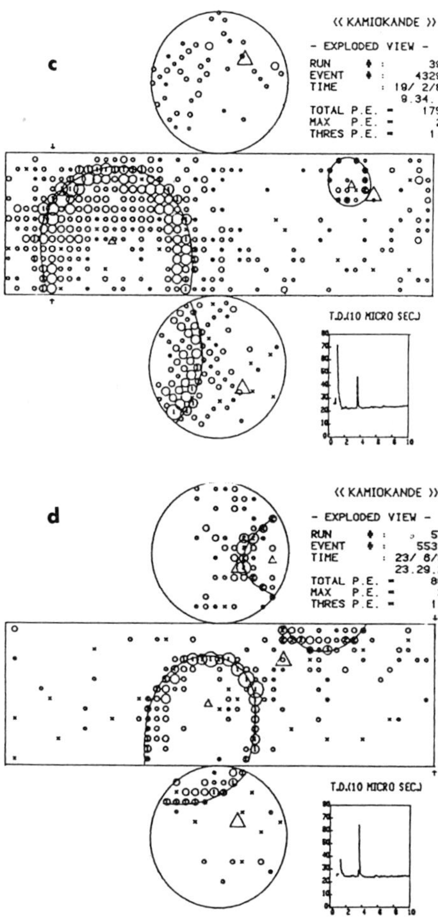

FIGURE 10c,d. KAMIOKANDE nucleon decay candidates. The small circles represent PMT hits projected onto an unfolded cylinder with the size proportional to light level. The large circles are drawn by the experimenters to guide the eye. Inset at lower right shows summed light output for the whole detector and indicates the presence or absence of stopping muon decays.

Detector Characteristics:

Fiducial mass = 3300 tonnes
Water Cherenkov type
2048 5-inch hemispherical phototubes on the surfaces of a parallelepiped
1.5% area coverage
Depth = 1670 mwe

Results:

3800 tonne-yr exposure
401 contained events
21 nucleon decay candidates

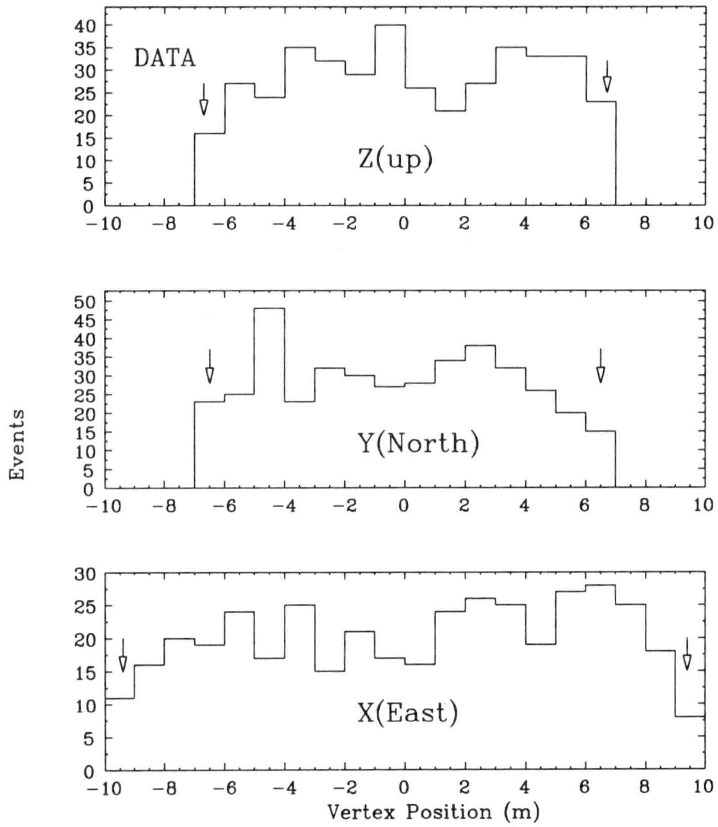

FIGURE 11. Projections of the vertices found in the IMB detector for 401 contained events. Arrows indicate the edges of the fiducial volume.

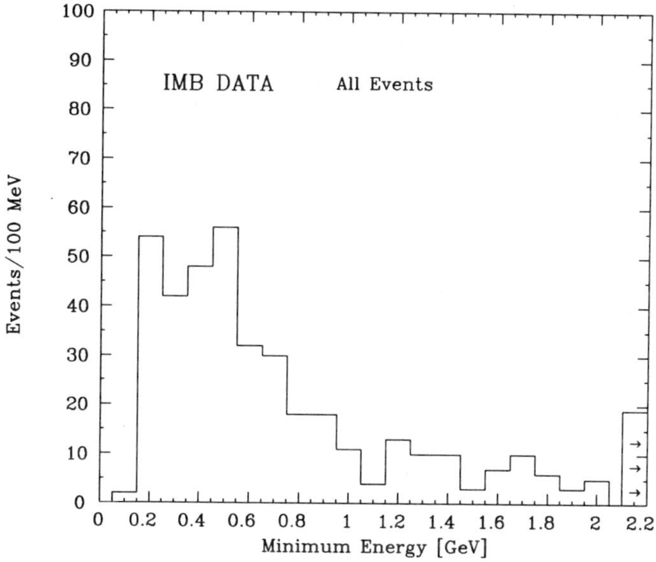

FIGURE 12. IMB energy distribution for 401 contained events.

The IMB collaboration has analyzed 417 live-days of data taken prior to the installation of the new wave-shifters around the tubes in the fall of 1984. The 401 contained events are distributed uniformly throughout the fiducial volume, which begins two meters in from the planes of tubes. FIGURE 11 shows the projections of the event vertices on the three orthogonal axes. Typical vertex resolutions are ~90 cm for single-track, or narrow-angle, events that produce light mainly in one hemisphere and ~50 cm for wide-angle events that have significant light recorded in two hemispheres.

The energy distribution for the 401 event sample is shown in FIGURE 12. Plotted here is the minimum visible energy (E_{min}). For events with all showering tracks, this is the total equivalent Cherenkov energy, E_c. When there is a muon decay detected, indicating the presence of a charged pion or muon, we define:

$$E_{min} = E_c + (200 \text{ MeV for each observed muon}).$$

The expected shape of the E_{min} distribution is shown simulated in FIGURE 13. The simulation is done by using neutrino and antineutrino events from the Gargamelle (Freon) bubble chamber data.[6] Only charged-current events with a final-state muon are used. The muon is replaced by an electron to simulate ν_e events and by a neutrino to simulate neutral currents. The events are chosen proportionate to their occurrence in the atmospheric spectrum. They are then used to generate simulated events in the detector. The simulated events are passed through the same analysis programs as are the real data. The absolute rates agree to within ±20% with estimates based on recent atmospheric flux calculations.[2,4]

In FIGURE 12, there is no evidence for a peak near the rest mass (940 MeV) of the nucleon. One can make use of event geometry in order to dig further into the neutrino background. Neutrino events tend to be single-track (or narrow-angle) two-track events at these energies, whereas nucleon decays should be wide-angle events with opening angles in the range of 150°–180°. We make use of an "anisotropy" variable (A) to differentiate between the two types of events. The value of A lies between 0 and 1 and is given by the mean vector sum of the unit vectors from the event vertex to each recorded phototube (PMT) hit. For a single short track that emits light only at the Cherenkov angle, A will be given by the cosine of the angle (0.75 for water).

A scatterplot of E_{min} versus A is shown in FIGURE 14. The region below and to the left of the dashed line is of interest for searching for nucleon decay. (The squares are events with two muon decays. Those and the three circled events will be discussed later.) The lower limit of 180 MeV in FIGURE 14 is due to the software threshold cut of 40 PMTs. One notices that the population of the regions below 1000 MeV are reasonable extrapolations of the regions above 1000 MeV for both the "isotropic" ($A < 0.5$) data and the "anisotropic" ($A > 0.5$) data. Hence, even without external knowledge of the neutrino background, one can judge from these extrapolations that there is a sizable neutrino background that must be dealt with in searching for nucleon decay, even in the "wide-angle" event sample.

If one uses just the two variables, E_{min} and A, to describe the events, then most of the events inside the dashed area of FIGURE 14 are "candidates" for one or more of the nucleon decays for which limits have recently been published.[7] (These published limits

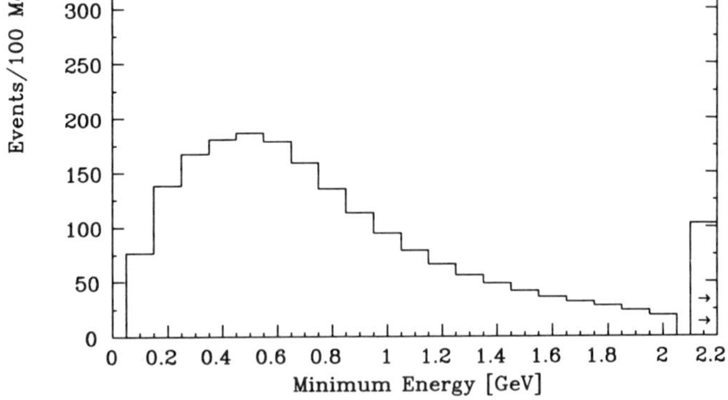

FIGURE 13. Energy distribution for IMB simulated events based on Gargamelle bubble chamber data. This represents the shape of the atmospheric neutrino background.

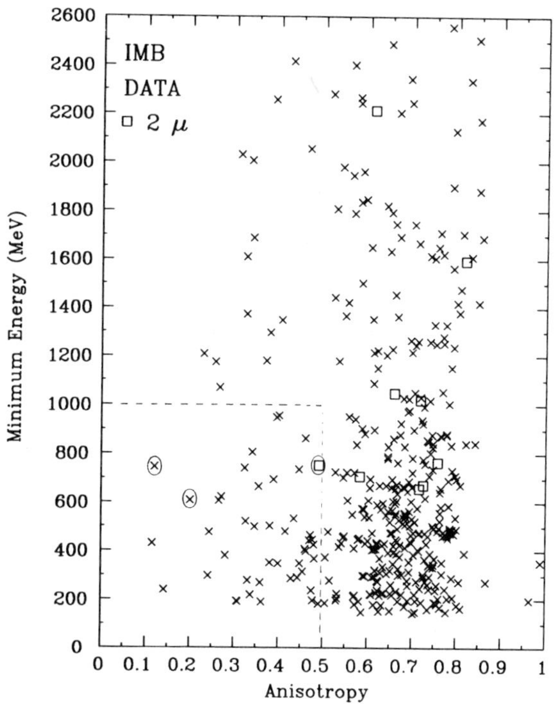

FIGURE 14. Scatterplot of E_{min} versus A for 401 IMB contained events. Open squares are events with two muon decays. (There is one additional 2μ event at 2700 MeV, $A = 0.46$.) Circled events are shown in FIGURES 15a, c, and e.

are conservative estimates calculated by taking the neutrino background to be negligible.)

In general, there is much more information in each event than just E_{min} and A. We have selected three particularly interesting events (circled on FIGURE 14) for closer inspection. The first event is shown in FIGURE 15a. Each lit PMT is indicated by one or more slashes, with each slash being proportional to the amount of light it recorded. The PMT hits have been projected onto a sphere whose center is at the event vertex. The sphere has been unfolded into a standard Mercator projection with the polar axis perpendicular to the track directions. The event is a nice example of a wide-angle two-track event, as indicated by two well-separated Cherenkov rings. The angle between the two tracks is measured as 178° with a typical error of ~10°.

There are two recorded muon decays (not shown) in the event. It is the only one of ten such events (see FIGURE 14) that is similar to the HPW events discussed earlier.

The left track in FIGURE 15a has $E_c = 301$ MeV and the right track has $E_c = 68$ MeV. [Typical errors in E_c are $\pm 100\%/\sqrt{\text{PMTs}} \pm 10\%$ (sys.).] The event is a candidate for the following decay modes:

$$N \rightarrow \mu^- \pi^+$$

$$P \rightarrow \mu^+\mu^+\mu^-$$

$$P \rightarrow \mu^+ + (K^\circ, \eta^\circ, \rho^\circ, \omega^\circ).$$

Taking the first interpretation, $N \rightarrow \mu^-\pi^+$, for example, one derives the following kinematics:

$$P_\mu = 490 \text{ MeV}/c$$

$$P_\pi = 297 \text{ MeV}/c$$

$$\Delta P = 193 \text{ MeV}/c$$

$$M = 806 \pm 120 \text{ MeV}.$$

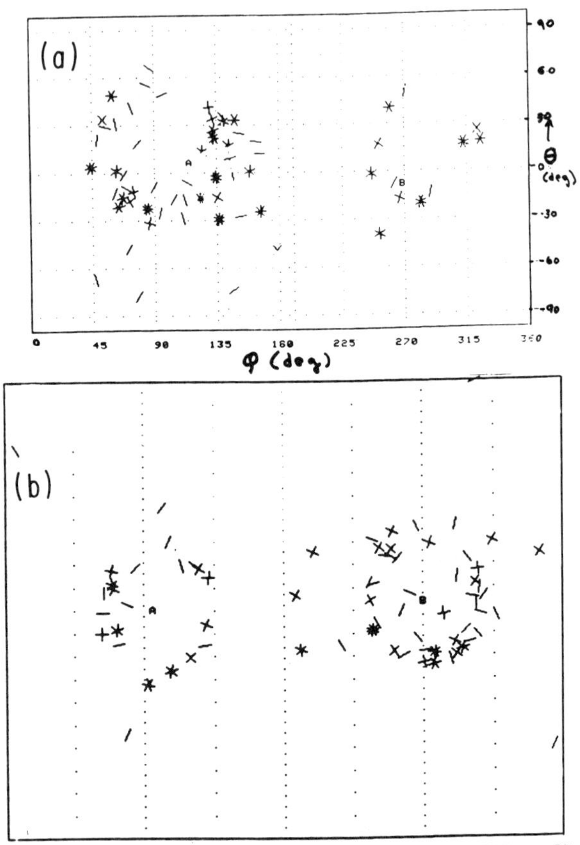

FIGURE 15a,b. (a) Wide-angle two-track events from the IMB detector (Event 8320; 2μ). (b) Simulated wide-angle two-track events from neutrino interactions ($\nu N \rightarrow \mu^-\Delta^+ \rightarrow \mu^-\pi^+N$). The points labeled "A" and "B" represent the reconstructed track directions as seen from the vertex. (See text.)

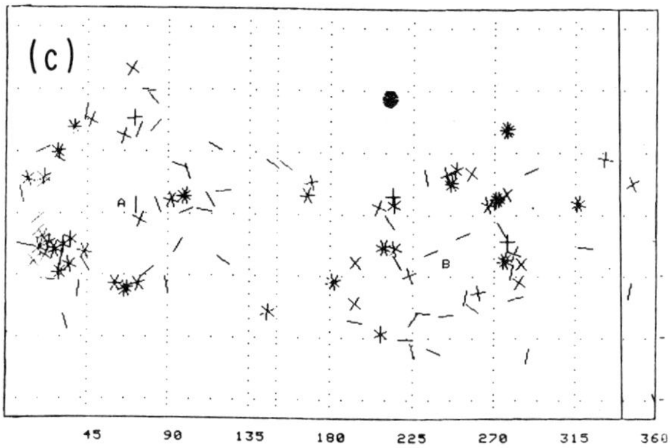

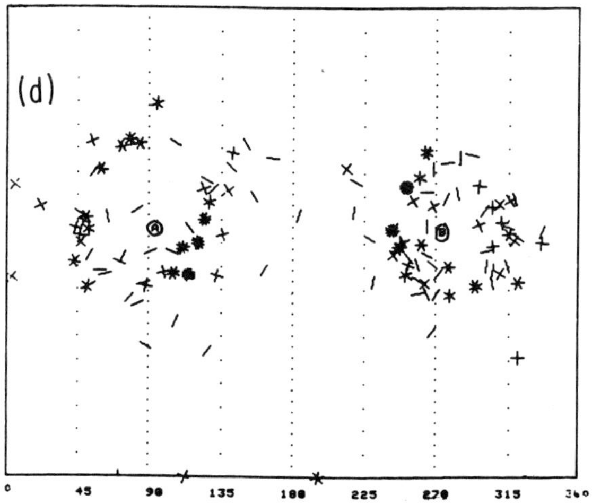

FIGURE 15c,d. (c) Wide-angle two-track events from the IMB detector (Event 17, 473; no μ). (d) Simulated wide-angle two-track events from neutrino interactions ($\nu N \rightarrow e^-\pi^+N$). The points labeled "A" and "B" represent the reconstructed track directions as seen from the vertex. (See text.)

Thus, although the invariant mass is somewhat low, it gives a reasonable fit to bound neutron decay.

Now, what about the background?

As mentioned earlier (FIGURE 2), atmospheric neutrinos do produce wide-angle two-track events. FIGURE 15b shows a simulated background event that has been passed through the standard analysis programs. It is quite similar to the real event of FIGURE 15a. The included angle is 174°. The background is $\nu N \rightarrow N\mu^-\pi^+$. We will estimate the frequency of such background events later in the discussion.

The second interesting real event is shown in FIGURE 15c. It has no recorded muon decays, but the total Cherenkov energy is 608 MeV, as opposed to 369 MeV for the first event. Its included angle is ~175° and it is a candidate for several nucleon decay modes, including $N \rightarrow e^+\pi^-, e^+\rho^-, e^-\rho^+$. The background calculation also, in this case, has a similar event, which is shown in FIGURE 15d. The background reaction is $\nu N \rightarrow Ne^-\pi^+$.

The third wide-angle event of interest is shown in FIGURE 15e. Its Cherenkov energy is still higher: $E_c = 722$ MeV. It also has no recorded muon decays and is a candidate for the same modes as the second event. It falls just below the acceptance cut

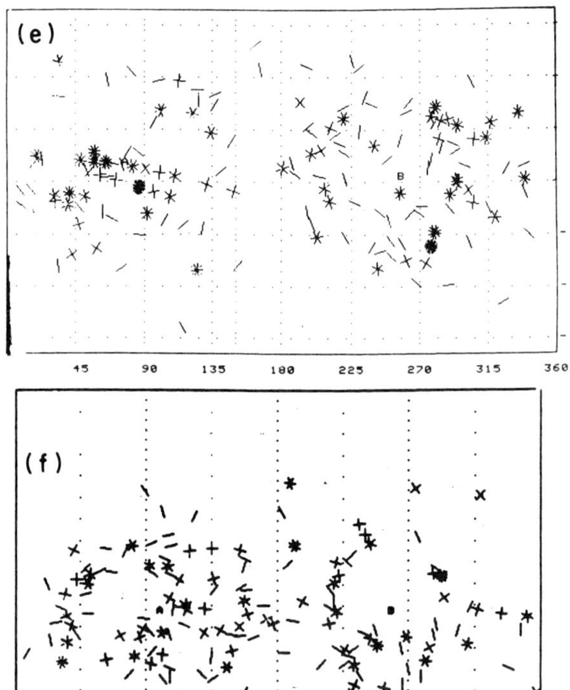

FIGURE 15e,f. (e) Wide-angle two-track events from the IMB detector (Event 42,652; no μ). (f) Simulated wide-angle two-track events from neutrino interactions ($\nu N \rightarrow e^-\pi^+N$). The points labeled "A" and "B" represent the reconstructed track directions as seen from the vertex. (See text.)

($E_c = 750$ MeV) for $p \rightarrow e^+\pi^0$. It also has a companion simulated background event, shown in FIGURE 15f, from the reaction of $\nu \rightarrow N e^-\pi^0$.

We see that wide-angle (>150°) two-track events consistent with nucleon decay definitely are present in the IMB data at a level of ~1% of the contained events. The question is, of course, are they nucleon decay? What is the expected background rate?

The cross sections for single pion production by neutrinos in the energy range of 1–3 GeV are fairly well known. I make the following estimate of the rate for wide-angle two-track events to pass the IMB selection criteria, using atmospheric neutrinos to generate single pions in charged-current interactions.

I expect 1.1 events/kilotonne-yr having the following properties:

(1) $0.2 < E_c < 1.0$ GeV;
(2) Two tracks both with $E_c > 50$ MeV;
(3) Included angle $> 150°$.

This estimate is in good agreement, given the statistics of a few events, with the rate seen by IMB.

CONCLUSIONS

(1) Atmospheric neutrino interactions have been recorded in several of the "modern" nucleon decay detectors. The rate of detected events is about as expected:

$$\sim 100 \text{ events/kt-yr with } E_{vig} > 0.2 \text{ GeV}.$$

(2) There are approximately five events/kt-yr that have $\geq$ two fairly wide-angle tracks and that fall into the general category of nucleon decay "candidates."
(3) Wide-angle ($>150°$) two-track events with $\Delta P \gtrsim 200$ MeV/c (the "classic" proton decay topology) do occur at a rate of $\sim$ one event/kt-yr. At this point, there is no good evidence that such events are inconsistent with the background; the rate is approximately as expected.
(4) Using the above numbers and adding some reasonable estimate of the background to be subtracted, one gets the following general limits:

$$\text{If } P \rightarrow \text{"classic" modes, then } \tau_p \geq 5 \times 10^{32} \text{ yr;}$$

$$\text{If } P \rightarrow \nu \text{ dominated modes, then } \tau_p \geq 5 \times 10^{31} \text{ yr.}$$

(5) If nucleon decays are hiding in the "neutrino curtain," then there is going to be a lot of hard work digging them out. One helpful clue should be the dominance of a single decay mode.

THE SCHEDULED FUTURE

We return to the fact that the existence of proton decay is a very important question and one that hopefully can be answered experimentally. The following experiments are being mounted to work on the question:

(1) The KFG group is building a new 300-tonne detector with better spatial resolution (6 mm plates). It will be ready in the summer of 1985.
(2) The Fréjus detector will be increased to 900 tonnes in the spring of 1985.
(3) The Kamioka detector will be surrounded with a water Cherenkov anti-

counter to increase the fiducial volume and lower the energy threshold. Timing information will be added to the tubes to improve spatial resolution.

(4) The IMB group has increased the light collection of its detector by a factor of two and will add another factor of two by the end of 1985.

(5) The Soudan-2 group (Minnesota, Argonne, Oxford, Rutherford, Tufts) will build a new detector with excellent (2 mm) spatial resolution and good ionization information. The group plans to have 500 tonnes operating in 1986 and 1000 tonnes in 1987.

Clearly the subject is alive and well; the question remains, is the proton? Time and patience, hopefully, will tell.

In closing, let me say that I'm very impressed with the way Martin Block and his cohorts have put together this conference. The physics, the hospitality, the recreation (including the fine dinners), and the camaraderie have all been first-rate. It's hard for a physicist not to find something wrong with an experiment, but this one has been nearly perfect; so good that it needs to be repeated.

ACKNOWLEDGMENTS

I am grateful to D. Ayres, R. Barloutaud, D. Cline, E. Fiorini, M. Marshak, V. S. Narasimham, and D. Perkins for providing much of the information on which this paper is based. Special thanks go to my colleagues on the IMB experiment (see reference 7) for their untiring efforts in producing and analyzing the data presented here, much of which is as yet unpublished.

NOTES AND REFERENCES

1. PERKINS, D. H. Proton decay experiments. To appear in Ann. Rev.
2. GAISSER, T. K. & T. STANEV. 1984. *In* Proceedings of the International Colloquium on Baryon Nonconservation, Salt Lake City, January, 1984. D. Cline, Ed.: 61. University of Wisconsin Press. Madison, Wisconsin; GAISSER, T. K. 1984. *In* Proceedings of the Eleventh International Conference on Neutrino and Astro Physics. K. Kleinknecht & E. A. Paschos, Eds.: 372. World Scientific. Singapore.
3. FIGURES 7 and 8 and TABLE 2 are taken from the Ph.D. Thesis of MORE, A. H. 1984. University of Wisconsin.
4. PERKINS, D. H. 1984. Atmospheric neutrino fluxes. Oxford University preprint 85/84.
5. KOSHIBA, M. 1984. Plenary Report, XXII International Conference on High Energy Physics, Leipzig. See FIGURE 15 of this report for a comparison of τ/B limits from various experiments. IMB limits since then have been updated in reference 7.
6. DEDEN, H. *et al.* 1975. Nucl. Phys. **B85:** 269. We are grateful to the Gargamelle Collaboration for supplying us with summary tapes of their data.
7. PARK, H. S., G. BLEWITT *et al.* 1985. Phys. Rev. Lett. **54:** 22. An updated table of IMB lifetime limits based on twice the livetime cited here will be available in the spring of 1985.

Monopoles, Cosmology, and Astrophysics—Update 1985[a]

MICHAEL S. TURNER

NASA/Fermilab Astrophysics Center
Fermi National Accelerator Laboratory
Batavia, Illinois 60510
and
Enrico Fermi Institute
University of Chicago
Chicago, Illinois 60637

INTRODUCTION

The superheavy magnetic monopoles predicted by unified theories remain extremely interesting and enigmatic objects. They offer one of the few windows to physics at energies much greater than 1 TeV and to the earliest moments of the universe ($t \lesssim 10^{-34}$sec). Theory still provides no useful or reliable prediction of their relic abundance, and the very stringent astrophysical bounds based upon the survival of astrophysical magnetic fields and catalysis of nucleon decay in neutron stars remain intact after very careful scrutiny (summarized in FIGURE 1). In this article, I present a brief review of the current situation, a situation that has not changed appreciably in the past few years. I refer the interested reader to the longer and more complete reviews in references 1–5.

WHAT IS A SUPERHEAVY MONOPOLE?

A superheavy magnetic monopole is a stable, nontrivial configuration of gauge and Higgs fields.[6,7] Such configurations exist in a gauge theory whenever a semisimple group (e.g., $SU(5)$, $SO(10)$, $E6$, etc.) is broken down to a group with an explicit $U(1)$ factor, e.g., $SU(3) \times SU(2) \times U(1)$. The three conspicuous features of the 't Hooft-Polyakov monopole are: (1) large mass, of order, M/α, where M = the symmetry breaking scale and α = the gauge coupling constant (for $SU(5)$, $m_M \simeq 10^{16}$ GeV); (2) hefty magnetic charge, $g = ng_D$, where n is an integer and $g_D = 1/2e \simeq 69e$ is the Dirac charge; (3) the remarkable ability to catalyze nucleon decay either via the Callan-Rubakov process[8,9] or via the weak anomaly.[10,11]

The cross section for catalysis via the Callan-Rubakov process is expected to be of the order of a strong interaction cross section (and velocity independent),

$$(\sigma v/c)_{\Delta B} = \sigma_{-28} 10^{-28}\text{cm}^2 , \tag{1}$$

[a]This work was supported in part by the DOE (at The University of Chicago and Fermilab), by the NASA (at Fermilab), and by an Alfred P. Sloan Fellowship.

while catalysis via the weak anomaly is thought to be suppressed by mixing angles and other kinematic factors, and is likely to be many orders of magnitude smaller ($\sigma_{-28} \ll 1$). Which monopoles catalyze nucleon decay via the Callan-Rubakov process and which via the weak anomaly? To oversimplify somewhat, if the physics at the core of the monopole does not conserve baryon number (as is the case in the simplest GUTs and simplest symmetry breaking schemes), then the monopole catalyzes nucleon decay

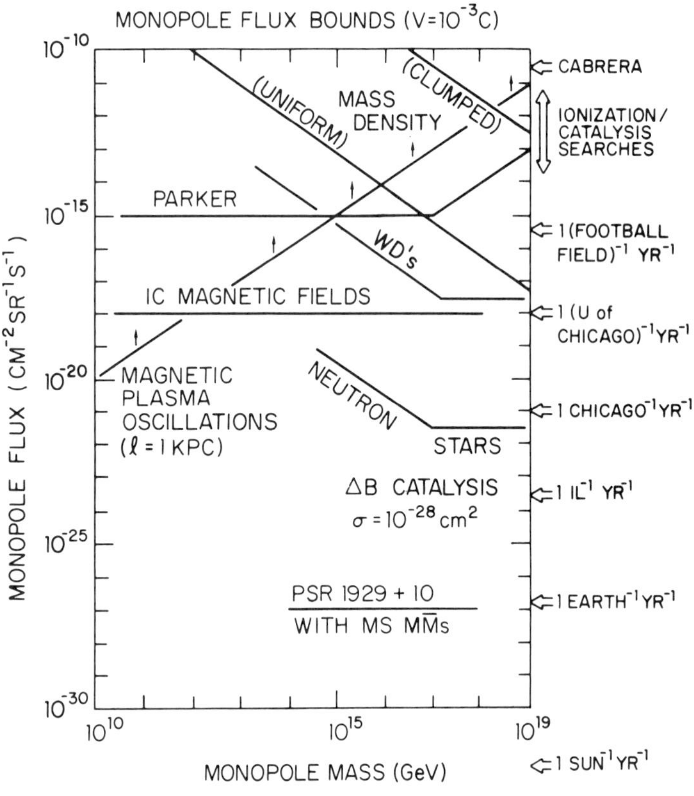

FIGURE 1. Summary of the astrophysical and cosmological monopole flux limits discussed in the text. Wherever necessary, a monopole velocity of $10^{-3}\,c$ was assumed. The line labeled "magnetic plasma oscillations" is the *minimum* flux required such that it is possible for coherent monopole effects to prevent the damping of the galactic magnetic field. Note that the strong magnetic fields of neutron stars may prevent monopoles less massive than about 10^{15} GeV from being captured.

via the Callan-Rubakov process. If the physics at the core of the monopole does not violate baryon number (as is the case in the more complicated SSB schemes, e.g., the Z_2 monopoles of $SO(10)$),[12] then the monopole only catalyzes nucleon decay via the weak anomaly.

These three properties lead to monopoles having interesting (and often conspicuous) astrophysical consequences, which in turn lead to the very stringent astrophysical consequences. I should emphasize that although superheavy magnetic monopoles are expected to have all three of these properties, the only property that a magnetic monopole must necessarily have is a magnetic charge that is a multiple of the Dirac charge.

BIRTH

There are certainly no contemporary sites, astrophysical or otherwise, where an object of a mass of 10^{16} GeV or so can be produced. Thus, we must look to the early universe ($t \lesssim 10^{-34}$sec, $T \gtrsim 10^{14}$GeV) as the birth site for these extremely interesting and important objects. There are two basic ways that monopoles can be produced in the early universe: (1) as topological defects in the SSB transition—the so-called Kibble process;[13] (2) thermal pair production in very energetic particle collisions.[14–16] Unless the SSB phase transition occurs at a very low temperature ($T \ll 10^{10}$GeV) or the symmetry breaking pattern is very complicated, process (1) leads to a severe overproduction of monopoles; in fact, for $M \simeq 10^{14}$ GeV, so many are produced by this process that the universe reaches a temperature of 3K at the very tender age of 30,000 years (references 1–5, 17–20). Subsequent annihilation is not efficient enough to reduce the number of monopoles to a safe level.[17] (However, Wasserman[21] has recently reexamined this question and raises the possibility that annihilations may be enhanced by the presence of the $e^{\pm}$ plasma, in which case they may be able to reduce the initial abundance to a safe level.)

If the SSB phase transition is inflationary,[22–24] then the overproduction is avoided as process (1) leads to less than one monopole in the observable universe. In more complicated inflationary scenarios, monopoles may be produced by process (1) during the inflationary epoch itself (toward the end)[25] or in a subsequent phase transition.[26] Whether or not an interesting number of monopoles can be produced in this way remains to be seen. By interesting, I mean large enough to be potentially observable, but not so as to be ruled out by the astrophysical constraints.

Process (2) occurs even if the universe underwent inflation. However, the number of monopoles produced this way is expected to be exponentially small (and exponentially uncertain). In fact, monopoles cannot be produced until SSB has occurred, $T < M$, and $m_M \simeq M/\alpha \simeq 100M$, and so the number expected is

$$n_M \propto \exp(-2m_M/M) \propto \exp(-\text{few } 100). \tag{2}$$

Although thermal production does not look too promising from a monopole hunter's point of view, the uncertainties are such that it is not impossible that an interesting abundance could have been produced this way.[14,27,28]

To summarize the present status of the theoretical predictions for the monopole abundance, pick a number, any number. In the absence of a meaningful prediction for the monopole abundance, all we can do is treat the monopole abundance as a free parameter. The following, though, are useful formulae:

$$\langle F\rangle \simeq 10^{10}(n_M/s)(v_M/10^{-3}c)\text{cm}^{-2}\,\text{sr}^{-1}\,\text{s}^{-1}, \tag{2a}$$

$$\simeq (n_M/n_B)(v_M/10^{-3}c)\text{cm}^{-2}\,\text{sr}^{-1}\,\text{s}^{-1}, \tag{2b}$$

$$\simeq 3\times 10^{-15}\Omega_M h^2(v_M/10^{-3}c)(10^{16}\text{GeV}/m_M)\,\text{cm}^{-2}\,\text{sr}^{-1}\,\text{s}^{-1}, \tag{2c}$$

where n_M, n_γ, n_B, and s ($\simeq 7.1\ n_\gamma$) are the average monopole number density in the universe, the photon number density, the baryon number density, and the entropy density, respectively; $<F>$ is the average monopole flux in the universe; $\Omega_M \equiv \rho_M/\rho_c$ is the fraction of critical density contributed by the monopoles; $\rho_c = 1.88h^2 \times 10^{-29}$g cm^{-3} is the critical density; $h = H_o/(100$ km sec^{-1} Mpc$^{-1})$; and v_M is the typical monopole velocity (to be discussed in more detail later).

The local monopole number density and flux (i.e., in the vicinity of the solar system) may be larger than the average values of these quantities in the universe if, for example, monopoles are concentrated in the galaxy. Denote the enhancement factor by f:

$$F_{GAL} \simeq f <F> . \tag{3}$$

Naively, we might expect $f \simeq \rho_{GAL}/\bar{\rho} \simeq 10^5$, which is a rather healthy factor. (Here $\bar{\rho}$ is the average density of the universe.) I will return to this issue later.

WHERE ARE THEY TODAY?

Because monopoles are so heavy, their internal thermal velocity dispersion is tiny: $\langle v_M^2 \rangle^{1/2} \ll 1$ cm sec^{-1}. Since they are effectively collisionless, they have only their velocity dispersion to support them against gravitational collapse. Due to their tiny velocity dispersion, they should be unstable to gravitational collapse on virtually all scales. However, as they are effectively collisionless, they cannot dissipate their gravitational energy and condense into tightly bound objects whose formation involves dissipation, such as the disks of spiral galaxies, stars, etc. (Later, after stars have formed, they can capture monopoles that impinge upon them, which is an issue to which I shall return.)

Naively then, we would expect to find monopoles in all structures whose formation does not involve dissipation, from galactic halos to clusters of galaxies to superclusters. The density contrasts for these objects are about 10^5, 100, and a few, respectively. Within these objects, we expect $f \simeq \rho/\bar{\rho}$. Since we live in a galaxy, this suggests a local enhancement of $f \simeq 10^5$ over the average monopole flux in the universe. However, the magnetic field of our galaxy will eject monopoles less massive than about 10^{20} GeV in less than the age of the galaxy.[32] Therefore, we should not expect to find a concentration of monopoles in our locality. The magnetic fields observed in clusters of galaxies are only potent enough to eject light monopoles, which are those lighter than about 10^{15} GeV.[37]

Since our galaxy is not a member of a cluster and is only on the outskirts of the Virgo supercluster where the density contrast is about equal to one, we would expect the monopole flux in our vicinity to be due primarily to monopoles that just happen to be passing through the galaxy (and due to perhaps an equal number that are bound to the Virgo supercluster). Thus, the local flux should be about equal to the average cosmic flux ($f \simeq 1$).

(It has also been suggested that the flux of monopoles in the solar system could be enhanced by a very large factor[29] (up to 10^6) due to a cloud of monopoles orbiting the sun. These monopoles are those that, over the life of the solar system, were captured by the sun. A more careful estimate[30] of this effect indicates that an enhancement of this kind might be at most a factor of a few.)

(It may be of interest to know that most monopoles should have become bound to protons about 10^5 s after the bang and if they do not catalyze nucleon decay, most monopoles should today still have protons bound to them. For a detailed discussion, see reference 54.)

HOW FAST ARE THEY MOVING?

From the point of view of detecting monopoles, their speed relative to the detector is a very important issue. As I discussed above, their thermal velocities are absolutely negligible. However, they will be accelerated by any gravitational and magnetic fields they encounter. The typical peculiar (meaning relative to the Hubble flow) velocities of objects in the universe are of the order of 10^{-3} c, thus implying typical monopole-galaxy velocities of this magnitude. Monopoles will be accelerated by the gravitational

TABLE 1. Sources of Monopole and Detector Motion

	Monopole	Detector
Gravitational		
—Virgo supercluster	$2 \times 10^{-3}\,c$	$2 \times 10^{-3}\,c$
—The galaxy	$10^{-3}\,c$	$7 \times 10^{-4}\,c$
—Solar system	few $\times\ 10^{-4}\,c$	$10^{-4}\,c$
—Earth	$3 \times 10^{-5}\,c$	—
Magnetic		
—Galactic B-field	$3 \times 10^{-3}\,c\,(10^{16}\mathrm{GeV}/m_M)^{1/2}$	
—Intragalactic B-field (upper limit)	$10^{-3}\,c\,(10^{16}\mathrm{GeV}/m_M)$	

fields of galaxies [to $0(10^{-3}c)$], of clusters [to $0(\text{few} \times 10^{-3}c)$], and of superclusters [to $0(10^{-2}c)$]. Monopoles will also be accelerated by magnetic fields. The intragalactic magnetic field strength is only known to be less than 3×10^{-11} G (reference 31) and will accelerate monopoles to velocities of

$$v_M \simeq 3 \times 10^{-4}c\,(B/10^{-11}G)(10^{16}\mathrm{GeV}/m_M). \tag{4}$$

The galactic magnetic field will accelerate monopoles otherwise at rest to velocities of[32]

$$v_M \simeq 3 \times 10^{-3}c\,(10^{16}\mathrm{GeV}/m_M)^{1/2}. \tag{5}$$

Of course, earth-based detectors are not at rest themselves. They have velocity components due to the earth's rotation (2×10^{-6}), the motion of the earth around the sun ($10^{-4}c$), and the orbital motion of the solar system through the galaxy ($7 \times 10^{-4}c$). I have summarized all of these velocity components for monopoles and monopole detectors in TABLE 1.

From the above discussion and the table, it is clear that one should expect monopole-detector relative velocities of at least a few $\times$ $10^{-3}c$. This is an important consideration that must be taken into account when designing the velocity acceptance of a detector. (In the very unlikely case that most of the monopole flux in our neighborhood is due to an orbiting monopole cloud, typical monopole velocities would be of an order of a few $\times$ $10^{-4}c$, thus implying monopole-detector relative velocities of an order of a few $\times$ $10^{-4}c$.)

Although monopoles would not be in objects such as stars, planets, etc., *ab initio* (because the formation of these objects clearly involves dissipation), they can be captured by these objects. Monopoles passing through matter lose energy by electronic interactions, hadronic interactions, magnetic interactions, and nuclear interactions. The electronic interactions (energy loss due to the eddy currents they induce) are usually the most important loss mechanism.[33] Monopoles less massive than about 10^{20}GeV will lose sufficient energy when passing through neutron stars and white dwarfs to become captured. Monopoles less massive than about 10^{18}GeV lose sufficient energy in main sequence stars to become captured. Jupiter-sized objects can stop monopoles as massive as 10^{16}GeV and the earth can stop very slow moving or light monopoles ($\lesssim 10^{15}$ GeV). The number of monopoles captured by an astrophysical object depends upon its gravitational capture cross, the monopole flux, and the time interval, τ:

$$N_{CAP} \simeq (4\pi R^2)(\pi - sr)(1 + 2GM/Rv_M^2)\, F\, \tau \qquad \textbf{(6a)}$$

$$\simeq 3.2 \times 10^{24}(F/10^{-16}\text{cm}^{-2}\text{sr}^{-1}\text{s}^{-1})(R/R_\odot)^2\, (\tau/10^9\text{yr})[(1 + 2GM/Rv_M^2)/5.3]\,, \qquad \textbf{(6b)}$$

where R and M are the radius and mass of the object, respectively, and $1 + 2GM/Rv_M^2$ is the ratio of the gravitational to the geometric cross section (whose value is 5.3 for a one solar mass star). Once they are captured, monopoles will sink to the center of the object and be supported against gravity by their thermal velocity dispersion or magnetic fields that may be present.[34] The number of monopoles residing in the object also depends upon the importance of monopole-antimonopole annihilations. Finally, I should mention that monopole capture does not deplete the cosmic stock of monopoles appreciably; even within the relatively high density environment of a galaxy, the mean free path of a monopole is 10^{42}cm.

ASTROPHYSICAL IMPLICATIONS—FLUX LIMITS

Because of their three extraordinary properties—macroscopic mass, hefty magnetic charge, and ability to catalyze nucleon decay—monopoles, if present even in small numbers, will make themselves astrophysically conspicuous. Their potential conspicuousness leads to stringent astrophysical bounds on their flux (and, if we can be very clever, astrophysics may provide us with very sensitive monopole detectors).

First, consider the mass they contribute. Although we do not have precise knowledge of Ω (the ratio of the cosmic mass density to the critical density), estimates of the deceleration parameter, q_0, and the age of the universe strongly suggest that; $\Omega h^2 < 1$. This is turn places a limit on the average flux on monopoles in the universe:

$$\langle F \rangle \lesssim 3 \times 10^{-15}(10^{16}\text{GeV}/m_M)\ \text{cm}^{-2}\,\text{sr}^{-1}\,\text{s}^{-1}. \qquad \textbf{(7)}$$

Recall that the local flux could be enhanced by a factor, f, which could be as large as 10^5.

Monopoles cannot contribute more mass density locally than we observe[35]:

$$\rho_{LOCAL} \simeq 10^{-23}\ \mathrm{g\ cm^{-3}}.$$

This leads to the flux limit of

$$F_{GAL} \lesssim 10^{-9}\ (10^{16}\mathrm{GeV}/m_M)\ \mathrm{cm^{-2}\ sr^{-1}\ s^{-1}}. \qquad \mathbf{(8)}$$

In fact, we can do better than this. By carefully modeling the galaxy, Bahcall[35] finds that the contribution of the halo material to the local mass density can be no more than about 1/30th of the total mass density, i.e., the disk component dominates the local mass density. (Remember, if monopoles are clustered in the galaxy, they must be in the halo since they have no mechanism for dissipating their gravitational energy and condensing into the disk.) This leads to the more stringent limit of

$$F_{GAL} \lesssim 3 \times 10^{-11}\ (10^{16}\mathrm{GeV}/m_M)\ \mathrm{cm^{-2}\ sr^{-1}\ s^{-1}}. \qquad \mathbf{(9)}$$

I should reemphasize that the bound based upon the average flux of monopoles in the universe is probably the relevant one as the galactic magnetic field will keep all but the most massive monopoles from clustering in the galaxy.

Next, consider the effect of monopoles on astrophysical magnetic fields. Because of their hefty magnetic charge, monopoles will respond to any magnetic fields they may encounter and in the process will usually gain KE. (Of course, this must occur at the expense of the magnetic field energy.) Parker pioneered the use of this simple principle to place limits on the flux of monopoles.[36] The basic idea is that monopoles tend to drain astrophysical magnetic fields with a damping time that is inversely proportional to the monopole flux. Survival of an astrophysical magnetic field requires that the damping time due to the monopoles be no shorter than the time required to regenerate the field. Parker then applied the argument to the galactic magnetic field (strength of $3 \times 10^{-6}\ G$ and coherence length of 300 pc), which is believed to be produced by dynamo action (which converts KE embodied in the differential rotation of the galaxy into magnetic field energy) and whose regeneration time is of the order of 100 Myr. The limit that follows is[32]

$$F_{GAL} \lesssim 10^{-15}\ \mathrm{cm^{-2}\ sr^{-1}\ s^{-1}},\ (m_M \lesssim 10^{17}\ \mathrm{GeV}), \qquad \mathbf{(10a)}$$

$$F_{GAL} \lesssim 10^{-15}\ (m_M/10^{17}\ \mathrm{GeV})\ \mathrm{cm^{-2}\ sr^{-1}\ s^{-1}},\ (m_M \gtrsim 10^{17}\ \mathrm{GeV}). \qquad \mathbf{(10b)}$$

Rephaeli and Turner[37] applied the same argument to the survival of the weaker magnetic fields that are observed in rich clusters ($10^{-7}\ G$ and coherence length $\simeq$ 100 kpc.) The result is a much more stringent bound of

$$\langle F \rangle \lesssim 10^{-18}\ \mathrm{cm^{-2}\ sr^{-1}\ s^{-1}}, \qquad \mathbf{(11)}$$

which, however, is somewhat less reliable than the Parker bound as our knowledge of intracluster magnetic fields is not as secure as that of galactic magnetic fields.

Other authors have applied Parker's logic to the magnetic fields in white dwarfs, neutron stars, and peculiar A stars.[38–40] Because of the additional assumptions involved, these bounds, while more stringent than the Parker bound, are very much less certain.

The implicit assumption made in deriving Parker-type bounds is that monopoles respond incoherently to the astrophysical magnetic field. If they can respond coherently (as in the case of magnetic plasma oscillations), then the KE they extract from the magnetic field energy will be returned a half cycle later.[32,41–43] In this case, the field energy is not drained, but only borrowed for a half cycle, and the bound in question can be circumvented. In fact, in some sense, the monopoles are participating in the maintenance of the magnetic field.

The key issue with regard to evading damping then is that of coherence. In general, this translates to a condition on the phase velocity of the oscillations:

$$v_{ph} \gtrsim \langle v_M^2 \rangle^{1/2}.$$

The phase velocity of the oscillations is just $v_{ph} = (\omega_p/2\pi)\ell$, where ℓ is the characteristic length scale of the mode and $\omega_p^2 = 4\pi g^2 n_M/m_M$ is the magnetic plasma frequency. The internal velocity dispersion of the monopoles depends upon the circumstances; for a galactic halo of monopoles, one expects it to be of the order of the virial velocity of the galaxy, which is $10^{-3}\,c$.

The condition that v_{ph} be greater than $\langle v_M^2 \rangle^{1/2}$ results in a *lower* bound to the monopole flux (if coherent effects are to be important):

$$F \gtrsim \tfrac{1}{4}\, m_M \langle v_M^2 \rangle^{3/2} (g\ell)^{-2}. \tag{12}$$

For the galaxy, that bound is

$$F_{GAL} \gtrsim 10^{-13}\,(m_M/10^{16}\text{GeV})\,(1\text{kpc}/\ell)^2\ \text{cm}^{-2}\,\text{sr}^{-1}\,\text{s}^{-1}. \tag{13}$$

While it is possible that the Parker bound can be circumvented by coherent effects, I think that it is an unlikely possibility. There are many issues to be addressed before this can be considered as a serious possibility: how such oscillations got set up in the first place; how the required spatial and temporal coherence is maintained in the face of the inhomogeneities known to exist in the galaxy; and how the numerous damping mechanisms can be avoided. The most difficult hurdle at present is the observational one: the present experimental limits on the monopole flux preclude the scenario unless the monopole mass is much less than 10^{16} GeV.

The most spectacular effect associated with superheavy magnetic monopoles is their apparent ability to catalyze nucleon decay at a prodigious rate.[8,9] The rate of energy release by this process is enormous:

$$dE/dt \simeq 10^{18}\ \text{ergs sec}^{-1}\, N_M\,(\rho_N/3\times 10^{14}\ \text{g cm}^{-3})\,\sigma_{-28}, \tag{14}$$

where ρ_N is the local nucleon density and N_M is the number of monopoles. In a neutron star, the energy release is 10^{18} ergs sec^{-1} per monopole. Thus, a single monopole catalyzing a nucleon in a neutron star produces almost as much power as is consumed by all the inhabitants of our planet. How does this process manifest itself in the cosmos?

Astrophysical objects such as stars, planets, etc., capture some or all of the monopoles that strike them. Once captured they sink to the center and accumulate there at a rate proportional to the monopole flux. There they catalyze nucleon decay and release about 1 GeV per catalysis event. The energy is thermalized and radiated—in the IR (for planets), visible (for main sequence stars), UV (for white dwarfs), or X ray (for neutron stars). In the process, some of the energy may also be released in neutrinos.

Using the observed photon flux from a variety of astrophysical objects, such as Earth,[44] Jupiter,[44] white dwarfs,[45] and neutron stars,[46–49] one can place very stringent limits on the flux of monopoles in the galaxy. The most stringent and I believe the most reliable limit comes from neutron stars:

$$F_{GAL} \lesssim 10^{-21}\,\sigma_{-28}^{-1}\,\mathrm{cm}^{-2}\,\mathrm{sr}^{-1}\,\mathrm{s}^{-1}. \tag{15}$$

I say most reliable because the various uncertainties, both due to astrophysics and particle physics, are better understood and believed to be the smallest for neutron stars. For example, when the catalysis process involves nuclei (as it would in planets or white dwarfs), there are angular momentum barriers and the question of whether the monopole gobbles the entire nucleus or just a few nucleons.[50] At small monopole-nucleon relative velocities, there may be threshold effects; however, in a neutron star, the relative velocities are of the order of the speed of light.

Monopole-antimonopole annihilations are almost certainly not important in neutron stars.[51] Owing to their high densities, neutron stars can capture monopoles at least as massive as 10^{20} GeV. The neutron star limit quoted above is actually based on a triad of observations: measured X-ray fluxes (or upper limits) from individual objects;[48] the negative results of serendipitous X-ray searches for bright, nearby neutron stars;[47] and the use of the measured intensity of the soft X-ray background to limit the integrated luminosity of all the old neutron stars in the galaxy.[49] In addition, all of the "astrophysical outs" have been carefully examined and found not to be important (e.g., the possibility that neutron stars would eject the monopoles they capture, that annihilations might be important, or that essentially all the catalysis energy would be released in neutrinos). A review of the neutron star catalysis limits is given in reference 49. I might mention that the limit based upon catalysis in white dwarfs is also very stringent,[45]

$$F_{GAL} \lesssim 3 \times 10^{-18}\sigma_{-28}^{-1}\,\mathrm{cm}^{-2}\,\mathrm{sr}^{-1}\,\mathrm{s}^{-1}. \tag{16}$$

The uncertainties are not too much worse than for neutron stars and they involve different physics and astrophysics.

Main sequence stars capture significant numbers of monopoles[34] (so long as they are less massive than 10^{18} GeV); see FIGURE 2. The main sequence progenitor of a neutron star captures 10^6 or so times as many monopoles as the neutron star itself does. If one takes into account the very likely presence of these monopoles in neutron stars, the limit quoted above improves by a factor of about 10^6:

$$F_{GAL} \lesssim 10^{-27}\,\sigma_{-28}^{-1}\,\mathrm{cm}^{-2}\,\mathrm{sr}^{-1}\,\mathrm{s}^{-1}. \tag{17}$$

However, it is only fair to mention the additional uncertainties involved—do the monopoles annihilate in the main sequence progenitor?, do they get ejected in the process of the formation of the neutron star?, etc.

EXOTIC ASTROPHYSICAL MONOPOLE DETECTORS

If monopoles do catalyze nucleon decay, the neutron star bound all but precludes building an earth-based detector to find them (which is good since a flux of 10^{-21} cm^2

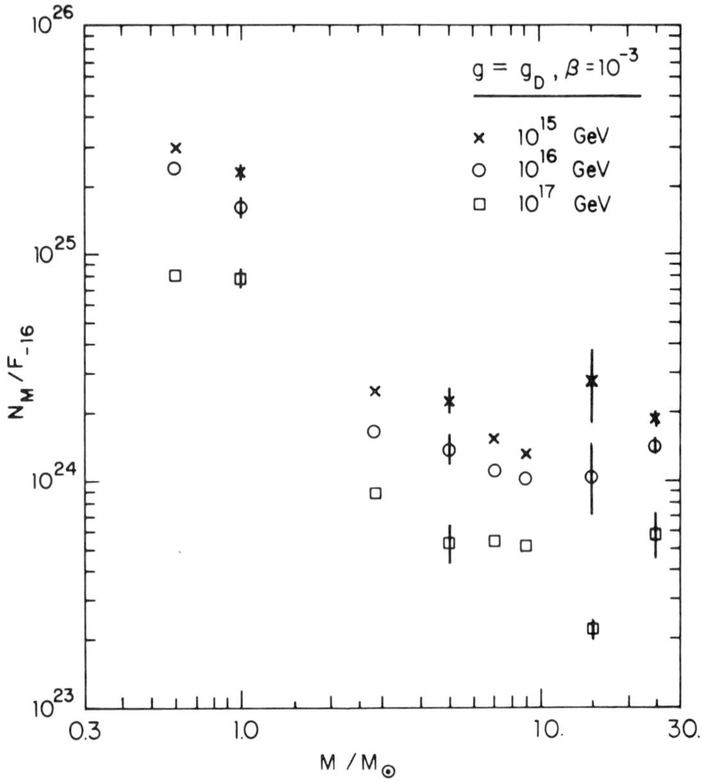

FIGURE 2. Number of monopoles (with Dirac charge) captured by main sequence stars during their lifetimes. Here, $v_M = 10^{-3}\,c$ and $F_{-16} = F_{GAL}/(10^{-16}\,\text{cm}^{-2}\,\text{sr}^{-1}\,\text{s}^{-1})$. Detailed stellar models were used and the monopole energy loss was assumed to be due to electronic interactions only; for more details, see reference 34.

$\text{sr}^{-1}\,\text{sec}^{-1}$ corresponds to one monopole through a large city per year.) We may then have to turn to astrophysical objects as detectors. In the next decade or so, there will be a number of soft X-ray and extreme UV satellite-based detectors operating. Soft X-ray and extreme UV are the optimal bands for detecting neutron stars heated by monopole catalysis. By observing a number of nearby, old neutron stars, it may be possible to "detect monopoles" by their heating effect on these objects, thereby providing us with extremely sensitive monopole detectors.

Another idea for an astrophysical detector is to detect the monopoles that are residing in the sun.[44,52,53] The sun captures a significant number of monopoles—about

$$N_{CAP} \simeq 10^{25}\,(F_{GAL}/10^{-16}\,\text{cm}^{-2}\,\text{sr}^{-1}\,\text{s}^{-1}).$$

If they avoid annihilation (for a discussion of this, see FIGURE 3 and reference 34), they will be catalyzing nucleon decays in the sun and will be producing neutrinos with energies of tens of MeV (from the decays of π's and μ's produced by catalysis). These neutrinos may be detectable in a large, underground earth-based detector such as

IMB, Kamiokande, or the Homestake solar neutrino detector. Preliminary estimates of the signal, though, suggest that even if all the monopoles captured avoid annihilation, the smallest detectable flux would be only about $10^{-20}\,\sigma_{-28}^{-1}\,\mathrm{cm}^{-2}\,\mathrm{sr}^{-1}\,\mathrm{s}^{-1}$, which is more than likely ruled out by the neutron star limits. The idea, however, warrants further study.

CONCLUDING REMARKS

What have we learned about GUT monopoles in the past five years or so? (1) They are exceedingly interesting objects, which if detected would provide us with a unique window to both the earliest moments of the universe and to physics at the very highest

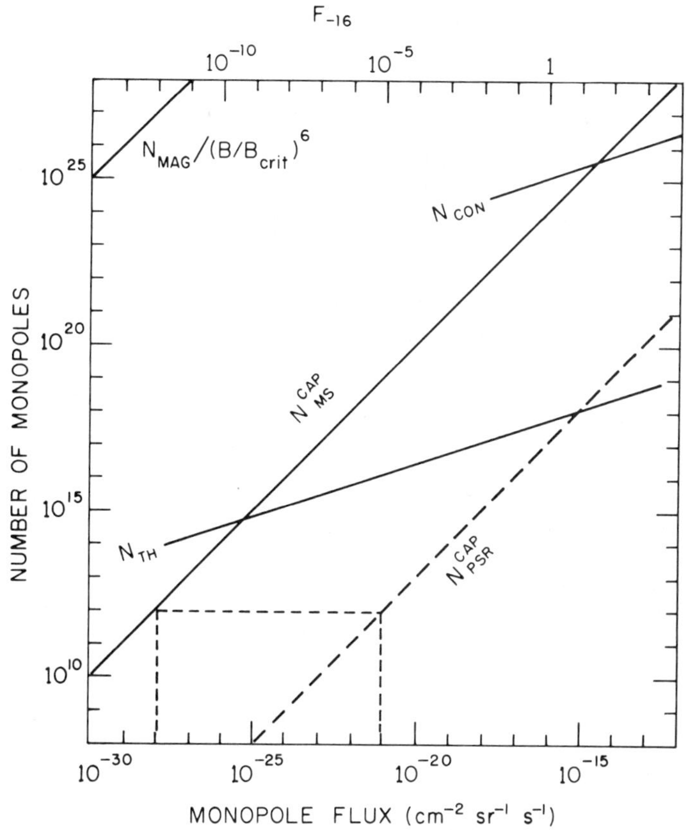

FIGURE 3. Number of monopoles captured by an $8M_{\odot}$ main sequence star in its lifetime (N_{MS}^{CAP}) and by a neutron star in 3×10^6 yrs (N_{PSR}^{CAP}). The effects of monopole-antimonople annihilations in the main sequence star are shown, i.e., the equilibrium number of monopoles for the case that monopoles are supported against gravity by: (1) their thermal velocity dispersion (N_{TH}); (2) magnetic field of order, 100 G (N_{MAG}); (3) convective motions of the core (N_{CON}). (When the equilibrium number is greater than N_{MS}^{CAP}, then annihilations are unimportant and the number surviving in the star is just N_{MS}^{CAP}.) For more details, see reference 34.

energies; (2) They remain one of the few generic predictions of GUTs that we can hope to verify in our low energy environment; (3) At present, there is no believable prediction for the flux of relic, superheavy monopoles; (4) Based upon astrophysical considerations, we can be reasonably certain that the monopole flux is small. Since it is not obligatory that monopoles catalyze nucleon decay at the prodigious rate, the most stringent and reliable upper bound to the monopole flux is the Parker bound: $F_{GAL} \lesssim 10^{-15}\ \mathrm{cm}^{-2}\ \mathrm{sr}^{-1}\ \mathrm{s}^{-1}$ (for $m_M \lesssim 10^{17}$ GeV). Note that this is *not* a prediction for the flux. In fact, it is very likely that the flux must be significantly smaller: $F_{GAL} \lesssim 10^{-18}\ \mathrm{cm}^{-2}\ \mathrm{sr}^{-1}\ \mathrm{s}^{-1}$, which is based on the survival of intracluster magnetic fields, and $F_{GAL} \lesssim 10^{-21}\ \mathrm{cm}^{-2}\ \mathrm{sr}^{-1}\ \mathrm{s}^{-1}$, which is based upon monopole-catalyzed nucleon decay; (5) There is every reason to believe that monopoles are moving (relative to earth) with typical velocities of *at least* a few $\times\ 10^{-3}\ c$.

In spite of the somewhat bleak theoretical situation (with respect to monopoles), I still believe that monopoles are worth hunting. The risks may be very great, but the potential payoffs are even greater.

REFERENCES

1. PRESKILL, J. 1984. Annu. Rev. Nucl. Part. Sci. **34:** 461–530.
2. PRESKILL, J. 1983. *In* The Very Early Universe. G. Gibbons, S. Hawking & S. Siklos, Eds. Cambridge University Press. Cambridge, England.
3. TURNER, M. S. 1983. *In* Magnetic Monopoles. R. A. Carrigan & W. P. Trower, Eds. Plenum. New York.
4. TURNER, M. S. 1984. *In* Monopole '83. J. Stone, Ed. Plenum. New York.
5. KOLB, E. W. 1984. Ann. N.Y. Acad. Sci. **422:** 33–44.
6. POLYAKOV, A. M. 1974. JETP Lett. **20:** 194.
7. 'T HOOFT, G. 1974. Nucl. Phys. **B79:** 276.
8. CALLAN, C. G. 1982. Phys. Rev. **D25:** 2141; 1982. Phys. Rev. **D26:** 2058.
9. RUBAKOV, V. A. 1981. JETP Lett. **33:** 644; 1982. Nucl. Phys. **B203:** 311.
10. SEN, A. 1985. Nucl. Phys. **B250:** 1–38.
11. GOLDHABER, A. 1983. *In* Fourth Workshop on Grand Unification. P. Langacker, P. J. Steinhardt & H. A. Weldon, Eds. Birkhäuser. Boston, Massachusetts; SCHELLEKENS, A. N. 1984. Phys. Rev. **D29:** 2378.
12. WEINBERG, E., D. LONDON & J. L. ROSNER. 1984. Nucl. Phys. **B236:** 90.
13. KIBBLE, T. W. B. 1976. J. Phys. **A9:** 1387.
14. TURNER, M. S. 1982. Phys. Lett. **115B:** 95.
15. LAZARIDES, G., Q. SHAFI & W. P. TROWER. 1982. Phys. Rev. Lett. **49:** 1756.
16. PRESKILL, J. 1983. *In* The Very Early Universe. G. Gibbons, S. Hawking & S. Siklos, Eds. Cambride University Press. Cambridge, England.
17. PRESKILL, J. 1979. Phys. Rev. Lett. **43:** 1365.
18. ZEL'DOVICH, YA. B. & M. YU. KHLOPOV. 1978. Phys. Lett. **79B:** 239.
19. WEINBERG, E. 1983. Phys. Lett. **126B:** 441.
20. LANGACKER, P. & S-Y. PI. 1980. Phys. Rev. Lett. **45:** 1.
21. WASSERMAN, I. 1984. Cornell University Report. Ithaca, New York.
22. GUTH, A. 1981. Phys. Rev. **D23:** 347.
23. LINDE, A. 1982. Phys. Lett. **108B:** 389.
24. ALBRECHT, A. & P. J. STEINHARDT. 1982. Phys. Rev. Lett. **48:** 1220.
25. OLIVE, K. A. & D. SECKEL. 1984. *In* Monopole '83. J. L. Stone, Ed. Plenum. New York.
26. LAZARIDES, G. & Q. SHAFI. 1983. *In* The Very Early Universe. G. Gibbons, S. Hawking & S. Siklos, Eds. Cambridge University Press. Cambridge, England.
27. COLLINS, W. & M. S. TURNER. 1984. Phys. Rev. **D29:** 2158.
28. LINDBLOM, P. R. & P. J. STEINHARDT. 1985. Phys. Rev. **D31:** 2151.
29. DIMOPOULOS, S., S. GLASHOW, E. PURCELL & F. WILCZEK. 1982. Nature **298:** 824.

30. Freese, K. & M. S. Turner. 1983. Phys. Lett. **123B:** 293.
31. Vallee, J. P. 1983. Astrophys. Lett. **23:** 85.
32. Turner, M. S., E. N. Parker & T. Bogdan. 1982. Phys. Rev. **D26:** 1296.
33. Ahlen, S. 1983. *In* Magnetic Monopoles. R. A. Carrigan & W. P. Trower, Eds. Plenum. New York.
34. Frieman, J., K. Freese & M. S. Turner. 1986. Astrophys. J. In press; Frieman, J. 1985. *In* Proc. of the Santa Fe Meeting. T. Goldman & M. M. Nieto, Eds. World Scientific. Philadelphia.
35. Bahcall, J. 1984. Astrophys. J. **276:** 169; 1984. Astrophys. J. **287:** 926.
36. Parker, E. N. 1970. Astrophys. J. **160:** 383.
37. Rephaeli, Y. & M. S. Turner. 1983. Phys. Lett. **121B:** 115.
38. Ritson, D. 1982. SLAC report 2977 (unpublished). Stanford, California.
39. Drukier, A. K. 1982. Unpublished report.
40. Harvey, J., M. A. Ruderman & J. Shaham. 1986. Phys. Rev. D. In press.
41. Arons, J. & R. Blandford. 1983. Phys. Rev. Lett. **50:** 544.
42. Salpeter, E. E., S. Shapiro & I. Wasserman. 1982. Phys. Rev. Lett. **49:** 1114.
43. Farouki, R., S. L. Shapiro & I. Wasserman. 1984. Astrophys. J. **284:** 282.
44. Turner, M. S. 1983. Nature **302:** 804.
45. Freese, K. 1984. Astrophys. J. **286:** 216.
46. Dimopoulos, S., J. Preskill & F. Wilczek. 1982. Phys. Lett. **119B:** 320.
47. Kolb, E. W., S. Colgate & J. Harvey. 1982. Phys. Rev. Lett. **49:** 1373.
48. Freese, K., M. S. Turner & D. N. Schramm. 1983. Phys. Rev. Lett. **51:** 1625.
49. Kolb, E. W. & M. S. Turner. 1984. Astrophys. J. **286:** 702–710.
50. Arafune, J. & M. Fukugita. 1983. Phys. Rev. Lett. **50:** 1901.
51. Harvey, J. 1984. Nucl. Phys. **B236:** 255; 1984. *In* Monopole '83. J. L. Stone, Ed. Plenum. New York.
52. Arafune, J., M. Fukugita & S. Yanagita. 1984. Unpublished report.
53. Dar, A. & S. P. Rosen. 1984. Los Alamos National Lab. Report (unpublished). Los Alamos, New Mexico.
54. Bracci, L., G. Fiorentini, G. Mezzorassi & P. Quarati. 1984. Phys. Lett. **143B:** 357.

The Status of Searches for Magnetic Monopoles

H.J. FRISCH

Department of Physics and Enrico Fermi Institute
University of Chicago
Chicago, Illinois 60637

I will summarize the status of terrestrial monopole searches. For a review of the limits that have been deduced from astronomical observations, I refer the reader to the discussion by Michael Turner at this conference.

My paper will present limits based on a hierarchy of possible properties of magnetic monopoles. These properties and the topics emphasized are:

(A) MAGNETIC CHARGE
(i) Faraday Induction Experiments
(ii) The Parker Bound from Galactic Magnetic Fields

(B) IONIZATION
(i) Expectations for monopole velocities
(ii) New results on the production of light by the excitation of scintillator by very slow protons from Ahlen *et al.*
(iii) New limits from the big Baksan and Texas A & M detectors
(iv) New detectors being planned

(C) CATALYSIS
(i) Dependence on the catalysis cross section (in particular, this requires a nonzero cross section)
(ii) Terrestrial searches in proton decay experiments
(iii) Strong limits from astrophysical observations

FARADAY INDUCTION EXPERIMENTS

On December 8, 1831, Michael Faraday put a magnetic pole through a loop of wire connected in series to a galvanometer (see FIGURE 1). His log for the day states: "When the marked pole of the magnet entered the helix *O* as in the figure, marked end of the needle went west; hence electrical currents as indicated by the arrows."[1] The change in flux through the hoop induced a current in the circuit, which Faraday saw as a transient.

Modern detectors function on exactly the same principles as Faraday's original. These detectors search for the current induced in a loop by the transit of a magnetic monopole through the loop. The use of a superconducting loop, first suggested by Alvarez,[2] allows one to look for a change in the DC current in the loop. The only assumptions that are used in these searches are that the charge is quantized in units of the Dirac charge,[3] $eg = n/2$, and Faraday's Law of Induction.

The strength of the induction technique is in its ability to be directly calibrated by the insertion of a magnetic pole. FIGURE 2 shows the effect of inserting the end of a long thin solenoid (called a pseudopole by our Chicago-Fermilab-Michigan group[4]) through a conducting loop. If the loop is superconducting or if the time constant is short with respect to the transit time of the pole, the field lines look as in FIGURE 2. A current is set up in the loop to make a flux equal and opposite to the pseudopole flux. FIGURE 3 shows the effect of a monopole—it looks exactly like the effect of a pseudopole, but with the thin solenoid removed.

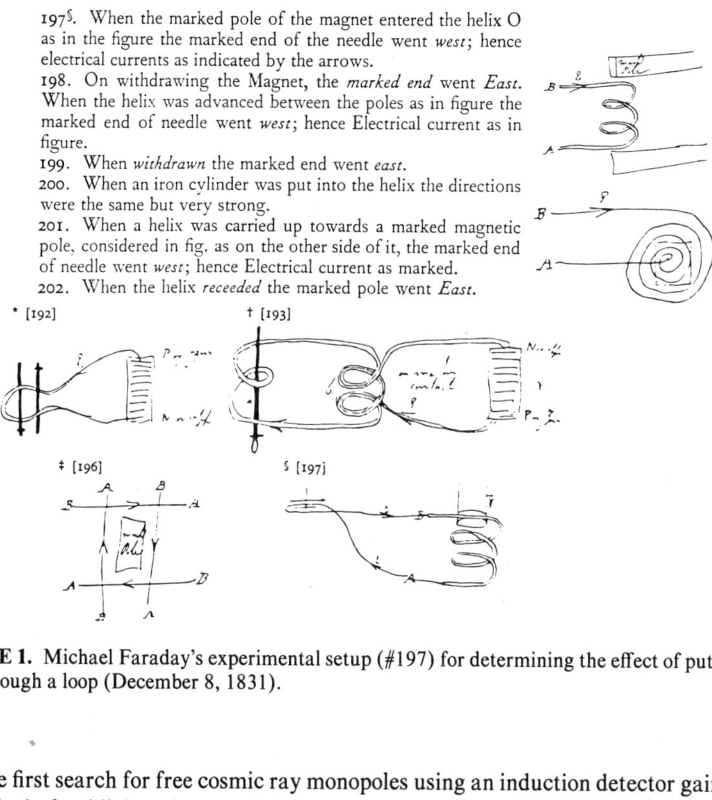
197§. When the marked pole of the magnet entered the helix O as in the figure the marked end of the needle went *west*; hence electrical currents as indicated by the arrows.
198. On withdrawing the Magnet, the *marked end* went *East*. When the helix was advanced between the poles as in figure the marked end of needle went *west*; hence Electrical current as in figure.
199. When *withdrawn* the marked end went *east*.
200. When an iron cylinder was put into the helix the directions were the same but very strong.
201. When a helix was carried up towards a marked magnetic pole, considered in fig. as on the other side of it, the marked end of needle went *west*; hence Electrical current as marked.
202. When the helix *receeded* the marked pole went *East*.

FIGURE 1. Michael Faraday's experimental setup (#197) for determining the effect of putting a pole through a loop (December 8, 1831).

The first search for free cosmic ray monopoles using an induction detector gained a great deal of publicity when, on February 14, 1982, B. Cabrera observed[5] a step in a single small (1″ diameter) loop that corresponded to an $n = 1$ Dirac charge. The step, shown in FIGURE 4, is very clean and is much larger than any of the other steps observed during the 151-day run.

Since that time, much larger and more redundant apparatuses have been built by several groups. The largest of the running detectors with their areas and approximate flux limits are shown in TABLE 1. The total exposure listed in this table is more than 600 times the exposure of Stanford I. In the original paper,[5] Cabrera pointed out that a

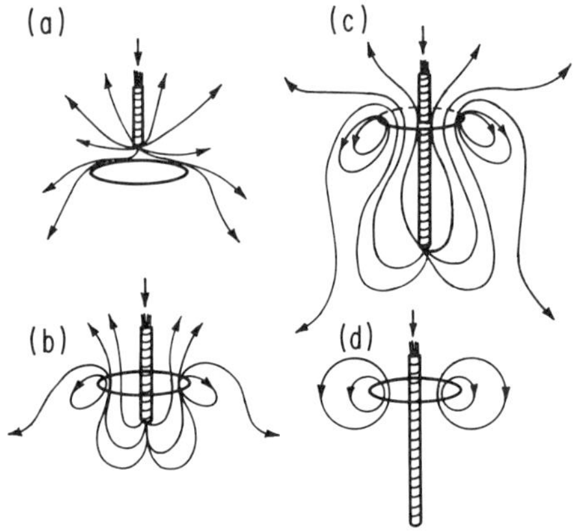

FIGURE 2. The field lines of a slender solenoid being inserted into a conducting coil.

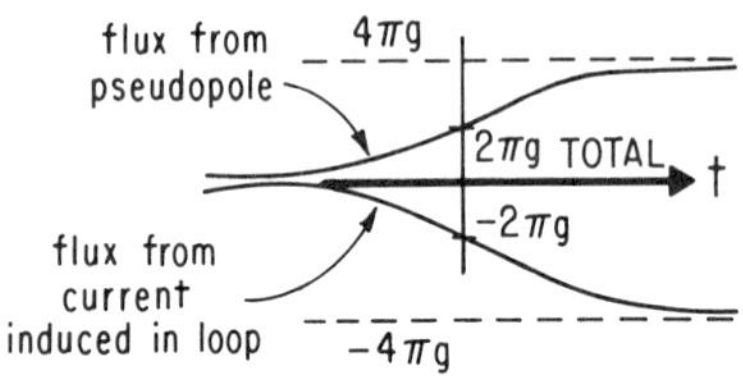

"spontaneous internal stress release mechanism" could be a source of such a step. I believe that everybody now should conclude that this was the case.

New, much larger, and much more redundant induction detectors are also being built. TABLE 2 lists the three largest that I know of.

Each of these detectors develops ideas that could be used to build a detector on a much larger (≥ 100 m^2) scale and each is redundant enough so that two or three events should be enough to establish a monopole signal. The use of mass-produced SQUIDs

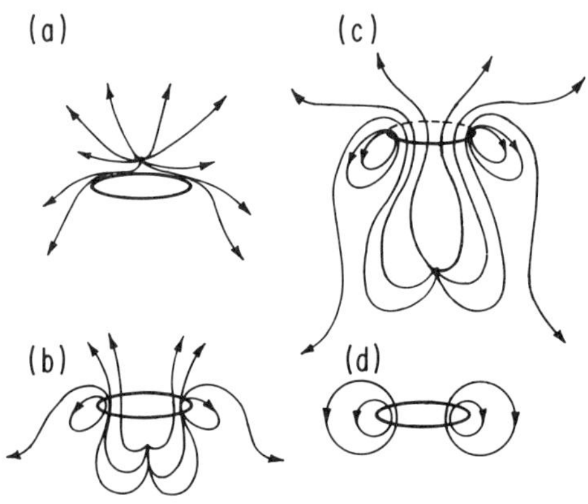

FIGURE 3. The field lines of a monopole being inserted into a conducting coil.

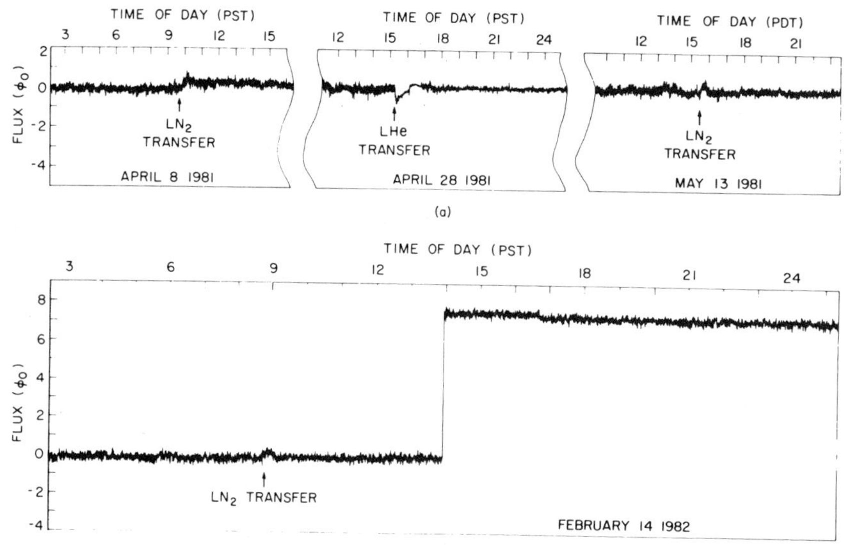

FIGURE 4. The "candidate" event of February 14, 1982 of B. Cabrera (reference 5).

by the IBM group is a big step toward a very large array; the superposed orthogonal gradiometer[12] and series-parallel techniques[11] are also ideas that help generate large area redundant detectors cheaply.

Induction detectors first were invented to look for monopoles bound to ferromagnetic materials, where monopoles could be expected to accumulate over geologic times as suggested by Goto[13] in 1958. A recent experiment by T. Ebisu and T. Watanabe of Kobe University treats iron ore in a furnace above a 14 cm diameter coil.[14] At the Curie point, monopoles of mass larger than 3×10^{15} GeV$/c^2$ are expected to fall through the loop. The Kobe group has now looked at 900 kg of iron ore and has put a limit on the monopole density of $\rho_m \leq 3.2 \times 10^{-6}$ monopoles/gm (95% C.L.). This result is to be compared with the limit of Eberhard and Alvarez *et al.*[15] of 1.7×10^{-4} monopoles/gm (also 95% C.L.). In addition, they now are setting up an experiment along the lines of a proposal by D. Cline, R. March *et al.*[16] to install a loop below the furnace of an operating blast furnace and examine orders of magnitude more material this way.

TABLE 1. The Largest Running Detectors

Detector	Area Averaged over Solid Angle	Flux Limit/Sensitivity (90% C.L. $cm^{-2}sr^{-1}s^{-1}$)
Chicago-Fermilab-Michigan[6]	0.21 m^2	6.7×10^{-12}
IBM II[7]	0.10 m^2	5.5×10^{-12}
Imperial College[8]	0.17 m^2	no limit yet
NBS (Boulder)[9]	0.02 m^2	$\sim 8 \times 10^{-11}$
Stanford II[10]	0.05 m^2	6.7×10^{-12}
Total		$F_m \leq 2 \times 10^{-12}$

TABLE 2. The Three Largest Induction Detectors Being Built

Group	$\langle \text{Area} \rangle_{4\pi}$	Description
IBM III	3.2 m^2	10 gradiometers; "homemade" DC SQUIDS
Stanford III	1.5 m^2	8 gradiometers; impedance matching by the series-parallel technique[11]
Chicago-FNAL-Michigan II	1.0 m^2	4 superposed orthogonal gradiometers;[12] impedance matching by the series-parallel technique.

IONIZATION DETECTORS

The detection of monopoles by the excitation or ionization of atoms leading to photons is generically termed "ionization detection." The signal of a monopole for such kinds of detectors depends on the monopole velocity, as the atomic electrons have to be excited.

Cosmic ray monopole velocities will depend on the magnetic and gravitational fields encountered by the monopole during its lifetime. We can make estimates of a lower bound on the typical expected velocities by ignoring the magnetic field and just considering the gravitational effects. In particular, if monopoles are heavy, they do not dissipate energy well and thus should have velocities comparable to the escape velocity from our galaxy.

From Mihalas and Binney,[17] a standard textbook on galactic astronomy, one learns that the rotational velocity of the sun in our galaxy is $\theta_c = 250 \pm 25$ km/sec. This is presumably a lower bound on the escape velocity.

To estimate escape velocity better, Mihalas and Binney look at the velocities relative to us of "spherical-component stars," which are those stars that are in the spherical halo of the galaxy, and not in the disk, where dissipative processes have slowed the material. Monopoles presumably would be halo objects if captured by galaxies and would have a range of speeds relative to us that are more comparable (and most likely greater than, due to acceleration by magnetic fields) than these stars. Mihalas and Binney tabulate the following kinds of objects, their average radial velocity relative to us, and the dispersions of the three components of velocity (TABLE 3).

TABLE 3. Parameters of Spherical-Component Stars[a]

Object	Number Observed	Radial Velocity $\langle\theta\rangle - \theta_c$ (km/sec)	Dispersions (km/sec) $\langle\pi^2\rangle^{1/2}$	$\langle(\theta - \langle\theta\rangle)^2\rangle^{1/2}$	$\langle z^2\rangle^{1/2}$
Globular Clusters	70	-165 ($5 \times 10^{-4}\,c$)	145	—	—
Extreme Subdwarfs	22	-184 ($6 \times 10^{-4}\,c$)	170	90	65
RR Lyrae Variables	33	-220 ($7 \times 10^{-4}\,c$)	210	120	90

[a] From Mihalas and Binney.[17]

As a β of $10^{-3}c$ is 300 km/sec, we see that the typical average value of β for these halo objects is $6 \times 10^{-4}c$, with dispersions that are of the same size. This puts a lower bound on the escape velocity on the order of 400–500 km/sec.

One, though, still does not know from the above what velocity distribution the monopoles would have. Monopoles are accelerated presumably by the galactic magnetic fields by the mechanisms discussed by Parker *et al.*[18] and those monopoles not ejected yet from the galaxy would have a velocity distribution shifted closer to the escape velocity than the distribution due to gravitational energy alone.

A conservative prescription toward setting limits on monopole fluxes using techniques that depend on the monopole velocity would be to use the observed velocity distributions for galactic halo objects ($\langle v \rangle \sim 8 \times 10^{-4}c$, $\sigma_v \sim 6 \times 10^{-4}c$) and convolute this distribution with the estimated detector response as a function of velocity (see below). Although this process still depends on many assumptions, such as that this limit is not "hard" in the way that the limits from induction detectors are, it is a better way to compare limits from different detectors. For example, we will see below that the Baksan detector trigger threshold is approximately $\beta = 10^{-3}$. The crude numbers for $\langle v \rangle$ and σ_v that I have taken above would then dilute their limit by a factor of approximately 2.4.

Given these expected monopole velocities, how does one expect a monopole to interact with a particle detector? There has been much hard theoretical work and lively discussion of this[19] for both scintillation and gas detectors. Several mechanisms for detection have been calculated: energy loss by electronic excitation and ionization, treating the electrons as a Fermi gas,[20] Zeeman excitation by level mixing by the magnetic field,[21] and the creation of lattice defects in crystalline materials by slow monopoles with attached nuclei.[22]

The largest detectors now running depend on detection by energy loss to the electrons of the material: scintillation in the cases of Baksan, Texas A & M, and Homestake detectors, and argon-ethylene in the case of the Frejus detector. The calculations of light-output from a scintillator have as their basis the scattering of electrons treated as a Fermi gas from a heavy particle. A partial check of the calculations is to measure the light output from very slow moving electrically charged particles.

This has been done very recently for protons in the range of $9 \times 10^{-4} < \beta < 6 \times 10^{-3}$ by S.P. Ahlen, T.M. Liss, G. Lane and G. Liu (Indiana, Chicago, and CalTech) in an experiment performed at the High Flux Beam Reactor at Brookhaven National Laboratory.[23] A beam of 24-keV neutrons impinged on a 3 mm thick target of NE110 scintillator as shown in FIGURE 5. Neutrons that scattered from protons in the scintillator were detected in two ^{6}Li-loaded glass scintillators. The recoil energy of the proton in the scintillator is determined by the kinematics of the elastic *n-p* collision: data were taken at three different angles (45°, 15°, and 7.5°), corresponding to recoil protons of energies, 12 keV, 1.6 keV and 0.41 keV, respectively. Time of flight was used between the pulse in the target scintillator caused by the recoil proton and the neutron ^{6}Li-detectors to discriminate between *n-p* and *n-C* scatterings and to eliminate γ-ray backgrounds.

What is measured is the average number of photoelectrons (<1) per stopping proton. To derive what is wanted, which is the light per unit track length, dL/dx,

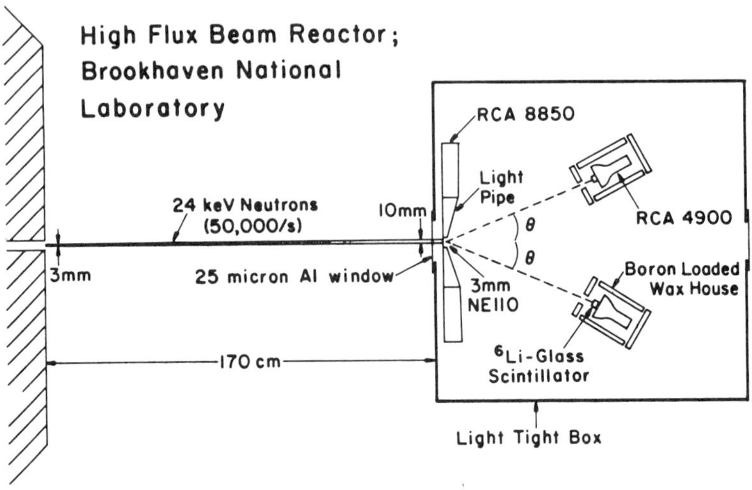

FIGURE 5. The experimental setup for the measurement of S. Ahlen and T. Liss and the light output by slow protons in the scintillator.

versus velocity, the experimenters write,

$$\frac{dL}{dx} = \frac{dL}{dE}\bigg/\frac{dE}{dx}_{electronic},$$

where $dE/dx(electronic)$ is the proton's electronic stopping power. (This can be calculated theoretically.) In addition, $dL/dE(electronic)$ is very close to what is measured directly, although it is averaged over the proton velocity.

The results of light output, dL/dx, versus velocity are shown in FIGURE 6 for two different calculations of the stopping power, $dE/dx(electronic)$.

One sees that for either of the two models for stopping power there is a signal expected at velocities at least as low as $9 \times 10^{-4}c$. For example, the Baksan detector's minimum detectable ionization is about $dL/dx = 4 \times 10^{-2}$ MeV/cm, which is below even the lowest of the points in FIGURE 6. (However, their trigger electronics limits them to $\beta \geq 10^{-3}$.)

Therefore, the conclusion is that the NE110 scintillator should certainly be efficient at detecting monopoles with bound protons at velocities around $10^{-3}c$, and it is reasonable that bare monopoles should also be detected in that velocity range.

The limits from ionization experiments are tabulated in TABLE 4, which is largely derived from Steven Errede's summary talk[24] at ICOBAN '84. The main changes from them are the Baksan limit, presented at the Leipzig Conference,[25] and the first results from the Texas A & M experiment, presented at the Santa Fe APS meeting.[26]

Several very large experiments are under way, under construction, or in planning. The Homestake detector[27] consists of a hollow box, 8m × 8m × 16m, consisting of 200 liquid scintillator filled modules. The acceptance is almost as large as the Baksan detector's, at 1200 m^2 sr. The electronics, however, has waveform digitization and is designed specifically for detecting and triggering on the long slow pulses made by a slowly moving monopole. The minimum ionization detectable is quite low (one photoelectron, where 500 corresponds to a minimizing ionizing particle). The threshold is limited by background radioactivity, but it will be below one-tenth of the minimum ionizing. The detector is now partially built and is triggering on cosmic ray muons. The experimenters expect to have the monopole electronics operating by late 1985.

The Frejus detector[28] has an effective area of 1000 m^2 sr, which consists of flash tubes and Geiger counters. A special pretrigger allows very slow ($\beta = 10^{-4}$) objects to trigger if they ionize the argon–ethyl-alcohol mix above about one-fiftieth of the minimum ionization. They are now taking data with muons while developing a monopole trigger.

The most ambitious detector is the Gran Sasso detector,[29] proposed by an Italy-US consortium. It would have both gaseous and liquid scintillator detectors for an effective area of 12,000 m^2 sr.

The detector would be a hollow box of scintillator tanks (like the Homestake detector) with an overall size of 12m × 5m × 110m (which is the size of a football field). The box would be filled with ten layers of limited streamer tubes filled with He-*n*-pentane to use the Drell mechanism.[21] The streamer tubes would thus have a threshold of $\beta \gtrsim 10^{-4}$ and the scintillation counters would have a threshold of something above $\beta \geq 6 \times 10^{-4}$, in principle.

One very clever experiment that is even more sensitive than the above future detectors (albeit heavily assumption dependent) is the search by B. Price and collaborators[30] for tracks of monopoles with attached nuclei in ancient mica. Although the area is small (~200 cm^2), the exposure is large (1.2×10^{16} sec). FIGURE 7 shows a representation of the experiment by J. Stone.[31] The assumed pickup of an Al nucleus,

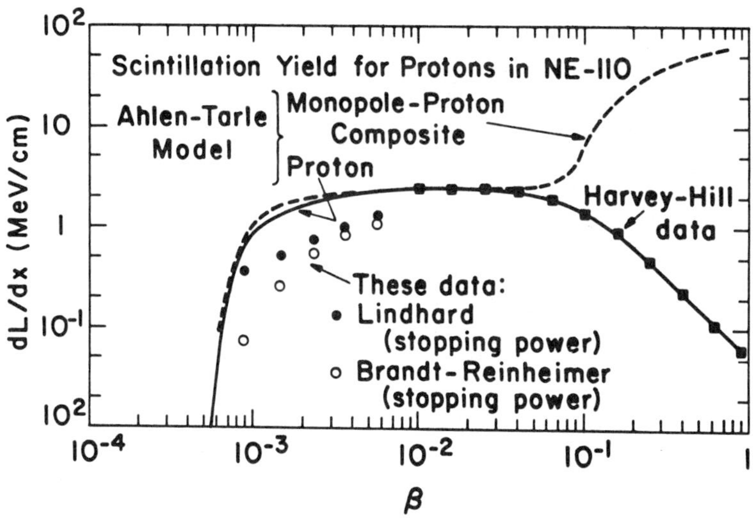

FIGURE 6. The light output per path length, dL/dX, versus velocity for recoil protons, as measured by S. Ahlen, T. Liss, G. Lane, and G. Liu (reference 22, preliminary).

TABLE 4. Summary of Ionization and Excitation (dE/dX) Monopole Search Experiments[a]

Experiment	Detector Type[b]	$A \cdot \Omega$ (m^2sr)	dE/dX (min) ($\times$ m.i.)	Velocity Range	Flux Limit (90% C.L.) ($cm^{-2}sr^{-1}s^{-1}$)
Baksan	LScint	1800	1/4	$1 \times 10^{-3} - 0.3$	$<3.0 \times 10^{-15}$
Bkly-Ind.	PScint	17.5	6/10	$6 \times 10^{-4} - 3 \times 10^{-3}$	$<8.2 \times 10^{-13}$
Bologna	PScint	33.0	20.0	$2 \times 10^{-3} - 6 \times 10^{-1}$	$<2.0 \times 10^{-13}$
BNL	PT(Ar)	1.9	2.0	$4 \times 10^{-4} - 1 \times 10^{-3}$	$<3.5 \times 10^{-11}$
BNL ν-expt.	PSc + DCH	14.5	1/3	$10^{-3} - 2 \times 10^{-1}$	$<5.2 \times 10^{-12}$
Kobe Univ.	PScint	20.0	1/10	$3 \times 10^{-3} - 1.0$	$<9.0 \times 10^{-13}$
KGF	PT(Ar)	250	2.5	$10^{-3} - 1.0$	$<1.9 \times 10^{-14}$
NUSEX	LST(Ar)	12.0	1/50	$3 \times 10^{-4} - 5 \times 10^{-1}$	$<6.4 \times 10^{-13}$
Soudan-I	PT(Ar)	71.6	1/3	$10^{-3} - 1.0$	$<1.5 \times 10^{-13}$
Texas A & M	PScint	300	1/2	$4 \times 10^{-4} - 10^{-1}$	$<2.4 \times 10^{-14}$
Tokyo-ICRR	PScint	11.0	1/20	$6 \times 10^{-4} - 10^{-1}$	$<1.8 \times 10^{-12}$
	PT(Ar)	24.7	1/20	$10^{-3} - 1.0$	$<1.6 \times 10^{-12}$
	PT(He)	24.7	3.0	$3 \times 10^{-4} - 1.0$	$<7.2 \times 10^{-13}$
Tokyo-Kam	PScint	22.0	1/16	$3 \times 10^{-4} - 1.0$	$<1.5 \times 10^{-12}$
Tokyo Univ.	PScint	2.0	5/4	$10^{-2} - 10^{-1}$	$<3.0 \times 10^{-12}$
Utah-Stnfd.	Pscint	2.7	1/8	$10^{-3} - 3 \times 10^{-2}$	$<8.2 \times 10^{-12}$
Berkeley	CR-39	94.0	$Z/\beta > 3$	$2 \times 10^{-2} - 1.0$	$<1.5 \times 10^{-13}$
Berkeley	Mica/Al	0.1	—	$2 \times 10^{-4} - 2 \times 10^{-3}$	$<2.0 \times 10^{-17}$
Kitami	NtrCll	1000	$Z/\beta > 6$	$4 \times 10^{-2} - 1.0$	$<5.2 \times 10^{-15}$
Future Detectors					
Caltech	PScint	58.3	1/10	$6 \times 10^{-4} - 10^{-1}$	$<1.2 \times 10^{-13}$ (expected in one years running)
Frejus	GT + FT	1000	1/50	$10^{-4} - 10^{-1}$	setting up
Homestake	LScint	1300	1/10	$6 \times 10^{-4} - 1.0$	setting up
Mt. Blanc-LSD	LScint	240	1/30	$4 \times 10^{-4} - 1.0$	taking data
Gran Sasso	PT LSci, TE	12,000	1/10	$1 \times 10^{-4} - 1.0$ (PT) $6 \times 10^{-3} - 1.0$ (LScint)	proposed

[a]Based on S. Errede, reference 24.

[b]PT(Xx) = Proportional Tubes (Gas); PScint = Plastic Scintillator = PSc; LScint = Liquid Scintillator; NtrCll = Nitrocellulose; GT + FT = Geiger Tubes + Flash Tubes; TE = Track Etch.

however, would not take place if the monopole is a positive dyon or if it picks up a proton or light nucleus. It also seems to be a little hard to calculate. The limit derived from the experiment is $Fm \leq 10^{-17} \text{cm}^{-2}\text{s}^{-1}\text{sr}^{-1}$ at a monopole velocity of $\sim 8 \times 10^{-4}c$.

CATALYSIS EXPERIMENTS

These experiments require monopoles, if they exist, to catalyze proton decay with a substantial cross section. The experiments to date are listed in TABLE 5, which has been updated from the ICOBAN '84 talk of S. Errede.[24]

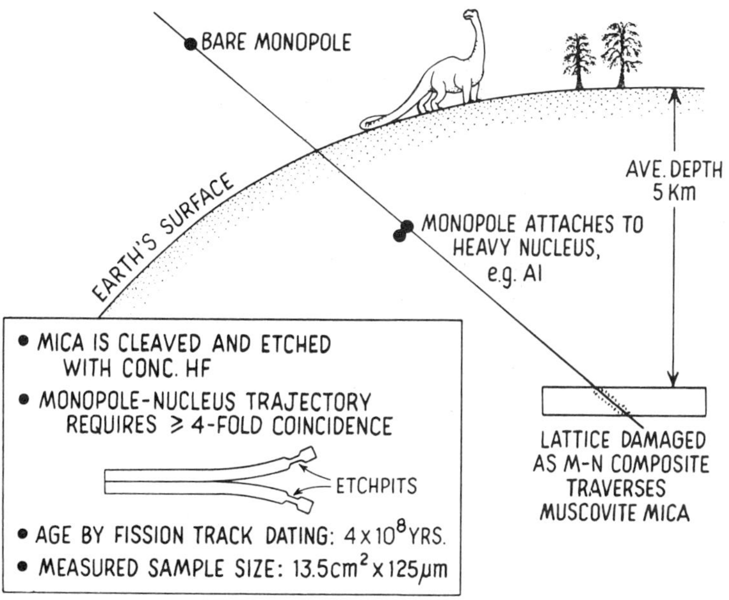

FIGURE 7. The monopole search by track-etch techniques in ancient mica of B. Price *et al.* (reference 29), as represented by J. Stone.

Suffice it to say that neither monopoles nor proton decays have been seen. The best limits are from the IMB and Kamiokande detectors and are listed with the assumed catalysis cross sections below:

IMB: $Fm < 8.5 \times 10^{-15} \text{cm}^{-2} \text{sr}^{-1} \text{s}^{-1}$, if $\sigma_c >= 1.7$ mb

$Fm < 3.4 \times 10^{-15} \text{cm}^{-2} \text{sr}^{-1} \text{s}^{-1}$, if $\sigma_c >= 17$ mb

Kamiokande: $Fm < 7.6 \times 10^{-15} \text{cm}^{-2} \text{sr}^{-2} \text{s}^{-1}$, if $\sigma_c > 10$ mb

$Fm < 6.4 \times 10^{-15} \text{cm}^{-2} \text{sr}^{-1} \text{s}^{-1}$, if $\sigma_c > 100$ mb

TABLE 5. Detector Parameters for Experiments Searching for Evidence of MCND[a]

	AHT	NUSEX	Soudan-I	KGF	Kamiokande	IMB
Depth (mwe)	0	5200	1800	7000	2700	1500
Detector Material	H_2O	Iron	Taconite Cement	Iron	H_2O	H_2O
Avg. Density (gm/cm^3)	1.0	3.5	1.9	1.5	1.0	1.0
Detector Mass (Ton) Tot/F. Vol	17 (17)	150 (100)	31 (16)	140 (100)	3000 (1000)	8000 (3300)
Detector Dim. (meters)	$3\phi \times 2.4h$	$(3.5)^3$	$3 \times 3 \times 2$	$4 \times 4 \times 6$	$15.5\phi \times 16h$	$23 \times 17 \times 18$
Eff. Area $\langle A \rangle$* of Detector (m^2)	7.0	18	10	32	220	550
Avg. Path Length $\langle L \rangle$ (m)	1.9	2.3	1.7	3.1	10.3	12.8
Electronic Live Time (μsec)	2.2	5.0	6.5	7.0	88	8.0
Electronic Dead Time (msec)	2.0	20	800	20	3.0	3.0

[a]Based on S. Errede, reference 24.

SUMMARY

The limits from each of the various types of detectors are shown in FIGURE 8. As pointed out in the discussion of velocity dependence, the limits from the scintillator ionization experiments, such as Baksan, may well be overstated by up to several factors

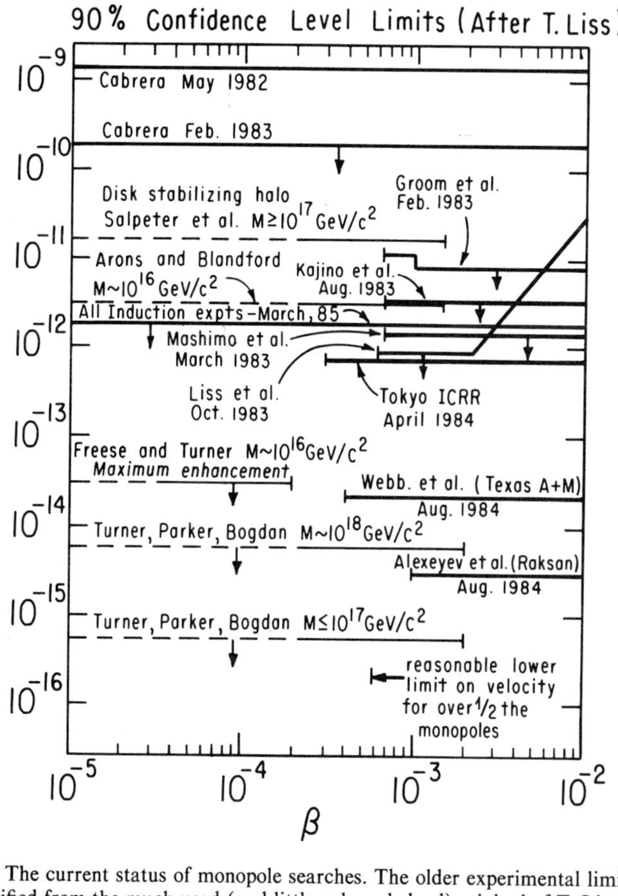

FIGURE 8. The current status of monopole searches. The older experimental limits have been slightly modified from the much used (and little acknowledged) original of T. Liss by changing them to be 90% C.L. limits.

of two in the velocity region near their threshold ($\beta = 10^{-3}$ for the Baksan detector). The region of the most likely monopole velocities is also shown in FIGURE 8.

In conclusion, a great deal of clever experimental work has gone on in monopole searches. The limits from the induction searches are now a factor of 600 below the flux corresponding to Cabrera's one candidate, which is most likely an experimental artifact. Large ionization detectors specifically designed for slow moving objects are running (Baksan, Texas A & M) or are being commissioned (Frejus, Homestake). A detector that is even an order of magnitude larger, using almost every trick so far invented, the Gran Sasso detector, has been proposed. The odds against finding monopoles seem long, but it is one of the few ways to try to do physics directly at energies 10^{13} orders of magnitude higher than available at accelerators.

ACKNOWLEDGMENTS

Thanks are due to many people who have helped me prepare this paper. In particular, I would like to thank Steve Errede for permission to quote and update the tables from his admirable ICOBAN '84 talk, James Stone for permission to use his picture of the mica (and dinosaur) experiment and for additional material, Peter Vandervoort for explaining galactic structure to me, and Tony Liss and Steve Ahlen for conversations and permission to present the results of their new BNL reactor experiment. I would also like to thank S. Berman, B. Cabrera, P. Eschstruth, F. Fickett, R. Steinberg, G. Tarlé and R.C. Webb for conversations and unpublished information on the status of their experiments. Finally, many thanks to Martin Block for the invitation to this excellent Aspen Conference, which also kept us away from Goldminer's Daughter in March.

NOTES AND REFERENCES

1. FARADAY, M. 1932. Faraday's Diary, Vol. I, December 8, 1831. Thomas Martin, Ed. G. Bell and Sons, Ltd. London. (See also October 17, 1831.)
2. ALVAREZ, L. W. 1963. LRL Physics Note 470. Unpublished.
3. DIRAC, P. A. M. 1931. Proc. R. Soc. London **A133:** 60.
4. KUCHNIR, M. Private communication.
5. CABRERA, B. 1982. Phys. Rev. Lett. **48:** 1378.
6. INCANDELA, J., M. CAMPBELL, H. FRISCH, S. SOMALWAR, M. KUCHNIR & H. R. GUSTAFSON. 1984. Phys. Rev. Lett. **53:** 2067.
7. BERMAN, S. Private communication.
8. GUY, C. N. Private communication.
9. FICKET, F. Private communication.
10. CABRERA, B., M. TABER, R. GARDNER & J. BOURG. 1984. Phys. Rev. Lett. **51:** 1933; also CABRERA, B. Private communication.
11. SOMALWAR, S., J. INCANDELA, H. FRISCH & M. KUCHNIR. 1984. Nucl. Instrum. Methods **226:** 341–343.
12. Two gradiometers can be superimposed with the cells overlapping, but with them offset by half a cell. They will then have zero mutual coupling and will be independent of each other. This can be done in an infinite number of ways corresponding to all kinds of orthogonal functions.
13. GOTO, E. 1958. J. Phys. Soc. Japan **10:** 1415; GOTO, E. *et al.* 1963. Phys. Rev. **132:** 307.
14. EBISU, T. & T. WATANABE. 1984. Kobe University preprints 84-07 and 84-08, October 1984.
15. EBERHARD, P. *et al.* 1971. Phys. Rev. **D4:** 3260.
16. See MARCH, R. 1984. *In* Monopole '83. J. Stone, Ed.: 511. Plenum Press. New York.
17. MIHALAS, D. & J. BINNEY. 1981. Galactic Astronomy. W. H. Freeman and Co., San Francisco.
18. PARKER, E. N. 1971. Astrophys. J. **163:** 225; TURNER, M. S., E. N. PARKER & T. J. BOGDAN. 1982. Phys. Rev. **D26.**

19. See AHLEN, S. 1984. *In* Monopole '83. J. Stone, Ed.: 629. Plenum Press. New York. This gives an excellent summary.
20. FERMI, E. & E. TELLER. 1947. Phys. Rev. **72:** 399; AHLEN, S. 1984. See reference 19.
21. DRELL, S., N. KROLL, M. MUELLER, S. PARKER & M. RUDERMAN. 1983. Phys. Rev. Lett. **50:** 644.
22. GOEBEL, C. 1984. *In* Monopole '83. J. Stone, Ed.: 333. Plenum Press. New York.
23. LISS, T. Private communication.
24. ERREDE, S. 1984. University of Michigan preprint UM/PDK 84-74. Invited talk at the ICOBAN '84 Conference, Park City, Utah, January 1984.
25. ALEXEYEV, E. N., N. M. BOLIEU, A. E. CHUDAKOV & S. R. MIKHEYEV. 1984. Paper 213 submitted to the XXII International Conference on High Energy Physics, Leipzig, July 19–25, 1984.
26. SHEPKO, M. J. & R. C. WEBB. 1984. Proceedings of the DPF meeting at Santa Fe, New Mexico, October 1984 (private communication from R. Webb, so far).
27. STEINBERG, R. 1984. Private communication and *In* Monopole '83. J. Stone, Ed.: 567. Plenum Press. New York.
28. ESCHSTRUTH, P. 1984. *In* Monopole '83. J. Stone, Ed.: 611. Plenum Press. New York; and private communication.
29. IAROCCI, E. *et al.* Proposal for a Large Area Detector Dedicated to Monopole Search, Astrophysics, and Cosmic Ray Physics at the Gran Sasso Laboratory (INFN, Frascati).
30. PRICE, B., S. L. GUO, S. P. AHLEN & R. L. FLEISCHER. 1984. Phys. Rev. Lett. **52:** 1265.
31. STONE, J. 1985. University of Michigan preprint UM HE 85-02. Invited talk at the Fermilab Conference Inner Space/Outer Space, May 2–5, 1984.

Production of W and Z in the UA2 Experiment at the CERN $\bar{p}p$ Collider

Presented by CLAUDIO CONTA

UA2 Collaboration[a]

Dipartimento di Fisica Nucleare e Teorica

Universitá di Pavia

and

INFN

Sezione di Pavia

Via Bassi, 6 I27100 Pavia, Italy

INTRODUCTION

We report here the results from a search for electrons with $p_T > 15$ GeV/c at the CERN $\bar{p}p$ collider ($\sqrt{s} = 540$ GeV) during its 1982 and 1983 periods of operation, which corresponded to an integrated luminosity of 131 nb^{-1}. In addition, we describe here some qualitative results obtained from the data collected during the last run from September to December 1984, which corresponded to an integrated luminosity of ~320 nb^{-1} at $\sqrt{s} = 630$ GeV.

Following a general discussion of the topology of the events containing an electron candidate, we shall compare the data with expectations in the framework of the Standard Model for the reactions:

$$\bar{p} + p \rightarrow W^{\pm} + \text{anything}, \quad W^{\pm} \rightarrow e^{\pm} + \nu\,(\bar{\nu})$$

$$\bar{p} + p \rightarrow Z^0 + \text{anything}, \quad Z^0 \rightarrow e^+ + e^- \text{ or } e^+ + e^- + \gamma .$$

We also report on the observation of high p_T electrons produced in association with hard jets and large missing transverse momenta. These events are found difficult to understand in terms of standard QCD production processes.

THE DETECTOR

The UA2 apparatus is described in detail elsewhere.[1] It comprises two calorimeters. The central part, covering the polar angles of 40° to 140°, consists of 240 towers, each segmented into one electromagnetic and two hadronic compartments.

The forward calorimeters (20° to 37.5° and 142.5° to 160°) consist of 240 electromagnetic cells, each split in depth in two compartments of 24 and 6 radiation lengths. They are located after a toroidal magnetic field that is followed by a set of drift chambers to allow for the measurement of the charged particle momentum.

[a]This UA2 Collaboration includes Bern, CERN, Copenhagen (NBI), Orsay (LAL), Pavia, and Saclay (CEN) Collaborations.

The interaction point is surrounded by a compact vertex detector to observe the charged particles produced in the $\bar{p}p$ collision and to measure the position of the event vertex. Both the central and the forward calorimeters are preceded by a preshower detector.

High p_T electrons are detected in the polar angular range of 20° to 160° and over the full azimuthal range. The electron signature requires an energy deposition in the calorimeters compatible with an electromagnetic shower, which has a matching charged track measured in the vertex detector that is aligned with a large signal observed in the preshower detectors. In the forward direction, agreement between the momentum and the energy of the electron is required. The detection efficiency is about 77% for electrons, while the rejection against jets is about 4×10^4.

RESULTS ON *W* AND *Z* DECAYS INTO LEPTONS

In recent publications,[3–5] we have described the observation in $\bar{p}p$ collisions of the W and Z^0 bosons via their electron decay mode.

Using the electron signature mentioned in the previous section, we have isolated a sample of 225 events with an electron candidate with $p_T > 15$ GeV/c. In addition to electrons, this sample still contains fake electrons resulting from misidentified hadrons or hadron jets; a jet is defined as an energy cluster with more than 3 GeV of transverse energy, with the cluster being obtained by grouping together adjacent cells with energy deposit. If the electron is genuine, the event must contain either another electron of opposite charge or a neutrino. However, if it is not genuine, we expect an additional jet at approximately opposite azimuth[2]; forty-five events contain no jet (FIGURE 1a). Events containing at least one jet fall into two classes: those with a jet back-to-back (FIGURE 1b), which is dominantly background, and those in which the dominant jet activity is not back-to-back (FIGURE 1c). Details of the selection criteria are described elsewhere,[5] as well as the determination of the shape and magnitude of the background of misidentified hadrons, which are shown as lines in FIGURE 1.

The combined p_T^e distributions of the 69 events from FIGURE 1a and FIGURE 1c are shown in FIGURE 2 together with the background estimate. There is a clear accumulation of events near $p_T^e = 40$ GeV/c, which is distinctive of the Jacobian peak expected from $W \rightarrow e\nu$ decay. Thirty-seven events have $p_T^e > 25$ GeV/c with a background contamination of less than 1.5 events. These events have been used to determine the mass of the W and its production cross section. The full sample of 37 events is contaminated by three background sources:

(a) misidentified electrons from two-jet background: 1.5 ± 0.1 events;
(b) electrons from $Z^0 \rightarrow e^+e^-$ decay with one electron undetected: 2.5 ± 0.9 events;
(c) electrons from $W \rightarrow \tau\nu$, $\tau \rightarrow e\nu\nu$ decay chain: 0.9 ± 0.1 events.

The eight events that are darker in FIGURE 1b are Z^0's decaying into e^+e^- pairs. They lead to the measurement of the Z^0 mass, width, and production cross section. TABLE 1 summarizes these measurements and the theoretical predictions.[8,9] This table shows the very good agreement of the masses with the theoretical predictions. Only the measured Z^0 cross section is too large (and hence the estimate of the number of

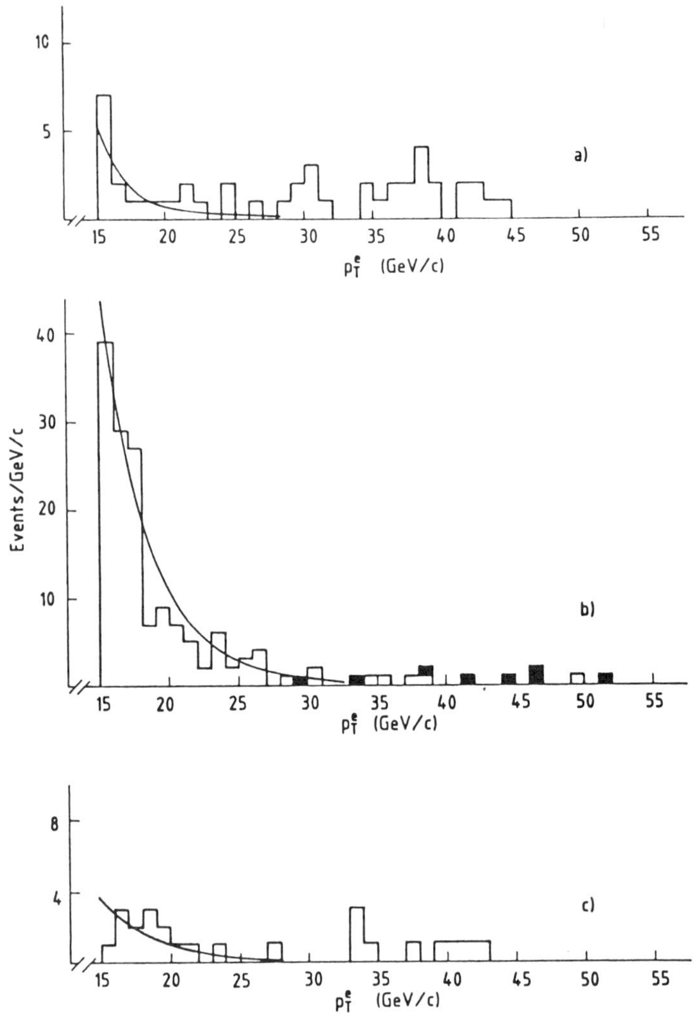

FIGURE 1. (a) Transverse momentum distribution of the electron candidates in events with no additional jets. (b) Transverse momentum distribution of the electron candidates in events with additional jets back-to-back in the transverse plane. The dark area corresponds to the eight Z^0 events for which only the electron with the higher p_T^e is plotted. (c) Transverse momentum distribution of the electron candidates in events with additional jets not back-to-back. In all three cases, the curves represent the background estimates.

neutrino families of mass $m_\nu \ll m_Z/2$ is tightly constrained). This feature is not statistically significant with the present limited sample. These results are also in good agreement with the UA1 measurements.[6,10]

FIGURE 3 shows the distribution of the measurement of the p_T of the W and Z^0, which can be as large as expected from a full QCD treatment.[9]

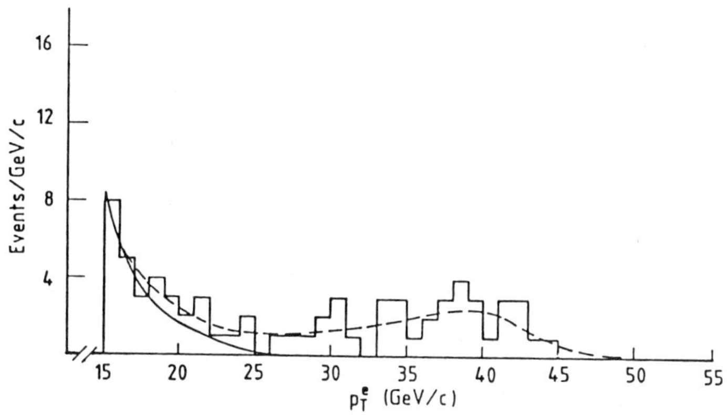

FIGURE 2. Transverse momentum distributions of the electron candidate in the $W \rightarrow e\nu$ event sample. Full curves: background estimates. Dashed curve: the sum of all expected contributions.

The observation of one Z^0 event compatible with the decay, $Z^0 \rightarrow e^+e^-\gamma$, and the observation of two similar events in the UA1 data[11] was unexpected. Using the UA2 event alone, we estimate a probability of 4% that at least one such event (with a configuration equal or less likely) should have resulted in our data sample from internal bremsstrahlung.[7] In a search for such events in the W decays, we have found one event with the electron and the photon collinear. It is not excluded that such an event results from the radiation of the electron in the vacuum chamber (4.5% probability).

A preliminary analysis of the 1984 data about the W and Z^0 production at $\sqrt{s}$ = 630 GeV shows the following qualitative features. From data corresponding to an integrated luminosity of ~210 nb^{-1} (about two-thirds of the full 1984 run), a sample of 122 events with a good electron candidate has been selected; sixty of them have $p_T^e > 25$

TABLE 1. $W^{\pm}$ and Z^0 Properties[a]

	$W^{\pm}$	Z^0
Mass(GeV/c^2)	83.1 ± 1.9 (stat.) ± 1.3 (syst.)	92.7 ± 1.7 ± 1.4
	$\left[83.0\ {}^{+2.9}_{-2.7}\right]$	$\left[93.8\ {}^{+2.4}_{-2.2}\right]$
Γ ($W^{\pm}$, $Z^0 \rightarrow$ all) (GeV/c^2)	2.7 (assumed)	<6.5 ± 0.6 (90% C.L.)
σ^e (nb)	0.53 ± 0.1 (stat.) ± 0.1 (syst.)	0.11 ± 0.04 ± 0.02
	$\left[0.37\ {}^{+0.11}_{-0.06}\right]$	$\left[0.042\ {}^{+0.012}_{-0.006}\right]$
$\langle p_T \rangle$ (GeV/c)	6.9 ± 1.0	4.1 ± 1.3
$\sin^2\theta_w$	0.216 ± 0.01 (stat.) ± 0.01 (syst.) (m_W = 38.65/sin θ_w) [0.217 ± 0.014]	

[a]Predicted quantities are in brackets.

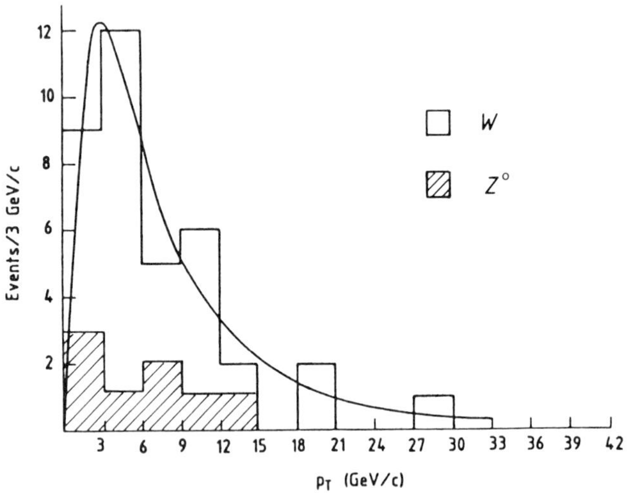

FIGURE 3. Distribution of the W and Z^0 transverse momentum. The curve represents a QCD prediction.[9]

GeV. Both the rate and the p_T distribution of these events are in good agreement with the results previously described. This behavior, therefore, qualitatively confirms both the $W \rightarrow e\nu$ cross section and the W mass values.

From the 1984 full data sample (~320 nb^{-1}), eight $Z^0 \rightarrow e^+e^-$ candidates have been selected with a mass distribution compatible with the previous results;[3,5] however, the rate is as much as ~2.5 less than before, thus leading to a cross section value in better agreement with theoretical predictions.[8,9]

Neither $Z^0 \rightarrow e^+e^-\gamma$ nor $W \rightarrow e\nu\gamma$ additional candidates have been found in the 1984 data.

LOW MASS ELECTRON PAIRS

Electron pairs originating from sources other than Z^0 decays are expected to result at lower mass from either the Drell-Yan mechanism or from the semileptonic decay of heavy flavor. The latter category is expected to have associated jet activity.

We use a sample of events in which at least two isolated electrons of $p_T^e > 5$ GeV/c are identified, with an invariant mass of 12 GeV/$c^2 < M_{ee} < M_Z$. Data from the 1982 run were excluded from this study because of incomplete azimuthal coverage of the UA2 detector. This leaves an integrated luminosity of 116 nb^{-1}.

Ten events are found in the mass range of 12 GeV/$c^2 < M_{ee} <$ 26 GeV/c^2 with an expected background due to fake electrons of 2.2 ± 0.6 events; the mass distribution of these events is shown in FIGURE 4 together with the estimated background curve.

The total event rate is consistent, after background subtraction, with the combined expectation of 2.5 ± 1.0 events from the Drell-Yan process[12] and of 3.0 ± 1.2 events from heavy flavor production.[13] However, because of kinematic constraints, the two event categories cannot be separated event-by-event.

An upper limit of 130 pb (90% C.L.) for the production of electron pairs with mass $M_{ee} > 26$ GeV/c^2 is also determined.

EVENTS WITH LARGE p_T^e AND LARGE p_T^ν

From the sample of 225 events previously described, we discard events collected in the 1982 period for which a wedge of 60° of the central calorimeter was removed to give room for a large angle magnetic spectrometer. We also discard ten events in which the electron candidate is observed near the interface between the central region and one of the forward regions of the detector and is associated with nearby calorimeter energies in each of the two regions. The initial sample is reduced to 190 events and the corresponding integrated luminosity to 116 nb^{-1}.

A large fraction of the 190 events have one or more jets. For each jet (j), we evaluate p_T^j and p_L^j, assuming the jet to be massless. We also evaluate $p_T(J)$ (the vector sum), $E_T(J)$ (the scalar sum), and $m(J)$ (the mass) for the system of all jets (J) or the system of jets and electrons (Je); $p_T(Je)$ measures the missing transverse momentum in the set of electrons plus detected jets with $E_T^J > 3$ GeV. The observation of large $p_T(Je)$ is generally equivalent to the presence of a neutrino, $p_T^\nu \sim p_T(Je)$. In this discussion, "neutrino" should be understood in the broad sense since the UA2 detector does not distinguish between a neutrino and any system of possible noninteracting particles, such as photinos predicted by supersymmetric models. The distribution of $p_T(Je)$ is shown in FIGURE 5a, where the superimposed background measures the probability that a multijet event contains a misidentified electron and/or an undetected jet(s) escaping the UA2 acceptance. The 35 events having $p_T(Je) > 25$ GeV/c have an estimated background of 3.4 ± 0.3 events.

For this sample of 35 events, FIGURE 5b shows $E_T(J)$, the level of hadron activity in the event. The 31 events having $E_T(J) < 30$ GeV all belong to the *W*-event sample mentioned in the third section of this paper (FIGURES 1a and 1c); four events (A–D) stand out above a background, in the [$p_T(Je)$, $E_T(J)$] plane, of 0.45 ± 0.04 events for $p_T(Je)$, $E_T(J) > 30$ GeV (FIGURE 6). We can evaluate the expected background

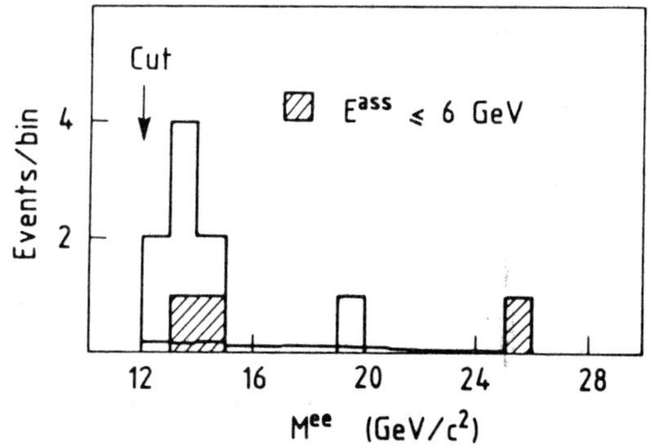

FIGURE 4. Electron pair mass distribution. The shaded area indicates three events with less than 6 GeV hadronic energy (E^{ass}) associated with each selected electron. The superimposed curve shows the estimated background of misidentified two-jet events.

contributions, B_1 and B_2, for the following configurations:

$$B_1 : p_T(Je) > p_0, E_T(J) > 30 \text{ GeV},$$

$$B_2 : p_T(Je) > 25 \text{ GeV/c}, E_T(J) > E_0.$$

In each of the four events, either B_1 or B_2 is always less than 0.02. No background

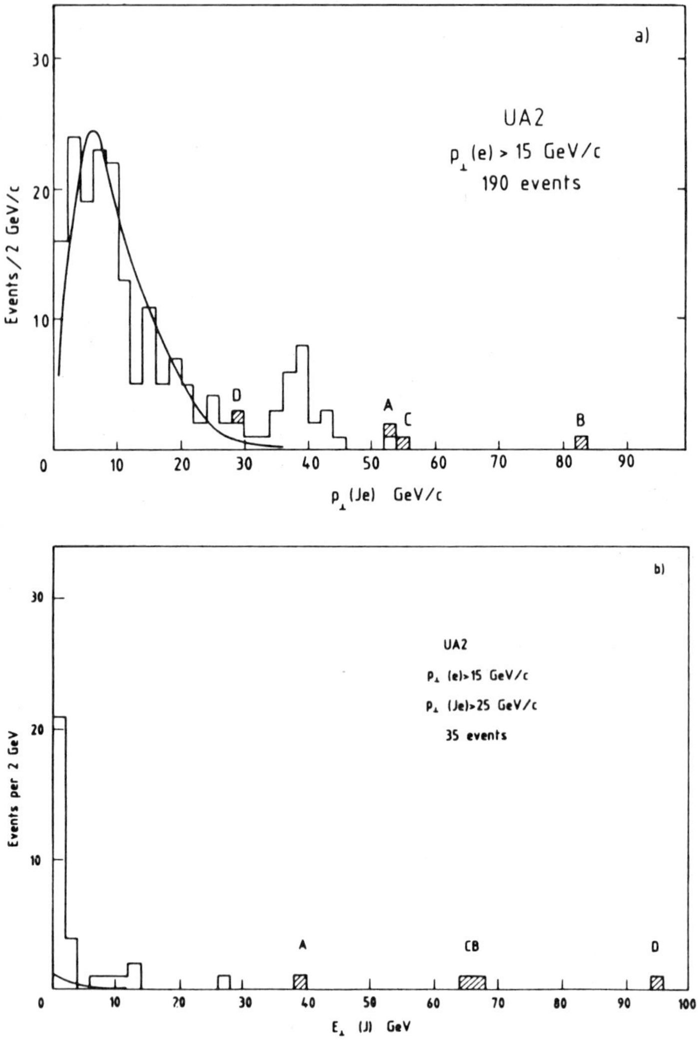

FIGURE 5. (a) Transverse momentum distribution of the system of electron and jets in the initial sample of 190 events. (b) Transverse energy distribution of the system of jets in the sample of 35 events having $p_T(Je) > 25$ GeV/c. In (a) and (b), the lines correspond to the calculated background contaminations. The four events having $p_T(Je) > 25$ GeV/c, $E_T(J) > 30$ GeV are crosshatched.

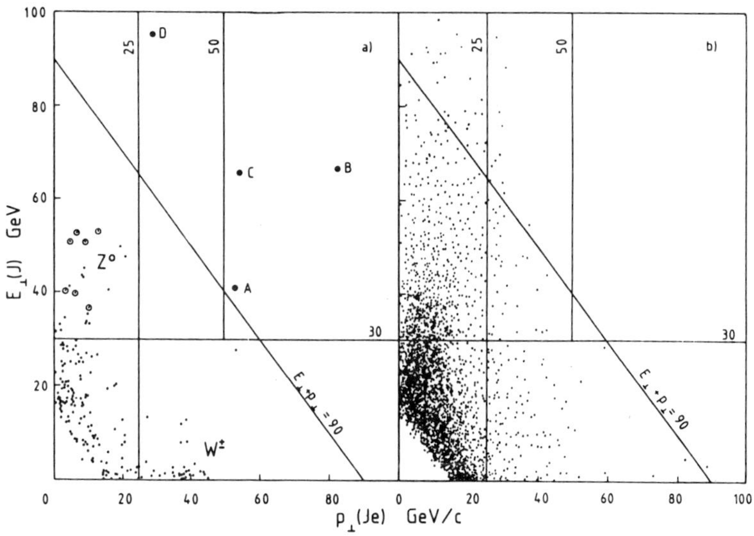

FIGURE 6. (a) Distribution of the 190 events of the initial sample in the $p_T(Je)$, $E_T(J)$ plane. Seven Z^0 events are circled (the eighth one was collected during the 1982 period; see reference 5). The W region is indicated. (b) Distribution of the background sample in the $p_T(Je)$, $E_T(J)$ plane. A reduction factor of 141 must be applied to infer from this sample the background contamination to the sample of FIGURE 6a. In both (a) and (b), lines of constant $p_T = 25$ and 50 GeV/c, $E_T(J) = 30$ GeV, and $p_T(Je) + E_T(J) = 90$ GeV are indicated.

event exists in the region of $p_T(Je) > 50$ GeV/c, $E_T(J) > 30$ GeV, which contains events A to C. We infer from this a background contamination of at most 0.02 events (90% confidence level) in this region.

In no event is there evidence of large p_T particles hitting passive detector elements, such as the magnet coils, in the azimuthal region where the neutrino is expected. However, event D has an azimuthal configuration similar to a two-jet event and this interpretation cannot be excluded.

In all events, the sharing of jet energies between the different calorimeter compartments is consistent with expectation. Also, each jet has several track pointing to the vertex, thus making interpretations such as beam-gas background or cosmic rays unlikely.

A muon can, in principle, simulate a neutrino in the UA2 detector. We searched without success in each event for a track near the neutrino azimuth that would also be associated with calorimeter energy consistent with the response of a minimum ionizing particle.

An event appearing at (p_0, E_0) in the $p_T(Je)$, $E_T(J)$ plane (FIGURE 6) may be interpreted as containing an undetected jet that would account, at least in part, for the measured value of p_0. If this jet had been detected, the same event would have appeared at (p_0^*, E_0^*) with $p_0^* < p_0$ and $p_0^* + E_0^* > p_0 + E_0$. However, the only events having $p_T(Je) + E_T(J) > 90$ GeV are the events of FIGURE 6. The absence of other events in this region, even with low values of $p_T(Je)$, makes that interpretation unlikely. (Details of each event are listed in table 1 of reference 14.)

While events A to C have a large $(e\nu)$ azimuthal opening angle, that of event D is small. Since its minimum $(e\nu J_1)$ mass is 23 ± 6 GeV/c^2 and its associated $(e\nu J_1 J_2)$ mass is 145 ± 15 GeV/c^2, event D cannot be easily interpreted as resulting from W decay. However, the configuration is suggestive of a $(\bar{q}q)$ pair where one quark decays semileptonically. Because of the absence of other events of similar topology and because of its similarity to a two-jet event, we defer such an interpretation until additional events of the same type will be observed.

The other events (A to C) have a transverse $(e\nu)$ mass between 56 and 82 GeV/c^2, suggestive of $W \rightarrow e\nu$ decay. Choosing the minimum $x_F(WJ)$ that forces $m(e\nu) = m(W)$, the mass, $m(WJ)$, of all three events is in the range of $160 < m(WJ) < 180$ GeV/c^2; this may be kinematic in origin since the background peaks in the same region (FIGURE 7).

If events A to C are interpreted as W-production with associated jet activity, their rate of occurrence should be predicted by known QCD processes. Integrating over all

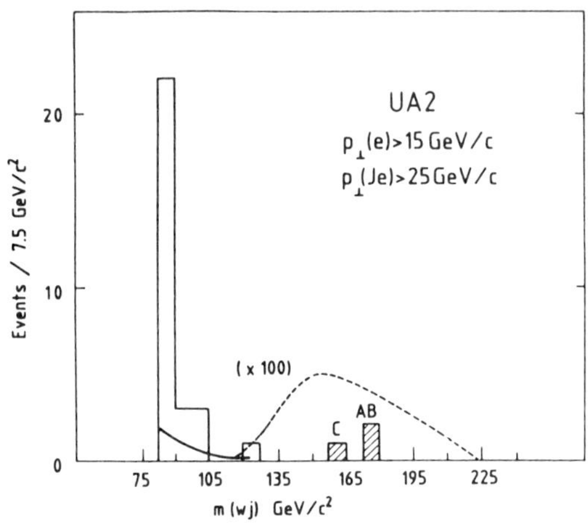

FIGURE 7. Distribution of $m(WJ)$ for the 35 events having $p_T(Je) > 25$ GeV/c. Events having in addition $E_T(J) > 30$ GeV are cross-hatched. Event D is omitted from the figure because its narrow $e\nu$ pair does not suggest an interpretation in terms of $W \rightarrow e\nu$. Such an interpretation would imply a very large value of $p_T(W)$ and $m(WJ) \simeq 230$ GeV/c^2. The smooth line corresponds to the calculated background. The dotted line is the background distribution (multiplied by 100) for events having $E_T(J) > 30$ GeV.

event topologies, Minkowski[15] predicts less than 0.14 events to satisfy $p_T(W) > 40$ GeV/c. A more realistic upper limit should take into account the topology of the events and thus we consider

$$\sigma(\bar{p}p \rightarrow W + J + \ldots)/\sigma(\bar{p}p \rightarrow W + \ldots)$$
$$\simeq \sigma(\bar{p}p \rightarrow j_1 j_2 + J + \ldots)/\sigma(\bar{p}p \rightarrow j_1 j_2 + \ldots),$$

where $j_1 j_2$ is a pair of jets having the same configuration as the $e\nu$ pair ascribed to a W decay. This relation holds approximately for processes in which $j_1 j_2$ couples via a gluon to a quark-antiquark pair (although the W-current, contrary to the gluon-current, is flavor changing). We expect the contribution of other QCD subprocesses and interferences among them to alter only slightly the right-hand side of the above relation. We evaluate such upper limits for events A to C from the sample of jet events.

We take as J any jet (jet pair) having a transverse momentum (invariant mass) at least as large as that of the corresponding jet (jet pair) in events A and B (C).

In our data sample, less than 0.4 events of the type of event A are expected; however, this event requires a W longitudinal momentum of ~150 GeV/c to be interpreted as a $W \rightarrow e\nu$ decay. The number of events like B and C, resulting from QCD processes, is less than 0.012 and 0.007, respectively, making their interpretation as W-jet(s) associated production from known processes unlikely. We should also expect events of the same type, with the ($e\nu$) pair replaced by a $W \rightarrow j_1 j_2$ decay, with a higher rate. From our own data, this event topology cannot be excluded.

It is important to note that the (unlikely) interpretation of events A to C in terms of $W \rightarrow e\nu$ decay is not mandatory. Indeed, the missing transverse energy may be shared by several undetected particles.

As far as 1984 data are concerned, the analysis, still in progress, shows the appearance of three more events with electrons associated with hard jets and large missing transverse momenta. However, in the [$p_T(Je)$, $E_T(J)$] plane, these events lie in a region around $p_T(Je) < 40$ GeV/c and $E_T(J) \sim 40$ GeV/c, where the background level is not negligible. Moreover, the analysis of the W + jet production, when completed, will have to show if these events have to be interpreted as the tail of the ordinary QCD W + jet production or as a new phenomena as indicated by the 1982–83 UA2 data.

CONCLUSIONS

We have summarized the results from the 1982–83 data on the production and decay into electron of the intermediate vector bosons, W and Z^0, and recalled the good agreement with the prediction of the $SU(2) \times U(1)$ theory.

We have found four events in which the ($e\nu$) pair is produced in association with large E_T jet(s) activity. Three of these events contain an ($e\nu$) pair of large transverse mass and we have attempted to describe them in terms of W-jet(s) associated production. Within the context of that hypothesis, at least two events are unlikely to result from known processes.

The existence of ($Z^0 \rightarrow e^+e^-\gamma$) decays was unexpected and more data are needed to ascertain whether the rate of this process is consistent with internal bremsstrahlung.

Preliminary and qualitative results, coming from the analysis of the 1984 data, have also been described.

REFERENCES

1. MANSOULIÉ, B. 1983. The UA2 apparatus at the CERN $\bar{p}p$ Collider. Proceedings 3rd Moriond Workshop on $\bar{p}p$ Physics. Editions Frontieres, p. 609; DIALINAS, M. *et al.* 1983. The vertex detector of the UA2 experiment. LAL-RT/83-14, ORSAY; CONTA, C. *et al.* 1984. The system of forward-backward drift chambers in the UA2 detector. Nucl. Instrum. Methods **224:** 65; BORER, K. *et al.* 1983. Multitube proportional chambers for the localization of electromagnetic showers in the UA2 detector. CERN-EP/83-177. Submitted to Nucl. Instrum. Methods; BEER, A. *et al.* 1984. The central calorimeter of the UA2 experiment at the CERN $\bar{p}p$ Collider. Nucl. Instrum. Methods **224:** 360.
2. BANNER, M. *et al.* UA2 COLLABORATION. 1982. Phys. Lett. **118B:** 203; BAGNAIA, P. *et al.*

UA2 COLLABORATION. 1983. Z. Phys. **C20:** 117; BAGNAIA, P. *et al.* UA2 COLLABORATION. 1984. Phys. Lett. **138B:** 430.
3. BANNER, M. *et al.* 1983. Phys. Lett. **122B:** 476.
4. BAGNAIA, P. *et al.* 1983. Phys. Lett. **129B:** 130.
5. BAGNAIA, P. *et al.* 1984. Z. Phys. **C24:** 1.
6. ARNISON, G. *et al.* UA1 COLLABORATION. 1983. Phys. Lett. **126B:** 398.
7. EADIE, W. *et al.* 1971. Statistical Methods in Experiment Physics, p. 231. North-Holland. Amsterdam.
8. For a review, see MARCIANO, W. J. 1983. Weak Interactions. Proc. Int. Symposium on Lepton and Photon Interactions at High Energies, p. 80 (Cornell University, Ithaca, New York).
9. ATARELLI, G., R. K. ELLIS, M. GRECO & G. MARTINELLI. 1984. Vector boson production at colliders: a theoretical reappraisal. Preprint CERN-TH 3851.
10. ARNISON, G. *et al.* 1983. Phys. Lett. **122B:** 103; ARNISON, G. *et al.* 1983. Phys. Lett. **129B:** 273.
11. ARNISON, G. *et al.* 1984. Phys. Lett. **135B:** 250.
12. DRELL, S. & T. YAN. 1970. Phys. Rev. **D25:** 316; For a review, see ALTARELLI, G. 1982. Phys. Rep. **81:** 1.
13. BERGER, E. & D. SOPER. 1984. CERN-TH/3850; FISHER, H. & W. GEIST. 1983. Z. Phys. **C19:** 159; PAKVASA, S. *et al.* 1979. Phys. Rev. **D20:** 2862; GLOVER, E., F. HALZEN & A. MARTIN. 1984. Phys. Lett. **141B:** 429.
14. BAGNAIA, P. *et al.* 1984. Phys. Lett. **139B:** 431.
15. MINKOWSKI, P. 1984. Phys. Lett. **139B:** 431.

Electron-Positron Electroweak Experiments

P. SÖDING

Deutsches Elektronen–Synchrotron DESY
Notkestrasse 85
2000 Hamburg 52, Federal Republic of Germany

INTRODUCTION

Electron-positron storage rings are now, after the latest upgrading of PETRA, reaching center-of-mass energies of $\sqrt{s} = 46.8$ GeV $= m_Z/2$. At such energies, the weak amplitudes expected from standard electroweak theory have sizes of up to 40% of the electromagnetic amplitude. Interference effects between weak and electromagnetic amplitudes are large and are measured in basically very simple two-particle to two-particle processes, with either lepton pairs or quark pairs in the final state (FIGURE 1). One already observes the tail of the Z^0 peak in the cross section for quark pair production at the higher cm energies, as well as up to 25% asymmetry in angular distributions induced by the weak force. The structure of the leptons is, under certain assumptions, being probed to a scale of a TeV or more.

In this talk, experimental results from e^+e^- storage rings on the electroweak interactions of the leptons and quarks will be surveyed and compared with the standard model.[1] A summary will be given at the end.

The basic process, $e^+e^- \rightarrow f\bar{f}$, is shown in FIGURE 1. For $e^+e^- \rightarrow e^+e^-$, t-channel diagrams also contribute. The ratio of the total cross section for $e\bar{e} \rightarrow f\bar{f}$ to the pointlike electromagnetic cross section for $e\bar{e} \rightarrow \mu\bar{\mu}$, from the diagrams of FIGURE 1, is

$$R_{f\bar{f}} = Q^{e^2} Q^{f^2} + 2Q^e Q^f g_v^e g_v^f Re\chi + (g_v^{e^2} + g_A^{e^2})(g_v^{f^2} + g_A^{f^2})\,|\chi|^2, \tag{1}$$

where

$$\chi = \frac{\rho G_F}{2\sqrt{2}\pi\alpha} \frac{m_Z^2 s}{s - m_Z^2 + i m_Z \Gamma_Z} \tag{2}$$

describes the ratio of the weak to the electromagnetic amplitude and

$$\rho = \frac{m_W^2}{m_Z^2 \cos^2 \theta_W} \tag{3}$$

measures the effective relative strength of the weak neutral current relative to the charged current interaction. (Q^f, g_v^f, and g_A^f are the charge and the weak neutral coupling constants of particle f.) The Z^0-photon interference contribution to $R_{f\bar{f}}$, which dominates the deviation from pure electromagnetism at low energies, is proportional to the vector couplings of the Z^0 to $e\bar{e}$ and $f\bar{f}$ since interference of axial amplitudes with the (vector) photon amplitude cancels on integration over the production angles, θ.

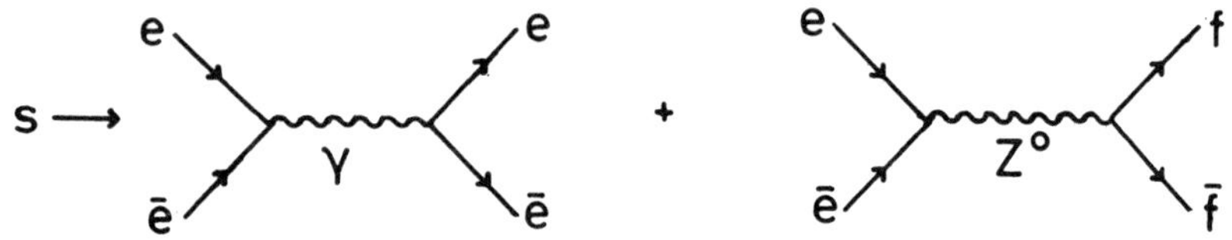

FIGURE 1. The basic processes of fermion (lepton or quark) pair production studied at electron-positron colliders.

In contrast to this, the angular distribution, which is of the general form,

$$1 + a \cos\theta + \cos^2\theta,$$

develops a forward-backward asymmetry,

$$A_{f\bar{f}} = \frac{\sigma\left(\theta < \frac{\pi}{2}\right) - \sigma\left(\theta > \frac{\pi}{2}\right)}{\sigma\left(\theta < \frac{\pi}{2}\right) + \sigma\left(\theta > \frac{\pi}{2}\right)}$$

$$= \frac{3}{2R_{f\bar{f}}}\left[Q^e Q^f g_A^e g_A^f Re\chi + 2 g_v^e g_A^e g_v^f g_A^f |\chi|^2\right], \tag{4}$$

which is dominated by interference of the axial Z^0 amplitude with the photon amplitude. Therefore, the vector and axial vector couplings of the Z^0 can be experimentally separated. The axial coupling is the easier one to measure due to the absence of a purely electromagnetic term in $A_{f\bar{f}}$, while for the determination of the weak vector coupling one has to rely on the absolute normalization of the measured cross section. Note that measurement of $R_{f\bar{f}}$ and $A_{f\bar{f}}$ does not give information on parity violation. To measure parity violation in e^+e^- annihilation, one needs a pseudoscalar observable, i.e., longitudinal polarization, which involves mixing of vector and axial couplings in the same Z^0 exchange amplitude. The expected effects are small and have not yet been observed.[2]

For later reference, the values of the weak neutral current couplings in the standard $SU(2) \times U(1)$ model are listed in TABLE 1.

TABLE 1. Weak Neutral Couplings in $SU(2) \times U(1)$

	Q^f	g_A^f	g_v^f	g_v^f for $\sin^2\theta_W = 0.22$
e, μ, τ	-1	$-\frac{1}{2}$	$-\frac{1}{2} + 2\sin^2\theta_W$	-0.06
u, c, t	$+\frac{2}{3}$	$+\frac{1}{2}$	$+\frac{1}{2} - \frac{4}{3}\sin^2\theta_W$	$+0.21$
d, s, b	$-\frac{1}{3}$	$-\frac{1}{2}$	$-\frac{1}{2} + \frac{2}{3}\sin^2\theta_W$	-0.36

TOTAL CROSS SECTION FOR $\ell\bar{\ell}$ AND $q\bar{q}$ PRODUCTION BY e^+e^- ANNIHILATION

By measuring the total cross section or the ratio, $R_{f\bar{f}}$, one expects to determine g_v^f and therefore, in the standard model, $\sin^2\theta_W$. However, for leptons, g_v^ℓ vanishes for $\sin^2\theta_W = 1/4$ and the electroweak interference term in $R_{\ell\bar{\ell}}$ is proportional to $(\sin^2\theta_W - 1/4)^2$. Consequently, the weak contribution is quite small for $\sin^2\theta_W \approx 0.22$ and the prediction of pure QED should hold to the percent level (FIGURE 2). For hadron production,

$$R_{had} = \sum_{flavors} R_{q\bar{q}} \propto \sum_{d,u,s,c,b} Q^q g_v^e g_v^q + \text{pure weak term} \tag{5}$$

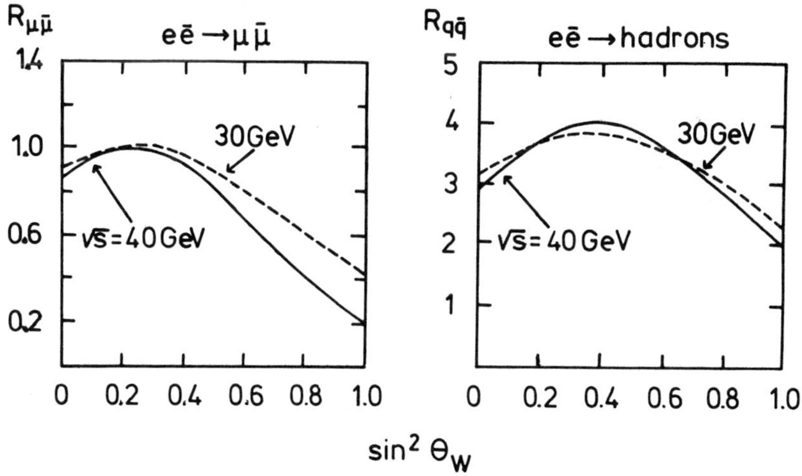

FIGURE 2. The $\sin^2\theta_W$ dependence of the muon pair production ($R_{\mu\mu}$) and the hadronic cross sections (R_{had}) normalized to the pointlike pure QED cross sections. (Adapted from reference 1.)

and the interference term vanishes only linearly at $\sin^2\theta_W = 1/4$, which results in some sensitivity to $\sin^2\theta_W$ (FIGURE 2).

The measurements[3] of lepton production, $R_{\mu\bar{\mu}}$ and $R_{\tau\bar{\tau}}$, from PEP and PETRA are indeed compatible with the QED prediction, as shown in FIGURE 3. The degree of agreement between the data and QED allows one to set limits on the validity of QED and on possible finite lepton size. A hypothetical lepton structure can be parametrized by form factors,

$$F_\pm(s) = 1 \mp \frac{s}{s - \Lambda_\pm^2}, \tag{6}$$

at the lepton-photon vertex. From the experiments, lower limits on the cutoff parameters, $\Lambda_\pm$, of typically 200 GeV or more for the muon and 120 GeV or more for

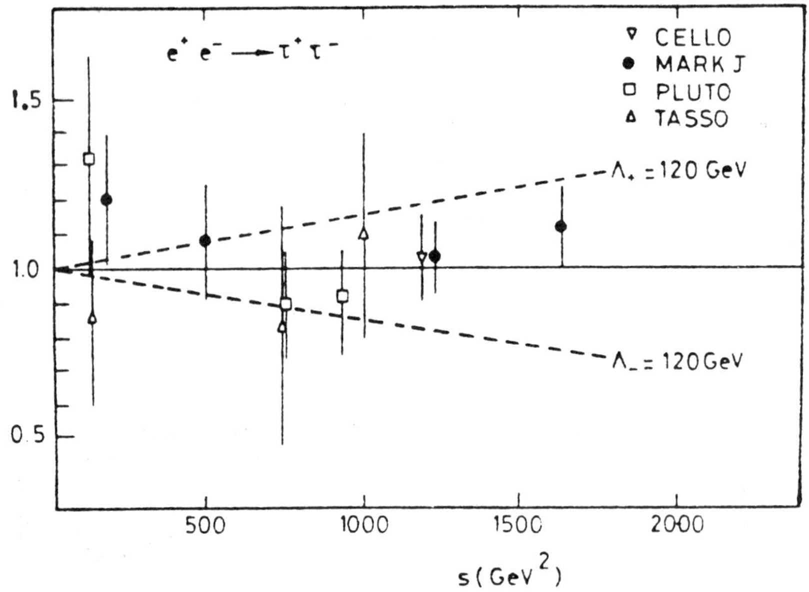

FIGURE 3a. Measured lepton pair production total cross sections divided by the QED prediction, as a function of cm energy squared. (Adapted from reference 1 and from Mark-J Collaboration. 1984. *Phys. Rep.* **109:** 131.)

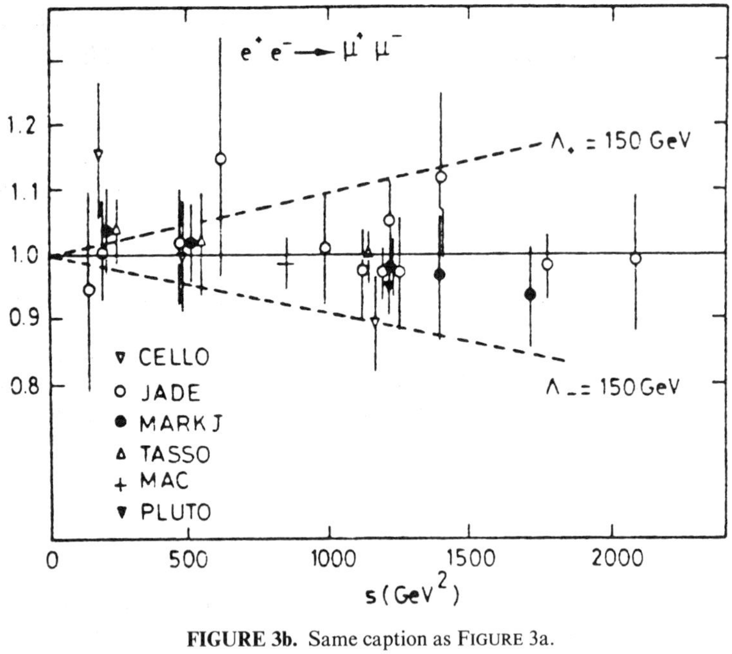

FIGURE 3b. Same caption as FIGURE 3a.

the τ, at 95% confidence level, are obtained. The form factor, $F_-(s)$, is equivalent to the exchange of a heavy vector particle with mass, Λ_-, and coupling constant, $\sqrt{2}e$.

Measurements of R_{had} averaged over PETRA experiments (and corrected for radiative effects) are compiled[4] in FIGURE 4. Above $\sqrt{s} = 35$ GeV, the values are rising with increasing energy. This is expected from the tail of the Z^0. In order to interpret the measurements quantitatively, in addition to the electroweak terms in equation (**1**), real and virtual gluonic radiative corrections [which modify $R_{q\bar{q}}$ by a factor of $1 + \alpha_s/\pi + 0(\alpha_s^2)$] also have to be taken into account; this contribution decreases

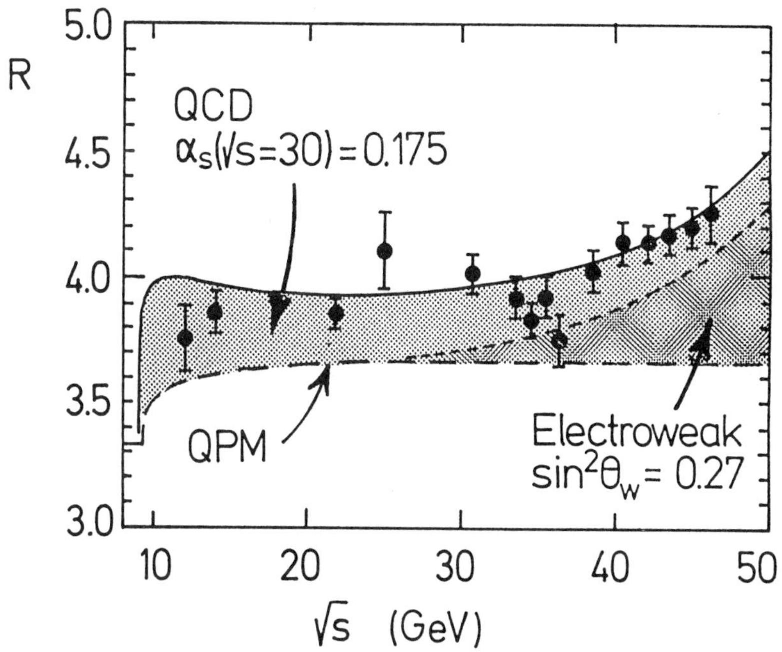

FIGURE 4. Measured R values for hadron production from PETRA experiments compared with the quark parton model (QPM) for the five flavors, d, u, s, c, and b, plus electroweak contribution and QCD (from reference 4).

logarithmically with increasing $\sqrt{s}$. A two-parameter fit[4] to the data gives

$$\alpha_s = 0.175 \pm 0.075 \qquad \text{at } \sqrt{s} = 30 \text{ GeV}$$

and

$$\sin^2 \theta_W = 0.27 \pm 0.04.$$

The weak contribution is shown shaded, with the QCD radiative corrections as the dotted part in FIGURE 4. Modification of the standard model amplitude by quark form factors in equation (6) leads to lower limits for the cutoff parameters, $\Lambda_{\pm}$, of 200 to 300 GeV.

ANGULAR ASYMMETRY IN LEPTON PAIR PRODUCTION

The asymmetries in the angular distributions for $\mu^+\mu^-$, $\tau^+\tau^-$, and $q\bar{q}$ production that are expected for standard model couplings (TABLE 1) are shown in FIGURE 5. Well below the Z^0 mass, they stem from the axial current and are therefore nearly independent of $\sin^2\theta_W$ (except for a weak dependence on $\sin^2\theta_W$ entering through the mass, m_Z, in the propagator). For $s \ll m_Z^2$, the asymmetry is largest for quarks of

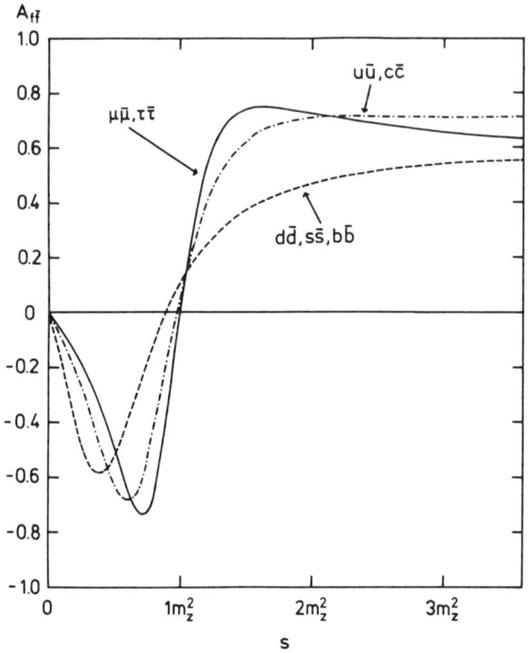

FIGURE 5. Expected asymmetry of the three types of charged fermions versus cm energy squared.

$Q^f = -⅓$ due to their smaller electromagnetic coupling, i.e., larger axial/vector ratio. Near the Z^0 pole, the weak amplitude dominates and leptons will in general show the strongest asymmetry. Examples of measured angular distributions[5] are shown in FIGURE 6.

Evidence for the cos θ term in the angular distribution is demonstrated in FIGURE 7. TABLES 2 and 3 summarize experimental values of $A_{\mu\bar{\mu}}$ and $A_{\tau\bar{\tau}}$. These values are shown plotted versus cms energy squared s in FIGURE 8. Systematic errors in the angular or charge measurements will lead to systematic shifts, $|\Delta\ A_{f\bar{f}}| \leq 1\%$. The radiative corrections depend on the selection criteria used to define the pair production process, e.g., the criteria differentiating $e\bar{e} \to \ell\bar{\ell}$ from $e\bar{e} \to \ell\bar{\ell}\gamma$. All the higher order electroweak effects have been calculated to the one-loop level.[6] They typically contribute

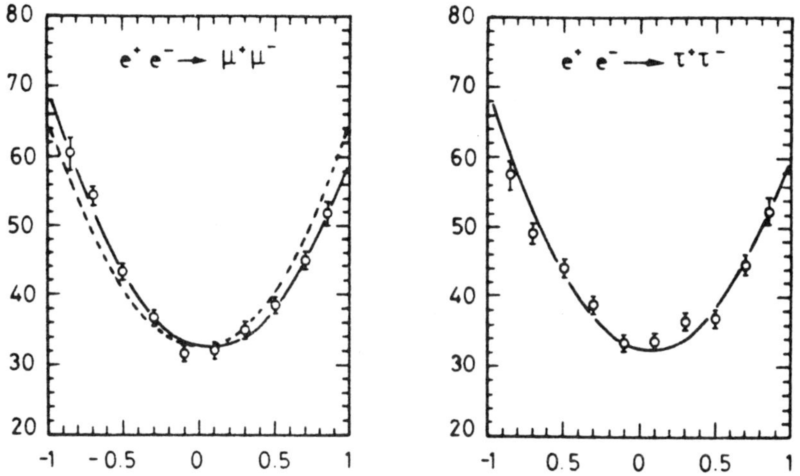

FIGURE 6. Differential cross section measurements, $sd\sigma/dcos\theta$, in nb GeV^2 from the MAC experiment. The dashed line shows QED and the solid one shows the standard model prediction (after $0(\alpha^3)$ radiative corrections).

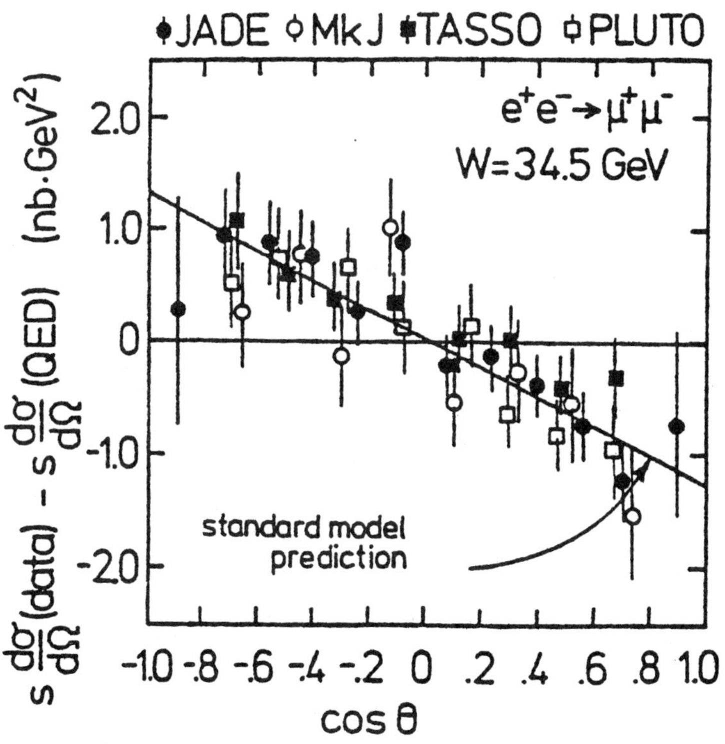

FIGURE 7. Difference between the measured cross section for $e\bar{e} \to \mu\bar{\mu}$ and QED, which shows the residual electroweak interference term that is linear in cos θ (from reference 4).

TABLE 2. Muon Pair Asymmetry

Expt.	$\sqrt{s}$ (GeV)	Asymmetry $A\mu\bar{\mu}$ (%) Measured	Predicted
MAC	29	−5.6 ± 0.8 ± 0.5	−5.9
MARK-2	29	−7.1 ± 1.7	−5.9
HRS	29	−5.0 ± 1.6	−5.9
Average	29	−5.9 ± 0.7	−5.9
CELLO	34.2	−6.4 ± 6.4	−8.5
MARK-J	34.6	−11.7 ± 1.9	−8.8
JADE	34.4	−11.6 ± 1.8 ± 1.0	−8.7
PLUTO	34.7	−13.4 ± 3.1 ± <1	−8.9
TASSO	34.5	−9.8 ± 2.3 ± 0.5	−8.7
Average	34.5	−11.3 ± 1.0	−8.7
CELLO	42.5	−13.4 ± 9.4	−14.4
MARK-J	41.1	−15.8 ± 5.3	−13.3
	44.1	−16 ± 10	−15.8
	46.0	−17 ± 10	−17.7
JADE	42.4	−20.1 ± 4.3 ± 1.7	−14.3
TASSO	43.0	−21 ± 7	−14.9
Average	42.5	−18.1 ± 2.7	−14.4

asymmetries of the order of $\Delta A_{ff} \approx +1.5\%$ (pure QED) and $+0.5\%$ (weak). The experimental data have been corrected for these effects so that they can be compared among each other as well as with leading order electroweak theory.

While the average of the $\mu^+\mu^-$ asymmetries measured at 29 GeV (PEP energy) agrees perfectly with the expectation from the standard model (given in TABLES 2 and 3, assuming $\rho = 1$), the PETRA averages of the asymmetry at 34.5 and 42.5 GeV exceed the expectations by 2.6 and 1.4 standard deviations, respectively. The $\tau\bar{\tau}$ asymmetry measurements all agree with the standard model expectations within their rather large errors. We will now discuss these results.

TABLE 3. Tau Pair Asymmetry

Expt.	$\sqrt{s}$ (GeV)	Asymmetry $A\tau\bar{\tau}$ (%) Measured	Predicted
MAC	29	−5.4 ± 1.2 ± 0.5	−5.9
MARK-2	29	−4.2 ± 2.0	−5.9
HRS	29	−4.5 ± 2.8	−5.9
Average	29	−5.0 ± 1.0	−5.9
CELLO	34.2	−10.3 ± 5.2	−8.5
JADE	34.6	−7.6 ± 2.7	−8.8
MARK-J	34.6	−7.8 ± 4.0	−8.8
TASSO	34.4	−4.9 ± 5.3 ± 1.2	−8.7
Average	34.5	−7.7 ± 1.9	−8.7
CELLO	42.5	−16.7 ± 9.0	−14.4
TASSO	43.0	−2.1 ± 15	−14.9
Average	42.6	−12.8 ± 7.7	−14.5

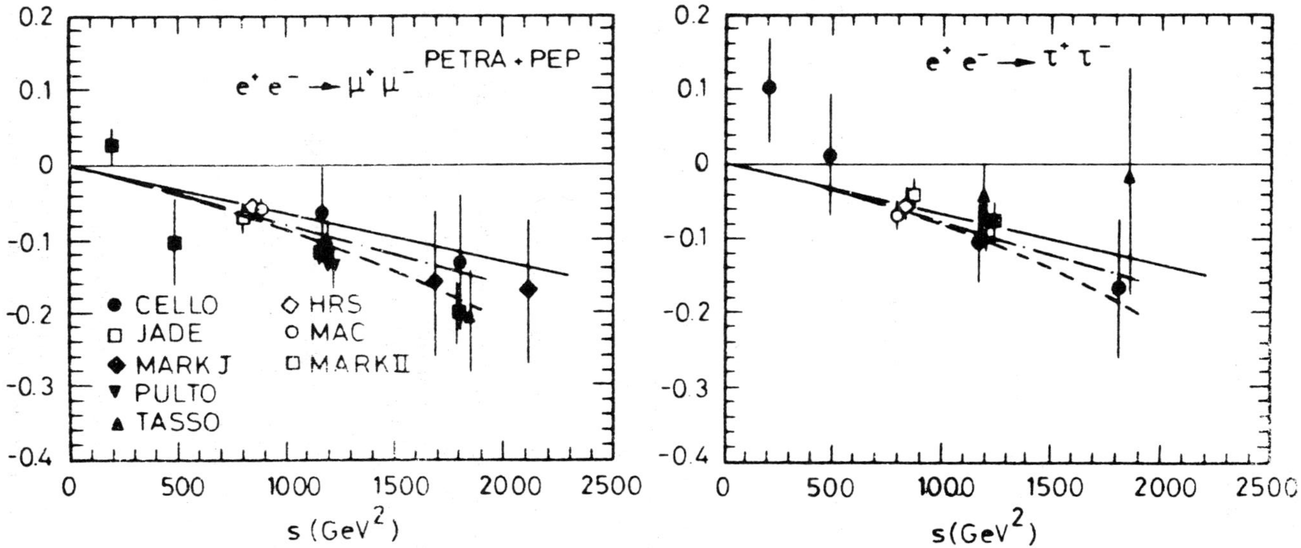

FIGURE 8. Measured angular asymmetries in lepton pair production. The solid line shows the standard model prediction for a pointlike interaction ($m_Z = \infty$). The dashed-dotted line shows the prediction for $m_Z = 93$ GeV, $\sin^2\theta_W = 0.22$ and the dashed line for $m_Z = 93$ GeV, $\sin^2\theta_W = 0.17$ (from reference 1).

Is There Evidence for a Finite Range of the Weak Force?

The dependence of the observed values of $A_{\mu\bar{\mu}}$ on s appears to be somewhat stronger than linear. The deviation from linearity contains information on m_Z, the propagator mass in the weak amplitude. A fit[1] to the s dependence of the combined $e\bar{e} \rightarrow \mu\bar{\mu}$, $\tau\bar{\tau}$ data gives

$$m_Z = 73 \pm {}^{10}_{6}\ \text{GeV}$$

or

$$63 < m_Z < 101\ \text{GeV}$$

at the 95% confidence level, thus establishing the finite range of the weak force that causes the asymmetry.

Value of the ρ Parameter

Assuming standard model couplings, g_v^ℓ and g_A^ℓ, and fixing the propagator mass at the value, $m_Z = 93 \pm 2$ GeV, observed in the UA1 and UA2 experiments,[7] the size of the observed asymmetries leads to[1]

$$\rho = 1.09 \pm 0.07.$$

In the minimal $SU(2) \times U(1)$ model in which Higgs particles are in doublets, one expects $\rho = 1$. The slight deviation from $\rho = 1$ is connected with the high values of the measured muon pair asymmetries at 34.5 and 42.5 GeV.

Value of the Z^0 Coupling Constants, g_v^ℓ, g_A^ℓ

If m_Z and ρ are fixed at their "standard" values ($m_Z = 93 \pm 2$ GeV, $\rho = 1$) and lepton universality is assumed such that only the squares of the two weak couplings, g_v^ℓ and g_A^ℓ, enter, then these can be determined by a common fit[1] to the data on $R_{\ell\bar{\ell}}$ (giving essentially $g_v^{\ell 2}$) and $A_{\ell\bar{\ell}}$ (giving $g_A^{\ell 2}$) from the three processes, $e\bar{e} \rightarrow e\bar{e}$, $\mu\bar{\mu}$, $\tau\bar{\tau}$. The result is shown in FIGURE 9; numerically,

$$g_v^{\ell 2} = -0.01 \pm 0.03 \qquad \text{(equivalent to } \sin^2\theta_W = 0.25 \pm 0.05\text{),}$$

$$g_A^{\ell 2} = 0.27 \pm 0.02 \qquad \text{(where 0.25 is expected from the standard model).}$$

The three processes, though, are not seen to give completely consistent results; there may be a slight difficulty with the normalization of $R_{\mu\bar{\mu}}$.

Determination of $\sin^2\theta_W$

We may altogether eliminate any reference to charged current couplings, i.e., to G_F and ρ, in equations (**1**) and (**4**).[8] Comparing the photon exchange amplitude with the axial Z^0 exchange amplitude, the former is proportional to e^2, the latter to $g^2_{axial}\ s/$

$(s - m_Z^2)$ with

$$g_{axial} = \frac{e}{2 \sin \theta_W \cos \theta_W} g_A^\ell.$$

The strength of the weak axial coupling is thereby expressed through the electromagnetic coupling and the electroweak mixing angle, θ_W. The expression for χ in equations (**1**) and (**4**) becomes

$$\chi = \frac{1}{4 \sin^2 \theta_W \cos^2 \theta_W} \cdot \frac{s}{s - m_Z^2 + i m_Z \Gamma_Z}$$

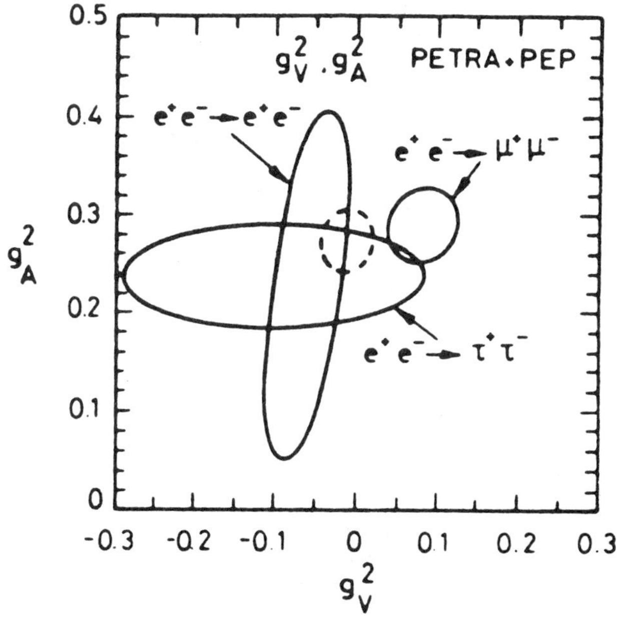

FIGURE 9. Determination of g_v^2 and g_A^2 from different leptonic processes. Shown are 95% confidence level contours for individual processes (solid lines) and for the combined result (dashed) (from reference 1).

and the relative size of the weak-electromagnetic interference effects depends only on the weak isospin and hypercharge, and the mixing angle, θ_W. Assuming the standard weak isospin assignments (which determine g_v^ℓ and g_A^ℓ), one can use the experimental data on $A_{\ell\bar{\ell}}$ and $R_{\ell\bar{\ell}}$ to solve for $\sin^2 \theta_W$ and m_Z. The result[1,8] of this is plotted in FIGURE 10. The ambiguity in the solution is due to the second order dependence of $R_{\ell\bar{\ell}}$ on $\sin^2 \theta_W - 1/4$, which is illustrated in FIGURE 2. Taking $m_Z = 93 \pm 2$ GeV from the CERN collider experiments[7] eliminates one of the solutions and gives

$$\sin^2 \theta_W = 0.18 \pm 0.03.$$

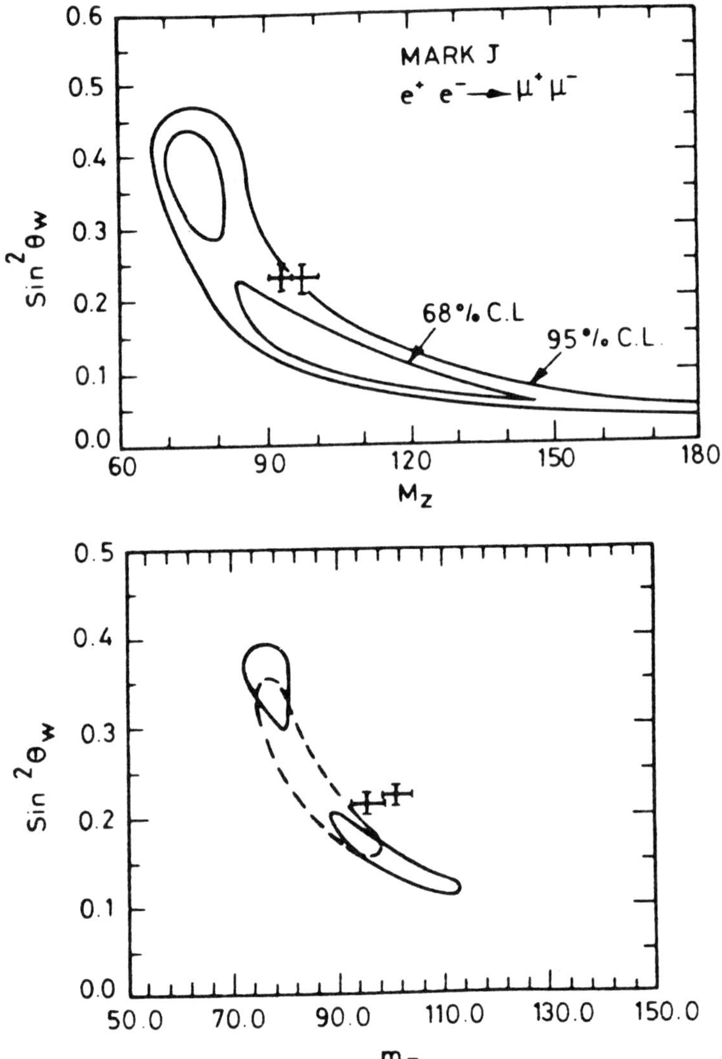

FIGURE 10. Determination of $\sin^2\theta_W$ and m_Z from measured $R_{\ell\bar{\ell}}$ and $A_{\ell\bar{\ell}}$ data. Upper part: MARK-J data, for $\mu^+\mu^-$ only. Lower part: All available data on muon and tau pair production (from reference 1). Solid line: 95% confidence level contours. Dashed line: contour after imposing a 2% change in normalization. Points with error bars: Results from the S$\bar{p}$pS collider.

Again, the slight deviation from the "standard" value of 0.22 is connected with the high values of the measured $A_{\mu\bar{\mu}}$ at the upper PETRA energies.

LIMITS ON THE SCALE OF LEPTON COMPOSITENESS

Bjorken has pointed out[9] that at energies well below the Z^0 pole the experimental information on the neutral current weak interaction can be described by an effective current-current Lagrangian:

$$\mathcal{L}_{eff} = \frac{4G_F}{\sqrt{2}}(J_\mu^{C^+}J_\mu^C + J_\mu^N J_\mu^N) - \frac{1}{2}\frac{e^2}{Q^2}J_\mu^{em}J_\mu^{em} + C\frac{4G_F}{\sqrt{2}}J_\mu^{em}J_\mu^{em}. \tag{7}$$

The first two terms are the low energy limit of the standard model; the extra C term is a parity conserving weak current coupling that would not have been detected in neutrino interactions. Such an extra term is absent if a single Z^0 dominates the low energy data.

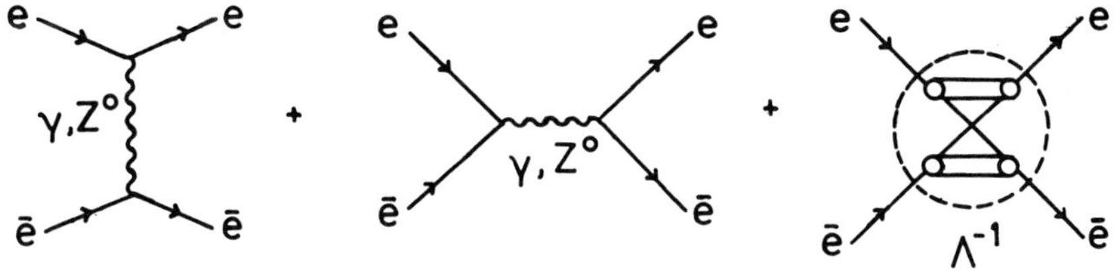

FIGURE 11. Diagrams describing Bhabha scattering for composite electrons, including a contact interaction.

The existence of more than one Z^0 of comparable mass or of a continuum of Z^0 states (as expected if the Z^0 boson is composite) would lead to $C > 0$. The extra weak term would interfere with the electromagnetic term and would consequently alter the normalization of R_{ff}. The best published experimental limit is $C < 0.012$ at the 95% confidence level.[10] This has to be compared with the factor, $\sin^4\theta_W = 0.05$, which is in front of the term of identical structure in the standard model (from equation (7), inserting $J_\mu^N = J_\mu^3 - \sin^2\theta_W J_\mu^{em}$). We conclude that in the presently accessible energy region, there is no possibility left for any sizable deviation from the standard model.

If the leptons have substructure, it may be possible to find excited states, ℓ^*. Searches for decays, $\ell^* \rightarrow \ell + \gamma$, and for ℓ^* exchange terms at PEP and PETRA have, however, excluded such states for masses up to $m_{e^*} \sim 70$ GeV and $m_{\mu^*} \sim 25$ GeV, provided that the couplings are not unreasonably small.[11]

Another consequence of lepton compositeness has recently been noted.[12] If quarks and leptons are composite at the energy scale, Λ, the strong forces binding their constituents induce flavor-diagonal contact interactions that have significant effects at reaction energies well below Λ. The effective "low energy" e^+e^- interaction (FIGURE

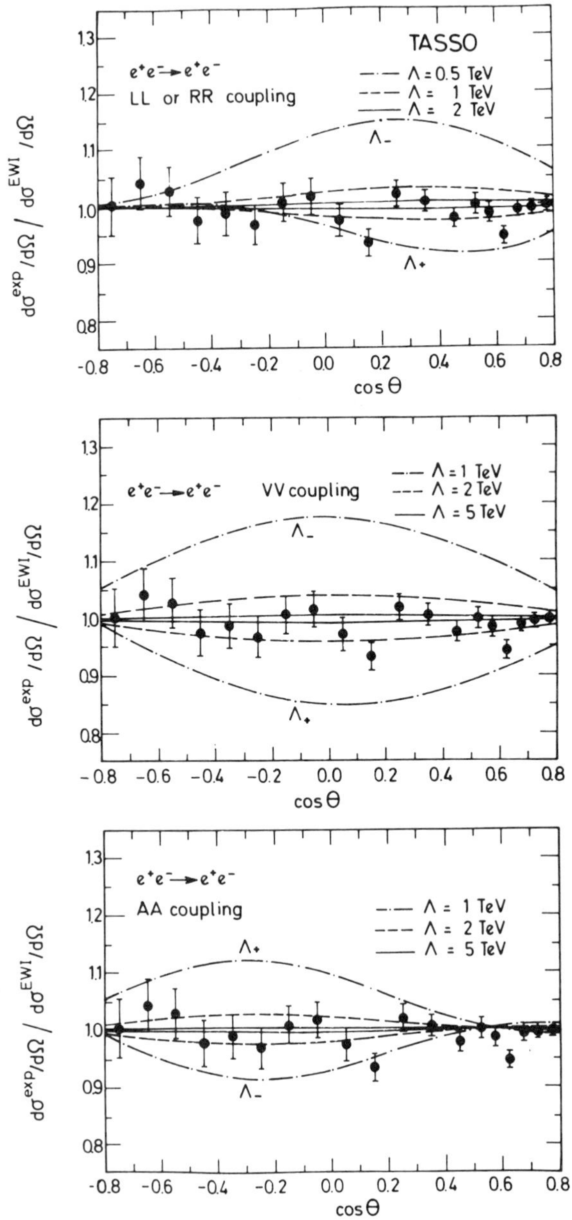

FIGURE 12. Bhabha scattering differential cross section at $\sqrt{s}$ = 34.5 GeV, divided by the standard electroweak interaction prediction. The dashed curves show predicted effects of electron substructure.

11) will be of the form,

$$\mathcal{L}_{eff} = \mathcal{L}_{std.mod.} \pm \frac{g^2}{2\Lambda_\pm^2} \sum_A C_A (\bar{e}\Gamma^A e)(\bar{e}\Gamma^A e), \tag{8}$$

where Γ^A can contain V − A, V + A, . . . terms describing the helicity structure of the constituent interchange process. For strong binding, $g^2/4\pi \sim 1$ and $\Lambda_\pm \sim 1/R_{lepton}$. Predictions from $\mathcal{L}_{eff}$ for different length scales of $\Lambda_\pm$ and different helicity structures of the contact interaction are compared with measurements of Bhabha scattering[13] in FIGURE 12. From the good agreement of the Bhabha scattering data with standard electroweak theory, one derives 95% confidence level lower limits for $\Lambda_\pm$ of 0.7 to 2.9 TeV, depending on the helicity structure of $\mathcal{L}_{eff}$. The sensitivity to an intrinsic length scale is much higher than in form factor determinations because the effect of the contact interaction is enhanced by interference effects with the standard model amplitude.

If electrons, muons, and tau leptons have constituents in common, then analogous effects are expected to show up in $e\bar{e} \to \mu\bar{\mu}$ and $e\bar{e} \to \tau\bar{\tau}$ measurements. The effects of $\mathcal{L}_{eff}$ can occur throughout the whole angular region in this case, while in elastic (Bhabha) scattering they are limited to wide angles. Muon pair production data have been analyzed and lower limits on $\Lambda_\pm$ of 2 to 4 TeV have been obtained.[13]

ANGULAR ASYMMETRIES AND WEAK NEUTRAL COUPLINGS OF *c* AND *b* QUARKS

Since a complete identification of quark flavors in the pair production process of $e\bar{e} \to q\bar{q}$ has not been achieved, the total cross section ratios, $R_{u\bar{u}}$, $R_{d\bar{d}}$, . . . , have not been measurable. One has learned, however, to identify samples of events,

$$e\bar{e} \to c\bar{c},\ b\bar{b},$$

that are unbiased with respect to the angular distribution and in which the quark and the antiquark jet can be distinguished. This allows a measurement of the angular asymmetries, $A_{c\bar{c}}$ and $A_{b\bar{b}}$.

The identification of c and b pair production events is possible owing to the fact that $c\bar{c}$ and $b\bar{b}$ production from the vacuum in the fragmentation process is negligible. Observation of the decay of a heavy hadron carrying charm or bottom flavor in a jet therefore identifies the primary quark. Fortunately, the fragmentation of c and b quarks into rank 1 mesons (i.e., those carrying the flavor of the primary quark) is characterized by a rather hard fragmentation function.[14] This makes "tagging" of jets from heavy quarks relatively easy.

The following techniques have been used:

(i) Direct identification of the decays, $D^+ \to K^-\pi^+\pi^+$, $D^0 \to K^-\pi^+$, as peaks in invariant mass distributions. Among the PEP and PETRA experiments, only the HRS detector has the momentum resolution required to acomplish this with a reasonable signal/background ratio (FIGURE 13).[15]

(ii) Identification of D^* decays by "Nussinov's trick"[16]: The decay chain,

$$D^{*+} \to D^0\pi^+,\ D^0 \to K^-\pi^+,\ K^-\pi^+\pi^0,\ K^-\rho^+,\ K^-\pi^+\pi^+\pi^-,$$

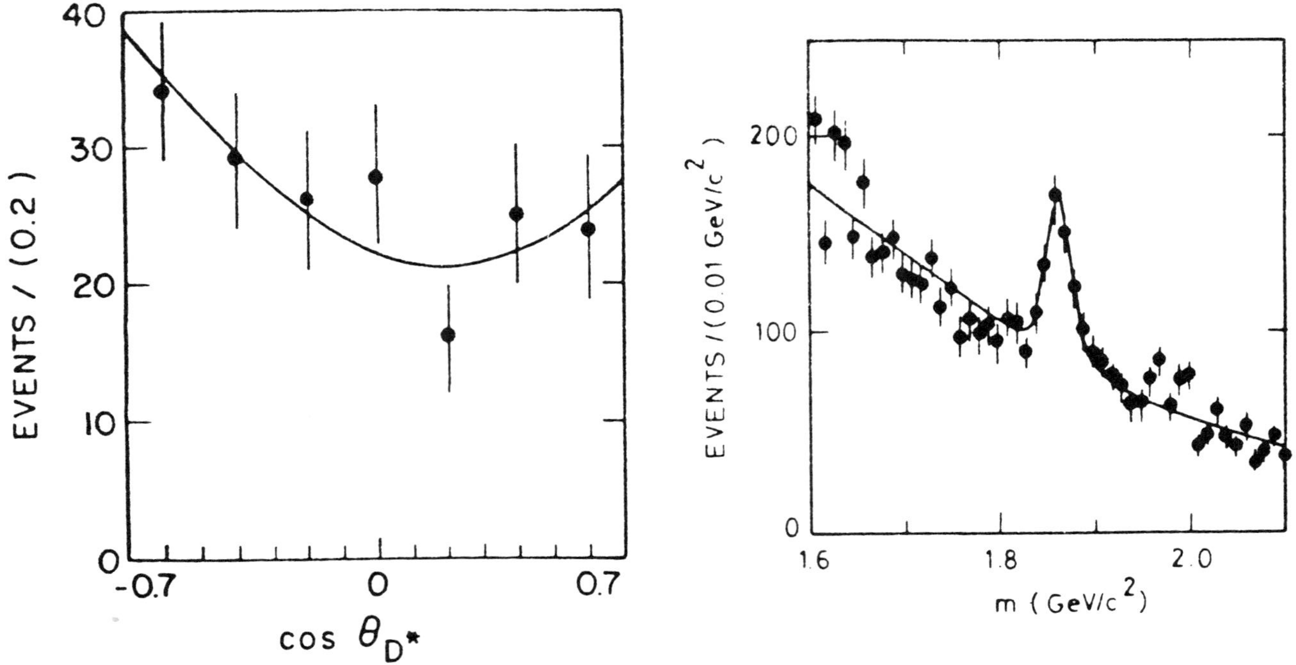

FIGURE 13. D^0 and D^+ mass spectrum ($x > 0.5$) and D^* production angular distribution, from the HRS experiment.

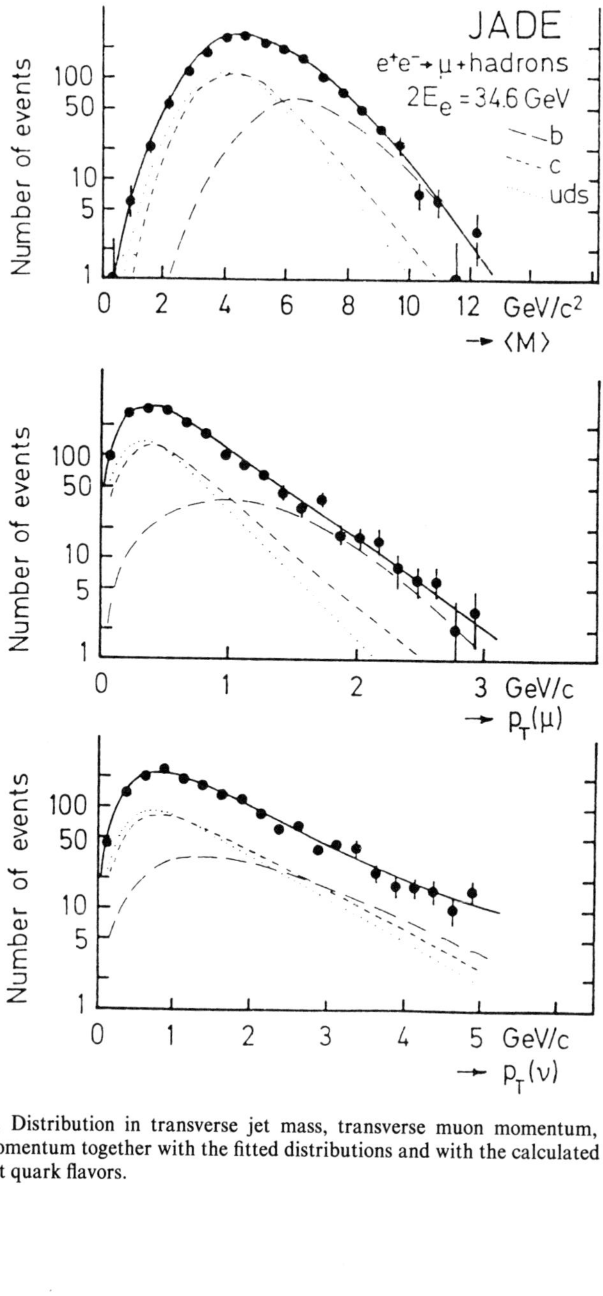

FIGURE 14. Distribution in transverse jet mass, transverse muon momentum, and missing transverse momentum together with the fitted distributions and with the calculated distributions from different quark flavors.

is detected by virtue of the small Q value of 6 MeV of the D^{*+} decay because it strongly restricts the kinematic configuration of the decay products and leads to a partial cancellation of the smearing effects from finite mass resolution.[17] The efficiency of identifying c jets by this method is not high, but the samples obtained are almost free of background.

(iii) Detection of the weak semileptonic decays,

$$c \longrightarrow qe\nu,\ q\mu\nu,$$
$$b \longrightarrow qe\nu,\ q\mu\nu.$$

The large masses of the decaying quarks lead to large transverse momenta, p_T^ℓ, of the charged lepton with respect to the jet axis. The p_T^ℓ spectra are peaking at ~0.6 GeV/c for c jets and at ~1.3 GeV/c for b jets. Therefore, selecting on

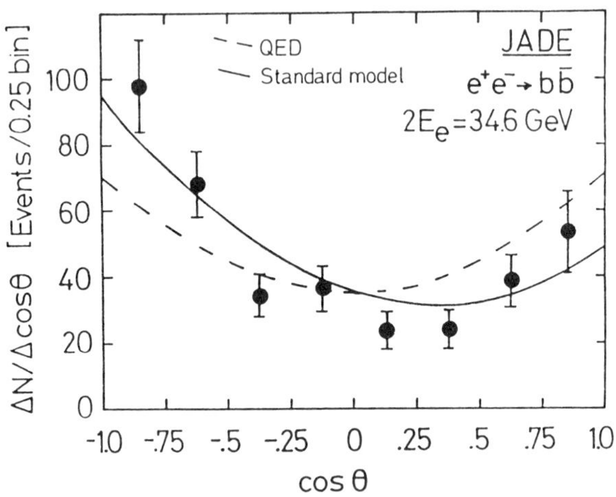

FIGURE 15. b quark production angular distribution resulting from fits in each bin of cos θ (from JADE).

high p_T^ℓ enriches a sample with b decays, while intermediate p_T^ℓ are strongly fed by c decays. The background from π and K decays and from hadrons misidentified as muons is peaking at low p_T^ℓ, but has a tail that extends out to a rather large p_T^ℓ. The difference of the shapes of the various contributions to the p_T^ℓ spectra allows a statistical unfolding from the measured data. To increase the reliability, other variables have been analyzed simultaneously with p_T^ℓ, such as transverse jet mass, total lepton momentum, or missing transverse momentum (FIGURE 14). In an overall fit of a superposition of the various decay spectra to the measured distributions of the different variables (including cos θ), the asymmetries, $A_{c\bar{c}}$ and $A_{b\bar{b}}$, are simultaneously determined. This technique is so far the only one used to extract $A_{b\bar{b}}$. While reliable for $A_{b\bar{b}}$ because some part of the kinematical region is uniquely associated with $b\bar{b}$

TABLE 4. Charm Results[a]

Expt.	$\sqrt{s}$ (GeV)	$A_{c\bar{c}}$ (%) Measured	Std. Model	g_A^c	Method
TPC	29			0.90 ± 0.46	$c \rightarrow e, \mu$
HRS	29	−12 ± 8	8.8	0.68 ± 0.91	D^0, D^+, D^*
MAC	29			0.8 ± 1.8	$c \rightarrow \mu$
MARK-J	35.5	−16 ± 9	−13.5	0.59 ± 0.33	$c \rightarrow \mu$
JADE	34.4	−14 ± 9	−12.9	0.54 ± 0.35	D^*
TASSO	34.4	−10 ± 9	−12.9	0.39 ± 0.35	$D^*, c \rightarrow e$

[a]Average: $g_A^c = 0.56 \pm 0.18$ (Std. model: 0.50).

events (FIGURE 14), the method is somewhat questionable for $A_{c\bar{c}}$, which is subject to a large background subtraction. This is particularly critical since $A_{b\bar{b}}$ and $A_{c\bar{c}}$ have the same sign such that after the leptonic decays, $b \rightarrow \ell^-$, $b \rightarrow c \rightarrow \ell^+$, and $c \rightarrow \ell^+$, the observed lepton asymmetries are partially canceling each other.

The most precise angular distribution for $b\bar{b}$ production measured so far[18] is shown in FIGURE 15. A strong asymmetry is seen. TABLES 4 and 5 show a summary of results from various experiments and a comparison with the standard model expectation. The values of the axial coupling constants, g_A^c and g_A^b, have been extracted from the measurements with the assumption of $g_A^e = -\frac{1}{2}$. The averages,

$$g_A^c = 0.56 \pm 0.18,$$

$$g_A^b = -0.50 \pm 0.10,$$

agree well with the expectations ($\frac{1}{2}$ and $-\frac{1}{2}$, respectively) from the standard model.

SUMMARY

The ratio of $R_{had} = \sigma(e^+e^- \rightarrow \text{hadrons})/\sigma_{\mu\bar{\mu}}$ is, at $\sqrt{s} \gtrsim 40$ GeV, beginning to show the tail of the Z^0 peak.

TABLE 5. Bottom Results[a]

Expt.	$\sqrt{s}$ (GeV)	$A_{b\bar{b}}$ (%) Measured	Std. Model	g_A^b	Method
TPC	29			−0.69 ± 0.42	$b \rightarrow \mu, e$
MARK-2	29			$-0.75 {+0.35 \atop -0.30}$	$b \rightarrow \mu, e$
MAC	29			−0.3 ± 0.35	$b \rightarrow \mu$
CELLO	34.0	−40 ± 17	−24.6	−0.81 ± 0.69	$b \rightarrow \mu, e$
JADE	34.4	−22.8 ± 6 ± 2.5	−25.2	−0.45 ± 0.12 ± 0.05	$b \rightarrow \mu$
MARK-J	34.6	−21 ± 19	−25.5	−0.41 ± 0.37	$b \rightarrow \mu$
TASSO	34.5	−30 ± 18	−25.4	−0.59 ± 0.35	$b \rightarrow \mu, e$

[a]Average: $g_A^b = -0.50 \pm 0.10$ (Std. model: −0.50).

The angular asymmetries, $A_{\mu\bar{\mu}}$ and $A_{\tau\bar{\tau}}$, observed in the pair production of leptons give evidence for a Z^0 propagator term, i.e., a finite range of the force responsible for the asymmetry, characterized by a mass of $63 < m_Z < 101$ GeV. If $m_Z = 93 \pm 2$ GeV is assumed, $\rho = 1.09 \pm 0.07$ follows from the combined data. The muon asymmetries, $A_{\mu\bar{\mu}}$, at 35 and 42 GeV, when all measurements are averaged, show a tendency of being slightly higher than expected from standard model couplings for $m_Z = 93$ GeV and $\rho = 1$; the deviation amounts to 2.6 and 1.4 standard deviations, respectively. The situation is expected to be clarified in the near future as more high energy measurements will be made at PETRA.

A common analysis of all the data in the lepton sector, $R_{\ell\bar{\ell}}$ and $A_{\ell\bar{\ell}}$, for $\ell = e, \mu, \tau$, leads to the following results: If $m_Z = 93 \pm 2$ GeV, $\rho = 1$, and the values of G_F from muon decay are input, then the data give Z^0-lepton couplings of $g_A^2 = 0.27 \pm 0.02$ and $g_v^2 = -0.01 \pm 0.03$, assuming lepton universality. This is consistent with the standard model expectations of $g_A^2 = 0.25$ and $g_v^2 = 0$. Alternatively, if both g_A and g_v^2 as given by the standard model and $m_Z = 93 \pm 2$ GeV are assumed and if the strength of the weak neutral coupling relative to the electromagnetic coupling is expressed by $\sin^2\theta_W$, then the data demand $\sin^2\theta_W = 0.18 \pm 0.03$.

Axial couplings of quarks have been extracted from the observed angular asymmetries of charm and bottom pair production. The results are $g_A^c = 0.56 \pm 0.18$ and $g_A^b = -0.50 \pm 0.10$ (assuming standard model couplings for the electron), which are consistent with the expectations of $g_A^c = 0.5$ and $g_A^b = -0.5$. This verifies the isodoublet nature of these quarks.

In the search for lepton and quark substructure manifested by form factor effects, lower limits for the scale parameters on the order of 100 to 300 GeV have been established. Assuming the leptons to be strongly bound states, searches for effects of compositeness through interference of the effective contact interaction on the constituent level, with the "soft" photon and Z^0 exchange amplitudes, give lower limits for the scale of compositeness of 1–4 TeV. This presumably represents the best spatial resolution so far obtained in a scattering experiment.

NOTES AND REFERENCES

1. For a comprehensive recent review, see SAXON, D. H. 1984. Electroweak Interference in e^+e^- Annihilation. Rutherford preprint RAL-84-094 *In* Proc. of Physics in Collision IV, UC Santa Cruz, August 1984. To be published. The present summary owes very much to this work.
2. A first attempt to observe parity violation effects has been made by measuring τ decays in which the angular distributions carry information on τ helicity. See BEHREND, H-J. *et al.* 1983. Phys. Lett. **127B:** 270. The effects expected according to the standard model are below the sensitivity limit of present experiments.
3. ADEVA, B. *et al.* 1982. Phys. Rev. Lett. **48:** 1701; 1984. Phys. Rep. **109:** 131; BEHREND, H-J. *et al.* 1983. Z. Phys. **C14:** 283; 1982. Phys. Lett. **114B:** 282; BARTEL, W. *et al.* 1984. Preprint DESY 84-078. To be published; 1983. Phys. Lett. **123B:** 353; FERNANDEZ, E. *et al.* 1983. Phys. Rev. Lett. **50:** 1238; 1983. SLAC-PUB-373; BERGER, C. *et al.* 1983. Z. Phys. **C21:** 53; 1981. Phys. Lett. **99B:** 489; BRANDELIK, R. *et al.* 1982. Phys. Lett. **110B:** 173; **117B:** 365; ALTHOFF, M. *et al.* 1984. Z. Phys. **C22:** 13; WU, S. L. 1984. Phys. Rep. **107:** 59.
4. MARSHALL, R. 1984. Rutherford preprint RAL-84-088 *In* Proc. of the 6th General Conference of EPS Trends in Physics, Prague, August 1984. To be published.

5. YOUNGMAN, C. 1984. Proc. XIX Recontre de Moriond; PREPOST, R. 1984. Proc. Leipzig Conference; LEVI, M. E. *et al.* 1983. Phys. Rev. Lett. **51:** 1941; KOLTICK, D. S. 1984. Proc. Symposium on High Energy e^+e^- Interactions, Vanderbilt, April 1984; FERNANDEZ, E. *et al.* 1984. Paper contributed to the Leipzig Conference; LOCKYER, N. *et al.* SLAC-PUB-3245. (See also references 1 and 3.)
6. BERENDS, F. A. & R. KLEISS. 1981. Nucl. Phys. **B177:** 237; BERENDS, F. A., R. KLEISS & S. JADACH. 1983. Nucl. Phys. **B202:** 63; WETZEL, W. 1983. Nucl. Phys. **B227:** 1; BROWN, R. W., R. DECKER & E. A. PASCHOS. 1984. Phys. Rev. Lett. **52:** 1192; BÖHM, M. & W. HOLLIK. 1982. Nucl. Phys. **B204:** 45; 1984. Phys. Lett. **139B:** 213.
7. See the articles by D. CLINE, C. CONTA, and J. ROHLF in these proceedings.
8. BÖHM, M. 1984. Aachen preprint PITHA 84-11; see also MARK-J COLLABORATION. 1984. Phys. Rep. **109:** 131.
9. BJORKEN, J. D. 1979. Phys. Rev. **D19:** 335.
10. ALTHOFF, M. *et al.* 1984. Z. Phys. **C22:** 13.
11. For a recent summary, see ADEVA, B. *et al.* 1984. MIT-LNS technical report no. 141. To be published.
12. EICHTEN, E. J., K. D. LANE & M. E. PESKIN. 1983. Phys. Rev. Lett. **50:** 811.
13. See reference 10 and the summary by YAMADA, S. 1984. Proc. Leipzig Conference. To be published.
14. SÖDING, P. 1983. Proc. Int. Europhysics Conference on High Energy Physics, Brighton, July 1983, p. 567.
15. AHLEN, S. *et al.* 1984. Phys. Rev. Lett. To be published.
16. NUSSINOV, S. 1976. Phys. Rev. Lett. **35:** 1672; FELDMAN, G. *et al.* 1977. Phys. Rev. Lett. **38:** 1313.
17. ALTHOFF, M. *et al.* 1983. Phys. Lett. **126B:** 493; 1984. Phys. Lett. **138B:** 317; BARTEL, W. *et al.* 1984. DESY 84-059. To be published; YELTON, J. M. *et al.* 1982. Phys. Rev. Lett. **49:** 430; DERRICK, M. *et al.* 1984. Phys. Rev. Lett. To be published.
18. BARTEL, W. *et al.* 1984. DESY 84-068. To be published.
19. ADEVA, B. *et al.* 1983. Phys. Rev. Lett. **51:** 443; ALTHOFF, M. *et al.* 1984. DESY 84-061. To be published; 1984. Z. Phys. **C22:** 219; AIHARA, H. *et al.* 1984. Tokyo Rep. UT-HE-84/03 and Proc. Leipzig Conference; BEHREND, H-J. *et al.* 1984. Z. Phys. **C19:** 291; NELSON, M. E. *et al.* 1983. Phys. Rev. Lett. **50:** 1542.

DISCUSSION OF THE PAPER

P. LANGACKER (*University of Pennsylvania, Philadelphia, PA*): Is there going to be more data on the muon asymmetry in the future?

P. SÖDING (*DESY, Hamburg, Germany*): PEP will presumably continue running at 29 GeV, while PETRA will work around 44 GeV during all of this year (1985) and maybe also in 1986, after which it will be converted into an injector for HERA. The question whether the high energy muon asymmetries are higher than the standard model predicts, will therefore presumably be clarified within the next year or so. After 1986 TRISTAN, the SLC and, eventually, LEP will open a new energy range for these measurements.

M. BALDO-CEOLIN (*Padua University, Italy*): In the measurements of the asymmetry in $b\bar{b}$ events, the results must be sensitive to the Monte Carlo simulation used to correct for the contamination of charm, which produces asymmetries of the opposite sign. Has this been taken into account in the systematic errors?

SÖDING: In the figure (FIGURE 14) showing the various distributions used in the JADE analysis, one notices that in the transverse mass and in the transverse momentum there is a small region where practically only $b\bar{b}$ production contributes. This is almost free of systematic errors since only kinematics enter, i.e., the large value of the b mass compared to the mass of all other quarks. Consequently, the systematic error to the asymmetry, $A_{b\bar{b}}$, estimated by the JADE group is only ~3%; due to the small number of decays that fall into this region, the statistical error is, however, substantial. For the $c\bar{c}$ asymmetry, the systematic errors from this analysis are, of course, much larger.

H. J. FRISCH (*University of Chicago, Chicago, IL*): What goes into the Monte Carlo curves for the heavy quark distribution in transverse jet mass, transverse lepton momentum, and missing momentum?

SÖDING: One simulates the jets produced by the various flavored quarks by a Field-Feynman type of cascade fragmentation process, including string fragmentation mechanism, gluon bremsstrahlung, etc. Of particular importance is the correct description of the semileptonic decays of the c and b quarks.

Physics at HERA

GÜNTER WOLF

Stanford Linear Accelerator Center (SLAC)
Stanford University
Stanford, California 94305
and
Deutsches Elektronen–Synchrotron (DESY)
Notkestrasse 85
D-2000 Hamburg 52, Federal Republic of Germany

THE HERA COLLIDER

DESY is presently constructing a new storage ring that will provide collisions between 30 GeV electrons and 820 GeV protons.[1] This collider is called HERA, which stands for Hadron-Elektron-Ring-Anlage. The layout of the machine is sketched in FIGURE 1. HERA intersects the PETRA storage ring that will serve as the injector for electrons and protons. The salient machine parameters are listed in TABLE 1.

An important aspect of HERA is the polarization of the electron beam. By the emission of synchrotron radiation, the circulating electrons become polarized perpendicular to the plane of the ring. From the physics point of view, this direction of polarization is not very attractive. What is needed are electrons of a given helicity. The electron spin will be rotated into the longitudinal direction by a spin rotator in front of each interaction region. The spin rotation is undone by a mirror symmetric spin rotator following the interaction region. The degree of longitudinal polarization is expected to be 85%.

The construction schedule is given in TABLE 2. Ground breaking took place in May 1984. The tunnel should be ready by the end of 1987. The installation of the machine components follows the tunnel boring machine, so by March 1988 the electron ring should be complete. Technically, the most demanding components are likely to be the superconducting magnets for the proton ring and the associated cryogenics. Series production of the proton magnets is expected to take from July 1986 until October 1988. The proton ring should be completed by the summer of 1989. The first *ep* collisions are foreseen for the beginning of 1990.

The total cost for HERA is estimated at 654 MDM using 1981 prices. Of this sum, 261 MDM are required for sites, buildings, and power installations, 112 MDM for the electron ring, and 281 MDM for the proton ring. The costs will be born by the German federal government and the city of Hamburg (in the ratio of 9:1) and by contributions from Canada, France, Holland, Israel, Italy, and England.

With respect to experiments the planning is as follows. A call has been issued for letters of intent by June 1985. After decisions on the proposed experiments by the end of 1985, the experimenters will have four years for the construction and installation of the detector.

HERA PHYSICS: GENERAL CONSIDERATIONS

The large momentum transfers possible between electron and proton, $Q^2_{max} = 10^5$ GeV^2, make HERA first and foremost an electron quark collider. The relevant diagram is shown in FIGURE 2. The incoming electron emits a lepton, ℓ, and exchanges a current, j, with one of the quarks of the incoming proton. This leads to the emission of a quark, q' (= current quark). Depending on whether a neutral current (γ, Z^0) or a charged current (W^-) is exchanged, the final state lepton is either an electron or a neutrino (FIGURE 3).

Of course, the scattering process is not confined to the known currents, quarks, and leptons. New currents may contribute and new quarks and leptons with masses up to the kinematic limit of 314 GeV may be produced. Any constituents that carry electromagnetic or weak charge and are within the kinematic limits can be produced at HERA. Thus, HERA is a hunting ground for the legions of new particles that have

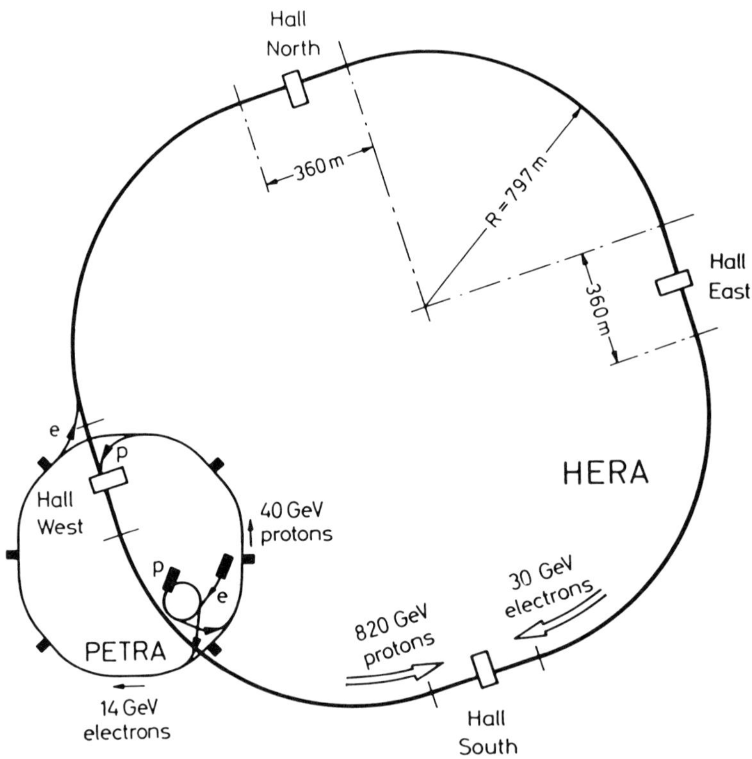

FIGURE 1. Schematic view of the HERA ring and the preaccelerators (two linear accelerators, two synchrotrons of 100 m diameter, and the PETRA storage ring). The proton ring is mounted on top of the electron ring.

TABLE 1. Parameters of HERA

	p-Ring		*e*-Ring	Units
Nominal energy	820		30	GeV
c.m. energy		314		GeV
Q^2 max		98.400		GeV^2
Luminosity		2.5×10^{31}		cm^2s^{-1}
Polarization time			25	min
Number of interaction points		4		
Crossing angle		0		
Free space for experiments		±5.5		m
Circumference		6336		m
Length of straight sections		360		m
Bending radius	603.8		540.9	m
Magnetic field	4.73		0.1849	T
Energy range	300–820		10–33	GeV
Injection energy	40		14	GeV
Circulating current	130		58	mA
Total number of particles	2.2×10^{13}		0.76×10^{13}	
Number of bunches		200		
Number of bunch buckets		220		
Time between crossings		95		nsec
Emittance (ϵ_x/ϵ_y)	0.47/0.24		1.6/0.16	10^{-8}m
Beta function (β_x^*/β_y^*)	10/1.0		2/0.7	m
Beam tune shift (Q_x, Q_y)	0.0026/0.0014		0.023/0.026	
Beam size at crossing σ_x^*	0.12 (0.43)		0.22	mm
Beam size at crossing σ_y^*	0.027		0.013	mm
Beam size at crossing σ_z^*	10		0.9	cm
Energy loss/turn	1.4×10^{-10}		142.3	MeV
Critical energy	10^{-6}		111	KeV
Max. circumf. voltage	0.7/2		165	MV
Total RF power	1		13.2	MW
RF frequency	52.033/208.13		499.667	MHz
Filling time	20		15	min
Heat loss at 4.2° K	13.2			kW
Available refrigeration power at 4.2° K	19.5 kW + 60 g/s			
Available refrigeration power at 40° K	60			kW
Cooldown time for one octant	30			hrs

been proposed: excited leptons and quarks, lepto quarks, supersymmetric particles, and many more.

Next to electron quark scattering, photon gluon fusion will play an important role at HERA. Photon gluon fusion, depicted in FIGURE 4, is presumably the dominant process for the production of heavy quarks, $Q = c, b, t, \ldots$

The following discussion of the physics potential that HERA offers is by no means exhaustive. Rather, a few processes that are archetypical for the different classes of physics will be considered. More detailed studies can be found in reference 2, in the 1979 ECFA study report, in the HERA proposal, and in the proceedings of the HERA meetings at Wuppertal, Amsterdam, and Genova[3] (and in the papers cited there).

TABLE 2. The HERA Construction Schedule

	1984	1985	1986	1987	1988	1989	1990
Tunnel and Buildings							
hall west	———	———	— 2				
south	———	———	— 8				
east		———	———	———	— 2		
north		———	———	— 8			
tunnel		———	———	— 4			
Electron Ring							
production of magnets		———	———	— 9			
installation of ring			———	———	— 3		
DESY II	———	———	— 3				
PETRA-*e*				— 1			
Proton Ring							
production of magnets		———	———	———	— 10		
installation of ring			———	———	———	— 6	
proton linac		———	———	— 4			
DESY III			———	— 12			
PETRA-p				— 9			

ELECTRON SCATTERING AND THE PROTON STRUCTURE FUNCTIONS

Neutral Current Processes

The diagram that describes *e*p scattering in lowest order is shown in FIGURE 2 (taking $j = \gamma$). The final state partons (quarks, gluons) and the "spectator" remnants from the incident proton are assumed to develop into jets of hadrons. Apart from the total center-of-mass energy squared,

$$s = (p_p + p_e)^2 = m_e^2 + m_p^2 + 2(E_e E_p + p_e p_p) \approx 4E_e E_p$$

(E_e, E_p—energies of electron and proton),

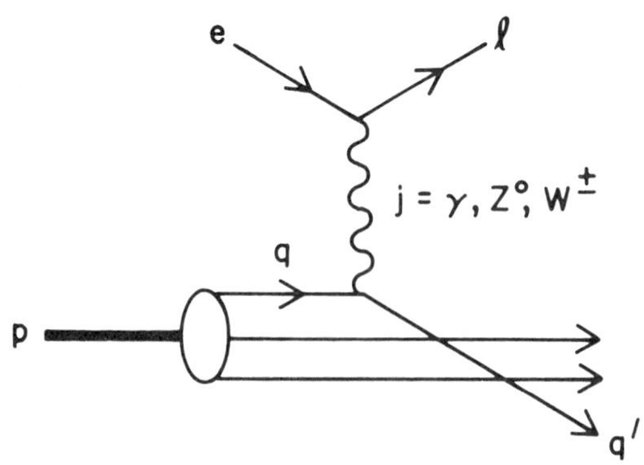

FIGURE 2. Lepton proton scattering.

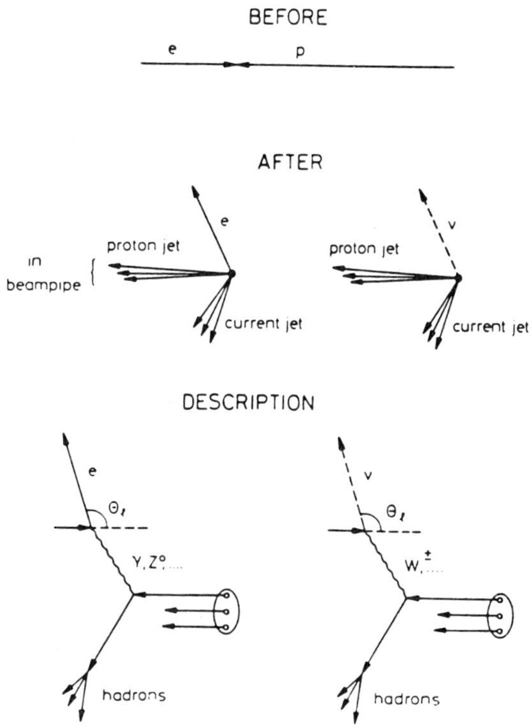

FIGURE 3. Diagrams for neutral and charged current scattering.

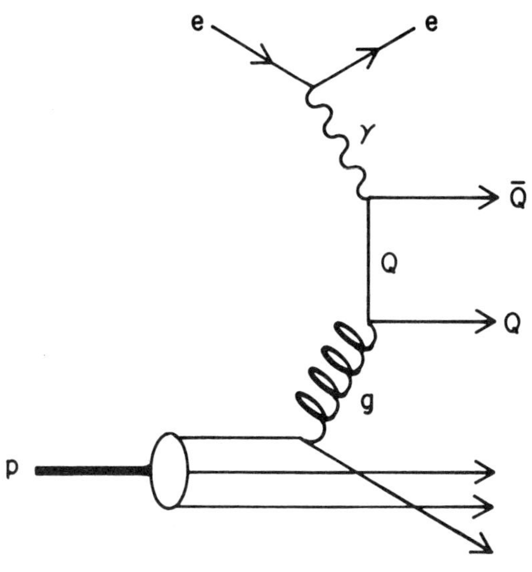

FIGURE 4. Quark pair production by photon gluon fusion.

there are two kinematic variables that describe the inclusive scattering process:

$q^2 = (p_e - p_e')^2 = -Q^2$, square of the four momentum transfer,

$W^2 = (q + p_p)^2$, square of the total mass of the final hadronic system;

or equivalently,

$$x = \frac{Q^2}{2(q \cdot p_p)} = \frac{Q^2}{2m_p\nu}, \quad \text{the Bjorken scaling variable,}$$

$$y = \frac{(q \cdot p_p)}{(q \cdot p_e)} = \frac{\nu}{\nu_{max}}. \quad \text{(Note that } Q^2 \simeq sxy.\text{)}$$

In the rest system of the incoming proton, ν measures the energy transferred by the current. The maximum value that ν can reach is

$$\nu_{max} = \frac{s - (m_e + m_p)^2}{2m_p} \approx \frac{s}{2m_p} = 2E_e E_p/m_p.$$

For $E_e = 30$ GeV and $E_p = 820$ GeV, $\nu_{max} = 52$ TeV. HERA therefore is equivalent to a fixed target experiment with an incident lepton beam of 52 TeV.

The one photon exchange cross section can be written in terms of two dimensionless structure functions, F_1 and F_2, which are functions of ν, Q^2 or x, y:

$$\frac{d^2\sigma(\gamma)}{dxdy} = \frac{4\pi\alpha^2}{s\,x^2y^2}[(1 - y)F_2(x, Q^2) + xy^2 F_1(x, Q^2)].$$

In the approximation where scattering on spin ½ partons dominates, F_1 can be expressed in terms of F_2 using the Callan-Gross relation: $2xF_1 = F_2$.

This leads to

$$\frac{d^2\sigma(\gamma)}{dxdy} \approx \frac{4\pi\alpha^2}{s\,x^2y^2}\left(1 - y + \frac{y^2}{2}\right)F_2(x, Q^2).$$

In the quark picture, ep scattering is described as incoherent eq scattering with x measuring the fractional momentum of the struck quark. The structure functions can then be expressed in terms of the quark distribution functions, $u(x)$, $d(x)$, . . . ,[a] where u, d, . . . , give the number of u, d, . . . quarks with fractional momentum between x and $x + dx$:

$$F_2(x) = 2xF_1(x) = x[q(x) + \bar{q}(x)],$$

where for the electromagnetic current,

$$q(x) = \frac{4}{9}u(x) + \frac{1}{9}d(x) + \frac{1}{9}s(x) + \frac{4}{9}c(x).$$

[a]The quark distribution functions, as well as the structure functions, depend on x and Q^2. For ease of writing, the Q^2 dependence is not shown.

At Q^2 values above 10^4 GeV^2, the contribution from Z^0 exchange becomes comparable to that from photon exchange. Formally, the cross section can be written as before,

$$\frac{d^2\sigma(\gamma + Z^0)}{dxdy} = \frac{4\pi\,\alpha^2}{s\,x^2y^2}\,[(1-y)F_2(x, Q^2) + y^2xF_1(x, Q^2)],$$

where now the structure functions, F_1 and F_2, receive contributions from γ exchange, from Z^0 exchange, and from the interference between the two processes. In terms of the quark distribution functions, the cross section reads for left–(right–)handed electrons[4]:

$$\frac{d^2\sigma}{dxdy}(e^-_{\{L/R\}} + \mathrm{p} \rightarrow e^-_{\{L/R\}}X)$$

$$= \frac{4\pi\alpha^2}{s(xy)^2}\,[Q_u^2(u(x) + u(x)) + Q_d^2(d(x) + \bar{d}(x))]x(1 - y + y^2/2)$$

$$+ \frac{\alpha}{xy}\frac{\sqrt{2}\,G_F\,M_Z^2}{sxy + M_Z^2}\begin{Bmatrix}1 - 2\sin^2\vartheta_W\\ -2\sin^2\vartheta_W\end{Bmatrix}$$

$$\cdot\,\{[Q_u(1 - 4\,Q_u\sin^2\vartheta_W)(u(x) + \bar{u}(x))$$

$$- Q_d(1 + 4\,Q_d\sin^2\vartheta_W)(d(x) + \bar{d}(x))]\cdot(1 - y + y^2/2)\}$$

$$+ \frac{\alpha}{xy}\frac{\sqrt{2}\,G_F\,M_Z^2}{sxy + M_Z^2}\begin{Bmatrix}1 - 2\sin^2\vartheta_W\\ 2\sin^2\vartheta_W\end{Bmatrix}$$

$$\cdot\,[Q_u(u(x) - \bar{u}(x)) - Q_d(d(x) - \bar{d}(x))]xy(1 - y/2)$$

$$+ \frac{s}{4\pi}\frac{G_F^2\,M_Z^4}{(sxy + M_Z^2)^2}\begin{Bmatrix}(1 - 2\sin^2\vartheta_W)^2\\ 4\sin^4\vartheta_W\end{Bmatrix}$$

$$\cdot\,\{[(1 - 4\,Q_u\sin^2\vartheta_W + 8Q_u^2\sin^4\vartheta_W)(u(x) + \bar{u}(x))$$

$$+ (1 + 4\,Q_d\sin^2\vartheta_W + 8\,Q_d^2\sin^4\vartheta_W)(d(x) + \bar{d}(x))]\cdot(1 - y + y^2/2)\}$$

$$+ \frac{s}{4\pi}\frac{G_F^2\,M_Z^4}{(sxy + M_Z^2)^2}\begin{Bmatrix}(1 - 2\sin^2\vartheta_W)^2\\ -4\sin^4\vartheta_W\end{Bmatrix}$$

$$\cdot\,\{[(1 - 4\,Q_u\sin^2\vartheta_W)(u(x) - \bar{u}(x))$$

$$+ (1 + 4\,Q_d\sin^2\vartheta_W)(d(x) - \bar{d}(x))]xy(1 - y/2)\},$$

with $Q_u = 2/3$ and $Q_d = -1/3$.

G_F is the fermi coupling constant, $G_F = 1.02 \times 10^{-5}/M_p^2$. To simplify the writing, only the u, $\bar{u}$ and d, $\bar{d}$ contributions are shown. The expected event rates have been calculated in reference 4 using the structure functions of reference 5 and assuming 10,000 h of data taking (about two years of operation) at an average luminosity of 0.5×10^{31} cm^{-2}s^{-1}. The results are shown in FIGURE 5. Note that Q^2 values less than 3

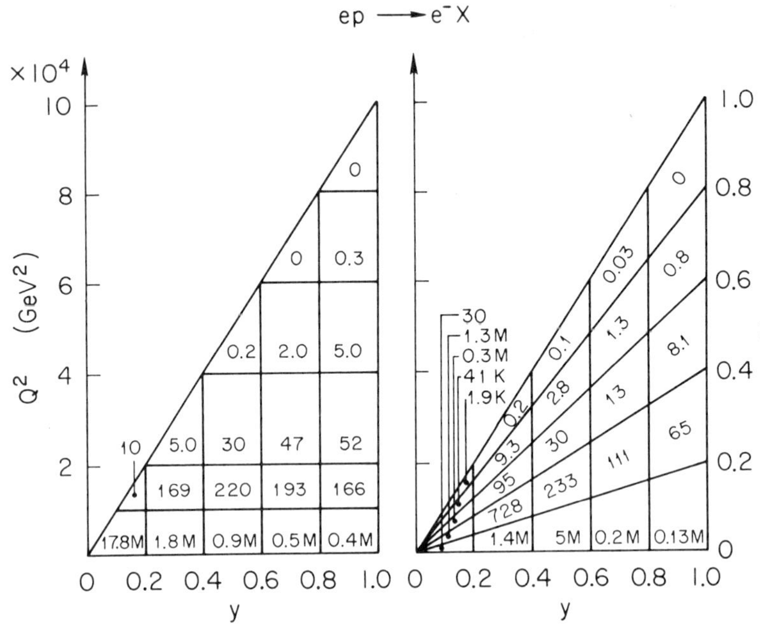

FIGURE 5. Expected neutral current event rates for $e^-p \rightarrow e^-X$ from $\gamma + Z^0$ exchange for a luminosity of 200 pb^{-1}.

GeV^2 have been suppressed. There are 20×10^6 events with Q^2 values between 3 and 10^4 GeV^2. Roughly a thousand events have Q^2 values above 10^4 GeV^2 where Z^0 effects dominate. If we demand a minimum of a hundred events, we see that neutral currents can be studied with sufficient statistics up to Q^2 values of 3×10^4 GeV^2. The dark corner with $Q^2 < 200$ GeV^2 and $y < 0.004$ marks the kinematical region that has so far been explored experimentally. HERA will extend the kinematic region by two orders of magnitude in either variable.

Charged Current Processes

Charged current reactions, $e^-p \rightarrow \nu X$, proceed through W^- exchange. The cross section for left-handed electrons can be written in terms of three structure functions, F_1, F_2, and F_3:

$$\frac{d^2\sigma}{dxdy}(e_L^- p \rightarrow \nu X) = \frac{G_F^2 s}{\pi} \frac{1}{(1 + Q^2/M_W^2)^2}$$

$$\left[(1 - y)F_2(x, Q^2) + y^2 x F_1(x, Q^2) + \left(y - \frac{y^2}{2}\right) x F_3(x, Q^2)\right].$$

The cross section for right-handed electrons vanishes since the neutrino is left-handed:

$$\frac{d^2\sigma}{dxdy}(e_R^- + \mathrm{p} \rightarrow \nu + X) = 0.$$

The quark-parton expression for the structure functions is

$$F_2(x) = 2xF_1(x) = x[q(x) + \bar{q}(x)],$$
$$xF_3(x) = x[q(x) - \bar{q}(x)],$$

where

$$q(x) = u(x) + c(x) + \ldots,$$
$$\bar{q}(x) = \bar{d}(x) + \bar{s}(x) + \ldots.$$

This leads to

$$\frac{d^2\sigma}{dxdy}(e_L^- \mathrm{p} \rightarrow \nu + X) = \frac{G_F^2 s}{\pi}\frac{1}{(1 + Q^2/M_W^2)^2}x\{u(x) + (1-y)^2\bar{d}(x)\},$$

where the higher mass quarks have been omitted. The expected event rates are shown in FIGURE 6. They have been calculated[4] under the same assumptions on luminosity and running time as above. Roughly 22,000 events are produced at $Q^2 < 10^4$ GeV^2 and

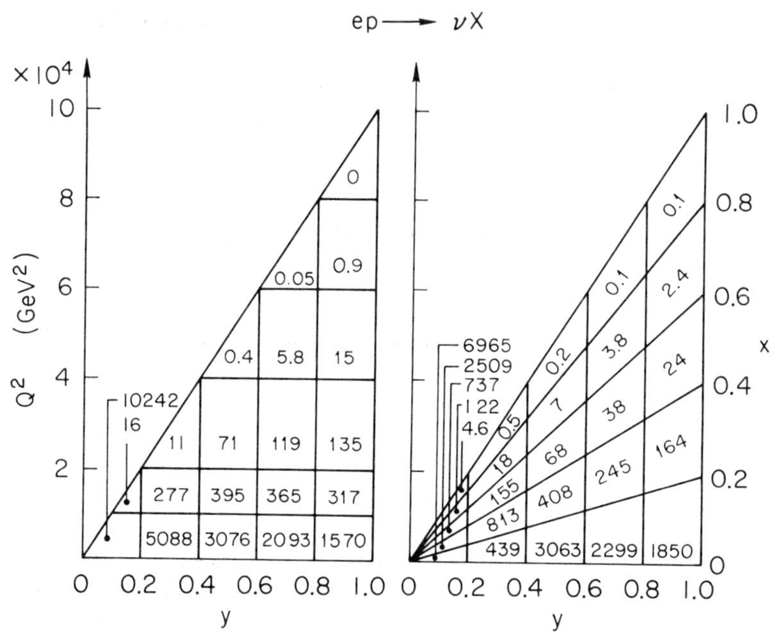

FIGURE 6. Expected charged current event rates for $e_L^- \mathrm{p} \rightarrow \nu X$ for a luminosity of 200 pb^{-1}.

1700 events at $Q^2 > 10^4$ GeV^2. FIGURE 6 shows that charged currents can also be studied with sufficient statistics up to $\sim 4 \times 10^4$ GeV^2.

The topology of neutral and charged current events is illustrated in FIGURE 7. The lepton and the current jet (several jets if gluons are also emitted) emerge on opposite sides of the beam axis; they balance each other in transverse momentum. The debris of the proton are emitted in a very narrow cone (of order, 10 mrad) around the proton beam direction, which renders their detection difficult. FIGURES 8 and 9 show neutral and charged current events. In the latter case, only the current jet is observable and Q^2, x have to be determined solely from the current jet properties.

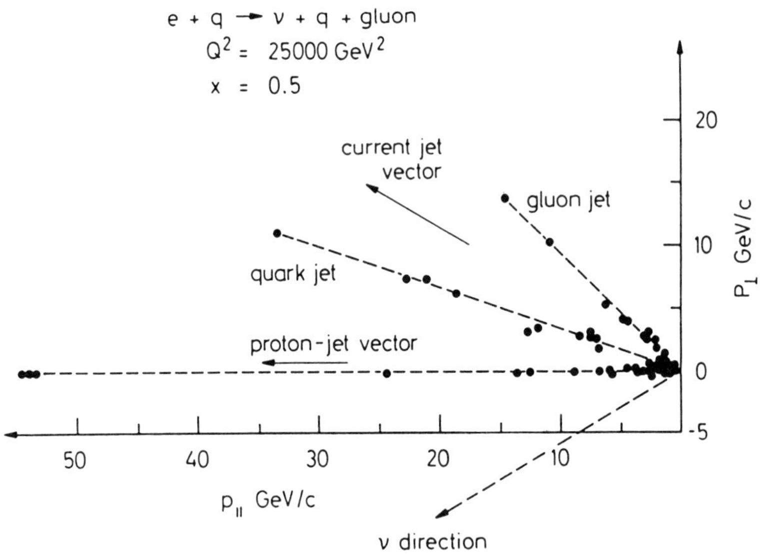

FIGURE 7. Topology of neutral and charged current events.

A precise determination of the structure functions will open the way to address several fundamental questions. (See below.)

QCD

The large Q^2 range accessible at HERA will allow a stringent test of QCD that predicts logarithmically falling structure functions. Mass corrections and higher twist contributions that affect present experiments should be negligible at HERA.

The accuracy with which one can hope to measure the QCD scale parameter is around ± 40 MeV for Λ_{QCD} = 200 MeV. The same data will also yield accurate measurements for the quark and gluon distribution functions.

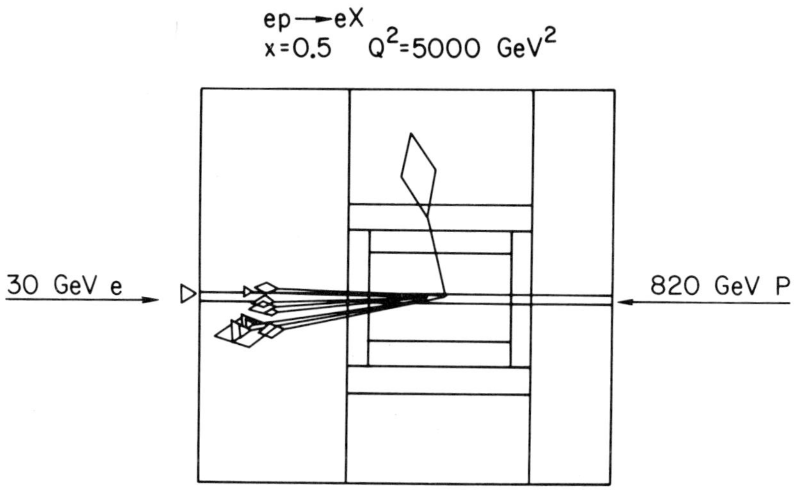

FIGURE 8. Example of a neutral event.

Structure of Quarks and Electrons

If quarks and/or electrons are extended objects, the structure functions will show power-law type deviations from their QCD predicted values. The sensitivity to possible structure of quarks and electrons has been estimated with the help of a model[6] that assumes quarks and electrons to be composites and to have common constituents. The

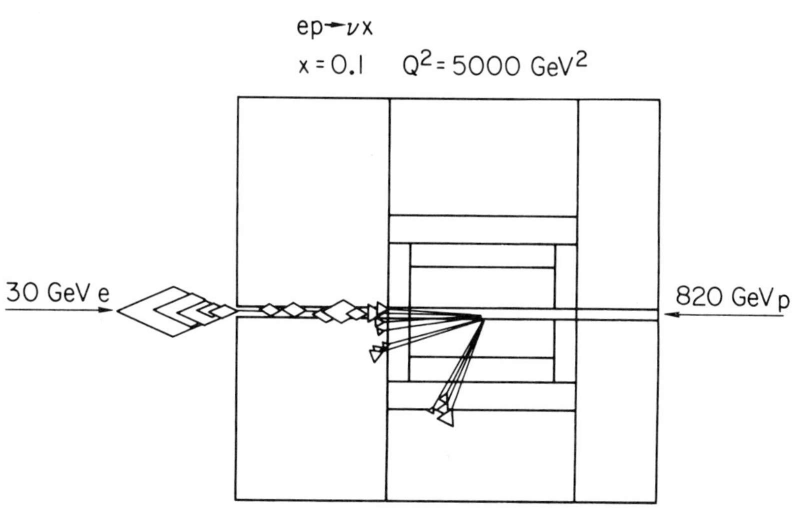

FIGURE 9. Example of a charged current event.

interchange of the constituents leads to a contact term of the form,

$$\mathcal{L} \sim \frac{g^2}{\Lambda^2} \Sigma\, \bar{e}_\alpha \gamma^\mu e_\alpha \bar{q}_\beta \gamma_\mu q_\beta,$$

where α, β denote states of a definite helicity ($\alpha = L, R$; $\beta = L, R$). The mass parameter, Λ, sets the compositeness scale; g measures the coupling strength. For the following, $g^2/4\pi = 1$ will be assumed. FIGURE 10 shows the ratio of the structure functions, F_2, obtained with the contributions from $\gamma + Z^0$ exchange and the contact interaction, and for γ exchange alone:

$$F_2(\gamma + Z^0 + C.I.)/F_2(\gamma).$$

The ratio is shown as a function of Q^2 and x for different Λ values. At Q^2 values above 10^4 GeV^2, a contact interaction with $\Lambda = 1$ TeV leads to large deviations (factors of five to ten) from the standard result. Given two years of data taking, HERA will be sensitive up to Λ values of ~7 TeV, corresponding to distances of 3×10^{-18} cm. The present lower limits on Λ are 2–4 TeV on e, μ from $e^+e^- \rightarrow \mu^+\mu^-$ and 300 GeV for quarks from the total cross section for e^+e^- annihilation into hadrons.[7] A theoretical analysis of some composite models has indicated that the energy of HERA would still be insufficient to see structure in the quarks and electrons.[8] It suggests a lower as well as an upper limit on Λ:

$$\lambda(20\text{–}30)\text{TeV} < \Lambda < 250 \text{ TeV}.$$

The lower limit is deduced from the limit on the decay, $K^+ \rightarrow \pi^+\mu e$; λ is a number of order unity. The upper limit stems from cosmological considerations.

New Currents

Altarelli *et al.*[3] have estimated the effect of a second W_2 on the rate of charged current events. The amplitude for the exchange of the standard $W(W_1)$ can be written as

$$A(Q^2) \sim \frac{g^2}{8(Q^2 + m_N^2)} \sim \frac{G_F/\sqrt{2}}{1 + Q^2/m_W^2}.$$

Under the assumption that the second W couples in the same way to leptons and quarks, the amplitude can be written as

$$A_{1+2}(Q^2) \sim \frac{g_1^2}{8(Q^2 + m_1^2)} + \frac{g_2^2}{8(Q^2 + m_2^2)}.$$

The coupling constants, g_1, g_2, have to be chosen such that the low Q^2 region remains unchanged, i.e.,

$$A(0) \equiv \frac{G_F}{\sqrt{2}} = A_{1+2}(0) = \frac{g_1^2}{8m_1^2} + \frac{g_2^2}{8m_2^2} \equiv \frac{g^2}{8m_W^2}.$$

Taking $m_1 = m_W$ and defining $r = g_2^2/g^2$, the ratio of the cross sections for two W's over one W is given by

$$\frac{\sigma(W_1 + W_2)}{\sigma(W_1)} = \left[1 - r\frac{m_1^2}{m_2^2} + r\frac{m_1^2}{m_2^2}\frac{1 + Q^2/m_1^2}{1 + Q^2/m_2^2}\right].$$

The ratio is shown in FIGURE 11 for different W_2 masses and assuming $r = 1$. For m_2 around 200 GeV, large deviations from the standard theory result will be observed. The deviations become smaller as m_2 increases. With two years of data taking, one will be sensitive to W_2 masses up to ~800 GeV. In addition, a similar result holds for neutral currents and additional Z^0's.

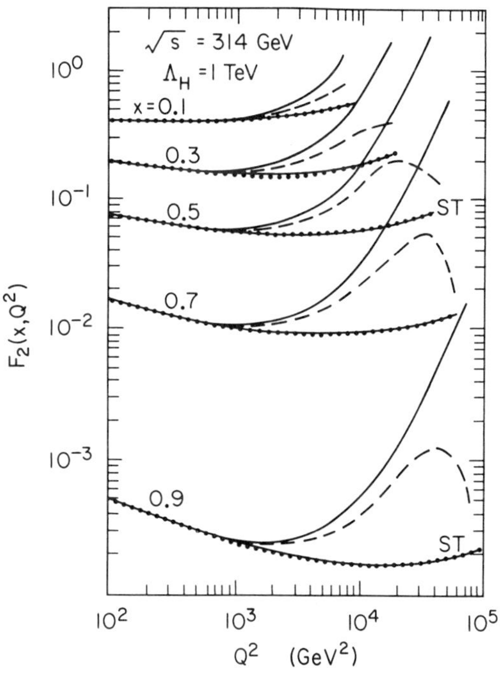

FIGURE 10. The structure function, $F_2(\gamma + Z^0 + C.I.)$.

Right-Handed Currents

One of the great mysteries of weak interactions is the asymmetry between left- and right-handedness; there are left-handed neutrinos and left-handed currents, but no right-handed counterparts. It has long been speculated that left-right symmetry is restored at some higher energy; namely, there exist heavy right-handed neutrinos, N_R, that couple to a right-handed W_R.[9] The mass of N_R has been related to the masses of the corresponding charged lepton, m_e, and the left-handed neutrino, m_ν[10]: $m_{N_R} \approx m_e^2/m_\nu$.

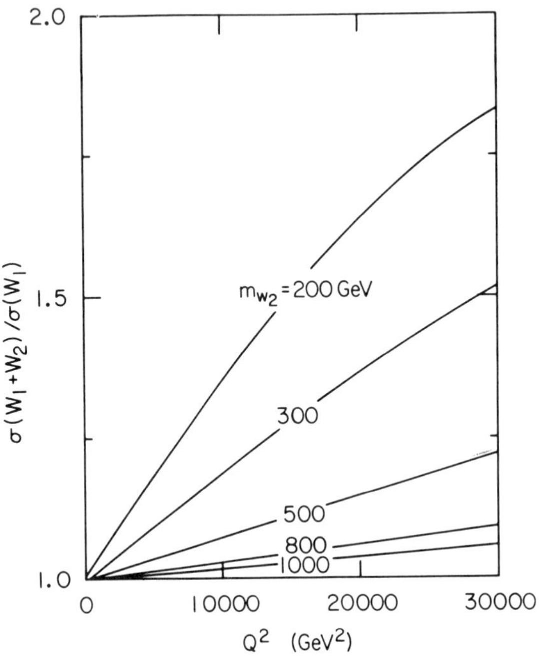

FIGURE 11. The ratio of the cross sections for two and one *W*.

For electrons, this leads to the following prediction for m_{N_R}:

m_ν(eV)	0.1	1	10
m_{N_R}(GeV)	2611	261	26.

Several experimental and theoretical limits have been put on the mass of W_R (see reference 11):

β decay of μ's	
if ν is a Dirac spinor:	$M_R > 380$ GeV
if ν is a Majorana spinor:	no limit
Nonleptonic decays	$M_R >$ 200–300 GeV
K_s^0–K_L^0 mass difference	
if the $(K–M)$ matrix elements are the same for W_L and W_R:	$M_R \geq$ 1–2 TeV
otherwise:	no limit.

The lower limit on a right-handed Z^0 is $M_{Z_R} \geq$ 150–200 GeV, provided $\sin^2\vartheta_W <$ 0.25.

Search for Right-Handed Currents with Longitudinally Polarized Electrons

Longitudinally polarized electrons make HERA ideally suited to search for right-handed currents. Any nonzero cross section contribution to $e_R^- \mathrm{p} \rightarrow \nu X$ or $e_L^+ \mathrm{p} \rightarrow \bar{\nu} X$ scattering signals the presence of a right-handed charged current (see CHARGED CURRENT PROCESSES). In the absence of a W_R contribution,

$$\frac{d^2\sigma}{dxdy}\begin{pmatrix} e_L^- \mathrm{p} \\ e_R^+ \mathrm{p} \\ e_R^- \mathrm{p} \\ e_L^+ \mathrm{p} \end{pmatrix} = \frac{G_F^2 s}{\pi} \frac{x}{(1 + Q^2/M_W^2)^2} \begin{pmatrix} (u + c) + (1 - y)^2(\bar{d} + \bar{s}) \\ (d + s) + (1 - y)^2(\bar{u} + \bar{c}) \\ 0 \\ 0 \end{pmatrix}.$$

For neutral current reactions, the situation is different. The photon is blind to the helicity of the electron, but the Z^0 gives different contributions to e_L and e_R scattering, as mentioned in NEUTRAL CURRENT PROCESSES. FIGURES 12a and 12b show the difference between the cross sections for $e_{L,R}^-$ and $e_{L,R}^+$ scattering. They amount to ~60% at Q^2 values of 10^4 GeV^2.

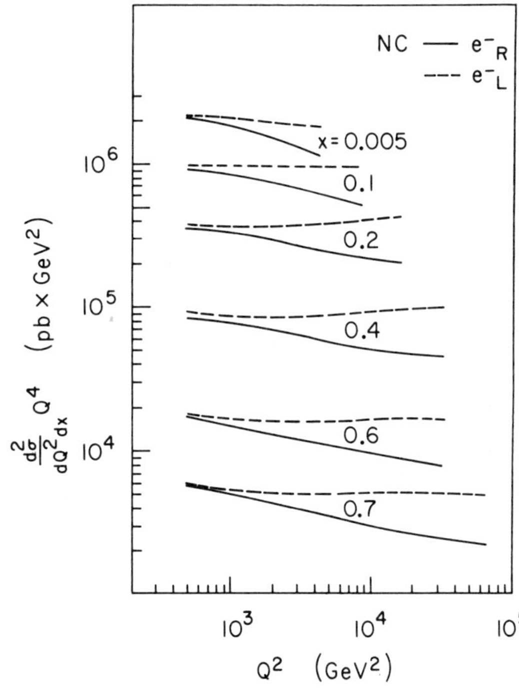

FIGURE 12a. The neutral current cross section for left- and right-handed e^-.

PRODUCTION OF NEW PARTICLES

Pair Production of Heavy Quarks

Photon-gluon fusion (FIGURE 4) will be a strong source of heavy quarks. A discussion of the process can be found in reference 12. The total cross section depends on the quark mass approximately as

$$\sigma(eq \rightarrow Q\bar{Q}X) \sim M_Q^{-4}.$$

Given a t quark mass of 40 GeV and a luminosity of 200 pb^{-1}, the expected number of $t\bar{t}$ events is a few hundred. Top-like quarks can be searched for up to masses of 100–120 GeV.

The heavy quarks will essentially be photoproduced ($Q^2 \approx 0$) and will be emitted in the direction of the incoming proton. Given the high mass, these heavy quarks will decay into many particles (~20–25 for $m_t = 40$ GeV) that are isotropically distributed in the plane perpendicular to the beams. Thus, the production of new heavy quarks should also be discernible from other processes. FIGURE 13 shows the production of a $t\bar{t}$ pair, assuming $M_t = 40$ GeV.

Another possible source of Q production might be an intrinsic $Q\bar{Q}$ component in the nucleon. It has been argued for instance[13] that the $c\bar{c}$ component could be as large as 1% and could become an important contributor once Q^2 exceeds $\sim 4M_c^2$.

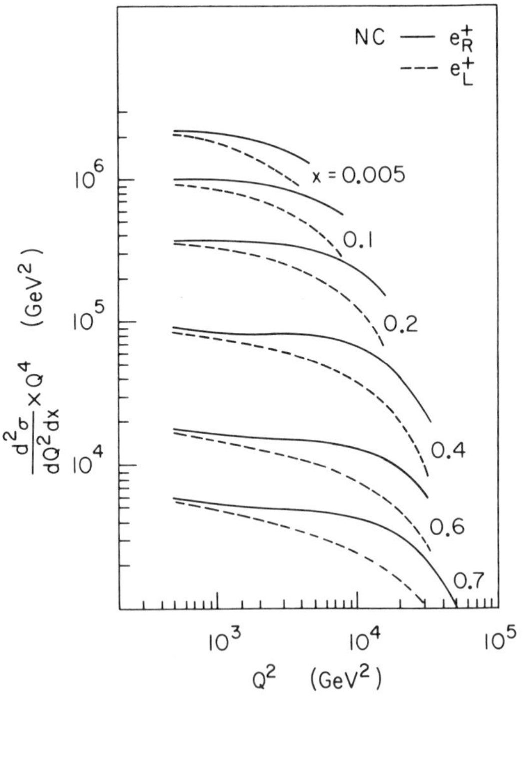

FIGURE 12b. The neutral current cross section for left- and right-handed e^+.

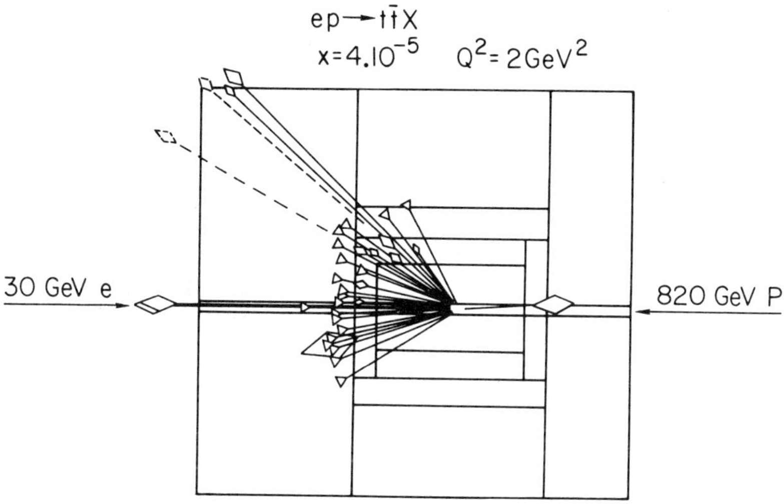

FIGURE 13. Example of a $t\bar{t}$ production event.

Vector Boson Production

Z^0 and $W^\pm$ production is expected to proceed predominantly through the diagrams shown in FIGURE 14. The total cross sections are predicted to be[14]

$$\sigma(ep \rightarrow Z^0X) \approx 0.1 \text{ pb},$$
$$\sigma(ep \rightarrow W^\pm X) \approx 0.05 \text{ pb},$$

leading to ~20 Z^0 events and ~10 $W^\pm$ events for 200 pb^{-1}.

In the (unlikely) case that Z^0 and $W^\pm$ have also a strongly interacting component, the time-honored pomeron exchange (FIGURE 15) may add to their production rate.

Higgs Production

Possible diagrams for production of Higgs particles are shown in FIGURE 16. For a total of 200 pb^{-1}, the expected yield of Higgs produced is not overwhelming[15]:

M_H =	10	20	50	100 Gev
number of events =	10	10	10	3.

Excited Quarks and Leptons

If quarks and leptons are composites, the existence of excited quarks and leptons, q^* and e^*, appears to be natural (remember p, Δ). Their production would follow the diagrams of FIGURE 17. The production rates are sufficient to search for q^*, e^* up to masses of 250 GeV.[16]

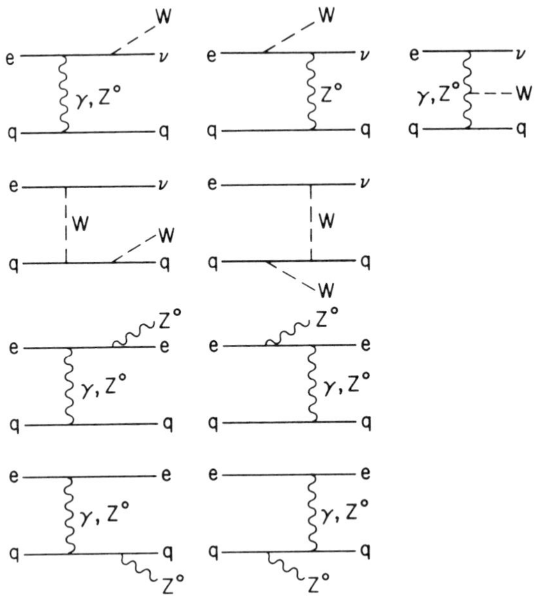

FIGURE 14. Diagrams for Z^0 and $W^{\pm}$ production.

New Currents, Quarks, and Leptons

Suppose that there are new currents, X^0, $X^{\pm}$, that link the known quarks and leptons to new quarks, $Q = U, D, S, C, B, T$, and leptons, L, and that have coupling strengths equal to that of the standard weak interactions. The cross section for the dominant associated production of D and L^0 as depicted in FIGURE 18 is given by

$$\frac{d\sigma}{dxdy} = \frac{G_F s}{2}\left(\frac{m_W}{m_X}\right)^4 \frac{1}{\left(1 + \dfrac{sxy}{M_x^2}\right)} \left\{xq(x)\left(1 - \frac{m_L^2 + m_D^2}{sx}\right) + x\bar{q}(x)\left(1 - y - \frac{m_L^2}{sx}\right)\left(1 - y - \frac{m_D^2}{sx}\right)\right\},$$

where

$$q(x) = u(x) + c(x) + \dots,$$
$$\bar{q}(x) = \bar{d}(x) + \bar{s}(x) + \dots.$$

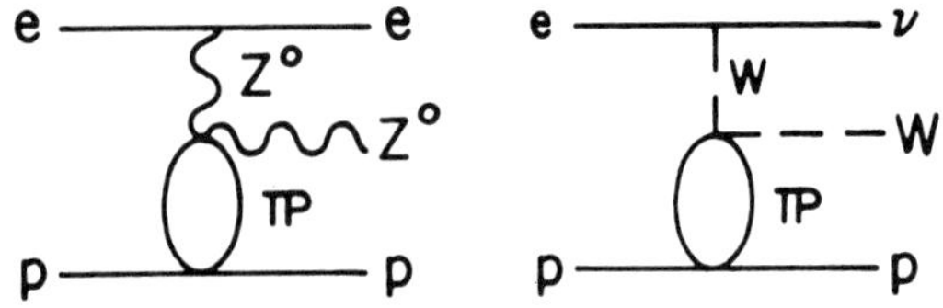

FIGURE 15. Diagrams for Z^0 and $W^{\pm}$ production by pomeron exchange.

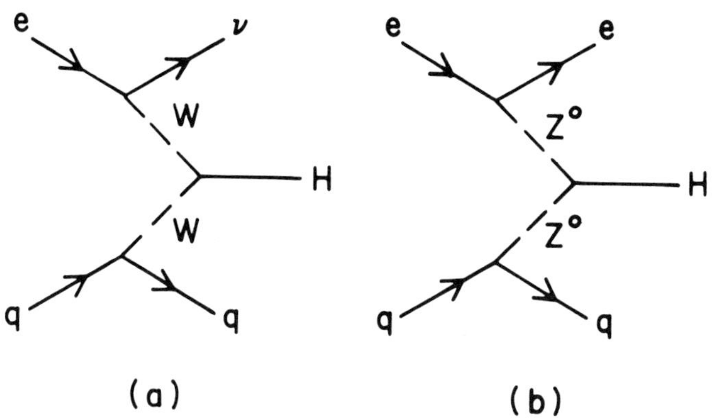

FIGURE 16. Diagrams for Higgs production.

The kinematically allowed regions for x and y are:

$$\frac{(m_L + m_D)^2}{s} < x < 1,$$

$$y_- < y < y_+,$$

$$y_\pm = \frac{1}{2sx}\left[sx - m_D^2 - m_L^2 \pm \sqrt{(sx - m_D^2 - m_L^2)^2 - 4m_D^2 m_L^2}\right].$$

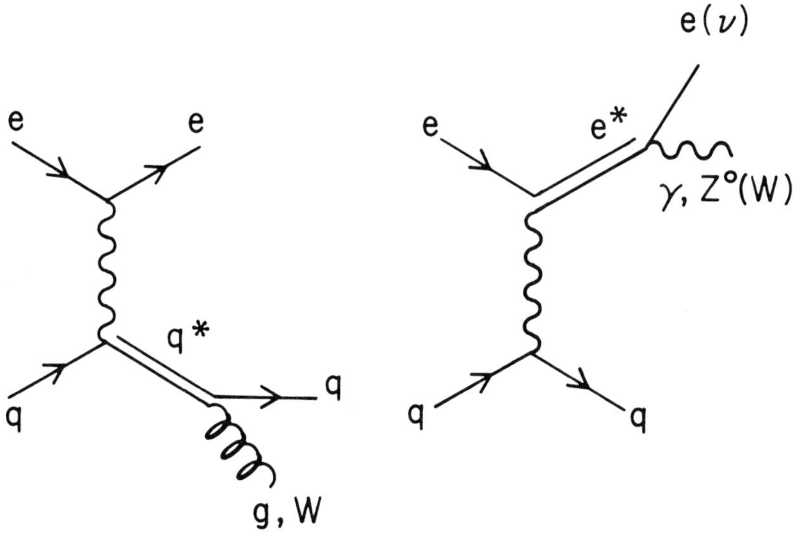

FIGURE 17. Diagrams for the production of excited quarks and leptons.

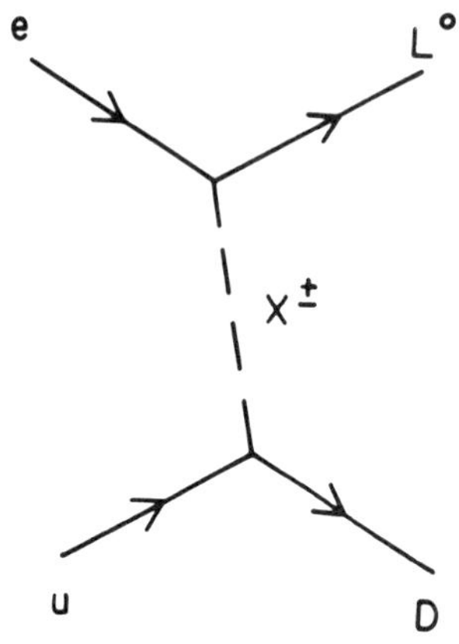

FIGURE 18. Diagram for quark and lepton production.

The process will lead to the event rates shown in FIGURE 19 as a function of m_D for different lepton masses, m_L (assuming $m_x = 83$ GeV and ten days of running at $L = 10^{31}\ \mathrm{cm^{-2}s^{-1}}$); 200 $\mathrm{pb^{-1}}$ would probe up to $m_L + m_D \approx 220$ GeV.

Leptoquarks

Technicolor offers an alternative, though not undisputed, explanation for the Higgs mechanism. In this theory, the Higgs is a pion-like composite: a state where a new quark and antiquark are bound together by the superstrong technicolor force. Besides new quarks, $U, D, \ldots$, new leptons, $E, M, \ldots$, are also predicted, which together lead to a new universe of particles. What is most exciting for HERA is the occurrence of leptoquarks, $TLQ \equiv \bar{u}E, \ldots$. The expected mass hierarchy is as follows:

	mass
—Technipions ($U, \bar{D}, \ldots$)	0 (10 GeV)
—Leptoquarks ($U, \bar{E}, \ldots$)	0 (160 GeV)
—Technicolor Octets ($U, \bar{D}, \ldots$)	0 (240 GeV).

Leptoquarks would be within the reach of HERA experiments. The diagrams expected to dominate the production are given in FIGURE 20. The cross section for $ep \rightarrow TLQ\,X$ is shown in FIGURE 21 as calculated in reference 17. Here, θ_{et} is the mixing angle. Given a $\sin^2\theta_{et}$ of ~ 0.05, leptoquarks with masses up to 180 GeV can be studied at HERA (FIGURE 21).

Technicolor theories already have difficulties explaining the small $K_L^0 \rightarrow \mu^+\mu^-$ branching ratio and no evidence has been found at PETRA or PEP for low-lying technipions.

Supersymmetric Particles

HERA is a hunter's paradise for supersymmetric particles; if, of course, they exist in the accessible mass range. The prototype reaction is the associated production of squark ($\tilde{q}$) and slepton ($\tilde{\ell}$) by photino ($\tilde{\gamma}$), zino ($\tilde{Z}$), or wino ($\tilde{W}$) exchange (FIGURE 22). The cross section has been calculated in reference 18. FIGURE 23 shows the cross section as a function of the squark mass for different slepton mass values. If we demand at least ten events per 200 pb^{-1}, then squark plus slepton production can be studied up to mass values of $m_{\tilde{q}} + m_{\tilde{\ell}} = 160$ GeV.

The production of scalar leptons has been studied in reference 14. The dependence on the $e^{\pm}$ polarization has been analyzed in reference 19.

DETECTION OF NEW PROCESSES

The mass range that is accessible to HERA for new particles was estimated in the previous sections by requiring the production of at least a few tens of these events a year. Since the yield of standard neutral and charged current events is many orders of magnitude larger, one might wonder whether those few exotic events will be discernible from the ordinary processes or whether their detection will be as difficult as, for example, that of the top quark at the S$\bar{p}$pS collider.

The events should be visible if the dominating background is ordinary neutral current scattering ($eq \rightarrow eq$). This is a consequence of two facts: (1) the kinematics in $eq \rightarrow eq$ is well defined and (2) the decay of a new heavy object will spread the

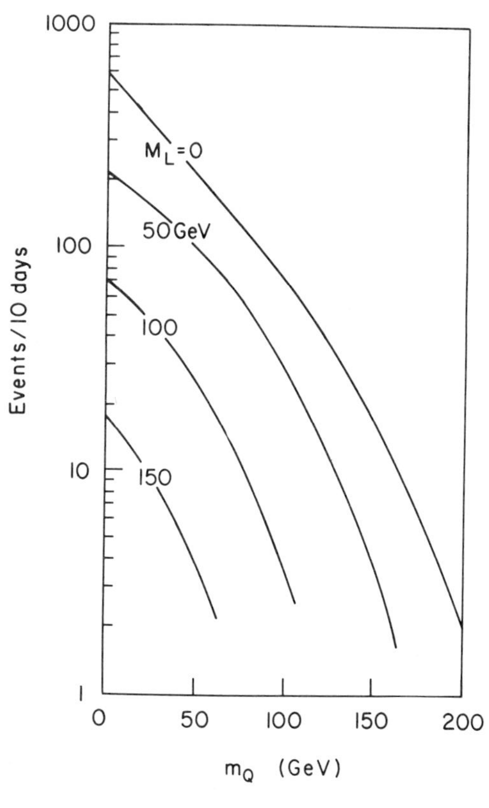

FIGURE 19. Production rates for $e^- p \rightarrow DL^0X$.

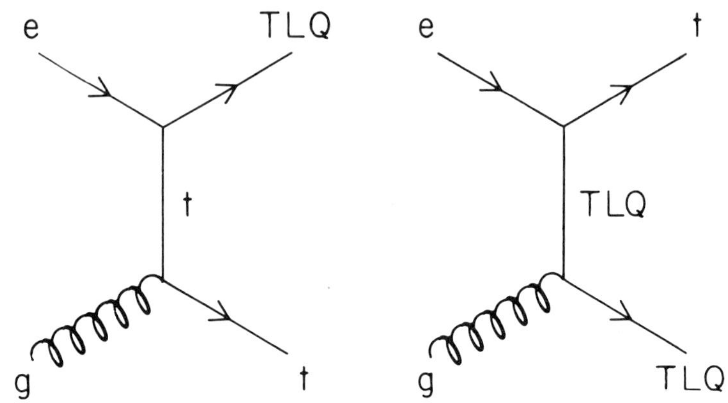

FIGURE 20. Diagrams for the production of leptoquarks.

emerging quark (lepton) over a large volume in phase-space. For definiteness, consider the supersymmetric reaction discussed before,

$$eq \longrightarrow \tilde{e}\tilde{q}.$$

The event rate relative to the standard neutral current reaction, $eq \longrightarrow eq$, is

$$\text{for } M_{\tilde{q}} + M_{\tilde{e}} = \begin{cases} 80 \text{ GeV} \\ 180 \text{ GeV} \end{cases} \quad \frac{\sigma(\tilde{e}\tilde{q})}{\sigma(eq)} \approx \begin{cases} 10^{-3} \\ 10^{-5} \end{cases}.$$

The expected decay chains for $\tilde{q}$ and $\tilde{e}$ are

$$\tilde{q} \longrightarrow q + \tilde{\gamma},$$

$$\tilde{q} \longrightarrow q + \tilde{g} \longrightarrow qg\tilde{G},$$

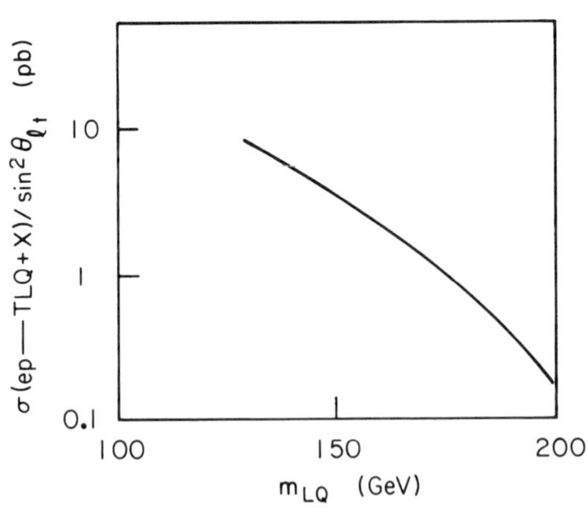

FIGURE 21. Cross section for the production of leptoquarks.

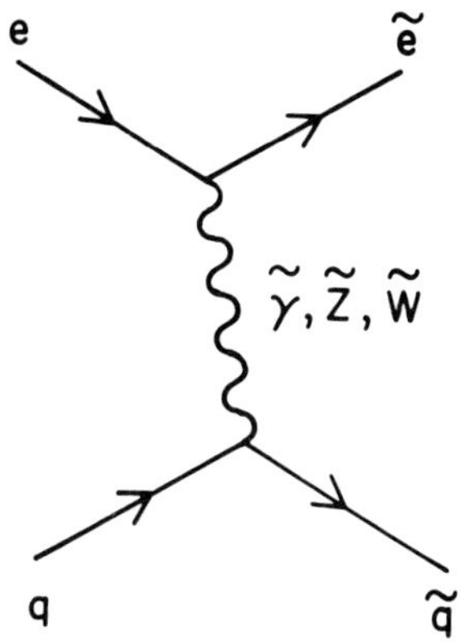

FIGURE 22. Diagram for $eq \rightarrow \tilde{e}\tilde{q}$.

and

$$\tilde{e} \rightarrow e + \tilde{\gamma} \text{ or } \tilde{e} \rightarrow e + \tilde{G}.$$

Under the assumption that $\tilde{\gamma}$ and $\tilde{G}$ are weakly interacting and leave the experiment undetected, then the final state consists of an electron and a quark jet as in the case of an ordinary neutral current event. The difference lies in the kinematics as seen in the plane perpendicular to the beams. This is illustrated in FIGURE 24.

In the standard case, the electron and the quark jet are anticollinear and balance each other in transverse momentum. For the supersymetric process, e and q are neither

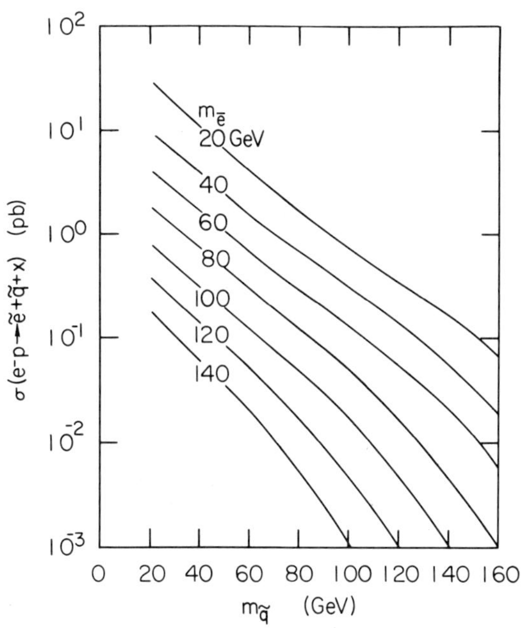

FIGURE 23. Cross section for the production of supersymmetic quarks and leptons.

anticollinear nor do they balance each other in transverse momentum. These two facts can be used to suppress the background. How efficiently that suppression can be done is demonstrated in FIGURE 25. Here, the acollinearity angle, $\Delta\varphi$, is plotted versus Δy, which measures the difference in y values computed from the current jet and the scattered electron:

$$\Delta_y = y_q - y_e,$$

$$y_q = \frac{\Sigma E_i - P_{/\!/i}}{2E_e},$$

$$y_e = 1 - \frac{E'_e - P'_{/\!/e}}{2E_e}.$$

E_i, $P_{/\!/i}$ are the energies and longitudinal momentum components of the current jet particles; E'_e, $P'_{/\!/e}$ are the energy and longitudinal momentum component (measured with respect to the incoming proton direction) of the scattered electron.

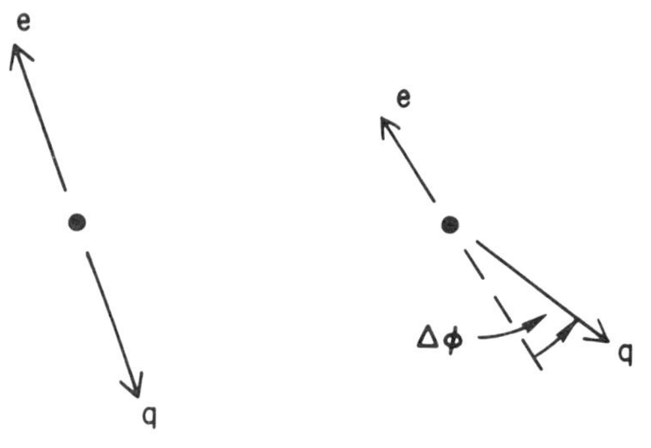

FIGURE 24. Comparison of the kinematics for the standard neutral current process, $eq \rightarrow eq$, and for the supersymmetric process, $eq \rightarrow \tilde{e}\tilde{q} \rightarrow eq$ + missing particles.

The distributions shown in FIGURE 25 have been computed by a Monte Carlo technique for a reasonably realistic detector assuming a finite hadronic energy resolution ($\Delta E/E = 60\%/\sqrt{E}$) and a blind angular region around the beam pipe. FIGURE 25c shows the result for $\sim 10^4$ standard neutral current events. As expected, the events cluster near $\Delta\varphi = \Delta y = 0$. In contrast, for the supersymmetric process (shown in FIGURE 25a for $m_{\tilde{q}} = m_{\tilde{e}} = 40$ GeV and in FIGURE 25b for $m_{\tilde{q}} = m_{\tilde{e}} = 100$ GeV), the events are spread all over the $\Delta\varphi$, Δy map. Thus, supersymmetric events can be well isolated from the standard ones without significant loss in efficiency.

SUMMARY AND CONCLUSIONS

Structure functions of neutral and charged current reactions can be studied with HERA for Q^2 values up to 30,000–40,000 GeV^2. The Q^2–ν space will be extended by two orders of magnitude in either variable over what can presently be reached. Precise structure function values will provide a stringent test of QCD, will probe for

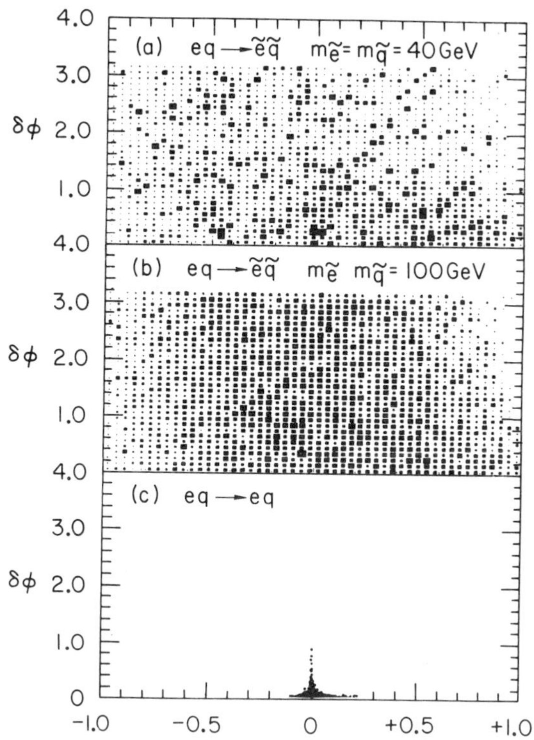

FIGURE 25. Distribution of $\Delta\varphi$ versus Δy for the supersymmetric process, $eq \rightarrow \tilde{e}\tilde{q}$, with $m_{\tilde{e}} = m_{\tilde{q}} = 40$ GeV (a) and $m_{\tilde{e}} = m_{\tilde{q}} = 100$ GeV (b), and for the standard process, $eq \rightarrow eq$ (c).

substructure of quarks and leptons down to distances of 3×10^{-18} cm, and could detect new neutral or charged pieces of the current. In the search for right-handed weak currents, longitudinally polarized electrons will be extremely useful.

HERA is potentially a rich source of new particles. Basically any state that has electric and/or weak charge can be produced up to masses of 200–250 GeV. The sensitivity in mass is shown in TABLE 3.

In many aspects, HERA is complementary to LEP. It probes primarily the spacelike region and gives access to charged currents over a unique kinematic range. For a number of the newly proposed particles, the mass range open to HERA experiments exceeds that of LEP II.

TABLE 3. Expected Sensitivity in Mass

New W	800 GeV
New Z^0	800 GeV
Right-handed W	500 GeV
New quark (t-like)	120 GeV
Excited q^*, e^*	250 GeV
New quarks, leptons	220 GeV
Leptoquarks	180 GeV
Supersymmetric quark plus lepton	160 GeV

The design of an optimal detector for HERA poses a veritable challenge. The large momentum imbalance between incident electron and proton and the nature of spacelike processes throw most final state particles into a narrow cone around the proton direction. On the other hand, the standard neutral current reaction in HERA, $ep \rightarrow eX$, offers an important advantage over $\bar{p}p/pp$ interactions. By tagging the scattered electron, the kinematics of the relevant subsystem is determined. It is this handle that provides HERA experiments with a special sensitivity to new exotic particles.

ACKNOWLEDGMENTS

I am very grateful to R. Peccei, P. Söding, and R. Taylor for a critical reading of the manuscript.

REFERENCES

1. WIIK, B. H. 1982. Electron-Proton Colliding Beams, the Physics Programme and the Machine. Proc. 10th SLAC Summer Institute. A. Mosher, Ed.: 233; 1983. DESY Internal Report, DESY F35-83-01; 1984. Proc. 1984 ICFA Seminar on Future Perspectives in High Energy Physics, May 14–20, 1984, KEK, Japan. S. Ozaki, S. Kurokawa & Y. Unno, Eds.: 123.
2. LLEWELLYN-SMITH, C. H. & B. H. WIIK. 1977. DESY Report 77/36.
3. Proc. Study of an *ep* Facility for Europe. 1979. U. Amaldi, Ed. DESY Report 79/48; HERA. 1981. Proposal report DESY-HERA 81/10; Résumé Discussion Meeting. 1981. Physics with *ep* colliders in view of HERA. Wuppertal, report DESY-HERA 81/18; Proc. Workshop Experimentation at HERA. 1983. NIKHEF, Amsterdam, June 9–11, 1983, report DESY-HERA 83/20; Proc. Discussion Meeting on HERA Experiments. 1985. Genova, Oct. 1–3, 1984, report DESY-HERA 85/01; ALTARELLI, G., B. MELE & R. RÜCKL. 1984. CERN report TH3932; BAGGER, J. A. & M. E. PESKIN. 1984. SLAC-PUB-3447; CASHMORE, J. R. *et al.* 1985. Exotic phenomena in high energy *ep* collisions. Phys. Rep.
4. See, for example, LOHRMANN, E. & K. H. MESS. 1983. DESY-HERA 83/08.
5. GLÜCK, M., H. HOFFMAN & E. REYA. 1982. Z. Phys. **C13:** 119.
6. EICHTEN, E. J., K. D. LANE, & M. E. PESKIN. 1983. Phys. Rev. Lett. **50:** 811; Also see RÜCKL, R. 1983. Phys. Lett. **129B:** 363.
7. FERNANDEZ, E. *et al.* MAC COLLABORATION. 1983. Phys. Rev. Lett. **50:** 123; ALTHOFF, M. *et al.* TASSO COLLABORATION. 1984. Z. Phys. **22:** 13; 1984. Phys. Lett. **138B:** 441; ADEVA, B. *et al.* MARK J COLLABORATION. 1984. Phys. Rep. **109:** 131; BENDER, D. *et al.* HRS COLLABORATION. ANL-HEP-PR-03,71; BERGER, C. *et al.* PLUTO COLLABORATION. 1985. Z. Phys. **27:** 341.
8. BARS, I., M. J. BOWICK & K. FREESE. 1984. Phys. Lett. **138B:** 159.
9. PATI, J. C. & A. SALAM. 1974. Phys. Rev. **D10:** 275; MOHAPATRA, R. N. & J. C. PATI. 1975. Phys. Rev. **D566:** 2559; MOHAPATRA, R. N. & R. E. MARSHAK. 1980. Phys. Rev. Lett. **44:** 1316.
10. MOHAPATRA, R. N. & G. SENJANOVIC. 1980. Phys. Rev. Lett. **44:** 912.
11. SENJANOVIC, G. 1984. Proc. SSC Fixed Target Workshop, The Woodlands, Texas; STOKER, D. P. *et al.* 1985. LBL–18935 and Phys. Rev. Lett.
12. LEVEILLE, J. P. & T. WEILER. 1979. Nucl. Phys. **B147:** 147.
13. BRODSKY, S. J. & J. F. GUNION. 1984. SLAC-PUB-3527.
14. ALTARELLI, G., G. MARTINELLI, B. MELE & R. RÜCKL. 1985. CERN TH 4094.
15. ELLIS, J., M. K. GAILLARD & D. V. NANOPOULOS. 1976. Nucl. Phys. **B106:** 292.
16. See, for example, KÜHN, J. H., H. D. THOLL & P. M. ZERWAS. 1985. CERN TH 4131.
17. RUDAZ, S. & J. A. M. VERMASEREN. 1981. CERN TH 2961.
18. JONES, S. K. & C. H. LLEWELLYN-SMITH. 1983. Nucl. Phys. **B217:** 145.
19. HARRISON, D. P. 1985. Nucl. Phys. **B249:** 704; MARLEAN, L. 1984. SLAC-PUB-3533.

CP Nonconservation

PAUL LANGACKER

Department of Physics
University of Pennsylvania
Philadelphia, Pennsylvania 19104

In this paper, I will describe the predictions of various models for CP violation and the implications of recent precise experiments. The first section consists of a brief review of the experimental numbers and of the kaon decay formalism. The second section is devoted to weak CP violation. Within the standard model, the recent measurements of a very long lifetime (1.11 ± 0.16 ps) for the b quark, the probable discovery of the t quark with a relatively small (30–50 GeV) mass, and the very small upper limit (≤ 0.04) on the ratio, $R = \Gamma(b \rightarrow u + X)/\Gamma(b \rightarrow c + X)$, have severely constrained the parameters of the Kobayashi-Maskawa-Cabibbo (KMC) matrix. It has been claimed that these results make it difficult to generate a large enough ϵ and a small enough ϵ'/ϵ, but it will be argued that the theoretical uncertainties are too large to conclude that there is a serious problem at present. Other CP or T violating effects, such as B^0 decay and the weak contribution, d_N^W, to the neutron electric dipole moment, will be briefly discussed, as will such alternatives to the standard model as supersymmetry, Higgs mediated CP breaking, $SU_{2L} \times SU_{2R} \times U_1$ models, and horizontal interactions.

The strong CP problem and possible solutions (fine tuning, the Peccei-Quinn mechanism, a massless quark, and spontaneous CP violation) are considered in the third section. There has been little progress (but some regression) in the last few years. Cosmological CP violation is discussed in the fourth section. I will assume CPT invariance throughout this paper.

THE EXPERIMENTAL SITUATION

CP violation has been observed only in neutral kaon decays and in the universe. The measured CP violating parameters in the kaon system are[1]

$$\eta_{+-} = \frac{A(K_L \rightarrow \pi^+\pi^-)}{A(K_S \rightarrow \pi^+\pi^-)} = |\eta_{+-}|e^{i\phi_{+-}}$$

$$\eta_{00} = \frac{A(K_L \rightarrow \pi^0\pi^0)}{A(K_S \rightarrow \pi^0\pi^0)} = |\eta_{00}|e^{i\phi_{00}} \qquad (1)$$

and the semileptonic charge asymmetries,

$$\delta_\ell \equiv \frac{\Gamma(K_L \rightarrow \ell^+\nu\pi^-) - \Gamma(K_L \rightarrow \ell^-\bar{\nu}\pi^+)}{\Gamma(K_L \rightarrow \ell^+\nu\pi^-) + \Gamma(K_L \rightarrow \ell^-\bar{\nu}\pi^+)} \qquad (2)$$

for $\ell = e$ or μ.

There are two origins for nonzero values of η_{+-}, η_{00}, and δ_ℓ: (a) $K^0 - \bar{K}^0$ mixing ($\Delta S = \pm 2$) and (b) the K^0 decay amplitude ($\Delta S = \pm 1$).

To discuss the $K^0 - \bar{K}^0$ mixing effects, it is useful to define the CP eigenstates:

$$|K_1> = \frac{1}{\sqrt{2}}(|K^0> + |\bar{K}^0>), \quad \mathrm{CP} = +1$$

$$|K_2> = \frac{1}{\sqrt{2}}(|K^0> - |\bar{K}^0>), \quad \mathrm{CP} = -1, \tag{3}$$

where I have adopted[2] the modern Kobayashi-Maskawa (KM) phase convention,

$$|K^0> = \mathrm{CP}\,|\bar{K}^0>. \tag{4}$$

The physical K_L and K_S states are then,

$$|K_L> = \frac{1}{\sqrt{1+|\epsilon_m|^2}}(|K_1> + \epsilon_m|K_2>)$$

$$|K_S> = \frac{1}{\sqrt{1+|\epsilon_m|^2}}(\epsilon_m|K_1> + |K_2>), \tag{5}$$

where the CP-violating mixing parameter is

$$\epsilon_m = \frac{e^{i\theta}}{\sqrt{2}}\frac{\mathrm{Im}\langle\bar{K}^0|H_{\Delta S=2}|K^0\rangle}{\Delta m}. \tag{6}$$

In equation (6),

$$\Delta m \equiv m_{K_L} - m_{K_S} = (3.521 \pm 0.014) \times 10^{-6}\,\mathrm{eV}, \tag{7a}$$

$$\theta \simeq \tan^{-1}\left(\frac{2\Delta m}{\Gamma_S}\right) \simeq 44°, \tag{7b}$$

and $H_{\Delta S=2}$ is the effective $\Delta S = 2$ interaction Hamiltonian. Δm is related by

$$\Delta m = -2\,\mathrm{Re}\,\langle\bar{K}^0|H_{\Delta S=2}|K^0\rangle. \tag{8}$$

To consider CP breaking in the decay amplitudes, one defines

$$A(K^0 \longrightarrow (\pi\pi)_I) = A_I\,e^{i\delta_I}, \tag{9}$$

where $I = 0$ or 2 is the isospin of the two-pion state and δ_I is the $(\pi\pi)_I$ S-wave phase shift (from final state interactions). Phases in A_I imply CP violation. Experimentally,

$$\frac{\mathrm{Re}A_2}{\mathrm{Re}A_0} \equiv \omega \simeq 0.045. \tag{10}$$

This tiny ratio, which is due to an unexplained enhancement of $\mathrm{Re}A_0$ rather than to a suppression of $\mathrm{Re}A_2$, is the famous $\Delta I = \frac{1}{2}$ rule.

The observed parameters in equations (1) and (2) can be reexpressed in terms of ϵ

and ϵ', where

$$\eta_{+-} = \epsilon + \frac{\epsilon'}{1 + \omega/\sqrt{2}} \simeq \epsilon + \epsilon',$$

$$\eta_{00} = \epsilon - \frac{2\epsilon'}{1 + \sqrt{2}\omega} \simeq \epsilon - 2\epsilon'. \tag{11}$$

Assuming the $\Delta S = \Delta Q$ rule for the weak charged currents (as well as CPT), the charge asymmetry parameters are then predicted to be

$$\delta_e = \delta_\mu = 2 \operatorname{Re} \epsilon/(1 + |\epsilon|^2). \tag{12}$$

In terms of ϵ_m and the A_I, one has

$$\epsilon = \epsilon_m + \frac{e^{i\theta}}{\sqrt{2}} \frac{\operatorname{Im} A_0}{\operatorname{Re} A_0} \tag{13}$$

and

$$\epsilon' \simeq \frac{e^{i\theta'}}{\sqrt{2}} \omega \left[\frac{\operatorname{Im} A_2}{\operatorname{Re} A_2} - \frac{\operatorname{Im} A_0}{\operatorname{Re} A_0} \right]. \tag{14}$$

The two terms in ϵ appear to be due to mixing and the decay amplitude, respectively. However, this interpretation is actually dependent on the KM phase convention. ϵ' is due entirely to CP breaking in the decay amplitude and would be zero in the superweak model.[3] It is proportional to the phase difference between A_2 and A_0 and is suppressed by a factor of ω. The phase of ϵ' is predicted to be

$$\theta' = \delta_2 - \delta_0 + \frac{\pi}{2} = (48 \pm 8)^\circ. \tag{15}$$

The predictions, $\theta \simeq \theta' \simeq 45^\circ$, follow only from CPT and experimental quantities.

The world average values,[4]

$$|\eta_{+-}| = (2.274 \pm 0.022) \times 10^{-3}$$
$$|\eta_{00}| = (2.33 \pm 0.08) \times 10^{-3}$$
$$\phi_{+-} = (44.6 \pm 1.2)^\circ$$
$$\phi_{00} = (54 \pm 5)^\circ, \tag{16}$$

are consistent with the CPT prediction of $\phi_{+-} \simeq \phi_{00} \simeq \theta \simeq \theta'$, but a new measurement of ϕ_{00} would be desirable. Equation (**16**) implies $\operatorname{Re} \epsilon = (1.621 \pm 0.088) \times 10^{-3}$, so the experimental values[4] of

$$\delta_e = (3.33 \pm 0.14) \times 10^{-3},$$
$$\delta_\mu = (3.19 \pm 0.38) \times 10^{-3} \tag{17}$$

are in excellent agreement with the CPT and $\Delta S = \Delta Q$ predictions of $\delta_\ell \simeq 2\operatorname{Re}\epsilon$.

The values of η_{+-} and η_{00} in equation (**16**) are consistent with the superweak predictions of $\epsilon' = 0$. Two recent high precision experiments (Chicago-Saclay) and (Yale-BNL) have greatly improved the limits of ϵ'/ϵ. The results are[5]

$$\frac{|\eta_{00}|^2}{|\eta_{+-}|^2} \simeq 1 - 6\frac{\epsilon'}{\epsilon} = \begin{cases} 1.028 \pm 0.032 \text{ (Chicago-Saclay)} \\ 0.99 \pm 0.05 \text{ (Yale-BNL)}, \end{cases} \quad \textbf{(18)}$$

implying

$$\frac{\epsilon'}{\epsilon} = \begin{cases} -0.0046 \pm 0.0053 \pm 0.0024 \text{ (Chicago-Saclay)} \\ 0.0017 \pm 0.0082 \text{ (Yale-BNL)}. \end{cases} \quad \textbf{(19)}$$

(The two uncertainties quoted for the Chicago-Saclay experiment are statistical and systematic, respectively.)

The only CP violation that has been observed outside of the kaon system is the baryon asymmetry of the universe. Observations indicate

$$\frac{n_B}{n_\gamma} \simeq (4\text{–}7) \times 10^{-10} (\text{nucleosynthesis}^{6}),$$

$$n_{\bar{B}} \ll n_B, \quad \textbf{(20)}$$

where $n_\gamma \simeq 400/\text{cm}^3$ is the number density of 2.7°K microwave photons in the present universe, and n_B ($n_{\bar{B}}$) is the average number density of baryons (antibaryons). The asymmetry, $(n_B - n_{\bar{B}})/n_\gamma \gtrsim 10^{-10}$, requires either CP violation in the dynamics relevant to the early universe or in the initial conditions.

Finally, a nonzero value for the neutron electric dipole moment, d_N, would imply time reversal violation and, therefore, CP violation (assuming that CPT is unbroken). The current experimental limit is

$$|d_N| < 6 \times 10^{-25} \text{e-cm} \quad \textbf{(21)}$$

from the Leningrad group.[7]

WEAK CP VIOLATION

The Standard Kobayashi-Maskawa-Cabibbo (KMC) Model

CP violation enters the standard model in explicit phases in the Yukawa couplings of Higgs bosons to fermions. After spontaneous symmetry breaking and the Higgs mechanism, the CP violation shows up in the charged current quark couplings. Physical CP violation in the standard model requires at least three fermion families; otherwise, all of the phases can be absorbed into the phases of the quark fields. For three families, the charged hadronic current is

$$J_W^{\mu\dagger} = (\bar{u}\,\bar{c}\,\bar{t})_L\, \gamma^\mu\, U_{\text{KMC}} \begin{pmatrix} d \\ s \\ b \end{pmatrix}_L, \quad \textbf{(22)}$$

where the KMC matrix is

$$U_{KMC} = \begin{pmatrix} c_1 & s_1c_3 & s_1s_3 \\ -s_1c_2 & c_1c_2c_3 - s_2s_3e^{i\delta} & c_1c_2s_3 + s_2c_3e^{i\delta} \\ -s_1s_2 & c_1s_2c_3 + c_2s_3e^{i\delta} & c_1s_2s_3 - c_2c_3e^{i\delta} \end{pmatrix}, \quad \textbf{(23)}$$

where $s_i = \sin\theta_i$, $c_i = \cos\theta_i$, $i = 1, 2$, or 3, and δ is the CP violating phase.

The KMC angle, θ_1, which is essentially the Cabibbo angle, θ_c, of the two family theory, can be determined from β decay and $\Delta S = 1$ kaon and hyperon decays. For example, a recent analysis[8] by Leutwyler and Roos yielded $|U_{us}| = s_1c_3 = 0.221 \pm 0.002$. Until recently, values of s_2 and s_3 as large as 0.5 could not be excluded.

The situation was changed rather dramatically by the measurement of a surprisingly long b lifetime by the MAC and MARK II groups. At this conference, W. Ford[9] has presented new results from the MAC and MARK II groups and has combined these with the DELCO, TASSO, and JADE values to obtain the average,

$$\tau_b = (1.11 \pm 0.16) \times 10^{-12}\,\text{s}. \quad \textbf{(24)}$$

This is consistent with, but somewhat smaller than, the older average of $(1.5 \pm 0.5) \times 10^{-12}$ s presented by Klanner[10] at the Leipzig Conference and tends to somewhat relax the pressure on the standard model. The value of τ_b in equation (**24**) is longer than the charmed hadron lifetimes, despite the expected $(m_b/m_c)^5$ enhancement of the b decay rate. Hence, the decay must be suppressed by very small mixing angles.

A second result is that the b quark decays preferentially into c rather than u. From the shapes of the lepton spectra, the CLEO[11] and CUSB[12] groups have obtained limits on $R_\ell = \Gamma(b \to u\ell\nu)/\Gamma(b \to c\ell\nu)$:

$$R_\ell < 0.04\ (\text{CLEO}^{11})$$

$$R_e < 0.055\ (\text{CUSB}^{12}). \quad \textbf{(25)}$$

Lee-Franzini[12] has argued that the CLEO and CUSB results can be combined to yield

$$R < 0.03\ (90\%\ \text{C.L.}). \quad \textbf{(26)}$$

From equations (**25**) and (**26**) and from the theoretical expression,

$$\tau_b^{-1} \simeq \frac{G_F^2\, m_b^5}{192\,\pi^3}\,[2.95\,|U_{cb}|^2 + 6.33\,|U_{ub}|^2], \quad \textbf{(27)}$$

where U is the Kobayashi-Maskawa-Cabibbo (KMC) matrix, one has[13]

$$|U_{ub}| < 0.0051$$

$$|U_{cb}| = 0.044 \pm 0.005. \quad \textbf{(28)}$$

That is, the b quark almost decouples from the first two families, with the couplings to the first family extremely weak. In spite of the small values in equation (**28**), the b decays are completely consistent with a sequential (doublet) quark.

The measurement of τ_b and the new limits on R have for the first time allowed reasonably precise determinations of s_2 and s_3. A number of groups[14] have studied the implications of the b measurements for the KMC matrix. For example, $\tau_b = 1$ psec and $R < 0.05$ leads (using $U_{ub} = s_1 s_3$ and $U_{cb} = c_1 c_2 s_3 + s_2 c_3 e^{i\delta}$) to the allowed region in the s_2–s_3 plane shown in FIGURE 1. It is seen that s_2 is very small and s_3 is consistent with zero. Clearly a measurement (rather than an upper limit) of R would be very useful since it would translate almost directly into a value for s_3. To proceed, additional input is needed. The dominant short-distance contribution to the $K_L - K_S$ mass difference [equation (8)] is given by the box diagram of FIGURE 2a. The original estimate of this diagram for the two family case by Gaillard and Lee[15] led to a correct prediction for the

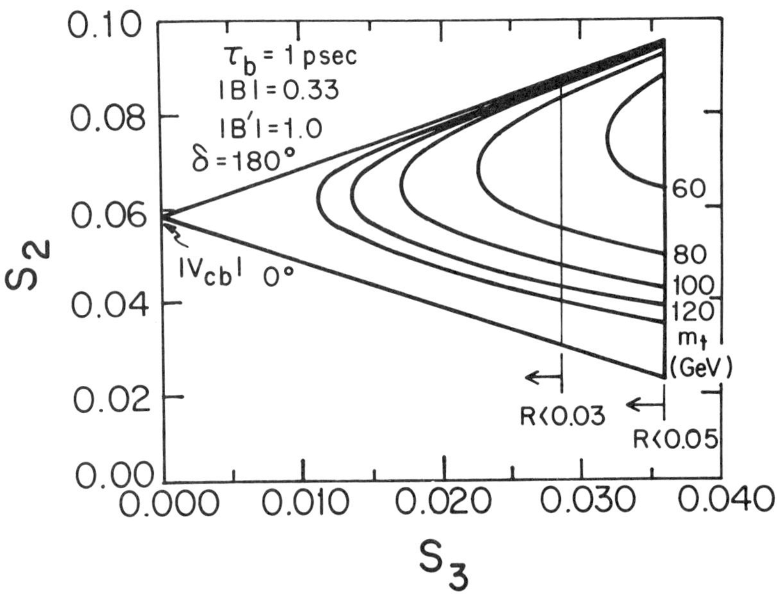

FIGURE 1. Allowed regions in s_2–s_3 for $\tau_b = 1$ psec and $R < 0.05$, from Chau, Cheng, and Keung.[14] For $R < 0.03$, one would have $s_3 \lesssim 0.028$. Each value of s_2 and s_3 corresponds to a definite $|\delta|$, with $|\delta|$ increasing from 0 to 180° as s_2 increases in the allowed region with fixed s_3. The solid curves are the additional constraints from the CP parameter, ϵ, for $B = 0.33$ and various values of m_t. Note that θ_1, θ_2 and θ_3 can be chosen to lie in the first quadrant by any appropriate choice of quark phases, while δ runs from $-\pi$ to π.

charm quark mass. However, this calculation should probably be regarded as only an order of magnitude estimate because of two major uncertainties. The first is the famous B parameter,

$$B \equiv \langle \bar{K}^0 | (\bar{s}_L \gamma_\mu d_L)^2 | K^0 \rangle / \frac{1}{3} m_K f_K^2, \tag{29}$$

that enters when one takes the relevant hadronic element of the four quark operator obtained from FIGURE 2. Gaillard and Lee used $B = 1$, obtained by vacuum saturation,

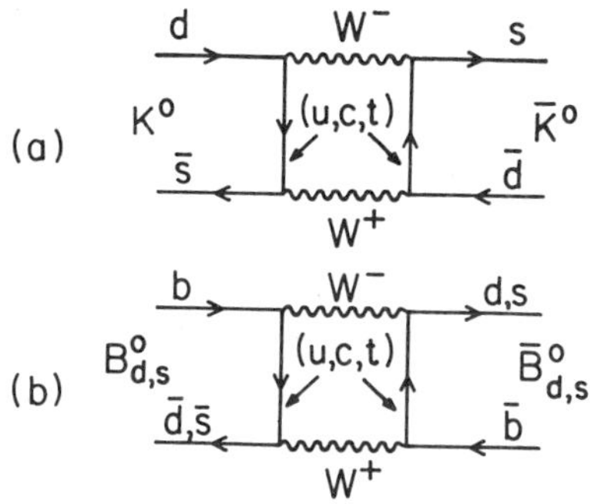

FIGURE 2. Box diagrams for (a) K^0–$\bar{K}^0$ and (b) B^0–$\bar{B}^0$ mixing.

but, as will be seen, this value is very uncertain. Secondly, several authors[16,17] have emphasized that there may be long-distance contributions to Δm (e.g., FIGURE 3) that could be larger than the box diagram.

The box diagram also gives the major short-distance contribution to the CP-violating mixing parameters, ϵ_m, of equation (**6**). ϵ_m is much more likely than Δm to be short-distance dominated because only diagrams involving the heavy t quark can contribute. Even this, though, is not absolutely certain. The η' contribution of FIGURE 3 could in principle yield[16,17] an important contribution to ϵ_m, but the observed small value of ϵ'/ϵ makes this unlikely. Also, double penguin diagrams could be important,[17] but this is controversial. In the following, I will neglect possible long-distance effects.

The box diagram (including QCD corrections) yields the approximate formula,[1]

$$\epsilon_m^{box} \simeq \frac{e^{i\pi/4}\,(0.81)}{2\sqrt{2}\,\mathrm{GeV}^2}\, B\, s_2 s_3 s_\delta \left[-0.7\, m_c^2 + 0.6\, s_2^2\, m_t^2 + 0.8\, m_c^2 \,\ell n \frac{m_t}{m_c}\right]. \qquad (\mathbf{30})$$

Unfortunately, neither the magnitude nor even the sign of B is reliably known, as can be seen from the range of theoretical estimates in TABLE 1. The original vacuum saturation estimate of Gaillard and Lee[15] ($B = 1$) is difficult to justify quantitatively. The SU_3 + current algebra estimate of Donoghue *et al.*[18] ($|B| = 0.33$) was until

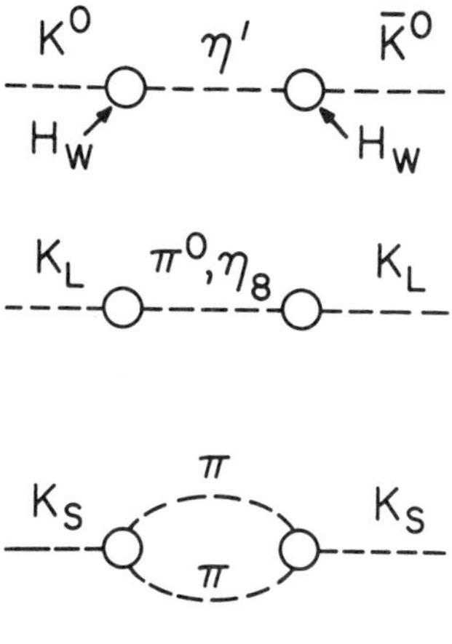

FIGURE 3. Long-distance contributions to the $\bar{K}^0$–K^0 mixing amplitude.

recently generally considered to be the most reliable. However, a recent calculation[19] by Bijnens *et al.* of the (formally leading) $0(m_K^2 \ell n\, m_K^2)$ terms yielded corrections larger than the zeroth order result, thus casting serious doubt on the Donoghue *et al.* value. It is still not clear why the corrections to the seemingly innocent SU_3 + current algebra calculations are so large. The value of B in the MIT bag model[20] is the difference of two much larger terms, leading to large uncertainties (even in the sign). From the estimates in TABLE 1, it seems likely that $0.2 < B < 1.5$, but a larger range cannot be rigorously excluded. More reliable estimates are urgently needed.

For a given B, m_t, and τ_b, the ϵ constraint determines a contour in the s_2–s_3 plane (recall that τ_b determines s_δ for a given s_2 and s_3), as can be seen in FIGURE 1. It was argued by Ginsparg *et al.*[27] that with the small value of $s_2 s_3$ implied by τ_b, it is difficult to generate a sufficiently large $\epsilon(\simeq\epsilon_m^{box})$ unless m_t is large. For example, for $B = 0.33$, $\tau_b = 1.4$ psec, and $R < 0.05$, one requires $m_t > 85$ GeV, which is considerably larger than the range of 30–50 GeV suggested[28] by the UA1 t-quark candidates. However, this limit is strongly dependent on B, as can be seen in FIGURE 4. For $B = 1$ and $\tau_b = 1.4$ psec, for example, one can have m_t as small as 30 GeV. The constraints are even weaker

TABLE 1. Estimates of the B Parameter

B	Method
$B = 1$	Vacuum saturation[15]
$\lvert B\rvert = 0.33$	Current algebra + SU_3[18,19]
$-0.4 < B < 0.34$	MIT bag[20]
$B = 0(1)$	Lattice[21]
$B = \begin{cases} 1.25\text{–}3.8 \\ 0.9\text{–}1.2 \end{cases}$	Vacuum + one particle state[22,23]
$\lvert B\rvert < 2.5$	QCD sum rule[24]
$B \sim 1$	Static quark[25,57]
$B \sim 1\text{–}3$	Harmonic oscillator[26]

for the new value of τ_b in equation (**24**). Until the theoretical issues of B and the long-distance effects are resolved, there does not seem to be any serious difficulty for the standard model. However, it would be difficult to generate the observed value of ϵ if R were smaller than $\simeq 0.01$.

Now let us turn to ϵ'/ϵ, which is experimentally very close to zero [equation (**19**)]. The only known standard model graph that can generate a significant contribution to ϵ'/ϵ is the penguin graph[29] of FIGURE 5, which yields an effective $\Delta S = 1$ Hamiltonian, H^P, involving the operator,

$$O_{LR} = \left(\bar{s}_L \gamma_\mu \frac{\lambda^a}{2} d_L\right)\left(\bar{q}_R \gamma_\mu \frac{\lambda^a}{2} q_R\right). \tag{31}$$

H^P contributes only to A_0 [equation (**9**)] because of the gluon line and it is expected[29] to have an enhanced matrix element between light meson states because of the chirality structure of O_{LR}. It was therefore proposed as a possible explanation of the $\Delta I = ½$ rule. Furthermore, H^P involves the t quark, so it can generate[30] a phase for A_0 (i.e., $\epsilon' \neq 0$).

Unfortunately, most later studies[31] concluded that the penguin diagram is too small

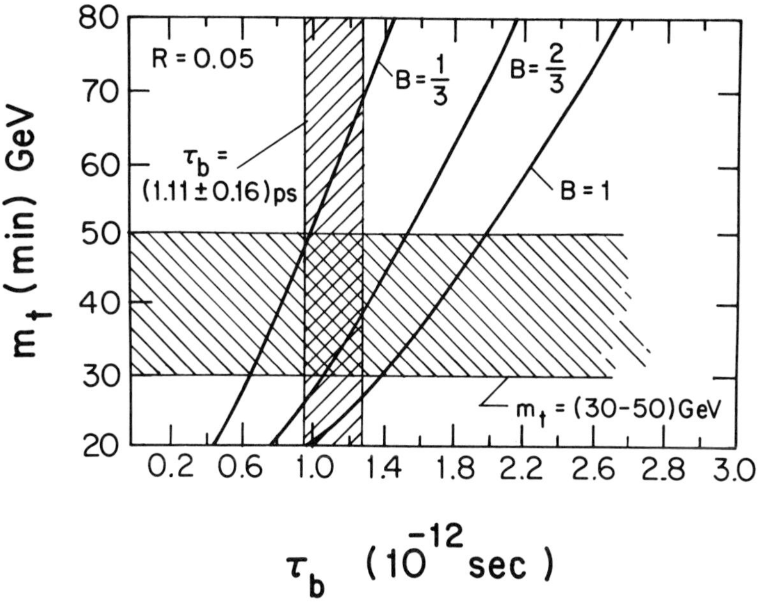

FIGURE 4. The minimum value of m_t that is compatible with ϵ in the standard model for $R = 0.05$ as a function of $\tau_b = B = 1/3$, $2/3$, and 1 (from Buras, Slominski and Steger[14]). Also shown is the range of $m_t \sim (30\text{–}50)$ GeV (UA1[28]) and $\tau_b = (1.11 \pm 0.16)$ psec.[9] The constraints become more stringent for smaller values of R.

by a factor of about five to account for the observed value of $|A_0|$, although there are large uncertainties from the matrix elements and QCD effects. This has been reinforced by two recent lattice calculations[21] that obtained results similar to vacuum saturation for the matrix elements of O_{LR}.

Another recent development is that several authors[32] have argued that many of the existing estimates of the penguin matrix element really included only the normal

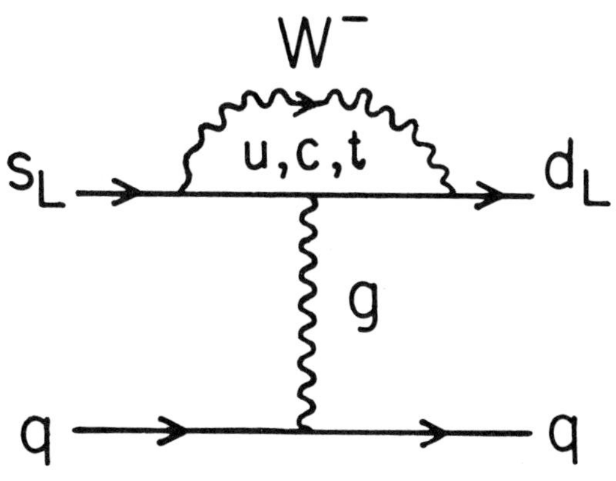

FIGURE 5. The penguin diagram, which contributes only to the $\Delta I = 1/2$ amplitude.

ordered part of O_{LR} (i.e., they ignored the possibility of closing the lower quark line in FIGURE 5 to form a loop). Including the extra term restores the correct soft pion limit for the $\langle\pi|O_{LR}|K\rangle$ matrix element,[33] but yields an even smaller $\Delta I = \frac{1}{2}$ amplitude ($\frac{1}{4}$–$\frac{1}{10}$ of the observed amplitude[32] when the nonpenguin contribution is included). However, again there are large uncertainties.

It therefore seems unlikely that the penguin can explain the $\Delta I = \frac{1}{2}$ amplitude enhancement. The origin could be messy long-distance effects or new physics.

Let us now turn to the imaginary part of the penguin diagram. Estimates[33,34] of ϵ'/ϵ in the KMC model have typically yielded results around 0.01 and Gilman and Hagelin[34] have given this value as a nonrigorous lower bound. The Chicago-Saclay result in equation (**19**) is 3σ below this value, so it is important to investigate the reliability of the theoretical estimate[35] and to consider whether a negative value for ϵ'/ϵ is possible.

To illustrate the basic issues, the expressions for ϵ' and ϵ (assuming it is short-distance dominated) are

$$\epsilon' \simeq C' e^{i\pi/4} \frac{\text{Re}A_2}{\text{Re}A_0} s_2c_2s_3s_\delta \frac{B'}{\text{Re}A_0}$$

$$\epsilon \simeq \epsilon^{box} \simeq C e^{i\pi/4} s_2c_2s_3s_\delta B\, F(m_t, m_c, s_2), \qquad (\mathbf{32})$$

where F is the function given in equation (**30**), C and C' are coefficients that are calculable in principle, and B' is a measure of the poorly known matrix element of the penguin operator. (See reference 36 for the exact definition.) The ratio, $\text{Re}A_2/\text{Re}A_0$, is known (and is positive) experimentally, so for a given m_t, the main uncertainties in

$$\frac{\epsilon'}{\epsilon} = \frac{C'}{C} \frac{\text{Re}A_2}{\text{Re}A_0} \frac{B'}{B} \frac{1}{F\,\text{Re}A_0} \qquad (\mathbf{33})$$

are in C', B, and B'. Although there are serious QCD uncertainties in the magnitude of C', the ratio, C'/C, is almost certainly positive. Also F is positive for allowed m_t. It is conceivable that B is negative (in which case, s_δ would be negative to ensure $\text{Re}\epsilon > 0$). However, Donoghue *et al.*[18] have used current algebra and SU_3 to relate B to $\text{Re}A_2$. This is the calculation for which Bijnens *et al.*[19] have found very large chiral corrections, but nevertheless it seems likely that the sign is correct. Finally, the various estimates of B' all give a positive sign. In particular, the sign in the bag model calculations is well defined (unlike B, for which there is a strong cancellation between two terms). It therefore seems reasonably certain (though not rigorously proven) that ϵ'/ϵ must be positive in the standard KMC model.

The magnitude is more difficult, though, because both B' and B are poorly known. The range of values of B is given in TABLE 1, while typical estimates of B' are

$$B' \sim \begin{cases} 0.33, \text{ new vacuum insertion estimates}^{37} \\ 1, \text{ bag}^{38} \\ 1.7, \text{ original vacuum insertion estimate}^{29}. \end{cases} \qquad (\mathbf{34})$$

($B' \sim 4$ would be needed to account for the $\Delta I = \frac{1}{2}$ amplitude via the penguin operator.) Most of the existing estimates of ϵ'/ϵ utilize definite values for B and B'. For example, the limit, $\epsilon'/\epsilon \gtrsim 0.01$, given in reference 34 was for $B = 1/3$ and $B' = 1.4$. However, smaller values are allowed if B' and B are varied over their possible ranges. For example, $B = 1$ and $B' = 0.33$ implies $\epsilon'/\epsilon = O\ (0.002)$. Given the theoretical uncertainties in B and B', it seems premature to conclude that the standard model is in trouble; all one can conclude is that the range, 0.001 to 0.015, is the most plausible (and at the level of $\epsilon'/\epsilon \simeq 0.001$, nonpenguin contributions may start to be important[39]). One can, however, interpret the new limits on ϵ'/ϵ as a strong indication that the penguin operator does not give a significant contribution to the $\Delta I = 1/2$ amplitude.

The box diagram of FIGURE 2b leads to B^0–$\bar{B}^0$ mixing,[40] where B^0 can be either $B_d^0 = b\bar{d}$ or $B_s^0 = b\bar{s}$. $B\bar{B}$ oscillation effects are determined by the ratio of the $B\bar{B}$ mixing amplitude (FIGURE 2b) and the total b decay width. The effects are expected to be much larger than for the $D^0\bar{D}^0$ system because (i) the t and c quarks in the loop for $B\bar{B}$ are heavier than the corresponding b, s quarks relevant to $D\bar{D}$, (ii) the relevant mixing angles are larger for B–$\bar{B}$ than for D–$\bar{D}$ (especially for B_s^0), and (iii) the B decay width is smaller.

B^0–$\bar{B}^0$ mixing leads to such effects as same-sign dilepton pairs via e^+e^-, pp, $\bar{\text{p}}$p $\to (B^0\bar{B}^0)_L + X \to \ell^\pm\ell^\pm + X$, where L is the orbital angular momentum of the $B^0\bar{B}^0$ pair. A convenient observable is

$$R_L = \frac{N_L(\ell^+\ell^+) + N_L(\ell^-\ell^-)}{N_L(\ell^+\ell^-)} \underset{L^{odd}}{\sim} \frac{(\Delta m_B/\Gamma_b)^2}{2 + (\Delta m_B/\Gamma_b)^2}. \tag{35}$$

Typical estimates[14,40] yield

$$R_{odd} \sim \begin{cases} 10^{-3}\text{–}0.02, & B_d^0 \\ 0.12\text{–}0.75, & B_s^0, \end{cases} \tag{36}$$

which is quite encouraging for B_s^0. (R_{odd} increases with B—the quoted range is for $1/3 < B < 1.5$.) Unfortunately, the CP-odd asymmetries such as

$$a_L = \frac{N_L(\ell^+) - N_L(\ell^-)}{N_L(\ell^+) + N_L(\ell^-)} \tag{37}$$

for e^+e^-, pp, $\bar{\text{p}}$p $\to (B^0\bar{B}^0)_L + X \to \ell^\pm + X$ are predicted to be discouragingly small [$\simeq$ (few) $\times 10^{-5}$ for both B_s^0 and B_d^0] in the standard model.

Finally, the electric dipole moment of the neutron [equation (**21**)] receives both strong and weak contributions. In the standard KMC model, the lowest order contribution to the quark weak dipole moment occurs at the three loop level and is very tiny ($\sim 10^{-34}$e-cm).[39,41] Diagrams involving two or more quarks (e.g., penguins) are more difficult to estimate,[41,42] but probably will yield $d_N^W \sim (10^{-30}\text{–}10^{-34})$ e-cm, which is too small to measure in the foreseeable future.

To summarize this section:

(a) The measurements of the b lifetime and $R = \Gamma(b \to u)/\Gamma(b \to c)$ have led to the first reasonably precise determination of the KMC parameters, s_2 and s_3.

(b) The values of s_2 and s_3 are very small, with s_3 consistent with zero. This implies that the third family is almost decoupled from the first two in the charged current weak interactions.

(c) The conflicts that have been claimed for the standard model with a small (30–50 GeV) value for m_t and the new limits on ϵ'/ϵ were dependent on specific assumptions for the poorly known matrix elements, B and B'. There is no real problem at present if B and B' are allowed to vary over reasonable ranges.

(d) The ranges of $R \gtrsim 0.01$ and $\epsilon'/\epsilon \sim (0.001$–$0.015)$ are very likely, but even these cannot be considered rigorous bounds.

(e) The small limits on ϵ'/ϵ imply that the penguin diagram does not explain the $\Delta I = \frac{1}{2}$ rule.

(f) Values (not limits) for R and ϵ'/ϵ are urgently needed and more precise values for m_t and τ_b would be very useful.

(g) More reliable estimates of B, B', and long-distance effects are also urgently needed.

Other Models

Most extensions of the standard model (e.g., supersymmetry or $SU_{2L} \times SU_{2R} \times U_1$) have several possible sources of CP violation, including the ordinary phase, δ, in U_{KMC} in most cases. Most studies of these models have ignored (often in a rather ad hoc way) all but one of the mechanisms.

Additional Fermion Families

If the mixing angles that couple the new fermion families to the light quarks are very small (as is suggested by the near decoupling of the third family), CP violation in the kaon system should be almost identical to the standard three family model. If the mixing angles are not negligible, then such models have many new free parameters; constraints on R, m_t, ϵ'/ϵ, etc., are weakened or lost entirely.[43]

Horizontal Interactions

It is possible to construct models[44] in which $\Delta S = 2$ processes are mediated by the tree-level exchange of a "horizontal" Higgs or gauge field, H (FIGURE 6a). Since this tree-level diagram should lead to values for ϵ_m and Δm comparable to the higher order box diagram of the standard model, the effective interaction must be extremely weak. One requires $G_H \equiv g_H^2/m_H^2 \simeq 10^{-14}$ GeV^{-2}, where g_H is the coupling of H to $\bar{s}d$ (any relevant mixing angles are included in g_H). For $g_H = 0(1)$, for example, one requires $m_H = 0$ (10^4 TeV). Because G_H is so small, the contributions of H exchange to ϵ'/ϵ and d_N^W are usually extremely small. Such models are, therefore, concrete realizations of the superweak model.[3] Many of the models utilizing spontaneous CP violation to solve the strong CP problem use horizontal interactions (see the section entitled THE STRONG CP PROBLEM).

Higgs Mediated CP Violation

CP violation may be associated with charged Higgs boson Yukawa couplings[45] in theories with additional Higgs multiplets. (The original CP breaking may be sponta-

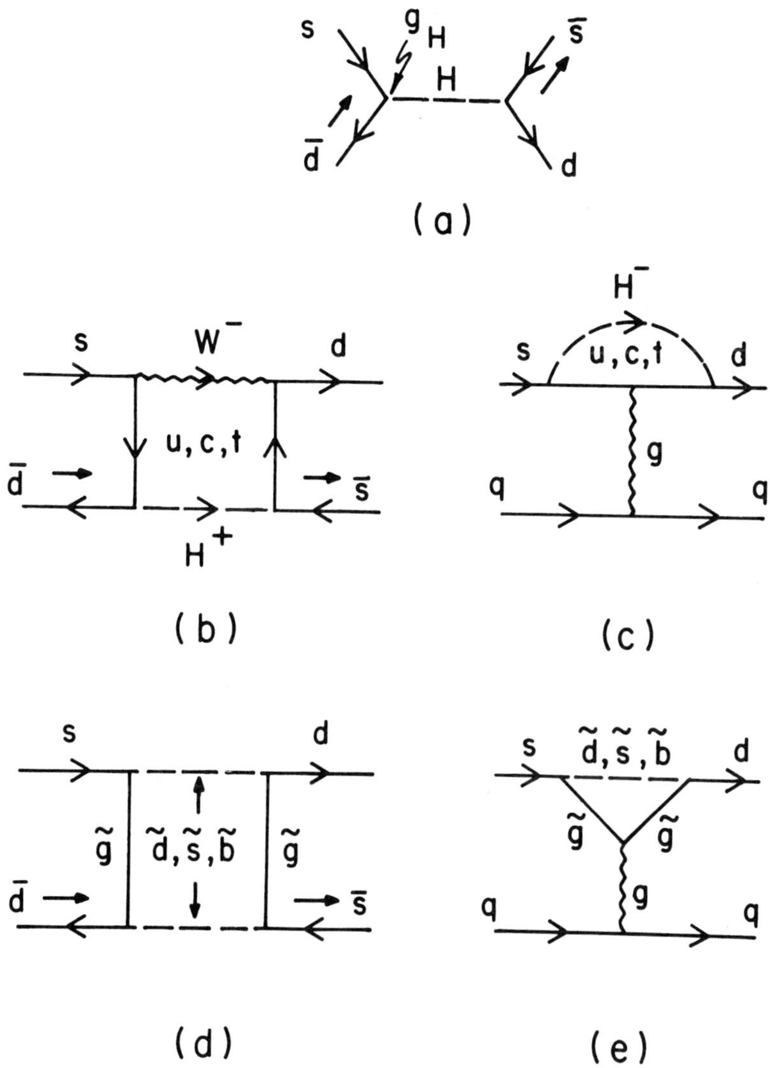

FIGURE 6. (a) Contribution of a horizontal gauge or Higgs boson, H, to $H_{\Delta S=2}$. (b) Charged Higgs contribution to ϵ_M. (c) The Higgs penguin. (d) A new $\Delta S = 2$ box diagram in supersymmetric models. ($\tilde{d}$, $\tilde{s}$, $\tilde{b}$) represent scalar quarks and $\tilde{g}$ is the gluino. (e) The superpenguin diagram: The gluon may also be attached to the scalar quark line.

neous or explicit in the Higgs potential.) ϵ_m may be generated by the box diagram of FIGURE 6b and $\mathrm{Im}A_0$ by the "Higgs penguin" of FIGURE 6c. Traditional estimates[46] have concluded that $\mathrm{Im}A_0/\mathrm{Re}A_0$ from the Higgs penguin is much larger than ϵ_m from the box; therefore [from equations (**13**) and (**14**)], $\epsilon'/\epsilon \sim -\omega \simeq -0.045$, which is much too large in magnitude. However, Dupont and Pham[47] have argued that there may be very important long-distance corrections to these estimates. Moreover, Donoghue and Holstein have recently claimed[48] that previous estimates of the Higgs penguin did not have the correct soft pion limit and that when this is corrected, $\mathrm{Im}A_0/\mathrm{Re}A_0$ is much smaller than the earlier calculations. They conclude that $\epsilon'/\epsilon \simeq -0.006$ with large (and hard to estimate) theoretical uncertainties. The Higgs mediated model therefore appears to be a viable possibility after all.

d_N^W has been estimated[49] to be $\simeq 10^{-25}$–10^{-26} e-cm in this class of models, which is close to the present limits.

$SU_{2L} \times SU_{2R} \times U_1$ *Models (Reference 50)*

Most studies of CP violation assume pseudomanifest left-right symmetry,

$$U_R = K_1 U_L{}^* K_2, \tag{38}$$

where U_L is the KMC matrix, U_R is the analogous matrix describing the W_R couplings to right-handed charged currents, and K_i are diagonal phase matrices. It is usually simply assumed[51] that U_L is real (i.e., $\delta = 0$ or π) so that all CP violation involves W_R exchange.

The box diagram analogous to FIGURE 2a, but with one W_L and one W_R, leads to an operator of the form, $(\bar{s}_L\gamma_\mu d_L)(\bar{s}_R\gamma^\mu d_R)$, which has a strongly enhanced matrix element between kaon states. One therefore needs $M_{W_R} \gtrsim (2$–$5)$ TeV to avoid conflicts with the K_L–K_S mass difference.[52,53] ϵ_m is generated by the same diagram.[53–57] However, the matrix elements of the operators relevant to ϵ'/ϵ and d_N^W do not have any analogous enhancements, so they are expected to be very small. In fact, ϵ'/ϵ requires W_L–W_R mixing in most models[51,54] with $\delta = 0$ and is expected to be 10^{-3}–10^{-4}. d_N^W has been estimated[57] to be 10^{-27} e-cm.

Supersymmetry

Supersymmetric models have several possible new sources of CP violation,[58] such as complex gaugino masses and trilinear scalar couplings. However, these new sources alone (with $\delta = 0$) cannot account for the observed CP effects.[59] They lead to either too large a value for d_N^W or too small an ϵ.

Several authors[59–61] have pointed out that even if these new phases are zero, there are still new $\Delta S = 2, 1$ diagrams involving the KMC phase, δ, as in FIGURES 6d and 6e. These come about because most viable supersymmetry and supergravity models require large radiative corrections to the Higgs potential in order to break $SU_2 \times U_1$. Essentially, the same radiative corrections (driven by a Yukawa coupling proportional to the t mass) also modify the scalar quark mass matrix, especially for the scalar partners, $\tilde{d}_L$, of the left-handed charge $-\frac{1}{3}$ quarks. It turns out that in a large class of models there are flavor-violating $\tilde{d}_L\, d_L\, \tilde{g}$ vertices[60,62] proportional to U_{KMC} (FIGURE 7), which can contribute to ϵ_m, ϵ, and ϵ'. These are supersymmetric QCD graphs: they involve only the strongly interacting particles and their SUSY parameters, and the only coupling constant is the strong gauge coupling.

The contribution of the superbox diagrams (FIGURE 6d) to Δm is sufficiently small[59–63] if the larger of the scalar quark and gluino masses is >30–40 GeV. However, because of the difference between fermion and boson propagators, the phase of the superbox is enhanced[59–61] with respect to that of the box by $\simeq (m_t^2/m_c^2)$. (The scalar quark mass splittings are related to the quark masses in most models.) Hence, for max $(m_{\tilde{g}}, m_{\tilde{q}}) \gtrsim 30$ GeV, the superbox can give a modest or small contribution to Δm, but it also gives the dominant contribution to ϵ_m. In such models, there is no significant lower bound on m_t (even for small B) and $\sin\delta$ can be much smaller than in the standard model (and even negative).

The superpenguin diagrams of FIGURE 6e can be nonnegligible. However for most parameter ranges, the ordinary penguin is still the most important contribution[59–61] to ϵ'. However, since s_δ is smaller than in the standard model, ϵ'/ϵ is also much smaller and can be negative.

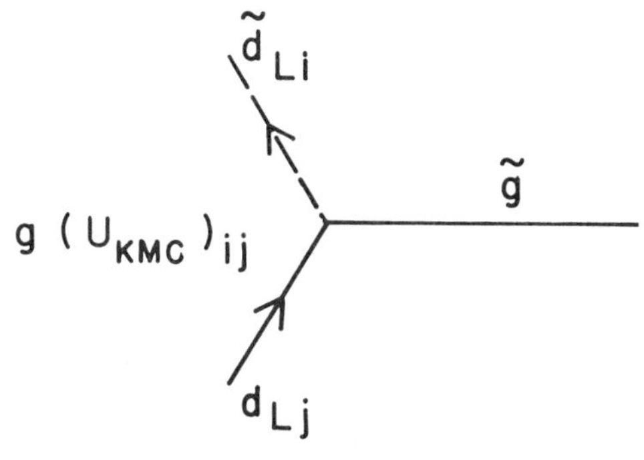

FIGURE 7. The flavor-changing scalar quark-quark-gluino vertex.

The B^0–$\bar{B}^0$ mixing parameters are generally enhanced in supersymmetric models[59] (e.g., R_{odd} (B_d^0) may be enhanced by $\simeq 10^2$). However, the CP-parameter, a_l, is still extremely small.

The gluino loop contribution to d_N^W is very small[64] ($<10^{-28}$ e-cm) in supersymmetric models, but other contributions (e.g., Higgsino exchange) may lead to larger effects,[65] possibly as large as 10^{-25} e-cm, but typically 10^{-27} e-cm.

Summary of Weak CP Violation

The predictions of various models for ϵ, ϵ'/ϵ, R_{odd}, a_l, and d_N^W are summarized in TABLE 2. It should be reemphasized that there are many variations on most of the models. I have presented the results for typical viable examples. Also, in most cases there are usually several possible sources of CP breaking; in most studies, all but one are ignored.

There are other possible experiments relevant to weak CP violation. These include

TABLE 2. Predictions of Typical Versions of Various Models for Weak, Strong, and Cosmological CP Violation[a]

Model	ϵ	ϵ'/ϵ	$R_{odd}\begin{pmatrix}B_d^0\\B_s^0\end{pmatrix}$	a_1	d_N^W	d_N^{QCD}	Baryon Asymmetry (GUT extension)
Standard KMC	need $R > 0.01$ $m_t > 30$	0.001–0.015	10^{-3}–10^{-2} 0.1–0.8	$0(10^{-5})$	10^{-34}–10^{-30}	arbitrary renorm small	ok (nonminimal Higgs)
≥ Four Families	similar	similar	similar	similar	similar	similar	ok
Horizontal Interaction	ok	very small	similar	similar	very small	can impose spontaneous CP	?
Higgs	ok	$-0.006 \pm$?	similar	similar	10^{-25}–10^{-26}	—arbitrary —renorm large	?
$SU_{2L} \times SU_{2R} \times U_1$ (real U_L)	ok	10^{-3}–10^{-4}	similar	similar	10^{-27}	arbitrary	problem in simple versions
SUSY (no new phases)	ok	—smaller than KMC —can be <0	enhanced	still small	10^{-27}–10^{-30}	easy to impose PQ in global	?

[a]The Peccei-Quinn or spontaneous CP violation solutions to the strong CP problem can be imposed on variants of most models.

T-odd asymmetries and correlations in β and μ decay, muon polarization[66] in $K_L \to \mu^+\mu^-$ ($P \gtrsim 10^{-3}$ in the standard model), asymmetries in $K^0(\overline{K}^0) \to 2\gamma$, which may be measurable[67] at LEAR, $K_S \to 3\pi$, which is especially useful if there are two sources of CP breaking with different Lorentz structures,[68] and enhanced electric dipole or magnetic quadrupole moments in nuclei.[69] Finally, several authors[70] have considered CP asymmetries in exclusive B decays, such as $(B^0, \overline{B}^0) \to f$, where $f = \psi K_S$, $\psi\eta$, $D\overline{D}K_S$, ... Asymmetries such as

$$\tilde{A}_f = \frac{\sigma(e^+e^- \to \ell^+ f + x) - \sigma(e^+e^- \to \ell^- f + x)}{\sigma(e^+e^- \to \ell^+ f + x) + \sigma(e^+e^- \to \ell^- f + x)} \tag{39}$$

may be substantial, but unfortunately the rates are expected to be small.

THE STRONG CP PROBLEM

The strong CP problem refers to the possibility of adding the P and T violating term,

$$L_\theta = \frac{\theta}{32\pi^2} g_s^2 F^a_{\mu\nu} \tilde{F}^a_{\mu\nu}, \tag{40}$$

to the QCD Lagrangian.[71] L_θ can be written as a 4-divergence, but its contribution to the action is nonvanishing because of nonperturbative instanton effects. θ is not directly observable, and because of anomalies, its value is changed by chiral transformations,

$$q_i \to e^{i\alpha_i\gamma_5} q_i$$
$$m_i \to e^{-2i\alpha_i} m_i, \tag{41}$$

where m_i is the current mass of quark, q_i (presumably generated by the Higgs mechanism in the weak sector of the theory). However, the combination,

$$\bar{\theta} = \theta - \arg(\det m), \tag{42}$$

where m is the quark mass matrix, is chiral invariant and observable. Therefore, it is not enough to simply set $\theta = 0$ in the original Lagrangian. Whatever mechanism generates weak CP violation generally yields phases of $0(10^{-3})$ or larger in m (these are conventionally removed by a subsequent chiral transformation), and therefore, $\bar{\theta} \geq 0(10^{-3})$.

On the other hand, $\bar{\theta}$ contributes a strong component,

$$d_N^{\text{QCD}} = C\bar{\theta}, \tag{43}$$

to the neutron dipole moment. Requiring $|d_N^{\text{QCD}}| < |d_N^{exp}|$, implies $\bar{\theta} < 10^{-9}$.[72] There are several possible solutions to this strong CP problem.

Fine Tuning

One can simply set $\theta = \arg(\det m)$ so that $\bar{\theta} = 0$. This is rather ugly because θ and m appear to be independent and unrelated. The fine tuning solution is implicit in the minimal version of the standard model.

Actually, $\bar{\theta}$ is a renormalized running parameter. Ellis and Gaillard[73] have argued that if $\bar{\theta}$ is chosen to vanish at the Planck scale, then the effective $\bar{\theta}$ at low energies is nonzero. However, it is of very high order, thus yielding $d_N^{QCD} \sim (10^{-31}$–$10^{-32})$ e-cm. Similarly, d_N^{QCD} may be of the order of $(10^{-24}$–$10^{-25})$ e-cm in Higgs mediated models. Finally, if the standard model is embedded in a grand unified theory, then the same diagrams that generate the baryon asymmetry contribute[74] to the renormalization of $\bar{\theta}$, suggesting $d_N^{QCD} > 3 \times 10^{-28}$ e-cm. All of these estimates are dependent on how the fine tuning is done, i.e., they assume that $\bar{\theta}$ vanishes at the Planck scale.

Global U(1) *Symmetry*

It is possible to impose a global chiral $U(1)$ symmetry on the standard model Lagrangian.

If the $U(1)$ symmetry is not spontaneously broken, there must be a quark, q_i, with $m_i = 0$. In this case, θ can be rotated to zero by an (unobservable) chiral transformation on q_i. Arg (det m) and therefore $\bar{\theta}$ are not well defined, so $\bar{\theta}$ must be unobservable [i.e., the constant C in (**43**) vanishes]. Unfortunately, phenomenological determinations[75] of the light quark current mass ratios yield

$$\frac{m_u}{m_d} = 0.47 \pm 0.11$$

$$\frac{m_d}{m_s} = 0.042 \pm 0.007, \tag{44}$$

which are not compatible with a massless quark.

Peccei and Quinn[76] also suggested the possibility that the U_1 symmetry is spontaneously broken by the vacuum expectation value, $\langle \Phi \rangle = V \exp(i\alpha)$, of a Higgs field, Φ. In this case, when instanton effects are included, the minimum of the Higgs potential occurs at a phase, α, such that $\bar{\theta} = 0$. Such models require extra Higgs multiplets.

The Peccei-Quinn solution is extremely elegant. However, Weinberg and Wilczek pointed out[77] that the spontaneous symmetry breaking implies the existence of an axion, a pseudo-Goldstone boson that acquires a small mass,

$$m_a \sim \frac{f_\pi m_\pi}{V}, \tag{45}$$

from instanton effects. The axion is expected to couple to fermions with a strength of m_f/V. Laboratory searches[78] rule out $V \sim 250$ GeV (the weak scale), which was assumed in the original model. Many authors[78] have suggested $V \gg 250$ GeV, which could occur if Φ is an SU_2 singlet. Such an "invisible" axion would have escaped direct searches because of its very weak couplings. However, the range, 250 GeV $< V < 10^8$ GeV, is excluded by stellar evolution (the axion would carry off too much energy). Also, $V > 10^{12}$ GeV (including the GUT scale) is excluded because relic axions would contribute too much energy density to the present universe unless fine tunings are assumed.

The range, 10^8 GeV $< V < 10^{12}$ GeV, is still possible, although somewhat ad hoc. One remaining problem is that some, but not all models lead to unacceptable domain walls in the universe. If all of the problems are overcome, then the relic axions would be a good candidate for the dark matter of the universe.[6] Sikivie[79] has suggested some experiments in which "invisible" axions might in fact be detectable.

Supersymmetric models require extra Higgs multiplets, so they are a good candidate for the Peccei-Quinn mechanism. In globally supersymmetric models, R invariance can be the necessary U_1. In supergravity, however, it is difficult to have both a Peccei-Quinn symmetry and a vanishing cosmological constant, at least in the simpler models.

Spontaneous CP Violation

The third possibility is to impose CP as an exact discrete invariance of the Lagrangian of the world, so that $\theta = 0$. In this case, weak CP violation requires spontaneous CP breaking.[45] One generally expects $\bar{\theta} = -\arg(\det m) = 0(\epsilon) \sim 10^{-3}$, which is much too large. A number of authors[80] have therefore proposed models in which not only CP, but also a second discrete, global, or gauge symmetry is imposed on the theory to force (det m) to be real at tree-level. A finite and calculable $\bar{\theta}$ is generated at loop level, but in many models it is sufficiently small.

Such models suffer many drawbacks. They tend to be ad hoc and complicated. In many cases, the extra symmetries suppress or forbid weak CP violation in the KMC matrix so that an extra horizontal interaction is often needed to account for kaon decays. Most of the models are hard to embed in grand unified theories, and it is hard to generate a cosmological baryon asymmetry[6] because spontaneously broken symmetries are generally restored at very high temperatures. Finally, such models may lead to unacceptable domain walls in the universe.

Many of these problems are avoided in an interesting class of models recently suggested by Nelson[81] and generalized by Barr.[82] These are grand unified theories in which the spontaneous CP breaking scale, v, is the unification scale, M_X. CP is therefore broken for temperatures, $T \lesssim M_X$, so that a baryon asymmetry can be generated. CP violating phases are generated in the effective low energy theory by fermion mixing effects; weak CP violation turns out to be exactly the same as the standard KMC model. A nonzero $\bar{\theta}$ is generated by higher order diagrams,[81,83] including those that generate the baryon asymmetry. Segrè has argued[83] that it may be possible to have a small enough $\bar{\theta}$, but it is somewhat marginal. However, these models may still have difficulties with domain walls.

Summary of Strong CP Violation

No completely satisfactory solution to the strong CP problem has been proposed. Given the many problems (and often ad hoc nature) of the models with global $U(1)$ symmetries or spontaneous CP breaking, we should perhaps consider the fine tuning solution (which may parametrize some so far unthought of mechanism) seriously.

COSMOLOGICAL CP VIOLATION

The baryon asymmetry [equation (**20**)] is the only CP violation that has been observed other than neutral kaon decays.

If baryon number, B, is absolutely conserved, as in the standard $SU_3 \times SU_2 \times U_1$ model (excluding tiny nonperturbative effects), then either: (a) $(n_B - n_{\bar{B}})/n_\gamma \geq 10^{-10}$ must be imposed as an asymmetric initial condition on the big bang. (The initial ratio may be larger because of photon generation.) This is possible, but not very elegant. Or (b), matter and antimatter are separated in domains on scales at least as large as galactic clusters. This possibility suffers from very serious difficulties.[84] In particular, the domains are expected to be separated by domain walls with unacceptably large energy densities except in very complicated models. Secondly, causality arguments imply that the domains should be much too small unless they are expanded by an inflationary phase. In the latter case, they will generally be much too large to be interesting. On the other hand, antimatter domains have been suggested[85] as one possible explanation for the large ($\bar{p}/p \sim 10^{-4}$) observed flux of low energy cosmic ray antiprotons.

If baryon number is not conserved, as in grand unified theories,[86] then the baryon asymmetry can be generated dynamically[6] in the first 10^{-35} s after the big bang, when the temperature is comparable to the unification scale, $M_X(\simeq 10^{14}$ GeV). In addition to B violation, the necessary ingredients are nonequilibrium (otherwise, $n_B = n_{\bar{B}}$ since B and $\bar{B}$ are degenerate by CPT) and C and CP violation (to distinguish baryons and antibaryons).

In most models, the actual mechanism is the decay of superheavy Higgs bosons, H, and their antiparticles, $\bar{H}$. CP violation allows the relative rates for $H \rightarrow qq$ and $\bar{q}\bar{\ell}$ to differ from those for $\bar{H} \rightarrow \bar{q}\bar{q}$ or $q\ell$. In the minimal SU_5 model with three fermion families, the asymmetry requires a three loop diagram and is too small ($\simeq 10^{-20}$). However, an adequate asymmetry can be generated for $\geq$ four families or if additional (superheavy) Higgs multiplets are added to the theory.

The baryon asymmetry is a very serious difficulty for almost all $SU_{2L} \times SU_{2R} \times U_1$ models with $M_{W_R} \ll M_X$ and for models with spontaneous CP breaking at a scale, $v \ll M_X$. Except for a few very complicated cases, CP or C invariance is generally restored at temperatures, $T \gg (M_{W_R}, v)$, so a baryon asymmetry cannot be generated during the GUT epoch. Supersymmetric theories may lead to very complicated cosmologies. It is not clear what the general implications are for the baryon asymmetry.

SUMMARY

The predictions of various models for weak, strong, and cosmological CP violation are summarized in TABLE 2.

Recent experimental values and limits for τ_b, R, m_t, and ϵ'/ϵ have strongly constrained the KMC matrix, but the theoretical uncertainties are still too large to allow a definitive test of the KMC model.

Values rather than limits for R, ϵ'/ϵ, and possibly d_N would be extremely useful, as would more accurate values for m_t and τ_b.

Much theoretical work is needed, including more reliable estimates of matrix elements and long-distance effects in the kaon system and more detailed estimates of alternative models.

CP violation is still one of the most mysterious aspects of nature and is potentially a very powerful probe of short-distance physics.

NOTES AND REFERENCES

1. For recent reviews, see CHAU, L. L. 1983. Phys. Rep. **95:** 3; WOLFENSTEIN, L. 1985. Comments Nucl. Part. Sci. **14:** 135; BURAS, A. 1984. MPI-PAE/PTh 46/84, 26/84; PAKVASA, S. 1984. Hawaii preprint UH-511-543-84; HOLSTEIN, B. R. 1984. University of Massachusetts preprint UMHEP-203; SEGRÈ, G. 1984. Pennsylvania preprint UPR-0258T; BIGI, I. I. 1984. Aachen preprint PITHA 84/29; BIGI, I. I. & A. I. SANDA. 1985. Comments Nucl. Part. Sci. **14:** 149.
2. The KM convention is more convenient for calculations than the older Wu-Yang convention, in which the phase of the K^0 state is chosen so that A_0 is real.
3. WOLFENSTEIN, L. 1964. Phys. Rev. Lett. **13:** 562.
4. PARTICLE DATA GROUP. 1984. Rev. Mod. Phys. **56:** S1.
5. CRONIN, J. 1985. These Proceedings.
6. For reviews, see SCHRAMM, D. 1985. These Proceedings; KOLB, E. W. & M. TURNER. 1983. ARNPS **33:** 645.
7. ALTAREV, I. S. *et al.* 1981. Phys. Lett. **102B:** 13; RAMSEY, N. F. 1981. ARNPS **32:** 211.
8. LEUTWYLER, H. & M. ROOS. 1984. Z. Phys. **C25:** 91.
9. FORD, W. 1985. This volume.
10. KLANNER, R. 1984. XXIIth Int. Conf. on High Energy Physics, Leipzig, GDR, July 1984.
11. CHEN, A. *et al.* 1984. Phys. Rev. Lett. **52:** 1084.
12. KLOPFENSTEIN, C. *et al.* 1983. Phys. Rev. Lett. **130B:** 444.
13. LEE-FRANZINI, J. XXIIth Int. Conf. on High Energy Physics, Leipzig, GDR, July 1984.
14. CHAU, L. L. & W-Y. KEUNG. 1984. Phys. Rev. **D29:** 592; CHAU, L. L., H-Y. CHENG & W-Y. KEUNG. 1985. Phys. Rev. **D32:** 1837; KLEINKNECHT, K. & B. RENK. 1984. Phys. Lett. **130B:** 459; PHAM, X. Y. & X. C. VU. 1984. Phys. Lett. **136B:** 209; An extremely thorough analysis is given by: BURAS, A. J., W. SLOMINSKI & H. STEGER. 1984. Nucl. Phys. **B238:** 529 and **B245:** 369; WOLFENSTEIN, L. 1983. Phys. Rev. Lett. **51:** 1945; WEBB, J. N. 1984. Phys. Lett. **138B:** 205; PASCHOS, E. A. & U. TÜRKE. 1984. Nucl. Phys. **B243:** 29.
15. GAILLARD, M. K. & B. W. LEE. 1974. Phys. Rev. **D10:** 897.
16. WOLFENSTEIN, L. 1979. Nucl. Phys. **B160:** 501; HILL, C. T. 1980. Phys. Lett. **97B:** 275; DONOGHUE, J. F. *et al.* 1984. Phys. Lett. **135B:** 481 and Phys. Rev. **D29:** 2088; BIGI, I. I. & A. I. SANDA. 1984. Phys. Lett. **148B:** 205; FRÈRE, J. M., J. HAGELIN & A. I. SANDA. 1985. Phys. Lett. **151B:** 161.
17. ECKER, G. 1984. Phys. Lett. **147B:** 369; HOCHBERG, D. & R. G. SACHS. 1983. Phys. Rev. **D27:** 606.
18. DONOGHUE, J. F., E. GOLOWICH & B. R. HOLSTEIN. 1982. Phys. Lett. **119B:** 412.
19. BIJNENS, J., H. SONODA & M. B. WISE. 1984. Phys. Rev. Lett. **53:** 2367.
20. SHROCK, R. E. & S. B. TREIMAN. 1979. Phys. Rev. **D19:** 2148; COLIC, P. *et al.* 1983. Nucl. Phys. **B221:** 141.
21. CABIBBO, N. *et al.* 1984. Nucl. Phys. **B244:** 381; BROWER, R. C. *et al.* 1984. Phys. Rev. Lett. **53:** 1318.
22. BILEC, N. & B. GUBERINA. 1984. Phys. Lett. **136B:** 440.
23. CEA, P., G. NARDULLI & G. PREPARATA. Phys. Lett. **148B:** 477.
24. GUBERINA, B., B. MACHET & E. DE RAFAEL. 1983. Phys. Lett. **128B:** 269.
25. WAGNER, F. 1984. MPI-PAE/PTh 89/88.
26. COLIC, P. *et al.* 1983. Nucl. Phys. **B221:** 141.
27. GINSPARG, P., S. GLASHOW & M. WISE. 1983. Phys. Rev. Lett. **50:** 1415.
28. ARNISON, G. *et al.* 1984. Phys. Lett. **147B:** 493.
29. SHIFMAN, M. A., A. T. VAINSHTEIN & V. T. ZAKHAROV. 1977. Nucl. Phys. **B120:** 316 and Zh. Eksp. Teor. Fiz. **72:** 1275 (Sov. Phys. JETP **45:** 670).
30. GILMAN, F. J. & M. B. WISE. 1979. Phys. Lett. **83B:** 83; Phys. Rev. **D20:** 2392; GUBERINA, B. & R. D. PECCEI. 1980. Nucl. Phys. **B163:** 289.
31. For a recent review, see DONOGHUE, J. F. 1983. Proc. of Phenomenology of Unified Theories; from the Standard Model to Supersymmetry. Dubrovnik (University of Massachusetts UM-HEP-186).
32. GAVELA, M. B. *et al.* 1984. Phys. Lett. **148B:** 225; ALFARO, J. *et al.* 1984. Phys. Lett.

147B: 357; DONOGHUE, J. F. 1984. Phys. Rev. **D30:** 1499; PHAM, T. N. Phys. Lett. **145B:** 113.

33. DUPONT, Y. & T. N. PHAM. 1984. Phys. Rev. **D29:** 1368.
34. GILMAN, F. J. & J. S. HAGELIN. 1983. Phys. Lett. **133B:** 443; **126B:** 111; GINSPARG, P. & M. B. WISE. 1983. Phys. Lett. **127B:** 265.
35. Recent articles include: CHAU, L. L., H. Y. CHENG & W. Y. KEUNG. See reference 14; BURAS, A. J. *et al.* See reference 14; FRÈRE, J. M. *et al.* See reference 16; HAGELIN, J. S. 1984. MIU-THP/84/010. See also references 1, 32, and 33.
36. CHAU, L. L., H. Y. CHENG & W. Y. KEUNG. See reference 14.
37. DONOGHUE, J. F. See reference 32.
38. DONOGHUE, J. F. *et al.* 1981. Phys. Rev. **D23:** 1213 and reference 31.
39. ELLIS, J., M. K. GAILLARD & D. V. NANOPOULOS. 1976. Nucl. Phys. **B109:** 213.
40. BIGI, I. I. & A. I. SANDA. 1984. Phys. Rev. **D29:** 1393 and references therein; PAKVASA, S. 1983. Phys. Rev. **D28:** 2915; ALI, A. & C. JARLSKOG. 1984. Phys. Lett. **144B:** 266.
41. SHABALIN, E. P. 1978. Sov. J. Nucl. Phys. **28:** 75; **31:** 864 (1980); **32:** 228 (1982).
42. GAVELA, M. B. *et al.* 1982. Phys. Lett. **109B:** 215; NANOPOULOS, D. V. *et al.* 1979. Phys. Lett. **87B:** 53; MOREL, B. F. 1979. Nucl. Phys. **B157:** 23.
43. See, for example, BIGI, I. I. in reference 1.
44. The earliest modern example is BÉG, M. A. B. 1973. Phys. Rev. **D8:** 665; Some recent examples include DECKER, R. *et al.* 1984. Phys. Lett. **137B:** 83; BROWN, T. *et al.* 1984. Phys. Lett. **141B:** 95; LIPARTELIANI, A. G. *et al.* 1982. Nucl. Phys. **B195:** 425 and Serpukhov preprint.
45. LEE, T. D. 1973. Phys. Rev. **D8:** 1226; WEINBERG, S. 1976. Phys. Rev. Lett. **37:** 657.
46. SANDA, A. I. 1981. Phys. Rev. **D23:** 2647; DESHPANDE, N. G. 1981. Phys. Rev. **D23:** 2653; DONOGHUE, J. F. *et al.* 1982. Phys. Rev. **D25:** 195; CHANG, D. 1982. Phys. Rev. **D25:** 1318.
47. DUPONT, Y. & T. N. PHAM. 1983. Phys. Rev. **D28:** 2169.
48. DONOGHUE, J. F. & B. R. HOLSTEIN. 1984. University of Massachusetts UMHEP-213.
49. BEALL, G. & N. G. DESHPANDE. 1983. Phys. Lett. **132B:** 427.
50. For a recent review, see MOHAPATRA, R. N. 1983. Nato Summer School on Particle Physics, Munich, September 1983.
51. BRANCO, G. G. *et al.* 1983. Nucl. Phys. **B221:** 317.
52. BEALL, G. *et al.* 1982. Phys. Rev. Lett. **48:** 848.
53. BIGI, I. I. & J. M. FRÈRE. 1983. Phys. Lett. **129B:** 469.
54. CHANG, D. 1983. Nucl. Phys. **B214:** 435.
55. HARARI, H. & M. LEURER. 1984. Nucl. Phys. **B233:** 221.
56. ECKER, G. E. *et al.* 1983. Nucl. Phys. **B229:** 421.
57. BEALL, G. & A. SONI. 1981. Phys. Rev. Lett. **47:** 552.
58. ELLIS, J. *et al.* 1982. Phys. Lett. **114B:** 231; BUCHMÜLLER, W. & D. WYLER. 1983. Phys. Lett. **121B:** 321; DEL AGUILA, F. *et al.* 1983. Phys. Lett. **126B:** 71 and **129B:** 77; POLCHINSKI, J. & M. B. WISE. 1983. Phys. Lett. **125B:** 393; NANOPOULOS, D. V. & M. SREDNICKI. 1983. Phys. Lett. **128B:** 61; FRÈRE, J. M. & M. B. GAVELA. 1983. Phys. Lett. **132B:** 107.
59. GERARD, J. M. *et al.* 1984. Phys. Lett. **140B:** 349; **141B:** 79; **145B:** 400; 1985. Nucl. Phys. **B253:** 93.
60. LANGACKER, P. & B. SATHIAPALAN. 1984. Phys. Lett. **144B:** 395, 401.
61. DUGAN, M., B. GRINSTEIN & L. HALL. 1985. Nucl. Phys. **B255:** 413.
62. DONOGHUE, J. F., H. P. NILLES & D. WYLER. 1983. Phys. Lett. **128B:** 55; DUNCAN, M. J. 1983. Nucl. Phys. **B221:** 285; DUNCAN, M. J. & J. TRAMPETIC. 1984. Phys. Lett. **134B:** 439.
63. BOUQUET, A. *et al.* 1984. Phys. Lett. **148B:** 69; GANGOPADHYAYA, A. 1984. CUNY preprint; GOLDBERG, H. 1984. Phys. Lett. **147B:** 289, where it has been suggested that gluinonium states may give a large contribution of the wrong sign to Δm.
64. DUNCAN, M. J. 1983. Nucl. Phys. **B224:** 289.
65. ENQVIST, K. *et al.* 1985. Phys. Lett. **151B:** 210.
66. HERCZEG, P. 1983. Phys. Rev. **D27:** 1512; CHANG, D. & R. N. MOHAPATRA. 1984. Phys. Rev. **D30:** 2005.
67. DECKER, R. 1984. CERN TH 3917.

68. LI, L. F. & L. WOLFENSTEIN. 1980. Phys. Rev. **D21:** 178; LANGACKER, P. & A. ZEE. Unpublished.
69. HAXTON, W. C. & E. M. HENLEY. 1983. Phys. Rev. Lett. **51:** 1937.
70. BIGI, I. I. & A. I. SANDA. See reference 40; CHAU, L. L. & H-Y. CHENG. 1984. Phys. Rev. Lett. **53:** 1037; WOLFENSTEIN, L. 1984. Nucl. Phys. **B246:** 45; BROWN, T., S. PAKVASA & S. F. TUAN. 1984. Phys. Lett. **136B:** 117; CARTER, A. B. & A. I. SANDA. 1981. Phys. Rev. **D23:** 1567; BIGI, I. I. See reference 1.
71. JACKIW, R. & C. REBBI. 1976. Phys. Rev. Lett. **37:** 172; CALLAN, C., R. DASHEN & D. GROSS. 1976. Phys. Lett. **63B:** 334.
72. BALUNI, V. 1979. Phys. Rev. **D19:** 2227; CREWTHER, R. J. *et al.* 1979. Phys. Lett. **88B:** 123; 1980. Phys. Lett. **91B:** 487(E); CEA, P. & G. NARDULLI. 1984. Phys. Lett. **144B:** 115.
73. ELLIS, J. & M. K. GAILLARD. 1979. Nucl. Phys. **B150:** 141.
74. ELLIS, J. *et al.* 1981. Phys. Lett. **99B:** 101.
75. For reviews, see LANGACKER, P. & D. SPARROW. 1982. Phys. Rev. **C25:** 1194; GASSER, J. & H. LEUTWYLER. 1982. Phys. Rep. **87:** 79.
76. PECCEI, R. D. & H. QUINN. 1977. Phys. Lett. **38:** 1440; Phys. Rev. **D16:** 1791.
77. WEINBERG, S. 1978. Phys. Rev. Lett. **40:** 223; WILCZEK, F. 1978. Phys. Rev. Lett. **40:** 279.
78. For a review, see SIKIVIE, P. 1983. Proc. of the Fourth Workshop on Grand Unification. H.A. Weldon *et al.*, Eds.: 214. Birkhäuser. Boston.
79. SIKIVIE, P. 1983. Phys. Rev. Lett. **51:** 1415.
80. BÉG, M. A. B. & H. S. TSAO. 1978. Phys. Rev. Lett. **41:** 278; BÉG, M. A. B. *et al.* 1979. Phys. Rev. **D19:** 2221; BARR, S. & P. LANGACKER. 1979. Phys. Rev. Lett. **42:** 1654; GEORGI, H. 1978. Had. J. **1:** 155; MOHAPATRA, R. N. & G. SENJANOVIC. 1978. Phys. Lett. **79B:** 283; SEGRÈ, G. & H. A. WELDON. 1979. Phys. Rev. Lett. **42:** 1191.
81. NELSON, A. 1984. Phys. Lett. **134B:** 422; **136B:** 387; **143B:** 165.
82. BARR, S. 1984. Phys. Rev. Lett. **53:** 329; 1984. Phys. Rev. **D30:** 1805.
83. SEGRÈ, G. 1984. Phys. Lett. **145B:** 231.
84. For the contrary point of view, see MOHANTY, A. K. & F. W. STECKER. 1984. Phys. Lett. **143B:** 351 and references therein.
85. STECKER, F. W. & A. W. WOLFENDALE. 1984. Nature **309:** 37.
86. For reviews, see LANGACKER, P. 1981. Phys. Rep. **72:** 185; 1981 Int. Symp. on Lepton and Photon Interactions at High Energies. W. Pfeil, Ed.: 823. Bonn University.

DISCUSSION OF THE PAPER

P. ROSEN: (*Los Alamos National Laboratory, Los Alamos, NM*): Why are you working so hard to save the standard model when there is a beautiful alternative to which Jim Cronin's talk is pointing, namely, the superweak model? Has not the time come to look at $\Delta S = 2$ decays, even though they are very hard, to check Lincoln's idea?

P. LANGACKER (*University of Pennsylvania, Philadelphia, PA*): I think that it is definitely worth exploring such alternatives to the standard model. However, given the theoretical uncertainties, it is simply not fair to say that the standard model is ruled out or is in serious trouble at the present time.

S. MESHKOV (*National Bureau of Standards, Washington, D.C.*): If I take what Cronin has told us at face value, it looks like ϵ'/ϵ can be zero. What are the simple consequences of taking $\epsilon'/\epsilon = 0$? Isn't that consistent with the decoupling of the third family that you described?

LANGACKER: As far as I know, only the superweak model yields ϵ'/ϵ exactly equal to zero naturally. However, both supersymmetric models and $SU_{2L} \times SU_{2R} \times U_1$ models can give very small values of ϵ'/ϵ, and they could give ϵ'/ϵ exactly equal to zero if there are unnatural cancellations.

Index of Contributors